M&E TECBOOKS

The primary need of students taking Technician Education Council (TEC) courses is for books which will reflect the new methodology and syllabus requirements. In presenting M&E TECBOOKS, we believe that the careful collaboration between subject editors and authors has resulted in the very best, "tailor-made" texts which could be devised for the student – and which are likely to be of guidance to the lecturer also. The aim has been to provide books based on TEC's own objectives, concisely and authoritatively presented, priced as closely as possible to the student budget.

The M&E TECBOOK Series

Design of Structural Elements
Level IV

BRIAN CURRIE
C Eng MIStructE
Senior Lecturer in Reinforced Concrete Design and Detailing,
Ulster Polytechnic

ROBERT A. SHARPE
C Eng MIStruct MIPHE
Senior Lecturer in Structural Steelwork Design and Detailing,
Ulster Polytechnic

Macdonald and Evans

Macdonald & Evans Ltd
Estover, Plymouth PL6 7PZ

First published 1984

British Library Cataloguing in Publication Data

Currie, Brian
Design of structural elements level IV.——
(The M & E TECBOOK series,
ISSN 0265–3109)
1. Structural design
I. Title II. Sharpe, Robert A.
624.1′771 TA658

ISBN 0–7121–0443–7

Set in 10 on 11pt Times by
Mathematical Composition Setters Ltd,
Salisbury and printed and bound in
Great Britain by J. W. Arrowsmith, Bristol

Preface

During the many years that the authors have been lecturing on the subject of structural design students have often asked the question "How is it done?"

There are many excellent books available that explain the theory associated with design and some that even try to explain the unexplainble (i.e. the logic behind some of the requirements of the British Standard Codes of Practice). However, there appears to be a gap in the available literature when it comes to illustrating how design is done (i.e. how available information is used in order to provide a structurally sound solution). It is hoped that this modest publication will help to fill this gap.

A word of warning, however, should be sounded in the ears of the student. This book is not intended to be a layman's guide to design. The uninitiated should tread warily, realising that little knowledge is a dangerous thing. It is expected that those using this volume have some competence in the theoretical behaviour of structures.

While primarily written to meet the needs of students who are studying the BTEC syllabus "Design of Structural Elements IV", the text has been enlarged to satisfy the needs of a wider readership and it is envisaged that the book will be useful to undergraduates and practising engineers.

To enable the reader to gauge his grasp of the text, self-assessment questions appear at the end of each chapter. These should be of value not only to the student but also to tutors who have to test students. It has been a conscious decision to exclude all structural detailing from this book. Some may find this strange, but the detailing requirements for most of the elements designed are to be found in the authors' companion book *Structural Detailing II* also published by Macdonald and Evans.

1984

BC
RAS

Acknowledgments

It is the desire of the authors to sincerely thank all who have helped in so many ways to make the publication of this book possible. It would not be feasible to name each individually, but special reference should be made to the following:

Professor R. B. Schofield, Director of Studies of the Department of Civil Engineering, Ulster Polytechnic, for his continuous support and encouragement; Mr B. R. Rodgers of the Cement and Concrete Association for his helpful and constructive criticism of the reinforced concrete section of the book.

Material has been utilised from many sources and extracts have been reproduced from CP3 Chapter V Part 1 *Dead and Imposed Loads*, BS648 *Schedule of Weights of Building Material*, CP110 *The Structural Use of Concrete*, CP112 Part 2 *The Structural Use of Masonry*, BS449 *The Use of Structural Steel in Building* and BS5628 Part 1 *The Structural Use of Masonry* all published by the British Standards Institution, from whom copies of the complete standards may be obtained; *Structural Steelwork Handbook* published by the British Constructional Steelwork Association Limited (BCSA) and the Constructional Steel Research and Development Organisation (CONSTRADO); *Structural Steelwork Handbook for Metric Angles* published by CONSTRADO; *Metric Practice for Structural Steelwork* and *Structural Fasteners and their Application*, both published by BCSA.

The typing of the manuscript, letters and all items associated with this volume has been done by Mrs Rea, Mrs McCauley, Mrs Currie, Mrs Wright and Mrs Convery. Our gratitude is expressed to them all. Finally we again must thank our respective families for their forbearance and understanding during the hours of manuscript preparation.

Contents

CHAPTER ONE

Limit State Design for Reinforced Concrete

CHAPTER OBJECTIVES

After studying this chapter you should be able to understand:

* the philosophy of limit state design;
* the basic properties of structural concrete and reinforcement;
* characteristic strengths of materials;
* characteristic loads;
* partial safety factors;
* partial load factors.

BRIEF HISTORY OF BRITISH CODES

It is desirable for engineers to have a guide to good design practice which is based on the results of research and observations from many years of experience. When all engineers use the same guide it means that each can understand and check the others' calculations. This guide to good design practice is called a code of practice.

The first attempt to produce such a document was in 1934 and through an evolutionary process by 1964 the structural concrete designer had to contend with three codes of practice. These were:

CP114 *The Structural Use of Reinforced Concrete in Buildings*
CP115 *The Structural Use of Prestressed Concrete in Buildings*
CP116 *The Structural Use of Precast Concrete*

While these British codes were being developed, various international committees were looking at the philosophical approach to structural design and in 1963 the European Concrete Committee (CEB) published its *Recommendations for an International Code of Practice for Reinforced Concrete* and sometime later its report dealing with prestressed concrete.

In 1964 a Structural Codes Advisory Committee was appointed by the Institution of Structural Engineers, with the job of trying to co-ordinate the codes to avoid any inconsistencies. As the work of this committee proceeded it became obvious that the three codes then in current use should be gathered together to form one code dealing with as many types of concrete structures as possible. The committee decided to adopt the CEB report as a guide in the formulation of this new British code. Perhaps the major implication of this decision was that the new British code would have for its basis limit state design. This new code was published in 1972 as CP110 *The Structural Use of Concrete*.

LIMIT STATE DESIGN

The former codes of practice permitted the strength of members to be calculated on the basis of either the elastic theory or the load factor theory. One of the ways that these differ from each other is with respect to the factor of safety. In the elastic theory the factor of safety is on the stresses with the actual loads being used, whereas in the load factor method the factor of safety is on the loads with the ultimate stresses used. Since the loads and the materials in a structure are quite different it was less than logical to have one global factor to assess the uncertainties in each. The limit state approach has introduced two partial safety factors, one for the loads and another for the materials which are based, as far as possible, on statistical data. This is explained later in this chapter.

Objectives of structural design

The objective of structural design is to provide a structure which complies with the clients' requirements with due attention paid to the following:

(*a*) economy;
(*b*) safety;
(*c*) serviceability;
(*d*) aesthetics.

It is the task of the structural engineer to produce a design which satisfies the appropriate safety and serviceability requirements, keeps within the budget and is pleasing to the eye. In order to do this the engineer needs guidance in these various areas. The economy aspect is normally determined by the client and is catered for in practice by a quantity surveyor. The aesthetics aspects is usually catered for by the architect, while the British Standard Code of Practice CP110 *The Structural Use of Concrete* gives guidance on the structural aspects.

After deciding on the most economical structural form which meets the other requirements, the engineer has to do a detailed design for the structure chosen. For this design the recommendations of CP110 (which is based on the limit state method of design) will normally be used.

Limit states

Section 2.1 of CP110 states:

> The purpose of design is the achievement of accepted probabilities that the structure being designed will not become unfit for the use for which it is required, i.e. that it will not reach a limit state.

There are many reasons why a structure may become unfit for use or reach a limit state. These are broadly divided into three categories which are listed below, with the code reference given in brackets:

(*a*) the ultimate limit state (2.2.2.);
(*b*) serviceability limit states (2.2.3.);
(*c*) other considerations (2.2.4.).

The ultimate limit state

This limit state is reached when the structure fails or collapses. This may be due to any part of the structure failing or buckling, or the whole overturning. The possibility of failure by accidental causes must also be considered, to ensure that damage should not be disproportionate to the original cause.

Serviceability limit states

In contrast to the ultimate limit state, a structure may become unfit for use while it is in service. When such a situation arises the structure is said to have reached a serviceability limit state. These can take many forms as outlined below.

(*a*) *Deflection*· the appearance or efficiency of the structure or any part thereof should not be adversely affected by deflection.

(*b*) *Cracking:* the appearance or durability of the structure should not be adversely affected by cracking.

(*c*) *Vibration:* special precautions must be taken where vibration produced by machinery, or other sources, is likely to cause alarm or discomfort.

(*d*) *Other limit states:* in unusual or special structures the engineer may deem it necessary to consider some other relevant limit state not included in (*a*)–(*c*).

Other considerations

A structure may become unfit for use for some reason not outlined in the sections on the ultimate limit state and serviceability limit states above. In CP110 these are entitled other considerations and are as follows:

(*a*) *Fatigue:* This should be considered where the load is likely to be cyclic.

(*b*) *Durability.* Where exposure conditions are exceptionally severe, precautions—beyond those normally given in the code—must be taken.

(*c*) *Fire resistance.* In section 10 of the code the minimum requirements for fire resistance of particular elements are given.

(*d*) *Lightning.* Protection ought to be provided in accordance with CP326 *The Protection of Structures Against Lightning*

As stated earlier the purpose of design is to ensure that the structure being designed will not reach a limit state. In normal reinforced concrete design the method usually adopted is to design for the ultimate limit state and to check that the other limit states are not reached.

MATERIAL STRENGTHS

It is obvious that before designing a structure the engineer must have a knowledge of the materials he is using and their behaviour under load. The two major materials that the reinforced concrete designer uses are concrete and steel reinforcement.

Concrete

The single property of concrete which is usually specified by an engineer is that of compressive strength. This is due to the fact that other properties of concrete such as durability and impermeability, are usually satisfactory when the concrete reaches its specified strength. There are, however, a number of factors which influence the strength of the concrete, the principal ones being:

(*a*) water–cement ratio;
(*b*) degree of compaction;
(*c*) age of the concrete when tested;
(*d*) type of curing;
(*e*) temperature of curing;
(*f*) type of cement used;
(*g*) type of aggregate used.

It is not the purpose of this book to deal specifically with these factors but where the student may require further information the references given at the end of the chapter may be helpful.

The majority of structural concrete in the UK is produced using ordinary Portland cement and in the UK its strength is usually measured by carrying out a compression test on a concrete cube. This test may be carried out on the concrete at various ages but it is usual to specify the strength required 28 days after production. Where an earlier indication of strength is required this may be found by doing the test at some earlier age and relating this to the 28-day strength using the graph shown in Fig. 1. (e.g. the 7-day strength is approximately 2/3 of the 28-day strength for samples cured at 21 °C).

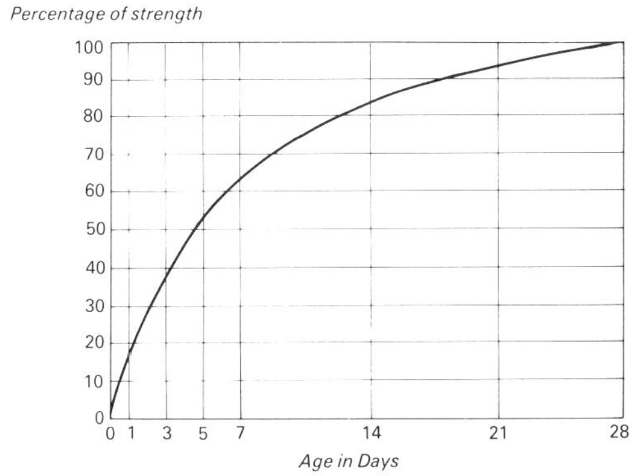

FIG. 1. *Development of compressive strenght of concrete cured at 21 °C.*

The tensile strength may also be related to the compressive strength and in this respect a ratio of 1 : 10 is often quoted (i.e. the tensile strength is about 1/10 of the compressive strength). Where a more accurate relationship is required the following equations may be used:

the direct tensile strength

$$= 0.45\sqrt{\text{compressive strength}} \quad \text{N/mm}^2$$

and

the bending tensile strength

$$= 0.7\sqrt{\text{compressive strength}} \quad \text{N/mm}^2$$

In reinforced concrete design, however, the concrete in tension is assumed to have cracked and therefore has no strength. The design is based on the strength of the concrete in compression, with the reinforcement carrying all the tensile stresses.

Characteristic strength

If a large number of compressive tests were done on particular concrete cubes it would be found that they would follow the normal frequency distribution curve shown in Fig. 2.

Rather than working to the mean strength, f_m, CP110 refers to the characteristic strength which is defined in section 2.3.2. as the value

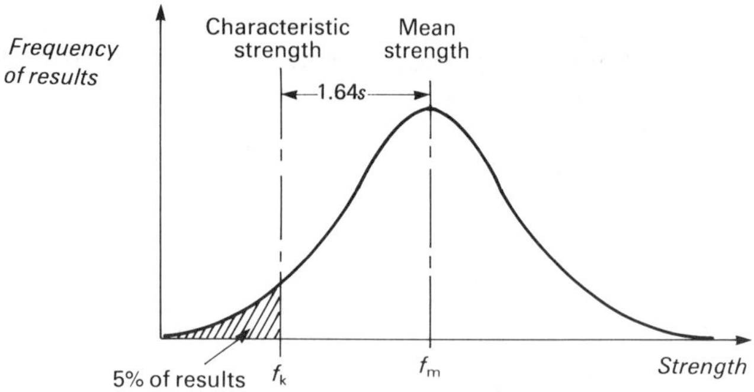

FIG. 2. *Characteristic strength of concrete cubes.*

"below which not more than 5 per cent of the test results fall". This characteristic strength is given the symbol f_k and is indicated in Fig. 2. If s is the standard deviation then

$$f_k = f_m - 1.64s$$

Concrete may be identified by giving it a "grade", e.g. a grade 30 concrete has a characteristic strength of 30 N/mm^2. Since the compressive strength of the concrete is related to its durability there must be conditions of use for which low grade concrete would not be suitable. This is shown in Table 1.

TABLE 1.
RECOMMENDED LOWEST GRADE OF CONCRETE

Grade	*Lowest grade for specified use*
7	Plain concrete
10	
15	Reinforced concrete with lightweight aggregate
20	Reinforced concrete with dense aggregate
25	
30	Concrete with post-tensioned tendons
40	Concrete with pre-tensioned tendons
50	
60	

Reinforcement

The reinforcement for concrete is produced in the form of bars or welded to give a mesh, either square or rectangular. The steel is factory produced to conform with BS449 *Specification for Hot Rolled Steel Bars for the Reinforcement of Concrete*; BS4461 *Specification for Cold Worked Steel Bars for the Reinforcement of Concrete;* or BS4482 *Hard Drawn Mild Steel Wire for the Reinforcement of Concrete*. The characteristic strengths of the more common reinforcements are shown in Table 2.

The same standard distribution as shown for concrete in Fig. 2 can be used for reinforcement and the definition for characteristic strength applies also to the reinforcement (i.e. the value below which not more than 5 per cent of the test results fall).

Throughout this book the main reinforcement will be in high-yield steel and the links in mild steel. This is because high-yield steels have a very significant stress advantage over mild steel while being only slightly more expensive, but mild steel can be bent to a tighter radius than high-yield steel, allowing more accurate placing of the main bars in the corner of the links.

TABLE 2.
STRENGTHS OF REINFORCEMENTS

Designation	*Nominal size (mm)*	*Specified characteristic strength (f_y) N/mm^2*
Hot-rolled mild steel (BS4449)	All sizes	250
Cold-worked high yield steel (BS4461)	Up to and including 16	460
	Over 16	425
Hard drawn steel wire (BS4482)	Up to and including 12	485

Partial factors of safety for materials

The characteristic strengths both for concrete and reinforcement are based on samples which are produced under ideal conditions. There is, however, a large difference between concrete produced and tested in a laboratory and that which is produced on site and tested in the laboratory. The same applies to the reinforcement stored and tested in a laboratory and that which is used on site. In the case of concrete, the material used on site may be segregrated, not properly compacted or badly cured while the reinforcement may be misshapen or corroded. To account for these effects partial safety factors, γ_m, are introduced and the characteristic strengths are divided by these partial safety factors to give the design strength of the material, i.e.

$$\text{design strength} = \frac{f_k}{\gamma_m}$$

The values for γ_m for each material are shown in Table 3.

TABLE 3.
γ_m VALUES FOR CONCRETE AND STEEL

	Values of γ_m	
Limit State	*Concrete*	*Steel*
Ultimate	1.5	1.15
Deflection	1.0	1.0
Cracking	1.3	1.0

Since the probabilities that can be accepted in each of the limit states vary, the γ_m values vary also.

Loads

Just as the characteristic strengths of the materials were obtained from a statistical analysis of the results, it should be possible to so assess the characteristic load on a structure using the same probability and standard deviation

$$F_k \text{ (characteristic load)} = F_m \pm 1.64\, s$$

However there is not sufficient information available to enable such an analysis to take place. Guidance for loading is obtained from BS648 *Schedule of Weights of Building Materials*; CP3 Chapter V Part 1 *Dead Load and Imposed Loads*; and CP3 Chapter V Part 2 *Wind Loads.*

The loads on a structure fall into one of three categories:

(*a*) dead loads G_k: the static weight of all permanent construction.

(*b*) imposed loads Q_k: the load produced by the intended occupancy or use; and

(*c*) wind loads W_k: the suction or pressure produced by wind forces.

Partial factors of safety for loads

The load on a structure or member may be inaccurate for a number of reasons:

(*a*) errors in calculation;
(*b*) inaccurate construction;
(*c*) unusual increases in load;
(*d*) the importance of the limit state being considered.

To take account of these effects partial safety factors γ_f are introduced and the characteristic loads are multiplied by these partial safety factors to give the design loading, i.e.:

$$\text{design load} = F_k \gamma_f$$

The values for γ_f depend on the load and the limit state being considered, and are shown in Tables 4 and 5.

TABLE 4.
ULTIMATE LIMIT STATE

	Dead load		*Imposed load*		
Load combination	*Max*	*Min*	*Max*	*Min*	*Wind load*
1 Dead plus imposed	1.4	1.0	1.6	0	0
2 Dead plus wind	1.4	0.9	0	0	1.4
3 Dead plus imposed plus wind	1.2	1.2	1.2	1.2	1.2

TABLE 5.
SERVICEABILITY LIMIT STATE

Load combination	*Dead load*	*Imposed load*	*Wind load*
1 Dead plus imposed	1.0	1.0	0
2 Dead plus wind	1.0	0	1.0
3 Dead plus imposed plus wind	1.0	0.8	0.8

In this book the usual combination will be dead plus imposed load for the ultimate limit state. Normally when checking the serviceability limit states the ultimate loading can be used since the checks required are given in terms deemed to satisfy rules which do not necessitate a recalculation of loading. The loads must be combined in order to produce the most severe conditions, as illustrated below for a simple beam.

Example
For the beam shown in Fig. 3 calculate the worst bending moments at B and mid span A–B for the ultimate limit state, when the dead load = 15/kN m and the imposed load = 10/kN m.

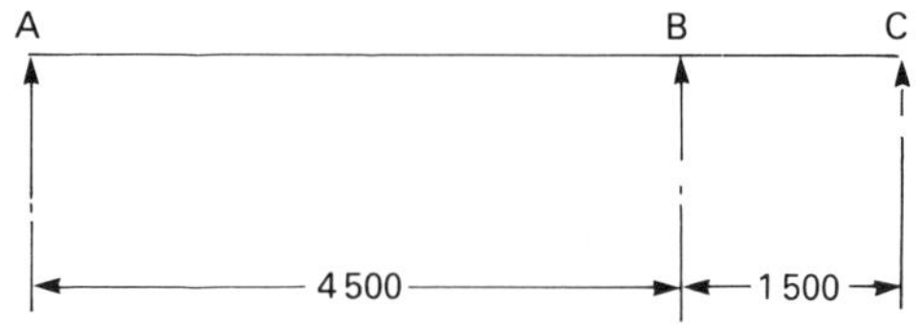

FIG. 3. *Method of support for beam.*

Solution
For the ultimate limit state, with dead and imposed load only, the γ_f values are:

Dead: maximum 1.4, minimum 1.0
Imposed: maximum 1.6, minimum 0

Load cases

(a) Full load across ABC, i.e. $1.4g_k + 1.6q_k$

$$\text{Design load } n = 1.4g_k + 1.6q_k = 1.4 \times 15 + 1.6 \times 10 = 37 \text{ kN m.}$$

$$\text{Bending moment at } B = \frac{37 \times 1.5^2}{2} = 41.625 \text{ kN m}$$

$$\text{“Free” bending moment in } A\text{–}B = \frac{37 \times 4.5^2}{8} = 93.66 \text{ kN m}$$

$$\therefore \text{ Maximum bending moment } A\text{–}B = 93.66 - \frac{41.625}{2} = 72.84 \text{ kN m}$$

(b) Minimum load ($1.0g_k$) on B—C, maximum load ($1.4g_k + 1.6q_k$) on A—B

$$\text{Bending moment at } B = \frac{15 \times 1.5^2}{2} = 16.875 \text{ kN m}$$

“Free” bending moment in

$$A\text{–}B = \frac{37 \times 4.5^2}{8} = 93.66 \text{ kN m}$$

∴ Maximum bending moment

$$A - B = 93.66 - \frac{16.875}{2} = 85.22 \text{ kN m.}$$

(c) Maximum load ($1.4g_k + 1.60q_k$) on B—C, minimum load ($1.0g_k$) on A—B

$$\text{Bending moment at } B = \frac{37 \times 1.5^2}{2} = 41.625 \text{ kN m}$$

“Free” bending moment in

$$A - B = \frac{15 \times 4.5^2}{8} = 37.97 \text{ kN m}$$

$$\therefore \text{ Maximum bending moment } A\text{–}B = 37.97 - \frac{41.625}{2} = 17.16 \text{ kN m}$$

Summarising the above results it is found that the maximum bending moment at B is

41.625 kN m under loading conditions (*a*) and (*c*) while the maximum bending moment in *A–B* is 85.22 kN m under loading condition (*b*). The above illustration is for a simple beam only, but the loads must be combined to produce the worst case in both frames (3.2.2.) and continuous beams (3.3.3.).

FURTHER READING

Properties of Concrete. A. M. Neville, Pitman, 1975.

Concrete Constituents and Mix Properties. B. W. Shacklock. Cement and Concrete Association, 1974.

Reinforced Concrete Design to CP110 Simply Explained. A. H. Allen. Cement and Concrete Association, 1974.

SELF-ASSESSMENT QUESTIONS

1. Explain the term "limit state" outlining various ways in which a limit state can be reached.
2. What is meant by the expression "characteristic strength" of a material?
3. Why are partial factors of safety necessary and how are they incorporated into CP110—the structural use of concrete?
4. Illustrate with an example of your own devising how loads should be combined to give the worst stresses.

CHAPTER TWO

Requirements for Beams and Slabs

CHAPTER OBJECTIVES

After studying this chapter you should be able to understand the requirements of CP110 with respect to simple beams and slabs for:

* bending;
* shear;
* bond; and
* cracking.

DESIGN PROCESS

The most common way to design a reinforced concrete member is to ensure that the member is safe for the ultimate limit state and then check that it also satisfies the serviceability limit states. This means that the design is firstly done to ensure that failure in either bending or shear will not take place and the section thus determined will satisfy the requirements of local bond, deflection and cracking. These are the common limit states but others, as outlined in Chapter One, may be checked as the engineer deems necessary.

Ultimate limit state

In this book the loads to be considered will always be dead loads and imposed loads. For simply supported members the worst bending moments and shear forces will be when the full span is loaded with the heaviest possible design load. The load factors are 1.4 × dead load and 1.6 × imposed load, and the design loading is:

$$f = 1.4G_k + 1.6Q_k \text{ (kN)}$$

or

$$n = 1.4g_k + 1.6q_k \text{ (kN/m) or (kN/m}^2\text{)}$$

The maximum bending moment will be at mid-span for a simply supported member and is either $Fl/8$ or $nl^2/8$, where l is the span of the member in metres. The span is defined in clause 3.3.1.1 as the smaller of either:

(*a*) the distance between the centres of bearings (supports); or
(*b*) the clear distance between supports plus the effective depth.

The maximum shearing force will be at the support and is either $F/2$ or $nl/2$.

Moment of resistance

Having obtained the maximum bending moment that the member has to support, this becomes the least moment of resistance that the member must develop in order to support the load without failure occurring. This moment of resistance is produced by an internal couple consisting, in a singly reinforced beam, of the compressive force of the concrete and the tensile force of the tension reinforcement, and in doubly reinforced beam of the compressive force in the concrete in conjunction with the force due to the compressive reinforcement and the tensile force of the tension reinforcement.

The bending moment is the product of loads and spans, while the moment of resistance is the product of stresses and section size. Thus the following relationship:

Externally	*Internally*
Loads	Stresses
+	+
Spans	Dimensions
‖	‖
Bending moment	Moment of resistance

The moment of resistance must at least balance the bending moment for the structure to be safe against bending failure.

From the above flow chart it can readily be appreciated that when the span, loading, cross-sectional dimensions and material strengths are known, then there is a direct link between the bending moment and the area of reinforcement required. This link, in CP110, can be formed in one of two ways, viz:

(*a*) using design formulae;
(*b*) using design charts.

In this book the reinforced concrete design examples done in the following chapters utilise the design charts in Part 2 of CP110. However, it would leave a large gap in the understanding of reinforced concrete if the design formulae were not understood, thus they are explained below.

DESIGN FORMULAE FOR BEAMS IN BENDING

These formulae are different for the various concrete sections but they are all based on the following assumptions:

(*a*) Plane sections before bending remain plane after bending.

(*b*) The concrete in tension has cracked and all the tensile stresses are taken by the reinforcement.

(*c*) There is good bond between the reinforcement and the concrete. Therefore the strain in the reinforcement is given by the theoretical strain in the adjacent concrete.

(*d*) The stresses in all bars are equal, therefore the resultant tensile force acts at the centroid of the reinforcement. In this book three types of simply supported beams are considered: (i) singly reinforced rectangular beams;

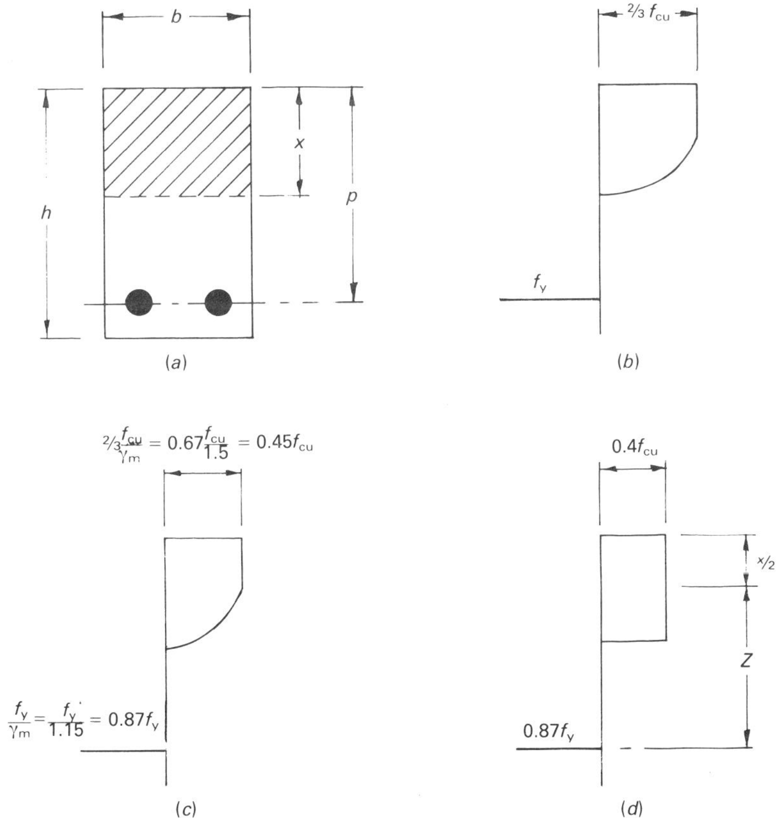

FIG. 4. *Singly reinforced beam.*

(ii) singly reinforced flanged beams; and (iii) doubly reinforced rectangular beams.

Singly reinforced beams

Consider the rectangular beam shown in Fig. 4 together with the stress distributions.

When the beam in Fig. 4(*a*) is subjected to pure bending, it has been determined by experiment that the stress distribution at failure is as shown in Fig. 4(*b*). When the partial safety factors are introduced the stress distribution equates to Fig. 4(*c*). Figure 4(*d*) shows the simplified stress distribution adopted by CP110 and used in the derivation of the design formulae.

For equilibrium the sum of the longitudinal forces in the beam must be zero (i.e. the force in the concrete = the force in the reinforcement):

$$0.4f_{cu}bx = 0.87f_yA_s$$

$$\therefore \quad x = \frac{0.87f_yA_s}{0.4f_{cu}b} \qquad (1)$$

The ultimate moment of resistance M_u is the result of the action of the couple formed by the tensile and compressive forces:

$$\therefore \quad M_u = \text{tensile force} \times \text{lever arm } (Z)$$

$$= \text{compressive force} \times \text{lever arm } (Z)$$

From the tensile force:

$$M_u = 0.87f_yA_s\left(d - \frac{x}{2}\right) \text{ where } Z = \left(d - \frac{x}{2}\right)$$

$$= 0.87f_yA_sZ \qquad (2)$$

From the compressive force

$$M_u = 0.4f_{cu}bx\left(d - \frac{x}{2}\right) \qquad (3)$$

Equation 2 corresponds to equation (1) in CP110, which stipulates the maximum value for $Z = 0.95d$, although Z may be calculated as follows using equation 1 above and substituting for x in the expression $Z = d - x/2$.

$$\therefore \quad Z = d - \left(\frac{0.87f_yA_s}{0.4f_{cu}b}\right)\frac{1}{2}$$

$$\therefore \quad Z = \left(1 - \frac{1.1f_yA_s}{f_{cu}bd}\right)d \qquad (4)$$

Equation 4 corresponds to equation (5) in CP110.

Failure

The type of bending failure observed in a reinforced concrete beam will largely depend on the amount of reinforcement present in the section. If the beam is under-reinforced it will deflect greatly as failure approaches and the reinforcement yields. This gives a visible warning of impending failure. Where it is over-reinforced the concrete fails in compression, prior to any significant deflection. This failure can occur explosively. To ensure that the beam gives some warning of approaching failure and does not produce an explosive failure it is necessary to limit the amount of tension reinforcement in the beam. It has been observed from experiment and can be shown by a strain compatibility approach that the steel yields before the concrete crushes when the ratio x/d is greater than 0.5 approximately.

Substituting this value in equation 3 above:

$$M_u = 0.4f_{cu}b\frac{d}{2}\left(d - \frac{d}{4}\right)$$

$$= 0.15f_{cu}bd^2 \qquad (5)$$

Equation 5 corresponds to equation (2) in CP110 and when used ensures that a singly reinforced beam will not be over-reinforced.

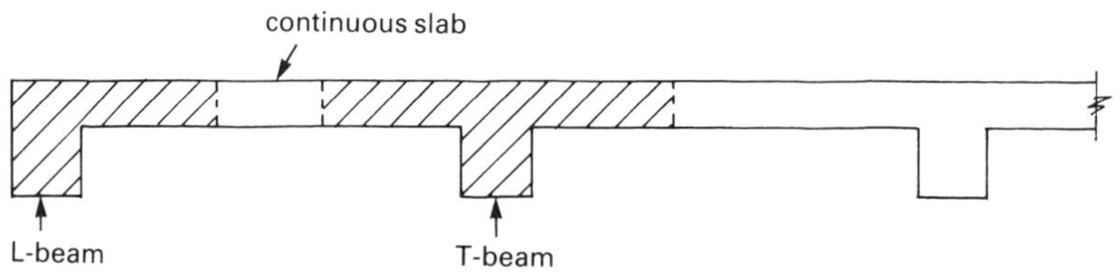

FIG. 5. *Structural section of flanged beam.*

Flanged beams

Where a concrete structure is poured *in situ*, the floor and the beam downstand combine to give the complete structural section as indicated in Fig. 5.

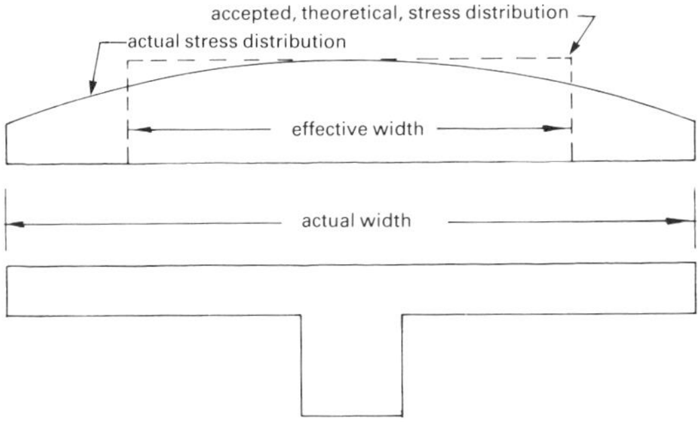

FIG. 6. *Difference between actual and theoretical stress distributions in flanged beam.*

In paragraph 3.3.1.2 CP110 gives guidance as to the effective width to be used in strength calculations. This accounts for the difference between the actual and theoretical stress distributions—(*see* Fig. 6).

There are two cases which must be considered:

(*a*) when the neutral axis lies within the flange;

(*b*) when the neutral axis lies below the flange.

Neutral axis within the flange

In this case the stress distribution is exactly as shown in Fig. 4(*d*) since the concrete below the neutral axis has cracked in tension and is ignored. Thus the calculations are as for a singly reinforced beam except that the breadth of the section is the effective width of the flange (*see* Fig. 7).

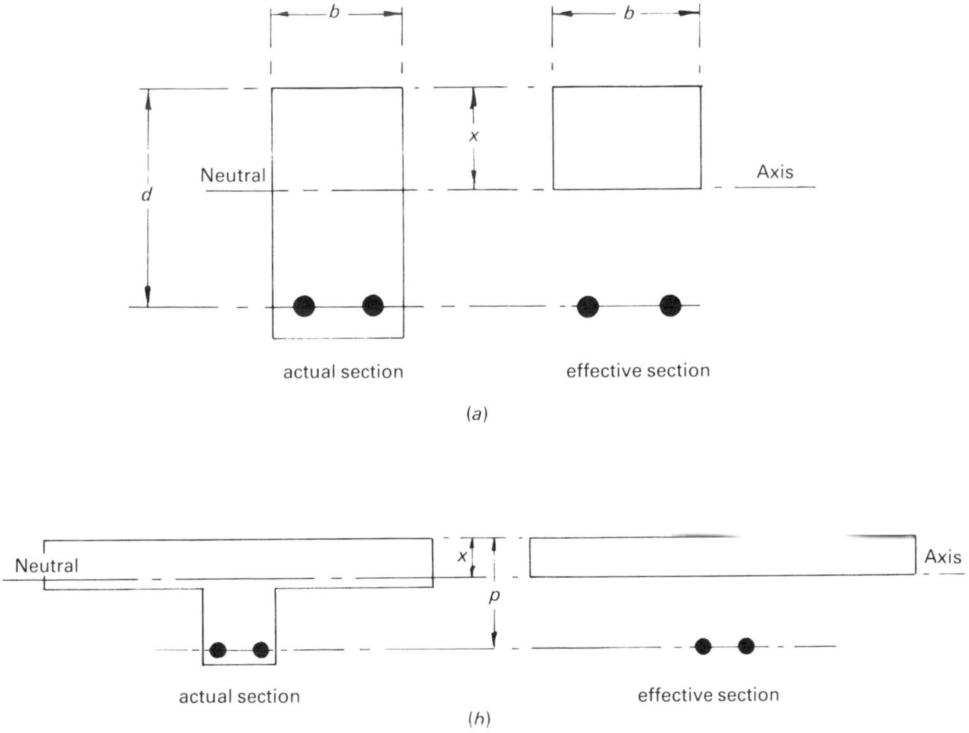

N.B. While the actual section in each case is different, the effective sections are similar in that each is a rectangular block of concrete plus the reinforcement.

FIG. 7. *Neutral axis within the flange.*

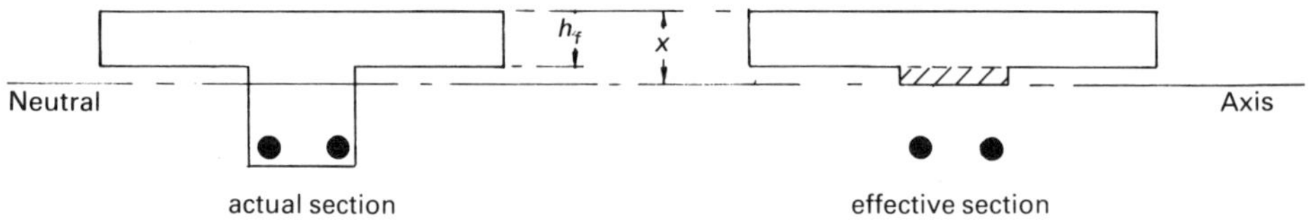

FIG. 8. *Neutral axis below the flange.*

To verify the position of the neutral axis the following equation can be used.

As with the singly reinforced beam the compressive force is equated to the tensile force, i.e.:

$$0.4 f_{cu} bx = A_s \frac{f_y}{1.15}$$

$$\therefore \quad x = \frac{0.87 f_y A_s}{0.4 f_{cu} b} \qquad (6)$$

Neutral axis below the flange
Where the neutral axis falls in the rib, the effective section includes a little rectangle of concrete shown shaded in Fig. 8.

However, it is usually assumed that the maximum moment of resistance occurs when $x = h_f$. This assumption reduces the calculations and is obviously conservative (*see* Fig. 9).

The ultimate moment of resistance M_u based on the compressive force is given by the equation:

$$M_u = 0.4 f_{cu} b h_f \left(d - \frac{h_f}{2}\right) \qquad (7)$$

Equation 7 corresponds to equation (7) in CP110.

If M_u is based on the tensile force then:

$$M_u = 0.87 f_y A_s \left(d - \frac{h_f}{2}\right) \qquad (8)$$

Equation 8 corresponds to equation (6) in CP110.

When designing flanged beams equation (7) can be used to calculate M_u and then A_s is found by using equation (8), i.e.:

$$A_s = \frac{M_u}{0.87 f_y (d - h_f/2)}$$

These equations allow the longitudinal reinforcement only to be designed, whereas there will also be bending transversely in the flange. In the case of an isolated beam the flange can be considered as two cantilever slabs, both cantilevering from the rib provided that the minimum amount of reinforcement, i.e. 0.3 per cent of the longitudinal cross sectional area of the flange, is provided, see clause 3.11.4.2 of CP110. However, in the more common case, the flange forms part of an *in situ* slab and is designed accordingly (*see* Fig. 10).

Doubly reinforced beams
Where the structural engineer has complete freedom to choose the dimensions of the beam, he will normally ensure that the beam is proportioned so that tension reinforcement only is required. If, however, there is some restriction on the constructional depth the section may have to be a depth such that the concrete available in compression will not be sufficient to withstand the applied stress. In this case it is necessary to introduce compression reinforcement. While the design of continuous beams is outside the scope of this book, it ought to be noted that a

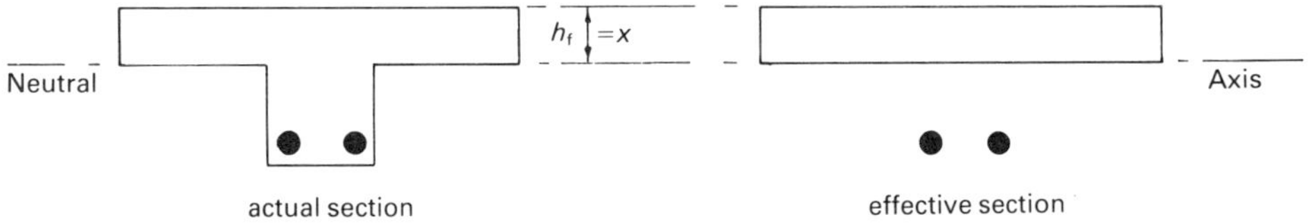

FIG. 9. *Assumed position of neutral axis for maximum moment of resistance.*

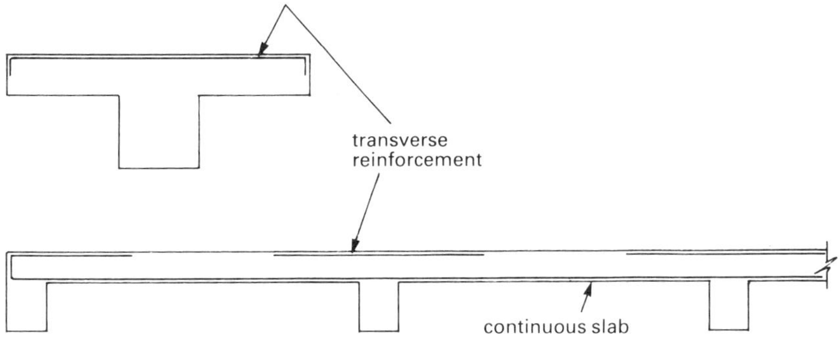

FIG. 10. *Design of flange as part of in situ slab.*

similar situation as described above often occurs at the supports of continuous flanged beams.

Moment of resistance

The moment of resistance of a doubly reinforced beam can be considered as consisting of two component parts.

As the load is applied the section acts as a singly reinforced beam until its ultimate moment of resistance is reached. As the load is increased the compressive stress exceeds that of the concrete alone which must be reinforced. This produces an additional compressive force which must be balanced by additional tensile reinforcement giving an equal and opposite tensile force. This is illustrated below in Fig. 11.

In Fig. 11(*a*) the singly reinforced beam is

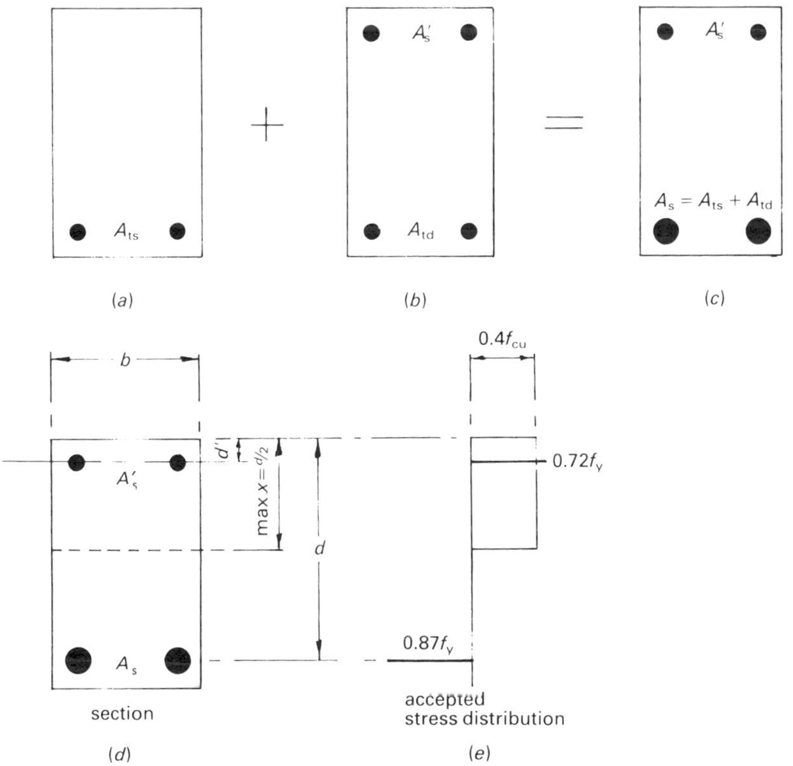

FIG. 11. *Moment of resistance in a doubly reinforced beam.*

shown with A_{ts} denoting the area of tension reinforcement in the singly reinforced beam. The ultimate moment of resistance of such a beam has been shown to be:

$$M_{us} = 0.15 f_{cu} b d^2 \quad \text{(9) as (5) above}$$

In Fig. 11(b) the compression reinforcement is added A_s' and the additional tensile reinforcement required to balance it A_{td}.

Figure 11(c) shows (a) and (b) combined to give the doubly reinforced section. Figures 11(d) and 11(e) show a typical section and the accepted stress distribution.

In order to derive the formulae for such a beam the compressive stress in the reinforcement must be known. CP110 gives the limiting stress as $(f_y/(\gamma_m + f_y/2000))$, *see* Fig. 2 in CP110.

With γ_m for reinforcement $= 1.15$ and $f_y = 460\ \text{N/mm}^2$, maximum:

$$\text{The stress in compression} = \frac{f_y}{1.15 + \dfrac{460}{2000}} = 0.72 f_y$$

This is obviously conservative for steels where $f_y < 460\ \text{N/mm}^2$.

The moment of resistance for the section in Fig. 11(b) can be found in two ways:

(a) Take moments about the compressive reinforcement:

$$M_{ud} = A_{td}\, 0.87 f_y (d - d') \tag{10}$$

(b) Take moments about the tensile reinforcement:

$$M_{ud} = A_s'\, 0.72 f_y (d - d') \tag{11}$$

Now combine equations (9) and (11):

$$M_u = 0.15 f_{cu} b d^2 + 0.72 f_y A_s'(d - d') \tag{12}$$

Equation 12 corresponds to equation (3) in CP110.

Also by equating the horizontal forces for Fig. 11(b):

$$0.72 f_y A_s' = 0.87 f_y A_{td} \tag{13}$$

And by equating the horizontal forces for Fig. 11(a):

$$0.4 f_{cu} b \frac{d}{2} = 0.87 f_y A_{ts} \tag{14}$$

Now combine equations (13) and (14):

$$0.72 f_y A_s' + 0.4 f_{cu} b \frac{d}{2} = 0.87 f_y A_{td} + 0.87 f_y A_{ts}$$

$$\therefore \quad 0.87 f_y A_s = 0.2 f_{cu} b d + 0.72 f_y A'_s \tag{15}$$

Equation 15 corresponds to equation (4) in CP110.

The equations developed above are for rectangular beams where the depth of concrete in compression is a maximum of $d/2$. For flanged beams this depth is h_f and similar equations can be found by simple substitution.

The equations can be used to design a doubly reinforced beam for bending used in the following steps:

1. Calculate the applied bending moment M.
2. Calculate the maximum moment of resistance which the beam can supply as singly reinforced with A_{ts}.
3. Calculate the areas A_s' and A_{td} needed to resist the moment which remains when (2) is subtracted from (1).
4. The areas of reinforcement required are A_s' in compression and $A_{ts} + A_{td} = A_s$ in tension.

Design charts for beams in bending

The above formulae have been derived using the simplified stress distribution shown in Fig. 4(d). A more economical use of the materials would be found if the formulae were to be derived using the rectangular parabolic stress distribution shown in Fig. 4(c). Obviously the algebra involved to derive corresponding expressions using Fig. 4(c) is much more complex. However, this more complex relationship giving slightly more economical results has been given in CP110, in the form of the design charts. Thus when dealing with simple cross sectional profiles the designer would naturally choose to obtain the area of reinforcement from the charts, which save him time and his client money.

It is therefore these charts which shall be

utilised in the remainder of the reinforced concrete portion of this book.

SHEAR IN BEAMS

Another ultimate limit state is that of shear. The analysis of shear in reinforced concrete beams is complex and has been the subject of research for many years. However, it is agreed that in a beam without shear reinforcement, the shear resistance is composed of three major parts. These are:

(*a*) the uncracked concrete in the compression zone (20–40%);
(*b*) the interlock of the aggregate across the crack zone (35–50%);
(*c*) the dowel action of the longitudinal reinforcement (15–25%).

The first and last of these can, to some extent, be quantified but obviously the second cannot. Thus Table 5 of CP110 which gives values of the permissible shear stress, ν, is calculated from the expression:

$$\nu = \frac{V}{bd} \qquad (16)$$

where V = the shear force due to ultimate loads; b = the breadth of the section which, for a flanged beam, should be taken as the rib width; and d = the effective depth. Where ν exceeds ν_c, shear reinforcement must be provided to account for the difference $(\nu - \nu_c)$. However, ν must never exceed the maximum value given in Table 6, even with shear reinforcement provided. If the values in Table 6 are used it will be found that the section will be congested with reinforcement. For practical purposes, therefore, it is better to have a maximum limit of say 0.5 times the values in Table 6 as a starting point.

The resistance to shear is normally provided in one of the following two ways:

(*a*) Vertical shear reinforcement, i.e. links or binders.
(*b*) Inclined shear reinforcement, i.e. bent up or inclined bars plus vertical shear reinforcement.

The total shear resistance is equal to the shear resistance of the plain concrete section and the shear resistance of the links plus the shear resistance of the inclined bars.

Shear resistance of links

When a reinforced concrete beam is broken in the laboratory under the action of shear it is difficult to determine precisely the angle of the shear crack. If it is assumed that the angle is 45° then the theoretical shear resistance calculated, as shown in Fig. 12, compares well with the experimental value.

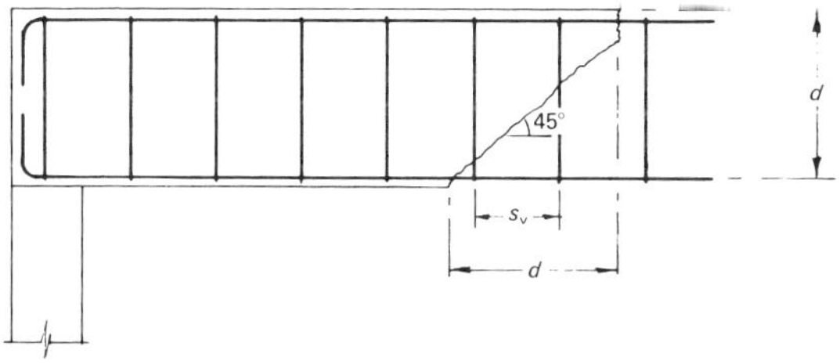

FIG. 12. *Calculation of shear resistance of links.*

The shear resistance due to vertical reinforcement is provided by the number of bars crossing the potential crack i.e. (d/s_ν).

Now, resolving the forces vertically, the shear resistance provided by the vertical links:

$$= \text{Area} \times \text{stress}$$

$$= \frac{d}{s_\nu} A_{sv} \frac{f_{y\nu}}{\gamma_m}$$

And with $\gamma_m = 1.15$ for reinforcement

$$= \frac{d}{s_\nu} A_{s\nu}\, 0.87 f_{y\nu}$$

For a beam, therefore, the total shear resistance will be provided by the addition of the shear resistance of the beam itself and that of the links, i.e.:

$$V = V \text{ for concrete} + V \text{ for links}$$

$$= \nu_c bd + \frac{d}{s_\nu} A_{s\nu}\, 0.87 f_{y\nu}$$

But

$$\nu = \frac{V}{bd} \qquad (16)$$

$$\therefore \quad \frac{V}{bd} = \nu = \nu_c + \frac{A_{s\nu}}{s_\nu b}\, 0.87 f_{y\nu}$$

$$\therefore \quad \frac{A_{sv}}{s_v} = \frac{(v - v_c)}{0.87 f_{yv}} b \qquad (17)$$

Equation 17 corresponds to equation (9) in CP110.

Notwithstanding the links calculated from equation 17 it must also be established that the minimum areas have been provided in accordance with clause 3.11.4.3. It should be noted that for the tension reinforcement to be considered in a shear calculation it must extend beyond the section a distance at least equal to the effective depth, except at an end support where all the reinforcement can be used provided the code requirements with regard to the detailing are met.

Generally throughout this book the links will be in mild steel having an $f_{yv} = 250\,\text{N/mm}^2$. Using mild steel ($f_y = 250\,\text{N/mm}^2$) allows the main reinforcement to be placed more accurately since the radius to which mild steel can be bent is two diameters as opposed to three diameters for high yield reinforcement.

Shear resistance of inclined or bent-up bars

It is not permissible, according to CP110, to have all the shear resistance provided by bent-up bars. Clause 3.3.6.1. states:

> Up to 50 per cent of the shear reinforcement may be in the form of bent up bars which should be assumed to form the tension members of one or more single systems of lattice girders in which the concrete forms the compression members.

This truss analogy is shown in Fig. 13.

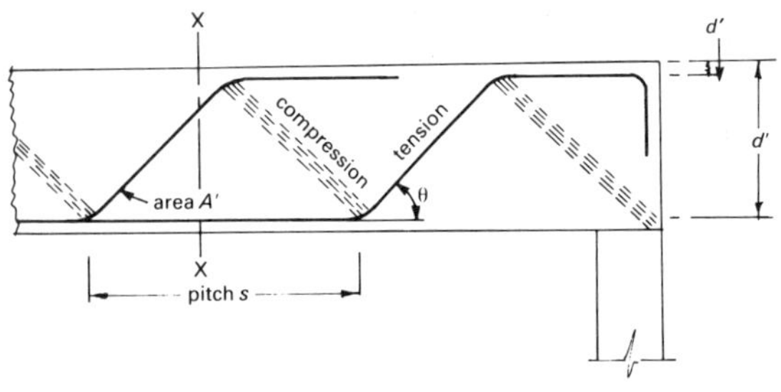

FIG. 13. *Truss analogy in inclined or bent-up bars.*

$$\text{The pitch } s = 2(d - d')\cot\theta \qquad (18)$$

Resolving vertically on the section X–X:

The shear resistance due to the inclined bars

$$= \frac{f_y}{\gamma_m} A \sin\theta$$

$$= 0.87 f_y A \sin\theta \qquad (19)$$

When the pitch is less than that in equation (18), the shear resistance is increased in inverse proportion (i.e. if the pitch is halved the shear resistance is doubled). The arrangement shown in Fig. 14 is called a single system, but if the spacing is halved it is called a double system. In a double system:

$$\text{Shear resistance} = 2 \times 0.87 f_y A \sin\theta$$

$$= 1.74 f_y A \sin\theta \qquad (20)$$

A common choice for θ is 45° and f_y would not be less than $425\,\text{N/mm}^2$ thus simplifying equations (19) and (20) to:

Shear resistance for a single system
$= 0.26A$ kN

and

The shear resistance for a double system
$= 0.52A$ kN

where A is the area of the reinforcement in mm^2.

The above formulae are for rectangular beams. However, the following points should be noted:

(*a*) For flanged beams only the rib is considered to resist the shear force.

(*b*) In a doubly reinforced beam the tension reinforcement only is considered when calculating ($100A_s/bd$) for Table 5.

Having dealt with the requirements for the ultimate limit states of bending and shear it remains to outline the requirements for the serviceability limit states of local bond, deflection and cracking. It is not intended to deal with these in the same detail as bending and shear but simply to state the requirements of CP110 which shall be illustrated in subsequent chapters.

BOND STRESSES

As stated previously one of the basic assumptions on which reinforced concrete theory is based is that there is good bond between the reinforcement and the concrete. In the present code, CP110, this must be checked. There are two locations which may give rise to problems (viz: the stress at a local position and the average stress along the bar).

Local bond stress

Local bond stresses are produced by large changes in tension over a short length of reinforcing bar. These changes in tension are produced by a change in bending moment and for equilibrium they must be balanced by the bond that exists between the reinforcement and the concrete. The rate of change of the bending moment is, in fact, the shear force. Thus CP110 gives the following as equation (42) simplified:

$$f_{bs} = \frac{V}{\Sigma U_s d} \qquad (21)$$

where f_{bs} = local bond stress; V = the shear force due to ultimate loads; ΣU_s = the sum of the effective perimeter of the tension reinforcement; and d = the effective depth of the section.

When f_{bs} is calculated from equation (21) it is compared with the permissible local bond stress from Table 21 in CP110. If the value from Table 21 is greater then the local bond stress is satisfactory.

Anchorage bond stress

The local bond stress discussed above checks that the bond between the concrete and the reinforcement is satisfactory at any section along the beam, whereas the anchorage bond stress checks that the average stress along the length of the bar is satisfactory (*see* Fig. 14).

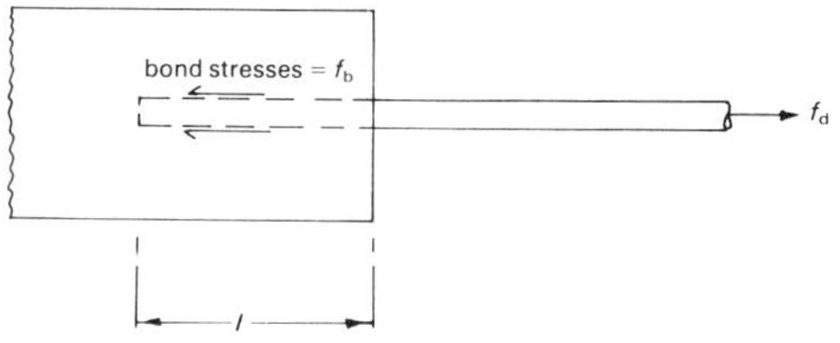

FIG. 14. *Use of anchorage bond stress.*

Considering the forces on the bar which must be in equilibrium:

The force causing

"pullout" = stress × cross-sectional area

$$= f_d \frac{\pi\phi^2}{4}$$

and with the ultimate tensile stress = $0.87 f_y$, this expression becomes:

$$\text{"pull-out" force} = 0.87 f_y \frac{\pi\phi^2}{4} \qquad (22)$$

The force resisting "pull out"

= anchorage bond stress × bond area

$$= f_b l \pi \phi \qquad (23)$$

For equilibrium, equation (22) = equation (23), i.e.

$$0.87 f_y \frac{\pi\phi^2}{4} = f_b l \pi \phi$$

$$\therefore \quad l = \frac{0.87 f_y \phi}{4 f_b}$$

Similarly for an anchorage in compression

$$l = \frac{0.72 f_y \phi}{4 f_b}$$

where in each case f_b is as given in Table 22 of CP110.

DEFLECTION

The method for checking deflection is as follows:

(*a*) Choose the basic (span/effective depth) ratio from Table 8 if the span is less than 10 m or Table 9 if the span is greater than 10 m and the engineer wishes to restrict the deflection to avoid damage to the finishes or partitions.

(*b*) This value is modified by a factor from Table 10, depending on the amount of tension reinforcement, and a factor from Table 11, depending on the amount of compression reinforcement.

(*c*) The modified ratio gives the permissible l/d ratio and when this is compared with the actual l/d ratio it can be seen whether or not deflection is satisfactory.

In the case of a flanged beam a further modification may have to be taken into account. This is because a flanged beam has a smaller second moment of area than a rectangular beam of the same overall dimensions. This modification factor is given in CP110 clause 3.3.8.3 and is shown graphically in Fig. 15.

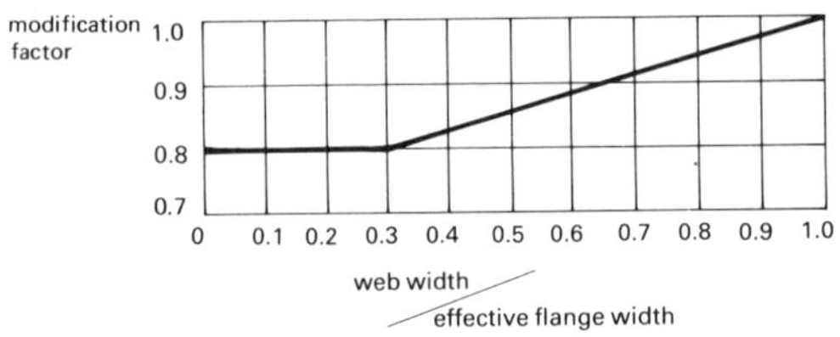

FIG. 15. *Modification factor for deflection in a flanged beam.*

Cracking

The method of controlling flexural cracking in a reinforced concrete beam is by ensuring that the spacing of the reinforcement in tension does not exceed stated maximum values. These values are given in clause 3.11.8.2 and Table 24 of CP110.

Occasionally there may be an advantage in calculating the actual crack width under service loads. A method for the calculation is given in Appendix A of CP110.

SOLID SLABS

The design of reinforced concrete solid slabs is basically the same as that outlined above for beams with respect to bending deflection and cracking. There are some differences in the rules for cracking but, as for beams, the requirements are based on bar spacing. However one difference does occur when a slab is subjected to shear.

Research has shown that wide, shallow members are relatively stronger in shear than the deep, narrow profiled beams. For this reason an enhancement factor ξ_s is applied to the permissible shear stress in slabs. Thus the permissible shear stress in a solid slab is given by the product of v_c from Table 5 and ξ_s from Table 14. Normal shear usually does not present a problem in solid slabs but where there is a heavy point load a problem may arise with punching shear. This is outlined in clause 3.4.5.2 of CP110 and will be further explained in Chapter Eight of this book when dealing with pad foundations.

SELF-ASSESSMENT QUESTIONS

1. Explain the relationship between bending moment and moment of resistance showing how the latter is developed in reinforced concrete.

2. Calculate the ultimate moment of resistance of the beam shown in Fig. 16. Use: (*a*) the concrete strength; and (*b*) the reinforcement strength.
Assume
$f_{cu} = 30\,\text{N/mm}^2$ and $f_y = 425\,\text{N/mm}^2$.

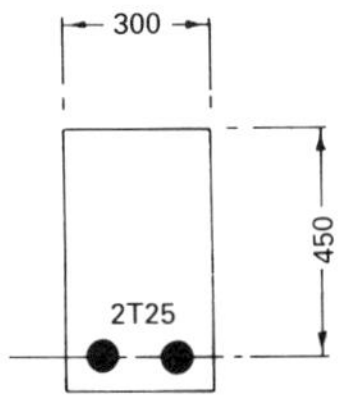

FIG. 16.

3. (*a*) Calculate the maximum bending moment that can be carried by the reinforced concrete beam shown in Fig. 17.
(*b*) How many 20 mm bars are required to support the bending moment calculated in (*a*)?
Assume
$f_{cu} = 25\,\text{N/mm}^2$ and $f_y = 425\,\text{N/mm}^2$.

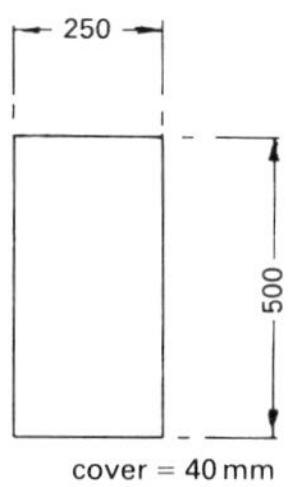

FIG. 17.

4. (*a*) Calculate the area of compressive reinforcement required to make the beam in Fig. 18 fully effective.
(*b*) Calculate its moment of resistance.

Assume
$f_{cu} = 30\,\text{N/mm}^2$ and $f_y = 425\,\text{N/mm}^2$.

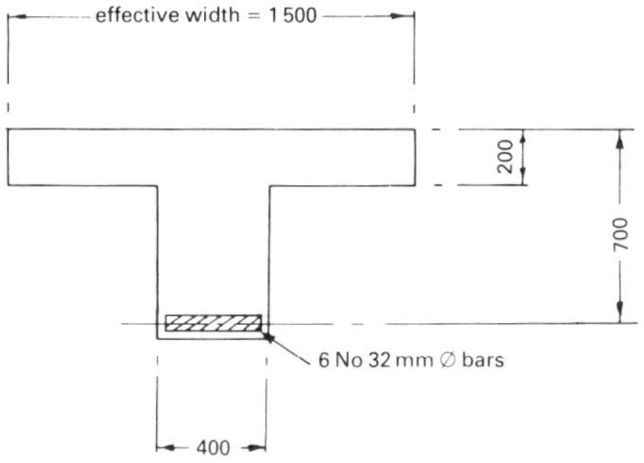

FIG. 18.

5. For the beam shown in Fig. 19 calculate the shear reinforcement necessary to resist a shear force of 550 kN. Assume $f_{cu} = 30\,\text{N/mm}^2$, $f_{yv} = 250\,\text{N/mm}^2$ and the tensile reinforcement consists of 4 T 25 bars, all fully anchored.

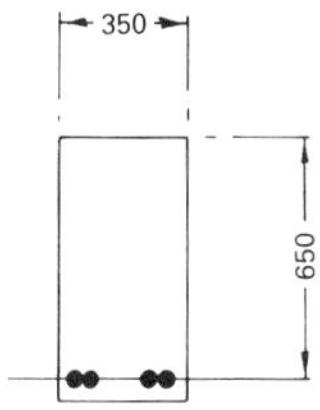

FIG. 19.

6. Outline the procedure for checking that the following are satisfactory in a slab: (*a*) deflection; (*b*) cracking; and (*c*) normal shear.

CHAPTER THREE

A Simply Supported, Rectangular, Reinforced Concrete Beam—Reinforced In Tension Only

CHAPTER OBJECTIVES

After studying this chapter you should be able to understand how a simply supported rectangular reinforced concrete beam, reinforced in tension only is:

* designed for the main reinforcement;
* checked for shear;
* checked for local bond;
* checked for deflection; and
* checked for cracking.

CURPE CONSULTANTS 46 Orburn Road DUNFIELD	Project DESIGN IV	Job ref CC/ME
		Sheet no 3/1
	Sub-section CHAPTER THREE S.S. BEAM - TENSION	Made by BC
		Date
	Structural Summary Sheet	Checked by
		Date

CHAPTER THREE

A SIMPLY SUPPORTED RECTANGULAR REINFORCED CONCRETE BEAM, REINFORCED IN TENSION ONLY

A reinforced concrete beam is required to span 6·50 m between the centres of supporting brick piers which are 300 mm wide. The beam carries a 1·75 m height of brickwork which consists of 2No. 100 mm skins with a 50 mm cavity between. The beam is to be designed to satisfy CP110 and the material strengths are as follows:-

f_{cu} (the characteristic strength of the concrete) = 25 N/mm²

f_y (the characteristic strength of the reinforcement)

= 425 N/mm² for the main reinforcement and

= 250 N/mm² for the links.

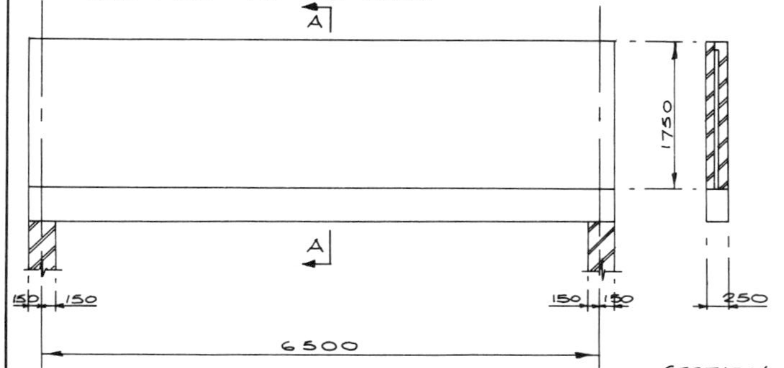

ELEVATION OF THE BEAM

SECTION A-A

CURPE CONSULTANTS
46 Orburn Road
DUNFIELD

Contract	Job ref
DESIGN IV	CC/ME
Part of structure: CHAPTER THREE S. S. BEAM - TENSION	Calc sheet No: 3/2
Drawing ref: — / Calculations by: B.C. / Checked by: —	Date: —

Code ref.	CALCULATIONS	OUTPUT
	TO DETERMINE THE BEAM DIMENSIONS	
table E	Depth : basic $\frac{\text{span}}{\text{eff. depth}} = 20$	
	$\therefore$ eff. depth $= \frac{6500}{20} = 325mm$	
table 19	Cover, for mild conditions of exposure, $= 20mm$	
	Assuming 20mm ϕ main bars and 8mm ϕ links, the overall depth	
	$'h' = 325 + 20 + 8 + \frac{20}{2}$	
	$= 363mm$	
	say $h = 375mm$	$h = 375mm$
	$\therefore d = 375 - (20 + 8 + 20/2)$	
	$= 337mm$	$d = 337mm$
	Let the breadth be determined by the brickwork dimension ie 250mm	$b = 250mm$
3.3.1.3	CHECK THE SLENDERNESS LIMITS	
	$60b_c = 60 \times 250 \times 10^{-3} = 15m$	
	$\frac{250\,b_c^2}{d} = \frac{250 \times 10^{-3}}{337} = 46.36m$	
	both greater than 6.20m	Slenderness Limits O.K.

CURPE CONSULTANTS
46 Orburn Road
DUNFIELD

Contract	Job ref
DESIGN IV	CC/ME
Part of structure: CHAPTER THREE S.S. BEAM - TENSION	Calc sheet No: 3/3
Drawing ref: — / Calculations by: B.C. / Checked by: —	Date: —

Code ref.	CALCULATIONS	OUTPUT
3.3.1.1.	THE EFFECTIVE SPAN	
	The calculations on page 3/2 were based on a span of 6.50m. Now that the effective depth is known this must be confirmed.	
	i.e. span is smaller of	
	(i) 6.50m	
	or (ii) $6.20 + 0.337 = 6.537m$	$\ell = 6.50m$
	LOADING - for a 1m length of beam	
BS648	Brickwork weighs say 55 kg/m^2 /25mm thickness	
	$\therefore$ Load from b'w'k $= \frac{75 \times 200 \times 55}{25} = 770\ kg/m$	
	Self weight $= 0.25 \times 0.375 \times 2400 = 225$	
	Total $g_k = 995\ kg/m$	
	$\therefore g_k = 995 \times 9.81 \times 10^{-3} = 9.76\ kN/m$	
	There is no imposed load $\therefore q_k = 0$	
2.3.3.1.	Design Load $F = 1.4\,G_k + 1.6\,Q_k$	
	$= (1.4 \times 9.76 + 1.6 \times 0)\,6.5$	
	$= 88.82kN$	$F = 88.82kN$
	BENDING MOMENT	
	$M_{max} = \frac{F\ell}{8} = \frac{88.82 \times 6.5}{8}$	
	$= 72.16\ kN.m$	$M = 72.16\ kN.m$

CURPE CONSULTANTS
46 Orburn Road
DUNFIELD

Contract	DESIGN IV	Job ref	CC/ME
Part of structure	CHAPTER THREE S.S. BEAM - TENSION	Calc sheet No	3/4
Drawing ref	Calculations by B.C.	Checked by	Date

Code ref.	CALCULATIONS	OUTPUT
	MAIN REINFORCEMENT	
	$\frac{M}{bd^2} = \frac{72.16 \times 10^6}{250 \times 337^2} = 2.54\ N/mm^2$	
chart 3	$\therefore \frac{100 A_s}{bd} = 0.81$	
	$\therefore A_s = \frac{0.81 \times 250 \times 337}{100}$	
	$= 682.43\ mm^2$	
	4T16 bars $= 804\ mm^2$	Main reinf = 4T16 bars
	When the effective depth was calculated the main bars were assumed to be 20mm ϕ. The fact that 16mm ϕ bars are being used changes the effective depth. The small difference of 2mm would be ignored in practice. The following calculation justifies this decision	
	New 'd' $= 375 - (20 + 8 + 16/2)$ $= 339\ mm$	
	$\frac{M}{bd^2} = \frac{72.16 \times 10^6}{250 \times 339^2} = 2.51\ N/mm^2$	
chart 3	$\frac{100 A_s}{bd} = 0.80$	
	$\therefore A_s = \frac{0.80 \times 250 \times 339}{100}$	
	$= 678\ mm^2$	
	ie 4T16 bars are still required	

CURPE CONSULTANTS
46 Orburn Road
DUNFIELD

Contract	DESIGN IV	Job ref	CC/ME
Part of structure	CHAPTER THREE S.S. BEAM - TENSION	Calc sheet No	3/5
Drawing ref	Calculations by B.C.	Checked by	Date

Code ref.	CALCULATIONS	OUTPUT
	Since 16 mm ϕ bars are to be used, the effective depth corresponding to these ie 339 mm, will be used in the remainder of this example.	
3.3.6.	CHECK SHEAR - Worst at support	
	Shear force $V = F/2$	
	$= \frac{88.82}{2} = 44.41\ kN$	
3.3.6.1.	Shear stress $v = \frac{V}{bd} = \frac{44.41 \times 10^3}{250 \times 339}$	
	$= 0.52\ N/mm^2$	$v = 0.52\ N/mm^2$
	The permissible shear stress v_c is obtained from table 5 depending on the concrete grade, ie f_{cu}, and the amount of tension reinforcement at the position under consideration.	
3.11.7.2	At the support only 50% of the mid span reinforcement will be available and anchored. ie 2T16 bars $= 402 mm^2$	
	At the support $\frac{100 A_s}{bd} = \frac{100 \times 402}{250 \times 339}$	
	$= 0.47$	
table 5	$f_{cu} = 25\ N/mm^2 \quad \therefore v_c = 0.48\ N/mm^2$	$v_c = 0.48\ N/mm^2$
	Since $v > v_c$ shear reinforcement must be designed.	
	SHEAR REINFORCEMENT	
eqn 9.	$\frac{A_{sv}}{s_v} \geq \frac{b(v - v_c)}{0.87 f_{yv}}$	

CURPE CONSULTANTS
46 Orburn Road
DUNFIELD

Contract	Job ref
DESIGN IV	CC/ME
Part of structure: CHAPTER THREE S.S. BEAM- TENSION	Calc sheet No: 3/6
Drawing ref: — Calculations by: B.C. Checked by: —	Date: —

Code ref.	CALCULATIONS	OUTPUT
	with 8 mm links $A_{sv} = 101 mm^2$	
	$\therefore\ S_v \leq \dfrac{101 \times 0.87 \times 250}{250\ (0.52 - 0.48)}$	
	$\leq 2196.75 mm$	
	max $S_v = 0.75\ d = 0.75 \times 339$	
	$= 254.25 mm$	
3.11.4.3	Nominal Links	
	$\dfrac{A_{sv}}{S_v} = 0.002\ b_t$	
	with $A_{sv} = 101 mm^2$	
	$S_v = \dfrac{101}{0.002 \times 250} = 202 mm$	
	$\therefore$ Use 8 mm ϕ links at 200 mm c/c	KB @ 200%
3.11.6.1	CHECK LOCAL BOND	
	The worst position will be at the support where the shear force is maximum and the number of bars a minimum.	
	$f_{bs} = \dfrac{V}{\Sigma U_s\ d} = \dfrac{44.41 \times 10^3}{2.\pi.16 \times 339}$	
	$= 1.30\ N/mm^2$	
table 2.	Permissible $f_{bs} = 2.5\ N/mm^2$	
	$\therefore$ SATISFACTORY	Local Bond O.K.

CURPE CONSULTANTS
46 Orburn Road
DUNFIELD

Contract	Job ref
DESIGN IV	CC/ME
Part of structure: CHAPTER THREE S.S. BEAM - TENSION	Calc sheet No: 3/7
Drawing ref: — Calculations by: B.C. Checked by: —	Date: —

Code ref.	CALCULATIONS	OUTPUT
	CHECK DEFLECTION	
table 8	basic $\ell/d = 20$	
	At mid span $\dfrac{100 A_s}{bd} = \dfrac{100 \times 804}{250 \times 339}$	
	$= 0.95$	
Eqn 13	Service stress $f_s = 0.58 \times \dfrac{f_y A_s req.}{A_s prov.} \times \dfrac{1}{\beta_b}$	
	$= 0.58 \times 425 \times \dfrac{678}{804} \times 1$	
	$= 207.87\ N/mm^2$	
table 10	By interpolation the Modification Factor	
	$= 1.14$	
	$\therefore$ Permissible $\ell/d = 20 \times 1.14$ $= 22.8$	
	Actual $\ell/d = \dfrac{6500}{339} = 19.17$	
	$\therefore$ SATISFACTORY	Deflection O.K
3.11.8.2.	CHECK CRACKING	
table 24	The maximum clear distance between the bars in tension = 180 mm	
	The worst tensile crack area will be in the centre of the span	

CURPE CONSULTANTS 46 Orburn Road DUNFIELD	Contract: DESIGN IV	Job ref: CC/ME
	Part of structure: CHAPTER THREE S.S. BEAM - TENSION	Calc sheet No: 3/8
	Drawing ref: / Calculations by: B.C. / Checked by:	Date:

Code ref.	CALCULATIONS	OUTPUT

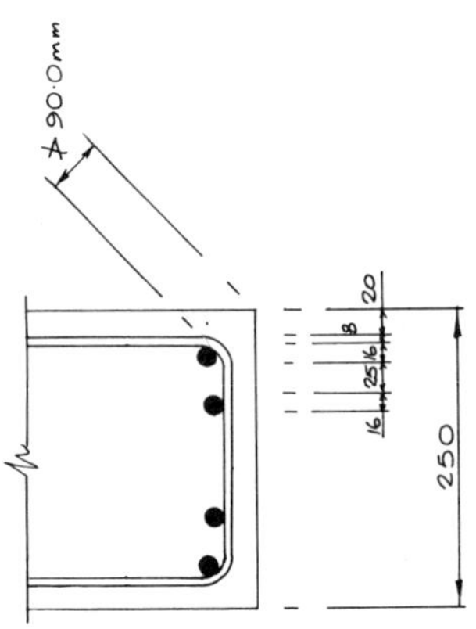

Actual clear distance

$= 250 - 2(16 + 25 + 16 + 8 + 20)$

$= 80\ mm$

i.e. $< 180\ mm$ ∴ SATISFACTORY

Also the maximum clear distance from the corner of the beam to the surface of the nearest longitudinal bar

$= 180/2 = 90{\cdot}0\ mm$

Actual distance $= \sqrt{28^2 + 28^2}$

$= 39{\cdot}60\ mm$

i.e. $< 90{\cdot}0\ mm$ ∴ SATISFACTORY

Cracking O.K.

SELF-ASSESSMENT QUESTIONS

1. Use the formula in CP110, outlined in Chapter Two, to design the main reinforcement for the beam in Chapter Three.

2. If the beam designed in Chapter Three spanned 7.0 m, use the same section to design the main reinforcement and check shear, deflection, local bond and cracking.

3. If a 15 kN point load, classified as dead load, was introduced in the centre of the span, re-design the beam in Chapter Three for bending and check for shear, deflection, local bond and cracking.

CHAPTER FOUR

A Simply Supported, Rectangular, Reinforced Concrete Beam—Reinforced In Tension and Compression

CHAPTER OBJECTIVES

After studying this chapter you should be able to understand how a simply supported rectangula- reinforced concrete beam, reinforced in tension and compression is:

* designed for the main reinforcement;
* checked for shear;
* checked for local bond;
* checked for deflection;
* checked for cracking.

CURPE CONSULTANTS 46 Orburn Road DUNFIELD	Project DESIGN IV	Job ref CC/ME
	Sub-section CHAPTER FOUR S.S. BEAM – TENS. & COMP.	Sheet no 4/1
		Made by BC
		Date
	Structural Summary Sheet	Checked by
		Date

CHAPTER FOUR

A SIMPLY SUPPORTED RECTANGULAR REINFORCED CONCRETE BEAM, REINFORCED IN TENSION AND COMPRESSION.

A reinforced concrete floor slab is introduced on top of the brickwork, in the beam designed in chapter three.

Section A-A on page 3/1 now becomes

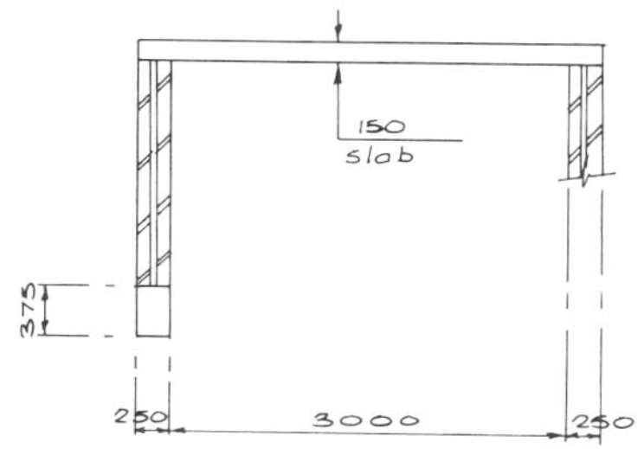

Assume : (i) There is a 25mm screed on top of the slab

(ii) The imposed load on the slab = 2.50 kN/m²

Project: DESIGN IV | Job ref: CC/ME
Part of structure: CHAPTER FOUR S.S. BEAM – TENS. & COMP. | Calc sheet no: 4 / 2 / rev
Drawing ref: | Calc by: B.C. | Date: | Check by: | Date:

Ref	Calculations	Output
	ADDITIONAL LOADING	
	Dead: Self weight of the slab	
	$= 2400 \times 0{\cdot}15 = 360\ kg/m^2$	
BS648	Screed weighs $29{\cdot}3\ kg/m^2/12{\cdot}7\ mm$ thickness	
	$\therefore$ 25mm screed $= \frac{29{\cdot}3 \times 25}{12{\cdot}7} = 57{\cdot}68$	
	Total $= 417{\cdot}68\ kg/m^2$	
	Convert to force units	
	$= 417{\cdot}68 \times 9{\cdot}81 \times 10^{-3} = 4{\cdot}10\ kN/m^2$	
	On a 1m span of the beam the additional dead load	
	$= 4{\cdot}10 \times (3/2 + 0{\cdot}25)$	
	$= 7{\cdot}18\ kN/m$	
	Imposed: On a 1m span of the beam	
	$= 2{\cdot}5\,(3/2 + 0{\cdot}25)$	
	$= 4{\cdot}375\ kN/m$	
2.3.3.1.	DESIGN LOAD	
	$F = 1{\cdot}4\,G_k + 1{\cdot}6\,Q_k$	
	$= [1{\cdot}4(7{\cdot}18 \times 9{\cdot}76) + 1{\cdot}6 \times 4{\cdot}375]\,6{\cdot}5$	
	$= 199{\cdot}65\ kN$	$F = 199{\cdot}65\ kN$

Project: DESIGN IV | Job ref: CC/ME
Part of structure: CHAPTER FOUR S.S. BEAM – TENS. & COMP. | Calc sheet no: 4 / 3 / rev
Drawing ref: | Calc by: B.C. | Date: | Check by: | Date:

Ref	Calculations	Output
	BENDING MOMENT	
	$M_{max} = \frac{Fl}{8} = \frac{199{\cdot}65 \times 6{\cdot}5}{8}$	
	$= 162{\cdot}22\ kN.m$	$M = 162{\cdot}22\ kN.m$
	MAIN REINFORCEMENT	
	$\frac{M}{bd^2} = \frac{162{\cdot}22 \times 10^6}{250 \times 337^2} = 5{\cdot}71\ N/mm^2$	
	This value cannot be read from chart 3 i.e. the beam must be doubly reinforced. In order to chose the design chart the value of d'/d must be known.	
	$d' = 20 + 8 + 20/2 = 38mm$	
	$\therefore \frac{d'}{d} = \frac{38}{337} = 0{\cdot}11$	
	$\therefore$ Use design chart No. 33.	
	It is possible to obtain an answer from charts 32 and 33 and to interpolate linearly for the required value of d'/d. However if the d'/d value is taken to the higher value obtainable, the answer will be conservative.	
chart 33	Assume $\frac{100A_s'}{bd} = 0{\cdot}9$	
	$\therefore A_s' = \frac{0{\cdot}9 \times 250 \times 337}{100}$	
	$= 758{\cdot}25\ mm^2$	
	Use 2T20 mm ϕ bars + 1T16 mm ϕ bar $= 829\ mm^2$	Top reinf.

Project	Job ref
DESIGN IV	CC/ME
Part of structure CHAPTER FOUR S.S. BEAM - TENS. & COMP.	Calc sheet no 4 / 4 rev /
Drawing ref — Calc by B.C. — Date	Check by — Date

Ref	Calculations	Output
	$\therefore \frac{100A_s'}{bd} = \frac{100 \times 829}{250 \times 337}$	
	$= 0.984$	
	From chart 33, with $\frac{100A_s'}{bd} = 0.984$	
	$\frac{100A_s}{bd} = 1.85$	
	$A_s = \frac{1.85 \times 250 \times 337}{100}$	
	$= 1558.63\ mm^2$	
	USE 2T25 mm ϕ bars + 2T20 mm ϕ bars = $1610 mm^2$	Btm. reinf.
3.3.6	CHECK SHEAR - Worst at the support	
	Shear force $V = F/2$	
	$= \frac{199.65}{2} = 99.83 kN$	
3.3.6.1.	Shear stress $v = \frac{V}{bd} = \frac{99.83 \times 10^3}{250 \times 337}$	
	$= 1.18\ N/mm^2$	$v = 1.18\ N/mm^2$
3.11.7.2.	Assume 2T25 bars are carried to the support	
	$\therefore \frac{100A_s}{bd} = \frac{100 \times 982}{250 \times 337}$	
	$= 1.17$	
table 5	with $f_{cu} = 25\ N/mm^2$	
	$v_c = 0.68\ N/mm^2$	$v_c = 0.68\ N/mm^2$

Project	Job ref
DESIGN IV	CC/ME
Part of structure CHAPTER FOUR S.S. BEAM - TENS. & COMP.	Calc sheet no 4 / 5 rev /
Drawing ref — Calc by B.C. — Date	Check by — Date

Ref	Calculations	Output
	Since $v > v_c$ shear reinforcement must be designed.	
eq^n 9	$\frac{A_{sv}}{s_v} \geq \frac{b(v - v_c)}{0.87 f_{yv}}$	
	with 8 mm ϕ links $A_{sv} = 101 mm^2$	
	$\therefore s_v = \frac{101 \times 0.87 \times 250}{250(1.18 - 0.68)}$	
	$= 175.74\ mm$	
	max $s_v = 0.75d = 0.75 \times 337$	
	$= 252.75\ mm$	
3.11.4.3.	Nominal Links	
	$\frac{A_{sv}}{s_v} = 0.002 b_t$	
	with $A_{sv} = 101 mm^2$	
	$s_v = \frac{101}{0.002 \times 250} = 202\ mm$	
	Also for compression reinforcement	
	minimum size of link = $\frac{1}{4} \times 20 \doteqdot 6 mm\ \phi$	
	maximum spacing $= 12 \times 16 = 192 mm$	
	$\therefore$ USE DESIGNED LINKS ie	
	8 mm ϕ links at 175 mm c/c	R8 @ 175 c/c

Project	Job ref
DESIGN IV	CC/ME
Part of structure CHAPTER FOUR SS. BEAM - TENS. & COMP.	Calc sheet no 4 / 6 / rev
Drawing ref / Calc by B.C. / Date	Check by / Date

Ref	Calculations	Output
3.11.6.1.	CHECK LOCAL BOND	
	The worst position will be at the support where the shear force is a maximum and the number of bars a minimum.	
	$f_{bs} = \frac{V}{\Sigma U_s d} = \frac{99.83 \times 10^3}{2.\pi.25 \times 337}$	
	$= 1.89\ N/mm^2$	
table 21	Permissible $f_{bs} = 2.5\ N/mm^2$	
	∴ SATISFACTORY	Local Bond O.K.
	CHECK DEFLECTION	
table 8	basic $l/d = 20$	
	At mid span $\frac{100 A_s}{bd} = \frac{100 \times 1610}{250 \times 337}$	
	$= 1.91$	
Eqⁿ 13	Service stress $f_s = 0.58 \times 425 \times \frac{1558}{1610} \times 1$	
	$= 238.54\ N/mm^2$	
table 10	By interpolation the Modification Factor $= 0.89$	
table 11	With $\frac{100 A_s'}{bd} = 0.984$ the Modification	
	Factor $= 1.25$	
	∴ Permissible $l/d = 20 \times 0.89 \times 1.25$	
	$= 22.25$	

Project	Job ref
DESIGN IV	CC/ME
Part of structure CHAPTER FOUR S.S. BEAM - TENS. & COMP.	Calc sheet no 4 / 7 / rev
Drawing ref / Calc by B.C. / Date	Check by / Date

Ref	Calculations	Output
	Actual $l/d = \frac{6500}{337} = 19.29$	
	∴ SATISFACTORY	Deflection O.K.
3.11.8.2.	CHECK CRACKING	
table 24	The maximum clear distance between the bars in tension $= 180\ mm$	
	At mid span the actual clear distance $=$	
	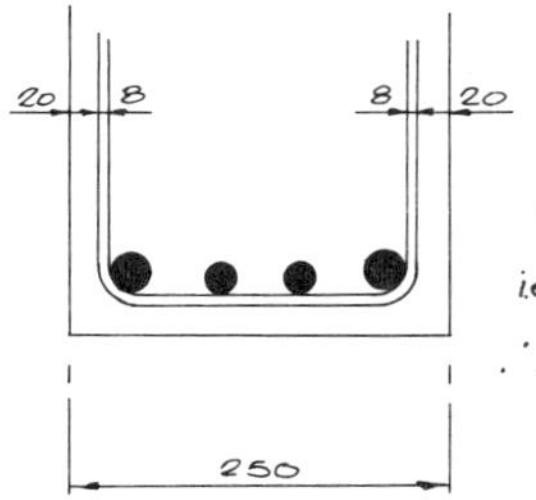	
	$\frac{1}{3}\left[250 - 2(20+8) - 2(20+25)\right]$	
	$= 34.67\ mm$	
	ie. $< 180mm$	
	∴ SATISFACTORY	
	Also the maximum clear distance from the corner of the beam to the surface of the nearest longitudinal bar $= 180/2 = 90mm$	
	The actual distance $= \sqrt{28^2 + 28^2}$	
	$= 39.60\ mm$	
	∴ SATISFACTORY	Cracking O.K.

SELF-ASSESSMENT QUESTIONS

1. Use the formula in CP110, outlined in Chapter Two, to design the main reinforcement for the beam in Chapter Four.

2. If the beam designed in Chapter Four spanned 7.0 m, use the same section to design the main reinforcement and check shear, deflection, local bond and cracking.

3. If two 15 kN point loads, classified as imposed load, were introduced at the third points, re-design the beam in Chapter Four for bending and check for shear, deflection, local bond and cracking.

CHAPTER FIVE

A Simply Supported Reinforced Concrete, Flanged Beam

CHAPTER OBJECTIVES

After studying this chapter you should be able to understand how a simply supported reinforced concrete flanged beam is:

* designed for the main reinforcement;
* checked for shear;
* checked for local bond;
* checked for deflection; and
* checked for cracking.

CURPE CONSULTANTS 46 Orburn Road DUNFIELD	Project: DESIGN IV Sub-section: CHAPTER FIVE S.S. FLANGED BEAM Structural Summary Sheet	Job ref: CC/ME Sheet no: 5/1 Made by: BC Date: Checked by: Date:

CHAPTER FIVE

A SIMPLY SUPPORTED FLANGED BEAM

The reinforced concrete floor slab introduced in chapter four is now positioned so as to come into the side of the beam.

The section shown on page 4/1 now becomes

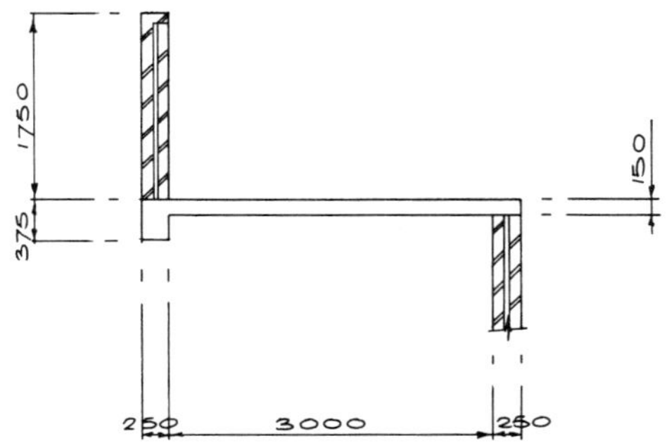

Project DESIGN IV			Job ref CC/ME	
Part of structure CHAPTER FIVE S.S. FLANGED BEAM			Calc sheet no 5/2	rev 1
Drawing ref	Calc by B.C.	Date	Check by	Date

Ref	Calculations	Output
	The loading and the maximum bending moment will be as calculated in chapter IV i.e. $F = 199 \cdot 65\,kN$	
	$M = 162 \cdot 22\,kN.m.$	
	Due to the position of the slab, there will be a part of the slab which will act in conjunction with the beam to form a flanged beam.	
3.3.1.2.	EFFECTIVE FLANGE WIDTH	
	lesser of (i) $250 + \frac{6500}{10} = 900mm$	
	or (ii) $3000/2 + 250 = 1750mm$	
	$\therefore b = 900mm$	$b = 900mm$
	MAIN REINFORCEMENT	
	$\frac{M}{bd^2} = \frac{162 \cdot 22 \times 10^6}{900 \times 337^2} = 1 \cdot 59\,N/mm^2$	
chart 3	$\frac{100A_s}{bd} = 0 \cdot 47$	
	$\therefore A_s = \frac{0 \cdot 47 \times 900 \times 337}{100}$	
	$= 1425 \cdot 5\,mm^2$	
	$\therefore$ USE 3T25mm ϕ bars $= 1470\,mm^2$	Btm reinf.

Project DESIGN IV			Job ref CC/ME	
Part of structure CHAPTER FIVE S.S. FLANGED BEAM			Calc sheet no 5/3	rev 1
Drawing ref	Calc by B.C.	Date	Check by	Date

Ref	Calculations	Output
	Typical Doubly Reinforced Beam Chart	

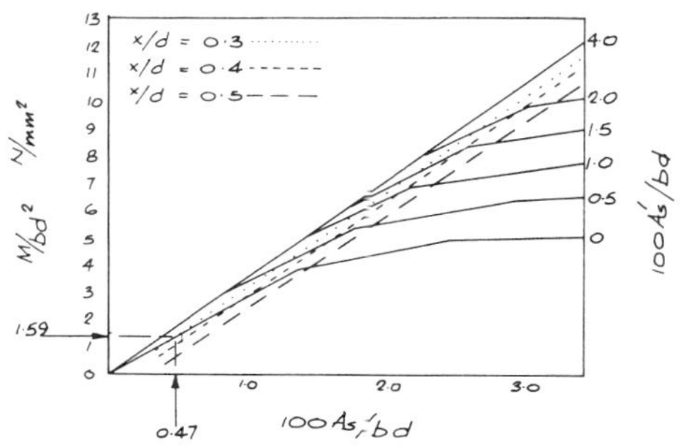

Ref	Calculations	Output
	From the above chart it can be clearly observed that the position read to obtain $100A_s/bd$ lies to the left of the line $x/d = 0 \cdot 3$	
	ie $x/d < 0 \cdot 3$	
	$\therefore x < 0 \cdot 3 \times 337$	
	$< 101 \cdot 1\,mm$	
	i.e. the neutral axis lies within the flange	
	Therefore the use of the charts, which are for rectangular beams only, is justified.	

Project: DESIGN IV		Job ref: CC/ME
Part of structure: CHAPTER FIVE S.S. FLANGED BEAM		Calc sheet no: 5 / 4 rev: /
Drawing ref:	Calc by: B.C. Date:	Check by: Date:

Ref	Calculations	Output
3.3.6	CHECK SHEAR - Worst at the support	
	Shear force $V = F/2$	
	$= \frac{199.65}{2} = 99.83\,kN$	
3.3.6.1.	shear stress $v = V/bd$	
	$= \frac{99.83 \times 10^3}{250 \times 337} = 1.18\,N/mm^2$	$v = 1.18\,N/mm^2$
	N.B. b for shear in a flanged beam is the rib width.	
3.11.7.2.	Assume 2T25 mm ϕ bars are carried to the support	
	$\therefore \frac{100A_s}{bd} = \frac{100 \times 982}{250 \times 337}$	
	$= 1.17$	
table 5	with $f_{cu} = 25\,N/mm^2$, $v_c = 0.68\,N/mm^2$	$v_c = 0.68\,N/mm^2$
	Since $v > v_c$ shear reinforcement must be designed.	
	The calculation for shear will be similar to that shown on page 4/5.	
	Also the calculation for nominal links will be similar except for the rules which apply to the compression reinforcement	
	$\therefore$ USE 8mm ϕ links at 175mm c/c	R8 @ 175 c/c

Project: DESIGN IV		Job ref: CC/ME
Part of structure: CHAPTER FIVE S.S. FLANGED BEAM		Calc sheet no: 5 / 5 rev: /
Drawing ref:	Calc by: B.C. Date:	Check by: Date:

Ref	Calculations	Output
3.11.6.1.	CHECK LOCAL BOND - Worst at the support	
	$f_{bs} = \frac{V}{\Sigma u_s d} = \frac{99.83 \times 10^3}{2.\pi.25 \times 337}$	
	$= 1.89\,N/mm^2$	
table 21	Permissible $f_{bs} = 2.5\,N/mm^2$	
	$\therefore$ SATISFACTORY	Local Bond O.K.
	CHECK DEFLECTION	
table 8	basic $l/d = 20$	
	At mid span $\frac{100A_s}{bd} = \frac{100 \times 1470}{900 \times 337}$	
	$= 0.48$	
Eq^n 13	Service stress $f_s = \frac{0.58 f_y A_{s\,req}}{A_{s\,prov}} \times \frac{1}{\beta_b}$	
	$= \frac{0.58 \times 425 \times 1425.5}{1470} \times 1$	
	$= 239.04\,N/mm^2$	
table 10	By interpolation the Modification	
	Factor = 1.26	
3.3.8.2.	Additional Modification Factor for a flanged beam	
	$0.3 \times$ effective flange width $= 0.3 \times 900$	
	$= 270\,mm$	

Project	DESIGN IV	Job ref	CC/ME
Part of structure	S.S. FLANGED BEAM CHAPTER FIVE	Calc sheet no	5/6 rev
Drawing ref		Calc by	B.C.
		Date	
		Check by	
		Date	

Ref	Calculations	Output
	i.e. > web width of 250 mm	
	∴ Additional factor = 0.8	
	∴ Permissible l/d = 20 × 1.26 × 0.8 = 20.16	
	Actual l/d = $\frac{6500}{337}$ = 19.29	
	∴ SATISFACTORY	Deflection O.K.
3.11.8.2	CHECK CRACKING	
table 24	The maximum clear distance between the bars in tension = 180 mm.	
	At mid span the actual clear distance	
	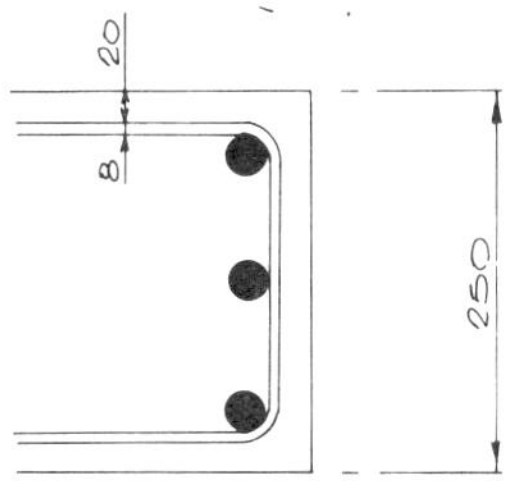	
	$= \frac{1}{2}[250 - 2(8+20) - (3 \times 25)]$	
	= 59.5 mm	
	ie < 180 mm	
	∴ SATISFACTORY	
	Also the maximum clear distance from the corner of the beam to the surface of the nearest longitudinal bar = 180/2 = 90 mm	
	As page 47 ∴ SATISFACTORY	Cracking O.K.

SELF-ASSESSMENT QUESTIONS

1. A floor slab is 160 mm thick and it spans 3.50 m from a brick wall to a beam which is 300 mm wide. The characteristic loads from the slab to the beam are:

(*a*) imposed = 6.50 kN/m;
(*b*) dead = 12.50 kN/m.

If high yield steel is to be used for the main reinforcement, mild steel for the links, and 30 grade concrete, use CP110 to design the beam, carrying out whatever checks are necessary.

CHAPTER SIX

A Simply Supported Reinforced Concrete Slab

CHAPTER OBJECTIVES

After studying this chapter you should be able to understand how a simply supported reinforced concrete slab is:

* designed for the main reinforcement;
* designed for secondary reinforcement;
* checked for shear;
* checked for local bond;
* checked for deflection; and
* checked for cracking.

CURPE CONSULTANTS 46 Orburn Road DUNFIELD	Project DESIGN IV	Job ref CC/ME
		Sheet no 6/1
	Sub-section CHAPTER SIX S. S. SLAB	Made by BC
		Date
	Structural Summary Sheet	Checked by
		Date

CHAPTER SIX

A SIMPLY SUPPORTED REINFORCED CONCRETE SLAB

A reinforced concrete slab is required to span between supporting brickwalls as shown below. The slab is to be designed to satisfy CP 110 with the loads and material strengths as follows:-

f_{cu} (the characteristic strength of the concrete) = 25 N/mm²

f_y (the characteristic strength of the reinforcement) = 460 N/mm²

Imposed loading on the slab = 2.50 kN/m²

Finishes and partitions on the slab = 1.75 kN/m²

The density of reinforced concrete = 2400 kg/m³

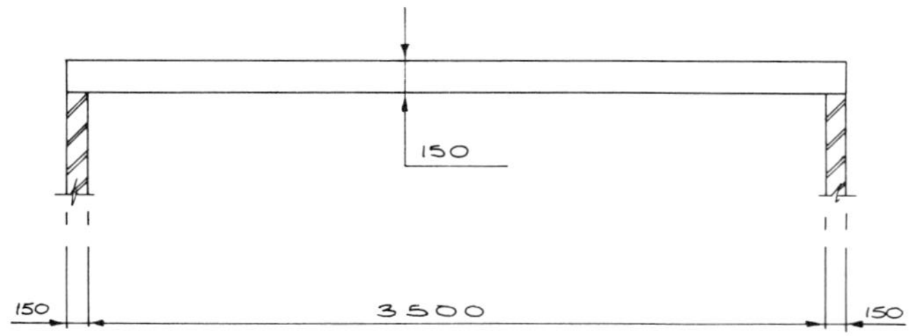

SECTION THROUGH THE SLAB

CURPE CONSULTANTS
46 Orburn Road
DUNFIELD

Contract	Job ref	Part of structure	Calc sheet No	Drawing ref	Calculations by	Checked by	Date
DESIGN IV	CC/ME	CHAPTER SIX S.S. SLAB	6/2		B. C.		

Code ref.	CALCULATIONS	OUTPUT
	EFFECTIVE DEPTH	
table 19	cover, for mild conditions of exposure,	
	= 20 mm	
	assume 12 mm ϕ main bars	
	$\therefore d = 150 - (20 + 12/2)$	
	= 124 mm	d = 124 mm
	EFFECTIVE SPAN	
3.3.1.1.	Smaller of 3500 + 150 = 3650 mm	
	or 3500 + 124 = 3624 mm	
	$\therefore$ span = 3624 mm	ℓ = 3624 mm
	LOADING	
	Dead: Finishes & partitions = 1·75 kN/m^2	
	S.Wt = $0{\cdot}15 \times 2400 \times 9{\cdot}81 \times 10^{-3}$ = 3·53 "	
	$\therefore$ Total g_k = 5·28 kN/m^2	
	Imposed: q_k = 2·50 kN/m^2	
2.3.3.1.	Design load $F = 1{\cdot}4\,G_k + 1{\cdot}6\,Q_k$	
	$= (1{\cdot}4 \times 5{\cdot}28 + 1{\cdot}6 \times 2{\cdot}50)\,3{\cdot}624$	
	= 41·28 kN	F = 41·28 kN
	BENDING MOMENT	
	$M_{max} = \frac{F\ell}{8} = \frac{41{\cdot}28 \times 3{\cdot}624}{8}$	
	= 18·70 kN.m.	M = 18·70 kN m

CURPE CONSULTANTS
46 Orburn Road
DUNFIELD

Contract	Job ref	Part of structure	Calc sheet No	Drawing ref	Calculations by	Checked by	Date
DESIGN IV	CC/ME	CHAPTER SIX S.S. SLAB	6/3		B.C.		

Code ref.	CALCULATIONS	OUTPUT
	MAIN REINFORCEMENT	
	$\frac{M}{bd^2} = \frac{18{\cdot}70 \times 10^6}{10^3 \times 124^2} = 1{\cdot}21\ N/mm^2$	
chart 4	$\frac{100A_s}{bd} = 0{\cdot}32$	
	$\therefore A_s = \frac{0{\cdot}32 \times 10^3 \times 124}{100} = 396{\cdot}8\ mm^2/m$	
	T12 mm ϕ bars at 250 mm c/c ($452\ mm^2/m$)	Main Reinf. T12 @ 250 c/c
3.11.4.1.	CHECK MINIMUM AREA OF MAIN REINFORCEM'T	
	$= 0{\cdot}15\%\ b_t d$	
	$= \frac{0{\cdot}15 \times 10^3 \times 124}{10^2} = 186\ mm^2/m$	
	$< 452\ mm^2/m$ provided	
	$\therefore$ SATISFACTORY	
3.11.4.2.	MINIMUM AREA OF SECONDARY ie DISTRIBUTION, REINFORCEMENT	
	$= 0{\cdot}12\%\ b_t h$	
	$= \frac{0{\cdot}12 \times 10^3 \times 150}{10^2} = 180\ mm^2/m$	
	T10 mm ϕ bars at 400 mm c/c ($196{\cdot}5\ mm^2/m$)	Sec. Reinf. T10 @ 400 c/c
	Spacing $\not>$ 5d ie 5 × 124 = 620 mm	
	$\therefore$ SATISFACTORY	

CURPE CONSULTANTS
46 Orburn Road
DUNFIELD

Contract: DESIGN IV | Job ref: CC/ME
Part of structure: CHAPTER SIX S.S. SLAB | Calc sheet No: 6/4
Drawing ref: | Calculations by: B.C. | Checked by: | Date:

Code ref.	CALCULATIONS	OUTPUT
3.4.6.	CHECK DEFLECTION - worst at mid span	
table 8	basic $l/d = 20$	
	$\frac{100A_s}{bd} = \frac{100 \times 452}{10^3 \times 124} = 0.36$	
eq^n 13	$f_s = 0.58 f_y \frac{A_s\,req}{A_s\,prov} \times \frac{1}{\beta_b}$	
	$= 0.58 \times 460 \times \frac{396.8}{452} \times \frac{1}{1}$	
	$= 234.22\ N/mm^2$	
table 10	Modification Factor, by interpolation, $= 1.47$	
	∴ Permissible $l/d = 20 \times 1.47 = 29.4$	
	Actual $l/d = \frac{3624}{124} = 29.2$	
	∴ SATISFACTORY	Deflection O.K.
3.4.5	CHECK SHEAR - worst at support	
	N.B. Under normal conditions of loading shear will seldom prove to be a problem in a solid slab. However the following illustrates the typical calculations to be done if deemed necessary. In the case of heavy point loads punching shear should be checked - this is dealt with in chapter eight under pad foundations.	

CURPE CONSULTANTS
46 Orburn Road
DUNFIELD

Contract: DESIGN IV | Job ref: CC/ME
Part of structure: CHAPTER SIX S.S. SLAB | Calc sheet No: 6/5
Drawing ref: | Calculations by: B.C. | Checked by: | Date:

Code ref.	CALCULATIONS	OUTPUT
	Maximum shear force $'V' = F/2$	
	$= 41.28/2$	
	$= 20.64\ kN$	
eq^n 18	Shear stress $v = \frac{V}{bd}$	
	$= \frac{20.64 \times 10^3}{10^3 \times 124}$	
	$= 0.17\ N/mm^2$	
3.11.7.3	At the support $\frac{100A_s}{bd} = \frac{100 \times 452}{10^3 \times 124} \times \frac{1}{2}$	
	$= 0.18$	
table 5	the value of $v_c = 0.35\ N/mm^2$, with $f_{cu} = 25$	
table 14	$\xi_s = 1.30$	
	Permissible shear stress $= v_c\,\xi_s$	
	$= 0.35 \times 1.30$	
	$= 0.46\ N/mm^2$	
	ie $> v = 0.17\ N/mm^2$	
	SATISFACTORY	shear O.K.
3.11.8.2	CHECK CRACKING	
	$h = 150mm$ ie $< 200mm$, therefore no check is required, but the max. clear distance between main bars $= 3 \times$ effective depth $= 3 \times 124 = 372mm > 238\ mm$ used	
	∴ SATISFACTORY	cracking O.K.

SELF-ASSESSMENT QUESTIONS

1. If the slab designed in Chapter Six spanned 3.80 m, use the same depth to design the main reinforcement in the slab, and check for deflection. If the slab failed in deflection, what remedial action could be taken?

2. If a slab spans 4.0 m between the centres of supporting brick walls, which are 250 mm thick, use CP110 and the following information to design all the reinforcement and carry out the necessary checks: Assume:

(*a*) the reinforcement is of high yield steel;
(*b*) the concrete is grade 30;
(*c*) the imposed loading on the slab = 4.00 kN/m^2;
(*d*) the dead load, excluding self weight = 2.00 kN/m^2; and
(*e*) the conditions of exposure are mild.

CHAPTER SEVEN

Short Reinforced Concrete Columns

CHAPTER OBJECTIVES

After studying this chapter you should know:
* how a column fails;
* the difference between a short and a slender column;
* the difference between a braced and an unbraced column;
* how the design formulae have been derived;
* how to use the design formulae to design a column;
* how to use the design charts to design a column;
* why links are required in a column; and
* how to design the links.

DEFINITION

CP110 does not define a column, but in clause 3.8.1.1 it defines a wall as a "vertical load bearing member whose greater lateral dimension is more than four times its least lateral dimension." By implication, therefore, a column may be defined as a vertical load-bearing member whose greatest lateral dimension is not more than four times its least lateral dimension.

FORMS OF FAILURE

One of three forms of failure may be observed in compression members (i.e. columns):

(*a*) If it is short and stubby, with respect to its cross section, it will fail by the compressive failure of the materials.
(*b*) If it is long and slender, with respect to its cross section it will fail by buckling.
(*c*) Failure is commonly due to a combination of (*a*) and (*b*).

Factors influencing failure

Of the many factors which influence the failure of a column, the following have the major effect:

(*a*) the strength of the materials;
(*b*) the elasticity of the materials;
(*c*) the second moment of areas of the cross section of the member;
(*d*) the length of the column;
(*e*) the restraints of the column ends.

Slenderness ratio

When the materials have been chosen (*a*) and (*b*) are known and the factors (*c*) (*d*) and (*e*) can be grouped together under the expression *slenderness ratio*. This may be defined as the effective height of the member divided by the breadth. In practice it is necessary to consider *two* slenderness ratios (*see* Fig. 20), namely those which correspond to the principal axes of cross-section i.e.:

$$\frac{l_{ex}}{h} \quad \text{and} \quad \frac{l_{ey}}{b}$$

These symbols are defined in clause 3.5.1.2.

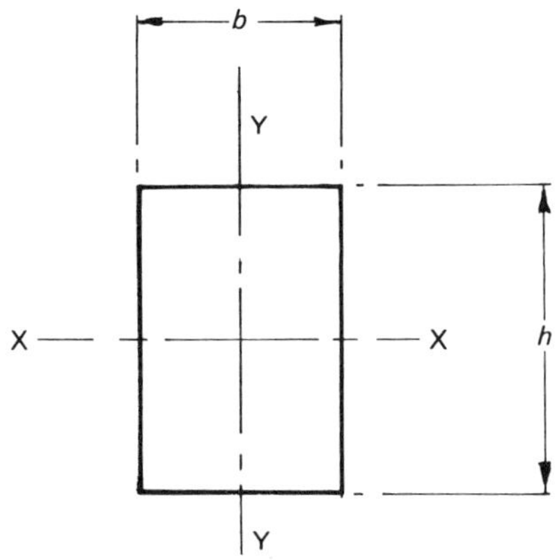

FIG. 20. *Two slenderness ratios.*

Short and slender columns (3.5.1.2)

The two ratios above, (l_{ex}/h) and (l_{ey}/b), are used to determine whether a column is short or slender. A column is said to be short when both ratios are less than 12. It should otherwise be considered as slender. This chapter deals only with short columns.

In order to determine whether a column is short or slender the effective length must be known. Guidance can be obtained from CP110 Table 15, but for the inexperienced designer this can be difficult to interpret. As an alternative use equations (20)–(23) in CP110. The former two will give the effective height of a braced column and the latter two that of an unbraced column. These equations are based on the relative beam and column stiffnesses.

Braced and unbraced columns (3.5.1.3)

It is important that the difference between braced and unbraced columns is appreciated, since if an unbraced column is designed as a braced column serious consequences and even failure, could result. This is illustrated as follows: The basic formula for short-column design is that of Euler which states that the load:

$$P = \frac{\pi^2 EI}{l^2}$$

where E = Young's modulus of elasticity; I = the second moment of area; and l = the effective length. For a given column E, I and π are constant. However, l for a braced column, which is properly restrained in direction at both ends, is equal to $0.75\,l_o$ (from CP110 Table 15). For a similar unbraced column the value is nearer $1.5\,l_o$. Since the Euler formula has l^2, the difference in the load carrying ability of an unbraced designed as a braced column is a factor of 4 (i.e. it will have about $\frac{1}{4}$ of the strength it should have and collapse is likely).

A *braced column* is one which is part of a braced frame (i.e. a frame which is not designed to resist lateral load). In such a frame all lateral loading is transmitted, by the floors, to shear walls, cores or bracing.

An *unbraced column* is one which is part of an unbraced frame (i.e. a frame which is designed to resist lateral loading). Such a frame is free to deflect laterally under load and resistance is supplied by the frame action of the structure.

SHORT COLUMN DESIGN

Short columns are divided into three categories by CP110. These are:

(*a*) short-braced axially loaded columns (3.5.3);

(*b*) short-braced columns supporting an approximately symmetrical arrangement of beams (3.5.4); and

(*c*) short columns resisting moments and axial forces (3.5.5 and 3.5.6).

Derivation of design formulae for (*a*) and (*b*) above

Consider a column having a cross-sectional area of concrete A_c with a total area of symmetrically placed reinforcement A_{sc} (*see* Fig. 21).

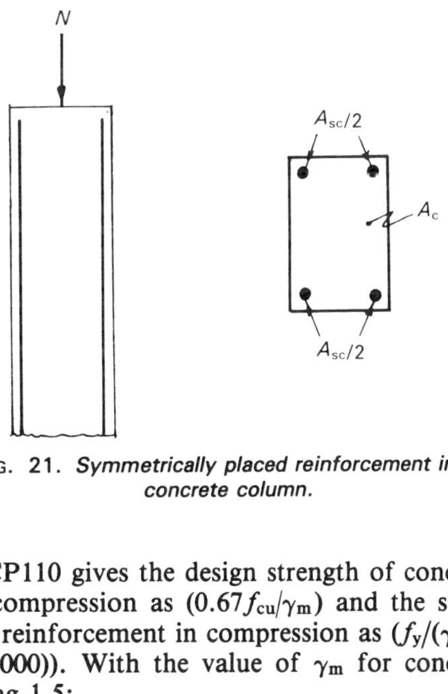

FIG. 21. *Symmetrically placed reinforcement in a concrete column.*

CP110 gives the design strength of concrete in compression as $(0.67 f_{cu}/\gamma_m)$ and the stress for reinforcement in compression as $(f_y/(\gamma_m + f_y/2000))$. With the value of γ_m for concrete being 1.5:

$$\text{Concrete stress} = \frac{0.67 f_{cu}}{1.5} = 0.45 f_{cu}$$

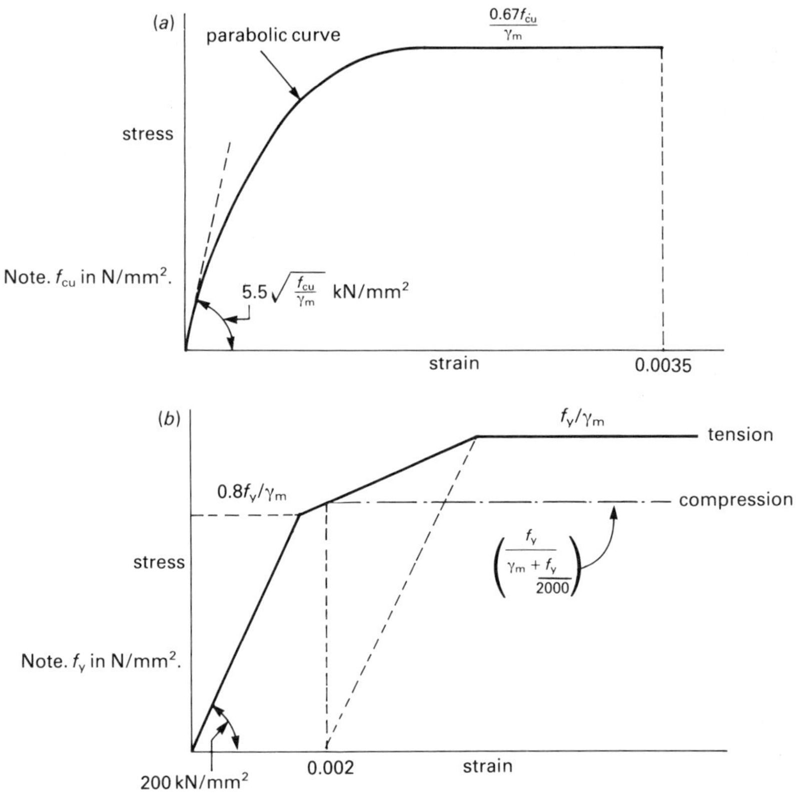

FIG. 22. *Short term design stress–strain curve:* (a) *for normal weight concrete; and* (b) *for reinforcement.*

With the value of γ_m for reinforcement being 1.15 and the value of f_y being 425 N/mm^2.

$$\text{Reinforcement stress} = \frac{f_y}{1.15 + \frac{425}{2000}} = 0.74f_y$$

Now the ultimate axial load N which can be carried by the column is the load carried by the concrete plus the load carried by the reinforcement i.e.:

$$N = 0.45f_{cu}A_c + 0.74f_yA_{sc} \qquad (24)$$

However, in practice—due to constructional inaccuracies—it is impossible to ensure an exact axial load. To cater for these inaccuracies clause 3.5.3 (i.e. (*a*) above) has reduced the stresses in equation (24) by 10% giving:

$$N = 0.4f_{cu}A_c + 0.67f_yA_{sc} \qquad (25)$$

Equation 25 corresponds to equation (25) in CP110.

It should be noted that such restrictions as are necessary to allow this formula to be used are rarely found and could only be expected where exceptional control and supervision are available.

Category (*b*) above lends itself to the internal columns of a multi-storey structure, clause 3.5.4 giving the following parameters:

(*a*) The beams are designed for uniformly distributed loads.

(*b*) The beam spans do not differ by more than 15 per cent of the longest.

These again can be difficult to achieve in practice but by reducing the stresses in equation (25) by a further 10 per cent a fair approximation can be made, i.e.:

$$N = 0.35 f_{cu} A_c + 0.60 f_y A_{sc} \qquad (26)$$

Equation 26 corresponds to equation (26) in CP110.

Use of design formulae for (*a*) and (*b*) above

Example 1

Calculate the ultimate axial load of a short, braced, 350 mm square, reinforced concrete column where the only eccentricity is due to constructional inaccuracies. The longitudinal reinforcement consists of four 32 mm dia. longitudinal bars. Also $f_{cu} = 30\,\text{N/mm}^2$ and $f_y = 425\,\text{N/mm}^2$.

Using equation (25) above:

$$\begin{aligned} N &= 0.4 f_{cu} A_c + 0.67 f_y A_{sc} \\ &= [0.4 \times 30 \times (350^2 - 3220) + 0.67 \times 425 \times 3220] \times 10^{-3} \\ &= 2348.26\,\text{kN} \end{aligned}$$

Example 2

Calculate the ultimate axial load for the column in Example 1, if it was an internal column in a multi-storey, equal bay, structure.

Using equation (26) above:

$$\begin{aligned} N &= 0.35 f_{cu} A_c + 0.60 f_y A_{sc} \\ &= [0.35 \times 30 \times (350^2 - 3220) + 0.60 \times 425 \times 3220] \times 10^{-3} \\ &= 2073.54\,\text{kN} \end{aligned}$$

Example 3

Design a square, short, braced, reinforced concrete column to support an ultimate axial load of 3000 kN if $f_{cu} = 30\,\text{N/mm}^2$ and $f_y = 425\,\text{N/mm}^2$. It can be assumed that control and supervision of construction are excellent.

Using equation (25) above:

$$N = 0.4 f_{cu} A_c + 0.67 f_y A_{sc}$$

$$3000 \times 10^3 = 0.4 \times 30 \times A_c + 0.67 \times 425 \times A_{sc} \qquad (27)$$

There are now two unknowns in equation (27), namely A_c and A_{sc}. However, minimum and maximum areas of reinforcement are given in CP110. Clause 3.11.4.1 gives the minimum area as 1 per cent of the cross section and clause 3.11.5 gives the maximum areas as 6 per cent in vertically cast columns and 8 per cent in horizontally cast columns, with 10 per cent at laps in each type.

It is common to have between 2–3 per cent reinforcement, which usually gives an amount of reinforcement which can be easily handled on site. We shall assume therefore, an amount of reinforcement equal to, say 2.7 per cent (i.e. $A_{sc} = 0.027 A_c$).

Now substitute this in equation (27):

$$\therefore \quad 3000 \times 10^3 = 0.4 \times 30 \times A_c + 0.67 \times 425 \times 0.027 A_c$$

$$\text{i.e. } A_c = 152375.1\,\text{mm}^2$$

For a square column the dimension of each side

$$= \sqrt{152375.1} = 390.35\,\text{mm}$$

Therefore use a column which is 400 mm square.

Now from equation (25):

$$3000 \times 10^3 = 0.4 \times 30 \times (400^2 - A_{sc}) + 0.67 \times 425 \times A_{sc}$$

$$\therefore \quad A_{sc} = 3959.5\ \text{mm}^2$$

Use 4 No 32 mm bars + 2 No 25 mm bars

(3200 + 982 = 4202 mm²)

Links

In order to prevent buckling of the longitudinal bars, links (or binders) are necessary (*see* Fig. 23).

These links are designed in accordance with clause 3.11.4.3. Their size is given as $\frac{1}{4} \times$ the largest compression bar with the spacing given as $12 \times$ the smallest compression bar.

For the longitudinal reinforcement above the links required are:

Size: $\frac{1}{4} \times 32 = 8\,\text{mm}$
Spacing: $12 \times 25 = 300\,\text{mm}$

In other words a summary of the column design is

Size: 400 mm square
Main reinforcement: 4T32 + 2T25
Links: T8 at 300 c/c

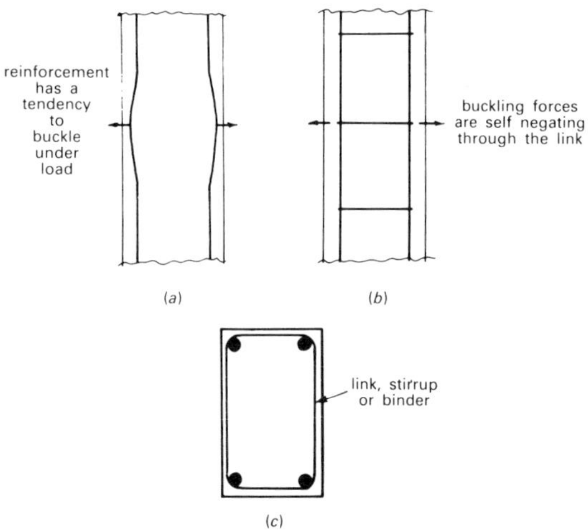

FIG. 23. *Use of links to prevent buckling of longitudinal bars:* (a) *column without links;* (b) *column with links; and* (c) *section through the column showing the shape of the link.*

It should also be noted that clause 3.11.4.3 states that:

> Links should be so arranged that every corner and alternative bar or group in an outer layer of reinforcement is supported by a link passing round the bar and having an included angle of not more than 135°. All other bars should be within 150 mm of a restrained bar (*see* Fig. 24).

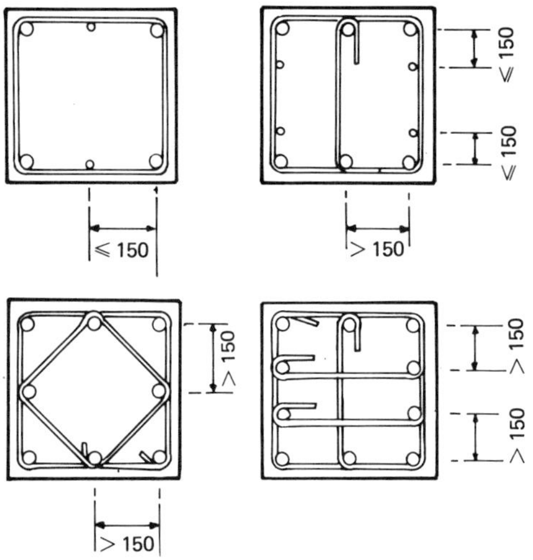

FIG. 24. *All bars or groups of bars within a compression zone should be within 150 mm of a retained bar.*

Example 4

An internal column in a braced three-storey building having equal bays carries an ultimate load of 2500 kN. It has cross-sectional dimensions of 450 mm × 300 mm and a height between end restraints of 4 m. These restraints are such that, from Table 15:

$$l_{ex} = 1.0 l_o \quad \text{and} \quad l_{ex} = 0.85 l_o$$

Calculate the main reinforcement required where $f_{cu} = 25\ \text{N/mm}^2$ and $f_y = 425\ \text{N/mm}^2$.

Check if the column is short or slender using the following method:

$$\frac{l_{ex}}{h} = \frac{1 \times 4 \times 10^3}{450} = 8.9$$

$$\frac{l_{ex}}{b} = \frac{0.85 \times 4 \times 10^3}{300} = 11.3$$

The column is short, since both values are less than 12.

Using equation (26):

$$N = 0.35 f_{cu} A_c + 0.60 f_y A_{sc}$$

$$2500 \times 10^3 = 0.35 \times 25 \times (450 \times 300 - A_{sc}) + 0.60 \times 425 \times A_{sc}$$

$$\therefore \quad A_{sc} = 5355.33\ \text{mm}^2$$

Use 6 No 32 mm bars and 2 No 20 mm bars
(4830 + 628 = 5458 mm²)

Links

Size: $\frac{1}{4} \times 32 = 8$ mm
Spacing: $12 \times 20 = 240$ mm

Reinforcement summary

Main reinforcement: 6T32 + 2T20
Links: T8 at 240 c/c

Check minimum and maximum areas

The minimum area is 1 per cent, i.e.:

$$1 \times \frac{450}{100} \times 300 = 1350\ \text{mm}^2$$

This is satisfied by the designed area.

The maximum area is 6 per cent, i.e.:

$$6 \times \frac{450}{100} \times 300 = 8100\ \text{mm}^2$$

This is not exceeded by the designed area.

Category (*c*)—Resisting moments and axial forces

These type of columns are common around the outside of a building where two situations occur:

(*a*) Eccentricity about one axis only (3.5.5).
(*b*) Eccentricity about both axes (3.5.6).

Eccentricity about one axis only

There are two methods for determining the area of reinforcement required in such a column:

(*a*) Design formulae 3.5.5.3.
(*b*) Design charts 3.5.5.2.

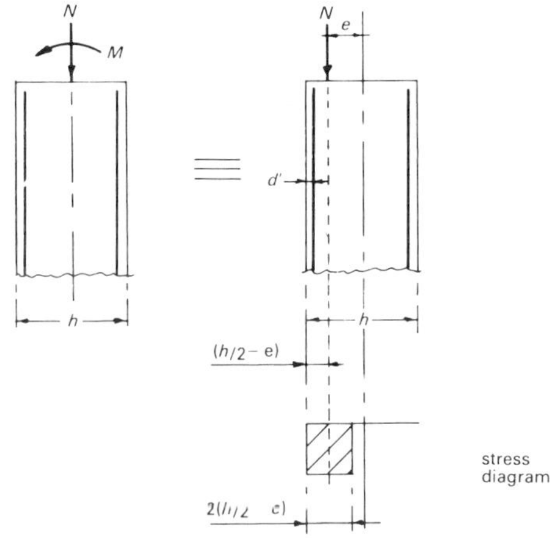

FIG. 25 *Stress block for a column with eccentricity about one axis—no tension occurring.*

Design formulae—compression only

Consider a column carrying an axial load N and a bending moment M, having cross-sectional dimensions $b \times h$. This is equivalent to a column carrying the load N at an eccentricity e where $e = M/N$ (*see* Fig. 25).

If the effects of the reinforcement are ignored the stress block which is available to resist the force N must be symmetrical about N. The width of the stress block is then determined by the outside edge of the column and is equal to $2(h/2 - e)$. With the column width being of dimension b the area resisting compression is $b[2(h/2 - e)]$. Substitute this value into equation (25), remembering that the reinforcement has been ignored, thus:

$$N = 0.40 f_{cu} b[2(h/2 - e)]$$
$$= 0.4 f_{cu} b(h - 2e) \qquad (28)$$

Equation 28 corresponds to equation (27) in CP110. It is obvious since the reinforcement has been ignored, that the column requires nominal reinforcement only.

When e is large enough to make N act at the face of the column, there is no width to the stress block resulting in an infinite stress. It is necessary, therefore, to place a limit on the use of this equation, and CP110 states that the eccentricity must not exceed $(h/2) - d'$ (3.5.5.3). This reduces the width of the stress block to $2d'$ which is very small. The use of this

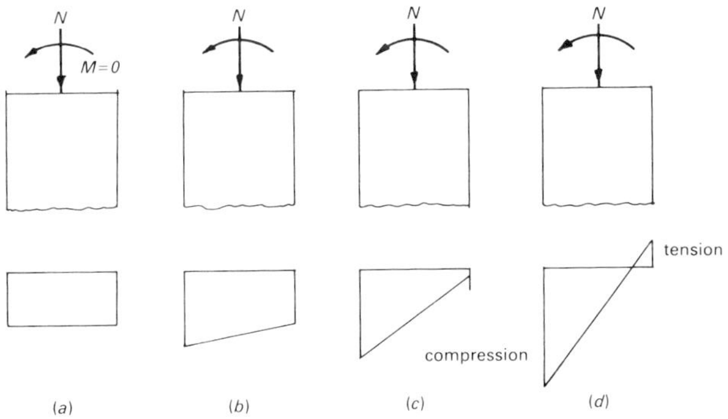

FIG. 26. *This shows how the stress distribution varies as the moment increases from zero in (a) to maximum in (d). As the moment increases the compression is increased on one face and decreased on the other face until it is so large that the more lightly compressed face eventually is in tension.*

formula implies that tension will not develop in the column.

Design formulae—tension and compression
Now consider the case of a symmetrically reinforced concrete column where there is no limit on the amount of eccentricity (i.e. tension can occur). In this case the reinforcement must be taken into account (*see* Fig. 26).

When N is very large with respect to M it is likely that the whole column, including all the reinforcement, will be in compression. As M increases on one side of the column the compression will reduce until it eventually becomes tension, while on the opposite side the compression will increase. When tension occurs the section will be as shown in Fig. 26 and will behave in a very similar way to a beam as analysed in Chapter Two. Consider the column shown in Fig. 27.

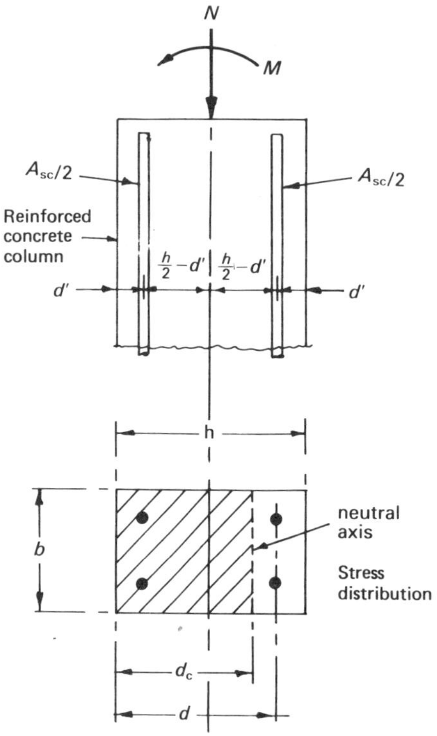

FIG. 27. *The distribution of stress when tension develops in a reinforced concrete column.*

Resolve all forces vertically:

(a) The concrete stress is $0.4f_{cu}$ and acts on an area bd_c, therefore the force is $0.4f_{cu}bd_c$.

(b) The compression bars are taken to be stressed at $0.72f_y$. This is to cover the use of $460\,N/mm^2$ steel in columns, which is rare. For the $425\,N/mm^2$ steel, which is more common, the stress could be taken as $0.74f_y$. However, to show the derivation of the formulae in CP110 the value of $0.72f_y$ is used. Therefore the force is $0.72f_y(A_{sc}/2)$.

(c) The bars on the right-hand side of the section may be in compression or tension depending on the position of the neutral axis. If they are in tension the stress will be $-0.87f_y$ or in compression $+0.72f_y$. Let f_{s2} denote whichever of these is applicable. Thus the force provided is $f_{s2}(A_{sc}/2)$. Now adding these three together:

$$N = 0.4f_{cu}bd_c + 0.72f_y\frac{A_{sc}}{2} + f_{s2}\frac{A_{sc}}{2} \quad (29)$$

If the amount of reinforcement is different on each face of the column, equation (29) can easily be modified to suit. Let the area of reinforcement in the more highly compressed face be A'_{s1} and in the other face A_{s2}, then:

$$N = 0.4f_{cu}bd_c + 0.72f_yA'_{s1} + f_{s2}A_{s2} \quad (30)$$

Equation 30 corresponds to equation (28) in CP110.

Now take moments of these forces about the centre line of the section

$$M = 0.4f_{cu}bd_c\left(\frac{h}{2} - \frac{d_c}{2}\right) + 0.72f_yA'_{s1}\left(\frac{h}{2} - d'\right) - f_{s2}A_{s2}\left(\frac{h}{2} - d'\right) \quad (31)$$

Where the reinforcement varies in diameter in each face this will affect the d' dimension since this is a function of half the bar diameter. Equation (31) can be modified by simply substituting the new value d_2 for d' in the last term i.e.:

$$M = 0.4f_{cu}bd_c\left(\frac{h}{2} - \frac{d_c}{2}\right) + 0.72f_yA'_{s1}\left(\frac{h}{2} - d'\right) - f_{s2}A_{s2}\left(\frac{h}{2} - d_2\right) \quad (32)$$

Equation 32 corresponds to equation (29) in CP110. The only restriction in the use of these formulae is that d_c must not be less than $2d'$.

Even with the common symmetrical arrangement of reinforcement and equal values of d' these equations are cumbersome and tedious to use since both d_c and A_{sc} are unknown. In a design utilising these formulae an answer is obtained by a process of trial and error, which is usually lengthy. Design charts reduce the amount of work considerably, but more of them later.

As an alternative to the above rigorous method of design CP110 permits an approximate method when the eccentricity is not less than $((h/2) - d_2)$. In this case (*see* Fig. 28) the moment M and the axial force N are replaced by an increased moment M_a plus a compressive force N acting through the tensile steel A_s, where:

$$M_a = M + N\left(\frac{h}{2} - d_2\right) \quad (33)$$

Equation 33 corresponds to equation (30) in CP110.

Using this simplified method there are two parts to the design:

(a) Treat the column as a doubly reinforced section resisting bending, then A'_s and A_s are found from the following equations:

$$M_a = 0.15f_{cu}bd^2 + 0.72f_yA'_s(d - d') \quad (34)$$

and

$$0.87f_yA_s = 0.2f_{cu}bd = 0.72f_yA'_s \quad (35)$$

Equation (34) corresponds to equation (12) on page 14 and equation (35) corresponds to equation (15) on page 14.

(b) The area of A_s calculated is reduced by an amount $N/0.87f_y$ to account for the compressive force N. Obviously this again allows for an unsymmetrical arrangement of reinforcement.

Eccentricity about both axes

This is called biaxial bending and is common at the corner columns of a building where moments are applied by beams framing into two adjacent faces of the column. The resulting stresses can be difficult to analyse by normal methods but CP110 gives the following simplified approach.

The code states that for a short column subjected to an ultimate load N and ultimate bending moments M_x and M_y about the major and minor axes respectively the following equation

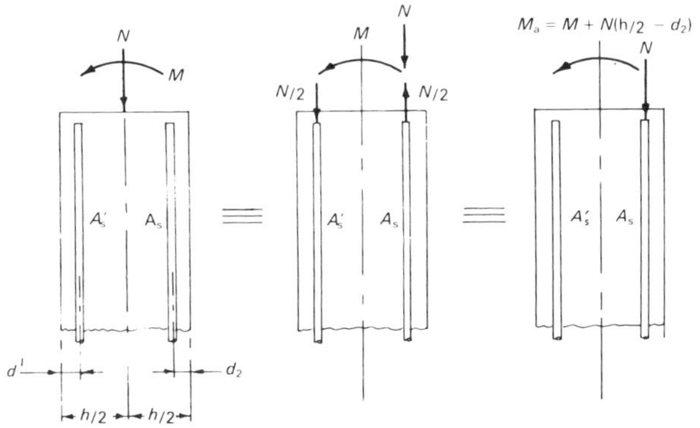

FIG. 28. M *and* N *replaced by* M_a *and* N.

must be satisfied:

$$\left(\frac{M_x}{M_{ux}}\right)^{\alpha_n} + \left(\frac{M_y}{M_{uy}}\right)^{\alpha_n} \leq 1.0 \qquad (36)$$

Equation 36 corresponds to equation (31) of CP110, where M_{ux} and M_{uy} are the maximum moment capacities assuming the ultimate axial load N and bending about the major and minor axes respectively. Also:

$$\alpha_n = 1.66 \frac{N}{N_{uz}} + 0.67 \qquad (37)$$

where

$$N_{uz} = 0.45 f_{cu} A_c + 0.75 f_y A_s' \qquad (38)$$

The relationship between α_n and (N/N_{uz}) is given in Table 16 of CP110 with α_n having limiting values of 1.0 and 2.0.

DESIGN CHARTS

As stated previously the formulae give rise to a long laborious design process. The use of the design charts considerably reduces the calculations, and hence the time involved. It should be noted that these graphs have been constructed using a symmetrical arrangement of reinforcement, the stress–strain graphs for concrete and steel as shown in Figs. 1 and 2 respectively of CP110, and a linear strain distribution.

Clause 3.5.5 of CP110 states that short columns must be designed for a minimum bending moment of $0.05 N h_{min}$, where h_{min} is the minimum depth of cross section. This value is built into the design charts and is the reason for the flat top to the graphs.

A typical column chart is shown in Fig. 29 together with the stress–strain graph for reinforcement which explains the "kinks" in the chart.

The following calculations illustrate the use of the charts in CP110 to design short columns with bending moment about one and two axes. It should be noted that three factors influence the choice of the design chart, *viz:* f_{cu}, f_y and d/h. The d/h values are given in 0.05 increments. Where a value lies between two charts, both may be read and the exact result found by linear interpolation. However, a conservative answer may be found by using the chart below the exact value necessary, e.g. if $(d/h) = 0.87$ use the chart where $(d/h) = 0.85$.

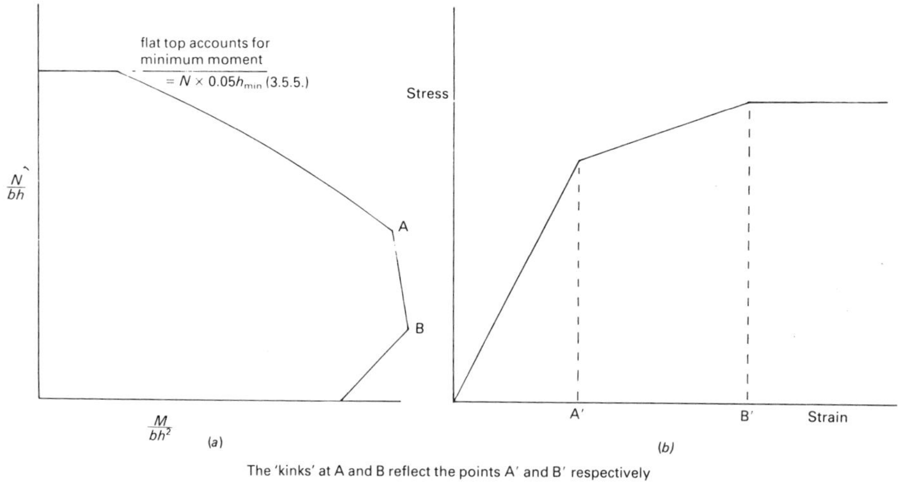

FIG. 29. *Typical column chart. The "kinks" at A and B reflect the points A′ and B′ respectively.*

CURPE CONSULTANTS 46 Orburn Road DUNFIELD	Project: DESIGN IV Sub-section: CHAPTER SEVEN SHORT COLUMNS	Job ref: CC/ME Sheet no: 7/1 Made by: B.C. Date: Checked by: Date:
	Structural Summary Sheet	

THE DESIGN OF A SHORT COLUMN CARRYING AXIAL LOAD AND BENDING ABOUT THE MAJOR AXIS ONLY

DESIGN DATA.

An internal column in a single storey building carries an ultimate axial load of 250 kN. and an ultimate bending moment of 160 kN.m. about the major axis. The column has cross-sectional dimensions of 450 mm × 300 mm and a height of 4.0 m between the end restraints. These restraints are such that $\ell_{ex} = 1.0\,\ell_o$ and $\ell_{ey} = 0.85\,\ell_o$.

Assume: (i) $f_{cu} = 25\ N/mm^2$

(ii) $f_y = 425\ N/mm^2$

(iii) ϕ of the main reinforcement = 25 mm

(iv) ϕ of the links = 8 mm.

Use the charts in C.P. 110 to design the column.

CURPE CONSULTANTS 46 Orburn Road DUNFIELD	Contract: DESIGN IV Part of structure: CHAPTER SEVEN SHORT COLUMNS Drawing ref: — Calculations by: B.C. — Checked by:	Job ref: CC/ME Calc. sheet No: 7/2 Date:

Code ref.	CALCULATIONS	OUTPUT
3.5.1.2	CHECK IF SHORT OR SLENDER	
	$\frac{\ell_{ex}}{h} = \frac{1 \times 4 \times 10^3}{450} = 8.9$	
	$\frac{\ell_{ey}}{b} = \frac{0.85 \times 4 \times 10^3}{300} = 11.3$	
	Since both are less than 12 the column is SHORT	Short Col.
Part 2	CHOSE THE APPROPRIATE DESIGN CHART	
table 19	Cover to the links = 40 mm; the conditions of exposure are moderate since the bottom of the column will be below ground level.	
	Assuming 25 mm diameter main bars and 8 mm diameter links	
	$d = 450 - (40 + 8 + 25/2)$	
	$= 389.5\ mm$	
	$\frac{d}{h} = \frac{389.5}{450} = 0.865$	
	'Round' this value down to 0.85 and Use chart no. 77	chart 77
	CALCULATE THE MAIN REINFORCEMENT	
	$\frac{N}{bh} = \frac{250 \times 10^3}{300 \times 450} = 1.85\ N/mm^2$	
	$\frac{M}{bh^2} = \frac{160 \times 10^6}{300 \times 450^2} = 2.64\ N/mm^2$	
chart 77	$\frac{100 A_{sc}}{bh} = 1.45$	

CURPE CONSULTANTS 46 Orburn Road DUNFIELD	Contract: DESIGN IV Part of structure: CHAPTER SEVEN SHORT COLUMNS	Job ref: CC/ME Calc sheet No: 7/3
Drawing ref:	Calculations by: B.C. / Checked by:	Date:

Code ref.	CALCULATIONS	OUTPUT
	$\therefore A_{sc} = \frac{1.45 \times 300 \times 450}{100}$	
	$= 1957.5\ mm^2$	
	USE 4 T25 mm ϕ bars = (1960 mm^2)	4T25
3.11.4.3	LINKS	
	Size = 1/4 × 25 ÷ 6 mm or 8mm.	
	Spacing = 12 × 25 = 300mm	
	USE 8mm links at 300mm c/c	8 @ 300 c/c.
3.11.4.1.	CHECK THE MINIMUM AREA OF MAIN REINFORCEMENT	
	Min. area = 1% < 1.45% designed	O.K.
3.11.5	CHECK THE MAXIMUM AREA OF MAIN REINFORCEMENT	
	Max. area = 6% for vertically cast	
	= 8% for horizontally cast	
	= 10% for lap positions	
	None of these values is exceeded	O.K

CURPE CONSULTANTS 46 Orburn Road DUNFIELD	Project: DESIGN IV Sub-section: CHAPTER SEVEN SHORT COLUMNS	Job ref: CC/ME Sheet no: 7/4 Made by: B.C. Date:
	Structural Summary Sheet	Checked by: Date:

THE DESIGN OF A SHORT COLUMN CARRYING AXIAL LOAD AND BENDING MOMENT ABOUT BOTH AXES.

DESIGN DATA

A short column carries an ultimate axial load of 650kN. and ultimate bending moments of 95 kN.m about the major axis and 75kN.m about the minor axis.

Assume:
(i) $f_{cu} = 25\ N/mm^2$
(ii) $f_y = 425\ N/mm^2$
(iii) $d/h = 0.90$
(iv) $h = 450mm$
(v) $b = 300mm$

CURPE CONSULTANTS
46 Orburn Road
DUNFIELD

Contract	DESIGN IV	Job ref	CC/ME
Part of structure	CHAPTER SEVEN SHORT COLUMNS	Calc. sheet No	7/5
Drawing ref	Calculations by B.C.	Checked by	Date

Code ref.	CALCULATIONS	OUTPUT
	A first estimate may be obtained by adding the two bending moments together and treating the sum as a bending moment applied about the major axis	
Part 2	CHOSE THE APPROPRIATE DESIGN CHART	
	With $f_{cu} = 25\ N/mm^2$, $f_y = 425\ N/mm^2$ and $d'/h = 0{\cdot}90$	
	USE CHART No 76	chart 76
	CALCULATE THE MAIN REINFORCEMENT	
	$\frac{N}{bh} = \frac{650 \times 10^3}{300 \times 450} = 4{\cdot}81\ N/mm^2$	
	$\frac{M}{bh^2} = \frac{(95+75) \times 10^6}{300 \times 450^2} = 2{\cdot}80\ N/mm^2$	
chart 76	$\frac{100 A_{sc}}{bh} = 1{\cdot}17$	
	$\therefore A_{sc} = \frac{1{\cdot}17 \times 300 \times 450}{100}$	
	$= 1579{\cdot}5\ mm^2$	
	TRY 4No 25mm ϕ bars $A_{sc} = 1960\ mm^2$	
	Actual $\frac{100 A_{sc}}{bh} = \frac{100 \times 1960}{300 \times 450} = 1{\cdot}45$	
	Now find the maximum moment capacity for each axis in turn assuming full axial load to be applied	

CURPE CONSULTANTS
46 Orburn Road
DUNFIELD

Contract	DESIGN IV	Job ref	CC/ME
Part of structure	CHAPTER SEVEN SHORT COLUMNS	Calc. sheet No	7/6
Drawing ref	Calculations by B.C.	Checked by	Date

Code ref.	CALCULATIONS	OUTPUT
	With $\frac{N}{bh} = 4{\cdot}81\ N/mm^2$ and	
	$\frac{100 A_{sc}}{bh} = 1{\cdot}45$	
chart 76	$\frac{M}{bh^2} = 3{\cdot}3\ N/mm^2$	
	hence $M_{ux} = 3{\cdot}3 \times 300 \times 450^2 \times 10^{-6}$	
	$= 200{\cdot}48\ kN.m$	
	$M_{uy} = 3{\cdot}3 \times 450 \times 300^2 \times 10^{-6}$	
	$= 133{\cdot}65\ kN.m$	
Eqn 32	$N_{uz} = 0{\cdot}45 f_{cu} A_c + 0{\cdot}75 f_y A'_s$	
	$= [(0{\cdot}45 \times 25 \times 300 \times 450)$	
	$+ (0{\cdot}75 \times 425 \times 1960)] \times 10^{-3}$	
	$= 2143{\cdot}50\ kN$	
	$\frac{N}{N_{uz}} = \frac{650}{2143{\cdot}5} = 0{\cdot}303$	
table 16	$\alpha_n = 1{\cdot}17$	
Eqn 31	$\left(\frac{M_x}{M_{ux}}\right)^{\alpha_n} + \left(\frac{M_y}{M_{uy}}\right)^{\alpha_n} \leq 1{\cdot}0$	
	$\left(\frac{95}{200{\cdot}48}\right)^{1{\cdot}17} + \left(\frac{75}{133{\cdot}65}\right)^{1{\cdot}17} = 0{\cdot}9261$	

CURPE CONSULTANTS, 46 Orburn Road, DUNFIELD	Contract: DESIGN IV	Job ref: CC/ME
	Part of structure: CHAPTER SEVEN SHORT COLUMNS	Calc sheet No: 7/7
Drawing ref	Calculations by: BC. / Checked by	Date

Code ref.	CALCULATIONS	OUTPUT
	This value of 0.9261 is less than 1.0 ie equation 31 and thus clause 3.5.6 is satisfied	
	∴ Use 4T25 mm φ bars.	4T25
	The links and the checks for the maximum and minimum areas of reinforcement are done as shown on page 7/3.	

SELF-ASSESSMENT QUESTIONS

1. Outline the principal factors which influence the ultimate load that a column can withstand.

2. Define the difference between:

(*a*) short and slender columns;
(*b*) braced and unbraced columns.

3. Calculate the ultimate axial load of a short-braced axially loaded column which is 400 mm square, has 6 No 25 mm dia. longitudinal bars, $f_{cu} = 25\,N/mm^2$ and $f_y = 425\,N/mm^2$ if;

(*a*) the column is internal in a multistorey structure;
(*b*) the only eccentricity is due to constructional inaccuracies.

4. A reinforced concrete column in a braced frame is required to support an ultimate axial load of 4500 kN. Use the following assumptions and the formulae in CP110 to design the column reinforcement, including the links, when:

(*a*) the control and supervision are excellent;
(*b*) the column is internal in a three-storey building.

Assume:

(*i*) the clear height between end restraints $= 4.75\,m$;
(*ii*) these restraints are such that $l_{ex} = 1.0 l_o$ and $l_{ey} = 0.85 l_o$;
(iii) the cross-sectional dimensions are 450 mm × 450 mm;
(*iv*) $f_{cu} = 30\,N/mm^2$;
(*v*) $f_y = 425\,N/mm^2$.

5. A column is 600 mm × 400 mm in cross section and is restrained such that $l_{ex} = 1.5 l_o$ and $l_{ey} = 0.85 l_o$ where $l_o = 4.50\,m$. The concrete grade is 30 and the characteristic strength of the reinforcement is $425\,N/mm^2$.

The column supports:

(*a*) an ultimate axial load of 400 kN and a bending moment about the major axis of 250 kNm;
(*b*) an ultimate axial load of 1000 kN and bending moments of 165 kNm about the major axis and 135 kNm about the minor axis.

Use the design charts in CP110 to design the columns.

CHAPTER EIGHT

Reinforced Concrete Pad Foundations

> CHAPTER OBJECTIVES
>
> After studying this chapter you should know:
> * the importance of foundations;
> * the various foundation types;
> * the sequence of design for a pad;
> * the load cases in the design of a pad;
> * the critical sections to be considered;
> * how to design a reinforced concrete pad foundation subjected to both direct load and bending.

FOUNDATIONS GENERALLY

Any structure can be divided into two broad groups, namely superstructure and substructure. The former is that part of the building that is above ground while the latter relates to the foundations that are below ground. The maxim "out of sight, out of mind" could well be applied to foundations and it is for this reason that they require special attention. The foundations are the first part of the structure to be constructed and their excavations are usually backfilled and hidden well in advance of the completion of the contract. Therefore it is impossible to inspect them when the full service loads are applied, nor can they display visually any subsequent signs of distress, as could more easily be detected in the superstructure. One of the responsibilities of the structural engineer is to ensure that the loads imposed on the structure are safely transmitted to the ground. The last link in the line of transfer is from the foundation to the soil (*see* Fig. 30).

Much time, labour and money can be spent on producing an attractive, well-admired and structurally sound superstructure, but if the foundations are not correct all else is wasted effort.

All this shows the unique importance of the foundations and thus it behoves any designer to give them the attention of which they are worthy.

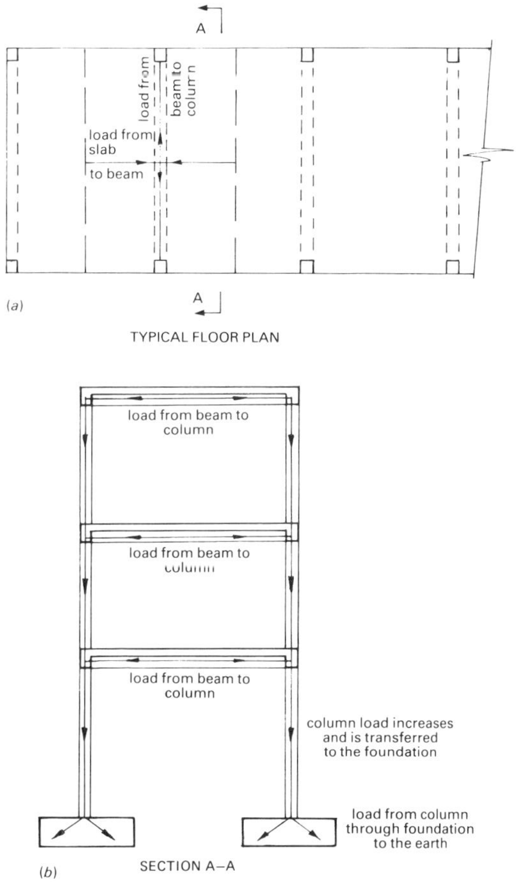

FIG. 30. *Reinforced concrete frame:* (a) *typical floor plan; and* (b) *section* A–A.

FOUNDATIONS IN PARTICULAR

It would be impossible to cover adequately the subject of foundations in such a short chapter as this, but the main types can be indicated.

Foundation types

It is obvious that, since the role of the foundation is to transfer the load to the ground, the type of ground and its strength must play an important role in the selection of the type of foundation. Thus prior to the engineer deciding on the foundation system he must have the results of a site investigation. This may vary from a trial hole, in a small contract, to a large number of deep boreholes in a large contract. This is the work of a specialist and is carried out in order to determine the nature of the ground and the contours of its strata. No matter how extensive the ground exploration is, there will always be the unknown factor and so special care must always be taken with foundations. This is reflected in the conservative factor of safety, usually three or more, depending on the job in hand.

Not only is the strength of the gound important, but the type of ground will also play a part in deciding the foundation system (e.g. it is difficult to drive piles through boulder clay, or to keep bored pile casings dry where the water table is high, and if the ground is "peaty" there may be sulphates existing which would attack the concrete).

Strip foundations

Where the ground is good, or the loads light and continuous (e.g. a block/brick wall), strip foundations are all that may be required. If dimension w is less than or equal to h then there is no transverse bending in the strip (*see* Fig. 31) and if any reinforcement is required it will only be a light mesh to enable the foundation to span across local soft spots. It should be noted that at large door openings there is the possibility in this foundation of a reversal of bending thus putting the top into tension. Here again reinforcement may be required (*see* Fig. 32).

Pad foundations

If the load is not continuous and the structure is supported on columns, the most economical foundation is often an isolated pad foundation, as shown in Fig. 33(*a*). One such is designed later in this chapter. However, if the ground properties to load relationship is such that the area covered by the pads is more than half the total area, then a strip foundation, designed as a continuous beam (*see* Fig. 33(*b*)) may be the answer but if this is unsatisfactory, a raft foundation may provide the solution (*see* Fig. 33(*c*)).

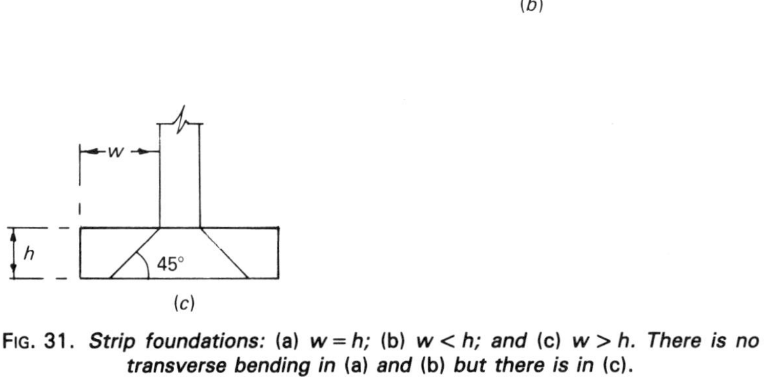

FIG. 31. *Strip foundations:* (a) $w = h$; (b) $w < h$; *and* (c) $w > h$. *There is no transverse bending in* (a) *and* (b) *but there is in* (c).

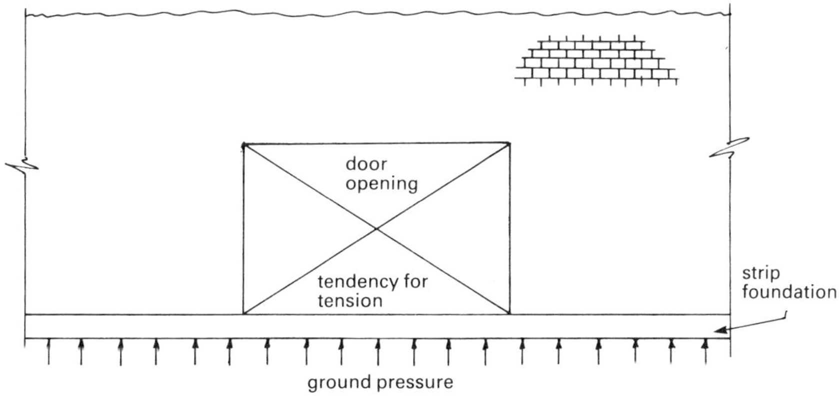

FIG. 32. *The ground reaction tends to produce tension in the top of the foundation at the door opening due to hogging.*

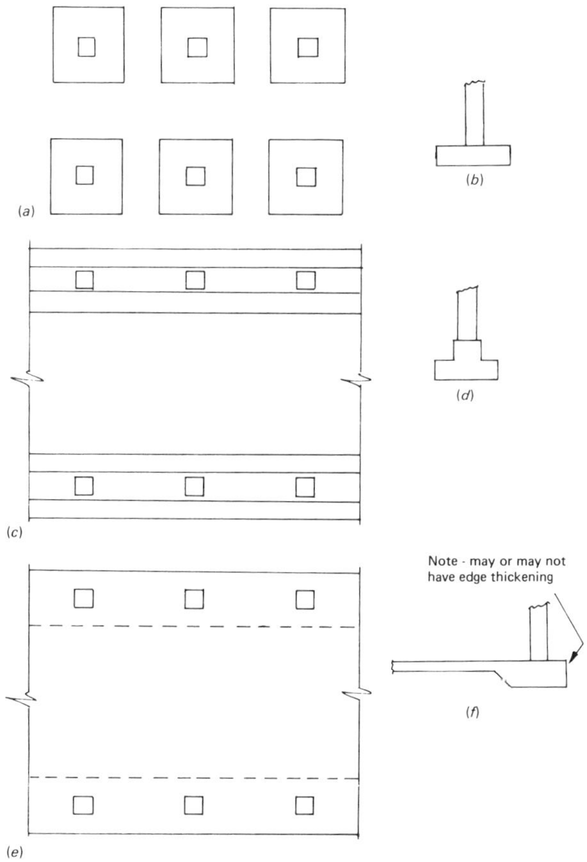

FIG. 33. Foundations: (a) *layout of pad foundations;* (b) *section through pad;* (c) *ground beams;* (d) *section through ground beam;* (e) *raft foundation;* (f) *section through raft foundation.*

Raft foundations

As stated above a raft can provide the answer where pads become so large that they are almost touching. If the ground is so poor that a raft could not spread the load sufficiently, thought can be given to a bouyant raft (*see* Fig. 34). This is a cellular raft, which in itself is lighter than the soil removed. This gives the supporting strata a relief of load which can thus be available to support the load from the superstructure.

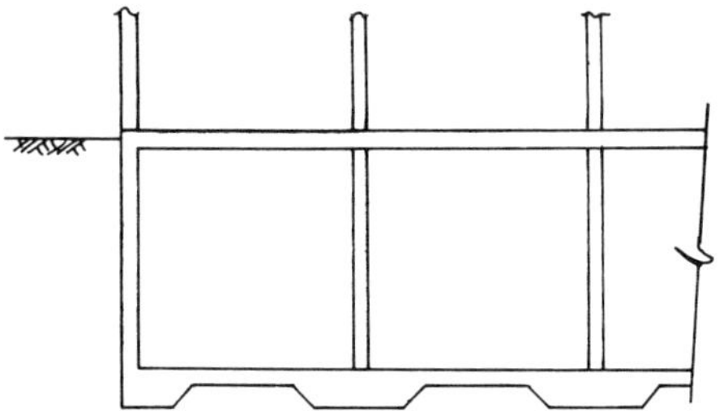

FIG. 34. *Bouyant raft.*

Piled foundations

Before going to raft foundations, if there was a good load-bearing strata at a reasonable depth piled foundations often provide a satisfactory solution (*see* Fig. 35). These are not usually subject to the settlement problems sometimes associated with the raft. This is due to the fact that the layers that would squeeze and give rise to settlement are bypassed as the pile transfers the load to the non-settling layers below.

Balanced or combined foundations

When developing a site where there are adjacent buildings it is often difficult to construct a foundation central on perimeter column. In this case a combined foundation which acts in a balancing fashion can provide a neat answer (*see* Fig. 36(*a*)). The balance may take place about a pad or piles as illustrated.

In foundation design it is essential that the ground-bearing pressure under all foundations is uniform. If it is otherwise, differential settlement will occur, eventually leading to unsightly cracks in the finished building. It would normally be expected that a perimeter column would carry less load than an internal one.

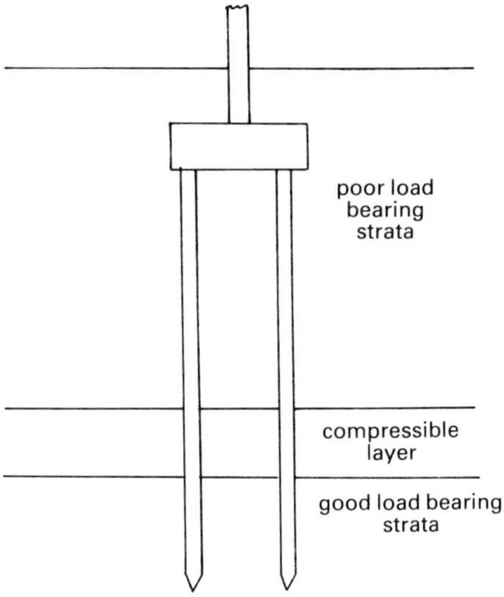

FIG. 35. *Piled foundations.*

Thus where the internal column is used to balance the perimeter one, the foundation is often shaped to have the centre of load corresponding to the centre of gravity of the shape (*see* Fig. 36(*b*)). A more simple solution to the problem is to position the perimeter columns back from the perimeter of the building, allowing the foundations to be easily constructed and cantilevering the superstructure to meet the existing buildings (*see* Fig. 36(*c*)).

However, within the confines of this simple book we shall consider in detail only the design of an isolated pad foundation.

Pad foundations

Sequence of design

In reinforced concrete, up to this point the sequence has been to design for bending initially and subsequently check for shear. However, when designing foundations, once the required foundation area has been established from the allowable bearing pressure and the service loadings, the thickness must be determined. In single footings the depth is normally governed by shear and it is generally undesirable to use shear reinforcement. The sequence of design for a pad foundation therefore is as follows:

(*a*) calculate the area of pad required;

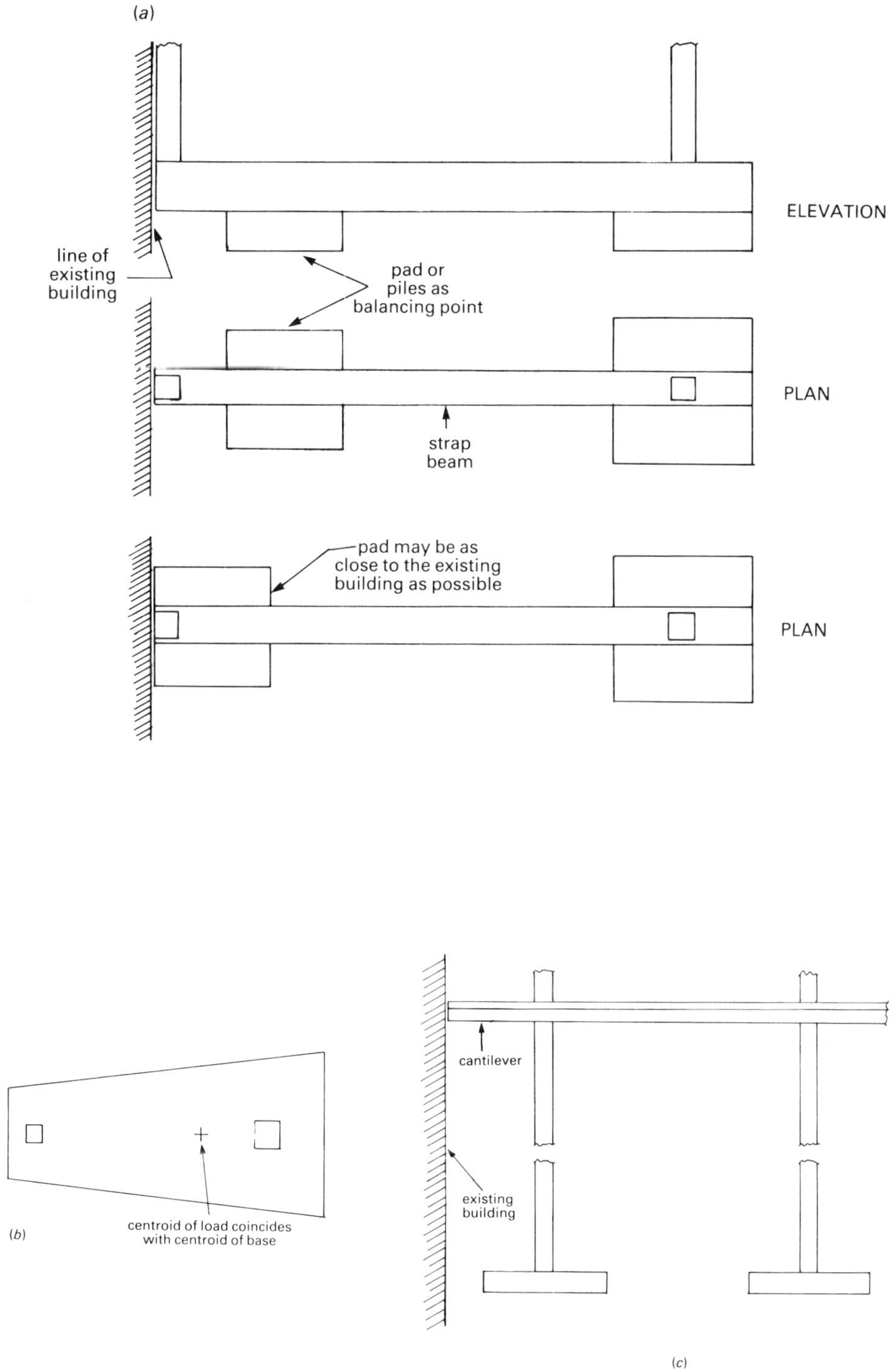

FIG. 36. *Various foundation systems for building against an existing structure.*

(*b*) design for shear, both punching and transverse;
(*c*) check for bending;
(*d*) check anchorage and cracking.

Load cases
There are two load cases which must be considered in the design of a foundation. These are:

(*a*) service loads and moments; and
(*b*) ultimate loads and moments.

(*a*) *Service loads and moments.* These are used in conjunction with the allowable ground-bearing pressure in order to calculate the base area required. The reasons for this are two-fold: (*i*) primarily because any settlement that takes place during the life of the building will be due to service and not ultimate loads; (*ii*) the allowable ground bearing pressure contains a factor of safety (i.e. it is *not* the ultimate pressure that the ground can withstand).

(*b*) *Ultimate loads and moments.* As stated in Chapter One the overall factor of safety is composed of two components, namely the partial factors of safety on the materials and the partial factors of safety on the loads. The ultimate loads and moments must be used to calculate the design pressure, which is used to design the section.

Critical sections
CP110 outlines the critical sections to be considered in the various stages of design:

(*a*) *punching shear* (*see* Fig. 37);
(*b*) *transverse shear* (*see* Fig. 38);
(*c*) *bending* (*see* Fig. 39).

In Figs. 37–39 the load on the shaded area is the load to be considered in the design.

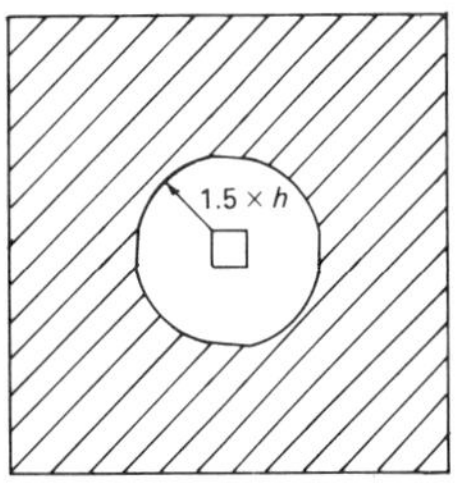

FIG. 37. *Critical section for punching shear.*

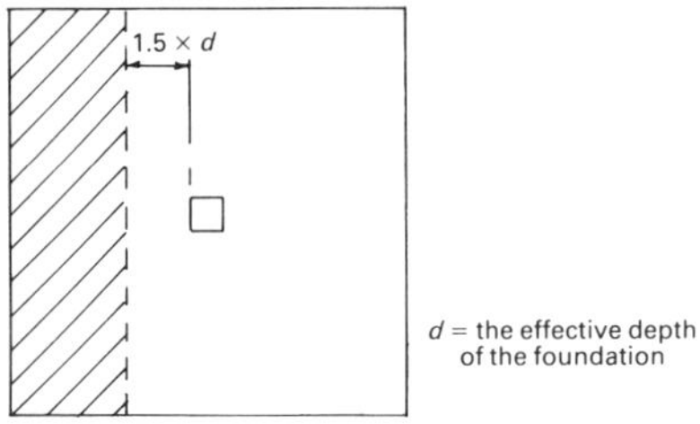

FIG. 38. *Critical section for transverse shear.*

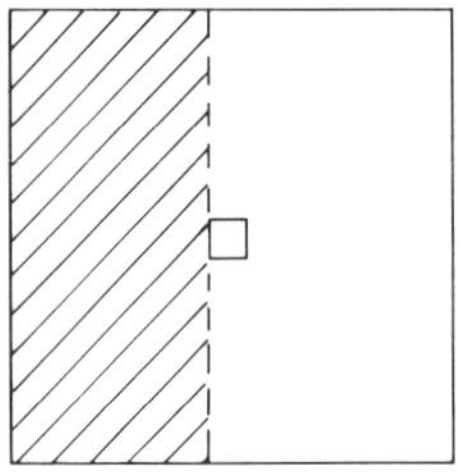

FIG. 39. *Critical section for bending.*

Bearing stress inside bends in reinforcement
This is a problem that is associated mainly with starter bars from a foundation to a column or wall. It may also occur at a fixed connection between a beam and column.

Where a bar is required to change direction while being fully stressed the nominal radii of 2 × the bar diameter for mild steel and 3 × the bar diameter for high yield steel are too tight to allow a free flow of stress around the bend. (The student may appreciate the rough analogy of water flowing freely through a pipe suddenly confronted with a right angle bend. This results in a large build up of pressure.) The stress built up at the bend position will be compressive and the result is that the concrete may crush. CP110 in clause 3.11.6.8. gives the following expressions which should be used to check the bearing stress:

$$\text{The bearing stress} = \frac{F_{bt}}{r\theta}$$

where F_{bt} = the tensile force due to ultimate loads in a bar or group of bars; r = the internal radius of the bend; and θ = the size of the bar or, in a bundle, the size of a bar of equivalent area.

This stress should not exceed the value given by $(1.5 f_{cu}/(1 + 2\phi/a_b))$ where a_b for a particular bar or group of bars in contact should be taken as the centre-to-centre distance between bars or groups of bars perpendicular to the plane of the bend; for a bar or group of bars adjacent to the face of the member, a_b should be taken as the cover plus ϕ as defined above.

Example 1

The main reinforcement in a reinforced concrete column consists of 4T32 mm ϕ longitudinal bars which are fully stressed in tension. The cover to the main reinforcement = 38 mm, $f_{cu} = 30\,\text{N/mm}^2$ and $f_y = 425\,\text{N/mm}^2$. Calculate the required radius of the bend in the starter bars.

The tensile force in one bar

$$= \text{area} \times \text{permissible stress}$$

$$= 804 \times \frac{425}{1.15} \times 10^{-3}$$

That is $\quad F_{bt} = 297.13\,\text{kN}$

The bearing stress $= \dfrac{1.5 f_{cu}}{1 + 2\phi/a_b}$

$$= \frac{1.5 \times 30}{1 + (2 \times 32/38 + 32)}$$

$$= 23.51\,\text{N/mm}^2$$

But the bearing stress also

$$= \frac{F_{bt}}{r\phi}$$

$$\therefore \quad r = \frac{F_{bt}}{\text{bearing stress} \times \phi}$$

$$= \frac{297.13 \times 10^3}{23.51 \times 32}$$

$$= 395.0\,\text{mm}$$

The normal radius would have been

$3\phi \quad = 96\,\text{mm}.$

It is unlikely that the full design stress of a bar would ever be developed in a bend, however even with the reduction due to the length of bar embedded in the concrete, the normal radius would not suffice.

Example 2

The reinforcement at the bottom of a reinforced concrete cantilever retaining wall is T16 mm ϕ bars at 200 mm c/c. If $f_{cu} = 30\,\text{N/mm}^2$ and $f_y = 425\,\text{N/mm}^2$ calculate the radius required in the bend of the starter bars.

The tensile force in one bar

$$= \text{area} \times \text{permissible stress}$$

$$= 201 \times \frac{425}{1.15} \times 10^{-3}$$

That is $\quad F_{bt} = 74.28\,\text{kN}$

The bearing stress $= \dfrac{1.5 f_{cu}}{1 + 2\phi/a_b}$

$$= \frac{1.5 \times 30}{1 + (2 \times 16/200)}$$

$$= 38.79\,\text{N/mm}^2$$

But the bearing stress also

$$= \frac{F_{bt}}{r\phi}$$

$$\therefore \quad r = \frac{F_{bt}}{\text{bearing stress} \times \phi}$$

$$= \frac{74.28 \times 10^3}{38.79 \times 16}$$

$$= 119.68\,\text{mm}$$

The normal radius would have been

$3\phi \quad = 48\,\text{mm}.$

DESIGN EXAMPLE

The following example illustrates the design of a typical pad foundation:

CURPE CONSULTANTS 46 Orburn Road DUNFIELD	Project DESIGN IV	Job ref CC/ME
	Sub-section CHAPTER EIGHT PAD FOUNDATIONS	Sheet no B/1
		Made by B.C.
		Date
	Structural Summary Sheet	Checked by
		Date

A PAD FOUNDATION SUBJECTED TO VERTICAL LOAD AND APPLIED MOMENT

A reinforced concrete pad foundation is required to meet the following conditions:

(i) Axial dead load = 1200 kN.
(ii) Axial imposed load = 550 kN.
(iii) Bending moment due to the dead load = 325 kN.m.
(iv) Bending moment due to the imposed load = 110 kN.m.
(v) The bending in (iii) & (iv) is about one axis only.
(vi) The column is 400 mm square and is central on the pad.
(vii) The safe ground bearing pressure = 150 kN/m^2
(viii) f_{cu} = 25 N/mm^2
(ix) f_y = 425 N/mm^2

CURPE CONSULTANTS 46 Orburn Road DUNFIELD	Contract DESIGN IV			Job ref CC/ME
	Part of structure CHAPTER EIGHT PAD FOUNDATIONS			Calc sheet No B/2
	Drawing ref	Calculations by B.C.	Checked by	Date

Code ref.	CALCULATIONS	OUTPUT
2.3.3.1.	Service loading = 1200 + 550 = 1750 kN.	
	Service Moment = 325 + 110 = 435 kN.m.	
	Ultimate loading = (1·4 × 1200) + (1·6 × 550) = 2560 kN.	
	Ultimate Moment = (1·4 × 325) + (1·6 × 110) = 631 kN.m.	

CALCULATE THE PAD SIZE

N.B. - The service loads are used in conjunction with the safe ground bearing pressure.

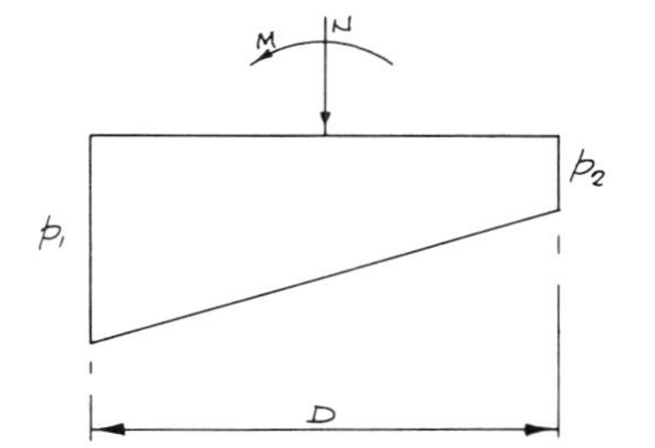

Typical pressure diagram

p_1 maximum = 150 kN/m^2

$$p_1 = \frac{N}{BD} + \frac{M.6}{BD^2}$$ where B = the breadth of the base

Let D = 2B.

CURPE CONSULTANTS
46 Orburn Road
DUNFIELD

Contract: DESIGN IV | Job ref: CC/ME
Part of structure: CHAPTER EIGHT PAD FOUNDATIONS | Calc sheet No: 8/3
Drawing ref: | Calculations by: B.C. | Checked by: | Date:

Code ref.	CALCULATIONS	OUTPUT

Since the ground has to withstand the base itself an allowance must be made for the self weight. Where soil has to be removed to construct the base, the only allowance to be made is for the difference between the weight of the soil removed and the weight of the concrete imposed. This can be done by subtracting the densities of the two materials.

Density of concrete = 2400 kg/m^3
Density of soil, say = 1750 "

Difference = 650 "

This is equivalent to 6.50 kN/m^3

For a base say 700 mm thick the difference

$$= 6.5 \times 0.7 = 4.55 \ kN/m^2$$

The safe ground bearing pressure is reduced by this amount,

$$ie \quad 150 - 4.55 = 145.45 \ kN/m^2$$

Now using the equation on page 8/2

$$p_1 = 145.45 = \frac{1750}{2B^2} + \frac{435 \times 6}{4B^3}$$

$$\therefore \quad B^3 - 6.015B - 4.486 = 0$$

$$\therefore \quad B \simeq 2.765 \ m$$

say $B = 2.80$ m

TRY A PAD 5.60 m × 2.80 m × 0.70 m — Output: 5.6 × 2.8 × 0.7

CURPE CONSULTANTS
46 Orburn Road
DUNFIELD

Contract: DESIGN IV | Job ref: CC/ME
Part of structure: CHAPTER EIGHT PAD FOUNDATIONS | Calc sheet No: 8/4
Drawing ref: | Calculations by: B.C. | Checked by: | Date:

Code ref.	CALCULATIONS	OUTPUT

THE DESIGN PRESSURES

$$p_1 = \frac{2560}{2.8 \times 5.6} + \frac{6 \times 631}{2.8 \times 5.6^2}$$

$$= 206.38 \ kN/m^2$$

Output: $p_1 = 206.38 \ kN/m^2$

$$p_2 = \frac{2560}{2.8 \times 5.6} - \frac{6 \times 631}{2.8 \times 5.6^2}$$

$$= 120.15 \ kN/m^2$$

Output: $p_2 = 120.15 \ kN/m^2$

THE EFFECTIVE DEPTHS

Since there will be longitudinal and transverse reinforcement, there will be two effective depths – one for each direction.

(table 19) cover = 40 mm.

Assuming 25 mm ϕ bars in the longitudinal direction and 20 mm ϕ bars in the transverse direction :-

For the longitudinal reinforcement

$$d = 700 - (40 + 25/2)$$

$$= 647.5 \ mm$$

For the transverse reinforcement

$$d = 700 - (40 + 25 + 20/2)$$

$$= 625 \ mm$$

Output: $d = 647.5$ mm. & 625 mm.

CURPE CONSULTANTS
46 Orburn Road
DUNFIELD

Contract	DESIGN IV
Job ref	CC/ME
Part of structure	CHAPTER EIGHT PAD FOUNDATIONS
Calc sheet No	8/5
Drawing ref	
Calculations by	B.C.
Checked by	
Date	

Code ref.	CALCULATIONS	OUTPUT
3.10.4.2	DESIGN FOR SHEAR	
3.3.6.1.	(i) shear along a vertical section	
	The critical section is 1½ d from the column face. ie 1·5 × 647·5 = 971·25 mm	

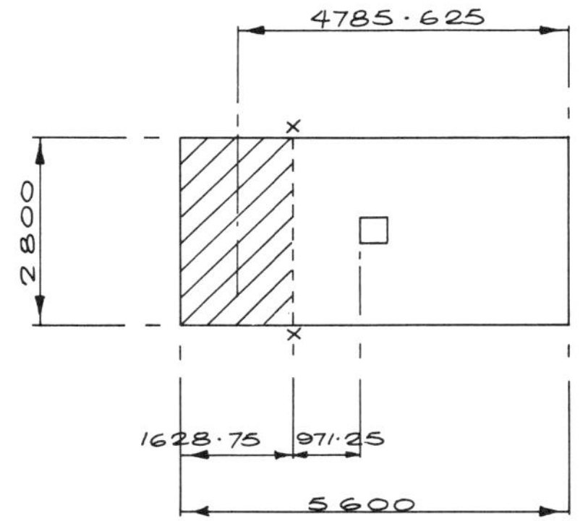

PLAN OF BASE

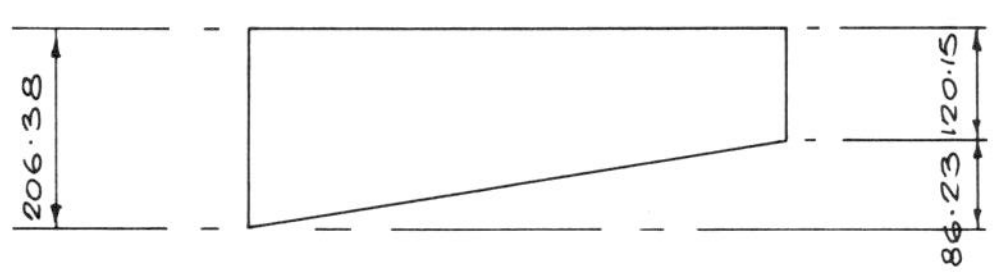

PRESSURE DIAGRAM (kN/m²)

CURPE CONSULTANTS
46 Orburn Road
DUNFIELD

Contract	DESIGN IV
Job ref	CC/ME
Part of structure	CHAPTER EIGHT PAD FOUNDATIONS
Calc sheet No	8/6
Drawing ref	
Calculations by	B.C.
Checked by	
Date	

Code ref.	CALCULATIONS	OUTPUT
	The force producing shear on section X-X is that on the shaded area. = area × average pressure.	
	Average pressure $= 120{\cdot}15 + \dfrac{86{\cdot}23 \times 4{\cdot}785}{5{\cdot}60}$	
	$= \underline{193{\cdot}84\ \text{kN/m}^2}$	
	Shear force = area × pressure	
	$= 2{\cdot}80 \times 1{\cdot}629 \times 193{\cdot}84$	
	$= \underline{884{\cdot}14\ \text{kN}}$	
3.4.5.1.	Shear stress $= \dfrac{884{\cdot}14 \times 10^3}{2800 \times 647{\cdot}5}$	
	$= \underline{0{\cdot}49\ \text{N/mm}^2}$	$v = 0{\cdot}49\ \text{N/mm}^2$
3.4.5.2	(ii) punching shear	
	The critical section is a perimeter 1½ h from the column face. ie. 1·5 × 700 = 1050 mm	

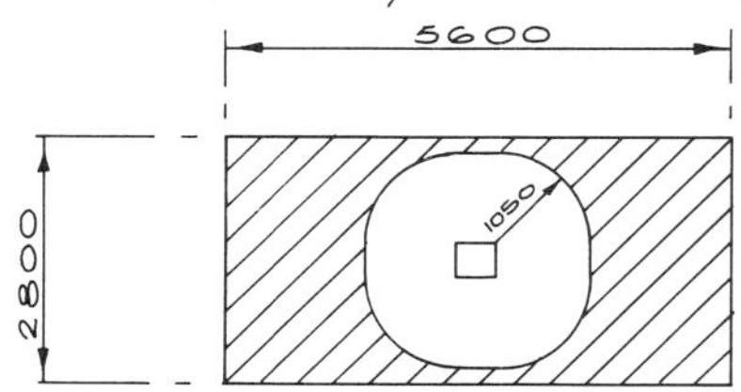

CURPE CONSULTANTS, 46 Orburn Road, DUNFIELD	Contract: DESIGN IV	Job ref: CC/ME
	Part of structure: CHAPTER EIGHT PAD FOUNDATIONS	Calc sheet No: 8/7
	Drawing ref: — Calculations by: B. C. Checked by: —	Date:

Code ref. | CALCULATIONS | OUTPUT

The force producing shear on the perimeter is that on the shaded area. This could be calculated as the area × average pressure.

However this would not compensate for the uneven distribution of pressure. When dealing with shear in flat slabs C.P. 110 states, in clause 3.6.2, 'the design shear force should be taken as 1.25 times the shear force calculated to allow for non symmetrical distribution of shear.' Notwithstanding this statement CP110 does not mention uneven distribution of pressure under bases.

In order to make a compensation it is suggested that the pressure used is not the average pressure, under the entire base, but the average of p_1 and p_3 as indicated below

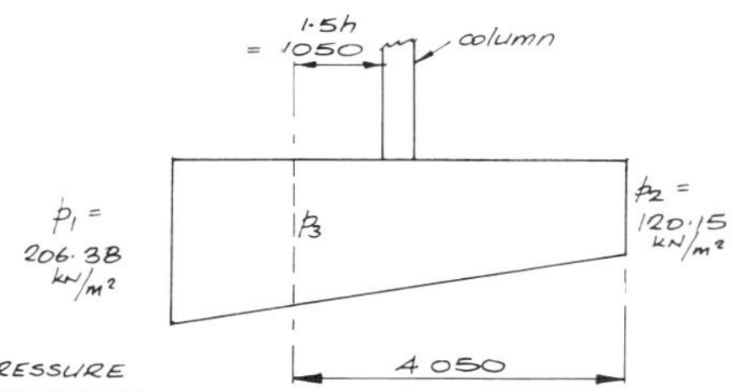

PRESSURE DISTRIBUTION

$$p_3 = \left[120{\cdot}15 + \frac{(206{\cdot}38 - 120{\cdot}15)}{5{\cdot}6} \times 4{\cdot}05\right]$$

$$= 182{\cdot}51\ \mathrm{kN/m^2}$$

CURPE CONSULTANTS, 46 Orburn Road, DUNFIELD	Contract: DESIGN IV	Job ref: CC/ME
	Part of structure: CHAPTER EIGHT PAD FOUNDATIONS	Calc sheet No: 8/8
	Drawing ref: — Calculations by: B. C. Checked by: —	Date:

Code ref. | CALCULATIONS | OUTPUT

The average of p_1 and p_3

$$= \frac{206{\cdot}38 + 182{\cdot}51}{2}$$

$$= 194{\cdot}45\ \mathrm{kN/m^2}$$

Area within the perimeter

= area of the column + area of the four rectangles corresponding to the sides of the column + the area of the four quadrants of a circle, making a complete circle

$$= (0{\cdot}4 \times 0{\cdot}4) + 4(0{\cdot}4 \times 1{\cdot}05) + \pi \times 1{\cdot}05^2$$

$$= 5{\cdot}304\ \mathrm{sq.m.}$$

The shaded area (pg 8/6) $= (5{\cdot}6 \times 2{\cdot}8) - 5{\cdot}304$

$= 10{\cdot}376$ sq.m.

The shear force $= 194{\cdot}45 \times 10{\cdot}376$

$= 2017{\cdot}56$ kN — **OUTPUT:** $V = 2017{\cdot}56$ kN

The length of the critical perimeter is 'b' in the calculation for shear stress

$b = 4 \times 0{\cdot}4 + 2 \times \pi \times 1{\cdot}05$

$= 8{\cdot}1973$ m — **OUTPUT:** $b = 8{\cdot}1973$ m

Since the punching shear perimeter embraces both the longitudinal and transverse reinforcement, the effective depth will be the average of these two :-

CURPE CONSULTANTS
46 Orburn Road
DUNFIELD

Contract: DESIGN IV	Job ref: CC/ME
Part of structure: CHAPTER EIGHT PAD FOUNDATIONS	Calc sheet No: 8/9
Drawing ref: — Calculations by: B.C. Checked by: —	Date: —

Code ref.	CALCULATIONS	OUTPUT
	i.e. $d = \dfrac{647.5 + 625}{2}$	
	$= \underline{636.25\text{mm}}$	$d = 636.25\text{mm}$
	shear stress $= \dfrac{2017.56 \times 10^3}{8197.3 \times 636.25} = \underline{0.39 \text{ N/mm}^2}$	$v = 0.39 \text{ N/mm}^2$
	The worst case is for (i) ie transverse shear where $v = 0.49 \text{ N/mm}^2$	$v_{max} = 0.49 \text{ N/mm}^2$
	<u>REINFORCEMENT FOR SHEAR</u>	
	Since it is undesirable to reinforce a pad for shear the situation of $v < \xi_s v_c$ must obtain.	
	i.e. $v = 0.49 < \xi_s v_c$	
table 14	$\xi_s = 1.00$	
	$\therefore v_c > 0.49 \text{ N/mm}^2$	
table 5	The value of $\dfrac{100A_s}{bd}$ required to ensure that $v_c > 0.49 \text{ N/mm}^2 = 0.48$	$\dfrac{100A_s}{bd} = 0.48$
	<u>CHECK BENDING</u>	
	The critical section for bending is at the face of the column	

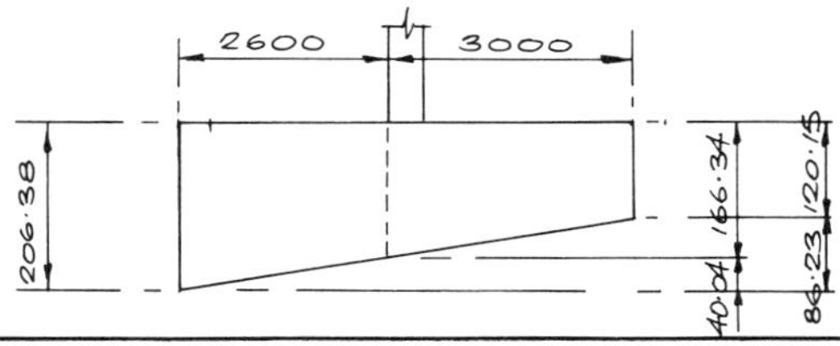

CURPE CONSULTANTS
46 Orburn Road
DUNFIELD

Contract: DESIGN IV	Job ref: CC/ME
Part of structure: CHAPTER EIGHT PAD FOUNDATIONS	Calc sheet No: 8/10
Drawing ref: — Calculations by: B.C. Checked by: —	Date: —

Code ref.	CALCULATIONS	OUTPUT
	Pressure at the face of the column	
	$= 120.15 + \dfrac{86.23 \times 3}{5.6}$	
	$= \underline{166.34 \text{ kN/m}^2}$	
	Bending moment at the face	
	$= 166.34 \times \dfrac{2.6^2}{2} \times 2.8$	
	$+ \dfrac{1}{2} \times 40.04 \times 2.6 \times 2.6 \times \dfrac{2}{3} \times 2.8$	
	$= \underline{1826.67 \text{ kN.m.}}$	B.M. = <u>1826.67kN.m</u>
	$\dfrac{M}{bd^2} = \dfrac{1826.67 \times 10^6}{2800 \times 647.5^2}$	
	$= 1.56 \text{ N/mm}^2$	
chart 3	$\dfrac{100A_s}{bd} = 0.46$	$\dfrac{100A_s}{bd} = 0.46$
	This value is insufficient for shear which requires 0.48	
	$\therefore A_s = \dfrac{0.48 \times 10^3 \times 647.5}{10^2}$	
	$= 3108 \text{ mm}^2/\text{m}$	
	<u>USE T25mm ϕ bars at 150mm c/c (3270)</u>	<u>T25 @ 150 c/c</u>

CURPE CONSULTANTS 46 Orburn Road DUNFIELD	Contract: DESIGN IV Part of structure: CHAPTER EIGHT PAD FOUNDATIONS Drawing ref: · Calculations by: B.C. · Checked by:	Job ref: CC/ME Calc sheet No: 8/11 Date:

Code ref.	CALCULATIONS	OUTPUT
3.11.6.1.	CHECK LOCAL BOND	
	This should be checked at the critical section for bending ie at the face of the column	
	The shear force = area × average press.	
	$= 2.6 \times 2.8 \times \left(166.34 + \frac{40.04}{2}\right)$	
	$= 1356.70$ kN	
	$f_{bs} = \frac{V}{\Sigma u_s d} = \frac{1356.70 \times 10^3}{19 . \pi . 25 \times 647.5}$	
	$= 1.40\ N/mm^2$	
table 21	Permissible $f_{bs} = 2.50\ N/mm^2$	
	∴ SATISFACTORY	Local Bond O.K.
3.11.8.2.	CHECK CRACKING	
	percentage of reinforcement	
	$= \frac{19 \times 491 \times 100}{2800 \times 700}$	
3.11.8.2. (3)	$= 0.48\% < 0.50\%$	
table 24	∴ the maximum clear distance between bars = 180 × 2 = 360 mm	
	the actual clear distance = 150 − 25 = 125 mm	
	∴ SATISFACTORY	Cracking O.K.

CURPE CONSULTANTS 46 Orburn Road DUNFIELD	Contract: DESIGN IV Part of structure: CHAPTER EIGHT PAD FOUNDATIONS Drawing ref: · Calculations by: B.C. · Checked by:	Job ref: CC/ME Calc sheet No: 8/12 Date:

Code ref.	CALCULATIONS	OUTPUT
	TRANSVERSE REINFORCEMENT	
	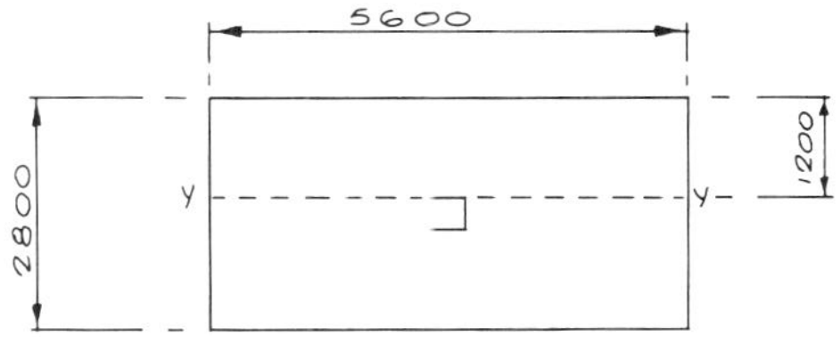	
	Bending Moment on Y-Y	
	$= \left(120.15 \times 5.6 \times \frac{1.2^2}{2}\right) + \left(\frac{86.23}{2} \times 5.6 \times \frac{1.2^2}{2}\right)$	
	$= 658.28$ kN m.	
	$\frac{M}{bd^2} = \frac{658.28 \times 10^6}{5600 \times 625^2} = 0.30\ N/mm^2$	
chart 3	$\frac{100A_s}{bd} < 0.15$ which is the minimum	
3.11.4.1.	$A_s = \frac{0.15 \times 10^3 \times 625}{10^2}$	
	$= 937.5\ mm^2/m$	
	Use T16 mm ϕ bars at 200 mm c/c (1010)	T16 @ 200 c/c
	N.B. This reinforcement should be distributed as outlined in clause 3.10.4.1. of CP110.	

SELF-ASSESSEMENT QUESTIONS

1. Outline at least four reasons why foundations are to be treated with special care.

2. Describe five types of foundations illustrating where each would be used.

3. Why is it necessary to consider two load cases when designing a pad foundation?

4. The main reinforcement in a concrete column consists of 6T25 mm dia. longitudinal bars which are fully stressed in tension. The cover to the main reinforcement is 36 mm, $f_{cu} = 30\,N/mm^2$ and $f_y = 425\,N/mm^2$. Calculate the radius required in the bend of the starter bars.

5. A reinforced concrete pad foundation is 3.25 m × 3.25 m × 0.5 m. The loads imposed from a 300 mm square, centrally placed column are:

(*a*) characteristic dead load = 1250 kN;
(*b*) characteristic imposed load = 750 kN.

If $f_{cu} = 30\,N/mm^2$ and $f_y = 425\,N/mm^2$, use CP110 to design the pad.

6. Design a pad foundation for the following design data:

(*a*) axial dead load = 1200 kN;
(*b*) axial imposed load = 575 kN;
(*c*) bending moment due to dead load = 375 kN m;
(*d*) bending moment due to imposed load = 25 kN m;
(*e*) the column is 450 mm square and is centrally positioned on the pad;
(*f*) safe ground bearing pressure = 200 kN/m²;
(*g*) $f_{cu} = 30\,N/mm^2$;
(*h*) $f_y = 425\,N/mm^2$.

CHAPTER NINE

Reinforced Concrete Cantilevered Retaining Walls

CHAPTER OBJECTIVES

After studying this chapter you should know:

* the common types of retaining walls;
* what pressures are on retaining walls;
* how these pressures are distributed;
* how to evaluate these pressures;
* how a retaining wall fails;
* how to design a reinforced concrete cantilever retaining wall.

INTRODUCTION

It is the purpose of this chapter to outline the types of retaining walls, pressures on retaining walls, their modes of failure and to illustrate the design of a cantilevered reinforced concrete retaining wall. The student will not find an exposition of the theories of soil mechanics since this is available in many textbooks.

TYPES OF RETAINING WALL

There are three main types of retaining wall:

(*a*) gravity;
(*b*) cantilevered;
(*c*) counterforted.

Gravity retaining walls

A gravity wall is designed on the simple principle that the mass of the wall must counteract, with a suitable factor of safety, the action of the retained material (*see* Chapter Sixteen). It can be constructed of mass concrete, brickwork, blockwork, masonary or rubble. In brickwork or blockwork a rough guide is that the thickness of the wall at the bottom should be about 1/3 of the height retained (*see* Fig. 40). Also it will be found that a reinforced concrete wall will be more economical for a retained height of around 2.0 m.

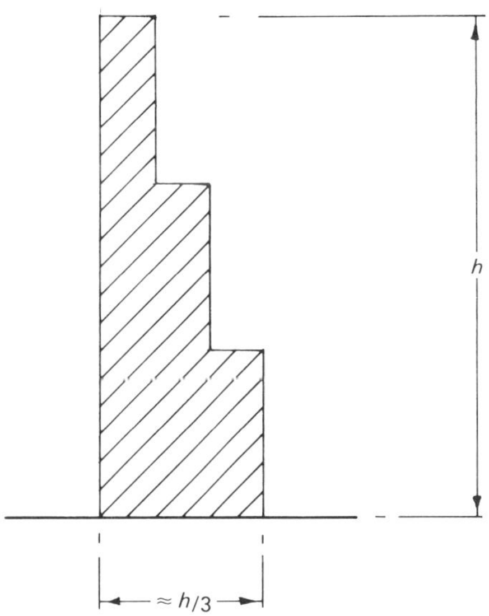

FIG. 40. Gravity retaining wall constructed in brickwork or blockwork.

Cantilevered retaining walls

This is perhaps the most common type of wall and may be constructed of reinforced concrete, precast concrete, driven or bored piles or steel sheet piles.

Counterforted retaining walls

A counterforted wall is used at heights that make a cantilever construction uneconomical (around 6–7 m for reinforced concrete). It is

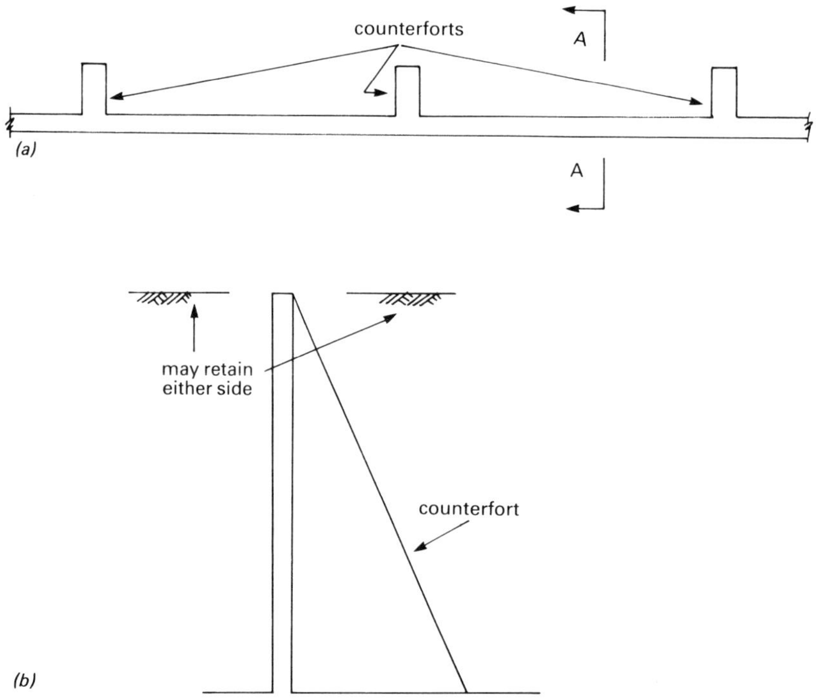

FIG. 41. *Counterforted retaining wall:* (a) *plan of the wall; and* (b) *Section* A–A.

designed as a slab spanning horizontally between the counterforts which are designed as cantilevers and are normally shaped to suit the cantilever moment (*see* Fig. 41).

PRESSURES ON RETAINING WALLS

The pressures on retaining walls can arise from one of three sources:

(*a*) earth;
(*b*) water;
(*c*) wind.

Earth pressure

The engineer normally uses the Rankine theory to estimate soil pressures. This is both easy to remember and simple to use. However, it by no means gives a finely accurate estimate of these pressures. In Rankine it is assumed that there is no friction between the wall and the adjacent soil. This is seldom true, but it results in an over-estimation of the pressure applied to the wall (the active pressure), and an under-estimation of the pressure resisting movement of the wall (the passive pressure). The Rankine theory is most accurate for cohesionless soils, and this is satisfactory since, for the majority of retaining walls, the backfill is a granular fill.

In general terms, Rankine states that the pressure per square metre at a depth h metres due to level fill of earth:

$$= \gamma h \left(\frac{1 - \sin\phi}{1 + \sin\phi}\right)$$

where γ = the unit weight of the soil in kN/m^3, and ϕ = the angle of internal friction of the soil.

The value of ϕ depends on whether total stress or effective stress conditions are being used (i.e. undrained or drained). In this chapter ϕ is taken in terms of total stress, with the exception of the values for clay and top soil in Table 6, which are in terms of effective stress.

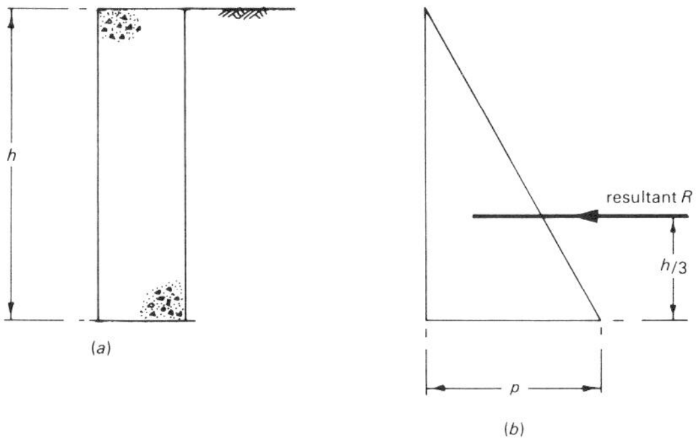

FIG. 42. (a) *Retaining wall and* (b) *pressure distribution.*

From Rankine (for Fig. 42):

$$p = \gamma h \left(\frac{1 - \sin\phi}{1 + \sin\phi}\right)$$

$$R = \frac{\gamma h}{2}\left(\frac{1 - \sin\phi}{1 + \sin\phi}\right)$$

These are usually calculated per metre length of wall.

Where there is an imposed load on top of the earth being retained, this can be considered as an equivalent depth of earth, as shown in Fig. 43, where:

$$h_1 = \frac{\text{intensity of imposed load}}{\text{unit weight of the earth}} = \frac{x}{\gamma}$$

Using Rankine:
the pressure at level A

$$= \gamma h_1 \left(\frac{1 - \sin\phi}{1 + \sin\phi}\right)$$

and the pressure at level B

$$= \gamma(h + h_1)\left(\frac{1 - \sin\phi}{1 + \sin\phi}\right)$$

The resultant pressure is the area of trapezoidal pressure distribution (shown shaded) and it acts at the centroid of the trapezoid. Some typical examples of γ and ϕ are shown in Table 6.

TABLE 6.
VALUES OF γ AND ϕ IN RANKINE EQUATIONS

Retained material	$\gamma (kN/m^3)$	ϕ
Common gravel	17–22	35°–40°
Dry sand	15–16	30°–35°
Crushed rock	12–22	35°–45°
Broken brick	15–16	35°–45°
Ashes	9–10	35°–45°
Clay	17–22	30°–45°*
Top soil	13–16	30°–40°*

*In terms of effective stress.

Water pressure

The pressure exerted by water varies directly with the depth and is therefore triangular (*see* Fig. 44):

$p = \omega h$, where ω = the specific weight of water = 9.81 kN/m^3

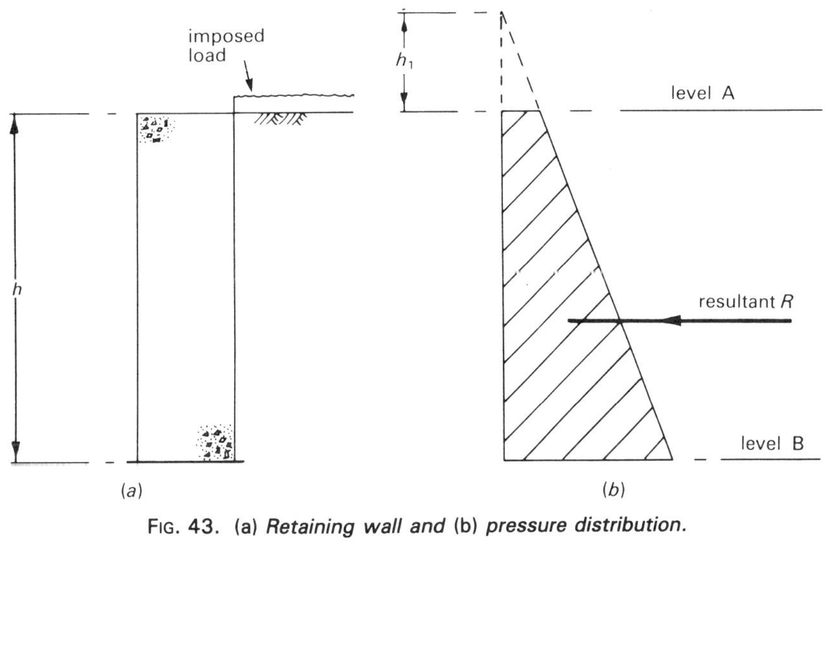

FIG. 43. (a) *Retaining wall and* (b) *pressure distribution.*

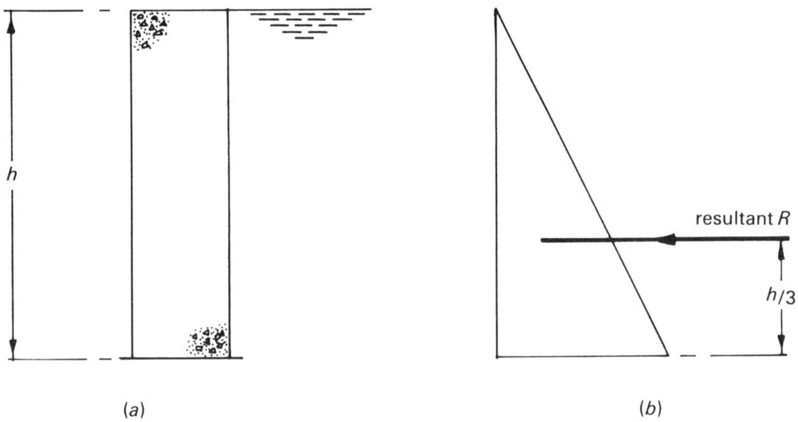

FIG. 44. *Water pressure.* (a) *Retaining wall.* (b) *Pressure distribution.*

R(per metre length of wall)

$$= \text{wetted area} \times \text{average pressure}$$

$$= (h \times 1) \times \left(\frac{\omega h}{2}\right)$$

$$\therefore \quad R = \frac{\omega h^2}{2}$$

Where the retaining face is non-vertical, the principle is the same, remembering that the pressure is always normal to the surface (*see* Fig. 45):

R(per metre length of wall)

$$= \text{wetted area} \times \text{average pressure}$$

$$= (L \times 1) \times \left(\frac{\omega h}{2}\right)$$

$$R = \frac{\omega h L}{2}$$

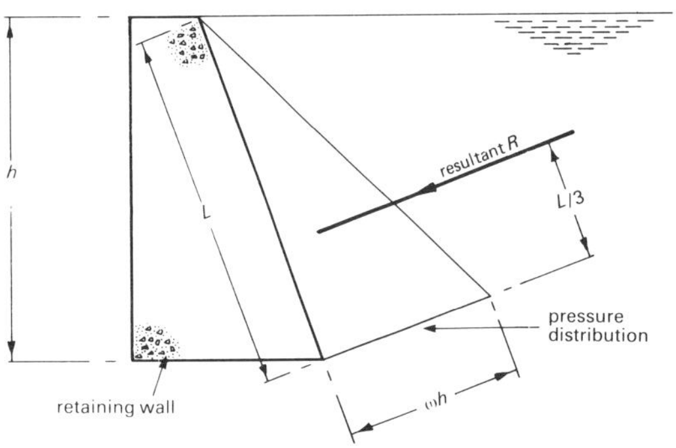

FIG. 45. *Water pressure for non-vertical retaining face.*

Wind pressure

The pressure from wind, unlike earth and water pressure, does not increase with depth. The pressure remains constant and thus the resultant force acts at half the height of the wall (*see* Fig. 46). This is true for relatively low walls, but in a tall structure the wind pressure increases uniformly with height.

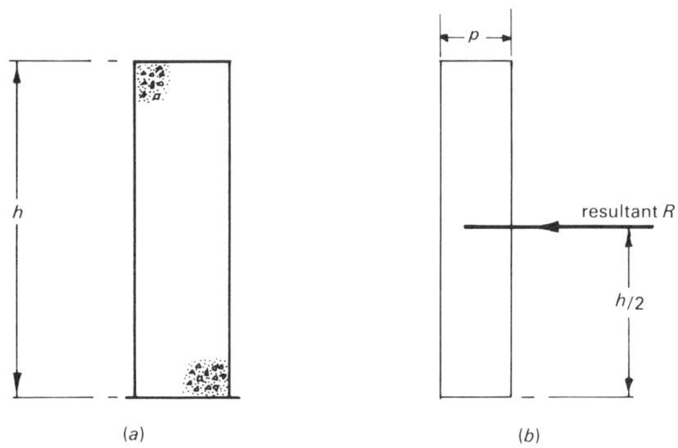

FIG. 46. *Wind pressure.* (a) *Retaining wall.* (b) *Pressure distribution.*

For a 1 m strip of wall $R = \text{ph}$. The pressure can be calculated using CP3 Chapter V: Part 2 *Wind Loads.*

MODES OF FAILURE

There are four main modes of failure:

(*a*) rotation of the wall;
(*b*) sliding of the wall;
(*c*) rotational failure of a soil mass;
(*d*) structural failure of the wall.

Failure by rotation or overturning

(*a*) *Due to the position of the resultant.* Figure 47 shows how the two forces acting on a free-standing wall, namely the active pressure P and the weight of the wall W, continue to produce an inclined resultant R.

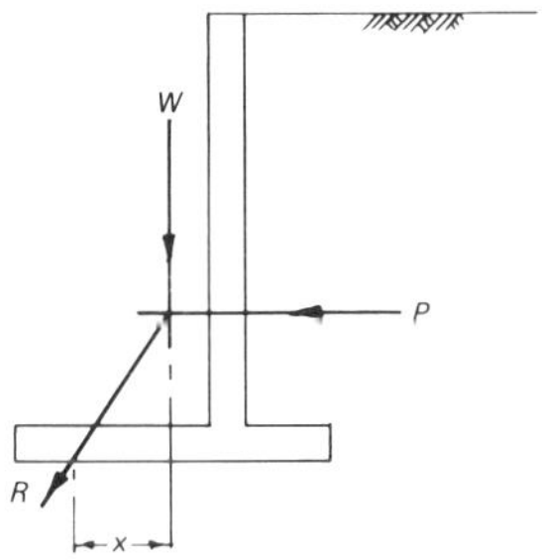

FIG. 47. *Failure by rotation or overturning due to the position of the resultant.*

If R falls outside the width of the base, the wall will fail immediately by overturning. However, if the base is wide enough and/or the resultant is back far enough, the wall will be stable against overturning.

(*b*) *Due to the bearing pressure being exceeded.* Even when the resultant is within the base, and by statistics it is proved that the wall will not overturn, there is the possibility that the bearing pressure under the toe of the wall will be exceeded, thus causing rotation (*see* Fig. 48).

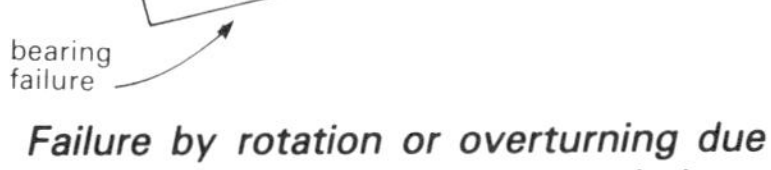

FIG. 48. *Failure by rotation or overturning due to the bearing pressure being exceeded.*

As stated above, the effect of the horizontal thrust and the vertical weight combine to produce an inclined resultant, the effect of which is to produce a base carrying an eccentric load (*see* Fig. 47). The magnitude of the eccentric load is the vertical component of the resultant (i.e. W). The distance x can easily be found from the triangle of forces:

$$x = \frac{h}{3} = P : W$$

$$\therefore \quad x = \frac{Ph}{3W}$$

The eccentricity (i.e. the distance from the centre of the base to point where the resultant cuts the base) can easily be calculated when x is known. This allows the pressures under the toe and the heel of the base to be calculated from the familiar formula:

$$p = \frac{W}{A} \pm \frac{M}{Z}$$

$$= \frac{W}{A} \pm \frac{We}{Z}$$

where A = the area of the base and Z = the section modulus of the base. These are usually calculated for a 1 m length of a wall.

Failure by sliding

Resistance to sliding is provided by two things—the passive resistance of the soil in front of the wall, and the friction between the soil and the underside of the base. The former of these is not often utilised by an engineer due to the difficulty in ensuring that the earth against which the base is cast is undisturbed at the time of construction and will remain so during the life of the wall. Also a cohesive soil is prone to seasonable movement due to shrinkage, and if a gap appears between the base and the earth, there is obviously no resistance to sliding. Thus it is usually the latter which is used in in design calculations.

For a cohesionless soil the base friction can be calculated as:

Resisting force = total vertical weight × coefficient of friction between the base and the earth.

That is $F_R = W \times \tan \phi$

For a cohesive soil the base friction can be

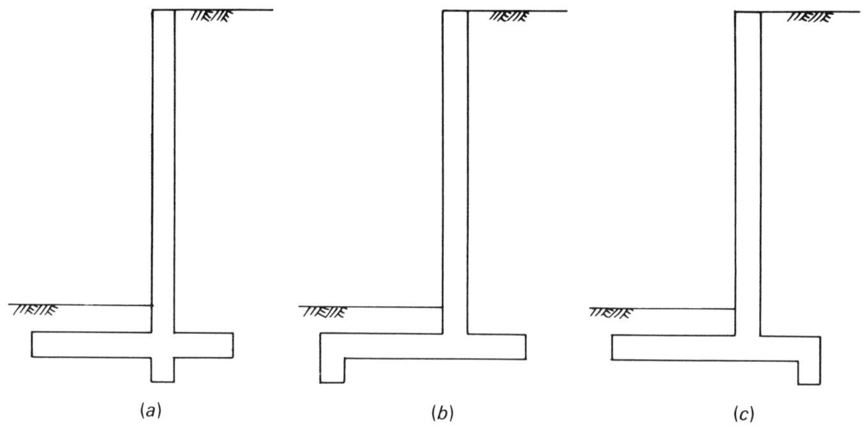

FIG. 49. *Keys to resist sliding.*

calculated as

$$F_R = cL$$

where c = the cohesion of the soil in terms of total stress, and L = the length of the base under pressure. Here c is usually limited to about $50\,kN/m^2$ because it is around this point that the adhesion between the concrete and the clay becomes the limiting factor, regardless of the shear strength of the clay.

If sliding is a problem, it is usual to provide a key as indicated in Fig. 49.

Failure by rotation of a soil mass (cartwheeling)

In this case it is not the ground close to the wall which fails, but it is a larger plane of failure which carries the wall with it (*see* Fig. 50).

Here the instability arises due to the out-of-balance moment caused by the moment of the weight of earth behind the wall being greater than the moment of the weight of earth in front of the wall, assisted by the shear resistance along the plane of failure. These moments are taken about some centre of rotation which is

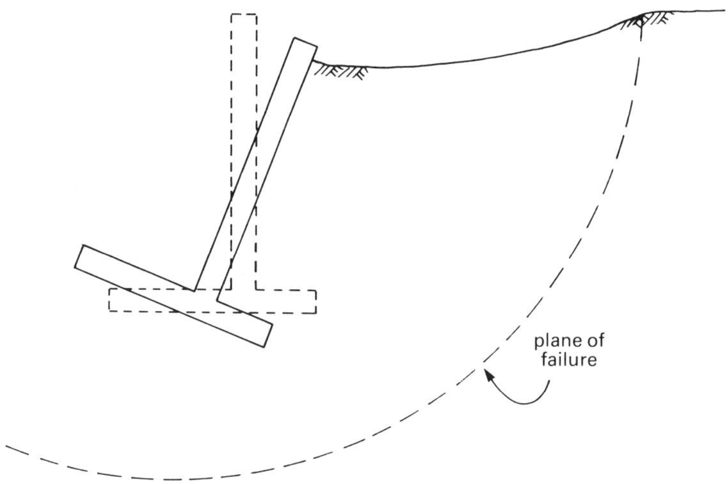

FIG. 50. *Failure by rotation of a soil mass.*

usually found from a trial and error procedure based on work by Fellenius.

This type of failure is most likely to occur in poor strength, non-frictional soils.

Failure of the wall itself

In this case the soil pressure exerts stresses on the wall which are beyond those that can safely be resisted by the materials of which the wall is constructed. The wall will therefore crack and fail. The following calculations illustrate how a reinforced concrete cantilevered retaining wall is designed to resist these pressures and stresses.

CURPE CONSULTANTS 46 Orburn Road DUNFIELD	Project DESIGN IV	Job ref CC/ME
	Sub-section CHAPTER NINE R.C. RETAINING WALL	Sheet no 9/1
		Made by B.C.
	Structural Summary Sheet	Date
		Checked by
		Date

A REINFORCED CONCRETE CANTILEVER RETAINING WALL

A reinforced concrete cantilever retaining wall is required to meet the following conditions:

(i) the wall and base have the dimensions shown below,
(ii) the characteristic strength of the concrete, $f_{cu} = 30\ N/mm^2$,
(iii) the characteristic strength of the reinforcement $f_y = 425\ N/mm^2$,
(iv) for the retained granular material the angle of internal friction = 35° and the density = 1800 kg/m³ (17.66 kN/m³),
(v) the maximum safe ground bearing pressure = 175 kN/m²,
(vi) the water table is below formation level

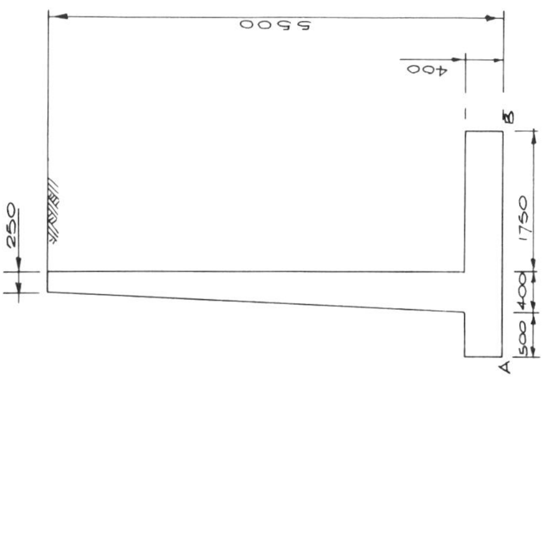

CURPE CONSULTANTS
46 Orburn Road
DUNFIELD

Contract	Job ref
DESIGN IV	CC/ME
Part of structure: CHAPTER NINE R.C. RETAINING WALL	Calc. sheet No: 9/2
Drawing ref: — Calculations by: B.C. Checked by: —	Date:

Code ref.	CALCULATIONS	OUTPUT

LOADING

There are two conditions of loading which must be considered.

(i) When the ground bearing pressure is being checked the service loads (i.e. unfactored) are used because the safe ground bearing pressure already contains a factor of safety

(ii) When the wall itself (i.e. the concrete and reinforcement) is being designed the loading must include the partial factors of safety on the loads, which when combined with the partial factors of safety on the stresses, give the overall factor of safety.

This concept has already been introduced in the previous chapter on pad foundations.

HORIZONTAL THRUST

From Rankine the horizontal thrust due to the retained earth

$$= \frac{\gamma h^2}{2}\left[\frac{1-\sin\phi}{1+\sin\phi}\right]$$

$$= \frac{17.66 \times h^2}{2}\left[\frac{1-\sin 35^\circ}{1+\sin 35^\circ}\right]$$

$$= 2.39\,h^2 \text{ kN}$$

CURPE CONSULTANTS
46 Orburn Road
DUNFIELD

Contract	Job ref
DESIGN IV	CC/ME
Part of structure: CHAPTER NINE R.C. RETAINING WALL	Calc. sheet No: 9/3
Drawing ref: — Calculations by: B.C. Checked by: —	Date:

Code ref.	CALCULATIONS	OUTPUT

FORCES AND MOMENTS - UNFACTORED

PART OF WALL	FORCE (kN)	DISTANCE FROM ℄ BASE TO c.g of PART (m)	MOMENT ABOUT ℄ OF BASE (kN. m) −	MOMENT ABOUT ℄ OF BASE (kN. m) +
Stem	5.10 × 0.25 × 23.6 = 30.09	0.550	16.55	
(triangle)	½ × 0.15 × 5.10 × 23.6 = 9.03	0.725	6.55	
Base	2.65 × 0.4 × 23.6 = 25.02	0	0	
Earth Backing	1.75 × 5.1 × 17.66 = 157.62	0.450		70.93
Horizontal Thrust	2.39×5.5^2 = 72.30	5.5/3	132.55	
	Σ vertical forces = 221.76		Σ moments: − 84.72	

CURPE CONSULTANTS 46 Orburn Road DUNFIELD	Contract: DESIGN IV Part of structure: CHAPTER NINE R.C. RETAINING WALL	Job ref: CC/ME Calc sheet No: 9/4
	Drawing ref: — Calculations by: B.C. Checked by: —	Date:

Code ref.	CALCULATIONS	OUTPUT
	the eccentricity $= \dfrac{\Sigma \text{ Moments}}{\Sigma \text{ vertical forces}}$	
	$= \dfrac{84.72}{221.76}$	
	$= 0.382m$	$e = 0.382m$
	ie within the middle third.	
	CHECK GROUND BEARING PRESSURE	
	$p = \dfrac{W}{A} \pm \dfrac{We}{Z}$	
	for a 1m length of wall $d = 1m$	
	$\therefore p = \dfrac{W}{b}\left[1 \pm \dfrac{6.e}{b}\right]$	
	$= \dfrac{221.76}{2.65}\left[1 \pm \dfrac{6 \times 0.382}{2.65}\right]$	
	$= 156.06 \ kN/m^2$	
	and $11.31 \ kN/m^2$	
	Since both values are less than $175 \ kN/m^2$ the ground bearing pressure is SATISFACTORY	S.B.P. - O.K.
	CHECK OVERTURNING	
	Take moments about the toe ie. point A as indicated on page 9/1	
	The moment causing overturning	
	$= 132.55 \ kN.m.$	

CURPE CONSULTANTS 46 Orburn Road DUNFIELD	Contract: DESIGN IV Part of structure: CHAPTER NINE R.C. RETAINING WALL	Job ref: CC/ME Calc sheet No: 9/5
	Drawing ref: — Calculations by: B.C. Checked by: —	Date:

Code ref.	CALCULATIONS	OUTPUT
	The restoring moment	
	$= 30.09 \times 0.775 + 9.03 \times 0.6$	
	$+ 25.02 \times 1.325 + 157.62 \times 1.775$	
	$= 320.68 \ kN.m$	
	The factor of safety against overturning	
	$= \dfrac{320.68}{132.55} = 2.42$	
	The factor of safety should be at least 1.5 although many engineers prefer 2.0. The value of 2.42 is SATISFACTORY	Overturning O.K
	CHECK SLIDING	
	The force causing sliding $= 72.30 \ kN$	
	The force resisting sliding	
	= the total vertical load × tan. ϕ	
	$= 221.76 \times \tan 35°$	
	$= 155.28 \ kN$	
	Factor of safety $= \dfrac{155.28}{72.30}$	
	$= 2.15$	
	This should be at least 1.5 and is therefore SATISFACTORY	Sliding O.K

CURPE CONSULTANTS 46 Orburn Road DUNFIELD	Contract: DESIGN IV	Job ref: CC/ME
	Part of structure: CHAPTER NINE R.C. RETAINING WALL	Calc sheet No: 9/6
	Drawing ref: — Calculations by: B.C. Checked by: —	Date:

Code ref. | CALCULATIONS | OUTPUT

TO DESIGN THE WALL & THE BASE

FORCES & MOMENTS - FACTORED

PART OF WALL	FORCE (kN)	DISTANCE FROM ℄ BASE TO c.g. OF PART (m)	MOMENT ABOUT ℄ OF BASE (kN.m) (– / +)
Stem	30·09 × 1·4 = 42·13	0·550	23·17
(triangle)	9·03 × 1·4 = 12·64	0·725	9·16
Base	25·02 × 1·4 = 35·03	0	0
Earth Backing	157·62 × 1·6 = 252·19	0·450	113·49
Horizontal Thrust	72·30 × 1·6 = 115·68	5·5/3	212·08

$\sum$ vertical forces = 341·99

$\sum$ moments = - 130·92

CURPE CONSULTANTS 46 Orburn Road DUNFIELD	Contract: DESIGN IV	Job ref: CC/ME
	Part of structure: CHAPTER NINE R.C. RETAINING WALL	Calc sheet No: 9/7
	Drawing ref: — Calculations by: B.C. Checked by: —	Date:

Code ref. | CALCULATIONS | OUTPUT

the eccentricity $= \dfrac{\sum \text{moments}}{\sum \text{vertical forces}}$

$= \dfrac{130 \cdot 92}{341 \cdot 99}$

$= 0 \cdot 383$ m. — OUTPUT: $e = 0{\cdot}383$m

DESIGN BASE PRESSURE

$= \dfrac{341 \cdot 99}{2 \cdot 65} \left[1 \pm \dfrac{6 \times 0 \cdot 383}{2 \cdot 65} \right]$

$= 240 \cdot 96$ kN/m²

and $17 \cdot 14$ kN/m²

THE EFFECTIVE DEPTH OF BASE & THE BOTTOM OF THE WALL

table 19 — the cover = 30mm

assuming 20mm ϕ bars

$d = 400 - (30 + 20/2)$

$= 360$mm — OUTPUT: $d = 360$mm

CURPE CONSULTANTS
46 Orburn Road
DUNFIELD

Contract	Job ref	Part of structure	Calc. sheet No	Drawing ref	Calculations by	Checked by	Date
DESIGN IV	CC/ME	CHAPTER NINE R.C. RETAINING WALL	9/8		B.C.		

Code ref.	CALCULATIONS	OUTPUT

TO FIND THE RESULTANT BASE PRESSURE

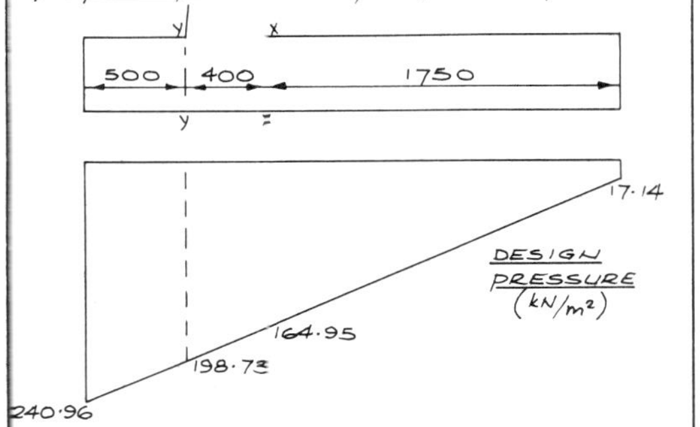

APPLIED PRESSURE (kN/m²)

EARTH $= 1.6 \times 5.10 \times 17.66$

$= 144.11$

CONC. BASE $= 1.4 \times 0.4 \times 23.6 = 13.22$

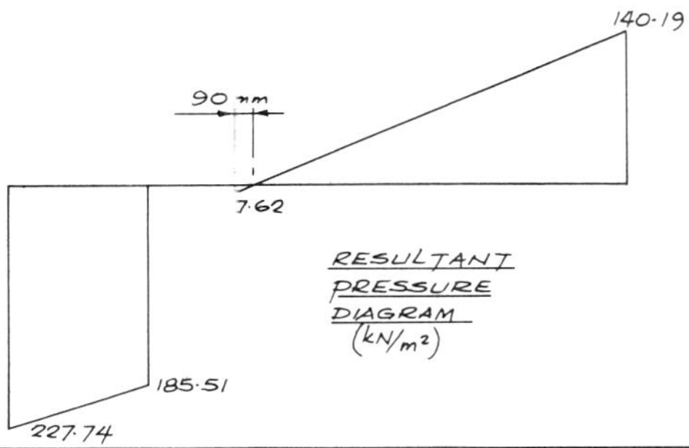

CURPE CONSULTANTS
46 Orburn Road
DUNFIELD

Contract	Job ref	Part of structure	Calc. sheet No	Drawing ref	Calculations by	Checked by	Date
DESIGN IV	CC/ME	CHAPTER NINE R.C. RETAINING WALL	9/9		B.C.		

Code ref.	CALCULATIONS	OUTPUT

DESIGN OF HEEL ie. SECTION X-X

Bending Moment

$$= (0.5 \times 140.19 \times 1.66) \times \left(1.66 \times \frac{2}{3} + 0.09\right)$$

$$- (0.5 \times 7.62 \times 0.09) \times \left(\frac{0.09}{3}\right)$$

$= 139.23$ kN.m. — OUTPUT: $M = 139.23$ kN.m

$$\frac{M}{bd^2} = \frac{139.23 \times 10^6}{10^3 \times 360^2} = 1.07 \text{ N/mm}^2$$

chart 3: $$\frac{100A_s}{bd} = 0.30$$

$$A_s = \frac{0.30 \times 10^3 \times 360}{10^2}$$

$= 1080$ mm²/m

Use T20 mm ϕ bars at 250 mm c/c (1260) — OUTPUT: T20 @ 250 or T16 @ 175

T16 mm ϕ bars at 175 mm c/c (1150) would also be sufficient and would give a slight increase in the effective depth.

The choice of bars may be left until the detailing stage when the most suitable would be chosen.

CURPE CONSULTANTS
46 Orburn Road
DUNFIELD

Contract	Job ref
DESIGN IV	CC/ME
Part of structure: CHAPTER NINE R.C. RETAINING WALL	Calc sheet No: 9/10
Drawing ref: — Calculations by: B.C. Checked by: —	Date: —

Code ref.	CALCULATIONS	OUTPUT
	DESIGN OF TOE i.e. SECTION Y-Y	
	Bending Moment	
	$= \left(185.51 \times \frac{0.5^2}{2}\right) + \left(\frac{227.74 - 185.51}{2}\right)$	
	$\times \left(0.5 \times 0.5 \times \frac{2}{3}\right)$	
	$= 26.71$ kN.m	M = 26.71 kN.m
	$\frac{M}{bd^2} = \frac{26.71 \times 10^6}{10^3 \times 360^2} = 0.21\ N/mm^2$	
chart 3	$\frac{100A_s}{bd}$ is less than the minimum of	
3.11.4.1.	0.15	
	$A_s = \frac{0.15 \times 10^3 \times 360}{10^2}$	
	$= 540\ mm^2/m$	
	USE T12 mm ϕ bars at 200mm c/c (566)	T12 @ 200

CURPE CONSULTANTS
46 Orburn Road
DUNFIELD

Contract	Job ref
DESIGN IV	CC/ME
Part of structure: CHAPTER NINE R.C. RETAINING WALL	Calc sheet No: 9/11
Drawing ref: — Calculations by: B.C. Checked by: —	Date: —

Code ref.	CALCULATIONS	OUTPUT
	DESIGN OF THE WALL	
	The maximum bending moment will occur at the bottom of the wall ie at a depth of 5.10m. This is the position for which the reinforcement is calculated below. Since the pressure and hence the bending moment reduces as the depth reduces, it is possible to calculate the reinforcement required at various levels and adjust the detailing to suit.	
	The maximum bending moment	
	$= 2.39 \times 5.10^2 \times \frac{5.10}{3} \times 1.6$	
	$= 169.09$ kN.m	M = 169.09 kN.m
	$\frac{M}{bd^2} = \frac{169.09 \times 10^6}{10^3 \times 360^2} = 1.30\ N/mm^2$	
chart 3	$\frac{100A_s}{bd} = 0.37$	
	$\therefore A_s = \frac{0.37 \times 10^3 \times 360}{10^2}$	
	$= 1332\ mm^2/m$	
	USE T20 mm ϕ bars at 200 mm c/c (1570)	T20 @ 200
		OR
	OR T16 mm ϕ bars at 150 mm c/c (1340)	T16 @ 150
	If it was to be decided to use T16 @ 175 in the heel of the wall (see page 9/9) and the spacing of 175mm was to be standardised then T16 bars at 350 mm c/c could be alternated with T20 bars at 350mm c/c, at 175 mm c/c, giving 1475 mm²/m.	OR T20 @ 350 alternating with T16 @ 350 at 175mm c/c.

CURPE CONSULTANTS 46 Orburn Road DUNFIELD	Contract: DESIGN IV	Job ref: CC/ME
	Part of structure: CHAPTER NINE R.C. RETAINING WALL	Calc sheet No: 9/12
	Drawing ref / Calculations by / Checked by	Date

Code ref.	CALCULATIONS	OUTPUT
	SECONDARY REINFORCEMENT	
3.11.4.2	$= \frac{0.12 \times 10^3 \times 400}{10^2}$	
	$= 480\ mm^2/m$	
	USE T10mm ϕ bars at 150mm c/c (523)	T10@150
	OR T12mm ϕ bars at 200mm c/c (566)	T12@200
	This is applicable to both the wall and the base.	
	TO CALCULATE THE RADIUS REQUIRED FOR THE STARTER BARS TO THE WALL - see also chapter eight.	
	The reinforcement is say T20 @ 200, see page 9/11.	
	The tensile force in the bar	
	$= 314 \times 0.87 \times 425 \times 10^{-3}$	
	$= 116.10\ kN$	
	The bearing stress	
	$= \frac{1.5 \times 30}{1 + 2 \times 20/200} = 37.50\ N/mm^2$	
	The radius $= \frac{116.10 \times 10^3}{37.50 \times 20} = 154.8\ mm$	For T20 bar r = 155mm
	A similar calculation for the T16 bars @ 150 mm c/c reveals a radius = 125 24 mm	For T16 bar r = 126mm

SELF-ASSESSMENT QUESTIONS

1. Outline the various types of retaining walls showing where each would be used.
2. Describe the various sources of pressure which a retaining wall may have to withstand. Illustrate the pressure distribution for each.
3. Indicate four ways in which a retaining wall may fail.
4. A reinforced concrete cantilever retaining wall is required to meet the following conditions:

(*a*) the wall and base have the dimensions shown in Fig. 51;
(*b*) $f_{cu} = 25\ N/mm^2$;
(*c*) $f_y = 425\ N/mm^2$;
(*d*) for the retained granular material the angle of internal friction = 30° and the unit weight = $16.5\ kN/m^3$;
(*e*) the safe ground bearing capacity = $150\ kN/m^2$;
(*f*) the water table is below formation level.
Design the wall using CP110.

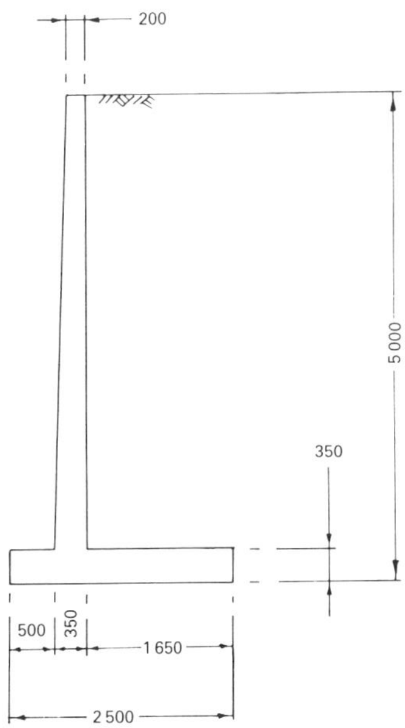

FIG. 51.

CHAPTER TEN

Structural Steelwork

> CHAPTER OBJECTIVES
>
> After studying this chapter you should know:
> * what a framework consists of;
> * the purpose of concrete casing;
> * the grades and sections of steel available;
> * the choice of design methods available;
> * the sources of information for loading.

STRUCTURAL DESIGN

Structural engineering covers the planning, design, detailing, fabrication and erection of structures using various materials. The purpose of structural design is to provide a safe and economical structure which also fulfills its intended function.

The architect usually produces the initial scheme for the structure to meet the requirements of the client or customer. An overall structural system scheme of adequate strength and stiffness is evolved taking aesthetic, economic and functional considerations also into account. This is generally the responsibility of the structural engineer. A decision also has to be made as to which structural material is to be used such as steel, concrete, timber or masonry.

Preliminary designs

Alternative designs for different structural systems are often carried out to enable appraisals and comparisons to be made during the planning stage and to arrive at the most suitable and economical scheme. Steel frame buildings will now be considered in a general fashion.

Steel frame buildings

Structural steelwork is suitable for single-bay, multi-bay, single-storey and multi-storey construction with flat or pitched roofs. It is also ideal for large clear span structures and the speed of erection on the site is an important factor in the choice of this material for many structures.

Roof and wall cladding can be readily fixed to the steel framing.

Steel framework

This consists of structural steel elements with ends prepared for connecting together by means of bolting or welding and designed to carry dead, imposed and wind loads, transmitting them safely to the foundations.

Elements

These include purlins, roof trusses, stanchions, bracing, framing rails, lattice girders, plate girders and portal frames.

Connections

Loads are transferred from one element to another by means of connections between them.

The more common type of connections are steel angle cleats, plates, T sections and section cuttings used in conjunction with bolts or welds.

Fabrication and erection

The steel framework is designed and detail drawings are prepared for fabrication purposes. The elements and the connections are fabricated in workshops, part assembled, and transported to the site for subsequent erection on prepared foundations. Examples of steel frameworks are shown in Fig. 52.

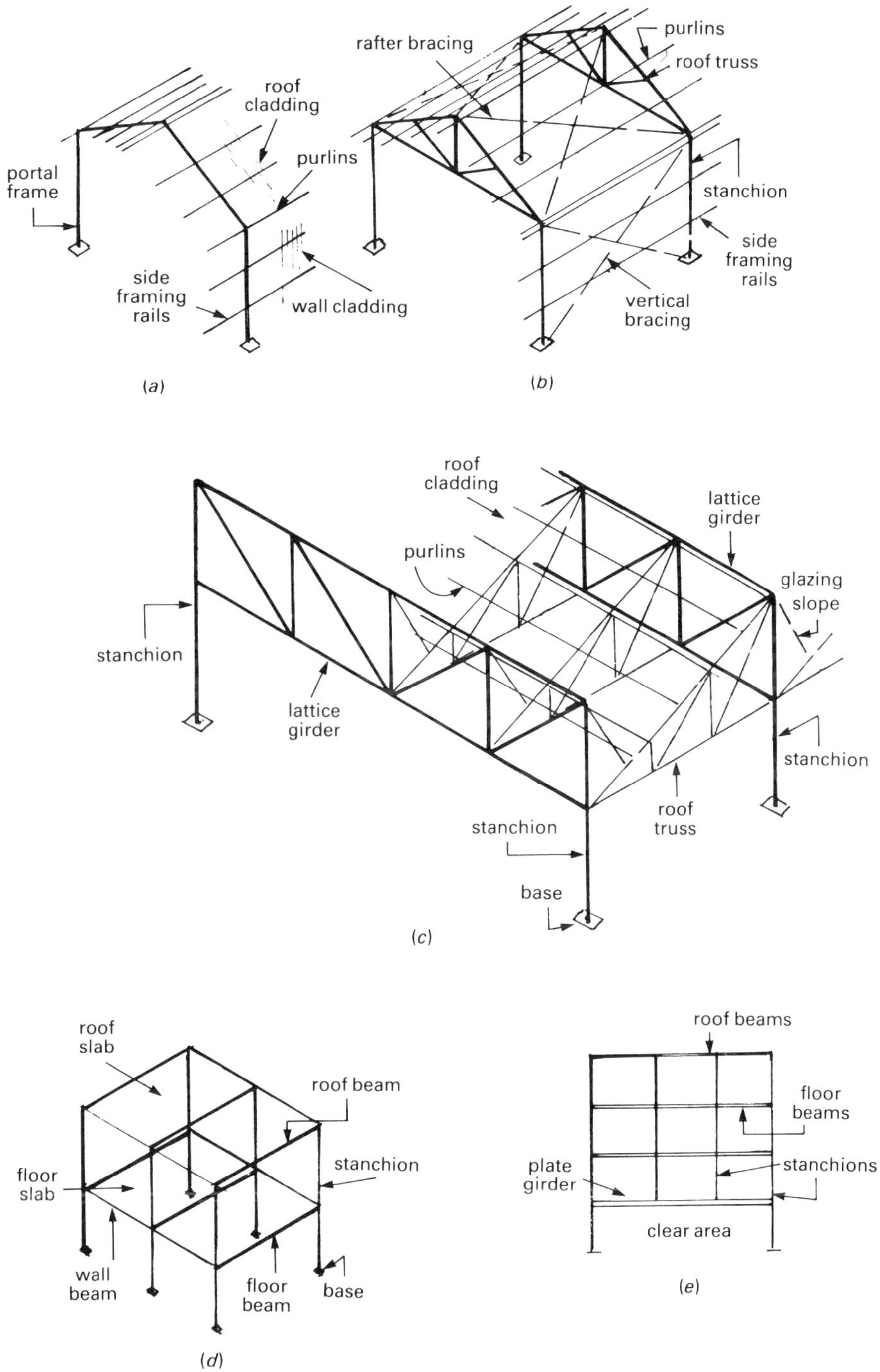

FIG. 52. *Fabrication and erection of steel work:* (a) *portal frame;* (b) *roof trusses on stanchions;* (c) *lattice girders on stanchions;* (d) *two-storey block; and* (e) *multi-storey building with plate girder.*

Concrete casing
Elements may be encased solidly in concrete on the site to provide fire protection and resistance to corrosion. The concrete may also provide additional strength to beams and stanchions if certain conditions are fulfilled.

Grades of steel
Structural steel is available in three basic grades, 43, 50 and 55. Grade 43 is mild steel with grades 50 and 55 being higher strength steels. Grade 43 is used for all examples throughout this book.

Standard steel sections
Hot rolled sections
A range of shapes called steel sections is produced at steel rolling mills. The shapes, dimensions, properties and safe loads for each section is given in:

(*a*) *Structural Steelwork Handbook* published by BCSA/CONSTRADO;
(*b*) *Structural Steelwork Handbook (Angles)* published by CONSTRADO.

A variety of steel sections is shown in Fig. 53.

Cold-formed sections
A range of shapes particularly suitable for roof purlins and framing or sheeting rails are available. Typical sections are shown in Fig. 54.

DESIGN METHODS

Analysis of the frames determines the loads, moments and shears in the elements, and design is carried out in accordance with BS449 Part 2, *The Use of Structural Steel in Building* and may be made by any of the different methods given in Clause 9. These are:

(*a*) simple design to the elastic theory;
(*b*) semi-rigid design;
(*c*) rigid design.

Simple design to elastic theory
The structure is to be assumed as pin jointed for the purpose of design. The following assumptions are made:

(*a*) Beams are simply supported.
(*b*) All connections of beams, girders or trusses are proportioned to resist the reaction shear forces, applied at the appropriate eccentricity.

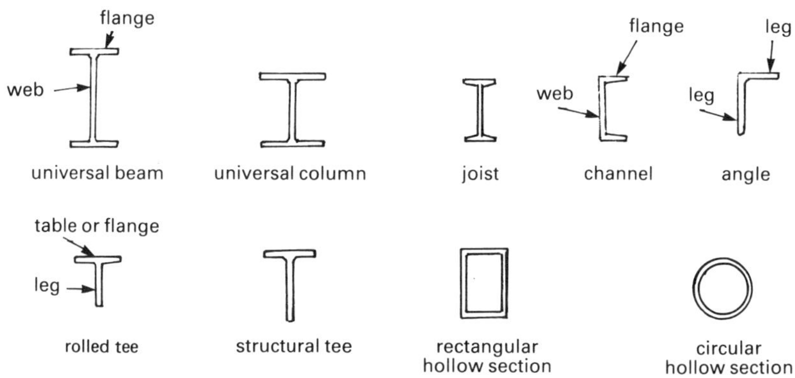

FIG. 53. *Typical structural sections.*

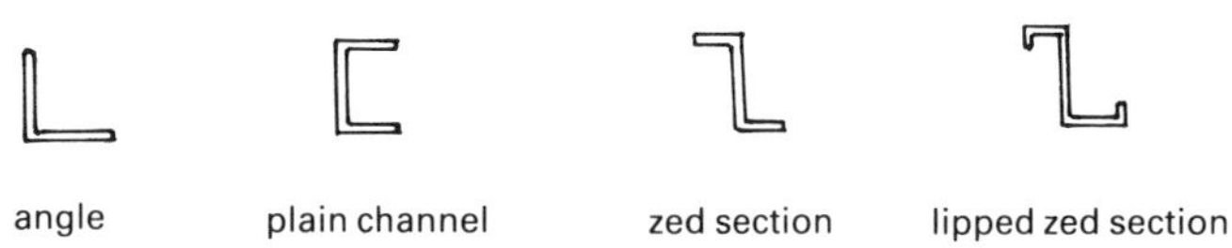

FIG. 54. *Typical cold-formed sections.*

(*c*) Members in compression are subjected to loads applied at the appropriate eccentricity (Clause 34) with effective lengths given in Clauses 30 and 31.

Members in tension are subjected to longitudinal loads applied over the net area of the section, as specified in Clause 42.

For a structure designed to this method, bracing, shear walls or core walls are necessary to provide resistance to lateral loading. Simple design is the method employed in this book.

Semi-rigid design permits a reduction in the maximum bending moment in beams suitably connected to their supports. It is seldom adopted. Fully rigid design will achieve the greatest rigidity and economy when used in appropriate cases. The design method may be based on elastic analysis or alternatively based on the principles of plastic design. Portal frames are designed to this method with welded and HSFG bolted joints.

Loading

Elements are designed to carry roof, floor, wall loads etc. For information regarding dead, imposed and wind loads reference should be made to:

(*a*) CP3 Chapter V Part 1 *Dead and Imposed Loads*;

(*b*) CP3 Chapter V Part 2 *Wind Loads*; and

(*c*) BS649 *Schedule of Weights of Building Materials.*

Another useful source of information is from manufacturers' catalogues which give dimensions, properties and safe loads for their range of products.

Scope

Selected structural elements and connections are treated as isolated components and designed as such in the following chapters.

The overall stability, strength and stiffness of the entire structural system produced by interaction of the elements and connections, however, requires investigation but this is outside the scope of this book.

CHAPTER ELEVEN

Bolted Connections

CHAPTER OBJECTIVES

After studying this chapter you should be able to:

* distinguish between single- and double-shear connections;
* calculate strengths of black bolts in shear, bearing, tension;
* calculate strength of HSFG bolts in shear and tension;
* design bolts for direct loaded connections;
* design eccentric bracket connections for: shear and torsion; shear and bending;
* design bolts for moment and shear connection at ends of beams.

INTRODUCTION

In structural steelwork, connections join individual steel elements or members together to form a structural framework. They are required to transmit various types of loading and must be designed to carry these loads without failure.

Bolts and welds in conjunction with angle cleats, plates, T- sections and section cuttings are used to make a variety of connections and joints between elements. Bolted connections are considered in this chapter. The two most common types of bolts used in structural steelwork are black bolts and high strength friction grip (HSFG) bolts.

BLACK BOLTS

These are available in grade 4.6 mild steel and grade 8.8 high strength steel and used in 2 mm clearance holes. They are commonly used for general purpose connections in simple design.

Typical shear, tension and moment connections are shown in Fig. 55. For modes of failure of bolted connections refer to Fig. 56.

Single shear connections

In a shear connection the design strength (criterion) of a bolt is either its shear or bearing capacity (value), whichever is the lower.

In single shear, the shearing value is given by:

$$\text{Shearing value} = P_s A$$

where P_s = allowable stress in shear and A = cross sectional area of the bolt shank.

In bearing, the bearing value is given by:

$$\text{Bearing value} = P_b D t$$

where P_b = allowable stress in bearing; D = diameter of bolt shank; and t = thickness of one plate (or thinner plate if appropriate).

Double-shear connections

In double shear, the shearing value of a bolt is twice that for single shear since two cross-sectional areas are involved in resisting failure. Thus:

$$\text{Shearing value} = 2P_s A$$

In bearing:

$$\text{Bearing value} = P_b D t$$

where t is the combined thickness of the two outer plates or the inner plate thickness, whichever is the smaller.

Tension connections

In a tension connection:

$$\text{Design strength of bolt} = P_t A_t$$

where P_t = allowable stress in tension and A_t = tensile stress area (i.e. the area at the root or bottom of the threaded section).

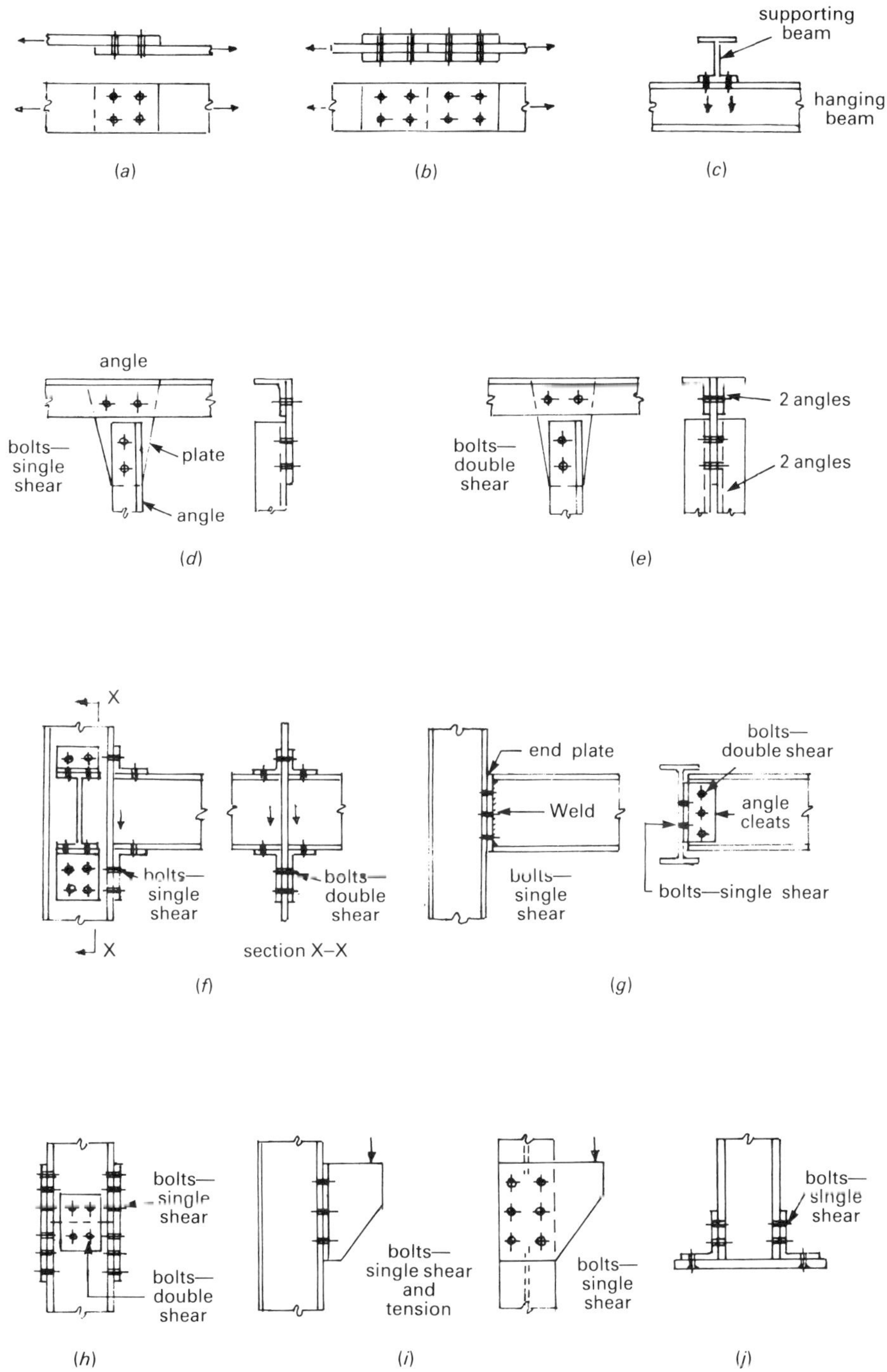

FIG. 55 *Black bolts:* (a) *single shear lap joint;* (b) *double-shear butt joint;* (c) *bolts in tension;* (d) *truss joint;* (e) *truss joint with two angles;* (f) *beam-to-stanchion connections;* (g) *beam-to-beam connection;* (h) *stanchion splice;* (i) *bracket connections; and* (j) *stanchion base.*

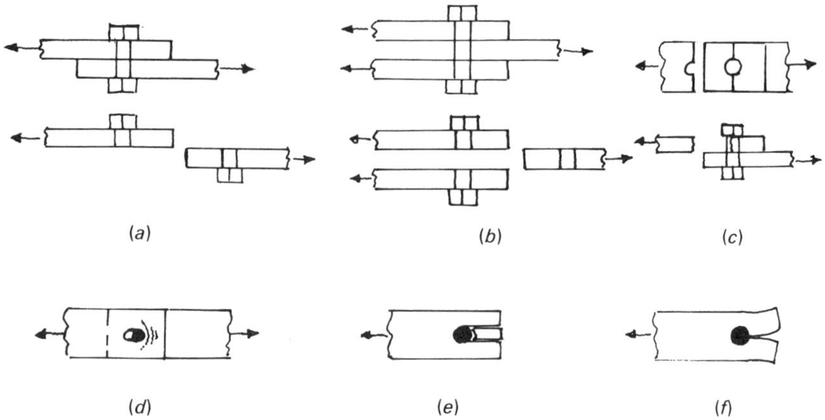

FIG. 56. *Modes of failure of bolted connections:* (a) *single shear;* (b) *double shear;* (c) *transverse tension tearing of plate;* (d) *plate bearing failure;* (e) *shear tearing failure; and* (f) *transverse tension failure.*

Combined shear and tension

Where a bolt is subjected to both shear and tension then the following stress conditions must be complied with:

$$f_s \not> P_s$$
$$f_t \not> P_t$$
$$\frac{f_s}{P_s} + \frac{f_t}{P_t} \not> 1.4$$

where f_s = actual shear stress; f_t = actual tensile stress; P_s = allowable shear stress; and P_t = allowable tensile stress.

Design strengths of grades 4.6 and 8.8 bolts

Before bolt strengths can be derived it is necessary to know:

(*a*) the allowable stresses in the bolts; and
(*b*) the allowable bearing stresses in the connected parts.

The allowable stresses in shear, bearing and tension for bolts used in clearance holes taken from Clause 50, BS449, are as follows (in N/mm^2):

(*a*) grade 4.6: $P_s = 80$; $P_b = 250$; $P_t = 120$.
(*b*) grade 8.8: $P_s = 187$; $P_b = 585$; $P_t = 280$.

The allowable bearing stresses in connected parts for bolts used in clearance holes taken from Table 20A, BS449 are as follows (in N/mm^2):

(*a*) grade 43: $P_b = 250$.
(*b*) grade 50: $P_b = 350$.
(*c*) grade 55: $P_b = 400$.

As an example, calculations will now be made to derive the design strengths of an M20 bolt subjected to various types of loading and conditions.

Bolts in single-shear connections

The following method is used to derive the design strength of an M20 bolt (for both grades 4.6 and 8.8), which connects together two plates each 6 mm thick, of grade 43 steel.

For a grade 4.6 bolt:

$$\text{Shearing value} = P_s A$$
$$= 80 \times \frac{\pi \times 20^2}{4} \times \frac{1}{10^3} = 25\,\text{kN}$$

$$\text{Bearing value} = P_b D t$$
$$= \frac{250 \times 20 \times 6}{10^3} = 30\,\text{kN}$$

Therefore the design strength

$$= 25\,\text{kN}$$

For a grade 8.8 bolt:

$$\text{Shearing value} = 187 \times \frac{\pi \times 20^2}{4} \times \frac{1}{10^3} = 59\,\text{kN}$$

$$\text{Bearing value} = \frac{250 \times 20 \times 6}{10^3} = 30\,\text{kN}$$

Therefore the design strength

$= 30\,\text{kN}$

Bolts in double-shear connections
The following method is used to derive the design strength of an M20 bolt (for both grades 4.6 and 8.8) with outer plates each 6 mm thick and inner plate 8 mm thick, of grade 43 steel.

For a grade 4.6 bolt:

Shearing value

$= 2P_sA$

$$= 2 \times 80 \times \frac{\pi \times 20^2}{4} \times \frac{1}{10^3} = 50\,\text{kN}$$

Bearing value

$= P_bDt$

$$= \frac{250 \times 20 \times 8}{10^3} = 40\,\text{kN}$$

Therefore the design strength

$= 40\,\text{kN}$

For a grade 8.8 bolt:

Shearing value

$$= 2 \times 187 \times \frac{\pi \times 20^2}{4} \times \frac{1}{10^3} = 117\,\text{kN}$$

Bearing value

$$= \frac{250 \times 20 \times 8}{10^3} = 40\,\text{kN}$$

Therefore the design strength

$= 40\,\text{kN}$

Bolts in tension connections
The following method is used to derive the design strength of an M20 bolt (for both grades 4.6 and 8.8):

$$\text{Design strength} = P_tA_t$$

$$\text{For a grade 4.6 bolt} = \frac{120 \times 245}{10^3} = 29.4\,\text{kN}$$

$$\text{For a grade 8.8 bolt} = \frac{280 \times 245}{10^3} = 68.6\,\text{kN}$$

In both cases the tensile stress area of 245 mm^2 is taken from the *Structural Steelwork Handbook*.

Bolts in combined shear and tension
Reference should be made to the examples on eccentric brackets given later in this chapter.

HIGH-STRENGTH FRICTION GRIP BOLTS

High strength friction grip (HSFG) bolts are made of high tensile steel tightened by a predetermined shank tension in order that the clamping force provided will transfer loads in the connected members by friction between the parts and not by shear in, or bearing on, the bolts or plies of connected parts. They are suitable for rigid connections as there is no slip or movement between the connected parts. It is important, however, that the bolts are tightened properly to provide the required tension. General grade bolts only are considered and the following information is given for design purposes.

Design strength of HSFG bolts

Shear connections
The design strength of one bolt is given by:

$$\frac{\text{slip factor}}{\text{load factor}} \times \text{number of interfaces} \times \text{proof load of one bolt}$$

where the slip factor is taken as 0.45; the load factor is taken as 1.4 generally (but as 1.2 for wind loads) and the number of interfaces is taken as one for single-shear connections and two for double-shear connections. Proof loads are given in the *Structural Steelwork Handbook* and for an M20 bolt is 144 kN.

Tension connections
The design strength of one bolt is given by:

$$0.6 \times \text{proof load of one bolt}$$

Combined shear and tension
The design strength of one bolt is given by:

$$\frac{\text{slip factor}}{\text{load factor}} \times \text{number of interfaces} \times (\text{proof load} - 1.7 \times \text{applied tension})$$

As an example, calculations will now be made to derive the design strengths of an M20 bolt subjected to different types of loading and conditions.

Bolts in shear connections
The following method is used to derive the design strength of an M20 bolt in single and double shear:

Design strength

$$= \frac{\text{slip factor}}{\text{load factor}} \times \text{number of interfaces} \times \text{proof load of one bolt}$$

Single shear

$$= \frac{0.45}{1.4} \times 1 \times 144 = 46.3\,\text{kN}$$

Double shear

$$= \frac{0.45}{1.4} \times 2 \times 144 = 92.6\,\text{kN}$$

Bolts in tension
The following method is used to derive the design strength of an M20 bolt in tension:

$$\text{Design strength} = 0.6 \times \text{proof load} = 0.6 \times 144 = 86.4\,\text{kN}$$

Bolts in combined shear and tension
Reference should be made to the examples on eccentric brackets given later in this chapter.

ECCENTRIC BRACKET CONNECTIONS

When a load P from a beam is applied on a line of action which does not pass through the centroid of a bolt group, the result is an eccentric loading effect. Forms of brackets fabricated from angles, section cuttings, etc. are extended out to support the beam. Connections may be subjected to shear and torsion or shear and bending.

Shear and torsion

Bolt groups may be analysed by using the traditional elastic method which is a generally conservative approach (*see* Fig. 54).

The load P applied at an eccentricity e, produces two components of shear stress at any bolt in the group. These are:

(*a*) stress due to direct shear;
(*b*) stress due to torsional moment.

The shear stresses are added vectorially to give the resultant shear stress at a point. The maximum resultant is found at the critical bolt A in the group. An adequate diameter of bolt can then be selected.

Considering direct shear, then for one bolt:

$$\text{Stress} = \frac{P}{NA}$$

where N = the number of bolts in the group; A = the cross sectional area of the bolt.

Since force = stress × area, then for one bolt, the direct shear force is given by:

$$F_s = \frac{P}{NA} \times A = \frac{P}{N}$$

Considering the torsional moment, at any bolt:

$$\text{Stress} = \frac{Mr}{\sum Ar^2}$$

where M = the torsional moment = Pe; r = the radial distance from the centroid to the bolt; and A = the cross sectional area of the bolt. This formula is derived as follows.

If F_1, F_2 etc. are the forces in the bolts then:

$$M = F_1r_1 + F_2r_2 + \ldots F_8r_8 = \sum Fr$$

The stresses are:

$$f_1 = \frac{F_1}{A} \qquad f_2 = \frac{F_2}{A} \qquad f_8 = \frac{F_8}{A}$$

where A is the cross-sectional area of the bolts.

The stress in each bolt is also proportional to its distance from the centroid:

$$\frac{f_1}{r_1} = \frac{f_2}{r_2} = \ldots \frac{f_8}{r_8}$$

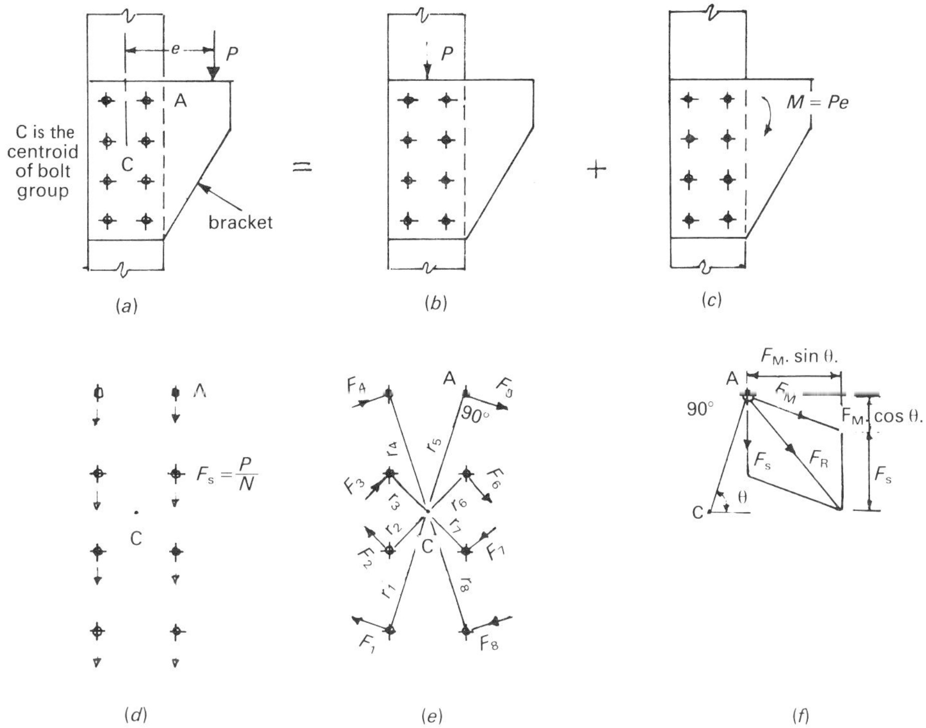

FIG. 57. *Eccentric bracket connections:* (a) *connection;* (b) direct forces; (c) *torsional moment;* (d) *direct shear forces;* (e) *torsional moment forces; and* (f) *forces vector diagram.*

Rewriting in terms of f_1 and r_1:

$$f_1 = \frac{f_1 r_1}{r_1} \quad f_2 = \frac{f_1 r_2}{r_1} \quad f_8 = \frac{f_1 r_8}{r_1}$$

Substituting:

$$M = f_1 A r_1 + f_2 A r_2 + \ldots f_8 A r_8$$

$$M = \frac{f_1 r_1}{r_1} A r_1 + \frac{f_1 r_2 A r_2}{r_1} + \ldots \frac{f_1 r_8 A r_8}{r_1}$$

$$M = \frac{f_1 r_1^2 A}{r_1} + \frac{f_1 r_2^2 A}{r_1} + \ldots \frac{f_1 r_8^2 A}{r_1}$$

$$= \frac{f_1 A}{r_1} [r_1^2 + r_2^2 + \ldots r_8^2]$$

$$= \frac{f_1}{r_1} \sum A r^2$$

The stress in bolt 1 is therefore:

$$f_1 = \frac{M r_1}{\sum A r^2}$$

Or in general terms:

$$f = \frac{Mr}{\sum A r^2}$$

Also from force = stress × area:

$$F_1 = f_1 A$$

$$= \frac{M r_1 A}{\sum A r^2}$$

$$= \frac{M r_1}{\sum r^2}$$

Or in general terms:

$$F_M = \frac{Mr}{\sum r^2} = \frac{Per}{\sum r^2}$$

The resultant shear force F_R is found by adding vectorially the direct shear force F_s and the shear force F_M due to the moment. The resultant force is given by:

$$F_R = \sqrt{(F_s^2 + F_M^2 + 2F_sF_M \cos\theta)}$$

This formula is derived using Pythagoras's theorem:

$$F_R = \sqrt{((F_M \sin\theta)^2 + (F_s + F_M \cos\theta)^2)}$$

$$= \sqrt{(F_M^2 \sin^2\theta + F_s^2 + 2F_sF_M \cos\theta + F_M^2 \cos^2\theta)}$$

$$= \sqrt{(F_M^2(\sin^2\theta + \cos^2\theta) + F_s^2 + 2F_sF_M \cos\theta)}$$

Since $\sin^2\theta + \cos^2\theta = 1$, then:

$$F_R = \sqrt{(F_s^2 + F_M^2 + 2F_sF_M \cos\theta)}$$

As an alternative F_R can be found graphically.

Shear and bending

Two methods of analysis may be used in this type of connection (*see* Fig. 58):

(*a*) *Method 1:* This method is conservative with the centre of rotation assumed to be at the bottom line of bolts with no compression taken by the contact surface area between the bracket and the face of the stanchion. The load P applied at an eccentricity e, produces two types of stress at any bolt in the group. These are (*i*) stress due to direct shear; and (*ii*) tensile stress due to bending.

Considering direct shear, then for one bolt:

$$\text{stress} = \frac{P}{NA}$$

For one bolt the direct shear force is given by:

$$F_s = \frac{P}{NA} \times A = \frac{P}{N}$$

where N = the number of bolts in the group; and A = the cross sectional area of the bolt.

Considering bending, the tensile stress at any bolt is given by:

$$\frac{MY}{I} = \frac{PeY}{I}$$

where Y = the distance from the centre of rotation to the bolt; and I = moment of inertia of the bolt group about the axis of rotation.

I is obtained as follows (with two vertical rows of bolts):

$$I = 2(AY_1^2) + 2(AY_2^2) + 2(AY_3^2) + 2(AY_4^2)$$

or in general terms:

$$I = 2\sum AY^2$$

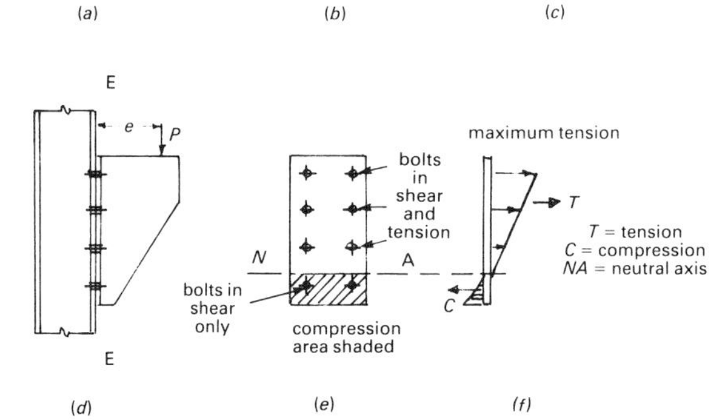

FIG. 58. *Shear and bending: Method 1* (a) *Connection;* (b) *direct shear forces and* (c) *moment forces; Method 2* (d) *connection,* (e) *section* E–E *and* (f) *Moment forces.*

and since force = stress × area then for one bolt, the tensile force due to bending is:

$$F_M = \frac{PeYA}{2\sum AY^2}$$
$$= \frac{PeY}{2\sum Y^2}$$

The bolt is designed for the combined forces F_s and F_M with the top bolts being critical.

(*b*) *Method 2:* In this method the load P applied at an eccentricity e produces (*i*) stress due to direct shear in each bolt in the group; (*ii*) tensile stress in bolts located above the neutral axis; and (*iii*) compression on part of the contact surface area between the bracket and the face of the stanchion.

The location of the neutral axis is found by taking area moments about the neutral axis assuming first of all a distance above the bottom of the bracket (e.g. above the bottom bolts). Calculations confirm or disprove this assumption and if necessary they are redone for a new location. Knowing the location of the neutral axis, the moment of inertia of the bolt areas in tension and the compression area of the bracket is calculated and the tensile stress in the top bolt determined from:

$$\frac{MY}{I} \quad \text{or} \quad \frac{PeY}{I}$$

The stress due to direct shear for one bolt is given by (P/NA). The bolt is designed for the combined stresses with the top bolts being critical.

This method is considered in further detail in *Metric Practice for Structural Steelwork* and *Structural Fasteners*.

MOMENT AND SHEAR AT ENDS OF BEAMS

Beam-to-stanchion connections may be subjected to both moment and vertical shear due to dead, imposed and wind loading (*see* Fig. 59).

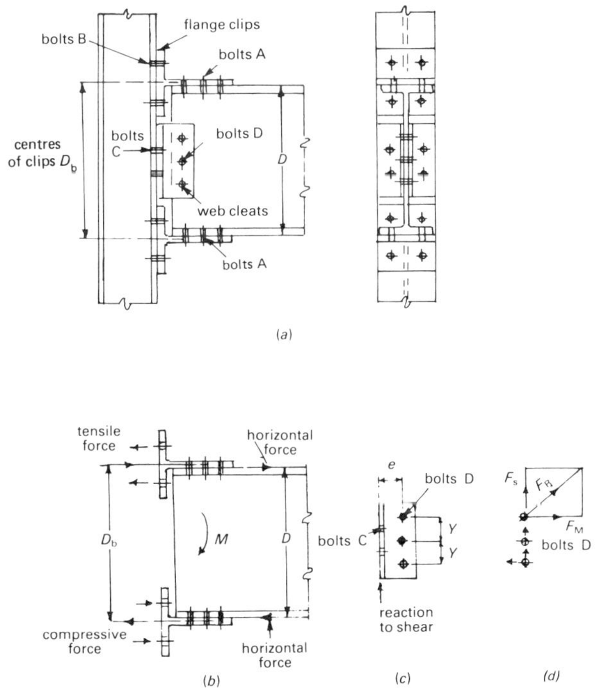

FIG. 59. *Moment and shear at ends of beams:* (a) *moment and shear connection;* (b) *moment connection;* (c) *web cleats; and* (d) *forces vector diagram.*

In this type of connection the moment is assumed to be resisted by the flange clips and vertical shear by the web cleats.

Bolts A
The single-shear force per bolt is given by:

$$F_s = \frac{M}{DN}$$

where N is the number of bolts A.

Bolts B
The tensile force per bolt is given by:

$$F_M = \frac{M}{D_b N}$$

where N is the number of bolts B.

Bolts C
The single-shear force per bolt is given by:

$$F_s = \frac{\text{vertical shear force}}{\text{number of bolts C}}$$

Bolts D
Bolts D are subjected to a direct shear force due to vertical shear and for one bolt this is given by:

$$F_s = \frac{\text{vertical shear force}}{\text{number of bolts D}}$$

The bolts are subjected also to a shear force due to moment and for one bolt this is given by:

$$F_M = \frac{PeY}{\sum Y^2}$$

The forces are added vectorially to determine the maximum resultant F_R at the top bolt.

The bolts A, B, C and D are designed individually.

The following examples illustrate the design of bolts used in structural connections.

CURPE CONSULTANTS
46 Orburn Road
DUNFIELD

Project: DESIGN IV | Job ref: CC/ME
Part of structure: CHAPTER ELEVEN BOLTED CONNECTIONS | Calc sheet no: 11/1 | rev
Drawing ref | Calc by: R.A.S. | Date | Check by | Date

Code ref. | Calculations | Output

EXAMPLE 11.1. Direct force acting in shear.

Seating cleats connected to stanchion.

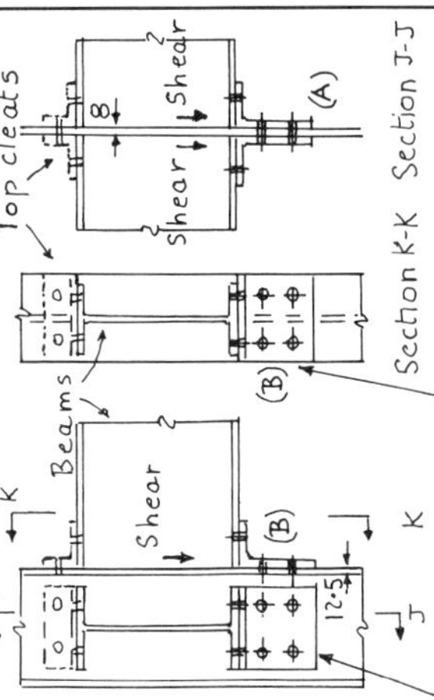

2 Cleats (A) 150×90×12 Angles M20 bolts

150×90×12 Angle cleat (B). M20 bolts

The seating cleats (A) and (B) resist the end shears for their respective beams, with the top cleats (or web cleats) steadying the top of the beam.

Cleats (A)

Determine the safe load for bolts. Using 4.6 grade black bolts, both shear and bearing have to be taken into account. For one bolt,

$$\text{Double shear value} = \frac{2 \times \pi \times 20^2 \times 80 \times 1}{4 \times 10^3}$$

$$= 50 \text{ kN}$$

CURPE CONSULTANTS
46 Orburn Road
DUNFIELD

Project: DESIGN IV	Job ref: CC/ME
Part of structure: CHAPTER ELEVEN BOLTED CONNECTIONS	Calc sheet no: 11/2 rev /
Drawing ref: — Calc by: R.A.S Date:	Check by: Date:

Code ref.	Calculations	Output
	EXAMPLE 11.1. continued.	
	Bearing value = $\frac{250 \times 20 \times 8}{10^3} = 40 kN$	
	Therefore, design strength of one bolt is 40kN i.e. the lesser value	
	And safe load for the 4 bolts is	
	4 × 40 = 160 kN.	Safe load 160kN Shear = 80kN
	Using HSFG bolts. For one bolt,	
	Double shear value = $\frac{0{\cdot}45 \times 2 \times 144}{1{\cdot}4} = 92{\cdot}6 kN$	
	Therefore safe load for the 4 bolts is	
	4 × 92·6 = 370·4 KN	Safe load 370·4kN Shear = 185·2
	Cleat (B)	
	Determine the safe load for bolts.	
	Using 4·6 grade black bolts. For one bolt	
	Single shear value = $\frac{\pi \times 20^2 \times 80}{4 \times 10^3} = 25 kN$	
	Bearing value = $\frac{250 \times 20 \times 12}{10^3} = 60 kN$	
	Therefore design strength of one bolt is 25 kN i.e. the lesser value	
	And safe load for the 4 bolts is	
	4 × 25 = 100 kN.	Safe load 100kN
	Using HSFG bolts	
	Safe load for the 4 bolts is 4 × 46·3 = 185·2	Safe load 185·2kN

CURPE CONSULTANTS
46 Orburn Road
DUNFIELD

Project: DESIGN IV	Job ref:
Part of structure: CHAPTER ELEVEN BOLTED CONNECTIONS	Calc sheet no: 11/3 rev /
Drawing ref: — Calc by: R.A.S. Date:	Check by: Date:

Code ref. | Calculations | Output

EXAMPLE 11.2. Design of bolt group in twist

Eccentric bracket connection subjected to shear and torsion.

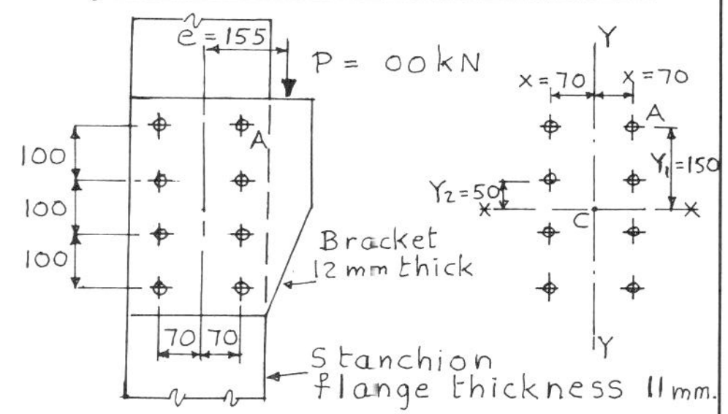

Connection Bolt group

Determine the size of bolts.

Assume cross sectional area of each bolt as unity.

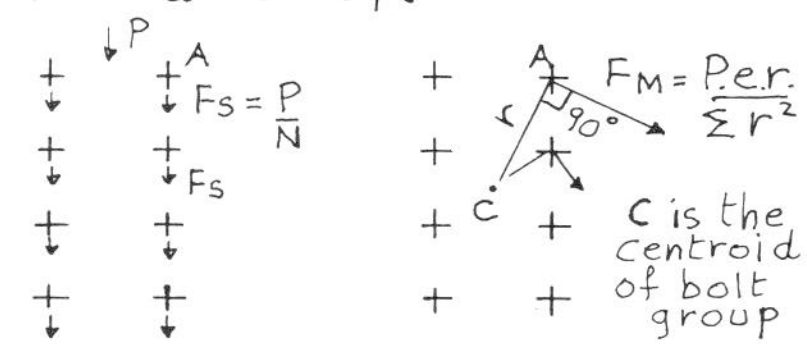

Direct shear Forces Torsional moment Forces

Design for critical bolt A

Force due to direct shear is

CURPE CONSULTANTS
46 Orburn Road
DUNFIELD

Project: DESIGN IV		Job ref: CC/ME	
Part of structure: CHAPTER ELEVEN BOLTED CONNECTIONS		Calc sheet no: 11/4	rev: 1
Drawing ref:	Calc by: R.A.S. Date:	Check by:	Date:

Code ref.	Calculations	Output

EXAMPLE 11.2. continued

$$F_S = \frac{P}{\text{Number of bolts}} = \frac{P}{N}$$

$$= \frac{100}{8} = 12{\cdot}5\ \text{kN}$$

Also force due to moment is

$$F_M = \frac{P.e.r}{\Sigma r^2}$$

$$r = \sqrt{70^2 + 150^2} = 165{\cdot}53\ \text{mm.}$$

$$\Sigma r^2 = \Sigma x^2 + \Sigma y^2$$

$$\Sigma x^2 = 8(1 \times x^2) = 8(1 \times 70^2) = 39200\ \text{mm}^4$$

$$\Sigma Y^2 = 4(1 \times Y_1^2) + 4(1 \times Y_2^2)$$

$$= 4(1 \times 150^2) + 4(1 \times 50^2) = 100000\ \text{mm}^4$$

Therefore $\Sigma r^2 = 39200 + 100000$

$$= 139200\ \text{mm}^4$$

$$F_M = \frac{100 \times 155 \times 165{\cdot}53 \times 1}{139200}$$

$$= 18{\cdot}43\ \text{kN}$$

Resultant F_R is found graphically or by calculation

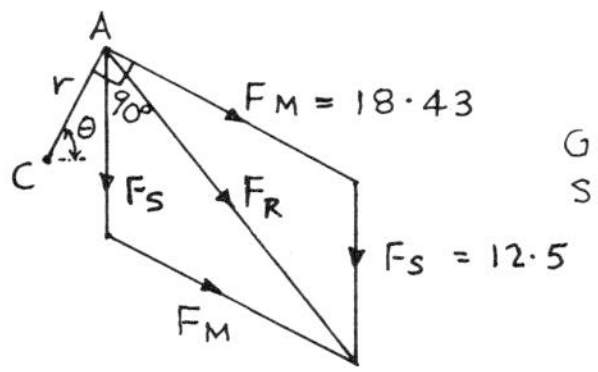

Graphical solution.

CURPE CONSULTANTS
46 Orburn Road
DUNFIELD

Project: DESIGN IV		Job ref: CC/ME	
Part of structure: CHAPTER ELEVEN BOLTED CONNECTIONS		Calc sheet no: 11/5	rev: 1
Drawing ref:	Calc by: R.A.S. Date:	Check by:	Date:

Code ref.	Calculations	Output

EXAMPLE 11.2. continued

Using the formula

$$F_R = \sqrt{F_S^2 + F_M^2 + 2 F_S . F_M . \cos\theta}$$

$$F_R = \sqrt{12{\cdot}5^2 + 18{\cdot}43^2 + 2\left(12{\cdot}5 \times 18{\cdot}43 \times \frac{70}{165{\cdot}53}\right)}$$

$$= 26{\cdot}28\ \text{kN}$$

Using HSFG bolts, say M16

Single shear value $= \frac{0{\cdot}45}{1{\cdot}4} \times 1 \times 92{\cdot}1 = 29{\cdot}6\ \text{kN}$

Since $26{\cdot}28 < 29{\cdot}6$ Bolt is adequate. — **Output:** M16 HSFG Adequate

Using M16, 8·8 grade bolts, both shear and bearing have to be accounted for.

Single shear value is

$$\frac{\pi \times 16^2}{4} \times 187 \times \frac{1}{10^3} = 37{\cdot}6\ \text{kN}$$

Bearing value is

$$\frac{250 \times 16 \times 11}{10^3} = 44\ \text{kN}$$

Therefore the safe load is 37·6 kN

Since $26{\cdot}28 < 37{\cdot}6$, Bolt is adequate. — **Output:** M16, 8·8 grade Adequate

CURPE CONSULTANTS
46 Orburn Road
DUNFIELD

Project: DESIGN IV | Job ref: CC/ME
Part of structure: CHAPTER ELEVEN BOLTED CONNECTIONS | Calc sheet no: 11/6 | rev: /
Drawing ref: | Calc by: R.A.S. | Date: | Check by: | Date:

Code ref. | Calculations | Output

EXAMPLE 11.3. Design of bolt group with moment at right angles

Eccentric bracket connection - Bolts subjected to shear and tension.

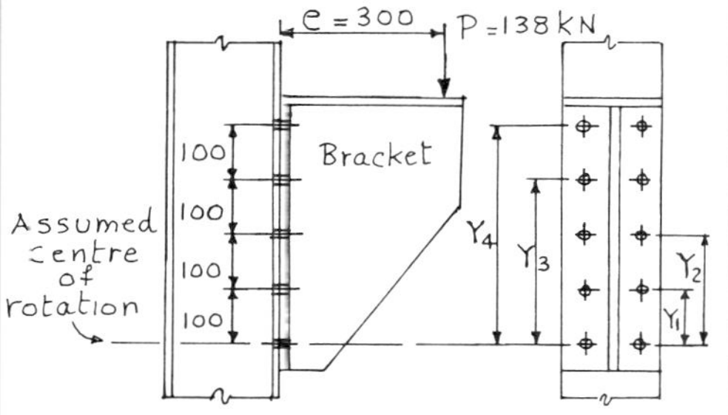

Connection

Determine the size of bolts.
Assume cross sectional area of each bolt as unity.

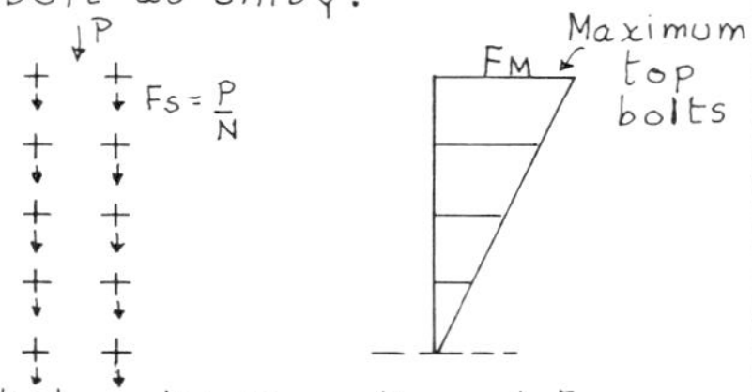

Direct shear Forces   Moment Forces

Design for critical bolt i.e. top bolts.

CURPE CONSULTANTS
46 Orburn Road
DUNFIELD

Project: DESIGN IV | Job ref: CC/ME
Part of structure: CHAPTER ELEVEN BOLTED CONNECTIONS | Calc sheet no: 11/7 | rev: /
Drawing ref: | Calc by: R.A.S. | Date: | Check by: | Date:

Code ref. | Calculations | Output

EXAMPLE 11.3. continued

Force due to direct shear is

$$F_S = \frac{P}{\text{Number of bolts}} = \frac{P}{N}$$

$$= \frac{138}{10} = 13{\cdot}8 \text{ kN}$$

Also force due to moment is

$$F_M = \frac{P.e.Y_4 \times 1}{2\sum Y^2}$$

$$2\sum Y^2 = 2(1 \times Y_1^2) + 2(1\times Y_2^2) + 2(1\times Y_3^2) + 2(1\times Y_4^2)$$

$$= 2(100^2) + 2(200^2) + 2(300^2) + 2(400^2)$$

$$= 600000 \text{ mm}^4$$

Therefore $$F_M = \frac{138 \times 300 \times 400 \times 1}{600000}$$

$$= 27{\cdot}6 \text{ kN}. \quad \text{Tension}$$

Using M20 HSFG bolts

$$\text{Single shear value} = \frac{0{\cdot}45 \times 1 \times 144}{1{\cdot}4} = 46{\cdot}3 \text{ kN}$$

For combined shear and tension this value is reduced to

$$\text{Single shear value} = \frac{0{\cdot}45 \times 1 \times (144 - 1{\cdot}7 \times 27{\cdot}6)}{1{\cdot}4}$$

$$= 31{\cdot}2 \text{ kN}.$$

Actual shear force $13{\cdot}8 < 31{\cdot}2$, therefore bolts adequate

Output: M20 HSFG Adequate

CURPE CONSULTANTS
46 Orburn Road
DUNFIELD

Project	Job ref
DESIGN IV	CC/ME
Part of structure: CHAPTER ELEVEN BOLTED CONNECTIONS	Calc sheet no: 11/8 rev: /
Drawing ref: — Calc by: R.A.S. Date: —	Check by: — Date: —

Code ref.	Calculations	Output

EXAMPLE 11.3. continued

Using M20, 4·6 grade black bolts

Cl. 50d. For combined shear and tension

$$\frac{f_s}{P_s} + \frac{f_t}{P_t} \not> 1{\cdot}4$$

$$f_s = \frac{13{\cdot}8 \times 10^3 \times 4}{\pi \times 20^2} = 43{\cdot}92 \text{ N/mm}^2$$

$$f_t = \frac{27{\cdot}6 \times 10^3 \times 4}{\pi \times 20^2} = 87{\cdot}85 \text{ N/mm}^2$$

Therefore $\frac{43{\cdot}92}{80} + \frac{87{\cdot}85}{120} = 1{\cdot}281 < 1{\cdot}4$

Bolt is adequate — **Output: M20, 4·6 grade Adequate**

The value of f_t above is determined assuming that the threaded portion of the bolt is kept sufficiently remote from the shear plane and the tensile force is taken by the shank of the bolt at that plane. Check for the threaded portion of the bolt resisting the tension.

$$f_t = \frac{27{\cdot}6 \times 10^3}{245} = 112{\cdot}65 \text{ N/mm}^2 < 120$$

Bolt is adequate — **Output: M20, 4·6 grade Adequate**

Note that the tensile stress area i.e. the area at the root of the threads is 245 mm^2 taken from the "Structural Steelwork Handbook

---

CURPE CONSULTANTS
46 Orburn Road
DUNFIELD

Project	Job ref
DESIGN IV	CC/ME
Part of structure: CHAPTER ELEVEN BOLTED CONNECTIONS	Calc sheet no: 11/9 rev: /
Drawing ref: — Calc by: R.A.S. Date: —	Check by: — Date: —

Code ref.	Calculations	Output

EXAMPLE 11.4. Moment and shear, beam to column connection

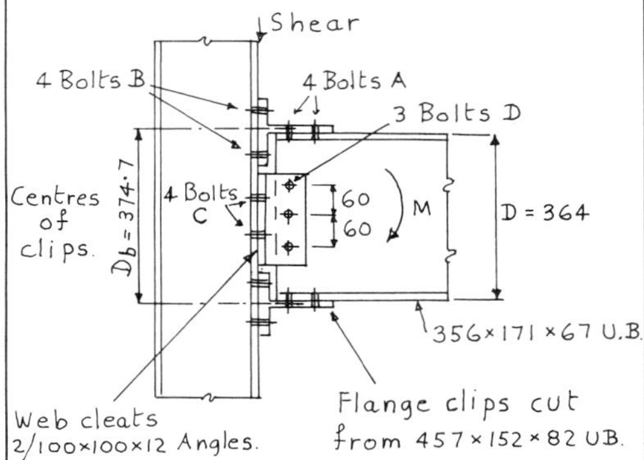

Loading details at end of beam.

Dead and imposed load = 48 kN

Wind load (Vertical) = 24 kN

Bending moment due to wind = 72 kN m.

Design the sizes of bolts (H.S.F.G).
Moment taken by flange clips.

Bolts A

Horizontal shear force due to the moment M is

$$\frac{M}{D} = \frac{72}{0{\cdot}364} = 197{\cdot}8 \text{ kN.}$$

CURPE CONSULTANTS
43 Orburn Road
DUNFIELD

Project	DESIGN IV
Part of structure	CHAPTER ELEVEN BOLTED CONNECTIONS
Drawing ref	
Calc by	R.A.S.
Date	
Job ref	CC/ME
Calc sheet no	11/10
rev	/
Check by	
Date	

Code ref.	Calculations	Output

EXAMPLE 11.4 continued

Try M20, H.S.F.G. bolts

For one bolt, single shear value is

$\frac{0{\cdot}45}{1{\cdot}2} \times 1 \times 144 = \underline{54\ kN}$

Note the load factor of 1·2 used for wind loading instead of 1·4 without wind.

Number of bolts required is

$\frac{197{\cdot}8}{54} = \underline{3{\cdot}66}$ ∴ Use 4 bolts. **Output:** Use 4 M20 HSFG.

Bolts B

Tensile force in bolts

$\frac{M}{D_b} = \frac{72}{0{\cdot}3747} = \underline{192\ kN}$

Try M20, H.S.F.G. bolts

For one bolt, tension value is

0·6 × Proof load

$= 0{\cdot}6 \times 144 = \underline{86{\cdot}4\ kN}$

Number of bolts required is

$\frac{192}{86{\cdot}4} = \underline{2{\cdot}22}$ ∴ Use 4 bolts. **Output:** Use 4 M20 HSFG

Bolts C

Since wind load of 24 kN exceeds

CURPE CONSULTANTS
46 Orburn Road
DUNFIELD

Project	DESIGN IV
Part of structure	CHAPTER ELEVEN BOLTED CONNECTIONS
Drawing ref	
Calc by	R.A.S.
Date	
Job ref	CC/ME
Calc sheet no	11/11
rev	/
Check by	
Date	

Code ref.	Calculations	Output

EXAMPLE 11.4 continued.

dead and imposed load of 48 kN by more than 25%, then design for dead + imposed + wind i.e. $48 + 24 = \underline{72\ kN}$

Try M16, H.S.F.G. bolts.

For one bolt, single shear value is

$\frac{0{\cdot}45}{1{\cdot}2} \times 1 \times 92{\cdot}1 = \underline{34{\cdot}5\ kN}$

Number of bolts required is

$\frac{72}{34{\cdot}5} = 2{\cdot}08$ ∴ Use 4 bolts. **Output:** Use 4 M16 HSFG.

Bolts D

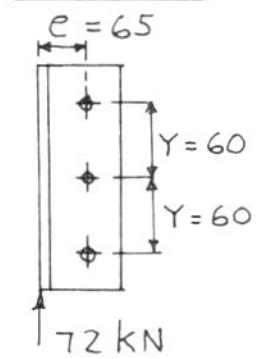

For top bolt

$F_S = \frac{72}{3} = \underline{24\ kN}$

$F_M = \frac{P.e.Y}{\Sigma Y^2}$

$F_M = \frac{72 \times 65 \times 60}{2 \times 60^2} = \underline{39\ kN}$

Resultant shear force $F_R = \sqrt{F_S^2 + F_M^2}$

$F_R = \sqrt{24^2 + 39^2} = \underline{45{\cdot}8\ kN}$

Try M16, HSFG bolts.

Bolts are in double shear.

For one bolt, double shear value is

$\frac{0{\cdot}45}{1{\cdot}2} \times 2 \times 92{\cdot}1 = \underline{69\ kN}$

Since $45{\cdot}8 < 69$, bolt is adequate **Output:** M16, HSFG Adequate

SELF-ASSESSMENT QUESTIONS

1. Check the adequacy of the 20 mm dia. bolts shown in Fig. 60 for the eccentric bracket connection using:

(*a*) black bolts, strength grade 4.6;
(*b*) HSFG bolts.

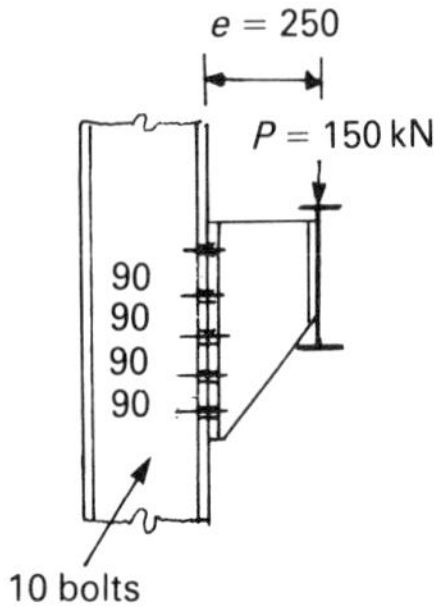

FIG. 60.

2. Check the adequacy of the 20 mm dia. bolts shown in Fig. 61 for the eccentric bracket connection using:

(*a*) black bolts, strength grade 8.8;
(*b*) HSFG bolts.

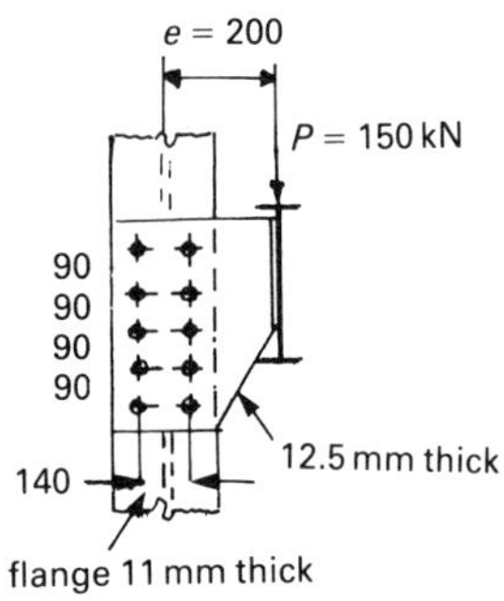

FIG. 61.

3. Determine the safe axial load W for the stanchion splice bolts 20 mm dia. shown in Fig. 62 using:

(*a*) black bolts, strength grade 8.8;
(*b*) HSFG bolts.

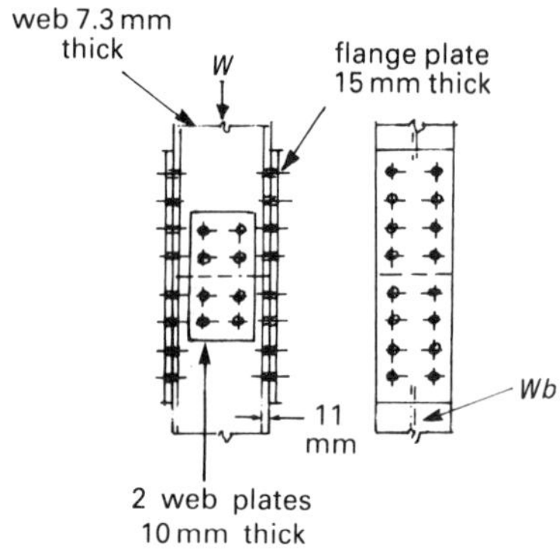

FIG. 62.

4. (*a*) Determine the capacity of both bolts A and bolts B used at the floor beam connection shown in Fig. 63.

(*b*) State the safe load for the connection.

All bolts are black bolts, strength grade 4.6.

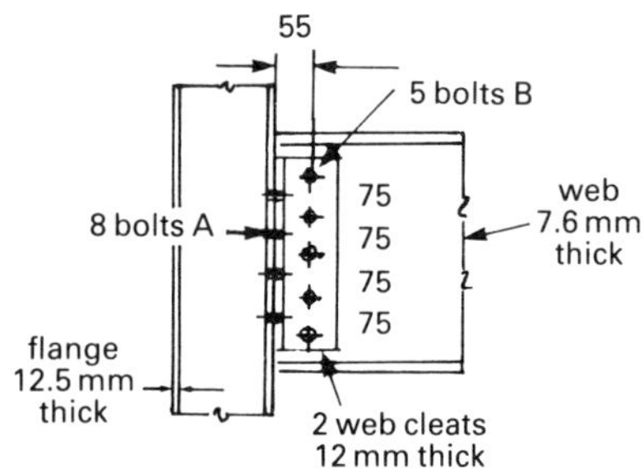

FIG. 63.

5. (*a*) Check the adequacy of the 20 mm dia. black bolts, strength grade 4.6 used for the seating cleat beam connections as shown in Fig. 64. End shears (reactions) are given.

(*b*) Could the bolts be reduced in diameter if HSFG bolts are substituted throughout instead of the black bolts.

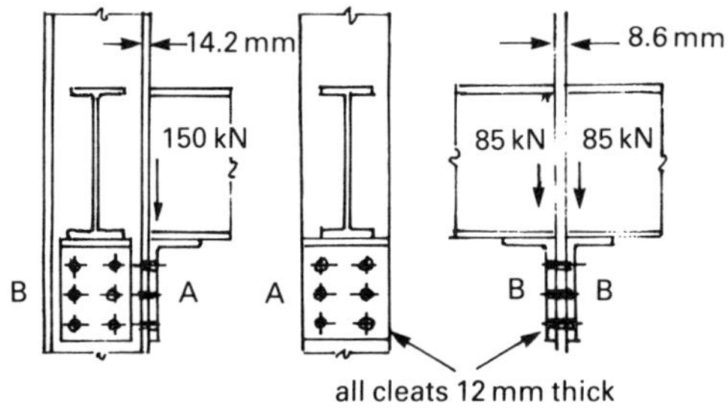

FIG. 64.

6. Determine the safe load P applied as shown in Fig. 65 using:

(*a*) 16 mm dia. black bolts, strength grade 4.6;
(*b*) 16 mm dia. HSFG bolts.

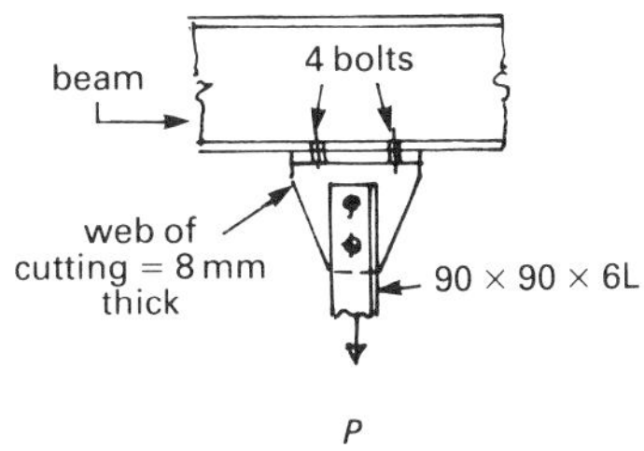

FIG. 65.

7. Determine the safe load loads for bolts A and bolts B in Fig. 66 using 16 mm dia. black bolts, strength grade 4.6.

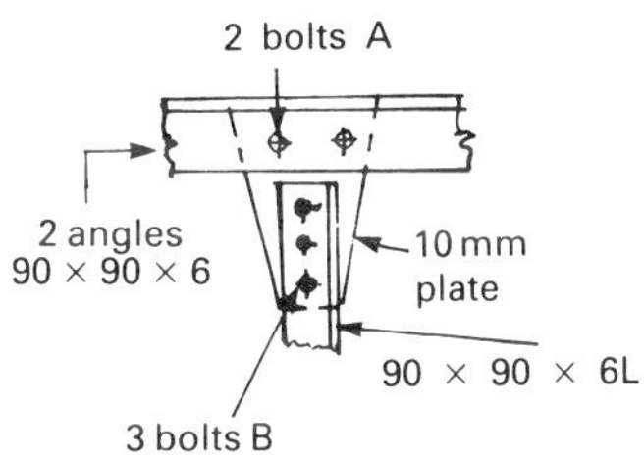

FIG. 66.

8. A moment and shear beam to stanchion connection is shown in Fig. 67. The following data is given:

Beam section 457 × 152 × 82 UB
Stanchion section 254 × 254 × 132 UB
End moment (wind) = 30 kN m
End shear (dead + imposed) = 50 kN
End shear (wind) = 25 kN
Flange clips Cut from 457 × 152 × 82 UB
Web cleats Two angles 100 × 100 × 12

Using HSFG bolts determine the number and minimum diameter of the bolts A, B, C and D required in flange clips and web cleat connections.

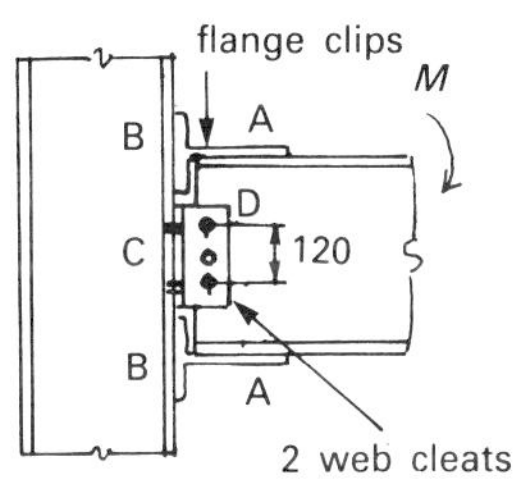

FIG. 67.

CHAPTER TWELVE

Welded Connections

> **CHAPTER OBJECTIVES**
>
> After studying this chapter you should be able to:
> * distinguish between butt and fillet welds;
> * appreciate some advantages and limitations of welding and determine the strength of welds by calculations;
> * design for direct loaded connections;
> * design for eccentric bracket connections, shear and torsion, shear and bending.

INTRODUCTION

Welding produces fusion by the application of heat from an electric arc struck between an electrode supplying filler metal and the material being joined, the base or parent metal. Welding offers some advantages over bolting.

(*a*) Economy is achieved by the elimination to a certain extent of connecting angle cleats, gusset plates and splice plates.

(*b*) A high degree of rigidity is obtained in a structural framework.

(*c*) It enables smoother lines to be obtained with the absence of protruding parts such as bolt heads and nuts and is more aesthetically pleasing to the eye.

(*d*) The effective area of tension members is not reduced by the presence of bolt holes.

(*e*) Full strength joints can be readily achieved.

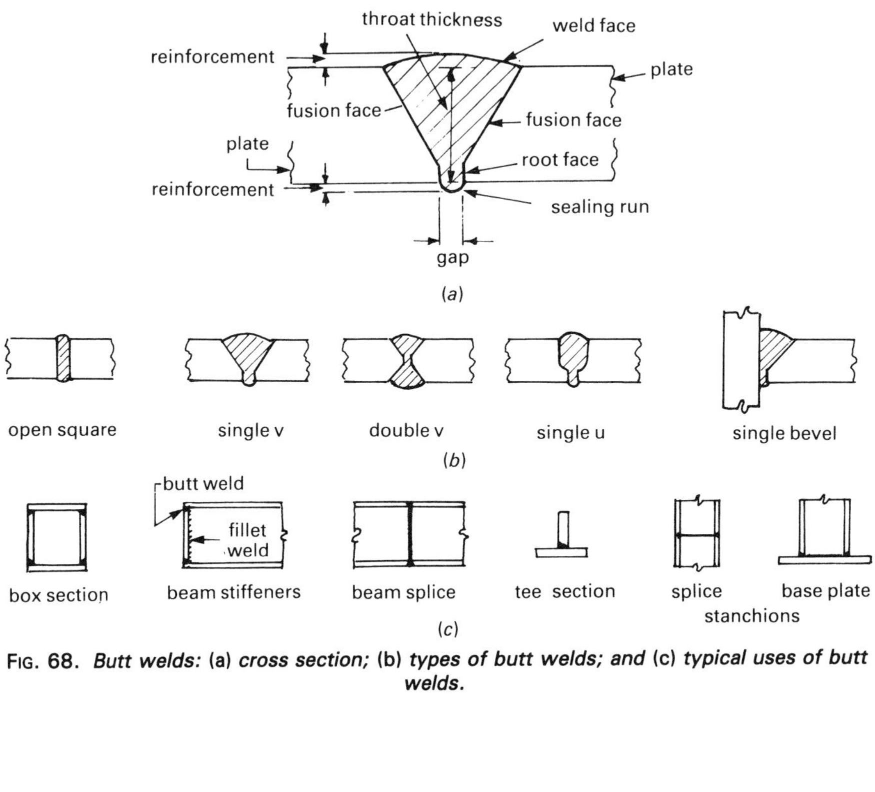

FIG. 68. *Butt welds:* (a) *cross section;* (b) *types of butt welds; and* (c) *typical uses of butt welds.*

Welding, however, has limitations—the major one being the problem of the quality of the welding and besides visual inspection, radiographic or ultrasonic inspection may be necessary. Distortion during welding, fatigue failures and brittle fractures are other problems. Site welding should be performed by highly skilled welders but it can be difficult and costly and so the site connections are often made using HSFG bolts, thus eliminating the problems of site weld inspection. The usual failure mode for a welded joint is not at the weld material but at some distance from it because of the higher strength of the weld metal than the parent metal.

Clauses 53 and 54, BS 449 cover the design and allowable stresses for welds. The two basic types generally used in structural steelwork are butt welds and fillet welds. Forms of welds and typical uses are shown in Figs 68 and 69.

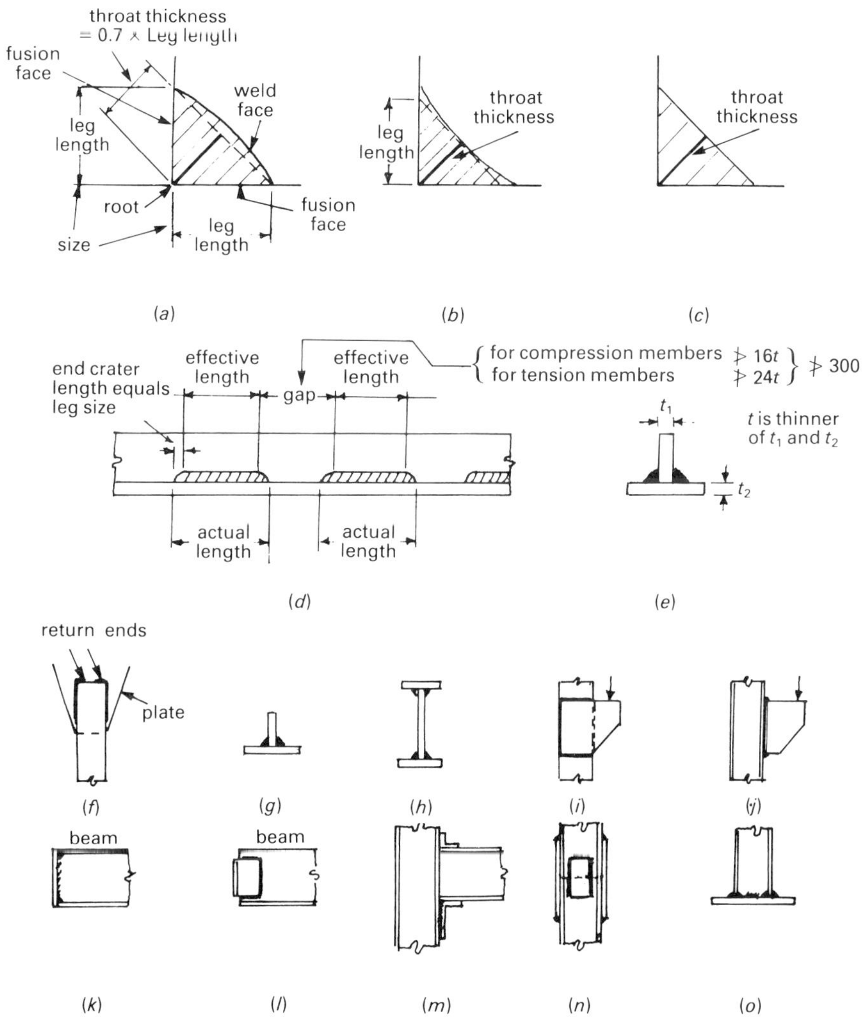

FIG. 69. *Types of fillet welds and their uses:* (a) *convex;* (b) *concave;* (c) *flat;* (d) *and* (e) *intermittent;* (f) *lap connection;* (g) *tee section;* (h) *plate girder;* (i) *and* (j) *eccentric brackets;* (k) *end plate;* (l) *web cleats;* (m) *seating cleat;* (n) *splice; and* (o) *base plate.*

BUTT WELDS

Butt welds are classified according to the shape into whch the plates or other structural sections are formed when the edges are prepared for welding.

They are treated as parent metal with a thickness equal to the throat thickness, or a reduced throat thickness. In order that the weld may develop the full strength of the members joined, it is necessary that complete penetration of weld metal into the joint is achieved. Incomplete penetration welds (e.g. unsealed single V, U, J or bevel butt welds), or any other butt welds which are welded from one side only, have a reduced throat thickness taken as not greater than $\frac{5}{8}$ of the thinner plate joined. The effective area of the weld for design strength is given by:

$$\text{Effective area} = \text{throat thickness} \times \text{effective length}$$

No design calculations are generally necessary for complete penetration butt welds as they usually provide a strength at the joint equivalent at least to that of the part or member itself.

FILLET WELDS

These are triangular in cross section with the face of the weld flat, concave or generally convex. They are used to connect overlapping or intersecting plates or members. The size of the weld is specified by its leg length. Before weld strength can be derived, it is necessary to know the:

(*a*) allowable stress in the weld;
(*b*) throat thickness;
(*c*) effective length.

The allowable stress in fillet welds based on the throat thickness is 115 N/mm^2 for grade 43 steel, Clause 53a, BS449.

The strength is based on the throat thickness and for a 90° angle between fusion faces:

$$\text{Throat thickness} = 0.7 \times \text{leg length}$$

For other angles between fusion faces:

$$\text{Throat thickness} = \text{an appropriate factor} \times \text{leg length}$$

To allow for end craters at the beginning and end of a run (length) of weld, for an open ended fillet weld:

$$\text{Effective length} = \text{actual length} - (2 \times \text{leg length})$$

The effective area of the weld is given by multiplying the throat thickness by the effective length.

The strength of welds in tension or compression is greater than that for shear and it is conservative to limit the strength to that controlled by shear alone:

$$\text{Allowable load} = \text{allowable stress} \times \text{effective length} \times \text{throat thickness}$$

Allowable load per mm length for a 6 mm fillet weld is therefore:

$$\frac{115 \times 0.7 \times 6 \times 1}{10^3} = 0.48\,\text{kN}$$

Intermittent fillet welds
The maximum distance between effective lengths of fillet welds is indicated on Fig. 69.

Return ends
Wherever practicable the ends shall be returned continuously around the corner for a distance of not less than twice the weld size for fillet welds terminating at the ends or sides of members. This provision applies in particular to side and top fillet welds in tension which connect brackets, beam seatings etc.

Lap joints
The minimum amount of lap shall be 4 times the thickness of the thinner part connected.

Side fillets
If side fillets alone are used in end connections, the length of each side fillet should not be less than the distance between the edges.

ECCENTRIC CONNECTIONS

The exact analysis of stresses in eccentric welded connections is impractical due to complex stresses. A general procedure for calculating stresses in weld configurations or groups for the purpose of design, is as follows:

(*a*) Draw the weld group or configuration showing effective lengths of welds.
(*b*) Determine the centroid of the welds.
(*c*) Determine the forces acting on the welds.
(*d*) Calculate the individual stresses in the welds at the critical position from direct shear, torsion and moment.
(*e*) Combine the stresses vectorially to obtain the resultant.

When designing weld groups the throat thickness t is usually unknown and it is convenient to assume unit thickness.

Eccentric bracket connections

When a load P from a beam is applied on a line of action which does not pass through the centroid of a weld group or arrangement, the result is an eccentric loading effect. Forms of brackets fabricated from angles, section cuttings, etc, are extended out to support the beam. Connections may be subjected to shear and torsion or shear and bending.

Shear and torsion

The load P applied at an eccentricity e, produces two components of shear stress at any point in the welds. These are:

(*a*) stress due to direct shear;
(*b*) stress due to torsional moment.

These shear stresses are added vectorially to give the resultant shear stress at a point. The maximum resultant is found at the critical point A in the weld group. An adequate weld size can then be selected (*see* Fig. 70).

Considering direct shear, the stress is given by:

$$\frac{\text{load}}{\text{area}} = \frac{P}{\text{effective length} \times \text{throat thickness}}$$

and since force = stress × area then for 1 mm length, the direct shear force is given by:

$$F_s = \frac{P \times \text{throat thickness} \times 1}{\text{effective length} \times \text{throat thickness}}$$

$$= \frac{P}{\text{effective length}} \text{ per mm length}$$

Considering the torsional moment, the stress is given by the torsion formula:

$$\text{stress} = \frac{Tr}{I_p}$$

where T = torsional moment = Pe; r = radial distance from the centroid to the point of stress; and F_p = polar moment of inertia of the group of welds; then stress due to torsional moment is also given by:

$$f_M = \frac{Per}{I_p}$$

and since force = stress × area then for 1 mm length, the force due to torsional moment is given by:

$$F_M = \frac{Per}{I_p} \times t \times 1$$

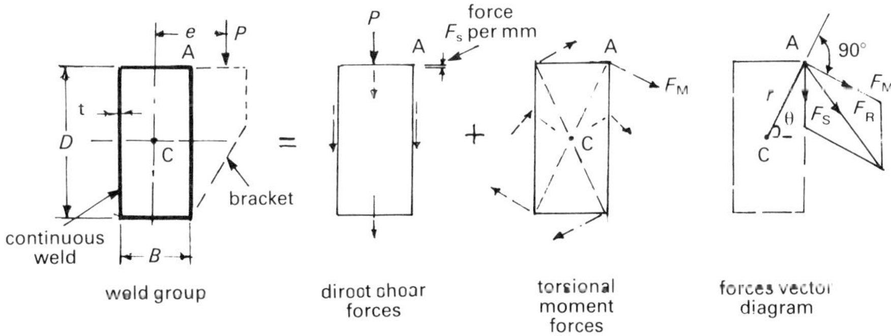

FIG. 70. *Shear and torsion.*

Now assuming $t = 1$ gives:

$$F_M = \frac{Per}{I_p} \text{ per mm length}$$

The resultant F_R can be found as for bolted connections using:

$$F_R = \sqrt{F_s^2 + F_M^2 + 2F_sF_M \cos\theta}$$

Shear and bending

Two methods of analysis (*see* Fig. 71) may be used in this type of connection:

(*a*) *Method 1:* The load P applied at an eccentricity e, produces two components of stress at any point in the welds. These are:

(*a*) stress due to direct shear;

(*b*) stress due to bending, considered for design as shear stress.

These shear stresses are added vectorially to give the resultant shear stress at a point. The maximum resultant is found at the critical level A in the weld group. An adequate weld size can then be selected (*see* Fig. 71).

As can be seen from the figure the actual maximum shear and bending stresses occur at different positions in the weld. Considering direct shear and since force = stress × area then for 1 mm length, the direct shear force is given by:

$$F_s = \frac{P}{\text{effective length}} \text{ per mm length}$$

Considering bending, then for 1 mm length, the force due to bending using the bending equation is given by

$$F_M = \frac{MY}{I} = \frac{PeY}{I} \text{ per mm length}$$

Forces F_s and F_M act at right angles one to the other and F_R is found using Pythagoras's theorem or graphically.

(*b*) *Method 2:* In this method it is assumed that the top flange weld resists the bending moment rotating about the bottom of the bracket and the vertical shear is resisted by the bottom flange weld. This method is considered in further detail in *Metric Practice For Structural Steelwork* published by BCSA.

The following examples illustrate the design of welds used in typical structural connections.

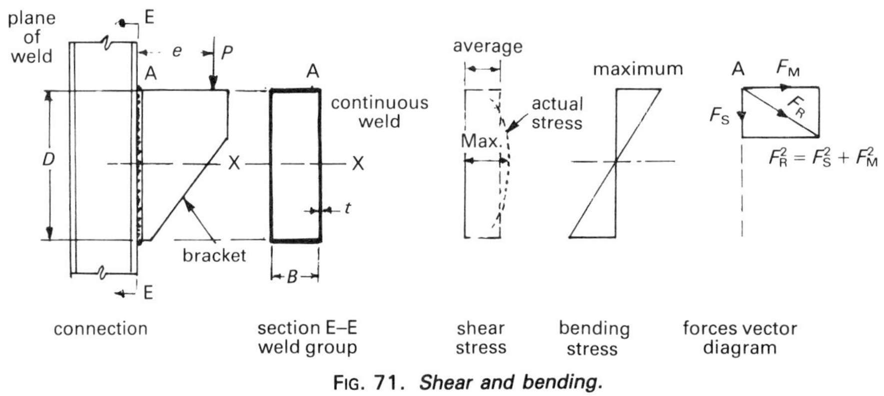

FIG. 71. *Shear and bending.*

CURPE CONSULTANTS, 46 Orburn Road, DUNFIELD	Project: DESIGN IV	Job ref: CC/ME
	Part of structure: CHAPTER TWELVE WELDED CONNECTIONS	Calc sheet no: 12/1 rev: /
	Drawing ref: — Calc by: R.A.S. Date: —	Check by: — Date: —

Code ref.	Calculations	Output
	EXAMPLE 12.1. Bottom bracket welds with shear only	
	Calculate the safe load for the welds (5 mm fillet) connecting the seating cleat to the stanchion.	

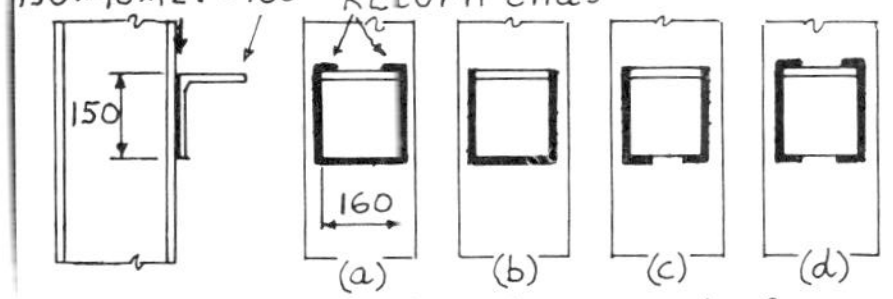

Code ref.	Calculations	Output
Cl.53a	Allowable stress in welds = $115\ N/mm^2$	
	Allowable load = allowable stress × area.	
	Allowable or safe load per mm length of weld is	
	$\frac{115 \times 0.7 \times 5 \times 1}{10^3} = 0.4\ kN$	
	Safe loads	
	(a) Length of weld = 150 + 160 + 150 = 460 mm.	
	Therefore safe load = 0.4 × 460 = 184 kN.	184 kN
	(b) Length of weld = 150 + 160 + 150 − (2×5) = 450 mm	
	Therefore safe load = 0.4 × 450 = 180 kN.	180 kN
	(c) Length of weld = 150 + 150 − (2×5) = 290 mm	
	Therefore safe load = 0.4 × 290 = 116 kN	16 kN
	(d) Length of weld = 150 + 150 = 300 mm.	
	Therefore safe load = 0.4 × 300 = 120 kN	120 kN
	The lengths of welds in the calculations are effective lengths allowing for end craters	

CURPE CONSULTANTS, 46 Orburn Road, DUNFIELD	Project: DESIGN IV	Job ref: CC/ME
	Part of structure: CHAPTER TWELVE WELDED CONNECTIONS	Calc sheet no: 12/2 rev: /
	Drawing ref: — Calc by: R.A.S. Date: —	Check by: — Date: —

Code ref.	Calculations	Output
	EXAMPLE 12.2. Connection resisting a direct force.	

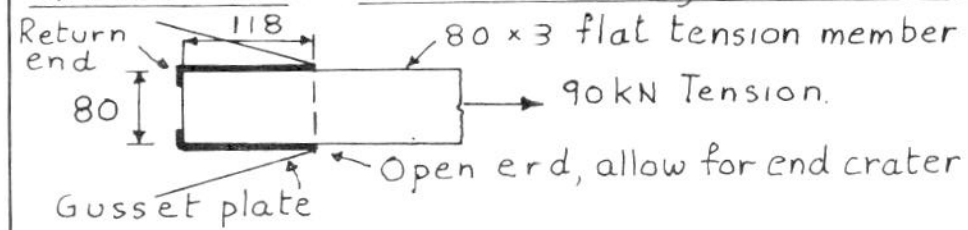

Code ref.	Calculations	Output
	(a) Determine the welding required (5 mm fillet)	
Cl.53a	Allowable stress in welds = $115\ N/mm^2$.	
	Allowable or safe load per mm length of weld is	
	$\frac{115 \times 0.7 \times 5 \times 1}{10^3} = 0.4\ kN$	
	Then effective length required is	
	$\frac{Load}{Safe\ load/mm} = \frac{90}{0.4} = 225\ mm.$	
Cl.54f	With welding along two edges of the flat requires a minimum of 80 mm each edge	
Cl.54e	Welding will include return ends and an allowance made for end craters at the open ends.	
	Then overall length of weld each side is	
	$\left(\frac{1}{2} \times 225\right)$ + end crater + return end	
	= 112.5 + 5 + return end	
	= Say 118 + return end.	Length = 118 mm
	(b) Check the strength of the 80×8 flat.	
Table 19	Allowable stress in tension $P_t = 155\ N/mm^2$	
	Actual stress in member $f_t = \frac{90 \times 10^3}{80 \times 8} = 140.6\ N/mm^2$	
	Since 140.6 < 155 ∴ Member adequate.	Member O.K.

CURPE CONSULTANTS — 46 Orburn Road, DUNFIELD

Project	Part of structure	Drawing ref	Calc by	Date	Job ref	Calc sheet no	rev	Check by	Date
DESIGN IV	CHAPTER TWELVE WELDED CONNECTIONS		R.A.S.		CC/ME	12/3	/		

Code ref.	Calculations	Output

EXAMPLE 12.3. End plate welded to beam and bolted to stanchion.

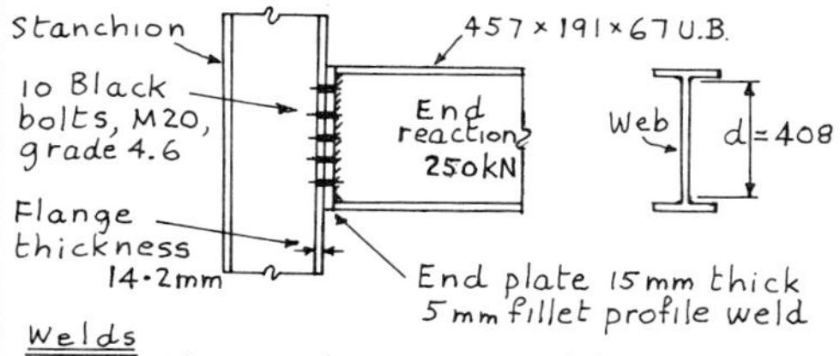

Welds

The weld must be capable of transmitting the end reaction to the web of the beam.

For the purpose of design the depth d between the fillets will be taken. i.e. 408 since shear is resisted mainly by the web.

Allowable or safe load per mm length of weld is

$$\frac{115 \times 0\cdot7 \times 5 \times 1}{10^3} = 0\cdot4\,\text{kN}$$

Effective length of weld is 2d with weld both sides of the web

Then the allowable or safe load is

Effective length × allowable load per mm length

$$= 2 \times 408 \times 0\cdot4 = 326\cdot4\,\text{kN}$$

Since the end reaction < Allowable load

The welding is adequate — **Output:** Welding O.K.

CURPE CONSULTANTS — 46 Orburn Road, DUNFIELD

Project	Part of structure	Drawing ref	Calc by	Date	Job ref	Calc sheet no	rev	Check by	Date
DESIGN IV	CHAPTER TWELVE WELDED CONNECTIONS		R.A.S.		CC/ME	12/4	/		

Code ref.	Calculations	Output

EXAMPLE 12.3 continued

Bolts

The bolts must be capable of transmitting the end reaction to the stanchion.

The design strength of one bolt is the lesser of its shear and bearing strengths.

Table 20 — Allowable stresses in bolts: Shear = 80 N/mm^2, Bearing = 250 N/mm^2

Table 20A — Allowable bearing stresses on connected parts = 250 N/mm^2.

Shear

Single shear strength of one bolt is

$$\frac{\pi \times D^2}{4} \times 80 \times \frac{1}{10^3}\ \text{kN}$$

$$= \frac{\pi \times 20^2}{4} \times 80 \times \frac{1}{10^3} = 25\,\text{kN}$$

Bearing

The thickness for bearing is the lesser of the end plate and stanchion flange i.e. 14·2 mm

Bearing strength of one bolt is

$$\frac{250 \times 20 \times 14\cdot2}{10^3} = 71\,\text{kN}.$$

Therefore the shear strength controls

Number of bolts required is

$$\frac{250}{25} = 10$$ AS SHOWN. — **Output:** Bolts O.K.

CURPE CONSULTANTS
46 Orburn Road
DUNFIELD

Project: DESIGN IV | Job ref: CC/ME
Part of structure: CHAPTER TWELVE WELDED CONNECTIONS | Calc sheet no: 12/5 | rev: /
Drawing ref: | Calc by: R.A.S. | Date: | Check by: | Date:

Code ref. | Calculations | Output

EXAMPLE 12.4. Design of weld group in twist.

Eccentric bracket connection - Welds subjected to Shear and Torsion.

e = eccentricity

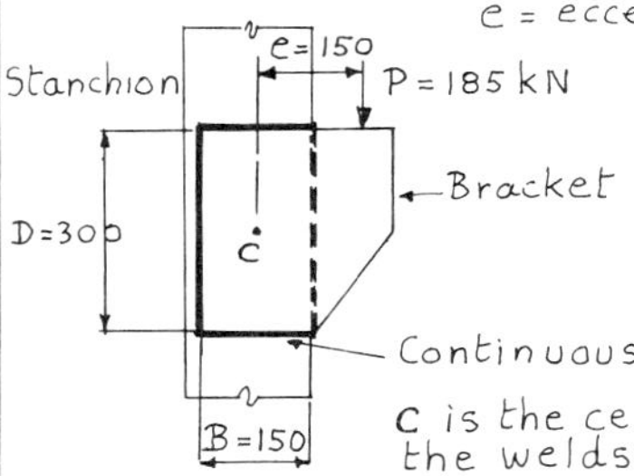

Determine size of welds

Allowable stress in welds = 115 N/mm².

Assume unit throat thickness

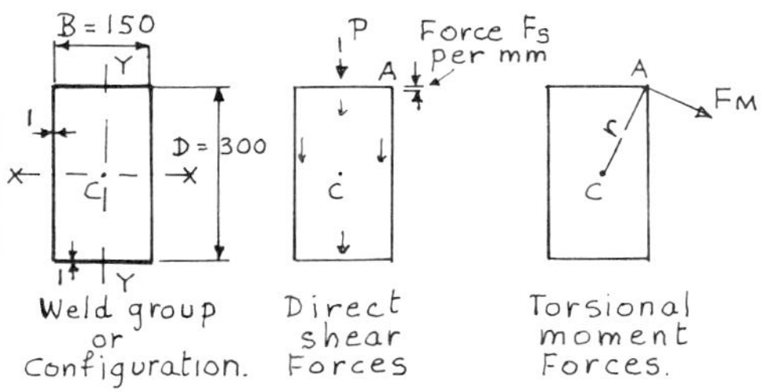

X-X and Y-Y are principal axes.

CURPE CONSULTANTS
46 Orburn Road
DUNFIELD

Project: DESIGN IV | Job ref:
Part of structure: CHAPTER TWELVE WELDED CONNECTIONS | Calc sheet no: 12/6 | rev: /
Drawing ref: | Calc by: R.A.S. | Date: | Check by: | Date:

Code ref. | Calculations | Output

EXAMPLE 12.4. continued

Design for critical position at A

Force due to direct shear is

$$F_S = \frac{P}{\text{Length of weld}} = \frac{185 \times 10^3}{2(150+300)}$$

$$= \underline{205 \cdot 5 \ \text{N/mm.}}$$

Also force due to moment is

$$F_M = \frac{P.e.r}{I_P}$$

$$r = \sqrt{150^2+75^2} = \underline{167 \cdot 7 \ \text{mm.}}$$

$$I_P = I_{XX} + I_{YY}$$

Neglect

$$I_{XX} = 2\left(\frac{1 \times 300^3}{12}\right) + 2\left[\frac{150 \times 1^3}{12} + (150 \times 1)150^2\right]$$

$$= 4500000 + 2[12 \cdot 5 + 3375000]$$

$$= 4500000 + 6750000 = 11250000 \ \text{mm}^2$$

$$I_{YY} = 2\left(\frac{1 \times 150^3}{12}\right) + 2\left[\frac{300 \times 1^3}{12} + (300 \times 1)75^2\right]$$

$$= 562500 + 2[25 + 1687500]$$

$$= 562500 + 3375000 = 3937500 \ \text{mm}^2$$

$$\therefore I_P = 11250000 + 3937500$$

$$= \underline{15187500 \ \text{mm}^4}$$

$$\therefore F_M = \frac{185 \times 10^3 \times 150 \times 167 \cdot 7}{15187500}$$

$$= \underline{306 \cdot 4 \ \text{N/mm.}}$$

Resultant F_R can be found either by calculation or graphically.

CURPE CONSULTANTS 46 Orburn Road DUNFIELD	Project: DESIGN IV	Job ref: CC/ME
	Part of structure: CHAPTER TWELVE WELDED CONNECTIONS	Calc sheet no: 12/7 rev: 1
	Drawing ref: — Calc by: R.A.S. Date: —	Check by: — Date: —

Code ref.	Calculations	Output

EXAMPLE 12.4. continued

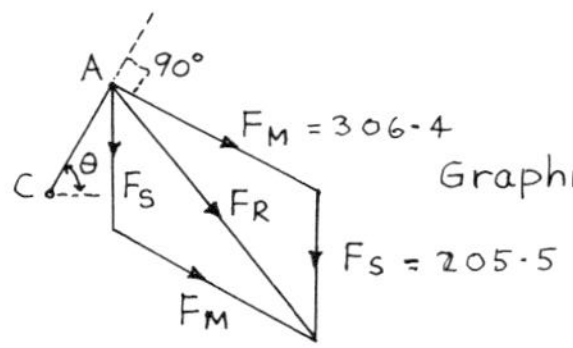

Graphical solution.

Using the formula derived for eccentric bolted connections

$$F_R = \sqrt{F_S^2 + F_M^2 + 2F_S.F_M.\cos\theta}$$

$$= \sqrt{205{\cdot}5^2 + 306{\cdot}4^2 + 2\left(205{\cdot}5 \times 306{\cdot}4 \times \frac{75}{167{\cdot}7}\right)}$$

$$= \underline{438{\cdot}6 \text{ N/mm}}.$$

Use 6 mm fillet weld; the allowable load is 480 N/mm length.

Output: Use 6 mm fillet weld

CURPE CONSULTANTS 46 Orburn Road DUNFIELD	Project: DESIGN IV	Job ref: CC/ME
	Part of structure: CHAPTER TWELVE WELDED CONNECTIONS	Calc sheet no: 12/8 rev: 1
	Drawing ref: — Calc by: R.A.S. Date: —	Check by: — Date: —

Code ref.	Calculations	Output

EXAMPLE 12.5.

Eccentric bracket connection - Welds subjected to shear and bending

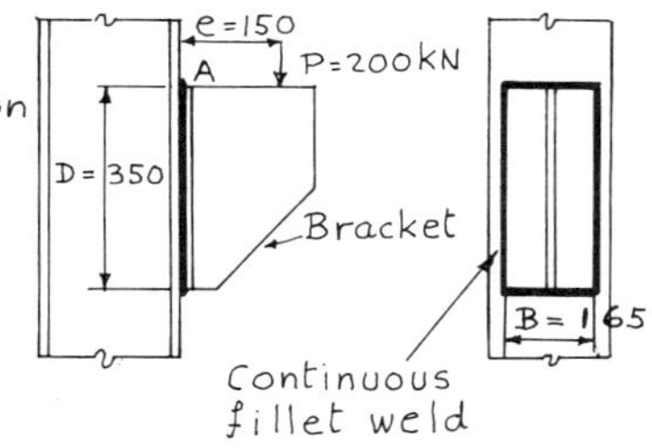

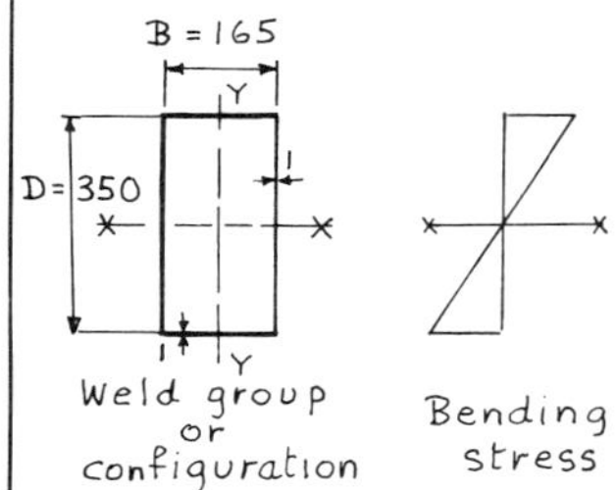

X-X and Y-Y are principal axes

Design for critical level A

Force due to direct shear is

$$F_S = \frac{P}{\text{Length of weld}} = \frac{200 \times 10^3}{2(165+350)}$$

$$= \underline{194{\cdot}2 \text{ N/mm}}.$$

CURPE CONSULTANTS 46 Orburn Road DUNFIELD	Project: DESIGN IV	Job ref: CC/ME
	Part of structure: CHAPTER TWELVE WELDED CONNECTIONS	Calc sheet no: 12/9 rev: 1
	Drawing ref: — Calc by: R.A.S. Date: —	Check by: — Date: —

Code ref.	Calculations	Output

EXAMPLE 12.5. continued

Also force due to moment is

$$F_M = \frac{P.e.Y}{I_{xx}}$$

$$Y = \frac{D}{2} = \frac{350}{2} = \underline{175\ mm}$$

$$I_{xx} = 2\left(\frac{1\times 350^3}{12}\right) + 2\left[\frac{165\times 1^3}{12} + (165\times 1)175^2\right]$$

Neglect $\frac{165\times 1^3}{12}$

$$= 7145833 + 2\left[13{\cdot}75 + 5053125\right]$$

$$= 7145833 + 10106250$$

$$= \underline{17252083\ mm^4}$$

$$F_M = \frac{200\times 10^3 \times 150 \times 175}{17252083}$$

$$= \underline{304{\cdot}3\ N/mm}$$

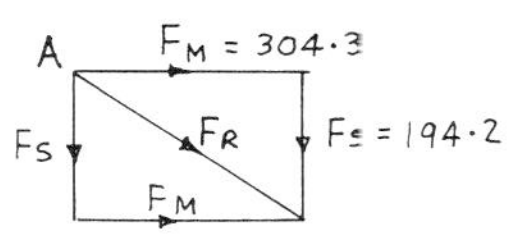

$$\text{Resultant } F_R = \sqrt{F_S^2 + F_M^2}$$

$$= \sqrt{194{\cdot}2^2 + 304{\cdot}3^2}$$

$$= \underline{361\ N/mm}.$$

Use 5 mm fillet weld; the allowable load is 402 N/mm length.

Output: Use 5 mm fillet weld

CURPE CONSULTANTS 46 Orburn Road DUNFIELD	Project: DESIGN IV	Job ref: CC/ME
	Part of structure: CHAPTER TWELVE WELDED CONNECTIONS	Calc sheet no: 12/10 rev: 1
	Drawing ref: — Calc by: R.A.S. Date: —	Check by: — Date: —

Code ref.	Calculations	Output

EXAMPLE 12.6.

Connection for single angle in tension

Disposition of welds to balance the forces about the centre of gravity of the angle. Use 6 mm fillet welds

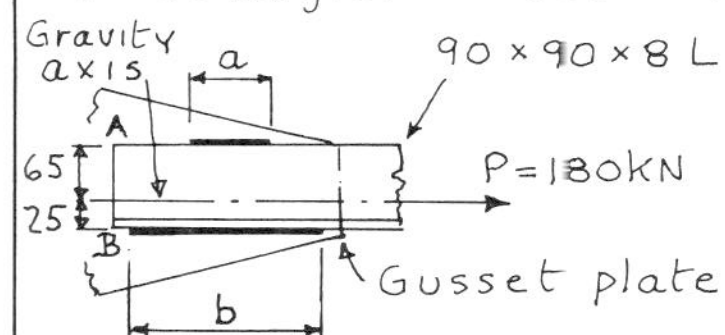

Strength of fillet weld is given as $\underline{0{\cdot}48\ kN/mm}$ length.

Total effective length required is

$$\frac{P}{\text{Safe load/mm}} = \frac{180}{0{\cdot}48} = \underline{375\ mm}$$

Weld length a

Taking moments about B

$$(0{\cdot}48 \times a \times 90) = 180 \times 25$$

$$a = 104{\cdot}2\ mm \quad \not< 90\ mm$$

Actual length = effective length + 2 end craters

$$= 104{\cdot}2 + (2\times 6) = 116{\cdot}2$$

Say $\underline{117\ mm}$

Output: a = 117 mm

CURPE CONSULTANTS 46 Orburn Road DUNFIELD	Project DESIGN IV		Job ref CC/ME	
	Part of structure CHAPTER TWELVE WELDED CONNECTIONS		Calc sheet no 12/11 rev /	
	Drawing ref	Calc by R.A.S. Date	Check by	Date

Code ref.	Calculations	Output
	EXAMPLE 12.6. continued.	
	Weld length b	
	Taking moments about A	
	$(0.48 \times b \times 90) = 180 \times 65$	
	$b = 270.8$ mm	
	Actual length $= 270.8 + (2 \times 6) = 282.8$	
	say 283 mm	b = 283 mm
	End and side fillet welds may be used where the available length is limited. as follows	
	Same example	
	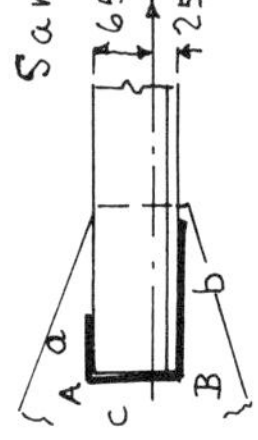	
	Moments about B	
	$(0.48 \times a \times 90) + (0.48 \times 90 \times \frac{90}{2}) = 180 \times 25$	
	$a = 59.2$ + 1 end crater	a = 66 mm
	Moments about A	
	$(0.48 \times b \times 90) + (0.48 \times 90 \times \frac{90}{2}) = 180 \times 65$	
	$b = 225.8$ + 1 end crater	b = 232 mm
	Check length $a + b + c$	
	$59.2 + 225.8 + 90 = 375$ mm As required.	

SELF-ASSESSMENT QUESTIONS

1. Determine an adequate economical size of fillet weld for the eccentric bracket connection shown in Fig. 72.

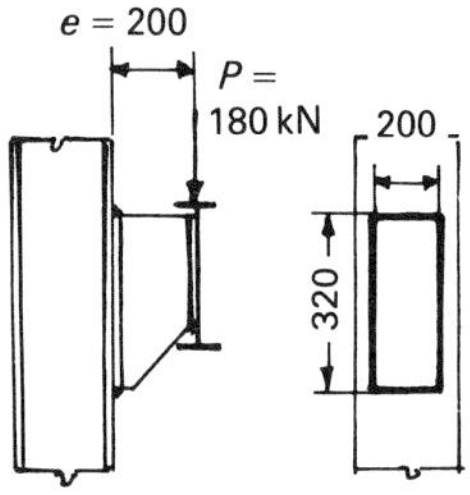

FIG. 72.

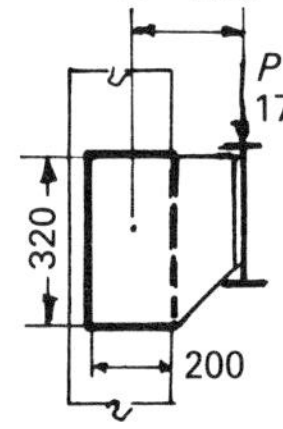

FIG. 73.

2. Determine an adequate economical size of fillet weld for the eccentric bracket connection shown in Fig. 73.

3. Calculate the safe load W for the 5 mm fillet welds used in a stanchion splice shown in Fig. 74.

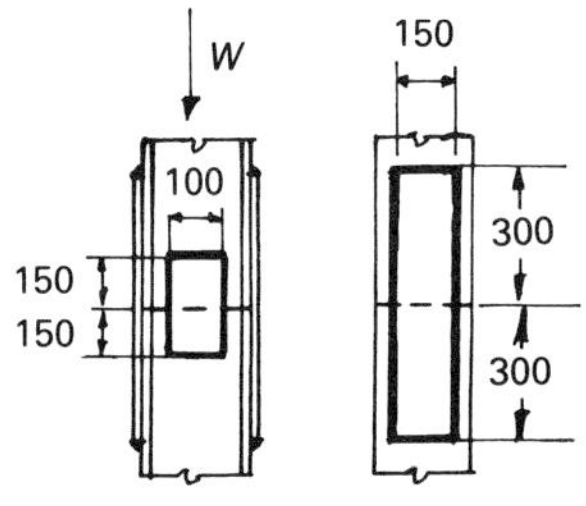

FIG. 74.

4. Determine the safe end shear (reaction) for the beam to stanchion connection in terms of the 5 mm fillet weld shown in Fig. 75.

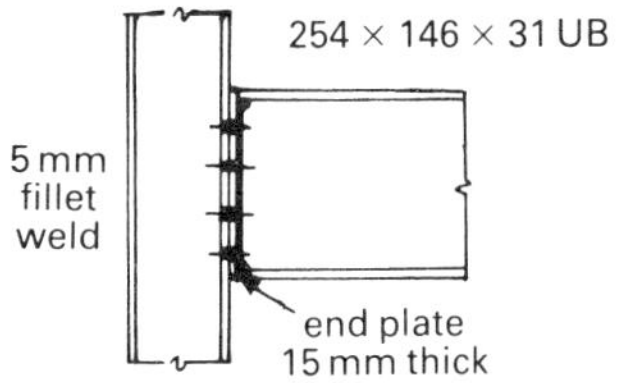

FIG. 75.

5. Check the adequacy of the 6 mm fillet welds connection seating cleats A and B to the stanchion shown in Fig. 76.

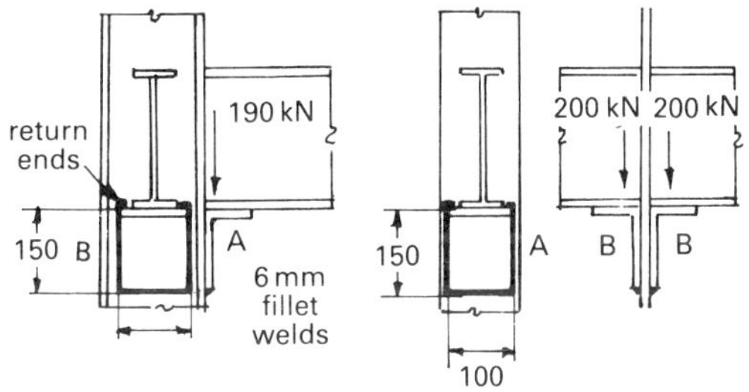

FIG. 76.

6. What maximum load P can be applied in terms of the 5 mm fillet weld shown in Fig. 77. The length shown is the actual length.

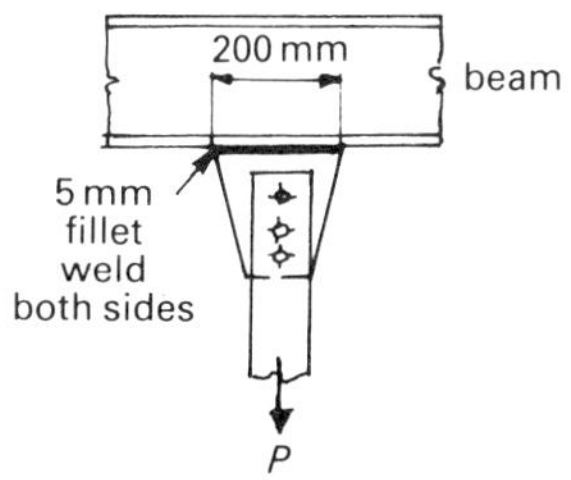

FIG. 77.

7. Determine the lengths of fillet welding required in connecting the angle to the T bar with the load of 170 kN assumed to act through the gravity axis of the angle as shown in Fig. 78. Assume that the T bar is deep enough to accommodate these lengths.

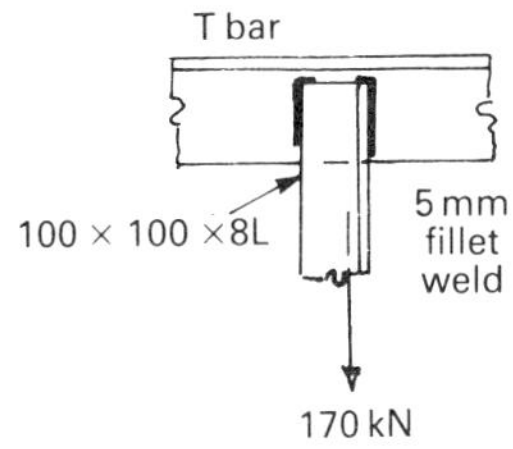

FIG. 78.

8. Calculate the safe axial load for the fillet welds shown at a stanchion base in Fig. 79.

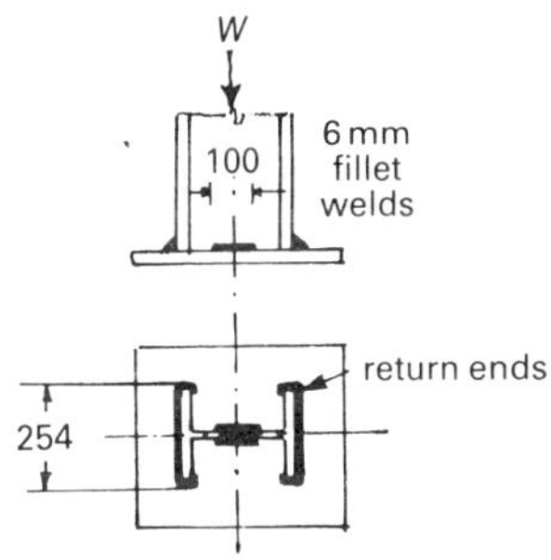

FIG. 79.

CHAPTER THIRTEEN

Steel Beams

> CHAPTER OBJECTIVES
>
> After studying this chapter you should be able to:
>
> * understand the effects of lateral restraint on beams;
> * design beams with and without lateral restraint;
> * design structurally cased beams;
> * check for shear, buckling and bearing stresses at supports;
> * appreciate the requirements for deflection of beams.

INTRODUCTION

The design of a beam involves a process of determining the different types of stresses resulting from applied loading and ensuring that specified allowable stresses are not exceeded at any point in the selected steel section.

Investigation of the following stresses should be made:

(*a*) bending;
(*b*) shear;
(*c*) web buckling;
(*d*) web bearing;
(*e*) combined bearing, bending and shear.

In addition to checking the stress values, the deflection of a beam should be limited so as not to impair the strength and efficiency of the structure, be unsightly, or lead to damage of the finishings. Bending, shear, web buckling, web bearing and deflection will now be considered in turn. Combined bearing, bending and shear may be critical in the case of cantilever or overhanging beams, neither of which is considered in the text.

BENDING

In the the majority of cases bending will determine the size of a steel beam. The theory of simple bending derives the equation which forms the basis of analysis of members subject to bending. This equation is:

$$\frac{M}{I}=\frac{f}{y}=\frac{E}{R}$$

Combining the first two ratios in the equation, then the bending stress is given by:

$$f=\frac{My}{I}$$

$$f=\frac{M}{Z}$$

where Z = elastic or section modulus of the beam = I/y.

The loading on the beam is determined and an estimated allowance for the self weight of the beam, and casing, if any, must be made before the total design load is established. This self weight can be checked later when the beam section is obtained.

The maximum bending moment is calculated and the compressive and tensile bending stresses determined. In the case of a simply supported beam, failure can occur with excessive compressive stress in the top flange or excessive tensile stress in the bottom flange. If the compression flange is not restrained against lateral buckling a lower allowable stress may have to be used. Lateral buckling and restraint will now be considered in some detail before an assessment of the allowable compressive stress in bending can be made.

Lateral buckling and restraint or support

When a beam is subject to bending, compressive and tensile stresses are produced at each cross section along the span of the beam. The resultant forces attain maximum value at the position of maximum bending moment.

The compression flange behaves as a strut and tends to buckle in the direction of least resistance, laterally or sideways, accompanied by a twisting action.

Figure 80 illustrates the buckled position of a simply supported laterally unrestrained beam. The deeper and narrower the beam section, the more susceptible it is to lateral buckling. When a beam is prevented from buckling laterally due to its surrounding construction, it is classified as laterally restrained or supported. Various degrees of restraint exist.

For simply supported beams there are three main categories of lateral restraint. These are:

1. Continuous or complete lateral restraint to the compression flange from the surrounding floor or roof construction, see Fig. 81.
2. Lateral restraint at intervals along the span from transverse restraint members which may be secondary beams or alternatively tie beams introduced at points along the span, see Figs 80 and 82.

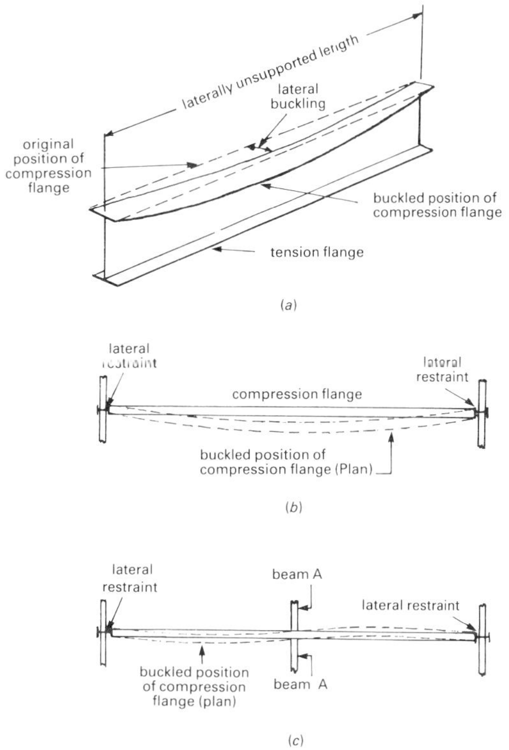

FIG. 80. *Buckling in compression flange* (a) *general instability;* (b) *lack of lateral restraint along span in category 3 beam; and* (c) *Lateral restraint provided at mid span by beams* A *in category 2 beam.*

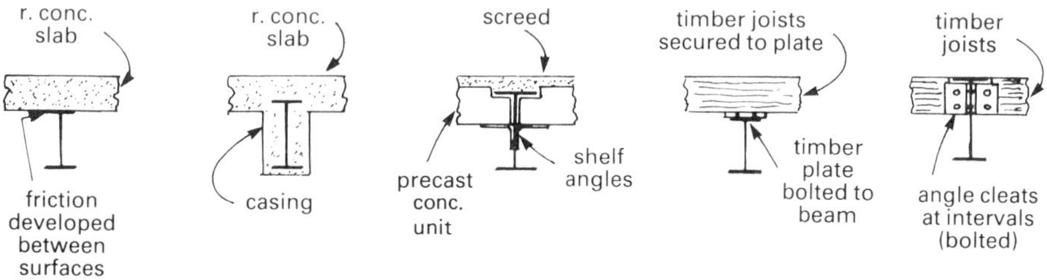

FIG. 81. *Continuous or complete lateral restraint.*

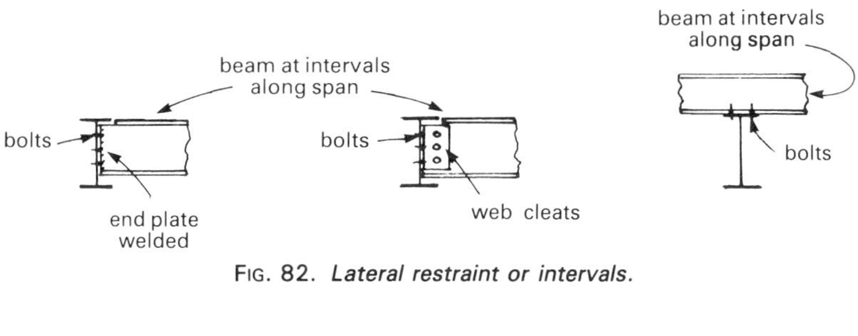

FIG. 82. *Lateral restraint or intervals.*

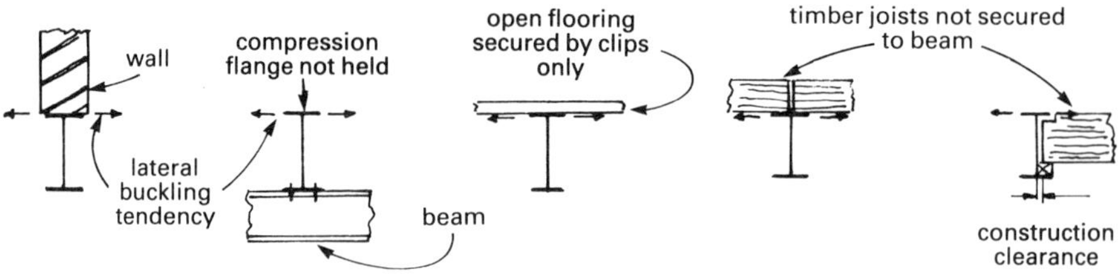

FIG. 83. *Complete lack of restraint.*

3. Complete lack of any form of lateral restraint along the span (i.e. laterally unsupported) with the ends of the beam connected to supporting stanchions or beams, *see* Figs 80 and 83.

It is necessary to establish into which category each beam falls in order to determine the allowable compressive bending stress P_{bc}.

Determing the allowable bending stresses P_{bc} and P_{bt}

Category 1
P_{bc} and P_{bt} are obtained directly from Table 2, BS 449.

Category 2
P_{bc} is obtained from Table 3(*a*), (*b*) or (*c*) depending on the grade of steel and considering two parameters l/r_y and D/T where

$$\frac{l}{r_y} = \frac{\text{effective length of the compression flange}}{\text{minimum radius of gyration}}$$

$$\frac{D}{T} = \frac{\text{overall depth of beam}}{\text{flange thickness}}$$

In this category:

$$\frac{l}{r_y} = \frac{\text{maximum distance, centre to centre of restraint members}}{r_y}$$

The values of r_y, D and T are constants for a given beam section. P_{bt} is obtained from Table 2, BS449.

Category 3
P_{bc} is obtained from Table 3(*a*), (*b*) or (*c*), BS449, again considering the two parameters l/r_y and D/T. In this category effective length l = a factor × actual length L where appropriate factors are given in Clause 26*a*, BS449, depending on the type of end connection provided.

Several types of end connection are illustrated in Fig. 84 with the appropriate factor and effective length given for each type.

SHEAR

Shear is most likely to become a controlling factor in design when heavy loads are carried on short spans or large concentrated loads are positioned near the end of the span.

Vertical shear stress is accompanied by an equal horizontal shear stress at the same point in a beam and together these produce diagonal compression on one plane at 45° and diagonal tension on the opposite plane again at 45°. The compression force tends to produce buckling of the web into a wave pattern and it is this phenomenon which influences the limit set for

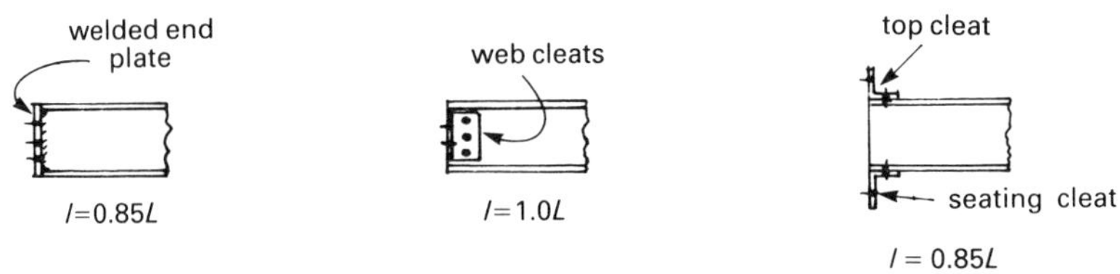

FIG. 84. *Types of end connection.*

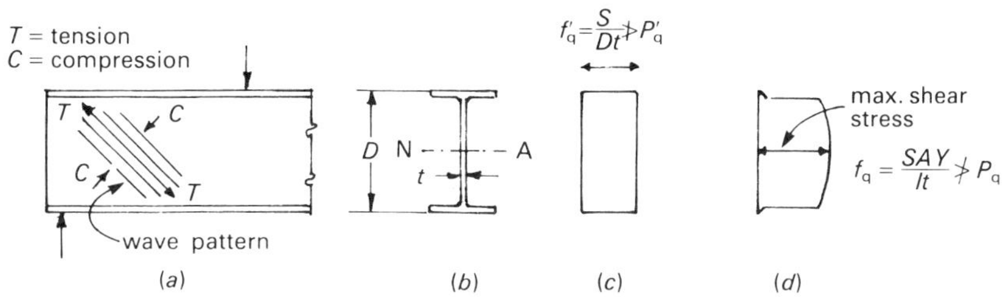

FIG. 85. *Shear distribution.*

the allowable shear stress in a steel beam, *see* Fig. 85.

To prevent web buckling due to shear, the ratio:

$$\text{Web slenderness} = \frac{\text{depth of web between flanges}}{\text{thickness of web}}$$

is limited and rolled beams are proportioned at manufacture accordingly to comply with the limits for each grade of steel.

Both maximum shear stress and average shear stress will now be considered, see Clause 23, BS449.

Maximum shear stress

The actual shear stress in a beam at any chosen section is given by

$$f_q = \frac{SAY}{It}$$

where S = the shear force at the section; A = the area of the section outside the level where the shear stress is required; Y = the distance from the centroidal axis of the beam to the centre of gravity of A; I = the moment of inertia of the beam; and t = the thickness of beam at the level considered.

The maximum shear stress occurs at the neutral axis of the section. The allowable maximum shear stresses P_q are given in Table 10, BS449 and $f_q \not> P_q$. The stress distribution is shown in Fig. 85.

Average shear stress

Referring to Fig. 85 the average shear stress for rolled I beams is given by

$$f_q^1 = \frac{S}{Dt}$$

where D = the overall depth of the beam and t = the web thickness.

The allowable average shear stresses P_q^1 are given in Table 11, BS449 for unstiffened webs and $f_q^1 \not> p_q^1$. For rolled I beams it is customary to check for shear strength using the average shear stress values.

If this is satisfactory then the requirements of (*a*) above are considered to be met.

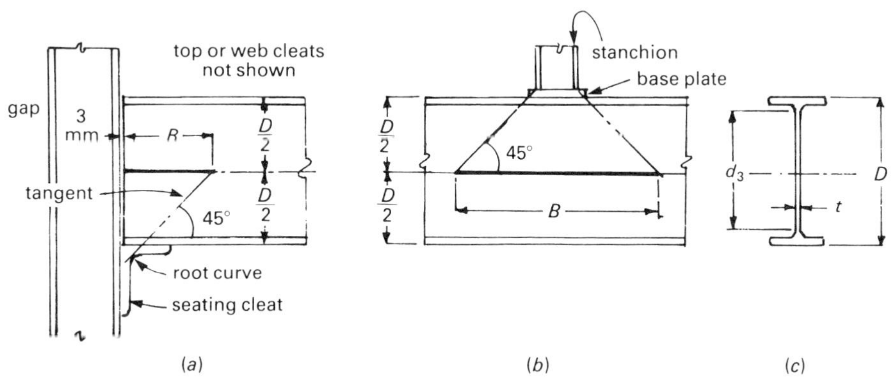

FIG. 86. *Web buckling:* (a) *end connection;* (b) *concentrated load; and* (c) *section.*

WEB BUCKLING

The web of a beam can also fail by buckling at its end support or under a concentrated load, due to excessive compressive stress.

The length of web B to be considered effective in resisting this compression is shown in Fig. 86.

The allowable buckling load is given by the formula:

$$\text{Load} = P_c t B$$

where P_c = the allowable compressive stress from Table 17; t = the web thickness; and B = the effective length of web using a 45° dispersal angle to the level of the neutral axis.

The value of P_c is obtained using a slenderness ratio of $d_3\sqrt{3}/t$, where d_3 is the clear depth of web between root fillets. The actual load or end reaction must not exceed the allowable buckling load, otherwise load bearing stiffeners are necessary. Lateral support should be provided at concentrated load positions.

WEB BEARING

The web of a beam can also fail by bearing or crushing of the material when the whole end reaction or concentrated load has to be carried on a small horizontal section of web area which may be overstressed.

The length of web B to be considered as effective in resisting the compression is shown in Fig. 87.

The allowable bearing load is given by $P_b tB$ where P_b = the allowable bearing stress from Table 9; t = the web thickness; and B = the effective length of web using a 30° dispersal angle to the level of the junction of the web and the root fillet.

The actual load or end reaction must not exceed the allowable bearing load otherwise bearing stiffeners are used to transmit the load through the web to the other flange of the beam. Lateral support should be proved at concentrated load positions.

CASED BEAMS

Steel beams may be solidly encased in *in situ* concrete to provide protection against fire and corrosion.

Beams of I section may be designed as cased beams if certain conditions are fulfilled. The conditions are illustrated in Fig. 88 and are in accordance with Clause 21.

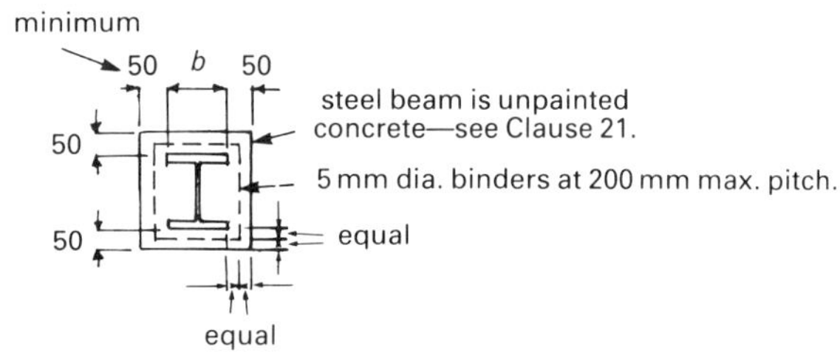

FIG. 88. *Cased beam.*

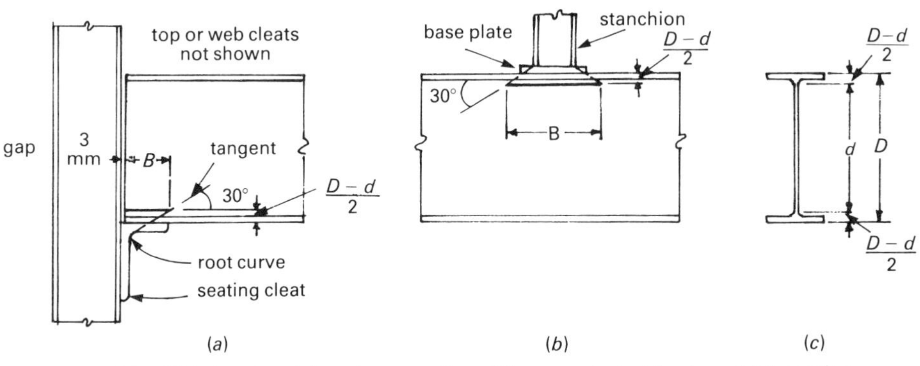

FIG. 87. *Web bearing:* (a) *end connection;* (b) *concentrated load; and* (c) *section.*

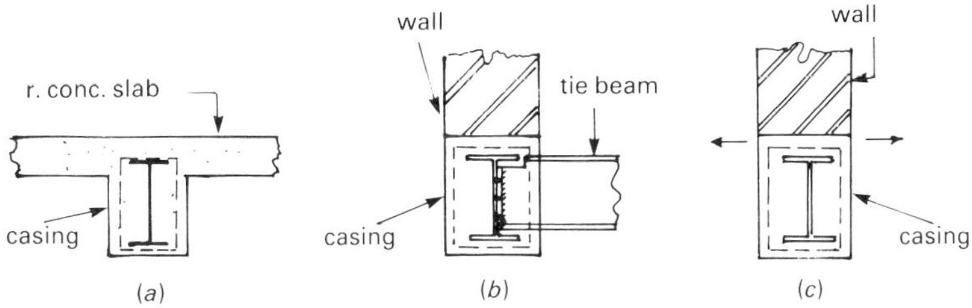

FIG. 89. *Bending:* (a) *Category 1;* (b) *Category 2; and* (c) *Category 3.*

Bending

Actual bending stresses are obtained using the section modulus of the steel section with no assistance assumed whatsoever from the concrete. When determining the allowable compressive bending stress P_{bc} the degree of lateral restraint existing should be assessed as for uncased beams (*see* Fig. 89).

For beams in category 1 with continuous lateral restraint no advantage is gained from the casing with P_{bc} taken from Table 2 as for uncased beams. For beams in categories 2 and 3 with lateral restraint only at intervals along the span or without lateral restraint along the span, P_{bc} is taken from Table 3 with advantage being taken of the concrete casing by using an increased value of r_y for the cased section, that is, $r_y = 0.2(b + 100)$. The value D/T is the same for both uncased and cased sections. The final design value of P_{bc} is restricted to $1\frac{1}{2} \times P_{bc}$ for the uncased section but also must not exceed the P_{bc} value from Table 2. The value of P_{bt} is obtained from Table 2.

Shear and deflection

The steel section alone is assumed to resist shear and deflection with no assistance whatsoever permitted from the concrete casing.

The following examples illustrate the design of beams with and without lateral support.

CURPE CONSULTANTS
46 Orburn Road
DUNFIELD

Project: DESIGN IV
Part of structure: CHAPTER THIRTEEN STRUCTURAL STEEL BEAMS
Drawing ref: | Calc by: R.A.S. | Date:
Job ref: CC/ME | Calc sheet no: 13/1 | rev
Check by: | Date:

Code ref.	Calculations	Output
	EXAMPLE 13.1. Bending stresses in chosen uncased and cased beams.	
	Consider a beam 533×210×82 UB of span L = 10 m	
	(a) Beam with top flange receiving continuous lateral restraint from concrete slab and carrying a uniformly distributed load of 200 kN.	
Table 2	Category I beam therefore $P_{bc} = P_{bt} = 165\ N/mm^2$	
	Maximum bending moment $= \frac{WL}{8} = \frac{200 \times 10}{8} = 250\ kN.m$	M = 250 kN.m
	Bending stresses $f_{bc} = f_{bt} = \frac{\text{Bending moment}}{\text{Section modulus}}$ $= \frac{250 \times 10^6}{1798 \times 10^3} = 139\ N/mm^2$	
	Since $f_{bc} < P_{bc}$ and $f_{bt} < P_{bt}$, beam is satisfactory for bending.	Bending O.K.
	(b) Beam with top flange unsupported between lateral restraints provided at centre and ends of span and carrying a uniformly distributed load of 190 kN	
Cl.26b	Category 2 beam with $l = 0.5L = 5$ metres	
	Maximum bending moment $= \frac{WL}{8} = \frac{190 \times 10}{8} = 237.5\ kN.m$	M = 237.5 kN.m

CURPE CONSULTANTS
46 Orburn Road
DUNFIELD

Project: DESIGN IV | Part of structure: CHAPTER THIRTEEN STRUCTURAL STEEL BEAMS | Drawing ref: | Calc by: R.A.S. | Date: | Job ref: CC/ME | Calc sheet no: 13/2 | rev: / | Check by: | Date:

Code ref.	Calculations	Output
	EXAMPLE 13.1. continued.	
	$f_{bc} = f_{bt} = \frac{237.5 \times 10^6}{1798 \times 10^3} = 132\ N/mm^2$	
	Now determine P_{bc}.	
	$\frac{\ell}{r_Y} = \frac{5 \times 10^3}{4.38 \times 10} = 114 \not> 300.$ $\quad \frac{D}{T} = 40$	
Table 3a	Therefore $P_{bc} = 142\ N/mm^2$ from Table 3a.	
Table 2	Also $P_{bt} = 165\ N/mm^2$ from Table 2	
	Since $f_{bc} < P_{bc}$ and $f_{bt} < P_{bt}$, beam is satisfactory for bending.	Bending O.K.
	(c) Beam with top flange lacking any lateral restraint along the span. Flange cleats are used for end connections to stanchions. The beam carries 90 kN uniformly distributed load.	
Cl. 26a	Category 3 beam with $\ell = 0.85L$	
	Maximum bending moment	
	$= \frac{WL}{8} = \frac{90 \times 10}{8} = 112.5$ kN.m.	M = 112.5 kN.m.
	$f_{bc} = f_{bt} = \frac{112.5 \times 10^6}{1798 \times 10^3} = 62.5\ N/mm^2$	
	Now determine P_{bc}	
	$\frac{\ell}{r_Y} = \frac{0.85 \times 10 \times 10^3}{4.38 \times 10} = 194 \not> 300$ $\quad \frac{D}{T} = 40$	
Table 3a	Therefore $P_{bc} = 66\ N/mm^2$ from Table 3a.	

CURPE CONSULTANTS
46 Orburn Road
DUNFIELD

Project: DESIGN IV | Part of structure: CHAPTER THIRTEEN STRUCTURAL STEEL BEAMS | Drawing ref: | Calc by: R.A.S. | Date: | Job ref: CC/ME | Calc sheet no: 13/3 | rev: / | Check by: | Date:

Code ref.	Calculations	Output
	EXAMPLE 13.1. continued.	
Table 2	Also $P_{bt} = 165\ N/mm^2$ from Table 2	
	Since $f_{bc} < P_{bc}$ and $f_{bt} < P_{bt}$, beam is satisfactory for bending.	Bending O.K.
	(d) Beam with top flange unsupported between lateral restraints provided at centre and ends of span. Beam is structurally cased in concrete and carries 200 kN uniformly distributed load.	
Cl. 26b.	Category 2 beam with $\ell = 0.5L = 5$ metres.	
	Maximum bending moment	
	$= \frac{WL}{8} = \frac{200 \times 10}{8} = 250$ kN.m.	M = 250 kN.m.
	$f_{bc} = f_{bt} = \frac{250 \times 10^6}{1798 \times 10^3} = 139\ N/mm^2$	
	Now determine P_{bc}	
	For uncased section $P_{bc} = 142\ N/mm^2$ from (b)	
	For cased section	
	$\frac{\ell}{r_Y} = \frac{5 \times 10^3}{6.17 \times 10} = 81$ $\quad \frac{D}{T} = 40$	
Table 3a	Therefore $P_{bc} = 165\ N/mm^2$ from Table 3a	
	Check for $1\frac{1}{2} \times$ uncased value of P_{bc}	
	$1\frac{1}{2} \times 142 = 213\ N/mm^2$	
	Therefore design $P_{bc} = 165\ N/mm^2$	
	Since $f_{bc} < P_{bc}$ i.e. $139 < 165$	

CURPE CONSULTANTS 46 Orburn Road DUNFIELD	Project: DESIGN IV Part of structure: CHAPTER THIRTEEN STRUCTURAL STEEL BEAMS	Job ref: CC/ME Calc sheet no: 13/4 rev: /
Drawing ref:	Calc by: R.A.S. Date:	Check by: Date:

Code ref.	Calculations	Output
	EXAMPLE 13.1. continued.	
	beam is satisfactory for bending.	Bending O.K.
	(e) Beam with top flange lacking any lateral restraint along the span. Flange cleats are used for end connections to stanchions. The beam is structurally cased in concrete and carries 130kN uniformly distributed load.	
Cl.26a	Category 3 beam with $\ell = 0{\cdot}85\,L$	
	Maximum bending moment	
	$= \frac{WL}{8} = \frac{130 \times 10}{8} = 162{\cdot}5$ kN.m.	M = 162·5 kN.m
	$f_{bc} = f_{bt} = \frac{162{\cdot}5 \times 10^6}{1798 \times 10^3} = 90{\cdot}3$ N/mm^2	
	Now determine P_{bc}	
	For uncased section $P_{bc} = 66$ N/mm^2 from (c)	
	For cased section	
	$\frac{\ell}{r_Y} = \frac{0{\cdot}85 \times 10 \times 10^3}{6{\cdot}17 \times 10} = 137{\cdot}8 \qquad \frac{D}{T} = 40$	
Table 3a	Therefore $P_{bc} = 117$ N/mm^2 from Table 3a.	
	Check for $1\frac{1}{2} \times$ uncased value of P_{bc}	
	$1\frac{1}{2} \times 66 = 99$ N/mm^2	
	Therefore design $P_{bc} = 99$ N/mm^2	
	Since $f_{bc} < P_{bc}$ and $f_{bt} < P_{bt}$	
	beam is satisfactory for bending	Bending O.K.

CURPE CONSULTANTS 46 Orburn Road DUNFIELD	Project: DESIGN IV Part of structure: CHAPTER THIRTEEN STRUCTURAL STEEL BEAMS	Job ref: CC/ME Calc sheet no: 13/5 rev: /
Drawing ref:	Calc by: R.A.S. Date:	Check by: Date:

Code ref.	Calculations	Output
	EXAMPLE 13.2. Shear stresses in a chosen section.	
	Consider a beam 533 × 210 × 82 U.B resisting a shear force of 500 kN.	
	(a) Maximum shear stress	
	This occurs at the neutral axis and is given by	
	$f_q = \frac{SAY}{I.t.}$	
	Use will be made of the properties of structural tees in the calculations.	
	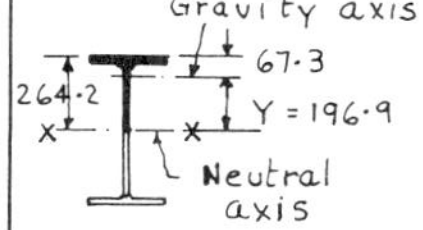 Dimension 67·3 is taken from properties of structural tees ∴ Y = 264·2 − 67·3 = 196·9	
	Area A is half the area of the U.B. section which equals the area of the structural tee and is shown shaded.	
	$f_q = \frac{500 \times 10^3 \times 52{\cdot}2 \times 10^2 \times 196{\cdot}9}{47491 \times 10^4 \times 9{\cdot}6} = 112{\cdot}7$ N/mm^2	
Table 10	Allowable maximum shear stress $P_q = 115$ N/mm^2	
	Since $f_q < P_q$, max. shear stress is satisfactory.	Max. shear stress O.K.
Table 11	(b) Average shear stress. Allowable $P'_q = 100$ N/mm^2	
	$f'_q = \frac{S}{D.t} = \frac{500 \times 10^3}{528{\cdot}3 \times 9{\cdot}6} = 98{\cdot}6 < 100$. Satisfactory.	Ave. shear stress O.K.

CURPE CONSULTANTS 46 Orburn Road DUNFIELD	Project: DESIGN IV Part of structure: CHAPTER THIRTEEN STRUCTURAL STEEL BEAMS Drawing ref:	Calc by: R.A.S.	Date:	Job ref: CC/ME Calc sheet no: 13/6 rev: / Check by: Date:

Code ref.	Calculations	Output
	EXAMPLE 13.3. Web buckling stress at a support in a chosen section	
	Consider a beam 533 × 210 × 82 U.B. with an end reaction of 265 kN carried by a seating cleat type connection.	
Cl. 28a	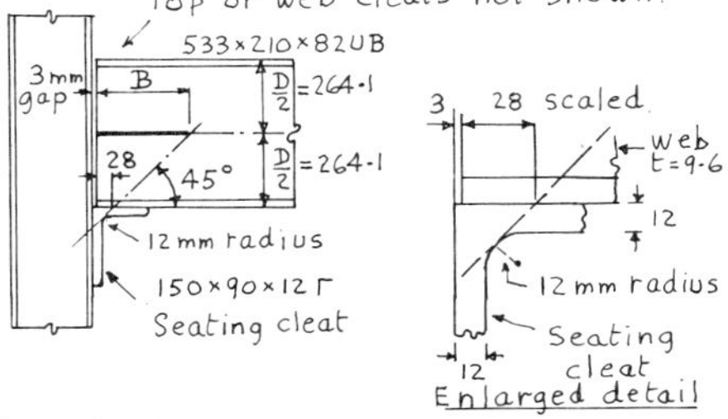	
	Length $B = 28 + \frac{D}{2} = 28 + 264{\cdot}1 = 292{\cdot}1$ mm	
Cl. 28a	Slenderness ratio $= \frac{d_3\sqrt{3}}{t} = \frac{476{\cdot}5\sqrt{3}}{9{\cdot}6} = 86$	
Table 17a	Therefore $P_c = 96$ N/mm² from Table 17a.	
	Actual buckling stress $f_c = \frac{\text{Load}}{\text{area}} = \frac{\text{Load}}{B.t}$	
	$f_c = \frac{265 \times 10^3}{292{\cdot}1 \times 9{\cdot}6} = 94{\cdot}5$ N/mm²	
	Since $f_c < P_c$, beam is satisfactory for web buckling. i.e. $94{\cdot}5 < 96$.	Web buckling O.K.

CURPE CONSULTANTS 46 Orburn Road DUNFIELD	Project: DESIGN IV Part of structure: CHAPTER THIRTEEN STRUCTURAL STEEL BEAMS Drawing ref:	Calc by: R.A.S.	Date:	Job ref: CC/ME Calc sheet no: 13/7 rev: / Check by: Date:

Code ref.	Calculations	Output
	EXAMPLE 13.4. Web bearing stress at a support in a chosen section	
	Consider a beam 533×210×82UB with an end reaction of 150 kN carried by a seating cleat type connection.	
Cl. 27e	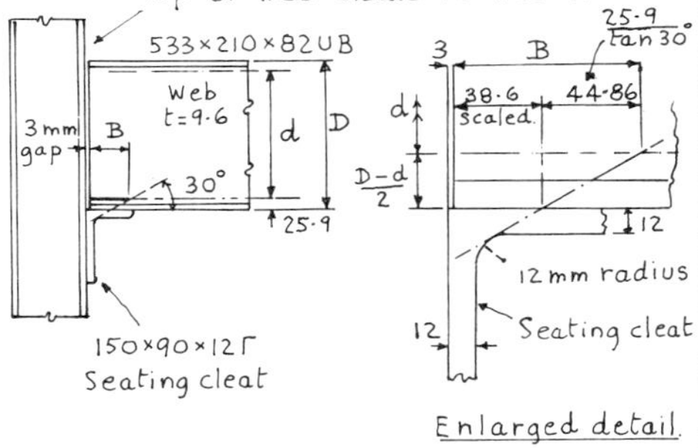	
	$B = 38{\cdot}6 + 44{\cdot}86 = 83{\cdot}46$ mm	
	Actual bearing stress $f_b = \frac{\text{Load}}{\text{area}} = \frac{\text{Load}}{B.t}$	
	$f_b = \frac{150 \times 10^3}{83{\cdot}46 \times 9{\cdot}6} = 187{\cdot}2$ N/mm²	
Table 9	Allowable bearing stress $P_b = 190$ N/mm² from Table 9.	
	Since $f_b < P_b$, beam is satisfactory for web bearing i.e. $187{\cdot}2 < 190$.	Web bearing O.K.
	The bearing strength of the cleat requires investigation.	

CURPE CONSULTANTS 46 Orburn Road DUNFIELD	Project: DESIGN IV	Job ref: CC/ME
	Part of structure: CHAPTER THIRTEEN STRUCTURAL STEEL BEAMS	Calc sheet no: 13/8 rev: /
	Drawing ref: — Calc by: R.A.S. Date: —	Check by: — Date: —

Code ref.	Calculations	Output

EXAMPLE 13.5. Design of roof beams.

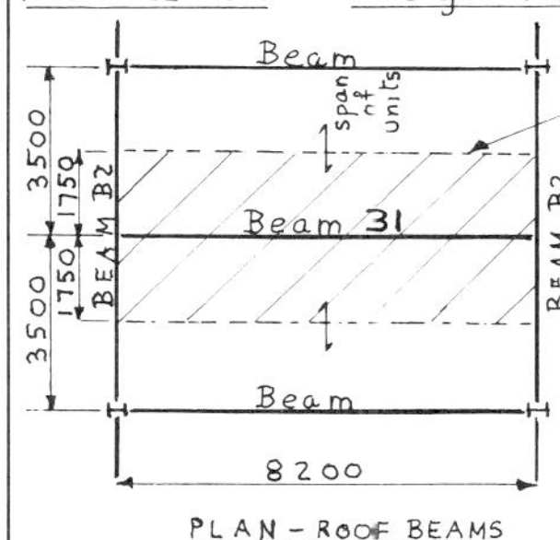

PLAN – ROOF BEAMS

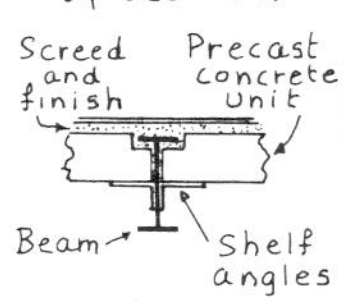

Section at beam B1.

Roof loading   Imposed $= 1{\cdot}5\ kN/m^2$

Dead including precast concrete units, screed and floor finish $= 3{\cdot}5\ kN/m^2$.

Assume 8 kN for self weight of beam B1 and 4 kN for self weight of beam B2.

Beam B1

Bending.

Floor load = loading/m^2 × area supported m^2

$= (1{\cdot}5 + 3{\cdot}5)\ 8{\cdot}2 \times 3{\cdot}5 = 143{\cdot}5\ kN$

Adding self weight then total load $= 143{\cdot}5 + 8 = 151{\cdot}5$

Max. bending moment $= \frac{WL}{8} = \frac{151{\cdot}5 \times 8{\cdot}2}{8} = 155{\cdot}2\ kNm$ **Output:** $M = 155{\cdot}2\ kN.m$

Beams are in Category I having continuous lateral restraint from precast concrete units

CURPE CONSULTANTS 46 Orburn Road DUNFIELD	Project: DESIGN IV	Job ref: CC/ME
	Part of structure: CHAPTER THIRTEEN STRUCTURAL STEEL BEAMS	Calc sheet no: 13/9 rev: /
	Drawing ref: — Calc by: R.A.S. Date: —	Check by: — Date: —

Code ref.	Calculations	Output

EXAMPLE 13.5. continued

Table 3a   therefore $p_{bc} = 165\ N/mm^2$ from Table 3a

Minimum Z required $= \dfrac{\text{Max. bending moment}}{\text{Allowable bending stress}}$

$= \dfrac{155{\cdot}2 \times 10^6}{165} = 940606\ mm^3$

457 × 152 × 52 U.B. Z given $= 949\ cm^3$.   **Output:** 457×152×52 UB Bending O.K

Shear

Shear at end of beam $= \frac{151{\cdot}5}{2} = 75{\cdot}75\ kN$

Average shear stress $f'_q = \dfrac{75{\cdot}75 \times 10^3}{449{\cdot}8 \times 7{\cdot}6} = 22{\cdot}15\ N/mm^2$

Table 11   Allowable average shear stress $P'_q = 100\ N/mm^2$ **Output:** Shear O.K.

Beam B2

Bending

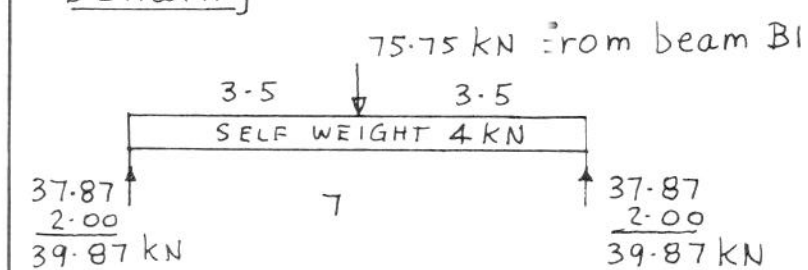

Max. bending moment $= (39{\cdot}87 \times 3{\cdot}5) - (2 \times 1{\cdot}75)$

$= 132{\cdot}54 - 3{\cdot}5 = 129{\cdot}04\ kNm$ **Output:** $M = 129{\cdot}04\ kN.m$

Beam is in Category 2 having lateral restraint provided by beam B1 connected at midspan therefore

Cl. 26b $\ell = 0{\cdot}5 \times \text{span} = 3500\ mm$.

CURPE CONSULTANTS 46 Orburn Road DUNFIELD	Project: DESIGN IV	Job ref: CC/ME
	Part of structure: CHAPTER THIRTEEN STRUCTURAL STEEL BEAMS	Calc sheet no: 13/10 rev: 1
	Drawing ref: — Calc by: R.A.S. Date:	Check by: Date:

Code ref.	Calculations	Output
	EXAMPLE 13.5. continued	
	To obtain a trial section assume $P_{bc} = 165\ N/mm^2$	
	then minimum Z required $= \frac{129.04 \times 10^6}{165} = 782060\ mm^3$	
	Try 356 × 171 × 51 UB Z given = $796.2\ cm^3$	
	$\frac{l}{r_y} = \frac{3500}{3.87 \times 10} = 90.4$	
Table 3a	Therefore $P_{bc} = 165\ N/mm^2$ from Table 3a	
	Beam is satisfactory for bending	356×171×51 UB Bending O.K.
	Shear	
	Shear = 39.87 kN	
	$f'_q = \frac{39.87 \times 10^3}{355.6 \times 7.3} = 15.35\ N/mm^2$	
Table 11	$p'_q = 100\ N/mm^2$	
	Beam is satisfactory for shear.	Shear O.K.
	Note - Beams should be checked for deflection	

Deflection

Clause 15 in BS449 states that the maximum deflection of a beam due to loads other than the weight of the structural floors or roof, steelwork and casing, if any, should not exceed 1/360 of the span.

Measures may be necessary to nullify the effects of the deflection due to dead loads by cambering or by adjustment in the casing of the beam.

Formulae for calculating the deflection of beams under various conditions of loading are given in structural handbooks. Two of the most commonly used formulae are for simply supported beams carrying a uniformly distributed load and a point load at the centre of the span. If the uniformly distributed load and the central point load occur simultaneously on a beam, then the total central deflection equals the sum of the deflections of the individual loads derived separately.

For a simply supported beam, the maximum uniformly distributed load allowed to meet deflection requirements is:

$$W_D = \frac{CI}{1000}$$

where C = the deflection coefficient tabulated for span L in m; E = the modulus of elasticity $= 21 \times 10^4\ N/mm^2$; and I = the moment of inertia about X–X axis in cm^4. Values of C and I are obtained from the *Structural Steelwork Handbook*.

SELF-ASSESSMENT QUESTIONS

1. Check the adequacy of a beam 457 × 191 × 74 UB of span 8 m:

(*a*) supporting a reinforced concrete slab and carrying a uniformly destributed Road (UDL) of 200 kN (including self weight);

(*b*) laterally restrained at mid span and ends of span and carrying a UDL of 170 kN (including self weight);

(*c*) lacking any lateral restraint along the

span with flange cleat connections at both ends and carrying a UDL of 80 kN (including self weight).

2. Check the adequacy of a beam 305 × 165 × 40 UB cased structurally and supporting a brick wall. The total UDL is 115 kN from brickwork, steel beam and casing. The span is 5 m.

3. Check the average shear stress and the maximum shear stress for a 457 × 191 × 67 UB subjected to a maximum shear force of 380 kN.

4. Check the web buckling stress in a beam at an end support given the following particulars. Beam section: 457 × 191 × 67 UB; seating cleat: 150 × 90 × 12; end clearance gap = 3 mm; beam end reaction = 190 kN.

5. Check the web bearing stress for the beam in Question 4 above, but with the beam end reaction 100 kN.

6. For the roof beam layout given in Ex. 13.5, design sections for beams B1 and B2 for bending given the following particulars: imposed loading = 1.5 kN/m^2; dead loading from timber joists, timber flooring and ceiling: = 0.8 kN/m^2. Assume that the timber joists span in same direction as that shown for the precast units and do not provide any lateral restraint to B1 along the span. Beams B1 are connected by end web cleats to the beams B2. For design purposes assume a self weight of 6 kN for both B1 and B2.

7. Design sections for beams B1 and B2 in Question 6 for bending assuming that the timber joists in this case provide continuous lateral restraint to beams B1. For design purposes assume a self weight of 4 kN for both B1 and B2.

8. A main beam 11.7 m span carries two concentrated point loads of 40 kN each from secondary beams positioned at third points along the span. For design purposes assume the self weight of the main beam as 8 kN. Check the adequacy of a 406 × 178 × 67 UB beam section for both bending and shear. Note that the main beam receives lateral restraint at intervals along the span from the secondary beams (i.e. $l = 3.9$ m).

CHAPTER FOURTEEN

Steel Compression Members

> CHAPTER OBJECTIVES
>
> After studying this chapter you should be able to:
> * obtain effective lengths of compression members;
> * calculate the slenderness ratio of cased and uncased compression members;
> * obtain allowable compressive stress;
> * identify type and amount of eccentric loading;
> * prepare calculations for single storey cased and uncased stanchions for axial and eccentric loading;
> * design and detail a slab base with concentric loading;
> * calculate tension and shear in holding down bolts.

STANCHIONS GENERALLY

Stanchions are the main vertical compression members in a structure. In this chapter only single-storey uncased and cased stanchions are considered. Compression members acting as struts in latticed frameworks such as roof trusses and girders are dealt with in the next chapter.

Generally it is found that short stanchions fail by crushing of the material and long slender stanchions fail by buckling. An important geometrical sectional property used in design is the radius of gyration.

Radius of gyration

In beam design the section or elastic modulus is an index of the strength of the member in bending. In the design of compression members the analogous sectional property is the radius of gyration which is given by:

$$r = \sqrt{\frac{I}{A}}$$

where I = the moment of inertia and A = the area of the section. It is a measure of the effectiveness of the compression member in resisting buckling. An understanding of effective length is also important in design.

Effective length

Other factors which affect the buckling tendency are:

(*a*) actual length L;
(*b*) end conditions or connections.

In practical design, end conditions are accounted for in assessing the effective length of compression members by making use of factors given in Clause 31, BS449. The effective length is given by:

$$l = \text{a factor} \times \text{actual length } L$$

The factors are summarised in Fig. 90 with deflected shapes shown.

Appendix D in BS449 gives typical cases of how the effective lengths of a stanchion, as affected by its end and beam connections, can be assessed to comply with BS449. It can be seen that a stanchion may have different effective lengths about the axes X–X and Y–Y.

Slenderness ratio

This is a measurement of the stanchion's tendency to buckle and is given by l/r. In practice it is necessary to consider two slenderness ratios corresponding to axis X–X and axis Y–Y; that is:

$$\text{Slenderness ratios} = \frac{l_{XX}}{r_{XX}} \quad \text{and} \quad \frac{l_{YY}}{r_{YY}}$$

where l_{XX} = the effective length, X–X axis; l_{YY} = the effective length, Y–Y axis; r_{XX} = the

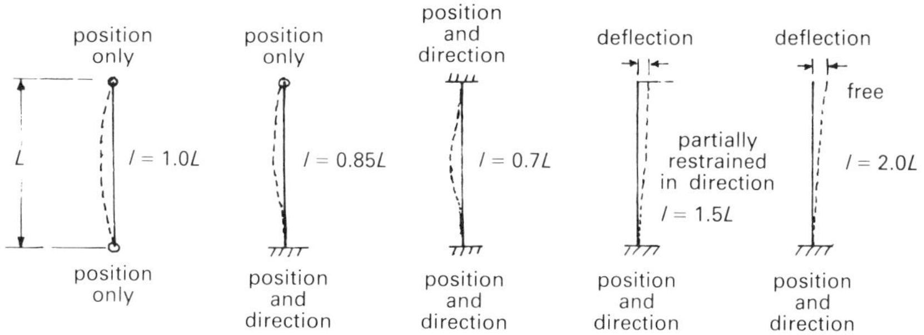

FIG. 90. *Effective length of compression members.*

radius of gyration, X–X axis; and r_{YY} = radius of gyration, Y–Y axis.

The maximum slenderness ratio permitted from Clause 33 is 180 for any member carrying dead loads with or without imposed loads.

Axial compressive stresses

The allowable compressive stress P_c is obtained from Table 17 using the larger value of the two calculated slenderness ratios.

The axial compressive stress is given by:

$$f_c = \frac{W}{A}$$

where W = the axial load and A = the area of the section.

Axially loaded stanchions

The allowable or safe load for uncased stanchions is given by P_cA and $W \not> P_cA$. Alternatively comparing stresses, $f_c \not> P_c$.

Eccentric loading

In practice most stanchions are eccentrically loaded with loads not balanced about one or both axes.

The magnitude of the eccentricity e to be used for the purpose of calculating bending moments and stresses may be determined by reference to Fig. 91.

The examples shown are interpreted from Clause 34, in BS449 and apply to single-storey stanchions. For examples (a) to (d) the bending moment is given by:

$$M = Pe$$

and the compressive bending stress:

$$f_{bc} = \frac{M}{Z}$$

where P = the beam end reaction; e = the eccentricity; and Z = appropriate section modulus of the stanchion.

If a stanchion supports several beams, the resultant or net bending moments are determined for each axis by the following method.

For bending about both axes the total compressive bending stress is given by:

$$f_{bc} = \frac{M_{xx}}{Z_{xx}} + \frac{M_{yy}}{Z_{yy}}$$

where M_{xx} = the resultant bending moment,

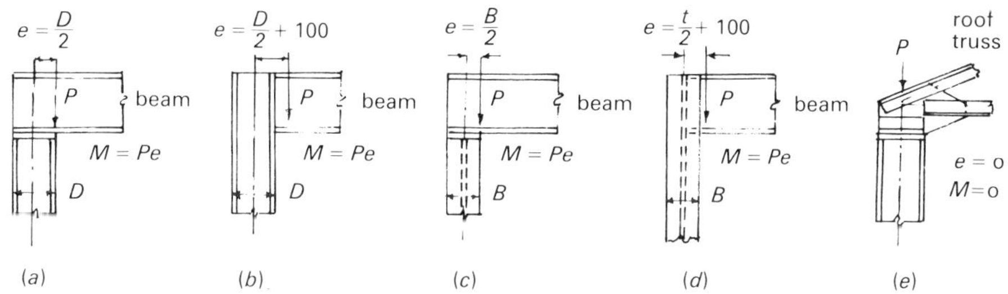

FIG. 91. *Magnitude of eccentricity e.*

X–X axis; M_{yy} = the resultant bending ment, Y – Y axis; Z_{xx} = the section modulus of stanchion, X – X axis; and Z_{yy} = the section modulus of stanchion, Y–Y axis.

The allowable bending stress P_{bc} is obtained from Table 2 or from Table 3 in BS449 using the parameters l/r_y and D/T for Table 3. Table 2 is applicable when a stanchion bends about its Y–Y axis and Table 3 when bending occurs about the X–X axis. For bending about both axes, the smaller value of P_{bc} derived from Tables 2 and 3 is used.

Eccentrically loaded stanchions

For uncased stanchions subjected to both axial compression and bending, then, from Clause 14a in BS449:

$$\frac{f_c}{P_c} + \frac{f_{bc}}{P_{bc}} \ngtr 1$$

STRUCTURALLY CASED STANCHIONS

Stanchions of single I section may be designed as cased struts when certain conditions are fulfilled all in accordance with Clause 30b in BS449. These conditions are shown in Fig. 92.

The concrete contributes to the strength of the composite section and also provides fire protection to the steel.

Slenderness ratio

For the uncased section the slenderness ratio measured over its full length, centre-to-centre of connections should not exceed 250. For the cased section the slenderness ratio should not exceed 180.

Axially loaded stanchions

Both concrete and steel share the load. The concrete casing increases the radius of gyration about the Y–Y axis and this results in a higher allowable compressive stress P_c.

The allowable or safe load in kN on the cased section is given by:

$$W = \frac{P_c}{10}\left(a_s + \frac{a_c}{0.19P_n}\right)$$

where P_c = the allowable compressive stress N/mm^2 for the cased steel section taken from Table 17; a_s = the area in cm^2 of steel core section; a_c = the gross area cm^2 of the concrete, i.e. ($B_c \times D_c/100$); and P_n = the numerical value of P_{bc} from Table 2 for the cased steel section.

The safe load W obtained from the above formula should in no case exceed twice that permitted for the uncased section.

Eccentrically loaded stanchions

Bending moments are resisted by the steel section alone. The casing increases the radius of gyration Y–Y axis which results in a smaller l/r_y value used in determining the value of P_{bc}.

The allowable bending stress P_{bc} is obtained from Table 2 or Table 3 as before using the cased radius of gyration. However, this should not exceed $1\frac{1}{2}$ times that permitted for the uncased section according to Clause 21 in BS449.

For cased stanchions subjected to both axial compression and bending, then, from Clause 14a in BS449:

$$\frac{\text{axial load}}{\text{allowable axial load}} + \frac{f_{bc}}{P_{bc}} \ngtr 1$$

Stanchion bases

Base plates of adequate dimensions and connected to the stanchion by either bolting or

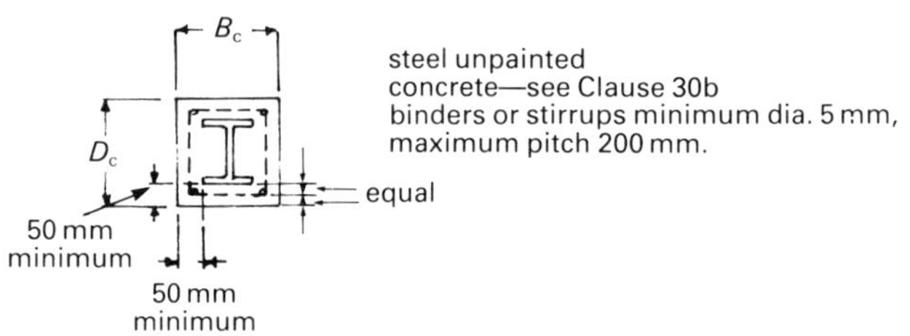

FIG. 92. *Structurally cased stanchion.*

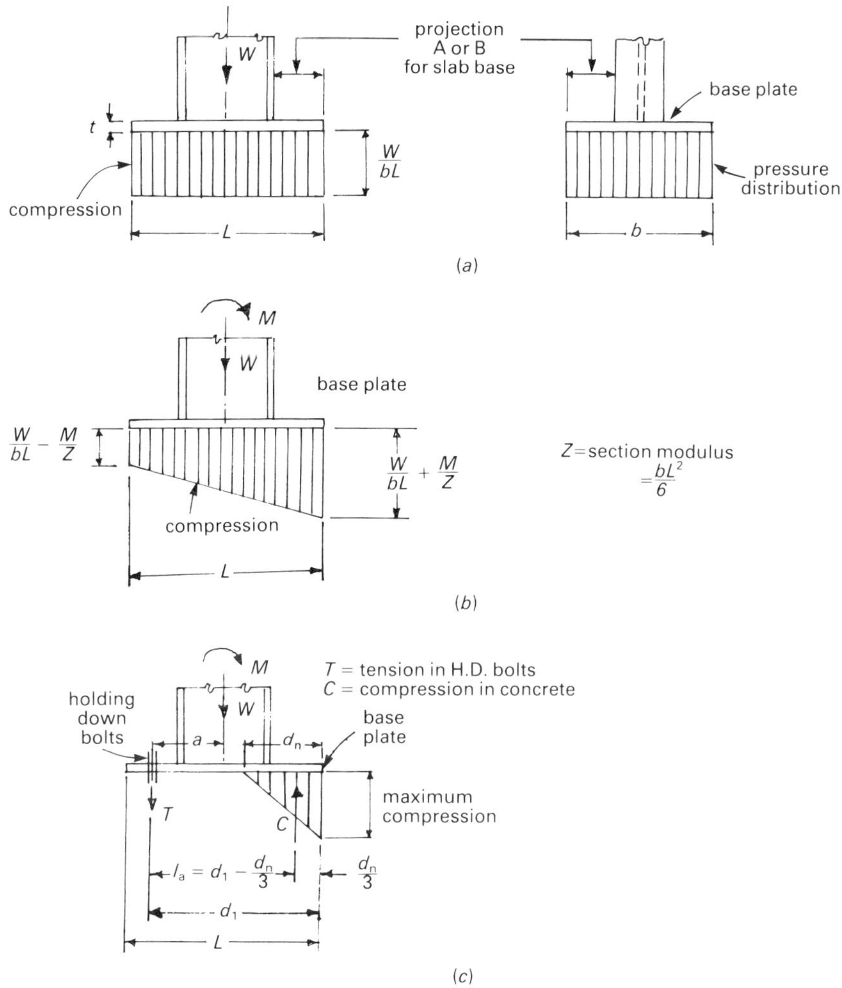

FIG. 93. *Stanchion bases:* (a) *axial loading; and* (b) *eccentric loading.*

welding, transmit the applied loading to the foundations. Holding down (HD) bolts connect the stanchions to the concrete foundations. Bases may be subjected to either axial (concentric) or eccentric loading as shown in Fig. 93.

Axially loaded bases

In this case compressive pressure under the whole base plate is uniform as shown in Fig. 93(*a*). The magnitude of this pressure is given by W/bL where W = the applied load; b = the breadth of the base plate; and L = the length of the base plate. This pressure must not exceed the allowable bearing pressure given for the concrete foundation.

Eccentrically loaded bases

Eccentric loading occurs when a vertical load is applied "off centre", (i.e. at a distance from either axis of the stanchion), thus producing a bending moment at the base.

A horizontal load (e.g. wind) applied at a distance above the base also produces a bending moment at the base which may be resolved into an "equivalent" eccentricity $e = M/W$. Depending on the relative magnitudes of the axial load and moment, two conditions arise:

(*a*) If M/W does not exceed $L/6$ then a compressive pressure exists under the whole base plate as shown in Fig. 93(*b*).

(*b*) If M/W exceeds $L/6$ then a compressive pressure exists under part of the base plate with a tensile force in the HD bolts as shown in Fig. 93(*c*). In this case it is not possible for tension to develop between the underside of the base plate and the concrete foundation so an analogy is drawn with a reinforced concrete section with the compression resisted by the concrete and the tension by the HD bolts giving an approximate and satisfactory solution.

Assuming that the maximum stresses in the concrete and steel occur simultaneously then the depth to the neutral axis is given by:

$$d_n = \frac{mP_{cc}d_1}{mP_{cc} + P_{st}}$$

where m = the modular ratio = 15; P_{cc} = the allowable compressive stress in the concrete; and P_{st} = the allowable tensile stress in the steel.

The distance of the centre of compression from the compression edge is $d_n/3$ for a triangular distribution and the lever arm is therefore $d_1 - d_n/3$, where d_1 is the effective depth.

C may be found by taking moments about the centre line of the HD bolts, that is:

$$C = \frac{M + Wa}{d_1 - \frac{d_n}{3}}$$

and since

$$C = T + W$$

then

$$T = C - W$$

Base types

The two basic base types are:

(*a*) gusseted bases;
(*b*) slab bases.

Gusseted bases consist of a base plate, gusset plates and stiffener plates usually of all welded construction suitable for bases subjected to large bending moments in addition to vertical loading. The stiffening effect of the plates results in a thinner base plate being required than that obtained using a slab base. Slab bases consist of a base plate welded directly to the stanchion or bolted to angle cleats connected to the stanchion. They are suitable for concentrically heavily loaded stanchions with the end of the stanchion and the top of the slab base machined to give complete bearing for the transmission of load.

The following examples illustrate the design of typical stanchions and bases.

CURPE CONSULTANTS 46 Orburn Road DUNFIELD	Project DESIGN IV		Job ref CC/ME
	Part of structure CHAPTER FOURTEEN COMPRESSION MEMBERS		Calc sheet no 14/1 rev /
	Drawing ref	Calc by R.A.S. Date	Check by Date

Code ref.	Calculations	Output

EXAMPLE 14.1. Single storey stanchion (uncased) with axial and eccentric loading

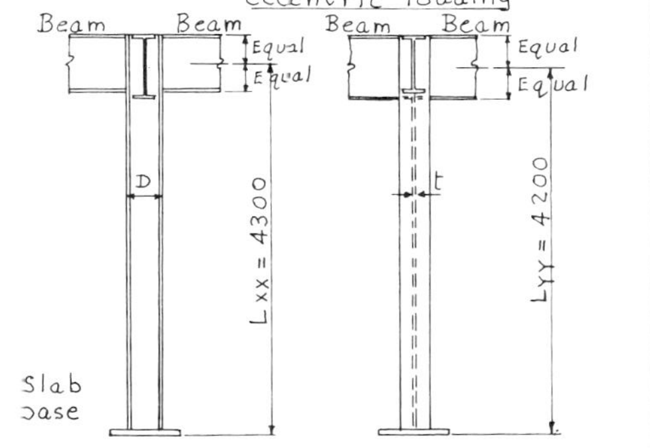

Elevations

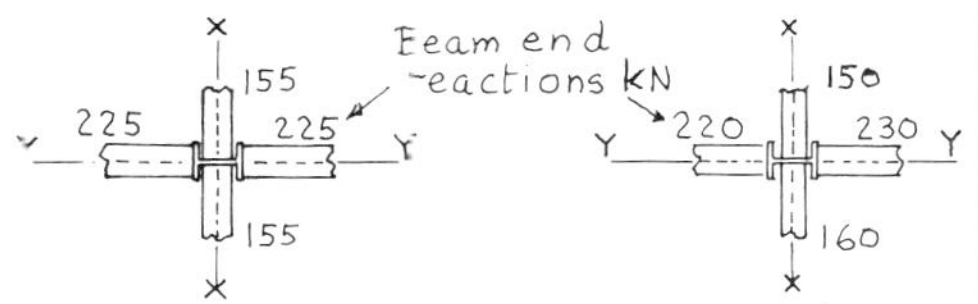

(a) Axial loading from beams.

(b) Eccentric loading from beams.

Plans at top of stanchion

Check the adequacy of a proposed trial section 203× 203× 52 UC

D = 206·2 mm, t = 8 mm.

CURPE CONSULTANTS 46 Orburn Road DUNFIELD	Project DESIGN IV		Job ref CC/ME
	Part of structure CHAPTER FOURTEEN COMPRESSION MEMBERS		Calc sheet no 14/2 rev /
	Drawing ref	Calc by R.A.S. Date	Check by Date

Code ref.	Calculations	Output

EXAMPLE 14.1. continued

(a) Axially loaded stanchion.

Load W = 225 + 225 + 155 + 155 = 760 kN

Cl. 31a — Effective lengths are assessed as 0·85 times the actual lengths for both axes, with both ends held in position and restrained in direction at one end (i.e. the top) about both axes.

Slenderness Ratio

$$\frac{\ell_{xx}}{r_{xx}} = \frac{0\cdot85\,L_{xx}}{r_{xx}} = \frac{0\cdot85\times4300}{89\times10} = 41$$

$$\frac{\ell_{YY}}{r_{YY}} = \frac{0\cdot85\,L_{YY}}{r_{YY}} = \frac{0\cdot85\times4200}{5\cdot16\times10} = 69\cdot2$$

Then using the larger value of 69·2

Table 17a — $p_c = 116\cdot6\ N/mm^2$ from Table 17a

Therefore allowable or safe load = $p_c.A$.

$$= \frac{116\cdot6\times66\cdot4\times10^2}{10^3} = 774\ kN$$

Since W < allowable load, i.e. 760 < 774

Stanchion is adequate for axial load. — Axial load OK.

(b) Eccentrically loaded stanchion

Load W = 220 + 230 + 150 + 160 = 760 kN as above.

CURPE CONSULTANTS 46 Orburn Road DUNFIELD	Project: DESIGN IV	Job ref: CC/ME
	Part of structure: CHAPTER FOURTEEN COMPRESSION MEMBERS	Calc sheet no: 14/3 rev: 1
	Drawing ref: — Calc by: R.A.S. Date: —	Check by: — Date: —

Code ref.	Calculations	Output
	EXAMPLE 14.1. continued	
Cl. 34a	Eccentricities	
	For x-x axis, $e = \frac{D}{2} + 100 = \frac{206.2}{2} + 100 = 203.1\,mm$	
	For y-y axis, $e = \frac{t}{2} + 100 = \frac{8}{2} + 100 = 104\,mm$	
	Bending moments	
	$M_{xx} = (230 - 220)\,203.1 = 2031\ kN.mm.$	
	$M_{YY} = (160 - 150)\,104 = 1040\ kN.mm.$	
	Axial compressive stress	
	$f_c = \frac{W}{A} = \frac{760 \times 10^3}{66.4 \times 10^2} = 114.4\ N/mm^2$	
	Compressive bending stress	
	$f_{bc} = \frac{M_{xx}}{Z_{xx}} + \frac{M_{YY}}{Z_{YY}} = \frac{2031 \times 10^3}{510.4 \times 10^3} + \frac{1040 \times 10^3}{173.6 \times 10^3}$	
	$f_{bc} = 3.98 + 5.99 = 9.97\ N/mm^2.$	
	Allowable compressive bending stress	
	$\frac{l}{r_Y} = \frac{0.85 \times 4200}{5.16 \times 10} = 69.2$ and $\frac{D}{T} = 16.5$	
Table 3a	$P_{bc} = 165\ N/mm^2$ from Table 3a	
	Combined axial compression + bending	
Cl. 14a	$\frac{f_c}{P_c} + \frac{f_{bc}}{P_{bc}} \ngtr 1.$	
	$\frac{114.4}{116.6} + \frac{9.97}{165} = 0.98 + 0.06 = 1.04 > 1$	Eccentric load. Inadequate
	A new section is chosen and the design procedure repeated until the result is less than 1.	

CURPE CONSULTANTS 46 Orburn Road DUNFIELD	Project: DESIGN IV	Job ref: CC/ME
	Part of structure: CHAPTER FOURTEEN COMPRESSION MEMBERS	Calc sheet no: 14/4 rev: 1
	Drawing ref: — Calc by: R.A.S. Date: —	Check by: — Date: —

Code ref.	Calculations	Output
	EXAMPLE 14.2. Single storey stanchion (cased) with axial and eccentric loading.	
	Stanchion used in example 14.1. will now be cased structurally in concrete. Although the stanchion was found to be adequate for axial load and therefore also adequate if cased, calculations will be produced for both axial and eccentric loading. to illustrate design procedure.	
	(a) Axially loaded stanchion (cased).	
	Load W = 760 kN (as before).	
	Effective lengths	
	$l_{xx} = 0.85\,L_{xx}$ for x-x axis (as before)	
	$l_{YY} = 0.85\,L_{YY}$ for y-y axis (as before)	
	Consider the uncased section	
	Slenderness Ratio	
	$\frac{l_{xx}}{r_{xx}} = \frac{0.85\,L_{xx}}{r_{xx}} = \frac{0.85 \times 4300}{8.9 \times 10} = 41$	
	$\frac{l_{YY}}{r_{YY}} = \frac{0.85\,L_{YY}}{r_{YY}} = \frac{0.85 \times 4200}{5.16 \times 10} = 69.2$	
Table 17a	$P_c = 116.6\ N/mm^2$ from Tale 17a.	
	Allowable or safe load = $P_c.A.$	
	$= \frac{116.6 \times 66.4 \times 10^2}{10^3} = 774\ kN$ (as before)	

CURPE CONSULTANTS 46 Orburn Road DUNFIELD	Project: DESIGN IV	Job ref: CC/ME
	Part of structure: CHAPTER FOURTEEN COMPRESSION MEMBERS	Calc sheet no: 14/5 rev: /
	Drawing ref: — Calc by: R.A.S. Date: —	Check by: — Date: —

Code ref.	Calculations	Output
	EXAMPLE 14.2. continued.	
	Now consider the cased section	
	Slenderness Ratio	
	$\frac{l_{xx}}{r_{xx}} = \frac{0.85 \times 4300}{8.9 \times 10} = 41.$	
	$\frac{l_{YY}}{r_{YY}} = \frac{0.85 \times 4200}{6.08 \times 10} = 58.7$	
	Then using the larger value of 58.7	
Table 17a	$P_c = 126.3\ N/mm^2$ from Table 17a.	
Cl.30b	Allowable load $W = \frac{P_c}{10}\left(a_s + \frac{a_c}{0.19 P_n}\right)$	
	$= \frac{126.3}{10}\left(66.4 + \frac{961}{0.19 \times 165}\right)$	
	$= 1226\ kN.$	
	The areas a_s and a_c are taken directly from the Structural Steelwork Handbook.	
	The allowable load ≯ 2 × allowable load for the uncased section.	
	2 × allowable load for the uncased section is	
	$2 \times 774 = 1548\ kN.$	
	The allowable or safe load is controlled therefore by the cased value which is 1226 kN.	Axial load ok
	(b) Eccentrically loaded stanchion (cased)	
	Load $W = 760\ kN$ (as before)	

CURPE CONSULTANTS 46 Orburn Road DUNFIELD	Project: DESIGN IV	Job ref: CC/ME
	Part of structure: CHAPTER FOURTEEN COMPRESSION MEMBERS	Calc sheet no: 14/6 rev: /
	Drawing ref: — Calc by: R.A.S. Date: —	Check by: — Date: —

Code ref.	Calculations	Output
	EXAMPLE 14.2. continued.	
	Effective lengths	
	$l_{xx} = 0.85\ L_{xx}$ for x-x axis (as before)	
	$l_{YY} = 0.85\ L_{YY}$ for y-y axis (as before)	
	Consider the uncased section	
	Slenderness Ratio	
	$\frac{l_{xx}}{r_{xx}} = \frac{0.85 \times 4300}{8.9 \times 10} = 4$	
	$\frac{l_{YY}}{r_{YY}} = \frac{0.85 \times 4200}{5.16 \times 10} = 69.2$	
Table 17a	$P_c = 116.6\ N/mm^2$ from Table 17a.	
	Allowable or safe load $= P_c.A.$	
	$= \frac{116.6 \times 66.4 \times 10^2}{10^3} = 774\ kN$ (as before)	
	Now consider the cased section	
	Slenderness Ratio	
	$\frac{l_{xx}}{r_{xx}} = \frac{0.85 \times 4300}{8.9 \times 10} = 41$	
	$\frac{l_{YY}}{r_{YY}} = \frac{0.85 \times 4200}{6.08 \times 10} = 58.7$	
Table 17a	$P_c = 126.3\ N/mm^2$ from Table 17a.	
	Allowable load $W = \frac{P_c}{10}\left(a_s + \frac{a_c}{0.19 P_n}\right)$	
	$= \frac{126.3}{10}\left(66.4 + \frac{961}{0.19 \times 165}\right)$	
	$\approx 1226\ kN$ (as before)	

CURPE CONSULTANTS 46 Orburn Road DUNFIELD	Project: DESIGN IV	Job ref: CC/ME
	Part of structure: CHAPTER FOURTEEN COMPRESSION MEMBERS	Calc sheet no: 14/7 rev: 1
	Drawing ref: — Calc by: R.A.S. Date: —	Check by: — Date: —

Code ref.	Calculations	Output
	EXAMPLE 14.2. continued.	
	Also 2 × allowable load for uncased section is	
	$2 \times 774 = 1548$ kN	
	The allowable or safe load is 1226kN (as before)	
	Eccentricities	
	For X-X axis, $e = \frac{D}{2} + 100 = \frac{206\cdot2}{2} + 100 = 203\cdot1$ mm	
	For Y-Y axis, $e = \frac{t}{2} + 100 = \frac{8}{2} + 100 = 104$ mm	
	Bending moments	
	$M_{xx} = (230 - 220)\,203\cdot1 = 2031$ kN.mm.	
	$M_{yy} = (160 - 150)\,104 = 1040$ kN.mm.	
	Compressive bending stress	
	$f_{bc} = \frac{2031 \times 10^3}{510\cdot4 \times 10^3} + \frac{1040 \times 10^3}{173\cdot6 \times 10^3} = 9\cdot97$ N/mm² (as before)	
	Allowable compressive bending stress	
	$\frac{\ell}{r_Y} = \frac{0\cdot85 \times 4200}{6\cdot08 \times 10} = 58\cdot7$ $\frac{D}{T} = 16\cdot5$	
Table 3a	$P_{bc} = 165$ N/mm² from Table 3a.	
	Since $P_{bc} = 165$ N/mm² for the uncased section then $1\frac{1}{2}$ × uncased value of P_{bc} will not apply.	
	Combined axial compression + bending	
Cl. 14a	$\frac{\text{Axial load}}{\text{Allowable axial load}} + \frac{f_{bc}}{P_{bc}} \not> 1.$	
	$\frac{760}{1226} + \frac{9\cdot97}{165} = 0\cdot62 + 0\cdot06 = 0\cdot68 < 1$	
	Stanchion adequate.	Eccentric load. Adequate

CURPE CONSULTANTS 46 Orburn Road DUNFIELD	Project: DESIGN IV	Job ref: CC/ME
	Part of structure: CHAPTER FOURTEEN COMPRESSION MEMBERS	Calc sheet no: 14/8 rev: 1
	Drawing ref: — Calc by: R.A.S. Date: —	Check by: — Date: —

Code ref.	Calculations	Output
	EXAMPLE 14.3. Steel base carrying a concentric load.	
	Concentric load = 1700kN	
	Stanchion section 254 × 254 × 73 UC.	
	Allowable bearing pressure on the concrete foundation = 5·3 N/mm².	
	Minimum area required	
	$= \frac{\text{Load}}{\text{Allowable bearing pressure}}$	
	$= \frac{1700 \times 10^3}{5\cdot3} = 320754$ mm².	
	Using a square base, sides are	
	$\sqrt{320754} = 566$ mm, but say 600mm	600mm × 600mm.
	Then the actual bearing pressure is	
	$\frac{1700 \times 10^3}{600 \times 600} = 4\cdot72$ N/mm² $< 5\cdot3$	
	The slab thickness is derived using the formula taken from BS449: Clause 38b.	
Cl. 38b	$t = \sqrt{\frac{3w}{P_{bct}}\left(A^2 - \frac{B^2}{4}\right)}$	
	where t = slab thickness (mm).	
	A = greater projection of plate (mm)	
	B = lesser projection of plate (mm)	
	w = pressure on underside of base N/mm²	
	$P_{bct} = 185$ N/mm²	

CURPE CONSULTANTS 46 Orburn Road DUNFIELD	Project DESIGN IV		Job ref CC/ME
	Part of structure CHAPTER FOURTEEN COMPRESSION MEMBERS		Calc sheet no 14/9 rev /
	Drawing ref	Calc by R.A.S. Date	Check by Date

Code ref.	Calculations	Output

EXAMPLE 14.3. Continued.

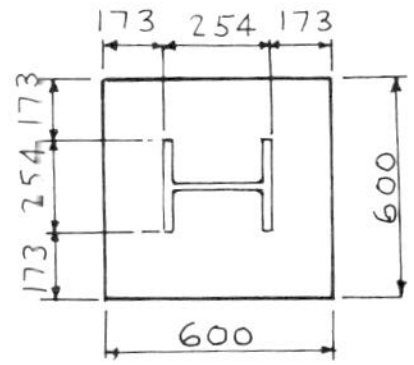

Overhang dimensions

$A = B = \frac{600 - 254}{2} = 173\text{mm}$

$t = \sqrt{\frac{3 \times 4{\cdot}72}{185}\left(173^2 - \frac{173^2}{4}\right)}$

$t = 41{\cdot}4$ mm, use 45 mm — Output: 45 mm thick

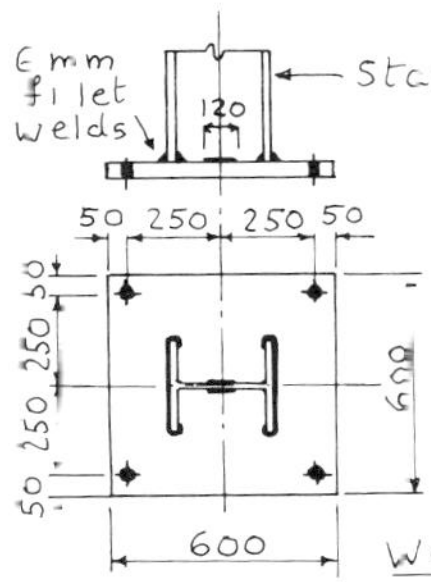

Base material

1 plate 600×45×600

4 holes 26 dia.

End of stanchion and top of plate machined.

Welded slab base

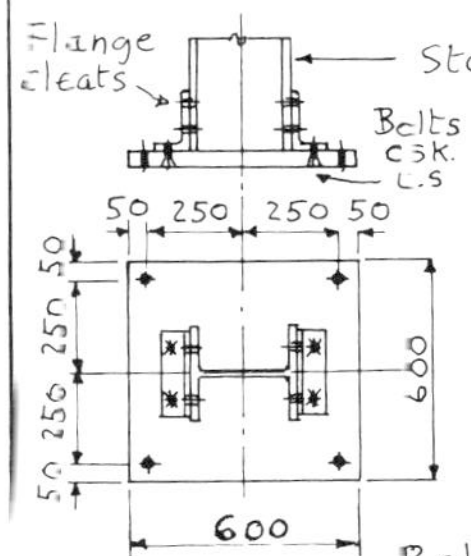

Base material

1 plate 600×45×600

4 holes 26 dia.

End of stanchion and top of plate machined.

Cleats 150×90×12L×250
M20, 4·6 grade Shopbolts

Bolted slab base

CURPE CONSULTANTS 46 Orburn Road DUNFIELD	Project DESIGN IV		Job ref CC/ME
	Part of structure CHAPTER FOURTEEN COMPRESSION MEMBERS		Calc sheet no 14/10 rev /
	Drawing ref	Calc by R.A.S. Date	Check by Date

Code ref.	Calculations	Output

EXAMPLE 14.4. Tension and shear in Holding Down (H.D) bolts.

W = 45 kN (Dead + Imposed + Wind)
20 kN (Dead + Wind)

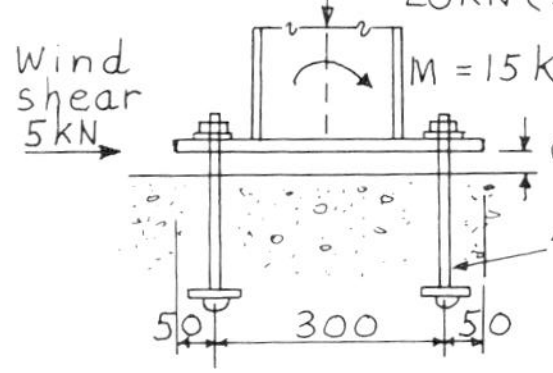

Two cases must be considered

Case (a) Dead + Imposed + Wind

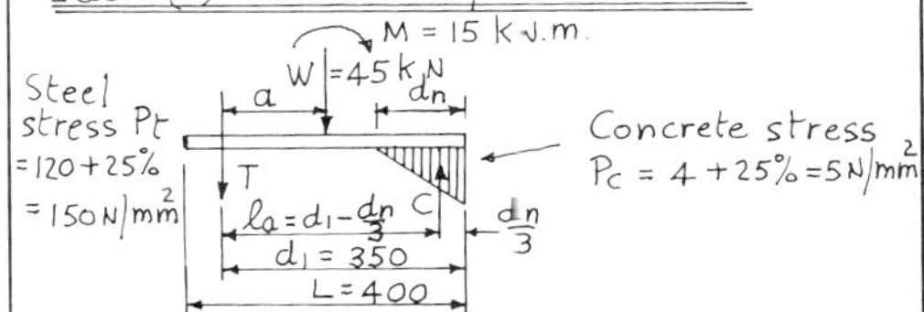

Cl.13 — The 25% increase is for wind loading.

As $\frac{M}{W} = \frac{15 \times 10^3}{45} = 333 > \frac{L}{6}$ then

tension occurs.

Depth to neutral axis

$d_n = \frac{m.P_c.d_1}{m.P_c + P_t} = \frac{15 \times 5 \times 350}{(15 \times 5) + 150} = 116{\cdot}66\text{ mm}$

Lever arm $l_a = d_1 - \frac{d_n}{3} = 350 - \frac{116{\cdot}66}{3}$

$l_a = 311{\cdot}12$ mm.

CURPE CONSULTANTS 46 Orburn Road DUNFIELD	Project DESIGN IV		Job ref CC/ME	
	Part of structure CHAPTER FOURTEEN COMPRESSION MEMBERS		Calc sheet no 14/11	rev /
	Drawing ref	Calc by R.A.S. / Date	Check by	Date

Code ref.	Calculations	Output
	EXAMPLE 14.4. continued.	
	Determine C and T	
	Take moments about centre line of bolts	
	$C = \frac{(15\times10^3)+(45\times150)}{350} = 62{\cdot}14\text{ kN}$	
	and $T = 62{\cdot}14 - 45 = 17{\cdot}14\text{ kN}$.	
	Breadth of plate b	
	$b = \frac{62{\cdot}14\times2\times10^6}{116{\cdot}66\times5\times10^3} = 213{\cdot}06\text{ mm}$	
Cl. 10b	Increase by 1·4, then $b = 213{\cdot}06\times1{\cdot}4 = 298$, say 300 mm	
	Case (b) Dead + Wind	
	Determine C and T	
	$C = \frac{(15\times10^3)+(20\times150)}{350} = 51{\cdot}42\text{ kN}$.	
	and $T = 51{\cdot}42 - 20 = 31{\cdot}42\text{ kN}$.	
Cl. 10b	Increase by 1·4, then $T = 1{\cdot}4\times31{\cdot}42 = 44\cdot\text{ kN}$	
	This case is the criterion for the design of the H.D. bolts with maximum tension.	
	Combined shear and tension in H.D. bolts	
Cl. 50d	$\frac{f_s}{P_s} + \frac{f_t}{P_t} \not> 1{\cdot}4$	
	For one bolt $f_s = \frac{5\times10^3\times4}{4\times\pi\times20^2} = 3{\cdot}98\text{ N/mm}^2$	
	and $f_t = \frac{44\times10^3}{2\times245} = 89{\cdot}8\text{ N/mm}^2$	
	$\frac{3{\cdot}98}{80+25\%} + \frac{89{\cdot}8}{150} = 0{\cdot}039 + 0{\cdot}598 = 0{\cdot}637$	
	Since $0{\cdot}637 < 1{\cdot}4$, bolts are adequate.	Bolts O.K.

SELF-ASSESSMENT QUESTIONS

1. A 254 × 254 × 89 UC stanchion of effective length 5 m carries an axial load of 1100 kN. Check the adequacy of the section.

2. The stanchion in Question 1 carries loads from four beams as follows: 300 kN connected to stanchion flange; 290 kN connected to stanchion flange; 260 kN connected to stanchion web; 250 kN connected to stanchion web. Check the adequacy of the section.

3. Calculate the safe loads for a 254 × 254 × 73 UC stanchion of 8 m actual length with:

(*a*) both ends fixed in position and direction;
(*b*) both ends held in position only.

4. A 305 × 305 × 97 UC stanchion of effective length 10 m is structurally cased in concrete. Calculate the safe axial load.

5. A 254 × 254 × 73 UC stanchion of effective length 8 m and structurally cased in concrete carries an axial load of 800 kN. Check the adequacy of the section.

6. The stanchion in Question 5 carries loads from four beams as follows: 200 kN connected to stanchion flange; 220 kN connected to stanchion flange; 200 kN connected to stanchion web; 180 kN connection to stanchion web. Check the adequacy of the section.

7. A slab base connected to a 305 × 305 × 97 UC carries an axial load of 3000 kN. Determine the dimensions of a square slab base given the allowable bearing pressure on the concrete foundation to be 5.3 N/mm^2.

8. To a scale of 1 : 10, prepare detailed elevations and a plan of the completed base designed in answer to Question 7 employing:

(*a*) welding;
(*b*) bolting to cleats.

9. A stanchion base is subjected to:

(*a*) an axial load of 25 kN (dead + wind)
(*b*) a moment of 18 kN m (wind)
(*c*) a horizontal shear of 10 kN (wind)

The length L of the base plate is 450 mm and 4 HD bolts 20 mm dia. are positioned 50 mm from the edges of the plate. The allowable stresses (including 25 per cent wind allowance) are: $P_{cc} = 5\text{ N/mm}^2$ and $P_{st} = 150\text{ N/mm}^2$. Check the adequacy of the HD bolts.

CHAPTER FIFTEEN

Steel Roof Trusses, Lattice Girders and Purlins

CHAPTER OBJECTIVES

After studying this chapter you should be able to design:

* simple-angle purlins;
* single- and double-angle compression members;
* single- and double-angle tension members;
* spacing of washer packs in double-angle members;
* simple roof trusses and lattice girders.

PURLINS

Purlins support the roof cladding or decking on flat and sloping roofs. In the case of roof slopes not exceeding 30° pitch, purlins may be designed in accordance with empirical rules which are based on a minimum imposed loading of 0.75 kN/m^2. See Clause 4.5 in BS449. For angle purlins of grade 43 steel:

$$\text{minimum depth of purlin} = \frac{L}{45}\ \text{mm}$$

$$\text{minimum width of purlin} = \frac{L}{60}\ \text{mm}$$

$$\text{minimum section modulus} = \frac{WL}{1.8 \times 10^3}\ \text{cm}^3$$

where L = the span of purlin (i.e. centre-to-centre in mm of the supporting trusses, beams or girders) and W = the total distributed load in kN from dead and imposed, assumed as acting normal to the roof.

Purlins are spaced to suit the type of roof covering employed but in each case must not exceed the spacing recommended in the manufacturer's specification.

ROOF TRUSSES

These are positioned at predetermined centres along the building to provide support for the purlins. The minimum angle of slope of the rafter depends on the particular type of roof covering used and should comply with that recommended in the manufacturer's specification. For economical reasons a height to the ridge equal to one fifth to one quarter of the span is often selected. The framework is generally triangulated using various arrangements of internal members for different spans of trusses. Typical examples of ridge and north light types are shown in Fig. 94.

The roof loading is applied to the truss by the purlins positioned along the rafters to suit the roof covering. When purlin positions coincide with node points as shown in Fig. 95(*a*) the most economical rafter section can be achieved, since all members are subjected to direct force only. When purlins are unavoidably positioned between node points as shown in Fig. 95(*b*) bending is produced in the rafter in addition to direct force. All the other members are subjected to direct forces only.

LATTICE GIRDERS

Where very large areas free of any obstructing stanchions are required, the use of lattice girders supporting roof trusses, purlins or beams, often provides a satisfactory solution to the problem.

The girder framework is normally triangulated and node points are arranged where possible to coincide with the positions of

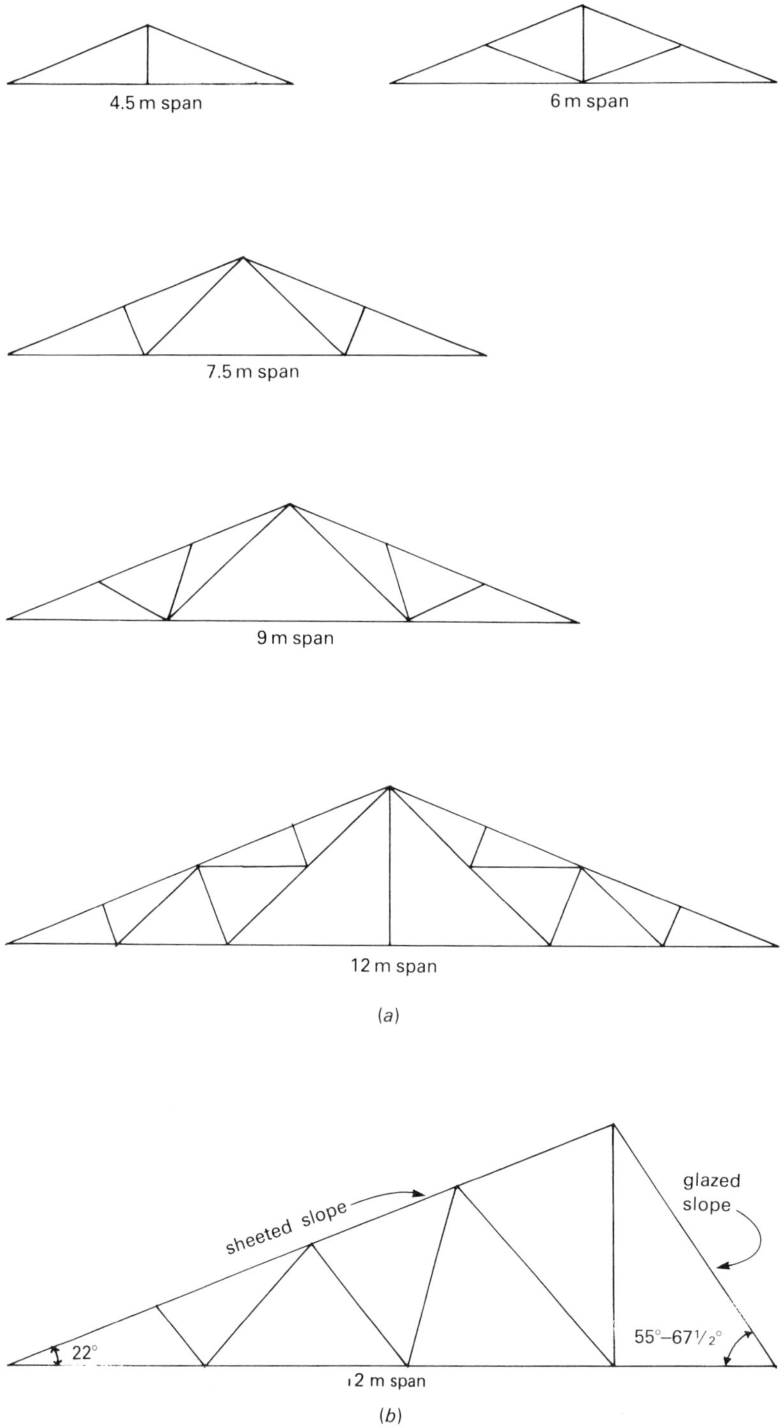

FIG. 94. *Examples of roof tissues:* (a) *ridge type; and* (b) *north light type.*

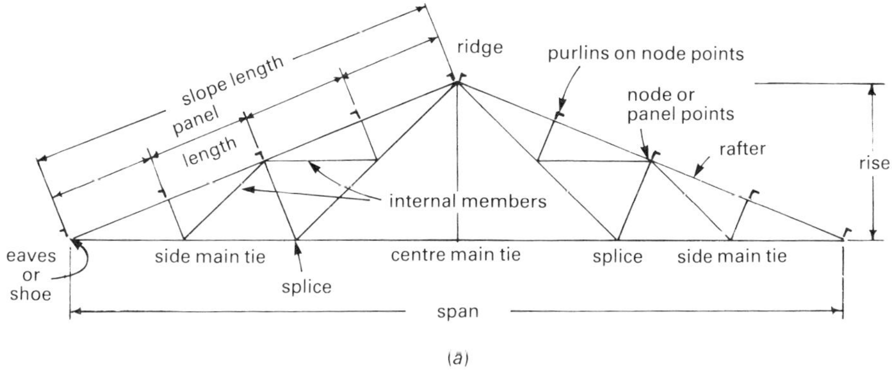

FIG. 95. *Purlins:* (a) *on node point; and* (b) *off node point.*

concentrated loads applied to the girder by roof trusses, purlins or beams. The economic depth of girders depends on the loading and may range from one tenth to one sixth of the span. Commonly used are the N and Warren girder types shown in Fig. 96.

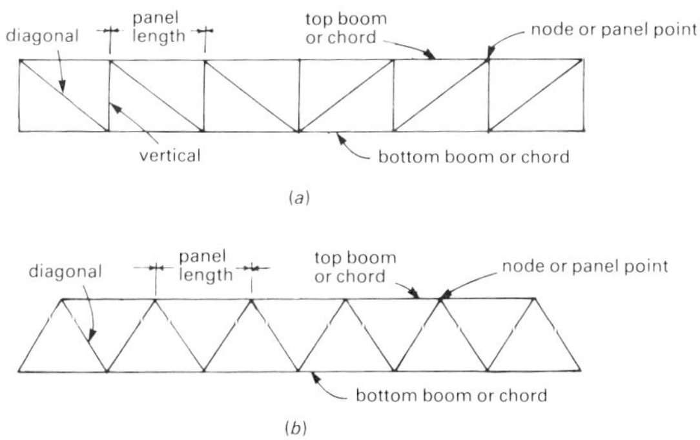

FIG. 96. *Lattice girders:* (a) *N type; and* (b) *Warren.*

Members and connections

Roof truss and lattice girder members may be fabricated from standard steel sections such as:

(*a*) angles;
(*b*) tees;
(*c*) channels;
(*d*) rectangular hollow;
(*e*) circular hollow.

The members may be connected together at the node points by:

(*a*) gusset plates and bolts;
(*b*) welding.

For hollow sections welding is the more obvious choice and for angles bolting or welding may be considered. In the roof truss and lattice girder design examples shown in this chapter, all members are fabricated from angle sections connected together by gusset plates and bolts. Dead and imposed loads are applied at the node points and include an estimated allowance for the self weight of the truss or girder.

Design of angle sections

Angle struts: Clause 30C in BS449
For the purpose of design these are considered as either discontinuous or continuous angle

struts:

(*a*) For single-angle discontinuous struts connected by not less than two bolts in line at each end or by the equivalent in welding, the slenderness ratio is given by:

$$\frac{l}{r} = \frac{0.85L}{r_{vv}}$$

where L = the length of strut, centre to centre of intersections at each end (this length is normally scaled from the drawing) and r_{vv} = the minimum radius of gyration of the angle (i.e. about V–V axis). The value of P_c is then obtained from Table 17 and the allowable load is given by P_cA where P_c = the allowable compressive stress and A = the gross cross sectional area of the angle.

(*b*) For double-angle discontinuous struts, back-to-back connected to both sides of a gusset or section by not less than two bolts in line at each end, or by the equivalent in welding, the effective length l is taken as between $0.7L$ and $0.85L$ depending on the degree of restraint. The slenderness ratio is given by l/r where r = the minimum radius of gyration i.e. lesser of the X–X or Y–Y axes. The value of P_c is then obtained from Table 17 and the allowable load is given by: P_cA where A = the gross cross sectional area of the pair of angles.

(*c*) For continuous struts such as rafters of trusses and top booms of lattice girders, design should be in accordance with Clause 26 and Table 17. All double-angle struts should be connected together in accordance with Clause 37. From Clause 33 the maximum slenderness ratio is 180 for compression members carrying dead loads with or without imposed loads.

Angle ties: Clause 42 in BS449

(*a*) Single-angle connected through one leg. It is necessary to make allowance for the eccentricity of the connection. Thus:

$$\text{Effective area} = a_1 + \left[a_2\left(\frac{3a_1}{3a_1 + a_2}\right)\right]$$

where a_1 = the net sectional area of the connected leg and a_2 = the sectional area of the unconnected leg. The allowable load is given by:

$$P_t \times \text{net or effective area of the angle}$$

where P_t = allowable stress in tension, Table 19.

If holes are required in both legs (e.g. at a splice in the main tie) holes must be staggered.

(*b*) Double angles placed back-to-back and connected to each side of a gusset or section. In this case:

$$\text{Allowable load} = P_t \times \text{net sectional area of the pair of angles}$$

where net sectional area = gross area less the sum of the sectional areas of the maximum number of holes in any cross section at right angles to the direction of stress in the member. The area to be deducted when the holes are staggered shall be that given above or, if greater, the sum of the sectional areas of all holes in any zig-zag line extending progressively across the member or part of the member, less $s^2t_1/4G$, where s = the staggered pitch of the holes; t_1 = the thickness of holed material; and G = the gauge of the holes. Members are connected together along their length as specified in Sub-clause 51e or 54.

The following examples illustrate the design of angles used as purlins and members of roof trusses and lattice girders.

CURPE CONSULTANTS 46 Orburn Road DUMFIELD	Project: DESIGN IV	Job ref: CC/ME
	Part of structure: ANGLE PURLINS	Calc sheet no: 15/1 rev: /
	Drawing ref: — Calc by: R.A.S. Date: —	Check by: — Date: —

Code ref.	Calculations	Output
	EXAMPLE 15.1. Design of angle purlin	
	Design an angle purlin to span 4·5 m.	
	Purlin spacing is 1·95 m. Roof slope = 1 in $2\frac{1}{2}$	
	Dead load measured on slope	
	Sheeting and insulation = 0·25 kN/m^2	
	Estimated purlin weight = 0·10 kN/m^2	
	0·35 kN/m^2	
	Imposed load measured on plan = 0·75 kN/m^2	
	Load on purlin	
	Dead load = 0·35 × 4·5 × 1·95 = 3·07 kN	
	Imposed load = $0{\cdot}75 \times 4{\cdot}5 \times 1{\cdot}75 \times \frac{2{\cdot}5}{2{\cdot}69} = 5{\cdot}49$ kN	
	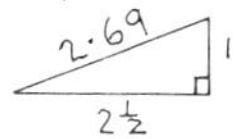	
	Total load = 8·56 kN.	W = 8·56 kN.
Cl.45	Minimum leg lengths	
	Depth = $\frac{L}{45} = \frac{4{\cdot}5 \times 10^3}{45} = 100$ mm.	Depth 100 mm
	Width = $\frac{L}{60} = \frac{4{\cdot}5 \times 10^3}{60} = 75$ mm.	Width 75 mm
Cl.45.	Minimum section modulus	
	Minimum $Z = \frac{WL}{1{\cdot}8 \times 10^3}$	
	$= \frac{8{\cdot}56 \times 4{\cdot}5 \times 10^3}{1{\cdot}8 \times 10^3} = 21{\cdot}4\ cm^3$	
	100 × 75 × 10 Γ Z_{xx} = 23·8 cm^3 13 kg/m	
	125 × 75 × 8 Γ Z_{xx} = 29·6 cm^3 12·2 kg/m	
	125 × 75 × 8 Γ lighter section	Use 125×75×8 Γ

CURPE CONSULTANTS 46 Orburn Road DUNFIELD	Project: DESIGN IV	Job ref: CC/ME
	Part of structure: ROOF TRUSS	Calc sheet no: 15/2 rev: /
	Drawing ref: — Calc by: R.A.S. Date: —	Check by: — Date: —

Code ref.	Calculations	Output
	EXAMPLE 15.2. Design of Roof Truss	
	Design dimensions of truss	
	Span = 14·5 m.	
	Rise = $\frac{1}{5}$ × span = 2·9 m.	
	Spacing = 4·5 m centres	
	Max. purlin spacing = 2 m.	
	Loading data	
	Dead load measured on slope	
	Sheeting and insulation = 0·25	
	Purlins = 0·10	
	Estimated self weight of truss = 0·12	
	0·47 kN/m^2	
	Imposed load measured on plan = 0·75 kN/m^2	
	Wind load will not be considered but in practice its effect would be investigated.	
	Loads on Roof Truss.	
	Dead load on one slope = 0·47 × 7·8 × 4·5 = 16·5 kN	
	Imposed " " " = 0·75 × 4·5 × 7·25 = 24·5 kN	
	41·0 kN	
	With the rafter slope divided into four equal panel lengths of 1·95 m then	
	Panel point loads are:	

CURPE CONSULTANTS 46 Orburn Road DUNFIELD	Project DESIGN IV	Job ref CC/ME
	Part of structure ROOF TRUSS	Calc sheet no 15/3 rev /
	Drawing ref / Calc by R.A.S. / Date	Check by / Date

Code ref.	Calculations	Output

EXAMPLE 15.2. continued

At intermediate points $\frac{41.0}{4} = 10.25\ kN$

At eaves and ridge $\frac{41.0}{8} = 5.125\ kN$

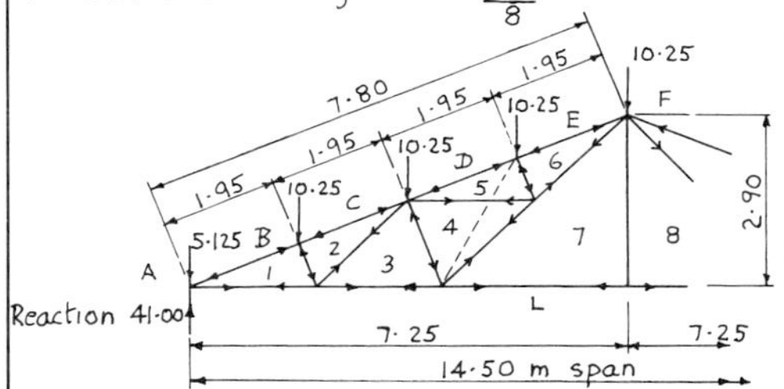

Frame diagram. Scale 1:100

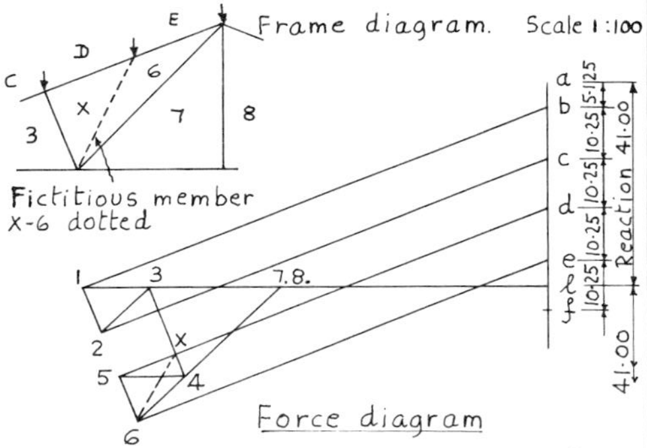

Force diagram

Dead + Imposed loads

Scale 1 mm = 1 kN.

Force diagram is drawn for half the truss only, because of the symmetry of

CURPE CONSULTANTS 46 Orburn Road DUNFIELD	Project DESIGN IV	Job ref CC/ME
	Part of structure ROOF TRUSS	Calc sheet no 15/4 rev /
	Drawing ref / Calc by R.A.S. / Date	Check by / Date

Code ref.	Calculations	Output

EXAMPLE 15.2. continued

both frame and loads. The members 4-5 and 5-6 are temporarily substituted by a fictitious member X-6, to enable the point 6 to be located and the force diagram to be completed.

Forces in members + compression − tension

Member	Dead + Imposed load kN	Strut or Tie
B-1	+99.0	Strut
C-2	+95.0	Strut
D-5	+91.0	Strut
E-6	+87.0	Strut
L-1	−92.0	Tie
L-3	−79.0	Tie
L-7	−53.0	Tie
1-2	+10.0	Strut
2-3	−13.0	Tie
3-4	+19.0	Strut
4-5	−13.0	Tie
5-6	+10.0	Strut
4-7	−26.0	Tie
6-7	−39.0	Tie
7-8	0	Sag Tie

CURPE CONSULTANTS, 46 Orburn Road, DUNFIELD	Project: DESIGN IV	Job ref: CC/ME
	Part of structure: ROOF TRUSS	Calc sheet no: 15/5, rev: /
	Drawing ref: — Calc by: R.A.S. Date: —	Check by: — Date: —

Code ref.	Calculations	Output
	EXAMPLE 15.2. continued	
	Design of members	
	Rafter - a continuous strut. $W = 99$ kN	
	These are formed with two angles, back to back, separated by gusset plates. Clause 30c (iv) states that truss rafters should be designed in accordance with Clause 26.	
Cl. 26a	Effective lengths	
	$\ell_{XX} = 0.7 \times$ panel length; assuming that the gusset plates provide full fixity at the node points each end.	
	$\ell_{YY} =$ purlin spacing; assuming a simple pinned purlin cleat connection at each end.	
	Try 2/50×50×6 ⅃⅂	
Cl. 33.	$\frac{\ell_{XX}}{r_{XX}} = \frac{0.7 \times 1.95 \times 10^3}{1.5 \times 10} = 91$; $\frac{\ell_{YY}}{r_{YY}} = \frac{1.95 \times 10^3}{2.38 \times 10} = 82$ $\Big\} \not> 180$	
Table 17a	Therefore $p_c = 90\ N/mm^2$ Table 17a.	
	Allowable or safe load $= p_c.A.$	
	$= \frac{90 \times 11.4 \times 10^2}{10^3} = 102.6$ kN.	
	Since $99 < 102.6$ Section satisfactory	Use 2/50×50×6 ⅃⅂

CURPE CONSULTANTS, 46 Orburn Road, DUNFIELD	Project: DESIGN IV	Job ref: CC/ME
	Part of structure: ROOF TRUSS	Calc sheet no: 15/6, rev: /
	Drawing ref: — Calc by: R.A.S. Date: —	Check by: — Date: —

Code ref.	Calculations	Output
	EXAMPLE 15.2. continued	
	Main tie Max $W = -92$ kN.	
	Single angle tie bolted to a gusset plate	
	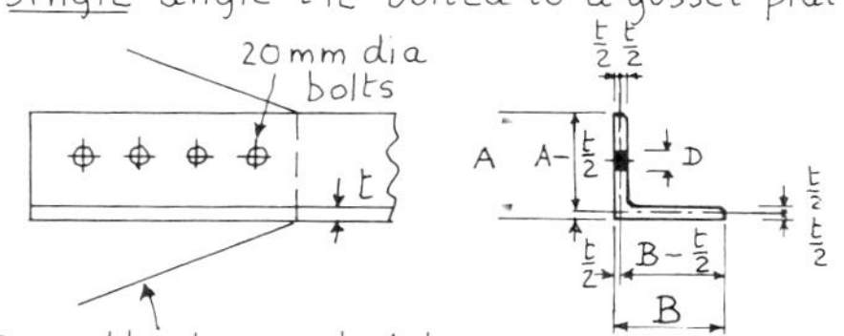	
	8 mm thick gusset plate	
Cl. 42a	Area of connected leg $c_1 = \left(A - \frac{t}{2} - D\right) t$	
	Area of unconnected leg $a_2 = \left(B - \frac{t}{2}\right) t$	
	Try 80×80×6 L	
	$a_1 = \left(80 - \frac{6}{2} - 22\right) 6 = 330\ mm^2$	
	$a_2 = \left(80 - \frac{6}{2}\right) 6 = 462\ mm^2$	
	Net or effective area $= a_1 + \left[a_2\left(\frac{3a_1}{3a_1 + a_2}\right)\right]$	
	$= 330 + \left[462\left(\frac{3 \times 330}{(3 \times 330) + 462}\right)\right]$	
	$= 645\ mm^2$	Net area $= 645\ mm^2$
	Allowable or safe load $= P_t \times$ net area	
Table 19	$= \frac{155 \times 645}{10^3} = 100$ kN	
	Since $92 < 100$ section satisfactory	Use 80×80×6 L

CURPE CONSULTANTS
46 Orburn Road
DUNFIELD

Project: DESIGN IV | Job ref: CC/ME
Part of structure: ROOF TRUSS. | Calc sheet no: 15/7 | rev: /
Drawing ref: | Calc by: R.A.S. | Date: | Check by: | Date:

Code ref.	Calculations	Output
	EXAMPLE 15.2. continued	
	Tie 4-7, 6-7 Max W = -39 kN	
	Try 50×50×6 L connected to gusset plate by 16 mm dia. bolts.	
Cl.42a	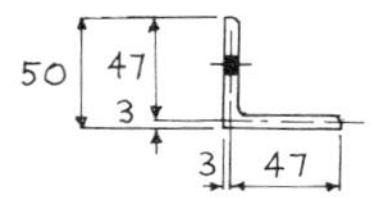 $a_1 = \left(50 - \frac{6}{2} - 18\right)6 = 174\,mm^2$	
	$a_2 = \left(50 - \frac{6}{2}\right)6 = 282\,mm^2$	
	Net or effective area $= 174 + \frac{282(3 \times 174)}{(3 \times 174) + 282}$	
	$= 357\ mm^2$	Net area = 357 mm²
Table 19	Allowable or safe load $= \frac{155 \times 357}{10^3} = 55.3\ kN$	
	Since 39 < 55.3, section satisfactory	Use 50×50×6 L
	Ties 2-3, 4-5 W = -13 kN	
	Considering 50×50×6 L as the minimum section suitable for this construction, then from previous calculation it is satisfactory.	Use 50×50×6 L
	Strut 3-4 W = 19 kN.	
Cl.30.c	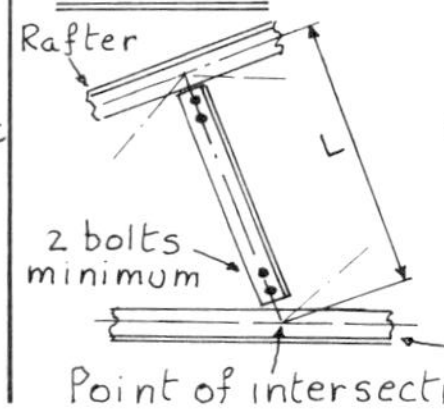 Scaled length L = 1·55 m Effective length $\ell = 0.85L$ This is a discontinuous strut.	

CURPE CONSULTANTS
46 Orburn Road
DUNFIELD

Project: DESIGN IV | Job ref:
Part of structure: ROOF TRUSS. | Calc sheet no: 15/8 | rev: /
Drawing ref: | Calc by: R.A.S. | Date: | Check by: | Date:

Code ref.	Calculations	Output
	EXAMPLE 15.2. continued.	
	Try 50 × 50 × 6 L	
Cl.33.	Slenderness ratio $\frac{\ell}{r_{vv}} = \frac{0.85 \times 1.55 \times 10^3}{0.97 \times 10} = 135.8 \not> 180$	
Table 17a	Therefore $p_c = 48.2\ N/mm^2$	
	Allowable or safe load = Pc. A.	
	$= \frac{48.2 \times 5.69 \times 10^2}{10^3} = 27.4\ kN.$	
	Since 13 < 27.4, section satisfactory	Use 50×50×6 L
	Struts 1-2, 5-6 W = +10 kN	
	Single angle discontinuous struts.	
	Scaled length L = 0·8 m.	
	Minimum section 50×50×6 L more than satisfactory.	Use 50×50×6 L
	Sag tie	Use 50×50×6 L
	Intermediate washer spacing in rafter	
Cl. 37.	Compression members composed of two angles back to back in contact or separated by a small distance shall be connected together by bolting or welding, so that the maximum slenderness ratio $\frac{\ell}{r}$ of each member between the connections is not greater than 40 or greater than 0·6 times the most unfavourable ratio of slenderness	

CURPE CONSULTANTS 46 Orburn Road DUNFIELD	Project: DESIGN IV	Job ref: CC/ME
	Part of structure: ROOF TRUSS	Calc sheet no: 15/9 rev: /
	Drawing ref: / Calc by: R.A.S. / Date:	Check by: / Date:

Code ref.	Calculations	Output

EXAMPLE 15.2 continued

of the strut as a whole, whichever is the less.

In this case:

0·6 × most unfavourable $\frac{\ell}{r}$ of the strut

$= 0{\cdot}6 \times 91 = 54{\cdot}6$

Thus $\frac{\ell}{r}$ of each angle $\not> 40$

Then maximum spacing $\ell = 40\, r_{vv}$

$\ell = 40 \times 0{\cdot}97 \times 10 = 388$

Number of spaces per panel length $= \frac{1950}{388} = 5$

Use 4 intermediate washers per panel length — Output: 4 per panel

Typical connections

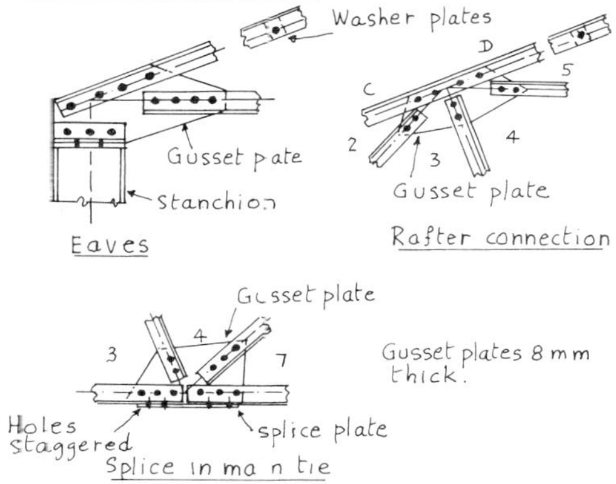

CURPE CONSULTANTS 46 Orburn Road DUNFIELD	Project: DESIGN IV	Job ref:
	Part of structure: LATTICE GIRDER	Calc sheet no: 15/10 rev: /
	Drawing ref: / Calc by: R.A.S / Date:	Check by: / Date:

Code ref.	Calculations	Output

EXAMPLE 15.3 Design of Lattice Girder

N type lattice girder supporting trusses

Span = 30 m

Depth = 3·2 m

Spacing = 7·5 m centres

Loading

Dead and Imposed loads including an estimated allowance for the self weight of the girder are applied at the panel points from trusses spaced at 5·0 m centres. Wind loads are not considered but in practice their effect would be investigated.

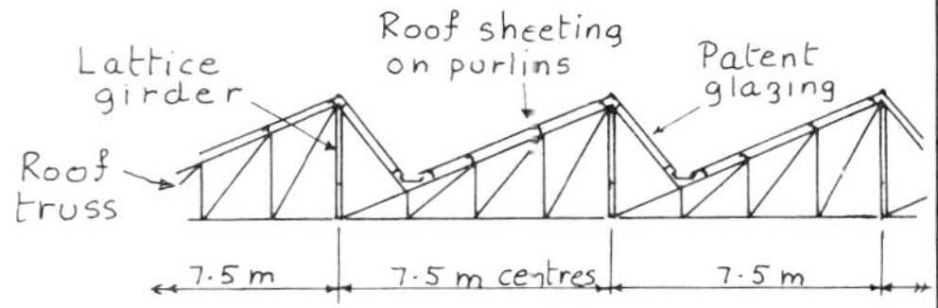

Section X-X

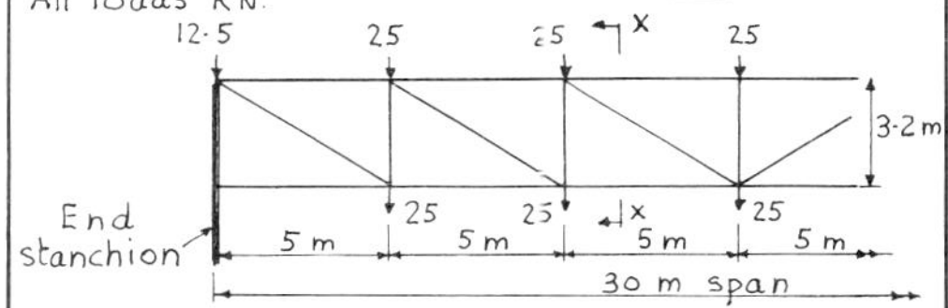

Part elevation of girder

CURPE CONSULTANTS
46 Orburn Road
DUNFIELD

Project: DESIGN IV | Job ref:
Part of structure: LATTICE GIRDER | Calc sheet no: 15/11 rev: /
Drawing ref: | Calc by: R.A.S. | Date: | Check by: | Date:

Code ref. | Calculations | Output

EXAMPLE 15.3. Continued

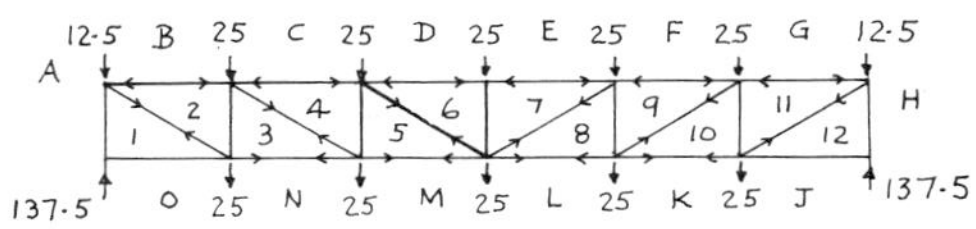

Frame diagram.

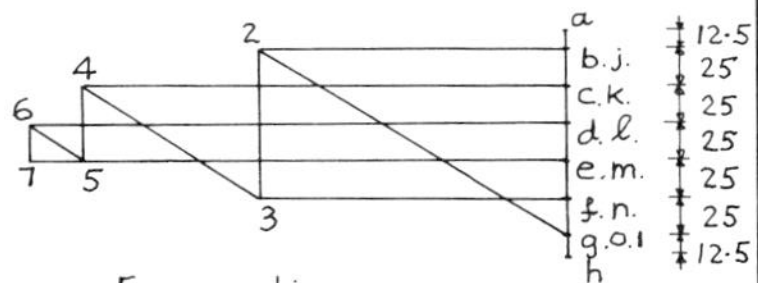

Force diagram

Force diagram is drawn for half the girder only, because of the symmetry of both frame and loads.

Forces in members + compression, − tension

Member	Dead+Imposed load kN	Strut or tie
B-2	+ 200	Strut
C-4	+ 320	Strut
D-6	+ 350	Strut
1-O	− 0	Tie
3-N	− 200	Tie
5-M	− 320	Tie
1-2	− 240	Tie
2-3	+ 100	Strut
3-4	− 140	Tie
4-5	+ 50	Strut
5-6	− 40	Tie
6-7	+ 25	Strut

As a check, determine the force in D-6 using the method of sections.

CURPE CONSULTANTS
46 Orburn Road
DUNFIELD

Project: DESIGN IV | Job ref:
Part of structure: LATTICE GIRDER | Calc sheet no: 15/12 rev: /
Drawing ref: | Calc by: R.A.S. | Date: | Check by: | Date:

Code ref. | Calculations | Output

EXAMPLE 15.3. continued.

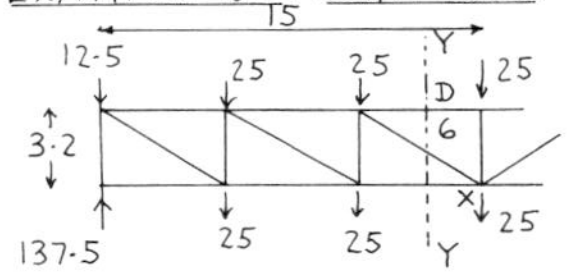

Section Y-Y cuts three members

Take moments about point X.

$(137{\cdot}5 - 12{\cdot}5)\,15 - (25+25+25+25)7{\cdot}5 = 3{\cdot}2\,F_{D\text{-}6}$

$F_{D\text{-}6} = \dfrac{1875 - 750}{3{\cdot}2} = 351{\cdot}5\ \text{kN}$

This force 351·5 kN agrees fairly well with 350 kN scaled from the force diagram

Design of members

Top Boom or Chord - a continuous strut

Max W = +350 kN.

These are formed with two angles, back to back, separated by gusset plates.

Cl. 30c — Clause 30 c(iv) states that girder flanges should be designed in accordance with Clause 26.

Cl. 26a — Effective lengths

ℓ_{XX} = 0·7 × panel length; assuming that the gusset plates provide full fixity at the node points each end.

ℓ_{YY} = truss spacing; assuming a simple

CURPE CONSULTANTS 43 Orburn Road DUNFIELD	Project: DESIGN IV		Job ref	
	Part of structure: LATTICE GIRDER		Calc sheet no: 15/13	rev: /
	Drawing ref	Calc by: R.A.S.	Date	Check by	Date

Code ref.	Calculations	Output
	EXAMPLE 15.3 continued	
	Pinned connection between truss and girder at each end.	
	Try 2/120×120×10 ⊤⊤	
	Slenderness ratio:	
	$\frac{\ell_{xx}}{r_{xx}} = \frac{0.7 \times 5 \times 10^3}{3.67 \times 10} = 95.3$	
	$\frac{\ell_{YY}}{r_{YY}} = \frac{5 \times 10^3}{5.37 \times 10} = 93.1$	
Table 17a	Therefore $P_c = 84.7\ N/mm^2$ Table 17a	
	Allowable or safe load = $P_c.A.$	
	$= \frac{84.7 \times 46.4 \times 10^2}{10^3} = 393\ kN.$	
	Since 350 < 393, Section satisfactory	Use 2/120×120×10⊤⊤
	Bottom Boom or Chord Max. W = −320 kN.	
	These are formed with two angles, back to back, separated by gusset plates.	
Cl.42b	Clause 42b states the net sectional area of the pair shall be taken.	
	Net area required = $\frac{W}{P_t} = \frac{320 \times 10^3}{155} = 2064\ mm^2$	
Table 19	$P_t = 155\ N/mm^2$	
	Try 2/80×80×8 ⅃L Gross area = $2450\ mm^2$	
	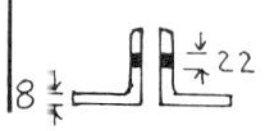 Assuming 20mm dia. bolts in 22 mm dia. holes	

CURPE CONSULTANTS 46 Orburn Road DUNFIELD	Project: DESIGN IV		Job ref	
	Part of structure: LATTICE GIRDER		Calc sheet no: 15/14	rev: /
	Drawing ref	Calc by: R.A.S.	Date	Check by	Date

Code ref.	Calculations	Output
	EXAMPLE 15.3 continued	
	Net area = gross area less holes.	
	$= 2450 - (2 \times 22 \times 8)$	
	$= 2098\ mm^2$	
	Allowable or safe load = $P_t.A.$	
	$= \frac{155 \times 2098}{10^3} = 325\ kN.$	
	Since 320 < 325, Section satisfactory	Use 2/80×80×8 ⅃L
	Alternatively the section can be arrived at as follows:	
	Net area required = $\frac{W}{P_t} = \frac{320 \times 10^3}{155} = 2064\ mm^2$	
	Assuming 8 mm thick angles then	
	Gross area required = net area + holes	
	$= 2064 + (2 \times 22 \times 8)$	
	$= 2416\ mm^2$	
	Gross area given for 2/80×80×8 ⅃L = $2450\ mm^2$	
	Therefore satisfactory.	
	Diagonal member 1-2 W = −240 kN.	
	Net area required = $\frac{240 \times 10^3}{155} = 1548\ mm^2$	
	Try 2/80×60×7 ⅃L Gross area = $1880\ mm^2$	
	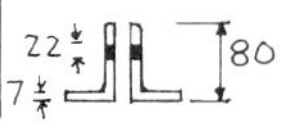 Assuming 20 mm dia. bolts in 22 mm dia. holes.	

CURPE CONSULTANTS 46 Orburn Road DUNFIELD	Project DESIGN IV			Job ref	
	Part of structure LATTICE GIRDER			Calc sheet no 15/15	rev /
	Drawing ref	Calc by R.A.S.	Date	Check by	Date

Code ref.	Calculations	Output

EXAMPLE 15.3. continued

Net area = gross area less holes

$= 1880 - (2 \times 22 \times 7)$

$= 1572\ mm^2$

Allowable or safe load $= \frac{155 \times 1572}{10^3} = 243\ kN$

Since $240 < 243$, Section satisfactory — **Output:** Use $2/80 \times 60 \times 7$ JL

Alternatively net area required $= 1548\ mm^2$

Assuming 7 mm thick angles then

Gross area required = net area + holes

$= 1548 + (2 \times 22 \times 7)$

$= 1856\ mm^2$

Gross area given for $2/80 \times 60 \times 7$ JL $= 1880\ mm^2$

Therefore satisfactory

Diagonal member 3-4 $\quad W = -140\ kN.$

Net area required $= \frac{140 \times 10^3}{155} = 903\ mm^2$

Try $100 \times 65 \times 8$ L $\quad$ Gross area $= 1270\ mm^2$

For a single angle

Cl. 42a — Net area $= a_1 + \left[a_2\left(\frac{3a_1}{3a_1 + a_2}\right)\right]$

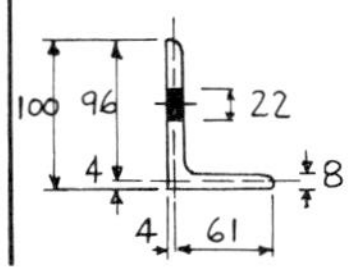

$a_1 = (96 - 22)8 = 592\ mm^2$

$a_2 = 61 \times 8 = 488\ mm^2$

CURPE CONSULTANTS 46 Orburn Road DUNFIELD	Project DESIGN IV			Job ref	
	Part of structure LATTICE GIRDER			Calc sheet no 15/16	rev /
	Drawing ref	Calc by R.A.S.	Date	Check by	Date

Code ref.	Calculations	Output

EXAMPLE 15.3. continued.

Net area $= 592 + \left[488\left(\frac{3 \times 592}{(3 \times 592) + 488}\right)\right]$

$= 975\ mm^2.$

Allowable or safe load = Pt. A.

$= \frac{155 \times 975}{10^3} = 151\ kN.$

Since $140 < 151$, Section satisfactory — **Output:** Use $100 \times 65 \times 8$ L

Diagonal member 5-6 $\quad W = -40\ kN.$

Net area required $= \frac{40 \times 10^3}{155} = 258\ mm^2$

Try $50 \times 50 \times 6$ L $\quad$ Gross area $= 5\ 690\ mm^2$

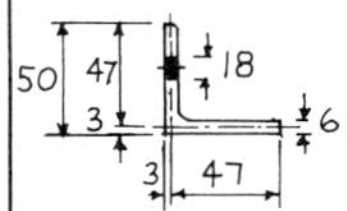

$a_1 = (47 - 18)6 = 174\ mm^2$

$a_2 = 47 \times 6 = 282\ mm^2$

Cl. 42a — Net area $= 174 + \left[282\left(\frac{3 \times 174}{(3 \times 174) + 282}\right)\right]$

$= 357\ mm^2$

Allowable or safe load $= \frac{155 \times 357}{10^3} = 55\ kN.$

Since $40 < 55$, Section satisfactory — **Output:** Use $50 \times 50 \times 6$ L

Vertical member 2-3 $\quad W = +100\ kN$

Length $L = 3{\cdot}2\ m$. A discontinuous strut.

Cl. 30c. — Assume effective length $\ell = 0{\cdot}85L$ for two angles, back to back, separated by gusset plates, with two bolts each end.

CURPE CONSULTANTS, 46 Orburn Road, DUNFIELD

Project	Part of structure	Drawing ref	Calc by	Date	Job ref	Calc sheet no	rev	Check by	Date
DESIGN IV	LATTICE GIRDER		R.A.S.			15/17	/		

Code ref.	Calculations	Output
	EXAMPLE 15.3. continued.	
	Try 2/80×80×6 ⅂⌐	
	Slenderness ratio:	
	$\frac{\ell_{XX}}{r_{XX}} = \frac{0\cdot85\times3\cdot2\times10^3}{2\cdot44\times10} = 111$	
	$\frac{\ell_{YY}}{r_{YY}} = \frac{0\cdot85\times3\cdot54\times10^3}{3\cdot54\times10} = 85$	
Table 17a	Using the larger value, $p_c = 68$ N/mm²	
	Allowable or safe load = Pc.A.	
	$= \frac{68\times18\cdot7\times10^2}{10^3} = 127$ kN	
	Since 100 < 127, Section satisfactory	Use 2/80×80×6 ⅂⌐
	Alternatively using a single angle	
	Try 120×120×8 L	
Cl. 30c	Slenderness ratio: $\frac{\ell}{r_{VV}} = \frac{0\cdot85\times3\cdot2\times10^3}{2\cdot37\times10} = 115$	
Table 17a	$p_c = 64$ N/mm²	
	Allowable or safe load = Pc.A.	
	$= \frac{64\times18\cdot7\times10^2}{10^3} = 119\cdot6$ kN	
	Since 100 < 119·6, Section satisfactory	Alternative 120×120×8L
	Vertical member 4-5 W = +50kN.	
	A discontinuous strut. L = 3·2 m.	
	Try 90×90×8 ⌐	
Cl. 30c	Slenderness ratio: $\frac{\ell}{r_{VV}} = \frac{0\cdot85\times3\cdot2\times10^3}{1\cdot76\times10} = 154\cdot5$	

CURPE CONSULTANTS, 46 Orburn Road, DUNFIELD

Project	Part of structure	Drawing ref	Calc by	Date	Job ref	Calc sheet no	rev	Check by	Date
DESIGN IV	LATTICE GIRDER		R.A.S.			15/18	/		

Code ref.	Calculations	Output
	EXAMPLE 15.3. continued	
Table 17a	$P_c = 38$ N/mm²	
	Allowable or safe load = Pc.A.	
	$= \frac{38\times13\cdot9\times10^2}{10^3} = 52\cdot8$ kN	
	Since 50 < 52·8, Section satisfactory	Use 90×90×8 ⌐
	Vertical member 6-7 W = +25 kN	
	A discontinuous strut. L = 3·2 m.	
	Try 80×80×6 L	
Cl. 30c	Slenderness ratio: $\frac{\ell}{r_{VV}} = \frac{3\cdot2\times0\cdot85\times10^3}{1\cdot57\times10} = 173$	
	Allowable or safe load = Pc.A.	
Table 17a	$P_c = 31$ N/mm²	
	Allowable load $= \frac{31\times9\cdot35\times10^2}{10^3} = 29$ kN.	
	Since 25 < 29, Section satisfactory	Use 80×80×6 ⌐
Cl. 37	Intermediate washer spacing in members composed of two angles back to back in contact or separated by a small distance	
	Top boom or chord 2/120×120×10 ⅂⌐	
	For this member	
	0·6 × most unfavourable $\frac{\ell}{r}$ of the strut	
	= 0·6 × 95·3 = 57	
	Thus $\frac{\ell}{r}$ of each angle ≯ 40	

CURPE CONSULTANTS
46 Orburn Road
DUNFIELD

Project DESIGN IV		Job ref
Part of structure LATTICE GIRDER		Calc sheet no 15/19 rev /
Drawing ref	Calc by R.A.S. Date	Check by Date

Code ref.	Calculations	Output
	EXAMPLE 15.3 continued	
	Then maximum spacing $\ell = 40\,r_{vv}$	
	$\ell = 40 \times 2{\cdot}36 \times 10 = 944$ mm	
	Number of spaces per panel length $= \frac{5000}{944} = 5$	Use 4 per panel
	Use 4 intermediate washers per panel length	
	Vertical member 2-3 2/80 × 80 × 6][	
Cl. 37.	$0{\cdot}6 \times 111 = 66{\cdot}6$	
	Thus $\frac{\ell}{r}$ of each angle ≯ 40	
	$\ell = 40 \times 1{\cdot}57 \times 10 = 628$ mm	
	Number of spaces $= \frac{3{\cdot}2 \times 10^3}{628} = 5{\cdot}1$	Use 4
	Use 4 intermediate washers	
Cl. 51e Cl. 42b	Bottom boom, Diagonal member 1-2	
	In the case of tension members the spacing should not exceed 1 m.	

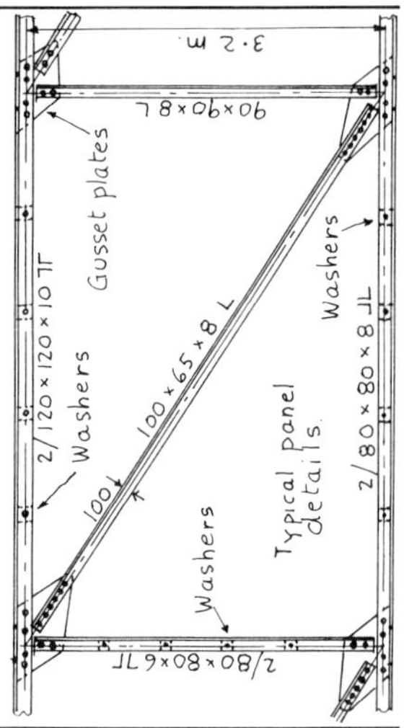

SELF-ASSESSMENT QUESTIONS

1. Design an angle purlin to span 5 m. Purlins are spaced at 2 m centres on a roof slope of 1 in 2. Dead load including self weight of the purlin is 0.4 kN/m² measured on slope. Imposed load is 0.75 kN/m² measured on plan.

2. For the roof truss shown in Fig. 97:

(*a*) Find the forces in the members.
(*b*) Check the adequacy of the proposed sections.

3. For the lattice girder shown in Fig. 98:

(*a*) Find the forces in the members.
(*b*) Check the adequacy of the proposed sections.

4. (*a*) Verify by calculation that the safe load is 91 kN for a discontinuous strut formed with two angles each of section 90 × 90 × 6, back-to-back separated by a gusset plate with an effective length of 4 m.
(*b*) Determine the intermediate washer spacing.

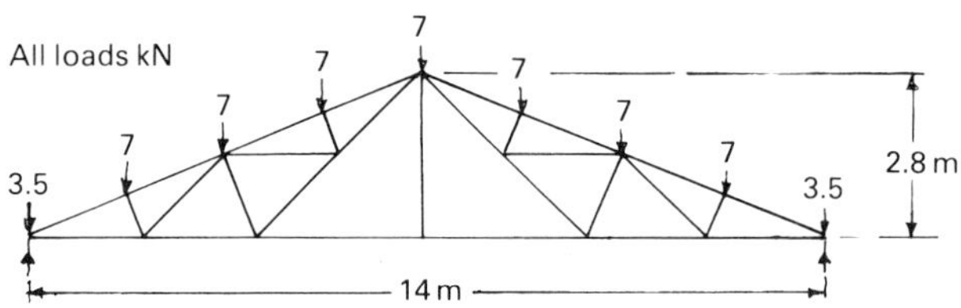

rafter 2/65 × 50 × 6][⌶65
main tie 80 × 80 × 8 L
internals 50 × 50 × 6 L
all bolts 16 mm dia.

FIG. 97.

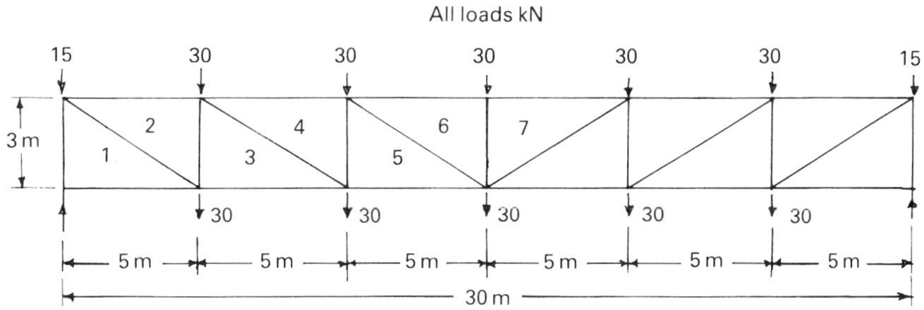

top boom 2/120 × 120 × 12 ┐┌
bottom boom 2/80 × 80 × 10 ┘└
1–2 2/100 × 65 × 7 ┐┌ ↕ 100
2–3 120 × 120 × 8 └
3–4 2/75 × 50 × 6 ┐┌ ↕ 75
4–5 90 × 90 × 8 ┌
5–6 70 × 70 × 6 └
6–7 90 × 90 × 6 └
all bolts 20 mm dia.

FIG. 98.

5. Verify by calculation that the safe load is 81 kN for a single angle discontinuous strut of section 100 × 100 × 8 and length between intersections is 3 m.

6. Verify by calculation that the safe load is 161 kN for a single angle tie of section 80 × 80 × 10 connected to a gusset plate with 20 mm dia. bolts.

7. (*a*) Verify that the safe load is 400 kN for a tie formed with two angles of section 80 × 80 × 10 separated by a gusset plate and connected to it by 20 mm dia. bolts.

(*b*) Determine the intermediate washer spacing.

CHAPTER SIXTEEN

Brickwork and Blockwork

CHAPTER OBJECTIVES

After reading this chapter you should be able to:

* explain the meaning and calculate the slenderness ratio, effective length, effective height, effective thickness of walls;
* understand the importance of lateral support;
* determine design loads and design strength;
* explain the importance of selecting suitable mortar mix;
* understand the effect of eccentricity and tension;
* design a brick or block wall, solid and cavity types;
* design a gravity retaining wall for overturning, sliding and bearing pressures.

INTRODUCTION

BS5628: Part 1 gives recommendations for the structural design of unreinforced masonry. It is intended ultimately to replace CP111.

In BS5628 the opportunity has been taken to use limit state philosophy in accordance with recent practice in other codes. All editions of CP111 are in permissible stress terms.

DESIGN REQUIREMENT

The design of load bearing masonry members should be undertaken primarily to ensure an adequate margin of safety against the ultimate limit state being reached. This is achieved by ensuring that:

$$\text{Design strength} \geqslant \text{design load}$$

Characteristic loads

The following characteristic loads should be used in design:

(*a*) Characteristic dead load G_k is the weight of the structure complete with finishes, fixtures and partitions defined in and calculated in accordance with CP3 Chapter V Part 1.

(*b*) Characteristic imposed load Q_k defined in and calculated in accordance with CP3 Chapter V Part I.

(*c*) Characteristic wind load W_k calculated in accordance with CP3 Chapter V Part 2.

Design loads

$$\text{Design load} = \text{characteristic load} \times \text{partial afety factor}$$

$$= (G_k \text{ or } Q_k \text{ or } W_k)\gamma_f$$

where γ_f is a partial safety factor for loads. Values are given in Clause 22.

In the case of combined dead and imposed loads:

$$\text{Design dead load} = 0.9G_k \text{ or } 1.4G_k$$

$$\text{Design imposed load} = 1.6Q_k$$

The numerical values 0.9, 1.4 and 1.6 are the γ_f factors. Where alternatives are shown, that producing the more severe condition should be selected.

Design strength

The design strength is a function of the characteristic strength of the masonry and the relevant partial safety factors on the materials and is given by:

$$\text{Design strength} = \frac{\text{characteristic strength}}{\gamma_m}$$

where γ_m is the partial safety factor for materials related to the quality control exercised during manufacture and construction. Values are given in Table 4 of the code, where normal and special categories are recognised.

TABLE 6.
CHARACTERISTIC COMPRESSIVE STRENGTH OF MASONRY f_k IN N/mm^2

(*a*) constructed with standard format bricks

Mortar designation	*Compressive strength of unit (N/mm²)* 5	10	15	20	27.5	35	50	70	100
(i)	2.5	4.4	6.0	7.4	9.2	11.4	15.0	19.2	24.0
(ii)	2.5	4.2	5.3	6.4	7.9	9.4	12.2	15.1	18.2
(iii)	2.5	4.1	5.0	5.8	7.1	8.5	10.6	13.1	15.5
(iv)	2.2	3.5	4.4	5.2	6.2	7.3	9.0	10.8	12.7

(*b*) *constructed with blocks having a ratio of height to least horizontal dimension of 0.6*

Mortar designtion	*Compressive strength of unit (N/mm²)* 2.8	3.5	5.0	7.0	10	15	20	35 or greater
(i)	1.4	1.7	2.5	3.4	4.4	6.0	7.4	11.4
(ii)	1.4	1.7	2.5	3.2	4.2	5.3	6.4	9.4
(iii)	1.4	1.7	2.5	3.2	4.1	5.0	5.8	8.5
(iv)	1.4	1.7	2.2	2.8	3.5	4.4	5.2	7.3

(*c*) *constructed with hollow blocks having a ratio of height to least horizontal dimension of between 2.0 and 4.0*

Mortar designation	*Compressive stength of unit (N/mm²)* 2.8	3.5	5.0	7.0	10	15	20	35 or greater
(i)	2.8	3.5	5.0	5.7	6.1	6.8	7.5	11.4
(ii)	2.8	3.5	5.0	5.5	5.7	6.1	6.5	9.4
(iii)	2.8	3.5	5.0	5.4	5.5	5.7	5.9	8.5
(iv)	2.8	3.5	4.4	4.8	4.9	5.1	5.3	7.3

(*d*) *Constructed from solid concrete blocks having a ratio of height to least horizontal dimension of between 2.0 and 4.0*

Mortar designation	*Compressive strength of unit (N/mm²)* 2.8	3.5	5.0	7.0	10	15	20	35 or greater
(i)	2.8	3.5	5.0	6.8	8.8	12.0	14.8	22.8
(ii)	2.8	3.5	5.0	6.4	8.4	10.6	12.8	18.8
(iii)	2.8	3.5	5.0	6.4	8.2	10.0	11.6	17.0
(iv)	2.8	3.5	4.4	5.6	7.0	8.8	10.4	14.6

Characteristic compressive strength f_k

Characteristic compressive strength f_k in terms of brick or block strength and mortor strength are presented in Table 2 of the Code. For compressive loading then:

$$\text{Design strength} = \frac{f_k}{\gamma_m}$$

Brickwork

Part (*a*) of Table 2 (reproduced at Table 6 below) applies to masonry built with standard format bricks. The value of f_k is modified if relevant to allow for the effects of narrow brick walls or small plan area as follows.

Narrow brick walls

Where solid brick walls or the loaded inner leaf of a cavity wall are constructed in standard format bricks with a thickness equal to the width of a single brick, the values of f_k obtained from Table 6(*a*) may be multiplied by 1.15.

Small plan area

Where the horizontal cross sectional area of a loaded wall or column is less than $0.2\,m^2$, the value of f_k should be multiplied by the factor $(0.7 + 1.5A)$ where A = the horizontal loaded cross sectional area of the wall or column (m^2).

Blockwork

The increased size of individual units means that fewer joints are required than for an equivalent wall built with standard bricks. The reduction in the number of joints results in an increased compressive strength for a given strength of unit or mortar.

For hollow block walls having a shape factor of between 0.6 and 2.0, the value of f_k should be obtained by interpolation between the values given in Tables 6(*b*) and 6(*c*). For solid concrete block walls having a shape factor of between 0.6 and 2.0, the value of f_k should be obtained by interpolation between the values given in Tables 6(*b*) and 6(*d*).

The shape factor of a unit is given by dividing the height of the unit by the least horizontal dimension of unit. As for brickwork, the value of f_k is modified, if relevant, to allow for the effects of plan area.

Mortars

A good mortar should possess the necessary "workability" in its fluid state and on setting and hardening obtain adequate strength and good bonding characteristics for structural stability and durability. However, whilst adequate strength is essential, the use of a stronger mortar does not necessarily result in a stronger wall, because mortar strength is not directly related to the strength of the brickwork or blockwork built with that mortar.

When a suitably matched mortar is used, any cracking will occur at the joints rather than in the masonry units. For any particular strength of unit, there is an optimum mortar strength and strength will not be increased by using a stronger than necessary mortar. Mortars are designated (i), (ii), (iii) and (iv) in order of decreasing strength but in increasing ability to accomodate movements caused by settlement and shrinkage.

To utilise the full capacity of high strength brick ($70\,N/mm^2$ or more) a 1 : 3 cement sand mix is necessary. For lower strength bricks or block, mortars with an increased proportion of lime can be used without any great loss in masonry strength.

Recommendations for mortar mixes and their suitability for various locations and conditions are given in CP121 Part I *Brick and Block Masonry*. Overall, the mortar used should contain no more cement than is necessary to provide adequate strength to the masonry.

Design for vertical loading

The basic principle can be expressed as:

$$\text{Design vertical loading} \leqslant \text{Design vertical load resistance}$$

Walls and columns are subjected to vertical loads arising from self weight of the masonry and any supported floors, roof, etc. A low squat wall fails by crushing of the material. Walls of a given thickness and material strength tend to fail by buckling at lower loads as their height increases. Eccentricity of loading also reduces the load-bearing capacity. In practice most loads are applied at some eccentricity to the centroidal axis of a wall. This eccentricity is usually expressed as a fraction of the wall thickness. The resultant eccentricity of the load at any level may be determined on the assumption that the total vertical load on a wall is axial immediately above a lateral support.

It is not necessary to consider the effects of eccentricity if:

$$e_x < 0.05t$$

where e_x = the eccentricity at the top of the wall and t = the thickness of wall.

A load transmitted to a wall by a single floor or roof is assumed to act at one third of the length of the bearing area from the loaded face, implying a triangular shaped pressure distribution. Where a uniform floor is continuous over a wall, each side of the floor may be taken as being supported individually on half the total bearing area.

The effects of slenderness ratio and eccentricity are combined and appear in the code as the capacity reduction factor β in Table 7.

TABLE 7.
CAPACITY REDUCTION FACTOR, β

Slenderness ratio h_{ef}/t_{ef}	*Eccentricity at top of wall,* e_x *Up to 0.05t* (see note 1)	*0.1t*	*0.2t*	*0.3t*
0	1.00	0.88	0.66	0.44
6	1.00	0.88	0.66	0.44
8	1.00	0.88	0.66	0.44
10	0.97	0.88	0.66	0.44
12	0.93	0.87	0.66	0.44
14	0.89	0.83	0.66	0.44
16	0.83	0.77	0.64	0.44
18	0.77	0.70	0.57	0.44
20	0.70	0.64	0.51	0.37
22	0.62	0.56	0.43	0.30
24	0.53	0.47	0.34	
26	0.45	0.38		
37	0.40	0.33		

NOTE 1. It is not necessary to consider the effects of eccentricities up to and including 0.05*t*.
NOTE 2. Linear interpolation between eccentricities and slenderness ratios is permitted.
NOTE 3. The derivation of β is given in appendix B of BS5628.

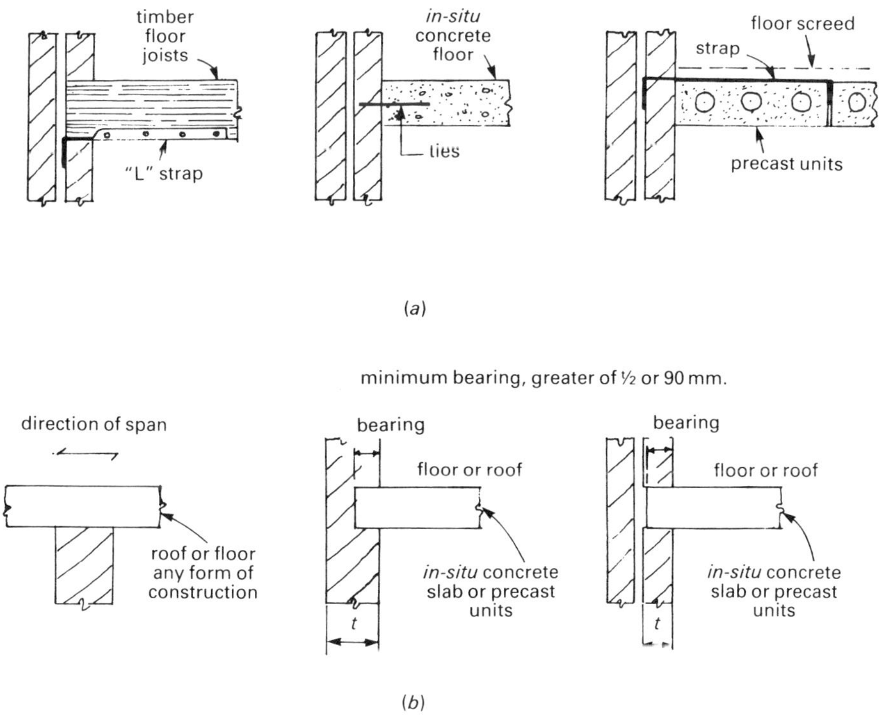

FIG. 99. *Horizontal lateral supports:* (a) *simple resistance; and* (b) *enhanced resistance.*

Slenderness ratio

This is defined as:

$$\frac{\text{effective height (or length)}}{\text{effective thickness}} = \frac{h_{ef} \text{ or } l_{ef}}{t_{ef}}$$

The effective length is used where this gives the lesser slenderness ratio.

Clause 28.1 states that the slenderness ratio should not exceed 27 except in the case of walls less than 90 mm thick in buildings of more than two storeys where it should not exceed 20.

Lateral support

A lateral support may be provided along either a horizontal or a vertical line, depending on whether the slenderness ratio is based on a vertical or horizontal dimension.

Horizontal lateral supports in the form of floors or roofs provide either "simple" or "enhanced" resistance to lateral movement of the wall. Examples are shown in Fig. 99.

Vertical lateral supports in the form of intersecting or return walls again provide either "simple" or "enhanced" resistance. Examples are shown in Fig. 100.

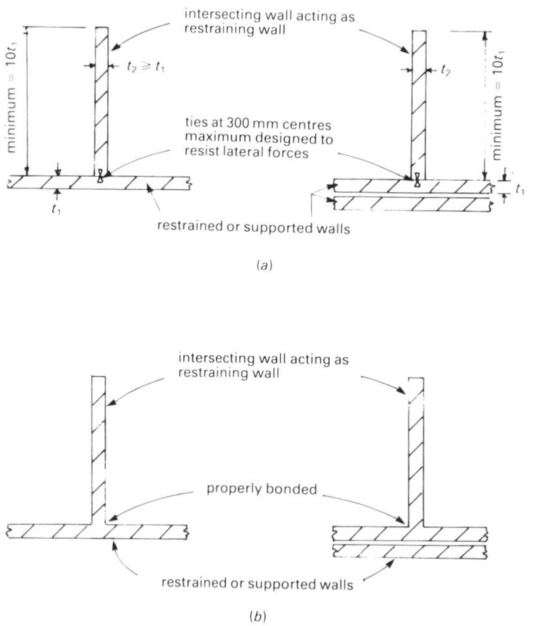

FIG. 100. *Vertical lateral supports:* (a) *simple resistance; and* (b) *enhanced resistance.*

Effective height

The effective height is related to the lateral support provided along horizontal lines.

The effective height of a wall may be taken as:

(*a*) 0.75 times the clear distance between lateral supports which provide "enhanced" resistance; or

(*b*) the clear distance between lateral supports which provide "simple" resistance.

Effective length

The effective length is related to the lateral support provided along vertical lines.

The effective length of a wall may be taken as:

(*a*) 0.75 times the clear distance between vertical lateral supports which provide "enhanced" resistance; or

(*b*) the clear distance between lateral supports which provide "simple" resistance.

Effective thickness

Case 1. Walls not stiffened by piers or intersecting walls.

For a single leaf wall, t_{ef} = actual thickness t.

For a cavity wall, t_{ef} is the greatest of:

(*a*)) $2/3(t_1 + t_2)$; or

(*b*) t_1; or

(*c*) t_2

where t_1 = the actual thickness of one leaf and t_2 = the actual thickness of the other leaf.

Case 2. Walls stiffened by piers or intersecting walls.

The effective thickness t_{ef} of a wall or leaf of a cavity wall is:

$$t_{ef} = tK$$

where t = the actual thickness of the wall or leaf and K = the appropriate stiffness coefficient taken from Table 5 of the code.

For a wall stiffened by intersecting walls, the appropriate stiffness coefficient may be determined from Table 5 of the code on the assumption that the intersecting walls are equivalent to piers of width equal to the thickness of the intersecting wall and thickness equal to three times the thickness of the stiffened wall.

Design vertical load resistance of walls

The design vertical load resistance of a wall per unit length is given by:

$$\frac{\beta t f_k}{\gamma_m}$$

where β = the capacity reduction factor allowing for the effects of slenderness and eccentricity and is obtained from Table 7; f_k = the characteristic strength of the masonry; γ_m = the partial safety factor for the material; and t = the thickness of wall.

Cavity walls

These are constructed with two leaves of masonry tied together at intervals by cavity ties.

Where the load from roof or floor is carried by one leaf only, the loadbearing capacity of the wall should be based on the horizontal cross section of that leaf alone, although the stiffening effect of the other leaf can be taken into account when calculating the slenderness ratio.

Tensile strength

Tensile strength is governed by the tensile bond at the brick/mortar interface.

In general, no direct tension should be allowed in masonry. However, at the designer's discretion half the values in Table 3 of the code may be allowed in direct tension when suction forces arising from wind loads on roofs are transmitted to masonry walls.

In no circumstances may combined direct and flexural tensile stresses exceed the values given in Table 3 of the code.

The following examples illustrate the design of solid, cavity and gravity retaining walls.

CURPE CONSULTANTS
46 Orburn Road
DUNFIELD

Project: DESIGN IV
Part of structure: WALL WITH AXIAL LOAD
Job ref: CC/ME
Calc sheet no: 16/1. rev
Drawing ref
Calc by: R.A.S. Date
Check by Date

Code ref. | Calculations | Output

Example 16.1 Wall with axial loading

Roof { Dead load = 5 kN/m²; Imposed load = 1.5 kN/m²

Reinforced concrete roof slab

Single storey internal load bearing walls at 5.5 m centres.

102.5 mm brick or 100 mm block.

Self weight of wall: 2.2 kN/m² (brick); 1.5 kN/m² (block)

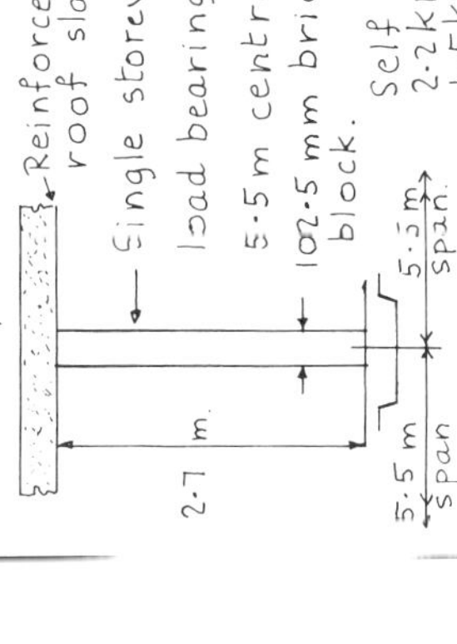

Partial safety factors:

γ_m = 3.5 for materials

γ_f = 1.4 for dead load

γ_f = 1.6 for imposed load

(1) Design for 102.5 mm thick bricks.

Calculate design load n_w (for 1 m length of wall)

Characteristic dead load G_k:

roof = $\frac{5.5 + 5.5}{2} \times 5 \times 1$ = 27.5 kN

wall = $2.2 \times 2.7 \times 1$ = 5.94 kN

33.44 kN

Output: Gk = 33.44

CURPE CONSULTANTS
46 Orburn Road
DUNFIELD

Project: DESIGN IV			Job ref: CC/ME	
Part of structure: WALL WITH AXIAL LOAD			Calc sheet no: 16/2	rev: /
Drawing ref:	Calc by: R.A.S.	Date:	Check by:	Date:

Code ref.	Calculations	Output
	Example 16.1. continued	
	Characteristic imposed load Q_k:	
	roof $= \frac{5\cdot5+5\cdot5}{2} \times 1\cdot5 \times 1 = 8\cdot25$ kN	$Q_k = 8\cdot25$
	Design load on wall	
	$n_w = 1\cdot4 G_k + 1\cdot6 Q_k$	
	$= 1\cdot4(33\cdot44) + 1\cdot6(8\cdot25)$	
	$= 46\cdot82 + 13\cdot2 = 60\cdot02$ kN/m	$n_w = 60\cdot02$
	Check slenderness ratio and determine β.	
	With enhanced resistance:	
CL. 28.3.1.1.	$h_{ef} = 0\cdot75 h$	
	The effective thickness:	
CL. 28.4.1.	t_{ef} = actual thickness	
	Slenderness ratio is then:	
	$\frac{h_{ef}}{t_{ef}} = \frac{0\cdot75 \times 2\cdot7 \times 10^3}{102\cdot5} = 19\cdot75 < 27$	
Table 7	Hence, from Table 7, $\beta = 0\cdot71$	$\beta = 0\cdot71$
	Calculate design vertical load resistance of wall.	
CL. 32.2.1. CL. 23.1.2.	$\frac{\beta . t . f_k}{\gamma_m}$ for 1 m length	
	$= \frac{0\cdot71 \times 102\cdot5(f_k \times 1\cdot15)10^3}{3\cdot5}$ N/m	

CURPE CONSULTANTS
46 Orburn Road
DUNFIELD

Project: DESIGN IV			Job ref: CC/ME	
Part of structure: WALL WITH AXIAL LOAD			Calc sheet no: 16/3	rev: /
Drawing ref:	Calc by: R.A.S.	Date:	Check by:	Date:

Code ref.	Calculations	Output
	Example 16.1. continued	
	Equate design load and resistance	
	$60\cdot02 \times 10^3 = \frac{0\cdot71 \times 102\cdot5(f_k \times 1\cdot15)10^3}{3\cdot5}$	
	f_k required $= 2\cdot51$ N/mm^2.	f_k required $2\cdot51$ N/mm^2
	Select brick strength and mortar	
Table 2a	From Table 2(a), using mortar designation (iv), a brick with compressive strength of 10 N/mm^2 giving $f_k = 3\cdot5$ N/mm^2, is satisfactory.	$f_k = 3\cdot5$
	(b) Design for 100 mm thick blocks	
	Calculate design load n_w (for 1 m length of wall).	
	Characteristic dead load G_k:	
	roof $= 27\cdot50$	
	wall $= 1\cdot5 \times 2\cdot7 \times 1 = 4\cdot05$	
	$31\cdot55$ kN	$G_k = 31\cdot55$
	Characteristic imposed load:	
	$Q_k = 8\cdot2$ kN	$Q_k = 8\cdot2$
	Design load on wall	
	$n_w = 1\cdot4(31\cdot55) + 1\cdot6(8\cdot2)$	
	$= 44\cdot17 + 13\cdot12 = 57\cdot29$ kN/m	$n_w = 57\cdot29$

CURPE CONSULTANTS 46 Orburn Road DUNFIELD	Project DESIGN IV		Job ref CC/ME
	Part of structure WALL WITH AXIAL LOAD		Calc sheet no 16/4 rev /
	Drawing ref	Calc by R.A.S. Date	Check by Date

Code ref.	Calculations	Output
	Example 16.1. continued	
	Check slenderness ratio and determine β	
	Slenderness ratio $= \dfrac{h_{ef}}{t_{ef}}$	
CL.28.1.	$= \dfrac{0.75 \times 2.7 \times 10^3}{100} = 20.25 < 27$	
Table 7	Hence, from Table 7, $\beta = 0.69$	$\beta = 0.69$
	Equate design load and resistance	
	$57.29 \times 10^3 = \dfrac{0.69 \times 100 \times f_k \times 10^3}{3.5}$	
	f_k required $= 2.91\ N/mm^2$	f_k required 2.91
	Select block size, strength and mortar	
	Block size say $440 \times 215 \times 100$ solid	
CL.23.	Shape factor $= \dfrac{\text{height}}{\text{thickness}} = \dfrac{215}{100} = 2.15$	
Table 2d	From Table 2(d), using mortar designation (iv), a block with compressive strength of $3.5\ N/mm^2$ giving $f_k = 3.5\ N/mm^2$, is satisfactory.	$f_k = 3.5$

CURPE CONSULTANTS 46 Orburn Road DUNFIELD	Project DESIGN IV		Job ref CC/ME
	Part of structure WALL WITH AXIAL LOAD		Calc sheet no 16/5 rev /
	Drawing ref	Calc by R.A.S. Date	Check by Date

Code ref.	Calculations	Output
	Example 16.2. Effect of small plan area	
	Consider the design of the wall in example 16.1. if its length is reduced to 1.79 m.	
	(a) 102.5 brick wall.	
	Plan area $= \dfrac{102.5}{10^3} \times 1.79 = 0.183 < 0.2\ m^2$	
CL. 23.1.1.	Therefore f_k must be multiplied by the factor:	
	$0.7 + 1.5A$	
	$= 0.7 + 1.5(0.183)$	
	$= 0.974$	
	Equate design load and resistance	
	$60.02 \times 10^3 = \dfrac{0.71 \times 102.5(f_k \times 1.15 \times 0.974)10^3}{3.5}$	
	f_k required $= 2.58\ N/mm^2$	f_k required 2.58
	(b) 100 block wall	
CL. 23.1.1.	Plan area $= \dfrac{100}{10^3} \times 1.79 = 0.179 < 0.2\ m^2$	
	Then factor is $0.7 + 1.5(0.179)$	
	$= 0.968$	
	Equate design load and resistance	
	$57.29 \times 10^3 = \dfrac{0.69 \times 100(f_k \times 0.968)10^3}{3.5}$	
	f_k required $= 3\ N/mm^2$	$f_k = 3$
	Previously selected bricks and blocks:	Adequate

CURPE CONSULTANTS
46 Orburn Road
DUNFIELD

Project: DESIGN IV		Job ref: CC/ME	
Part of structure: WALL WITH ECCENTRIC LOADING		Calc sheet no: 16/6	rev: /
Drawing ref	Calc by: R.A.S. / Date	Check by	Date

Code ref.	Calculations	Output

Example 16.3. Wall design with eccentric loading

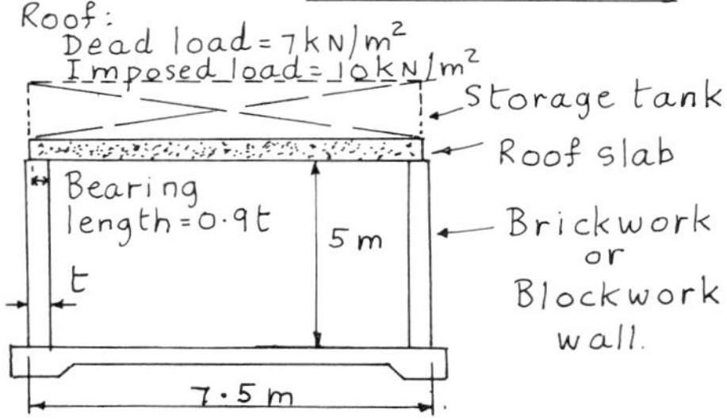

Partial safety factors:

γ_m = 3·5 for materials.

γ_f = 1·4 for dead load

γ_f = 1·6 for imposed load.

Self weight of walls:

4·5 kN/m² for brickwork (215 mm thick)

3·0 kN/m² for blockwork (190 mm thick)

(a) Design of brick wall

Try 215 mm thick

Calculate design load n_w for 1 m length

Characteristic dead load Gk:

CURPE CONSULTANTS
46 Orburn Road
DUNFIELD

Project: DESIGN IV		Job ref: CC/ME	
Part of structure: WALL WITH ECCENTRIC LOADING		Calc sheet no: 16/7	rev: /
Drawing ref	Calc by: R.A.S. / Date	Check by	Date

Code ref.	Calculations	Output
	Example 16.3. continued	
	slab $= \frac{7\cdot5}{2} \times 7 \times 1 = 26\cdot25$	
	wall $= 4\cdot5 \times 5 \times 1 = 22\cdot5$	
	$48\cdot75$ kN/m	$G_k = 48\cdot75$
	Characteristic imposed load Qk:	
	storage $= \frac{7\cdot5}{2} \times 10 \times 1 = 37\cdot5$ kN/m	$Q_k = 37\cdot5$
	Design load $n_w = 1\cdot4G_k + 1\cdot6Q_k$	
	$= 1\cdot4(48\cdot75) + 1\cdot6(37\cdot5)$	
	$= 68\cdot25 + 60 = 128\cdot25$ kN/m	$n_w = 128\cdot25$
	Calculate slenderness ratio, eccentricity, and determine β.	
	Slenderness ratio $= \frac{h_{ef}}{t_{ef}}$	
CL.28.1.	$= \frac{0\cdot75 \times 5 \times 10^3}{215} = 17\cdot5 < 27$	
	$e_x = \frac{t}{2} - \frac{0\cdot9t}{3} = 0\cdot5t - 0\cdot3t = 0\cdot2t$	
CL.31.	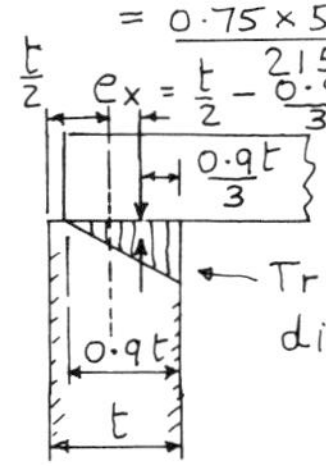 Triangular pressure distribution	
Table 7	From Table 7, with slenderness ratio = 17·5 and $e_x = 0\cdot2t$, by interpolation β = 0·58	β = 0·58

CURPE CONSULTANTS
46 Orburn Road
DUNFIELD

Project: DESIGN IV | Job ref: CC/ME
Part of structure: WALL WITH ECCENTRIC LOADING. | Calc sheet no: 16/8 | rev: 1
Drawing ref: | Calc by: R.A.S. | Date: | Check by: | Date:

Code ref.	Calculations	Output
	Example 16.3. continued	
	Equate design load and resistance of wall	
	$128 \cdot 25 \times 10^3 = \frac{0 \cdot 58 \times 215 \times f_k \times 10^3}{3 \cdot 5}$	
	f_k required $= 3 \cdot 6$ N/mm^2	f_k required $= 3 \cdot 6$
	From Table 2(a), using mortar designation (iii), standard format bricks with compressive strength 10 N/mm^2 giving $f_k = 4 \cdot 1$ N/mm^2,	
Table 2a	are satisfactory	$f_k = 4 \cdot 1$
	(b) Design of block wall	
	Try 190 mm thick solid blocks.	
	Calculate design load n_w for 1 m length.	
	slab = as before 26·25	
	wall = 3 × 5 × 1 = 15	
	G_k = 41·25 kN/m	$G_k = 41 \cdot 25$
	As before Q_k = 37·5 kN/m	$Q_k = 37 \cdot 5$
	Design load $n_w = 1 \cdot 4(41 \cdot 25) + 1 \cdot 6(37 \cdot 5)$	
	$= 57 \cdot 75 + 60$	
	$= 117 \cdot 75$ kN/m	$n_w = 117 \cdot 75$
	Calculate slenderness ratio, eccentricity, and determine β	

CURPE CONSULTANTS
46 Orburn Road
DUNFIELD

Project: DESIGN IV | Job ref: CC/ME
Part of structure: WALL WITH ECCENTRIC LOADING. | Calc sheet no: 16/9. | rev: 1
Drawing ref: | Calc by: R.A.S. | Date: | Check by: | Date:

Code ref.	Calculations	Output
	EXAMPLE 16.3. continued.	
CL.28.1.	Slenderness ratio $= \frac{0 \cdot 75 \times 5 \times 10^3}{190}$	
	$= 19 \cdot 74 < 27$	
	Eccentricity $e_x = 0 \cdot 2t$ as before	
	From Table 7, with slenderness ratio = 19·74 and $e_x = 0 \cdot 2t$,	
Table 7	by interpolation $\beta = 0 \cdot 517$	$\beta = 0 \cdot 517$
	Equate design load and resistance of wall	
	$117 \cdot 75 \times 10^3 = \frac{0 \cdot 517 \times 190 \times f_k \times 10^3}{3 \cdot 5}$	
	f_k required $= 4 \cdot 19$ N/mm^2	f_k required $4 \cdot 19$
	Select block size, strength & mortar	
	Block size say 440 × 190 × 190 solid	
CL.23.	Shape factor $= \frac{\text{height}}{\text{thickness}} = \frac{190}{190} = 1 \cdot 0$	
	Try solid blocks with compressive strength 10 N/mm^2 set in mortar designation (iv)	
Table 2b	From Table 2(b), $f_k = 3 \cdot 5$	
Table 2d	From Table 2(d), $f_k = 5 \cdot 6$	
	By interpolation, $f_k = 4 \cdot 5$ N/mm^2	$f_k = 4 \cdot 5$
	This is greater than 4·19 required	Satisfactory

CURPE CONSULTANTS 46 Orburn Road DUNFIELD	Project DESIGN IV			Job ref CC/ME	
	Part of structure CAVITY WALL			Calc sheet no 16/10	rev /
	Drawing ref	Calc by R.A.S.	Date	Check by	Date

Code ref.	Calculations	Output
	EXAMPLE 16.4. Design of cavity wall.	
	Roof { Dead load = 7 kN/m²; Imposed load = 1.5 kN/m²	

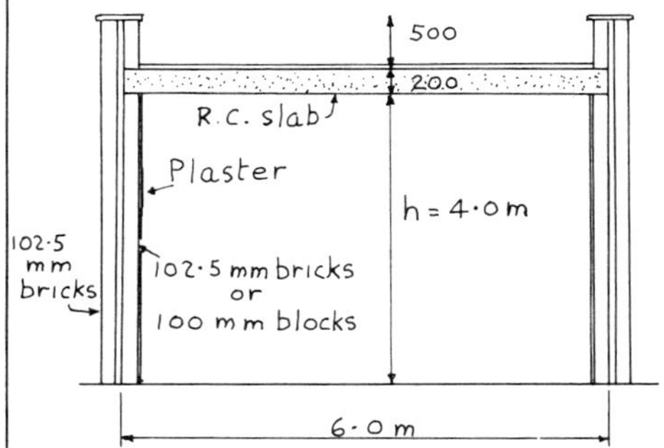

Code ref.	Calculations	Output
	Partial safety factors:	
	γ_m = 3.5 for materials.	
	γ_f = 1.4 for dead loads.	
	γ_f = 1.6 for imposed loads.	
	Self weight of wall: Inner leaf	
	102.5 mm thick brick + plaster = 2.5 kN/m²	
	100 mm thick block + plaster = 1.8 kN/m²	
CL. 29.1.1.	Outer leaf is self supporting.	
	(a) Design for 102.5 mm thick bricks.	
	Calculate design load n_w for 1 m	

CURPE CONSULTANTS 46 Orburn Road DUNFIELD	Project DESIGN IV			Job ref CC/ME	
	Part of structure CAVITY WALL			Calc sheet no 16/11.	rev /
	Drawing ref	Calc by R.A.S.	Date	Check by	Date

Code ref.	Calculations	Output
	EXAMPLE 16.4. continued.	
	length of wall.	
	Characteristic dead load Gk:	
	roof slab = $\frac{6}{2} \times 7 \times 1$ = 21.00	
	wall, inner leaf = $(4+0.5)2.5\times1$ = 11.25	
	32.25 kN/m	Gk = 32.25
	Characteristic imposed load Qk:	
	$\frac{6}{2} \times 1.5 \times 1 = 4.5$ kN/m	Qk = 4.5
	Design load on wall	
	$n_w = 1.4\,Gk + 1.6\,Qk$	
	$= 1.4(32.25) + 1.6(4.5)$	
	$= 45.15 + 7.2 = 52.35$ kN/m	n_w = 52.35
	Calculate slenderness ratio, eccentricity, and determine β	
CL. 28.3.1.1.	With enhanced resistance:	
	$h_{ef} = 0.75\,h$	
CL 28.4.1.	Effective thickness:	
	$t_{ef} = \frac{2}{3}(102.5 + 102.5) = 136.66$ mm.	
CL 28.1.	Slenderness ratio = $\frac{h_{ef}}{t_{ef}} = \frac{0.75\times4\times10^3}{136.66} = 22 < 27$	
CL. 31.	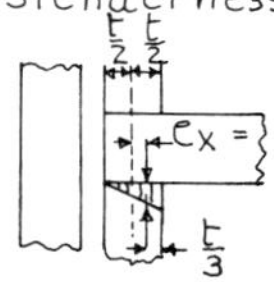 $e_x = \frac{t}{2} - \frac{t}{3} = \frac{t}{6} = 0.167t$	

CURPE CONSULTANTS
46 Orburn Road
DUNFIELD

Project: DESIGN IV | Job ref: CC/ME
Part of structure: CAVITY WALL | Calc sheet no: 16/12 | rev
Drawing ref | Calc by: R.A.S. | Date | Check by | Date

Code ref.	Calculations	Output
	EXAMPLE 16.4. Continued	
	From Table 7, with slenderness ratio = 22	
Table 7	and $e_x = 0.167t$, $\beta = 0.473$	$\beta = 0.473$
	Equate design load and resistance of wall	
	$52.35 \times 10^3 = \frac{0.473 \times 102.5 \times (f_k \times 1.15) \times 10^3}{3.5}$	
	f_k required $= 3.28\ N/mm^2$	f_k required 3.28
	Select brick strength and mortar	
	From Table 2(a), using mortar designation	
	(iv), standard format bricks with	
	compressive strength of $10\ N/mm^2$ giving	
Table 2a	$f_k = 3.5\ N/mm^2$, are satisfactory.	$f_k = 3.5$
	(b) Design for 100 mm thick blocks	
	Calculate design load n_w for 1 m length of wall	
	Characteristic dead load G_k:	
	roof slab $= 21.0$	
	wall, inner leaf $= (4 + 0.5)\,1.8 \times 1 = 8.1$	
	29.1 kN/m	$G_k = 29.1$
	Characteristic imposed load:	
	$Q_k = 4.5$ kN/m	$Q_k = 4.5$
	Design load on wall:	
	$n_w = 1.4(29.1) + 1.6(4.5)$	
	$= 40.74 + 7.2 = 47.94$ kN/m	$n_w = 47.94$

CURPE CONSULTANTS
46 Orburn Road
DUNFIELD

Project: DESIGN IV | Job ref: CC/ME
Part of structure: CAVITY WALL | Calc sheet no: 16/13 | rev
Drawing ref | Calc by: R.A.S. | Date | Check by | Date

Code ref.	Calculations	Output
	EXAMPLE 16.4. Continued.	
	Calculate slenderness ratio, eccentricity and determine β	
CL. 28.3.1.1.	$h_{ef} = 0.75 \times 4 \times 10^3$ mm	
CL. 28.4.1.	$t_{ef} = \frac{2}{3}(102.5 + 100) = 135$ mm	
	Slenderness ratio $= \frac{0.75 \times 4 \times 10^3}{135} = 22 < 27$	
CL. 31.	Eccentricity $e_x = \frac{t}{6} = 0.167t$	
Table 7	Hence, $\beta = 0.473$ as before	$\beta = 0.473$
	Equate design load and resistance of wall	
	$47.94 \times 10^3 = \frac{0.473 \times 100 \times f_k \times 10^3}{3.5}$	
	f_k required $= 3.55\ N/mm^2$	f_k required 3.55
	Select block size, strength and mortar	
	Using 440 × 215 × 100 solid blocks then	
	Shape factor $= \frac{215}{100} = 2.15$	
	From Table 2(d), using mortar	
	designation (iv), blocks with compressive	
Table 2d	strength of $5.0\ N/mm^2$ giving $f_k = 4.4\ N/mm^2$	$f_k = 4.4$
	are satisfactory.	

CURPE CONSULTANTS 46 Orburn Road DUNFIELD	Project: DESIGN IV	Job ref: CC/ME
	Part of structure: GRAVITY RETAINING WALL	Calc sheet no: 16/14 rev: 1
	Drawing ref: — Calc by: R.A.S. Date: —	Check by: — Date: —

Code ref.	Calculations	Output

Example 16.5. Gravity retaining wall.

A wall of overall height 750 mm retains cohesionless soil of weight $17 kN/m^3$ and angle of internal friction $\phi = 30°$. Weight of brickwork is given as $20 kN/m^3$.

Coefficient of friction between concrete/masonry and concrete/soil is 0·6.

The weight of the concrete foundation (base) is $24 kN/m^3$.

For this example limit state design will not be considered.

Factors of safety against overturning and sliding will be taken as 1·5 minimum.

Determine minimum wall thickness.

This will be based on the "middle-third" rule for no tension in the brickwork.

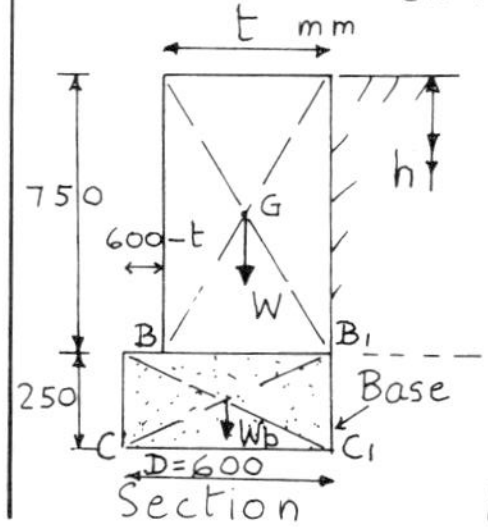

Section

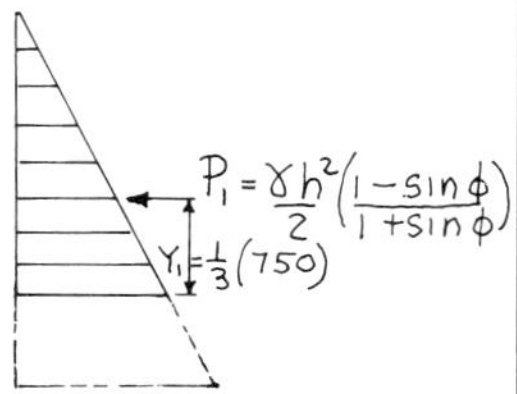

Horizontal pressure

CURPE CONSULTANTS 46 Orburn Road DUNFIELD	Project: DESIGN IV	Job ref: CC/ME
	Part of structure: GRAVITY RETAINING WALL	Calc sheet no: 16/15 rev: 1
	Drawing ref: — Calc by: R.A.S. Date: —	Check by: — Date: —

Code ref.	Calculations	Output

Example 16.5. continued

Weight of wall $W = 20 \times 0{\cdot}75 \times \frac{t}{10^3} \times 1 = 0{\cdot}015t$ kN

Horizontal force P_1 for depth $h = 0{\cdot}75$ m

$$= \frac{17 \times 0{\cdot}75^2}{2}\left(\frac{1 - \sin 30}{1 + \sin 30}\right)$$

$$= 1{\cdot}593 \text{ kN}$$

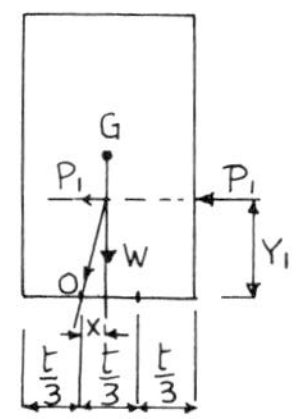

G is the centre of gravity of the wall.

Moments about O give:

$P_1 . Y_1 = W . x$.

For no tension $x = \frac{t}{6}$ mm

where x = "shift" of W.

$P_1 . Y = W . x$.

$$1{\cdot}593 \times \frac{1}{3} \times 0{\cdot}75 = 0{\cdot}015t \times \frac{t}{6} \times \frac{1}{10^3}$$

$t^2 = 159300$

$t = 399$ mm (minimum).

Then design for 440 mm thick. (215+10+215)

Assuming foundation (base) as 250 mm deep and 600 mm wide, then weight per m length is:

$W_b = 24 \times 0{\cdot}25 \times 0{\cdot}6 \times 1 = 3{\cdot}6$ kN.

Weight of wall $W = 20 \times 0{\cdot}75 \times 0{\cdot}44 \times 1 = 6{\cdot}6$ kN

CURPE CONSULTANTS 46 Orburn Road DUNFIELD	Project DESIGN IV			Job ref CC/ME	
	Part of structure GRAVITY RETAINING WALL			Calc sheet no 16/16	rev /
	Drawing ref	Calc by R.A.S.	Date	Check by	Date

Code ref.	Calculations	Output

Example 16.5. continued

Check for overturning

For the wall, in this case, calculations are unnesessary, since the resultant falls within the middle third and not outside the face of the wall. Nevertheless calculations will be made to demonstrate procedure.

Level B-B₁

Take moments about B.

Overturning moment (O.T.M) = $P_1 . Y_1$.

$= 1.593 \times \frac{0.75}{3} = 0.398$ kN.m.

Restoring moment (R.M) = $W . \frac{t}{2}$

$= 6.6 \times \frac{0.44}{2} = 1.452$ kN m.

Factor of safety = $\frac{R.M}{O.T.M} = \frac{1.452}{0.398} = 3.64 > 1.5$ — Satisfactory

Level C-C₁

Take moments about C

O.T.M = $P_2 . Y_2$.

where $P_2 = \frac{17 \times (0.75 + 0.25)^2}{2} \left(\frac{1 - \sin 30}{1 + \sin 30}\right)$

$= 2.833$ kN.

CURPE CONSULTANTS 46 Orburn Road DUNFIELD	Project DESIGN IV			Job ref CC/ME	
	Part of structure GRAVITY RETAINING WALL.			Calc sheet no 16/17	rev /
	Drawing ref	Calc by R.A.S.	Date	Check by	Date

Code ref.	Calculations	Output

Example 16.5. continued.

$Y_2 = (0.75 + 0.25) \frac{1}{3} = \frac{1}{3}$

Then O.T.M. $= 2.833 \times \frac{1}{3} = 0.944$ kN m.

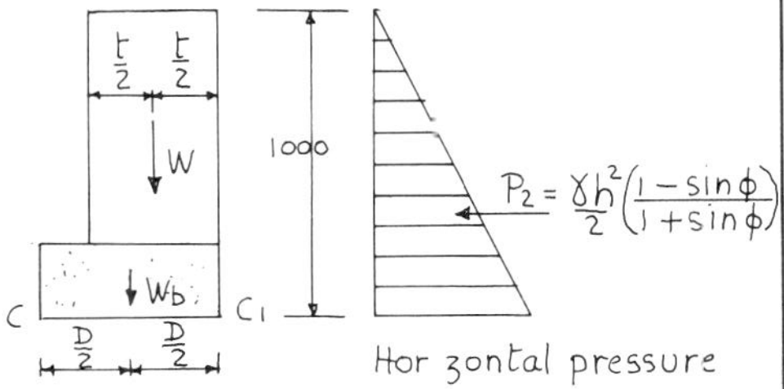

R.M. $= W\left(D - \frac{t}{2}\right) + \left(W_b \times \frac{D}{2}\right)$

$= 6.6\left(0.6 - \frac{0.44}{2}\right) + \left(3.6 \times \frac{0.6}{2}\right)$

$= 2.508 + 1.08 = 3.588$ kN m

Factor of safety $= \frac{3.588}{0.944} = 3.8 > 1.5$ — Overturning satisfactory

Check for sliding

Level B-B₁

Force causing sliding $= P_1 = 1.593$ kN.

Frictional resisting force $= \mu . R$

$= 0.6\ W = 0.6 \times 6.6 = 3.96$ kN.

Factor of safety $= \frac{3.96}{1.593} = 2.48 > 1.5$ — Satisfactory.

CURPE CONSULTANTS 46 Orburn Road DUNFIELD	Project: DESIGN IV Part of structure: GRAVITY RETAINING WALL. Drawing ref:	Calc by: R.A.S.	Date:	Job ref: CC/ME Calc sheet no: 16/18 rev: / Check by: Date:

Code ref.	Calculations	Output
	Example 16.5. continued	
	Level $C\text{-}C_1$	
	Force causing sliding $= P_2 = 2{\cdot}833$ kN.	
	Frictional resisting force $= \mu . R.$	
	$= 0{\cdot}6(W + W_b) = 0{\cdot}6(6{\cdot}6 + 3{\cdot}6) = 6{\cdot}12$ kN	
	Factor of safety $= \frac{6{\cdot}12}{2{\cdot}833} = 2{\cdot}16 > 1{\cdot}5$	Sliding - satisfactory.
	Determine maximum stress in brickwork	
	Moment about centroidal axis of the wall is $P_1 . Y_1. = 1{\cdot}593 \times \frac{0{\cdot}75}{3} = 0{\cdot}398$ kN.m.	
	Stresses are given by $f = \frac{W}{A} \pm \frac{M}{Z}$ $Z = \frac{BD^2}{6}$	
	$f = \frac{6{\cdot}6 \times 10^3}{1 \times 10^3 \times 440} \pm \frac{0{\cdot}398 \times 10^6 \times 6}{1 \times 10^3 \times 440^2}$	
	$f = 0{\cdot}015 \pm 0{\cdot}012 = 0{\cdot}027$ and $0{\cdot}003$	
	To comply with CP121: Part I: 1973, Table 4.1.	
	select special quality bricks, designation (i) mortar.	Special quality desig (i) mortar
	Bearing pressures on soil. Resultant moment about centroidal axis of base $= P_2 . Y_2 - W(\frac{D}{2} - \frac{t}{2})$	
	$= (2{\cdot}833 \times \frac{1}{3}) - 6{\cdot}6(\frac{0{\cdot}6}{2} - \frac{0{\cdot}44}{2})$	
	$= 0{\cdot}944 - 0{\cdot}528 = 0{\cdot}416$ kN.m.	
	$f = \frac{(6{\cdot}6 + 3{\cdot}6)}{1 \times 0{\cdot}6} \pm \frac{0{\cdot}416 \times 6}{1 \times 0{\cdot}6^2} = 17 \pm 6{\cdot}93$	
	$f = 23{\cdot}92$ kN/m² (Maximum)	$23{\cdot}92$ kN/m² maximum
	$f = 10{\cdot}06$ kN/m² (Minimum)	
	Maximum $f \not> $ safe bearing pressure.	

SELF-ASSESSMENT QUESTIONS

1. Determine the design load kN/m² given:

(a) $G_k = 3.0\,\text{kN/m}^2$;
(b) $Q_k = 1.5\,\text{kN/m}^2$.

2. Determine the characteristic compressive strength of masonry f_k constructed with standard format bricks of compressive strength 70 N/mm² and a $1:\frac{1}{4}:3$ mortar.

3. Determine the shape factor for a concrete block size 440 mm long × 190 mm high × 150 mm thick.

4. Determine the design strength of the brickwork given:

(a) Characteristic compressive strength $f_k = 19.2\,\text{N/mm}^2$;
(b) Partial safety factors for material $\gamma_m = 3.1$.

5. Determine the slenderness ratio of a solid wall given:

(a) Effective height = 2 m;
(b) Effective length = 3 m;
(c) Actual thickness = 215 mm.

6. Determine the effective thickness of a cavity wall constructed of

(a) Inner leaf 215 mm;
(b) Outer leaf 102.5 mm.

7. An internal loadbearing wall 2.8 m high required to support a design load $n_w = 150$ kN/m length (including self weight). If the wall is constructed with 102.5 mm thick standard format bricks set in mortar designation (111), determine a suitable unit strength. Assume "enhanced" resistance to lateral movement and partial safety factor for materials $\gamma_m = 3.5$.

8. Figure 101 shows the plan and section of a single storey building. The following data is given: Roof load: dead = 5.5 kN/m² and imposed = 1.5 kN/m². Walls are 102.5 mm thick brick or alternatively 100 mm thick block. Self weight to be taken as 2.2 kN/m² and 1.5 kN/m² respectively. Partial safety factors: $\gamma_m = 3.5$; γ_f for dead = 1.4; and γ_f for imposed = 1.6.

Design walls A and B.

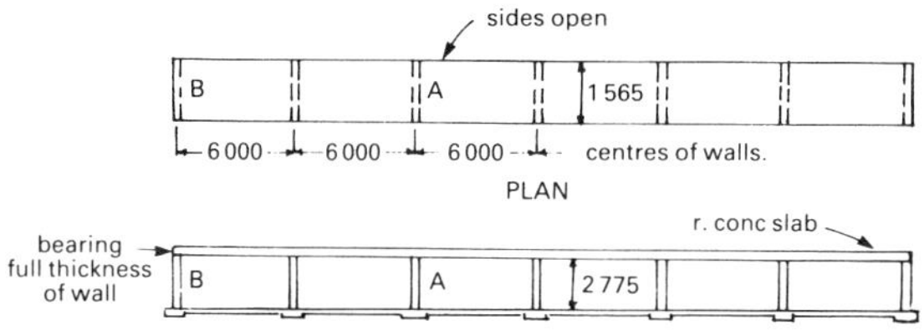

FIG. 101.

9. Figure 102 shows the plan and section of a single storey building. The following data is given. Roof load: dead = 5 kN/m^2 and imposed = 1.5 kN/m^2. All internal walls are 102.5 mm thick brick or alternatively 100 mm thick block. Self weight to be taken as 2.2 kN/m^2 and 1.5 kN/m^2 respectively. External walls are 102.5 mm thick brick outer leaf with inner leaf 102.5 mm thick brick or alternatively 100 mm thick block. Partial safety factors are as for previous question.

Design walls A and B.

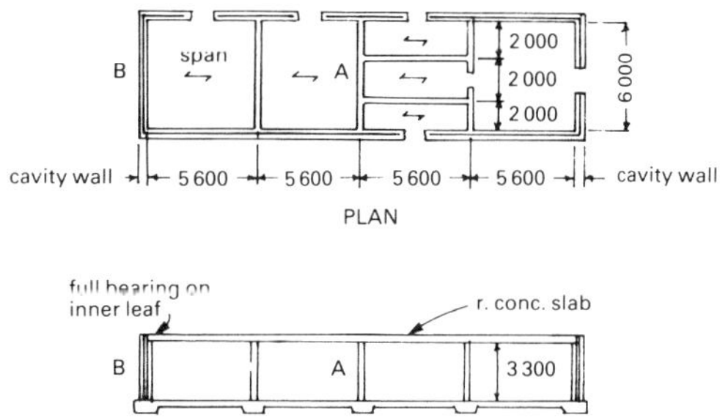

FIG. 102.

CHAPTER SEVENTEEN

Structural Timber

CHAPTER OBJECTIVES

After reading this chapter you should be able to:

* explain the concepts of duration of load and load sharing;
* appreciate the strength of timber related to species grouping and stress grading;
* apply modification factors if relevant;
* understand the necessity of limiting deflection in flexural members;
* determine slenderness ratio for compression members;
* design timber compression members subject to vertical loading;
* design simply supported timber beams and joists;
* design timber support work for a trench.

SCOPE

The Code of Practice CP112: Part 2 deals in general with the structural use of timber. The code includes recommendations on the quality of materials, permissible stresses and modification factors applicable to timber when used as simple solid members or as parts of built up components and structures incorporating other materials.

Definitions

For the purposes of the code the following definitions apply.

Basic stress
The stress that can safely be permanently sustained by timber containing no strength reducing characteristics.

Dry stress
A stress applicable to timber having a moisture content not exceeding 18 per cent.

Grade stress
The stress which can safely be permanently sustained by timber of a particular grade.

Green stress
A stress applicable to timber having a moisture content exceeding 18 per cent.

Permissible stress
The stress that can safely be sustained by a structural component under the particular condition of service and loading.

Species groups

Softwood species having similar strength and stiffness properties are grouped together for convenience in design procedure. The species groups are presented as S1, S2 and S3 in Table 8 below and grade stresses for SS and GS grades are given in Tables 9 and 10 below.

TABLE 8.
SOFTWOOD SPECIES GROUPS (TABLE 9 OF CP112)

Species group	*Standard name*	*Origin*
S1	Douglas fir Pitch pine	Imported
S2	Western hemlock (unmixed) Western hemlock (commercial) Parana pine Redwood Whitewood Canadian spruce	Imported
	Douglas fir Larch Scots pine	Home-grown
S3	European spruce Sitka spruce	Home-grown
	Western red cedar	Imported

TABLE 9.
GREEN STRESSES AND MODULI OF ELASTICITY FOR GROUPED SOFTWOODS (TABLE 10A OF CP112)

Species group	*Grade*	*Bending* (N/mm^2)	*Tension* (N/mm^2)	*Compression parallel to grain* (N/mm^2)	*Compression perpendicular to grain* (N/mm^2)	*Shear parallel to grain* (N/mm^2)	*Modulus of elasiticity* *Mean* (N/mm^2)	*Minimum* (N/mm^2)
S1	SS	6.9	4.8	6.3	1.29	0.70	9 200	4 400
	GS	4.8	3.4	4.4	1.15	0.70	8 200	4 000
S2	SS	5.5	3.8	5.4	1.03	0.70	8 000	4 300
	GS	3.8	2.7	3.7	0.92	0.70	7 000	3 900
S3	SS	3.8	2.7	3.6	0.77	0.55	6 100	3 100
	GS	2.7	1.9	2.5	0.69	0.55	5 500	2 800

NOTE. These stresses apply to timber having a moisture content exceeding 18%.

TABLE 10.
DRY STRESSES AND MODULI OF ELASTICITY FOR GROUPED SOFTWOODS (TABLE 11A of CP112)

Species group	*Grade*	*Bending* (N/mm^2)	*Tension* (N/mm^2)	*Compression parallel to grain* (N/mm^2)	*Compression perpendicular to grain* (N/mm^2)	*Shear parallel to grain* (N/mm^2)	*Modulus of elasticity* *Mean* (N/mm^2)	*Minimum* (N/mm^2)
S1	SS	8.6	6.0	8.6	1.93	0.87	10 600	5 100
	GS	6.0	4.2	6.0	1.72	0.87	9 500	4 600
S2	SS	6.9	4.8	7.0	1.55	0.79	8 900	4 800
	GS	4.8	3.3	4.8	1.38	0.79	8 000	4 300
S3	SS	5.2	3.6	5.0	1.16	0.66	7 500	3 700
	GS	3.6	2.5	3.5	1.03	0.66	6 800	3 300

NOTE. These stresses apply to timber having a moisture content not exceeding 18%

Stress graded timber

Timber may be stress graded in accordance with BS4978 by visual inspection or by the use of a stress grading machine. Visual stress grading is based on the physical characteristics of the timber (e.g. the presence, number and size of knots, shape of the grain fissures and wane). Two visual grades are described:

(*a*) GS (General structural); and
(*b*) SS (Special structural).

Mechanical stress grading is based on a non-destructive test that assesses the modulus of elasticity of the timber. Stress grades are specified as:

(*a*) MGS (General structural);
(*b*) MSS (Special structural);
(*c*) M50;
(*d*) M75.

The grade stresses for timber visually graded as SS or GS are given in Tables 11 and 12 below.

The grade stresses for timber machine graded as MSS, MGS, M75 or M50 are given in Tables 13 and 14 below.

Permissible stresses

These are governed by the conditions of service and loading. They are given by:

grade stress × appropriate modification factor

TABLE 11.
GREEN STRESSES AND MODULI OF ELASTICITY (TABLE 3A OF CP112)

	Bending		*Tension*		*Compression parallel to grain*		*Compression perpendicular to grain*		*Shear parallel to grain*		*Modulus of elasticity*			
											Mean		*Minimum*	
Standard name	*SS* (N/mm^2)	*GS* (N/mm^2)	*SS* (N/mm^2)	*GS* (N/mm^2)	*SS* (N/mm^2)	*GS* (N/mm^2)	*SS* (N/mm^2)	*GS* (N/mm^2)	*SS* (N/mm^2)	*GS* (N/mm^2)	*SS* (N/mm^2)	*GS* (N/mm^2)	*SS* (N/mm^2)	*GS* (N/mm^2)
(*a*) *Imported*														
Douglas fir-larch(1)	7.6	5.3	5.3	3.7	7.2	5.0	1.35	1.20	0.86	0.86	11 300	10 200	6 600	6 000
Hem-fir(1)	5.9	4.0	4.0	2.9	5.9	4.0	1.03	0.92	0.70	0.70	9 500	8 500	5 900	5 200
Parana pine	5.9	4.0	4.0	2.9	6.7	4.6	1.14	1.01	0.76	0.76	9 500	8 500	5 400	4 800
Pitch pine	7.6	5.3	5.3	3.7	7.2	5.0	1.35	1.20	0.86	0.86	11 900	10 700	7 800	7 000
Redwood	5.9	4.0	4.0	2.9	5.4	3.7	1.14	1.01	0.70	0.70	8 200	7 000	4 700	4 000
Whitewood	5.9	4.0	4.0	2.9	5.4	3.7	1.03	0.92	0.70	0.70	8 200	7 000	4 700	4 000
Spruce-pine-fir(1)	5.5	3.8	3.8	2.7	5.4	3.7	1.03	0.92	0.70	0.70	8 000	7 200	4 300	3 900
Western red cedar	4.5	3.2	3.2	2.2	4.0	2.8	0.77	0.69	0.62	0.62	7 400	6 700	5 000	4 600
(*b*) *Home-grown*														
Douglas fir, Larch, Scots pine	5.9	4.0	4.0	2.9	5.4	3.7	1.41	1.01	0.70	0.70	8 200	7 000	4 700	4 000
European spruce	4.2	2.9	2.9	2.0	4.0	2.8	0.82	0.73	0.55	0.55	6 100	5 500	3 200	2 900
Sitka spruce	3.8	2.7	2.7	1.9	3.6	2.5	0.82	0.73	0.55	0.55	6 500	5 900	3 100	2 800

NOTE. These stresses apply to timber having a moisture content exceeding 18%. These stresses also apply respectively to Douglas fir, western hemlock (commercial) and Canadian spruce as designated in Table 1 when graded to the visual grades in BS4978. Where appropriate (see 3.6.1) the basic stresses in compression perpendicular to grain, given in Tables 3 and 4, for these species may be used for the above species groups.

TABLE 12.

DRY STRESSES AND MODULI OF ELASTICITY (TABLE 4A OF CP112)

	Bending		Tension		Compression parallel to grain		Compression perpendicular to grain		Shear parallel to grain		Modulus of elasticity			
											Mean		Minimum	
Standard name	SS (N/mm^2)	GS (N/mm^2)	SS (N/mm^2)	GS (N/mm^2)	SS (N/mm^2)	GS (N/mm^2)	SS (N/mm^2)	GS (N/mm^2)	SS (N/mm^2)	GS (N/mm^2)	SS (N/mm^2)	GS (N/mm^2)	SS (N/mm^2)	GS (N/mm^2)
(*a*) *Imported*														
Douglas fir-larch(1)	9.3	6.5	6.5	4.6	9.6	6.7	2.00	1.80	0.96	0.96	12 500	11 300	7 300	6 600
Hem-fir(1)	7.3	5.1	5.1	3.5	7.9	5.5	1.55	1.38	0.80	0.80	10 700	9 600	6 600	5 900
Parana pine	7.3	5.1	5.1	3.5	9.1	6.3	1.71	1.52	0.90	0.90	10 500	9 400	6 000	5 300
Pitch pine	9.3	6.5	6.5	4.6	9.1	6.3	2.00	1.80	0.94	0.94	12 900	11 600	8 500	7 600
Redwood	7.3	5.1	5.1	3.5	8.0	5.6	1.71	1.52	0.86	0.86	10 000	8 600	5 700	4 900
Whitewood	7.3	5.1	5.1	3.5	8.0	5.6	1.55	1.38	0.86	0.86	10 000	8 600	5 700	4 900
Spruce-pine-fir (1)	6.9	4.8	4.8	3.3	7.0	4.8	1.55	1.38	0.79	0.79	8 900	8 000	4 800	4 300
Western red cedar	5.2	3.6	3.6	2.5	5.9	4.1	1.16	1.03	0.70	0.70	8 000	7 000	5 500	4 900
(*b*) *Home-grown*														
Douglas fir, Larch, Scots pine	7.3	5.1	5.1	3.5	8.0	5.6	1.71	1.52	0.86	0.86	10 000	8 600	5 700	4 900
European spruce	5.5	3.9	3.9	2.7	5.6	3.9	1.24	1.10	0.71	0.71	7 500	6 800	4 000	3 600
Sitka spruce	5.2	3.6	3.6	2.5	5.0	3.5	1.24	1.10	0.66	0.66	7 800	7 000	3 700	3 300

NOTE. These stresses apply to timber having a moisture content not exceeding 18%. These stresses also apply respectively to Douglas fir, western hemlock (commercial) and Canadian spruce as designated in Table 1 when graded to the visual grades in BS4978. Where appropriate (see 3.6.1) the basic stresses in compression perpendicular to grain, given in Tables 3 and 4, for these species may be used for the above species groups.

TABLE 13.

GREEN STRESSES AND MODULI OF ELASTICITY FOR MACHINE STRESS-GRADED SOFTWOOD (TABLE 3B OF CP112)

		Standard name					
Property	*Grade*	*Western hemlock (commercial) (N/mm²)*	*Redwood and Whitewood (N/mm²)*	*Douglas fir (home-grown) (N/mm²)*	*Scots pine (home-grown) (N/mm²)*	*Sitka spruce (home-grown) (N/mm²)*	*Western white spruce (N/mm²)*
Bending	M75	8.1	8.1	8.1	8.1	4.9	7.6
	M50	5.3	5.3	5.3	5.3	3.3	5.0
	MSS	5.9	5.9	5.9	5.9	3.8	5.5
	MGS	4.0	4.0	4.0	4.0	2.7	3.8
Tension	M75	5.7	5.7	5.7	5.7	3.4	5.3
	M50	3.7	3.7	3.7	3.7	2.3	3.5
	MSS	4.0	4.0	4.0	4.0	2.7	3.8
	MGS	2.9	2.9	2.9	2.9	1.9	2.7
Compression parallel to grain	M75	7.9	7.3	7.3	7.3	4.6	7.3
	M50	5.2	4.8	4.8	4.8	3.1	4.8
	MSS	5.9	5.4	5.4	5.4	3.6	5.4
	MGS	4.0	3.7	3.7	3.7	2.5	3.7
Compression perpendicular to grain	M75	1.20	1.20(1)	1.32	1.32	0.97	1.20
	M50	1.03	1.03(1)	1.13	1.13	0.82	1.03
	MSS	1.03	1.03(1)	1.13	1.13	0.82	1.03
	MGS	0.92	0.92(1)	1.01	1.01	0.73	0.92
Shear parallel to grain	M75	1.03	1.03	1.03	1.03	0.82	1.03
	M50	0.70	0.70	0.70	0.70	0.55	0.70
	MSS	0.70	0.70	0.70	0.70	0.55	0.70
	MGS	0.70	0.70	0.70	0.70	0.55	0.70
Mean modulus of elasticity	M75	10 600	8 700	8 700	8 700	7 500	9 200
	M50	9 300	7 300	7 300	7 300	6 400	7 700
	MSS	9 600	8 300	8 300	8 300	6 800	8 300
	MGS	8 700	7 200	7 200	7 200	6 000	7 400
Minimum modulus of elasticity	M75	7 100	5 500	5 500	5 500	4 600	5 800
	M50	6 200	4 500	4 500	4 500	4 000	4 900
	MSS	6 400	5 200	5 200	5 200	4 200	5 200
	MGS	5 800	4 400	4 400	4 400	3 700	4 600

NOTE. These stresses apply to timber having a moisture content exceeding 18%.

(1) Where redwood is used separately these stresses may be increased by a factor of 1.1.

TABLE 14.

DRY STRESSES AND MODULI OF ELASTICITY FOR MACHINE STRESS-GRADED SOFTWOODS (TABLE 4B OF CP112)

		Standard name					
Property	*Grade*	*Western hemlock (commercial) (N/mm²)*	*Redwood and Whitewood (N/mm²)*	*Douglas fir (home-grown) (N/mm²)*	*Scots pine (home-grown) (N/mm²)*	*Sitka spruce (home-grown) (N/mm²)*	*Western white spruce (N/mm²)*
Bending	M75	10.0	10.0	10.0	10.0	6.6	9.7
	M50	6.6	6.6	6.6	6.6	4.5	6.2
	MSS	7.3	7.3	7.3	7.3	5.2	6.9
	MGS	5.1	5.1	5.1	5.1	3.6	4.8
Tension	M75	7.0	7.0	7.0	7.0	4.6	6.7
	M50	4.6	4.6	4.6	4.6	3.2	4.3
	MSS	5.1	5.1	5.1	5.1	3.6	4.8
	MGS	3.5	3.5	3.5	3.5	2.5	3.3
Compression parallel to grain	M75	10.8	10.8	10.8	10.8	6.4	9.5
	M50	7.1	7.1	7.1	7.1	4.3	6.3
	MSS	7.9	8.0	8.0	8.0	5.0	7.0
	MGS	5.5	5.6	5.6	5.6	3.5	4.8
Compression perpendicular to grain	M75	1.80	1.80(1)	1.98	1.98	1.45	1.80
	M50	1.55	1.55(1)	1.70	1.70	1.24	1.55
	MSS	1.55	1.55(1)	1.70	1.70	1.24	1.55
	MGS	1.38	1.38(1)	1.52	1.52	1.10	1.38
Shear parallel to grain	M75	1.19	1.28	1.28	1.28	0.98	1.19
	M50	0.80	0.86	0.86	0.86	0.66	0.79
	MSS	0.80	0.86	0.86	0.86	0.66	0.79
	MGS	0.80	0.86	0.86	0.86	0.66	0.79
Mean modulus of elasticity	M75	12 000	10 700	10 700	10 700	9 000	10 200
	M50	10 500	9 000	9 000	9 000	7 700	8 600
	MSS	10 800	10 200	10 200	10 200	8 200	9 200
	MGS	9 800	8 800	8 800	8 800	7 200	8 200
Minimum modulus of elasticity	M75	8 000	6 700	6 700	6 700	5 500	6 400
	M50	7 000	5 500	5 500	5 500	4 800	5 400
	MSS	7 200	6 400	6 400	6 400	5 100	5 800
	MGS	6 500	5 400	5 400	5 400	4 500	5 100

NOTE. These stresses apply to timber having a moisture content not exceeding 18%.
(1) Where redwood is used separately these stresses may be increased by a factor of 1.1.

Modification factors

Modification factor for flexural members—duration of loading

Basic and grade stresses are applicable to permanent loading. Timber has the characteristic of being able to sustain a very much greater load for short periods of time than it can for longer periods. Basic stresses are obtained on a statistical basis from short term tests on clear specimens.

Long, medium and short term periods of duration of loading are shown in Table 12 of the Code giving the following factors:

(*a*) 1.00 for long term (e.g. dead + permanent imposed);
(*b*) 1.25 for medium term (e.g. dead + snow, dead + temporary loads);
(*c*) 1.5 for short term (e.g. dead + imposed + wind, dead + imposed + snow + wind).

Permissible stress = grade stress in bending × modification factor.

Modification factor for slenderness and various conditions of loading on compression members

For compression members formed from SS, GS, MSS, MGS or M50 grade softwoods, Table 15 below gives the modification factor to be used accounting for the combined effect of slenderness ratio and duration of loading. The slenderness ratio of a solid rectangular member is calculated from l/b where l = the effective height or length given in Table 18 of the code, and b = the least lateral dimension of the member.

Permissible stress = grade stress in compression parallel to grain × modification factor

TABLE 15.
MODIFICATION FACTOR K_{18} FOR SLENDERNESS RATIO AND DURATION OF LOADING ON COMPRESSION MEMBERS OF 40 GRADE AND 50 GRADE SOFTWOODS (TABLE 15 OF CP112)

Slenderness ratio		*Values of K_{18}*		
l/r	l/b	*Long-term loads*	*Medium-term loads*	*Short-term loads*
Less than 5	1.4	1.00	1.25	1.50
5	1.4	0.99	1.24	1.49
10	2.9	0.98	1.23	1.47
20	5.8	0.96	1.20	1.44
30	8.7	0.94	1.17	1.40
40	11.5	0.91	1.13	1.34
50	14.4	0.87	1.08	1.27
60	17.3	0.83	1.00	1.16
70	20.2	0.77	0.90	1.01
80	23.0	0.70	0.79	0.86
90	26.0	0.61	0.68	0.72
100	28.8	0.53	0.58	0.60
120	34.6	0.40	0.42	0.44
140	40.4	0.31	0.32	0.33
160	46.2	0.24	0.25	0.25
180	52.0	0.20	0.20	0.20
200	57.7	0.16	0.16	0.17
220	63.5	0.13	0.14	0.14
240	69.2	0.11	0.12	0.12
250	72.2	0.10	0.11	0.11

Modification factor for tension members

The modification factor for duration of loading (Table 12 of CP112) applies also to members in tension.

Modification factor for load sharing systems

Where four or more are considered to act together to support a common load, the grade stress should be multiplied by 1.1. This applies to rafters, joists, trusses and studwalling. These should be spaced at 610 mm or closer and the load distributed by means of purlins, boarding, etc.

Other modification factors not applicable in this book are given in the Code.

Flexural members

Effective span

The span is the distance between the centres of bearing.

Stiffness and deflection

Dimensions of flexural members should be adequate to resist deflection within limits. Excessive deflection may cause damage to ceilings, partitions and finishings.

For floors, the deflection of joists when fully loaded should not exceed 0.003 of the span. For the purpose of calculating the deflection the mean value of the modulus of elasticity should be used for rafters, floor joists and other systems where transverse distribution of applied load is achieved. The minimum value of the modulus of elasticity should be used for principals, binders and other components acting along and also where the stress induced by the dead or permanent load is more than 60 per cent of the permissible stress.

Lateral support
Maximum depth to breadth ratios for solid and laminated members with rectangular cross sections are given in Table 16 below. These ratios depend on the degree of lateral support provided.

It is possible for a beam or joist to buckle laterally if it is not restrained in position at the ends and held in line by decking and adequate bridging or blocking.

Compression members

Effective length
For the purpose of calculating the slenderness ratio the values of effective length l in terms of actual length L and depending on the condition of end restraint are obtained from Table 18 of the Code.

Slenderness ratio
The maximum slenderness ratio calculated in terms of l/r should be 180 for a member carrying dead and imposed load and 250 for a member carrying wind loads only.

If calculated in terms of l/b the corresponding values should be 52 and 72.2 respectively.

Bending and axial compression
In this case the quantity:

$$\frac{f_{apar}}{f_{ppar}} + \frac{C_{apar}}{C_{ppar}} \not> \begin{cases} 1.0 \text{ for } l/r \text{ not exceeding } 20 \\ 0.9 \text{ for } l/r \text{ exceeding } 20 \end{cases}$$

where f_{apar} = the applied flexural stress parallel to the grain; f_{ppar} = the permissible stress parallel to the grain; C_{apar} = the applied compression stress parallel to the grain; and C_{ppar} = the permissible stress parallel to the grain.

Spaced columns

These consist of two equal rectangular shafts spaced apart by end and intermediate blocking pieces suitably glued, bolted, screwed or otherwise adequately connected together.

The following examples illustrate the design of flexural and compression members consisting of structural timber:

TABLE 16.
MAXIMUM DEPTH TO BREADTH RATIOS (SOLID AND LAMINATED MEMBERS) (TABLE 17. OF CP112)

Degree of lateral support	*Maximum depth to breadth ratio*
No lateral support	2
Ends held in position	3
Ends held in position and member held in line, as by purlins or tie rods	4
Ends held in position and compression edge held in line, as by direct connection of sheathing, deck or joists	5
Ends held in position and compression edge held in line, as by direct connection of sheathing, deck or joists, together with adequate bridging or blocking spaced at intervals not exceeding 6 times the depth	6
Ends held in position and both edges firmly held in line	7

CURPE CONSULTANTS
46 Orburn Road
DUNFIELD

Project	DESIGN IV
Part of structure	Timber floor joists
Drawing ref	
Calc by	R.A.S
Date	
Job ref	CC/ME
Calc sheet no	17/1
rev	/
Check by	
Date	

Code ref.	Calculations	Output
	EXAMPLE 17.1. <u>Design of timber floor joists</u>	
	Timber joists: Species Group S2, G.S. grade	
	Section 50 mm × 200 mm	
	Span = 4 m	
	Spacing = 400 mm centres.	
	Loading: Imposed = 1·5	
	Boarding, joists = 0·25	
	Ceiling = 0·15	
	Σ = 1·90 kN/m^2.	
	Long term loading is assumed.	
Table 11A	Grade stress for bending f_g par = 4·8 N/mm^2	
"	Grade stress for shear parallel to the grain = 0·79 N/mm^2	
"	Modulus of elasticity (Mean) = 8000 N/mm^2.	
	Check for bending, deflection, shear	
	<u>Bending</u>	
	Load to one joist = 1·9 × 0·4 × 4 = 3·04 kN.	
	Bending moment $M = \frac{WL}{8} = \frac{3{\cdot}04 \times 4}{8} = 1{\cdot}52$ kNm.	M = 1·52 kNm
	Bending stress = $\frac{M}{Z}$ where $Z = \frac{bd^2}{6}$	
	$= \frac{1{\cdot}52 \times 10^6 \times 6}{50 \times 200^2} = 4{\cdot}56$ N/mm^2	
	Permissible bending stress =	
	f_p par = f_g par × load sharing K × K_{12}	

CURPE CONSULTANTS
46 Orburn Road
DUNFIELD

Project	DESIGN IV
Part of structure	Timber floor joists
Drawing ref	
Calc by	R.A.S.
Date	
Job ref	CC/ME
Calc sheet no	17/2
rev	/
Check by	
Date	

Code ref.	Calculations	Output
	EXAMPLE 17.1 <u>continued</u>	
	= 4·8 × 1·1 × 1·0 = 5·28 N/mm^2	
	Since 4·56 < 5·28, bending satisfactory	Bending O.K.
	<u>Defection</u>	
	Permissible deflection = 0·003 × span.	
	= 0·003 × 4 × 10^3 = 12 mm.	
	Actual deflection = $\frac{5\,WL^3}{384\,EI}$ where $I = \frac{bd^3}{12}$	
	$= \frac{5 \times 3{\cdot}04 \times 10^3 \times 4^3 \times 10^9 \times 12}{384 \times 8000 \times 50 \times 200^3} = 9{\cdot}5$ mm	
	Since 9·5 < 12, deflection satisfactory	Deflection O.K.
	<u>Shear</u>	
	Maximum shear force = $\frac{W}{2} = \frac{3{\cdot}04}{2} = 1{\cdot}52$ kN.	
	Maximum shear stress = $1\frac{1}{2}$ × average	
	$= \frac{1{\cdot}5 \times 1{\cdot}52 \times 10^3}{50 \times 200} = 0{\cdot}228$ N/mm^2	
	Since 0·228 < 0·79 × 1·1 shear satisfactory	Shear O.K.
Table 17	<u>Note</u>. that the maximum depth to breadth ratio is $\frac{d}{b} = \frac{200}{50} = 4 < 5$	
	This is satisfactory and enables the the full grade <u>bending</u> stress to be used in this example.	Lateral support O.K.

CURPE CONSULTANTS 46 Orburn Road DUNFIELD	Project DESIGN IV			Job ref CC/ME	
	Part of structure Timber beam			Calc sheet no 17/3	rev /
	Drawing ref	Calc by R.A.S	Date	Check by	Date

Code ref.	Calculations	Output
	EXAMPLE 17.2 Design of timber beam	
	Loads kN from timber joists & floor	

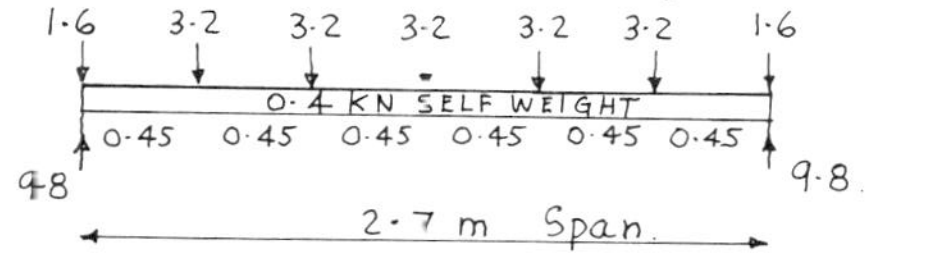

Code ref.	Calculations	Output
	Check beam for bending, deflection and shear assuming medium term loading	
	Bending	
	Maximum bending moment at centre of span is then	
	$M = (9\cdot8 - 1\cdot6)1\cdot35 - (3\cdot2 \times 0\cdot9) - (3\cdot2 \times 0\cdot45) - (0\cdot2 \times \frac{2\cdot7}{4})$	
	$= 11\cdot07 - 2\cdot88 - 1\cdot44 - 0\cdot135$	
	$= 6\cdot615$ kN.m.	$M = 6\cdot615$ kNm
	Using a whitewood, grade G.S. section,	
Table 4A	Grade stress in bending $= 5\cdot1$ N/mm^2	
Table 12	Modification factor for duration of loading $K_{12} = 1\cdot25$ for medium term.	
	Permissible bending stress	
	$=$ grade stress $\times K_{12}$	
	$= 5\cdot1 \times 1\cdot25 = 6\cdot375$ N/mm^2.	
	Z required $= \frac{6\cdot615 \times 10^6}{6\cdot375} = 1037647$	
	Assuming a section 75 mm breadth	$b = 75$ mm

CURPE CONSULTANTS 46 Orburn Road DUNFIELD	Project DESIGN IV			Job ref CC/ME	
	Part of structure Timber beam			Calc sheet no 17/4	rev /
	Drawing ref	Calc by R.A.S.	Date	Check by	Date

Code ref.	Calculations	Output
	EXAMPLE 17.2. continued	
	and since $Z = \frac{bd^2}{6}$ then $d^2 = \frac{6Z}{b}$	
	Minimum $d = \sqrt{\frac{6 \times 1037647}{75}} = 288\cdot1$ mm	$d = 300$ mm
Table 17	$\frac{d}{b} = \frac{300}{75} < 5$ Satisfactory	Lateral support O.K.
	Deflection	
	Treat closely spaced point loads as U.D.L.	
	E "minimum" is used in the formula.	
	Allowable deflection $= 0\cdot003 \times$ span	
	$= 0\cdot003 \times 2\cdot7 \times 10^3 = 8\cdot1$ mm	
	Actual deflection $= \frac{5}{384} \frac{WL^3}{EI}$ for U.D.L.	
	$= \frac{5 \times 19\cdot6 \times 10^3 \times 2\cdot7^3 \times 10^9 \times 12}{384 \times 4900 \times 75 \times 300^3} = 6\cdot07$ mm	
	Since $6\cdot07 < 8\cdot1$ Satisfactory	Deflection O.K.
	Shear	
	Maximum value of shear force at ends	
	$= 9\cdot8 - 1\cdot6 = 8\cdot2$ kN	
	Maximum shear stress $= 1\cdot5 \times$ average	
	$= \frac{1\cdot5 \times 8\cdot2 \times 10^3}{75 \times 300} = 0\cdot54$ N/mm^2	
Table 4A	Grade stress $= 0\cdot86$ N/mm^2 which is also the permissible stress.	
	Since $0\cdot54 < 0\cdot86$ Satisfactory	Shear O.K.

CURPE CONSULTANTS
46 Orburn Road
DUNFIELD

Project	DESIGN IV	Job ref	CC/ME
Part of structure	Timber strut	Calc sheet no / rev	17/5 / 1
Drawing ref	Calc by R.A.S. / Date	Check by	Date

Code ref.	Calculations	Output
	EXAMPLE 17.3. Strut subject to eccentric load	
	The column or strut consists of a 75mm × 150mm redwood SS grade member.	
	Actual length is 3460 mm.	
	Effective lengths: $\ell_{xx} = \ell_{YY} = 3460$ mm	
	A long term load of 10kN is applied at an eccentricity of 15mm perpendicular to X-X axis. i.e. $M_{xx} = 10 \times 15$ kNmm.	
	Slenderness ratio and compression	
	$\frac{\ell_{xx}}{D} = \frac{3460}{150} = 23$	
	$\frac{\ell_{YY}}{B} = \frac{3460}{75} = 46$	
Table 15	$K_{18} = 0.24$	
Table 4A	$C_g\,par = 8\ N/mm^2$ for redwood SS grade	
	$C_p\,par = C_g\,par \times K_{18}$	
	$= 8.0 \times 0.24 = 1.92\ N/mm^2$.	$C_p\,par = 1.92\ N/mm^2$
	$C_a\,par = \frac{W}{A} = \frac{10 \times 10^3}{75 \times 150} = 0.88\ N/mm^2$	$C_a\,par = 0.88\ N/mm^2$
	Bending	
	$f_a\,par = \frac{M_{xx}}{Z_{xx}} = \frac{10 \times 15 \times 10^3 \times 6}{75 \times 150^2} = 0.53\ N/mm^2$	$f_a\,par = 0.53\ N/mm^2$
	For bending the column must be restrained about the Y-Y axis.	

CURPE CONSULTANTS
46 Orburn Road
DUNFIELD

Project	DESIGN IV	Job ref	CC/ME
Part of structure	Timber strut	Calc sheet no / rev	17/6 / 1
Drawing ref	Calc by R.A.S. / Date	Check by	Date

Code ref.	Calculations	Output
	EXAMPLE 17.3. continued.	
	$\frac{D}{B} = \frac{150}{75} = 2$	
Table 17	From Table 17 the maximum $\frac{D}{B}$ may be 2 to permit use of the full grade bending stress.	
Table 4A	$f_g\,par = 7.3\ N/mm^2$ for redwood SS grade	
	$f_p\,par = f_g\,par \times K_{12}$	
	$= 7.3 \times 1 = 7.3\ N/mm^2$	$f_p\,par = 7.3\ N/mm^2$
CL. 3.14.3 Table 19	$\frac{f_a\,par}{f_p\,par} + \frac{C_a\,par}{C_p\,par} \not> 0.9$	
	$\frac{0.53}{7.3} + \frac{0.88}{1.92} = 0.072 + 0.458 = 0.53$	
	Since $0.53 < 0.9$ Column satisfactory	Column satisfactory
	Note. The permissible axial long term load is equal to:	
	$C_p\,par \times area$	
	$= \frac{1.92 \times 75 \times 150}{10^3} = 21.6$ kN	Permissible axial load 21.6 kN.

CURPE CONSULTANTS 46 Orburn Road DUNFIELD	Project: DESIGN IV	Job ref: CC/ME
	Part of structure: Timber spaced column	Calc sheet no: 17/7 rev: /
	Drawing ref: — Calc by: R.A.S. Date: —	Check by: — Date: —

Code ref.	Calculations	Output

EXAMPLE 17.4. Spaced Column design

(a) Check that the spaced column complies with the design requirements of CP112.

The column consists of two 75 mm × 150 mm redwood GS grade members spaced at 75 mm apart. All joints are glued.

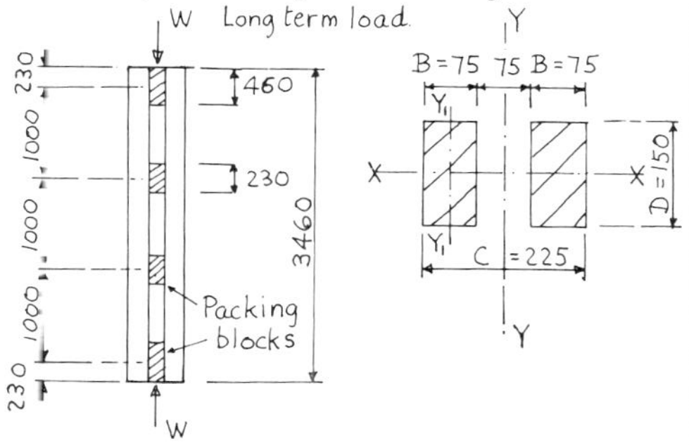

When subjected to a compressive load the column may buckle about any of the axes; X-X, Y-Y or Y_1-Y_1

Axes X-X and Y-Y are related to the composite section and Y_1-Y_1 to each of the individual sections.

CL.3.14.7 The permissible load is taken as the least of the following:

CURPE CONSULTANTS 46 Orburn Road DUNFIELD	Project: DESIGN IV	Job ref: CC/ME
	Part of structure: Timber spaced column	Calc sheet no: 17/8 rev: /
	Drawing ref: — Calc by: R.A.S. Date: —	Check by: — Date: —

Code ref.	Calculations	Output

EXAMPLE 17.4 continued.

(1) that for a solid column bending about the axis X-X.

(2) that for a solid column whose section is that of one member of the built-up column, with effective length equal to spacing of packing blocks, multiplied by the number of shafts

(3) that for a column bending about the Y-Y axis of built-up column cross section, but whose effective length is multiplied by modification factor K20.

Calculate radii of gyration for both individual and built-up sections.

$$I_{X_1X_1} = \frac{BD^3}{12} = \frac{75 \times 150^3}{12} = 21093750 \text{ mm}^4$$

$$I_{Y_1Y_1} = \frac{DB^3}{12} = \frac{150 \times 75^3}{12} = 5273437{\cdot}5 \text{ mm}^4$$

Area $A_1 = 75 \times 150 = 11250 \text{ mm}^2$ — Output: $A_1 = 11250 \text{ mm}^2$

$$r_{X_1X_1} = \sqrt{\frac{I_{X_1X_1}}{A_1}} = \sqrt{\frac{21093750}{11250}} = 43{\cdot}3 \text{ mm}$$ — Output: $r_{X_1X_1} = 43{\cdot}3$ mm

$$r_{Y_1Y_1} = \sqrt{\frac{I_{Y_1Y_1}}{A_1}} = \sqrt{\frac{5273437{\cdot}5}{11250}} = 21{\cdot}65 \text{ mm}$$ — Output: $r_{Y_1Y_1} = 21{\cdot}65$ mm

CURPE CONSULTANTS, 46 Orburn Road, DUNFIELD
Project: DESIGN IV — Part of structure: Timber spaced column — Drawing ref: — Calc by: R.A.S. — Date: — Job ref: CC/ME — Calc sheet no: 17/9 — rev: / — Check by: — Date:

Code ref.	Calculations	Output
	EXAMPLE 17.4 continued	
	For the composite section:	
	$A = 2 \times 75 \times 150 = 22500\ mm^2$	$A = 22500\ mm^2$
	$I_{XX} = \frac{2BD^3}{12} = \frac{2 \times 75 \times 150^3}{12} = 42187500\ mm^4$	
	$I_{YY} = \frac{DC^3}{12} - \frac{D(C-2B)^3}{12}$	
	$= \frac{150 \times 225^3}{12} - \frac{150(225-(2\times75))^3}{12}$	
	$= 142382810 - 5273437.5$	
	$= 137109372.5\ mm^4$	
	$r_{XX} = \sqrt{\frac{42187500}{22500}} = 43.3\ mm$ Same as $r_{X_1X_1}$	$r_{XX} = 43.3\ mm$
	$r_{YY} = \sqrt{\frac{137109372.5}{22500}} = 78.06\ mm$	$r_{YY} = 78.06\ mm$
	<u>Determine slenderness ratios and permissible loads for cases (1), (2) and (3)</u>	
	(1) $\frac{\ell}{r_{XX}}$ for composite section $= \frac{3460}{43.3} = 79.9$	
	Then from Table 15, and assuming long	
Table 15	term load; modification factor $K_{18} = 0.7$.	
	For redwood GS grade; compression	
Table 4A	parallel to grain $= 5.6\ N/mm^2$ (Table 4a)	C_g par $= 5.6\ N/mm^2$
	Permissible stress = grade stress $\times K_{18}$	
	$= 5.6 \times 0.7 = 3.92\ N/mm^2$	C_p par $= 3.92\ N/mm^2$

CURPE CONSULTANTS, 46 Orburn Road, DUNFIELD
Project: DESIGN IV — Part of structure: Timber spaced column — Drawing ref: — Calc by: R.A.S. — Date: — Job ref: CC/ME — Calc sheet no: 17/10 — rev: / — Check by: — Date:

Code ref.	Calculations	Output
	EXAMPLE 17.4 continued.	
	Permissible load = permissible stress × area	
	$= C_p$ par × area	
	$= \frac{3.92 \times 2 \times 75 \times 150}{10^3} = 88.2\ kN$	88.2 kN.
	(2) $\frac{\ell}{r_{Y_1Y_1}} = \frac{1000}{21.65} = 46.2$, for individual section	
Table 15	$K_{18} = 0.89$	
Table 4A	C_g par $= 5.6\ N/mm^2$	
	C_p par $= 5.6 \times 0.89 = 4.98\ N/mm^2$	C_p par $= 4.98$
	Permissible load $= \frac{4.98 \times 2 \times 75 \times 150}{10^3} = 112\ kN$	112 kN
	(3) Slenderness ratio $= \frac{\ell}{r_{YY}} \times K_{20}$ for	
	Composite section.	
Table 20	$\frac{\text{Space between members}}{\text{member thickness}} = \frac{75}{75} = 1$	
	Then $K_{20} = 1.1$ for glued connections	
	$\frac{\ell}{r_{YY}} \times K_{20} = \frac{3460 \times 1.1}{78.06} = 48.7$	
Table 15	$K_{18} = 0.87$	
	C_p par $= 5.6 \times 0.87 = 4.87\ N/mm^2$	C_p par $= 4.87$
	Permissible load $= \frac{4.87 \times 2 \times 75 \times 150}{10^3} = 109.5\ kN$	109.5 kN.
	Permissible load = 88.2 kN; the least value obtained from (1), (2) and (3)	Permissible load = 88.2 kN.

CURPE CONSULTANTS
46 Orburn Road
DUNFIELD

Project: DESIGN IV | Job ref: CC/ME
Part of structure: Timber spaced column | Calc sheet no: 17/11 | rev
Drawing ref | Calc by: R.A.S | Date | Check by | Date

Code ref.	Calculations	Output
	EXAMPLE 17.4 continued	
CL 3.14.5	Maximum clear space between individual shafts	
	$= 3 \times$ thickness of shaft $= 3 \times 75 = 225$ mm	
	Since space $= 75$ mm Satisfactory	Satisfactory
CL.3. 14.6.1.2.	Minimum length end packing blocks	
	$= 6 \times$ thickness of individual shaft	
	$= 6 \times 75 = 450$ mm	460 mm satisfactory
CL.3. 14.6.2.	Minimum length intermediate packings	
	$= 230$ mm. Since 230mm is provided	Satisfactory
CL.3. 14.6.3.	Minimum number of intermediate packings	
	$= 2$. Since 2 are provided	Satisfactory
CL.3 14.6.3.	In addition, sufficient packings should be provided to ensure that:	
	the greatest local slenderness of an individual shaft is limited to either 70 or $0.7 \times$ greatest $\frac{\ell}{r}$ of the whole column, whichever is the east. Thus:	
	Local slenderness ratio $= \frac{1000}{21.65} = 46.2$ and	
	$0.7 \times$ greatest $\frac{\ell}{r} = 0.7 \times 79.9 = 55.9$	
	Local slenderness is limited to 55.9	
	Since $46.2 < 55.9$ Satisfactory	Satisfactory
	No calculations are made here for glued joints.	

CURPE CONSULTANTS
46 Orburn Road
DUNFIELD

Project: DESIGN IV | Job ref: CC/ME
Part of structure: Timber spaced column | Calc sheet no: 17/12 | rev
Drawing ref | Calc by: R.A.S. | Date | Check by | Date

Code ref.	Calculations	Output
	EXAMPLE 17.4. continued	
	(b) Now check the adequacy of the column if subjected to a long term load of 50kN applied at an eccentricity of 15 mm perpendicular to X-X axis. i.e. $M_{xx} = 50 \times 15$	
CL.3.14.3 Table 19	$\frac{f_{a\,par}}{f_{p\,par}} + \frac{C_{a\,par}}{C_{p\,par}} \not> 0.9$	
	$C_{a\,par} = \frac{W}{A} = \frac{50 \times 10^3}{75 \times 150 \times 2} = 2.22\ N/mm^2$	
	$C_{p\,par} = 3.92\ N/mm^2$ as before	
	$f_{a\,par} = \frac{M_{xx}}{Z_{xx}} = \frac{50 \times 15 \times 10^3 \times 6}{(2 \times 75) 150^2} = 1.33\ N/mm^2$	
	For bending about the x-x axis the column must be restrained about the Y-Y axis	
	The ratio $\frac{D}{B}$ for one joist $= \frac{150}{75} = 2$	
Table 17	From Table 17 the maximum $\frac{D}{B}$ may be 2 to permit use of the full grade bending stress	
Table 4A	$f_{g\,par} = 5.1\ N/mm^2$	
	$f_{p\,par} = f_{g\,par} \times K_{12}$	
	$= 5.1 \times 1.0 = 5.1\ N/mm^2$	
	Then $\frac{1.33}{5.1} + \frac{2.22}{3.92} = 0.261 + 0.566 = 0.827 < 0.9$	
	Since $0.827 < 0.9$ Column satisfactory	Column satisfactory

CURPE CONSULTANTS 46 Orburn Road DUNFIELD

Project	DESIGN IV	Job ref	CC/ME
Part of structure	STUD WALLING	Calc sheet no / rev	17/13 /
Drawing ref	Calc by R.A.S. / Date	Check by	Date

Code ref.	Calculations	Output

EXAMPLE 17.5 Design of stud walling

Loading 15 kN/m length (long term)

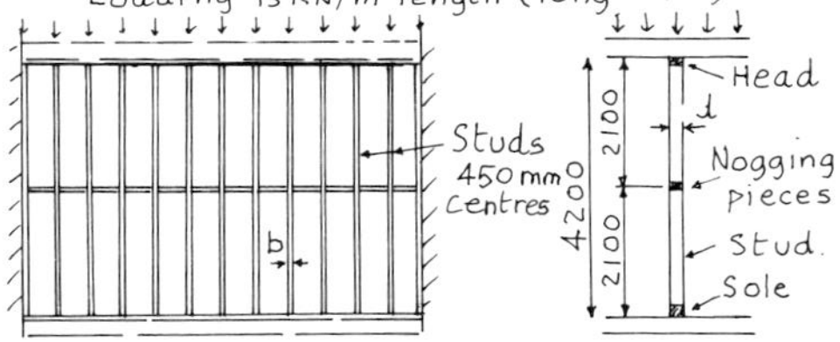

Elevation Section

Framing for stud wall consists of whitewood GS grade members spaced at 450 mm centres. No openings are reqd.

Determine minimum dimensions of stud to meet slenderness ratio requirements.

CL. 3.14.2 Maximum slenderness ratio $\frac{\ell}{r} = 180$ or if calculations are based on length and cross section dimensions then

Table 15 $\frac{\ell_{YY}}{b}$ or $\frac{\ell_{XX}}{d} \ngtr 52$

Effective length X-X axis = actual height

Effective length Y-Y axis = $\frac{1}{2}$ actual height with restraint provided by adequate securely fixed nogging pieces.

$\frac{\ell_{YY}}{b} = \frac{2100}{b} = 52$ $\therefore$ minimum $b = \frac{2100}{52} = 40.38$

CURPE CONSULTANTS 46 Orburn Road DUNFIELD

Project	DESIGN IV	Job ref	CC/ME
Part of structure	STUD WALLING	Calc sheet no / rev	17/14 /
Drawing ref	Calc by R.A.S. / Date	Check by	Date

Code ref.	Calculations	Output

EXAMPLE 17.5 continued

$\frac{\ell_{XX}}{d} = \frac{4200}{d} = 52$ $\therefore$ minimum $d = \frac{4200}{52} = 80.76$

Try 44 mm × 125 mm section for studs

Slenderness ratio:

$\frac{\ell_{XX}}{d} = \frac{4200}{125} = 33.6$

$\frac{\ell_{YY}}{b} = \frac{2100}{44} = 47.7$

Table 15 Then $K_{18} = 0.24$

Table 4A $C_g\,par = 5.6\ N/mm^2$

$C_p\,par = C_g\,par \times K_{18} \times$ load sharing K

CL.3.12.4 $= 5.6 \times 0.24 \times 1.1$

$= 1.478\ N/mm^2$

Permissible load = $C_p\,par \times$ area

$= \frac{1.478 \times 44 \times 125}{10^3}$

$= 8.13\ kN$ — Output: 8.13 kN.

Actual load to one stud

$= 15 \times 0.45 = 6.75\ kN$ — Output: W = 6.75 kN

Since $6.75 < 8.13$ kN section satisfactory — Output: 44 mm × 125 mm Satisfactory

Nogging pieces

Use same section as studs — Output: 44 mm × 125 mm.

CURPE CONSULTANTS 46 Orburn Road DUNFIELD	Project: DESIGN IV	Job ref: CC/ME
	Part of structure: FLOOR SHUTTER SYSTEM	Calc sheet no: 17/15, rev: 1
	Drawing ref: — / Calc by: R.A.S. / Date: —	Check by: — / Date: —

Code ref.	Calculations	Output

EXAMPLE 17.5 Floor shutter system.

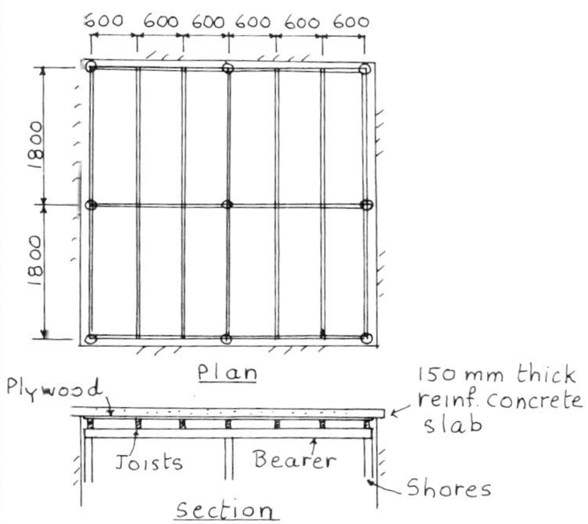

Design data:

Loading		
	Imposed	$2 \cdot 0\ kN/m^2$
	Plywood	$0 \cdot 1$
	slab	$3 \cdot 6$
	joists	$0 \cdot 1$
	Σ =	$5 \cdot 8\ kN/m^2$

Permissible bending stress for joists and bearers allowing for increase of 33% for duration of loading is $7 \cdot 31\ N/mm^2$ (green stress).

Modulus of elasticity $E = 4100\ N/mm^2$

CURPE CONSULTANTS 46 Orburn Road DUNFIELD	Project: DESIGN IV	Job ref: CC/ME
	Part of structure: FLOOR SHUTTER SYSTEM	Calc sheet no: 17/16, rev: 1
	Drawing ref: — / Calc by: R.A.S / Date: —	Check by: — / Date: —

Code ref.	Calculations	Output

EXAMPLE 17.5. continued

Allowable deflection = $\frac{1}{270}$ of the span

Assume joists and bearers simply supported

Design of joists

Load $W = 5 \cdot 8 \times 0 \cdot 6 \times 1 \cdot 8 = 6 \cdot 264\ kN$.

$M = \frac{WL}{8} = \frac{6 \cdot 264 \times 1 \cdot 8}{8} = 1 \cdot 41\ kNm$

Z required $= \frac{M}{f_{ppar}} = \frac{1 \cdot 41 \times 10^6}{7 \cdot 31} = 192886\ mm^3$

If $b = 50$ mm then $\frac{50 d^2}{6} = 192886$

$d^2 = 23146 \quad \therefore d = 152$ mm (minimum)

Deflection Allowable $= \frac{L}{270}$

Actual $= \frac{5}{384}\frac{WL^3}{EI}$

Equating: $\frac{L}{270} = \frac{5}{384}\frac{WL^3}{EI}$

I required $= \frac{5 \times W \times L^2 \times 270}{384 \times E}$

$= \frac{5 \times 6 \cdot 264 \times 10^3 \times 1 \cdot 3^2 \times 10^6 \times 270}{384 \times 4100}$

$= 17402652$ mm (minimum)

If $b = 50$ mm then $\frac{5 c d^3}{12} = 17402652$

$d^3 = 4176636 \quad \therefore d = 161$ mm (minimum)

Use 50×175 joists. | Output: Use 50×175

CURPE CONSULTANTS 46 Orburn Road DUNFIELD	Project DESIGN IV			Job ref CC/ME	
	Part of structure FLOOR SHUTTER SYSTEM			Calc sheet no 17/17	rev /
	Drawing ref	Calc by R.A.S.	Date	Check by	Date

Code ref.	Calculations	Output

EXAMPLE 17.5. Continued

Design of bearers

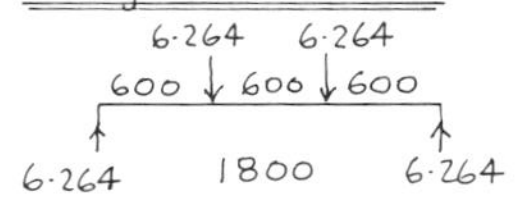

$M = 6{\cdot}264 \times 0{\cdot}6 = 3{\cdot}758$ kNm

Z required $= \dfrac{M}{f_{p\,par}} = \dfrac{3{\cdot}758 \times 10^6}{7{\cdot}31} = 514145$ mm^3

If $b = 50$ mm then $\dfrac{50d^2}{6} = 514145$

$d^2 = 61697 \quad \therefore d = 248$ mm (minimum)

Deflection Allowable $\dfrac{L}{270}$

Actual $= \dfrac{23\,WL^3}{648\,EI}$

Equating: $\dfrac{L}{270} = \dfrac{23\,WL^3}{648\,EI}$

I required $= \dfrac{23 \times 6{\cdot}264 \times 10^3 \times 1{\cdot}8^2 \times 10^6 \times 270}{648 \times 4100}$

$= 47438341$ mm^4 (minimum)

If $b = 50$ mm then $\dfrac{50d^3}{12} = 47438341$

$d^3 = 11385201 \quad \therefore d = 225$ (minimum)

Use 50 × 250 joist bearers — Output: Use 50×250

NOTE. The thickness of the plywood soffit shuttering (not designed here) is selected to resist bending and restrict deflection

CURPE CONSULTANTS 46 Orburn Road DUNFIELD	Project DESIGN IV			Job ref CC/ME	
	Part of structure Timber support for trench			Calc sheet no 17/18	rev /
	Drawing ref	Calc by R.A.S.	Date	Check by	Date

Code ref.	Calculations	Output

EXAMPLE 17.6. Timber support for trench

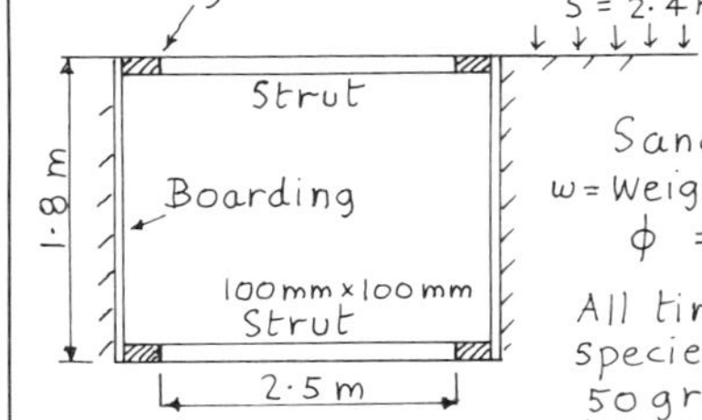

Sand
w = Weight = 15 kN/m^3
$\phi = 30°$

All timber to be Species group S2, 50 grade.
Use green stresses

Horizontal forces per m length

Sand:

$P = \dfrac{wh^2}{2}\left(\dfrac{1-\sin\phi}{1+\sin\phi}\right) = \dfrac{15 \times 1{\cdot}8^2}{2}\left(\dfrac{1-\sin 30°}{1+\sin 30°}\right)$

$= 24{\cdot}3\left(\tfrac{1}{3}\right) = 8{\cdot}1$ kN. — Output: $P_{sand} = 8{\cdot}1$ kN

Surcharge:

$P = sh\left(\dfrac{1-\sin\phi}{1+\sin\phi}\right) = 2{\cdot}4 \times 1{\cdot}8 \times \tfrac{1}{3} = 1{\cdot}44$ kN. — Output: $P_{surcharge} = 1{\cdot}44$ kN

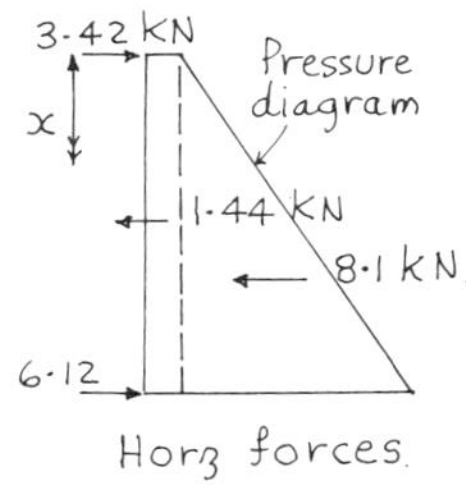

Horz forces

Reactions per m

At top:
$\dfrac{1{\cdot}44}{2} + \dfrac{8{\cdot}1}{3} = 3{\cdot}42$ kN

At bottom:
$\dfrac{1{\cdot}44}{2} + \dfrac{2}{3} \times 8{\cdot}1 = 6{\cdot}12$ kN

CURPE CONSULTANTS 46 Orburn Road DUNFIELD

Project	DESIGN IV	Job ref	CC/ME
Part of structure	Timber support for trench.	Calc sheet no / rev	17/19 /
Drawing ref	Calc by R.A.S. / Date	Check by / Date	

Code ref.	Calculations	Output

EXAMPLE 17.6. continued.

Consider pressure due to sand (Hor3)

At 1·8 m depth pressure $= wh\left(\frac{1}{3}\right)$

$= 15 \times 1{\cdot}8 \times \frac{1}{3} = 9\ \text{kN/m}^2$

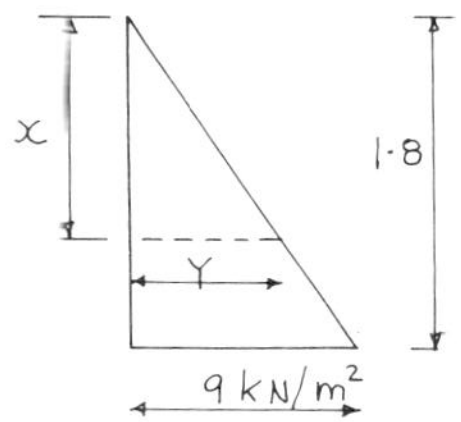

$9 : 1{\cdot}8$

$Y : x$

$\therefore Y = \frac{9x}{1{\cdot}8} = 5x$

Determine position of zero shear i.e. position of maximum bending moment for total loading applied to boarding.

At distance x from top

$3{\cdot}42 = \frac{1{\cdot}44x}{1{\cdot}8} + 5x \times \frac{x}{2}$

$3{\cdot}42 = 0{\cdot}8x + 2{\cdot}5x^2$

$2{\cdot}5x^2 + 0{\cdot}8x - 3{\cdot}42 = 0$

$x^2 + 0{\cdot}32x - 1{\cdot}368 = 0$ Quadratic equat

From which $x = 1{\cdot}02$ m. — Output: $x = 1{\cdot}02$ m

Then maximum bending moment

$M = (3{\cdot}42 \times 1{\cdot}02) - (0{\cdot}81 \times 0{\cdot}51) - \left(2{\cdot}60 \times \frac{1{\cdot}02}{3}\right)$

CURPE CONSULTANTS 46 Orburn Road DUNFIELD

Project	DESIGN IV	Job ref	CC/ME
Part of structure	Timber support for trench.	Calc sheet no / rev	17/20 /
Drawing ref	Calc by R.A.S. / Date	Check by / Date	

Code ref.	Calculations	Output

EXAMPLE 17.6. continued.

$M = 2{\cdot}19$ kNm — Output: $M = 2{\cdot}19$ kNm.

Table 10 — Grade stress (green) for bending $= 5{\cdot}5\ \text{N/mm}^2$

Modification factor for duration of loading to be taken as 1·25

Therefore permissible bending stress

$f_{ppar} = 5{\cdot}5 \times 1{\cdot}25 = 6{\cdot}875\ \text{N/mm}^2$

Z required $= \frac{M}{f_{ppar}} = \frac{2{\cdot}19 \times 10^6}{6{\cdot}875} = 318545\ \text{mm}^3$

Z of boarding $= \frac{bd^2}{6} = \frac{10^3 \times d^2}{6}$

$\frac{10^3 d^2}{6} = 318545 \qquad d^2 = \frac{318545 \times 6}{10^3}$

Minimum $d = 43{\cdot}7$ mm — Output: Use 50 mm boarding.

Check shear: Maximum force $= 6{\cdot}12$ kN

Maximum shear stress $= \frac{1{\cdot}5 \times 6{\cdot}12 \times 10^3}{10^3 \times 50} = 0{\cdot}183$

Table 10 — Since $0{\cdot}183 < 0{\cdot}69$, where $0{\cdot}69\ \text{N/mm}^2$ is grade stress (green) for shear, this is satisfactory. — Output: Shear O.K.

Design of bottom waling 200×100 section

Applied uniformly distributed load per metre length $= 6{\cdot}12$ kN. See reactions

Let spacing of struts $= L$ metres.

CURPE CONSULTANTS 46 Orburn Road DUNFIELD	Project: DESIGN IV		Job ref: CC/ME
	Part of structure: Timber support for trench		Calc sheet no: 17/21.1 rev
	Drawing ref	Calc by: R.A.S. Date	Check by / Date

Code ref.	Calculations	Output
	EXAMPLE 17.6 continued	
	Maximum bending moment $M = \frac{WL}{8}$	
	$= \frac{6.12L}{8} \times L$ kNm	
	$M = f_{ppar} \times Z$ $\quad Z = \frac{bd^2}{6} = \frac{100 \times 200^2}{6}$	
	$\frac{6.12L^2}{8} \times 10^6 = \frac{6.875 \times 100 \times 200^2}{6}$	
	$L^2 = \frac{8 \times 6.875 \times 100 \times 200^2}{6 \times 6.12 \times 10^6} = 5.99$	
	$L = \sqrt{5.99} = 2.45$ metres	200 × 100 Section span = 2.45 m
	Design of bottom strut 100×100 section.	
	Applied load = 6.12 × 2.45 = 15 kN.	
	Slenderness ratio $= \frac{l}{b} = \frac{2.5 \times 10^3}{100} = 25$	
Table 15	$K_{18} = 0.70$ for medium term loads.	
	Permissible stress = grade stress × K_{18}	
Table 10	Grade stress (green) for compression	
	perpendicular to grain = 4.1 N/mm²	
	Then permissible stress = 4.1 × 0.7 = 2.87 N/mm²	
	Permissible load $= \frac{2.87 \times 100 \times 100}{10^3} = 28.7$ kN.	
	Since 15 < 28.7 Section satisfactory	100 × 100 section O.K.

SELF-ASSESSMENT QUESTIONS

1. Timber joists of species group S2, G.S. grade and of section 200 mm × 50 mm, span 3.8 m. If the total loading, including the self weight is 2 kN/m² (medium term loading) determine the maximum spacing at which the joists could be placed given that the:

(*a*) grade stress in bending = 4.8 N/mm²;
(*b*) grade stress for shear parallel to the grain = 0.79 N/mm²; and
(*c*) mean modulus of elasticity E = 800 N/mm².

2. A timber beam of 2.5 m span supports a uniformly distributed load of 16.0 kN including the self weight (long term loading). Determine a suitable section for the beam given that the:

(*a*) grade stress in bending = 5.1 N/mm²;
(*b*) minimum modulus of elasticity E = 4900 N/mm²; and
(*c*) grade stress for shear parallel to the grain = 0.86 N/mm².

3. A timber column consists of a 100 mm square section having an effective length of 3.50 m.

(*a*) Determine the safe axial long term load if the grade stress in bending parallel to the grain = 8.0 N/mm².
(*b*) If a long term load of 12.0 kN is applied at an eccentricity of 12 mm (i.e. $M_{xx} = 12 \times 12$ kN/mm), check the adequacy of the section for this load only.

4. A spaced column of redwood GS grade which complies with the design requirements of CP112, consists of 2 No. 100 mm square members spaced 75 mm apart. All joints are glued. Determine the safe axial long term if the effective length of the column is 5.0 m and the spacing of packing blocks is 1.10 m. The grade stress in compression parallel to the grain may be taken as 5.60 N/mm².

5. Design a typical stud walling member carrying a long term loading of 20.0 kN/m length. The overall height is 3.80 m with nogg-

ing pieces at mid height. The studs consist of whitewood, GS grade timber. The grade stress in compression parallel to the grain = 5.60 N/mm^2.

6. A floor shuttering system consists of plywood with joists spaced at 700 mm centres, beams at 2.0 m centres and shores spaced to a grid 2.10 m × 2.00 m. Design suitable joists and bearers assuming that the:

(*a*) total loading is 6 kN/m^2;
(*b*) permissible bending stress = 7.3 N/mm^2;
(*c*) modulus of elasticity $E = 4100$ N/mm^2; and
(*d*) allowable deflection = span/270.

7. Design suitable boarding, wallings and struts for a trench support given that the:

(*a*) depth of excavation = 2.0 m;
(*b*) length of strut = 3.0 m;
(*c*) weight of retained soil = 18.0 kN/m^3;
(*d*) angle of internal friction $\theta = 30°$;
(*e*) surcharge = 2 kN/m^2;
(*f*) grade stress in bending = 5.50 N/mm^2;
(*g*) modification factor for duration of loading = 1.25;
(*h*) grade stress for shear = 0.69 N/mm^2;
(*i*) grade stress in compression perpendicular to the grain = 4.10 N/mm^2.

Appendix I
Summary of symbols used from CP110-'The Structural Use of Concrete.'

A_c Area of concrete
A_{cf} Area of effective concrete flange
A_s Area of tension reinforcement
A'_s Area of compression reinforcement
A_{sc} Area of longitudinal reinforcement (for columns)
$A_{s.prov.}$ Area of tension reinforcement provided
$A_{s.req.}$ Area of tension reinforcement required
A_{sv} Cross-sectional area of the two legs of a link
b Width of section
b_c Breadth of compression face
b_t Breadth of section at level of tension reinforcement
b_w Breadth of web or rib of a member
d Effective depth of tension reinforcement
d' Depth of compression reinforcement
d_c Depth of concrete in compression
d_2 Depth to reinforcement
e Eccentricity
F Ultimate load
F_k Characteristic load
f_{bs} Bond stress
f_{cu} Characteristic concrete cube strength
f_k Characteristic strength
f_s Service stress
f_y Characteristic strength of reinforcement
G_k Characteristic dead load
g Distributed dead load
g_k Characteristic dead load per unit area
h Overall depth of section in plane of bending
h_f Thickness of flange
I Second moment of area
l_e Effective height of a column or wall
l_{ex} Effective height for bending about the major axis
l_{ey} Effective height for bending about the minor axis
l_o Clear height of column between end restraints
M Bending moment due to ultimate loads
M_a Increased moment in column
M_t Total moment in a column due to ultimate loads
M_u Ultimate resistance moment
M_{ux} Maximum moment capacity in a short column assuming ultimate axial load and bending about the major axis only
M_{uy} Maximum moment capacity in a short column assuming ultimate axial load and bending about the minor axis only
M_x,M_y Moments about the major and minor axes of a short column due to ultimate loads
N Ultimate axial load at section considered
n Total ultimate load per unit area ($1.4\ g_k + 1.6\ q_k$)
S_v Spacing of links along the member
u Perimeter
u_{crit} Length of a critical perimeter
V Shear force due to ultimate loads
v Shear stress
v_c Ultimate shear stress in concrete
W_k Characteristic wind load
x Neutral axis depth
z Lever arm
γ_f Partial safety factor for load
γ_m Partial safety factor for strength
μ Coefficient of friction
ϱ $\varrho = \dfrac{A_s}{bd}$
ΣA_{sv} Area of shear reinforcement
ΣU_s Sum of the effective perimeters of the tension reinforcement
ϕ Bar size

Appendix II
Reinforcement Area Charts

STEEL AREA CHART FOR NUMBER OF BARS

Bar diameter (mm)	Areas in mm^2 for numbers of bars									
	1	*2*	*3*	*4*	*5*	*6*	*7*	*8*	*9*	*10*
6	28	57	85	113	142	170	198	226	255	283
8	50	101	151	201	252	302	352	402	453	503
10	79	157	236	314	392	471	550	628	707	785
12	113	226	339	452	566	679	792	905	1,020	1,130
16	201	402	603	804	1,010	1,210	1,410	1,610	1,810	2,010
20	314	628	943	1,260	1,570	1,890	2,200	2,510	2,830	3,140
25	491	982	1,470	1,960	2,450	2,950	3,440	3,930	4,420	4,910
32	804	1,610	2,410	3,220	4,020	4,830	5,630	6,430	7,240	8,040
40	1,260	2,510	3,770	5,030	6,280	7,540	8,800	10,100	11,300	12,600

STEEL AREA CHART FOR SPACING OF BARS

Bar diameter (mm)	Areas in mm^2 for spacings in mm								
	50	*75*	*100*	*125*	*150*	*175*	*200*	*250*	*300*
6	566	377	283	226	189	162	142	113	94
8	1,010	671	503	402	335	287	252	201	168
10	1,570	1,050	785	628	523	449	393	314	262
12	2,260	1,510	1,130	905	745	646	566	452	377
16	4,020	2,680	2,010	1,610	1,340	1,150	1,010	804	670
20	6,280	4,190	3,140	2,510	2,090	1,800	1,570	1,260	1,050
25	9,820	6,550	4,910	3,930	3,270	2,810	2,450	1,960	1,640
32	16,100	10,700	8,040	6,430	5,360	4,600	4,020	3,220	2,680
40	25,100	16,800	12,600	10,100	8,380	7,180	6,280	5,030	4,190

Index

Electrical and Electronic Principles Level II
F. GOODALL & D.K. RISHTON

This book is specifically designed to cover in detail the BTEC unit of the same name, though it will also be useful in any study of basic electrical engineering including all electrical courses in City and Guilds, GCE "O" Level and some CSE topics.
The authors have assumed no background knowledge, and by simple analogies taken from everyday experience the aim is to outline the fundamental principles which are necessary to understand how electrical circuits function. Written in a clear explanatory style, key principles are developed from very simple concepts so that the student can easily follow the text. New terminology is explained in full and where mathematical principles have needed to be included, these are explained as part of the chapter without the need to turn to a second textbook. A wide variety of self-assessment questions has been included both to help consolidate the student's understanding and to help develop a knowledge of the different types which are likely to be part of BTEC assignments in colleges.
Amply illustrated and including many worked examples, the book deals first with the basic units and elements involved in simple a.c. and d.c. circuits. This is followed by a description of measuring techniques and instruments, and concludes with a study of semiconductors and junction transistors. ". . . a good introduction to the topics covered. . . ." *Journal of the Institution of Electronics and Telecommunication Engineers*

Illustrated

Fluid Mechanics Level III
W. MADILL

This book has been written specifically to cover the BTEC Level II Manufacturing Technology unit in the Mechanical/Production Engineering Programme, although it will also be useful for those following the BTEC General Engineering Programme or those on a BTEC Diploma course as well as to undergraduates in their first year of a BSc course.

An entirely new work based on the author's industrial, commercial and teaching experience, the text is fully in accord with the most efficient modern practice and the appropriate British Standards. Clearly written in a straightforward simple style incorporating many diagrams to illustrate the points under discussion, each chapter commences with the objectives to be covered and concludes with a summary of the contents and a series of self-assessment questions by which the student may test the knowledge he has gained.

Illustrated

Manufacturing Technology Level II

K.J. PARSLEY

This new book has been specifically written for the BTEC Level III Unit of the same name, which is in turn part of the syllabus for students preparing for the examinations of the Higher Certificate and Higher Diploma in Mechanical and Production Engineering. It covers the principles involved in common industrial pressure and flow measurements and also in the determination of forces on immersed surfaces as well as the application of the Bernoulli equation. Unlike many existing texts the subject matter has been restricted to that which is directly relevant to this unit, and the emphasis has been placed on developing an understanding of fundamental principles. Numerical exercises are included throughout in order to reinforce and illustrate the theory, and each chapter concludes with a selection of both multiple choice and conventional questions.

Some degree of prior knowledge is assumed, as provided, for example, by the BTEC Units Engineering Science Level II and Mathematics Level II. Although specifically designed for BTEC students the book will also be suitable as supplementary reading for any course including elementary fluid mechanics such as first year degree courses in mechanical and production engineering and civil engineering, and also for the relevant sections of the BTEC Mechanical Engineering Science Level III and the Applied Mathematics Level III Units.

Illustrated

Mechanical Science Level III

P.A.N. LILLEY

This book covers the new BTEC Standard Unit U82/040, Mechanical Science Level III, whose aim is to extend the student's understanding of Mechanical Engineering and to develop an analytical approach to the solution of problems associated with deformation of materials, dynamics and fluid flow. The emphasis of the text is on the clear exposition of basic principles and the book contains continuous worked examples that enhance comprehension and illustrate application. The plentiful line drawings and explanatory style serve to lead the student through the text with a confident understanding of the basic concepts. Continuity is maintained between the various sections which include: stress and strain, bending, torsion, angular motion, energy, vibration and fluids. The scope of the book, although aimed at BTEC students in particular, would make it a highly appropriate course text for adoption by students following similar level courses involving mechanical engineering. As well as the usual self-assessment questions, each chapter concludes with a summary of essential formulae and information dealt with in that chapter. The book assumes that the student has completed his/her study of the Level 2 Unit, and includes some peripheral work slightly outside the Level 3 syllabus. As such the book will also appeal to "O" and "A" level students and students preparing for the Ordinary National Certificate and Diploma and Higher National Diploma.

Illustrated

Soil Mechanics Level IV
ALAN HARTLEY

Soil is an extremely complex and variable substance, yet its behaviour as an engineering material is in the main governed by the few basic properties of strength, permeability, compressibility and stress transfer. This book, provides an introduction to the engineering characteristics of soils and the methods used to determine their properties in the laboratory and on site. It covers the syllabus of the BTEC Soil Mechanics (Properties) IV unit, an essential component of Higher Technician Diploma and Certificate courses in Civil Engineering. The text will also be useful for students taking soil mechanics as part of a building degree, for civil engineering degree students or as part of the CEI examination studies.

Written in a simple, straightforward style and in a manner particularly suitable for BTEC study purposes, the text begins with a discussion of the historical background to the subject and the importance of site investigation, followed by an examination of the nature of soils and the methods used to group and classify them for engineering purposes. The effect of water in the flow and non-flow conditions is then discussed, followed by the shearing strength of soils and their compressibility in terms of both compaction and consolidation. Lateral earth pressures are then considered as is their effect on retaining structures. A broad view of the laboratory testing techniques then follows, and finally an introduction to the stability of earth slopes is given, together with relevant British Standards and other information sources concerned with this important aspect of soil mechanics.
Worked examples are included throughout to illustrate the use of the principles discussed and each chapter concludes with a series of questions which the student can use to appraise his understanding of that section of study.

Illustrated

Structural Detailing Level II
B. CURRIE & R.A. SHARPE

This book has been specifically written to cover the syllabus requirements of the BTEC Level II Unit, Structural Detailing.
For many years information concerning the detailing of reinforced concrete and structural steelwork has been available only from numerous scattered sources, and in many cases such information has been handed down to the newcomer in the structural drawing office by those experienced in the field. With the introduction of standard BTEC units, a carefully constructed textbook designed to correspond to the relevant unit is clearly a necessity.
The book, covers a wider field than that dealt with in this BTEC unit, and as such it will prove useful to all students involved in structural detailing, including those following degree, diploma and certificate courses in both structural and civil engineering and

building. It will assist the site supervisor having to interpret the technical drawings provided by the structural detailer, and may also be adopted by structural consultants as an office standard. "It has been written specifically to cover the requirements of the BTEC Level II unit of the same name. ". . . it covers these requirements admirably." *NATFHE Journal*

Illustrated

CW01638927

Biología y Geología 1

BACHILLERATO

ÍNDICE

CONTENIDOS ESPECÍFICOS

Para recordar

Contenidos para reforzar conceptos estudiados en cursos o unidades anteriores.

Para ampliar

El objetivo de estos contenidos es profundizar en detalles de interés, conocer las biografías de científicos y ampliar conceptos de mayor complejidad relacionados con los contenidos de la unidad.

Para conocer

Se trata de contenidos curriculares complementarios a los que se tratan en el cuerpo central de la unidad.

Actividades de Selectividad

Se encuentran en las páginas de actividades de síntesis de algunas unidades. Basándose en los contenidos de la unidad correspondiente, se plantean ejercicios del mismo tipo que los que se encuentran en las pruebas de Selectividad.

Actividades de Internet

Se hallan en las páginas de actividades de síntesis de todas las unidades. La resolución de estas actividades debe realizarse con información disponible en Internet.

1

El universo

El estudio del universo y sus componentes: galaxias, estrellas, planetas, satélites..., nos proporciona información sobre el origen y la dinámica de nuestro planeta así como de los fenómenos astronómicos que nos afectan.

Mediante el análisis del Sol y su actividad es posible determinar cuándo se producirán las denominadas *tormentas solares* y mitigar sus efectos sobre los circuitos eléctricos, los transformadores y los sistemas de comunicación.

Además del conocimiento científico adquirido directamente gracias al estudio del universo, el avance tecnológico asociado a la investigación espacial ha aportado destacados beneficios a nuestra calidad de vida. De este modo, nuevos materiales y tecnologías creados y utilizados para las misiones espaciales son ahora parte de nuestro entorno cotidiano, como las prendas ignífugas de los bomberos, las antenas de televisión vía satélite o los sistemas de generación y análisis de imágenes utilizados para diagnóstico médico.

ORGANIZACIÓN DE LOS CONTENIDOS

1. El conocimiento del universo

2. La exploración del universo

3. Origen y evolución del universo

4. Las galaxias

4.1. Las estrellas

4.2. Las nebulosas

5. El Sistema Solar

5.1. El Sol

5.2. Los planetas

5.3. Los satélites

5.4. Asteroides, cometas y meteoritos

6. Origen y evolución de la Tierra

6.1. Métodos de datación y eras geológicas

OBJETIVOS

- Comprender la evolución que ha sufrido el conocimiento sobre el universo a lo largo de la historia y valorar la aportación de los diferentes científicos a su conocimiento.
- Comprender la teoría del big bang sobre el origen del universo y la posible evolución de éste.
- Conocer las características de las galaxias y de sus componentes.
- Mostrar interés por conocer las características del Sistema Solar y de sus componentes.

Observatorio maya dedicado al planeta Venus en Chichén Itzá (México)

1. El conocimiento del universo

El universo es un espacio inmenso que incluye todo cuanto nos rodea, las galaxias, las estrellas, el Sistema Solar, la Tierra y a nosotros mismos, sus minúsculos habitantes.

Desde su aparición en la Tierra, el ser humano ha contemplado el cielo nocturno buscando datos que le ayuden a conocer el universo: su composición, su origen, su evolución, el movimiento de los astros... Las pinturas sobre constelaciones que decoran las cuevas habitadas por los primeros seres humanos y los calendarios elaborados por antiguas civilizaciones son una muestra de este interés.

A lo largo de la historia han existido distintas concepciones sobre el universo. Estas concepciones se han ido modificando a partir de las observaciones de numerosos científicos.

Las modificaciones de los diferentes modelos sobre el universo se han producido al mismo tiempo que se perfeccionaban los instrumentos de observación y se avanzaba en el esclarecimiento de las leyes de la física.

s. IV aC	Aristóteles propone el modelo geocéntrico, en el cual la Tierra es el centro del universo, en torno a la cual giran los demás astros unidos a las esferas celestes, bajo la bóveda de las estrellas fijas.	Sistema geocéntrico propuesto por Aristóteles 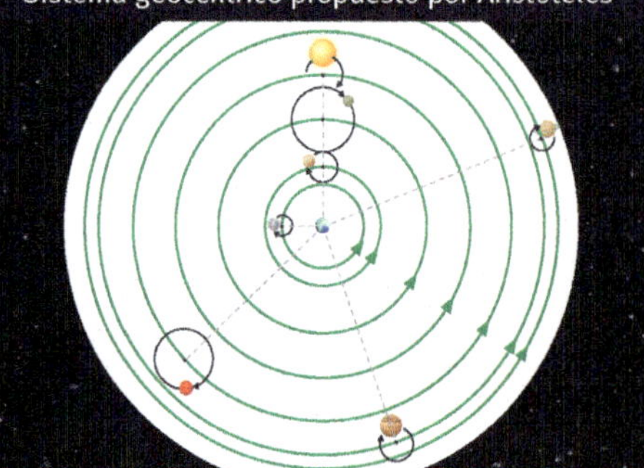
s. II dC	Claudio Ptolomeo introduce el concepto de epiciclo: los planetas no están fijos a las esferas, sino que realizan un giro, lo que justifica su aparente cambio de trayectoria en el cielo.	
s. XV-XVI	Nicolás Copérnico propone que la Tierra no es el centro del universo, sino que gira alrededor del Sol, al igual que los otros planetas, en órbitas circulares.	Sistema heliocéntrico copernicano 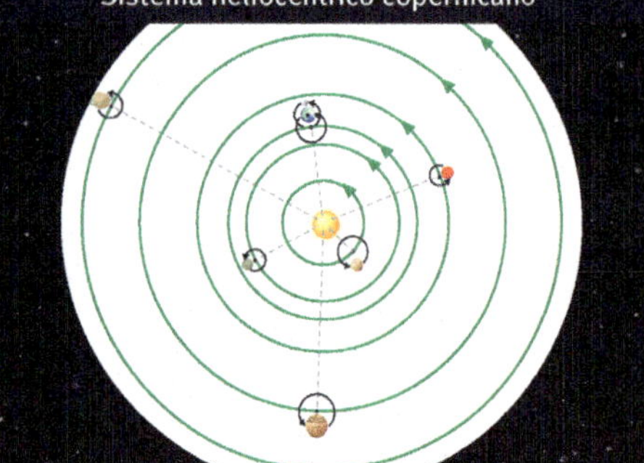
	Galileo Galilei descubre los satélites de Júpiter: el Sol y la Tierra no son los únicos astros alrededor de los cuales giran otros cuerpos.	
s. XVI-XVII	Tycho Brahe modifica el sistema copernicano: el Sol gira alrededor de la Tierra, y los demás planetas alrededor del Sol.	Rotación elíptica de los planetas según la segunda ley de Kepler 
	Johannes Kepler, apoyándose en los datos obtenidos por Brahe, propone sus tres leyes: 1. Los planetas se mueven en elipses con el Sol en uno de sus focos. 2. Los radios que unen el planeta con el Sol barren áreas iguales en tiempos iguales. 3. Los cuadrados de los períodos de los planetas son proporcionales a los cubos de sus distancias al Sol.	

s. XVII	Isaac Newton enuncia la ley de la gravitación universal: la fuerza de la gravedad es la que mantiene a la Luna en la órbita de la Tierra, al igual que cualquier par de masas se atraen entre ellas. Estas leyes son aplicables a todo el universo.
s. XVIII	William Herschel afirma que el Sol es una estrella en una galaxia, la Vía Láctea, y que ocupa el centro de ésta.
s. XX 1900	Max Planck inicia la física cuántica, que permitirá unir el mundo atómico con el mundo estelar.
1905	Albert Einstein propone la teoría de la relatividad, trastocando los conceptos de espacio y tiempo, y permitiendo la interpretación de fenómenos no explicables por las leyes de Newton.
1915	Harlow Shapley propone que el Sol se halla en uno de los brazos de la galaxia, no en su centro.
1934	Edwin Hubble identifica a Andrómeda como una galaxia más allá de la Vía Láctea. Nuestra galaxia no es más que una entre millones. Las galaxias se separan entre ellas: el universo está en expansión.
1964	Arno Penzias y Robert Wilson corroboran la teoría del big bang mediante el estudio de la radiación de fondo.

Ley de la gravitación universal

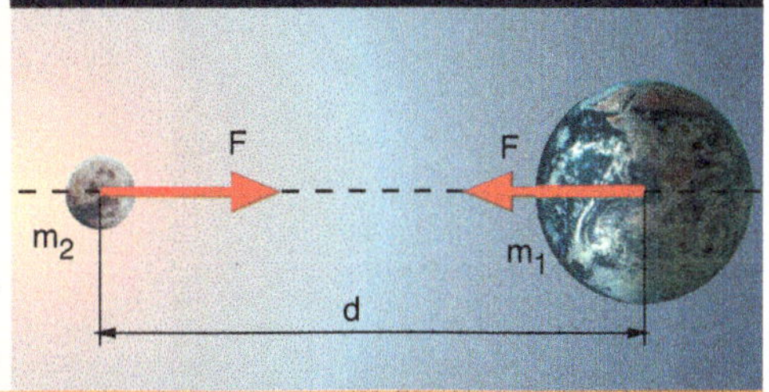

Representación de la deformación del espacio-tiempo por la masa

Ubicación del Sol en la galaxia

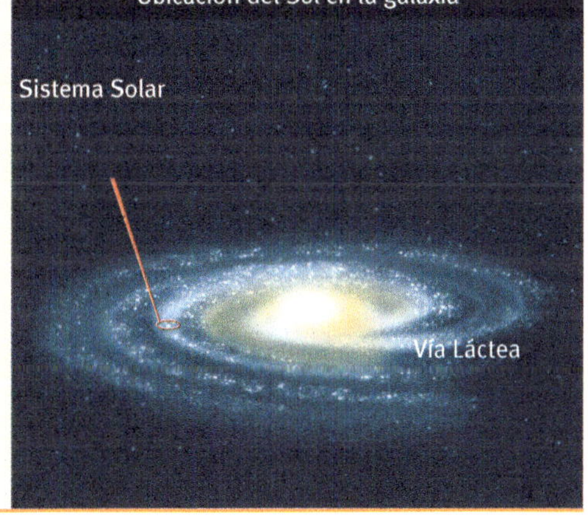

ACTIVIDADES

1. Realiza un cuadro comparativo entre los modelos geocéntrico y heliocéntrico del universo, destacando sus diferencias fundamentales así como sus principales valedores.
2. ¿Crees que los cambios sociales que se dieron en los siglos XVI y XVII influyeron en el desarrollo de los nuevos sistemas cosmológicos? Justifica tu respuesta.
3. ¿Qué diferencia existe entre las órbitas de la teoría heliocéntrica copernicana de las del modelo de Kepler?
4. ¿Qué fuerza provoca que la Luna se mantenga en la órbita terrestre? ¿Y que la Tierra orbite alrededor del Sol? ¿Qué característica han de tener los cuerpos para verse sometidos a esa fuerza?
5. ¿Podrías decir cuáles son las dos grandes teorías enunciadas a principio del s. XX que cambiaron radicalmente la manera de ver el universo físico que nos rodea, y quiénes fueron los científicos que las enunciaron?

2. La exploración del universo

Telescopio refractor en el observatorio Lowell, Arizona, Estados Unidos

El desarrollo de aparatos e instrumentos cada vez más precisos ha permitido ampliar el conocimiento acerca del universo. Los más utilizados son: los *telescopios ópticos,* los *radiotelescopios,* las *sondas espaciales* y las *naves espaciales tripuladas*.

Los telescopios ópticos

Captan la luz procedente de los astros y producen imágenes lo bastante grandes y nítidas para que puedan ser examinadas con detalle.

Las principales características que definen un buen telescopio son:

- El **poder de recogida,** que representa el área de luz que capta el telescopio.
- El **poder de resolución,** que es la capacidad para distinguir separadamente dos puntos muy próximos. A mayor poder de resolución, más capacidad para diferenciar los detalles en una imagen.

Los telescopios pueden ser *refractores* o *reflectores*, según estén constituidos por lentes o por espejos.

- Los telescopios **refractores** se basan en la refracción de la luz, producida por las **lentes** del objetivo y del ocular.
- Los telescopios **reflectores** se fundamentan en el fenómeno de la reflexión de la luz y, por tanto, el objetivo y el ocular son **espejos**.

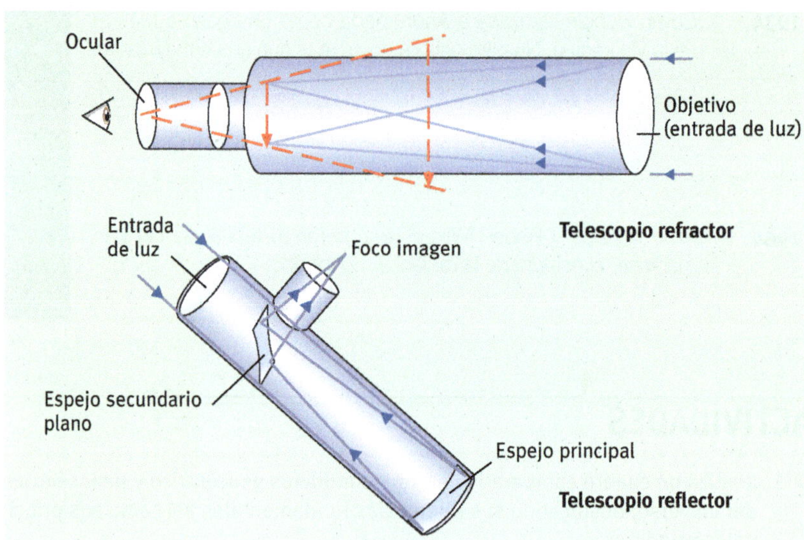

La atmósfera absorbe una parte de la luz que llega del espacio, especialmente si hay nubes o el aire está contaminado. Por esta razón, los observatorios donde se encuentran los telescopios están situados lejos de las áreas urbanas, en lugares altos y de clima seco, lo que permite llevar a cabo mejores observaciones del espacio.

En las últimas décadas se han lanzado al espacio **telescopios espaciales** que, al estar fuera de la atmósfera terrestre, nos ofrecen mejores imágenes. Estos telescopios están automatizados y responden a órdenes enviadas por los científicos desde la Tierra.

Los radiotelescopios

Son aparatos que captan las ondas radioeléctricas que tienen una frecuencia aproximada de 100 megahercios. Estos aparatos disponen de una gran antena parabólica que recibe las ondas y las concentra sobre otra pequeña antena. La señal procedente de ésta es amplificada y procesada por un receptor, que la transforma en una imagen de la zona observada.

En comparación con los telescopios ópticos, los radiotelescopios ofrecen la ventaja de que las ondas electromagnéticas penetran a través de la atmósfera independientemente de cuáles sean las condiciones meteorológicas.

Sin embargo, su poder de resolución es mucho más bajo que el de un telescopio, motivo por el cual es necesario procesar simultáneamente la información procedente de varios radiotelescopios.

Radiotelescopio de antena fija en Arecibo (Puerto Rico)

Las sondas espaciales

Se trata de pequeñas naves que se lanzan desde la Tierra para que realicen un recorrido determinado. Estas naves son capaces de detectar diferentes tipos de radiaciones y envían a la Tierra la información que captan.

Desde 1957, año en que la antigua Unión Soviética lanzó al espacio la sonda espacial *Sputnik 1,* este tipo de naves ha sido muy utilizado. La *Voyager 2* es la primera sonda que ha recorrido todo el Sistema Solar y que ha salido de él.

La *Mars Pathfinder* ha sido uno de los dispositivos enviados a Marte para recopilar información sobre el planeta rojo.

Las naves espaciales tripuladas

Son vehículos espaciales que permiten al ser humano la observación astronómica in situ. La Luna es el único astro que ha sido visitado hasta el momento por el ser humano. El primer alunizaje fue realizado por una nave de Estados Unidos, el 20 de julio de 1969. Dentro del programa *Apolo*, entre los años 1969 y 1972 se realizaron seis alunizajes con éxito.

El coste de los viajes tripulados es mayor que el de las sondas espaciales. Además, un viaje por el espacio siempre entraña riesgos. Por ello, en la actualidad, los viajes tripulados son trayectos alrededor de la órbita terrestre.

ACTIVIDADES

6. ¿Cuáles son los dos tipos principales de telescopios ópticos y en qué se diferencian?

7. ¿A qué es debido que los telescopios en órbita nos proporcionen mejores imágenes que aquellos que se encuentran en la superficie terrestre?

8. ¿Qué ventaja presenta un radiotelescopio sobre un telescopio óptico?

9. Últimamente Marte ha sido objeto de varias misiones de reconocimiento con sondas espaciales. A pesar de ello, aún no se ha enviado ninguna nave espacial tripulada al planeta rojo. ¿Qué inconvenientes y dificultades crees que presenta una misión tripulada en comparación con las sondas espaciales?

3. Origen y evolución del universo

En el primer apartado hemos descrito algunos de los estudios y descubrimientos sobre el universo que se han sucedido a lo largo de la historia. Aunque desde la perspectiva actual, algunos parezcan alejados de la realidad, todos han contribuido al conocimiento del universo.

Este conocimiento ha permitido la elaboración de un modelo sobre la formación y la evolución del universo que ha sido aceptado hasta el momento por la comunidad científica, pero que, sin duda, puede verse modificado con nuevos descubrimientos.

Según las últimas teorías sobre el origen del universo, éste se habría originado hace entre 15 000 y 20 000 millones de años a partir de la expansión de un único punto de temperatura y densidad infinitas en lo que conocemos como el **big bang.** Este punto o singularidad inicial contendría toda la materia y la energía que constituyen nuestro universo, y su expansión marcaría el inicio del tiempo y el espacio.

A continuación vamos a ver los fenómenos que estas teorías suponen que tuvieron lugar desde lo que se considera el inicio del tiempo y del universo.

C

La conocida fórmula $E=mc^2$, descubierta a principios del s. XX por **Albert Einstein**, relaciona dos conceptos que hasta el momento habían estado completamente separados: la energía y la materia. La materia no es más que una forma condensada de energía, y, por lo tanto, puede generarse a partir de ella.

0	Singularidad de la gran explosión inicial. Big bang. Leyes de la física exóticas y desconocidas.
10^{-43} segundos	Era de la teoría de la gran unificación. El equilibrio entre materia y antimateria se decanta a favor de la materia.
10^{-35} segundos	Era electrodébil, dominada por quarks y antiquarks.
10^{-10} segundos	Era de los hadrones y leptones. Los quarks se asocian formando protones, neutrones y otras partículas.
1 segundo	Los protones y neutrones se combinan formando núcleos de hidrógeno, helio, litio y deuterio.
3 minutos	La materia y la radiación se acoplan y se forman los primeros átomos estables.
1000 millones de años	Cúmulos de materia forman quásares, estrellas y protogalaxias. En el interior de las estrellas comienzan a sintetizarse núcleos más pesados.
15000 millones de años	Se forman nuevas galaxias con sistemas solares alrededor de las estrellas. Los átomos se enlazan entre ellos para formar moléculas.

+

Los **quarks** son partículas subnucleares, que existen asociados formando compuestos llamados hadrones. Los protones y los neutrones, que son un tipo de hadrones, están formados por la asociación de tres quarks.

El otro tipo de partículas subnucleares que constituye la materia son los **leptones**, como el electrón o los neutrinos.

A medida que se producía la expansión la temperatura disminuyó, hasta alcanzar los 100 000 millones de grados aproximadamente tres minutos después del inicio de la expansión. Este descenso abrió la posibilidad de la condensación de la energía en partículas subatómicas, quarks en primer lugar, que acabaron asociándose en protones, neutrones, mesones y bariones. Unos tres minutos después del big bang, estas partículas dieron lugar a los núcleos atómicos más ligeros.

Cientos de miles de años más tarde, con temperaturas de unos miles de grados, los electrones fueron frenados lo suficiente como para que ser capturados por los núcleos atómicos, originando los primeros átomos completos. La materia así creada se asoció 1000 millones de años después, formando protogalaxias, compuestas por estrellas que en su interior sintetizaron núcleos atómicos pesados, como carbono, nitrógeno, hierro... A partir de los productos de estas protogalaxias se formaron nuevas generaciones de galaxias, como la Vía Láctea.

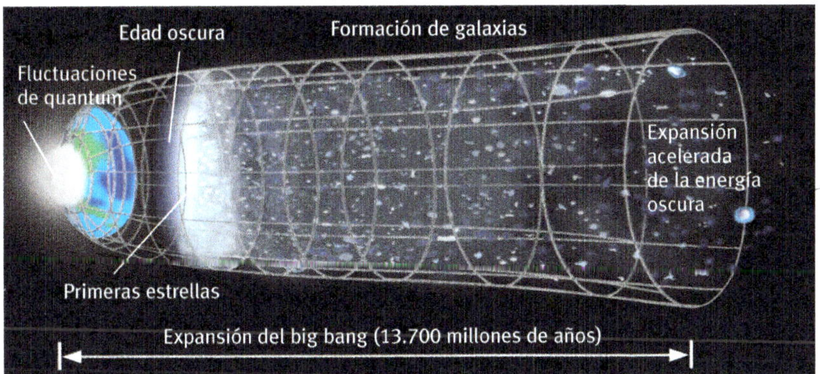

Representación del big bang

En la actualidad, la expansión y el enfriamiento del universo continúan, aunque no se ha podido precisar si este fenómeno ocurrirá indefinidamente. La evolución futura del universo depende de su masa total y de la fuerza gravitatoria que se ejerce entre la materia que lo forma:

- Si la masa del universo no es suficientemente elevada, no generará una fuerza gravitatoria capaz de detener la expansión, y el universo continuará expansionándose indefinidamente. En tal supuesto, el **universo** sería **abierto.**
- Si la masa del universo es suficientemente elevada, ejercerá una gran fuerza gravitatoria sobre sus constituyentes y, como consecuencia, en un momento futuro se detendrá la expansión. Cuando llegue ese momento, es posible que se inicie la contracción del universo, en un proceso denominado **big crunch,** y se origine un nuevo superátomo. En este caso, el **universo** sería **cerrado**. El modelo de universo cerrado nos plantea dos posibilidades:
  - El superátomo originado por el *big crunch* da lugar a otro *big bang*, que iniciará otra expansión del universo, y así sucesivamente. En este caso se trataría de un universo cíclico.
  - No se produce un nuevo *big bang* y se pasa a una situación en la que el universo dejaría de existir.

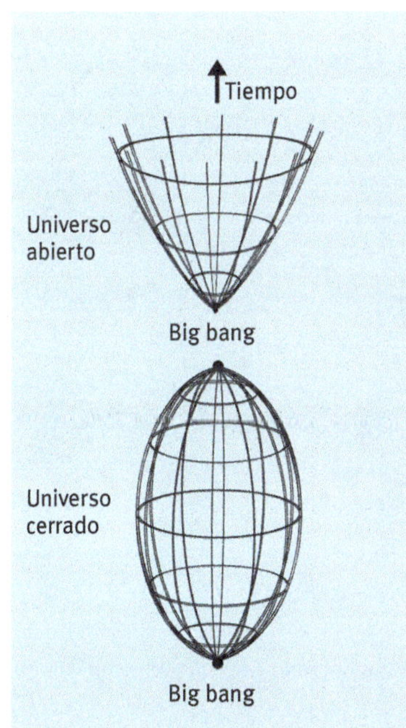

Representación de un universo abierto y de un universo cerrado

ACTIVIDADES

10. ¿Cuál se cree que es la edad del universo hoy día?

11. En las centrales nucleares la materia de los átomos radiactivos es convertida en energía. En las etapas tempranas del universo, la energía se condensó para dar lugar a partículas materiales. ¿Qué teoría nos explica estas transformaciones?

12. ¿Cuáles son los átomos más ligeros? ¿De qué están compuestos?

13. ¿Cómo podrías argumentar que la Vía Láctea no se originó inmediatamente después del big bang, sino que forma parte de una generación posterior de galaxias?

Galaxia espiral barrada NGC 1300

4. Las galaxias

Durante el siglo XX se definió la galaxia como la unidad básica del universo. Las galaxias están formadas por *agregados de estrellas* y *nebulosas*.

- Los **agregados de estrellas** son conjuntos de centenares o incluso miles de millones de estrellas.
- Las **nebulosas** son inmensas nubes de gas y polvo, de densidad variable.

Las galaxias se agrupan dando lugar a **cúmulos**, los cuales, a su vez, forman grupos mayores denominados **supercúmulos**.

Hasta el momento, en la zona del universo que ha podido ser estudiada, se han detectado más de mil millones de galaxias, que se encuentran separadas entre sí por grandes extensiones de espacio vacío.

La primera clasificación de las galaxias se basó en criterios morfológicos: según su forma fueron agrupadas en irregulares, elípticas, lenticulares y espirales, y dentro de éstas en normales y barradas.

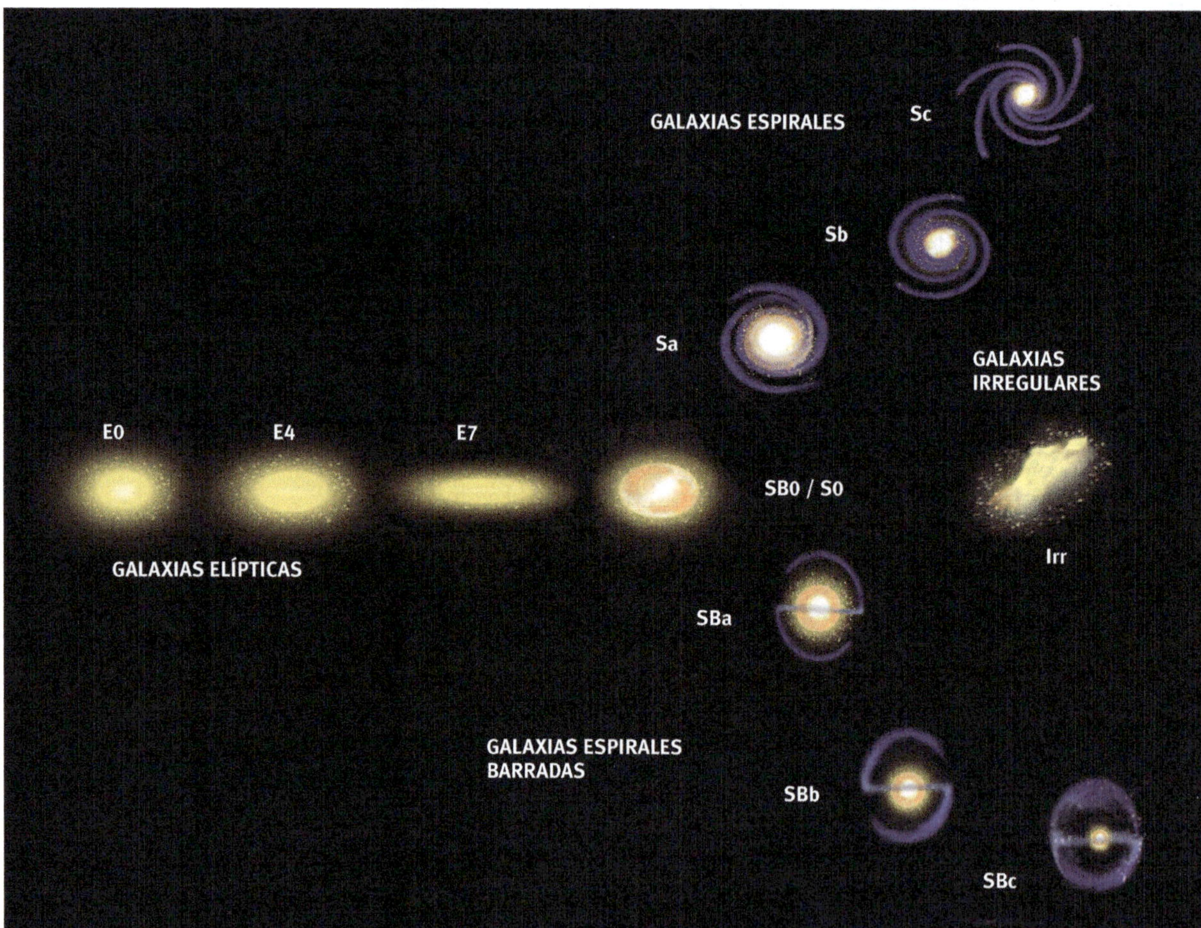

Clasificación morfológica de las galaxias

En la actualidad esta clasificación únicamente se aplica a las llamadas galaxias luminosas, habiéndose detectado muchos otros tipos galácticos, como las galaxias enanas, las de bajo brillo o las galaxias peculiares: las galaxias Seyfert, las radiogalaxias o los quásares.

Las galaxias no son elementos estáticos, sino que presentan tres movimientos característicos. La mayor parte de ellas presentan una rotación alrededor de un eje central. También tienen un movimiento de traslación respecto a otras galaxias de su entorno, que puede llegar a producir colisiones galácticas. Por último, las galaxias se desplazan alejándose las unas de las otras. Este desplazamiento fue una de las primeras indicaciones de un posible origen expansivo del universo.

Galaxias colisionando en El Ratón

Las extraordinarias dimensiones de las galaxias y de las distancias entre ellas obligan a utilizar como unidad de longitud el **año luz**. Un año luz equivale a 9 460 800 000 000 km.

Nuestro planeta se encuentra en la galaxia de la **Vía Láctea,** una galaxia espiral que tiene un diámetro de 100 000 años luz y un espesor de 20 000 años luz en su núcleo central.

La Vía Láctea presenta cuatro brazos que reciben los nombres de *Brazo de Sagitario-Carina, Brazo de Orión, Brazo de Perseo* y *Brazo del Cisne.*

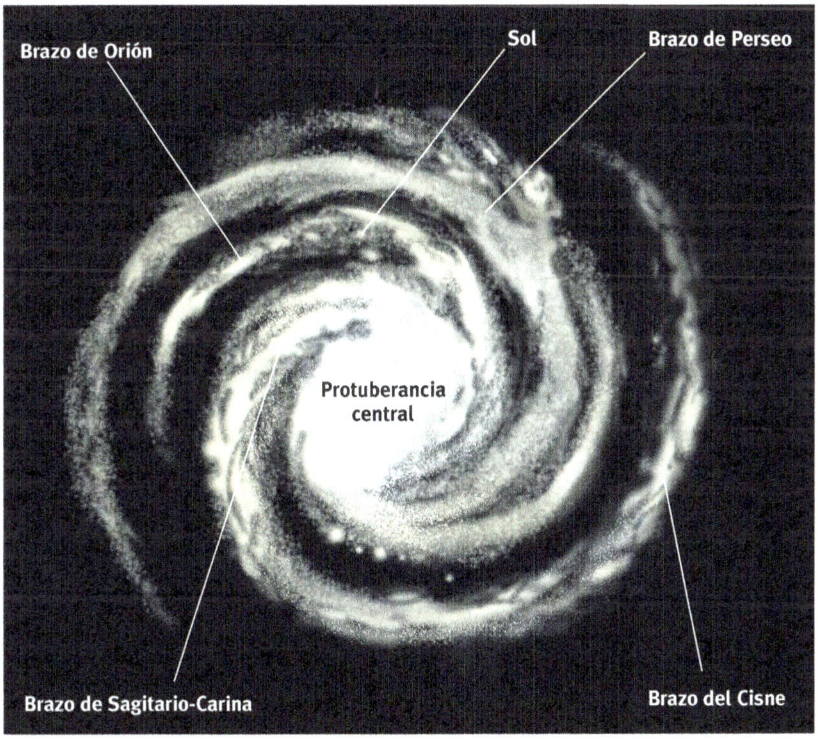

La Vía Láctea forma parte de un cúmulo que está constituido por aproximadamente 30 galaxias y que se conoce con el nombre de **Grupo Local.** En esta región se localizan galaxias irregulares, como la *Gran Nube de Magallanes,* que se encuentra a unos 170 000 años luz de la nuestra, o como la *Pequeña Nube de Magallanes* a 190 000 años luz. Además también se localizan diversas galaxias espirales, como *Andrómeda,* situada a 2 000 000 años luz.

El Grupo Local se agrupa con otros cúmulos de galaxias, como por ejemplo el *Quinteto de Stephan,* que juntos forman el **Supercúmulo de Virgo.**

+

Edwin Powell Hubble (1889-1953) fue un astrónomo estadounidense que demostró en 1923 que la nebulosa de Andrómeda era una galaxia diferenciada de la Vía Láctea, a diferencia de lo que se creía hasta el momento.

Continuando su estudio de las galaxias, llegó a medir el movimiento de expansión, que las aleja las unas de las otras y puso las bases de la teoría del big bang, calculando de forma aproximada el tamaño y la edad del universo. En su honor se bautizó al telescopio espacial Hubble.

4.1. *Las estrellas*

Las estrellas son enormes concentraciones de materia, brillantes y calientes. En la clasificación de los cuerpos celestes realizada por la Unión Astronómica Internacional, quedarían incluidas en la categoría de los objetos primarios, juntamente con las enanas marrones y las subenanas marrones. Los objetos primarios están formados por el colapso gravitatorio de la materia de una nube interestelar.

Vamos a ver en la siguiente tabla las principales características de los objetos primarios:

DENOMINACIÓN	MASA (EN MASAS DE JÚPITER)	CARACTERÍSTICAS
Estrella	m>80	Mantiene reacciones estables de fusión nuclear del hidrógeno.
Enana marrón	13<m<80	Presentan reacciones de fusión nuclear de deuterio durante un período breve de tiempo.
Subenana marrón	m<13	

+

La fuerza gravitatoria es la responsable de la atracción que ejercen los cuerpos con masa entre ellos.

Si dos partículas se atraen y colisionan, la masa resultante, mayor, actuará sobre otras partículas más lejanas, repitiendo el proceso y conduciendo a la formación de núcleos masivos. Todo ello comporta la reducción de tamaño de la nebulosa, el incremento de su densidad y el inicio de la rotación. Es el **colapso gravitatorio.**

La reacción de fusión nuclear genera una enorme cantidad de energía, que será la responsable de la temperatura y el brillo de la estrella.

Por lo tanto, la estrella se verá sometida a dos fuerzas fundamentalmente: la fuerza gravitatoria, que la comprimirá hacia su centro y generará las altas presiones que permitirán la fusión nuclear, y la fuerza expansiva originada por la energía liberada. En una estrella como el Sol estas dos fuerzas están equilibradas.

C

La **fusión nuclear** es un proceso en el cual dos núcleos preexistentes chocan y se unen dando lugar a un núcleo mayor. Para que este proceso pueda realizarse, son necesarias unas condiciones de presión y temperatura extremadamente altas, de forma que pueda vencerse la repulsión entre ellos. La reacción fundamental es la que une dos núcleos de hidrógeno, dando lugar a uno de helio.

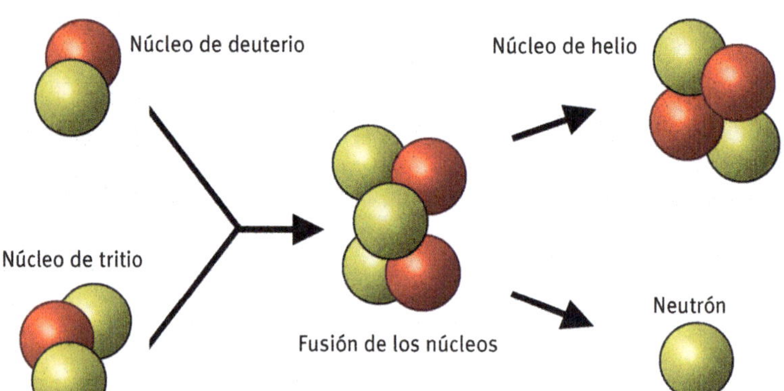

$^{2}_{1}H$ (deuterio) + $^{3}_{1}H$ (tritio)→ $^{4}_{2}He$ (helio) + $^{1}_{0}n$

Cuando el combustible nuclear de la estrella se agota, la estrella inicia un nuevo período de reacciones utilizando los nuevos núcleos generados anteriormente. En esta fase, la relación entre las fuerzas gravitatorias y expansivas varía, la estrella aumenta de volumen y su color cambia al rojo. También cambiará su luminosidad, que corresponde a la cantidad de energía que emite por unidad de tiempo.

La temperatura superficial determina el color de las estrellas. Así, si es elevada, desprenden una luz azulada y, si es baja, una luz rojiza. Según su color, las estrellas se clasifican en siete tipos, que se distinguen con una letra: azul (O), blanco azulado (B), blanco (A), amarillo blanquecino (F), amarillo (G), naranja (K) y rojo (M). En el **diagrama de Hertzsprung-Rusell** aparecen clasificadas según su color y su luminosidad.

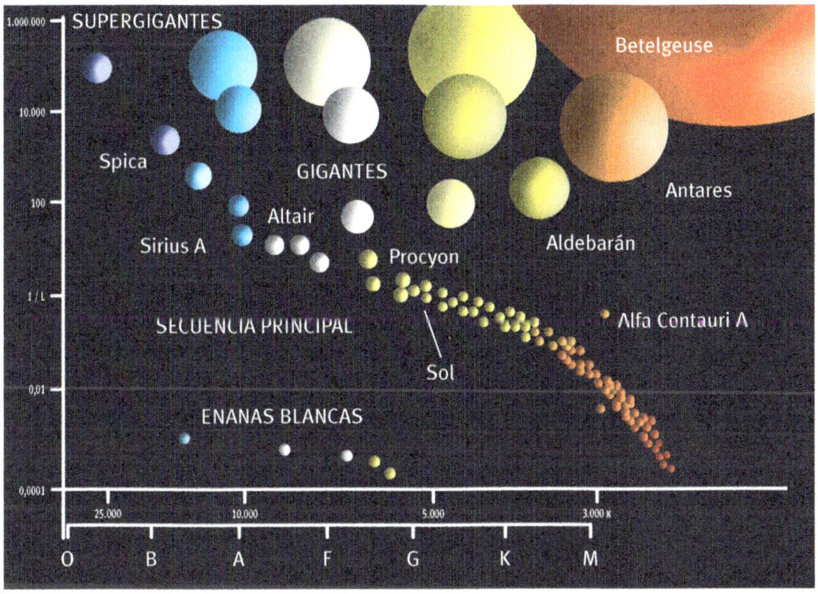

El diagrama de Herzsprung-Rusell relaciona la temperatura y la luminosidad de las estrellas. La luminosidad se indica dando el valor de 1 a la luminosidad del Sol.

+

Cuando una estrella de gran masa se colapsa sobre ella misma por efecto de la fuerza gravitatoria no existe una presión capaz de frenar esta caída hacia su interior: continuará comprimiéndose hasta tamaño cero y densidad infinita, formando lo que se llama una singularidad.

La región del espacio-tiempo que queda a su alrededor es un **agujero negro**. Nada, ni siquiera la luz, puede escapar de su interior debido a la intensidad de su gravedad.

En sus fases finales, si la estrella es relativamente pequeña, menor de 1,4 veces la masa del Sol, el fin de las reacciones nucleares provoca que se condense por acción de la gravedad hasta dar lugar a una **enana blanca;** en el proceso se libera una burbuja de gas, la nebulosa planetaria. El enfriamiento del núcleo estelar la convertirá en una **enana negra,** sin brillo.

Las estrellas mayores consumen rápidamente el hidrógeno, tras lo cual utiliza el helio para dar lugar a elementos más pesados en reacciones que liberan menos energía. Como consecuencia, comienzan a colapsarse tras una violenta explosión denominada **supernova.** La mayor parte de los materiales sintetizados a lo largo de su vida son expulsados al espacio, quedando sólo un núcleo muy denso, que puede ser una **estrella de neutrones** o un **agujero negro,** dependiendo de la masa inicial.

Residuos expulsados al espacio por una supernova

En el siguiente esquema podemos ver las fases de la evolución estelar en los dos casos: cuando la masa estelar es inferior a 1,4 veces la masa del Sol y cuando es superior.

4.2. *Las nebulosas*

Son nubes de gas, principalmente de hidrógeno y de helio. También contienen polvo formado por partículas de carbono, hierro y silicio, recubiertas por amoníaco, metano o agua.

El polvo procede del material que se desprende de las estrellas en las últimas etapas de su actividad.

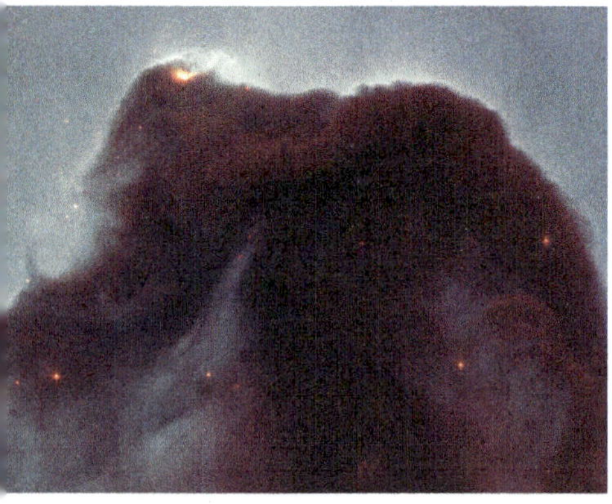

La nebulosa de la Cabeza del Caballo es una nebulosa de absorción.

Existen tres tipos de nebulosas:

- **De absorción.** No emiten luz, pero podemos observar sus siluetas oscuras gracias a que absorben la luz de las estrellas situadas detrás. Es el caso de la nebulosa del Saco de Carbón.
- **De reflexión.** Tampoco emiten luz, pero reflejan la luz de las estrellas más cercanas y por ello se observan tenuemente, como sucede con la nebulosa de las Pléyades.
- **De emisión.** Emiten una luz rosada procedente de alguna estrella muy activa que se encuentra en su interior. La nebulosa del Pelícano es un ejemplo.

ACTIVIDADES

- **14.** ¿Cuáles son los componentes de las galaxias?
- **15.** ¿En qué categorías morfológicas se dividen las galaxias?
- **16.** Situa nuestro Sol en el diagrama Hertzsprung-Russell, sabiendo que tiene una temperatura superficial de 5800 K y un magnitud visual absoluta de 4,79 mag. ¿A qué tipo estelar corresponde?
- **17.** ¿Podrías decir cuál sería la evolución de una estrella con una masa equivalente a la de 15 soles?
- **18.** ¿Cuál es la diferencia principal entre una nebulosa de emisión y una de reflexión?

5. El Sistema Solar

Hasta 1995, en que Didier y Queloz identificaron por vez primera un planeta orbitando alrededor de la estrella 51 Pegasi, el Sistema Solar era el único sistema planetario conocido. Hoy en día han sido descubiertos más de 200 mundos que orbitan alrededor de multitud de estrellas, la mayor parte similares al Sol, lo que nos hace pensar que estos sistemas son fenómenos asociados comúnmente a este tipo estelar.

C

Para medir la distancia entre un planeta y su estrella primaria se hace servir la **ua** o **unidad astronómica.** Una unidad astronómica de distancia equivale a $1{,}495 \cdot 10^{11}$ m, o lo que es lo mismo, la distancia media entre la Tierra y el Sol.

ALGUNOS PLANETAS EXTRASOLARES

Estrella a la que pertenecen	Masa (masas de Júpiter)	Semieje orbital (ua)	Período orbital
51 Pegasi	0,44	0,05	4,2 días
55 Cancri	0,85	0,11	14,6 días
τ Bootis	3,0	0,05	3,3 días
ρ Coronae Borealis	1,13	0,23	39,6 días
47 Ursae Majoris	2,3	2,11	3 años
Gliese 876	2,1	0,2	61,1 días

El Sistema Solar está formado por una estrella y un grupo de astros que giran a su alrededor. Tiene una estructura aplanada y se encuentra en uno de los brazos de la Vía Láctea, el Brazo de Orión, situado a unos 28 000 años luz del centro de la galaxia. El **Sol** es la estrella alrededor de la cual giran los astros que integran este sistema.

En la actualidad se cree que la formación de los sistemas planetarios se da paralelamente a la formación de la estrella entorno a la cual orbitan. Mientras la estrella se colapsa sobre ella misma, una cierta cantidad de gas y polvo cósmico forma un disco aplanado en torno a su ecuador: el disco protoplanetario. Muchos de estos discos han sido observados en fecha reciente alrededor de estrellas en las proximidades de Orión.

Formación del disco protoplanetario alrededor del protosol. A partir de una distribución homogénea de la materia en el disco protoestelar, aparecen concentraciones de masa que darán lugar a la formación de los planetas.

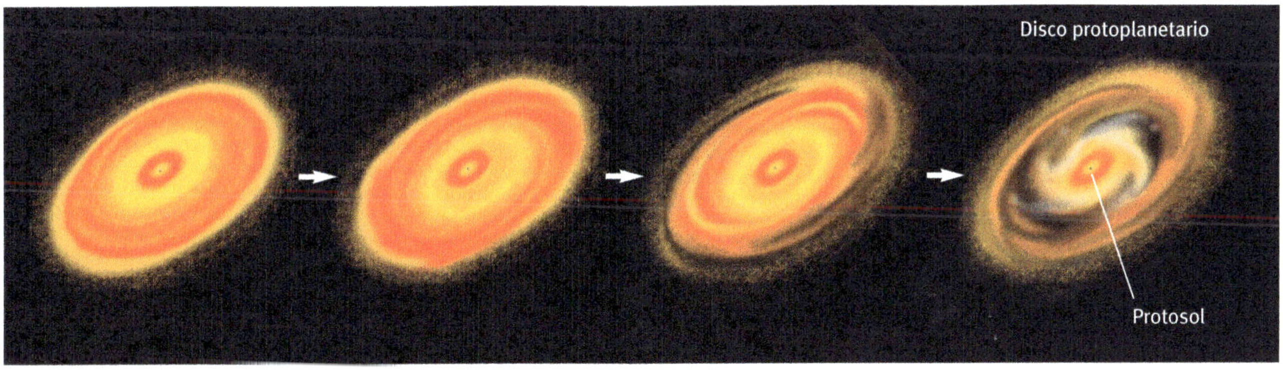

La materia que forma el disco colisiona entre ella formando agregados que crecen en tamaño y masa hasta dar lugar a pequeños cuerpos, los planetesimales. Éstos crecen uniéndose entre ellos y forman un núcleo de roca y metal. Si su masa es suficiente y se hallan relativamente alejados de su estrella, pueden capturar gas y envolverse de una atmósfera.

Los sistemas planetarios como el Sistema Solar están formados, por tanto, por los elementos que la Unión Astronómica Internacional ha caracterizado como objetos **secundarios, terciarios** y **escombros.**

OBJETOS SECUNDARIOS **Planetas**	Se forman por acreción de granos de polvo en un disco que gira alrededor de una estrella. Capturan la mayor parte del material de su entorno en la fase de crecimiento, dejando libre su órbita. Tienen forma esférica, y a partir de una cierta masa pueden atraer una atmósfera densa. El Sistema Solar se compone de 8 planetas: Mercurio, Venus, Marte, Tierra, Júpiter, Saturno, Urano y Neptuno
OBJETOS TERCIARIOS **Satélites**	Son cuerpos que orbitan en torno a los planetas, o bien porque se formaron allí o bien porque fueron capturados por la gravedad de éstos. Hay multitud de satélites en el Sistema Solar: sólo Saturno ya tiene 18. La Luna es el satélite del planeta Tierra.
ESCOMBROS **Asteroides, cometas y objetos del cinturón de Kuiper**	Inician su formación como los objetos secundarios, pero su crecimiento se detiene. Los asteroides están formados por material rocoso, mientras que los objetos del cinturón de Kuiper, más allá de Neptuno, son pequeñas masas de hielo. Cuando sus órbitas los llevan cerca del Sol, se convierten en cometas.

5.1. *El Sol*

El Sol es la estrella más cercana a la Tierra y, por esta razón, es la más estudiada y conocida. Se formó hace 5000 millones de años y, a partir de los datos que se conocen sobre él, se ha calculado que seguirá brillando unos 5000 millones de años más.

Es una estrella amarilla, de tipo G2, y, por tanto, es de mediana edad. Su temperatura superficial es de 6000 °C aproximadamente y en su interior puede alcanzar los 15 millones de grados. La luz y el calor que emite hacen posible la vida en nuestro planeta.

El siguiente cuadro nos muestra algunas de sus características.

DISTANCIA A LA TIERRA	149 600 000 km	**MASA**	1,989 · 10^{30} kg
DISTANCIA AL CENTRO DE LA VÍA LÁCTEA	28 000 años luz	**PERÍODO DE ROTACIÓN**	25,38 días terrestres
DIÁMETRO	1 393 000 km	**PERÍODO DE TRASLACIÓN**	220 millones de años terrestres

El Sol se compone de varias capas concéntricas que, ordenadas de más interna a más externa, son: *núcleo, zona radiativa, zona convectiva, fotosfera, cromosfera* y *corona.*

INTERIOR SOLAR	**NÚCLEO** Tiene un radio aproximado de 500 000 km y contiene aproximadamente el 60% de la masa solar. En esta zona se dan las reacciones de fusión nuclear. **ZONA RADIATIVA** Ocupa desde los 500 000 km hasta 1 000 000 de km aproximadamente. La energía generada en el núcleo solar atraviesa esta capa en forma de radiación, en un proceso que tarda 10 000 000 de años. **ZONA CONVECTIVA** Abarca desde el final de la zona radiativa hasta la superficie del Sol. En ella la energía se transmite hacia el exterior por la circulación de la materia solar.
SUPERFICIE SOLAR	**FOTOSFERA** Constituye la superficie visible. Está formada por gas muy tenue.Tiene un espesor de unos 400 km y en ella se produce un descenso de temperatura desde los 8000 K a los 4500 K. En ella se localizan las manchas solares.
ATMÓSFERA SOLAR	**CROMOSFERA** Está compuesta de gas muy tenue y tiene un grosor de unos 500 km. A medida que se aleja de la fotosfera se produce un incremento de temperatura. En esta capa se originan las protuberancias solares, erupciones de brillo muy intenso que pueden llegar a medir más de 14 000 km. **CORONA** Es la región más externa y menos densa del Sol. Está formada por gases muy poco densos altamente ionizados. Alcanza hasta una decena de radios solares desde la superficie.

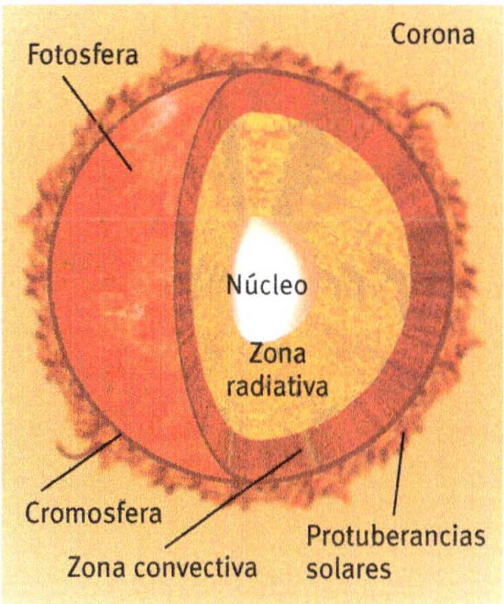

El tamaño del Sol permite predecir que, cuando llegue a su fase final de actividad, se convertirá primero en una estrella roja, después en una enana blanca y, posteriormente, en una enana negra. El fin del Sol implicará la desaparición del Sistema Solar.

5.2. *Los planetas*

Los planetas del Sistema Solar suelen clasificarse en dos grupos, según su proximidad al Sol, sus características físicas y el número de satélites que tengan.

- **Los planetas interiores o terrestres.** Son los más próximos al Sol. Son Mercurio, Venus, la Tierra y Marte. Tienen un tamaño pequeño; son sólidos y de aspecto rocoso. Poseen pocos o ningún satélite.

  Están formados por elementos como el hierro o el silicio, y poseen poco hidrógeno y helio, por lo que sus densidades son superiores o iguales a 4 g/cm^3.

C

Todos los astros del Sistema Solar presentan dos tipos de movimiento:

- **Rotación.** Es el giro de los astros sobre sí mismos.
- **Traslación.** Es el recorrido que describen los astros alrededor de otro astro.

Si se miran desde el polo Norte celeste, los planetas presentan en su mayoría movimientos de traslación y rotación en sentido antihorario. Si el movimiento se da en sentido opuesto, se denomina retrógrado y se identifica con un signo negativo.

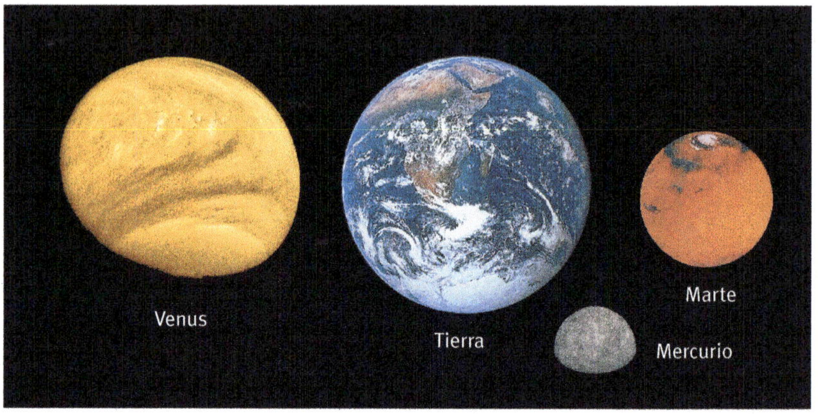

	MERCURIO	VENUS	TIERRA	MARTE
Distancia al Sol (ua)	0,3871	0,7233	1	1,5237
Diámetro ecuatorial (km)	4879,4	12 103,8	12 756,3	6794
Masa (kg)	$3{,}30 \cdot 10^{23}$	$4{,}87 \cdot 10^{24}$	$5{,}97 \cdot 10^{24}$	$6{,}42 \cdot 10^{23}$
Gravedad superficial (G)	0,37	0,88	1	0,38
Temperatura media (°C)	260	255	22	- 23
Período de rotación	58,65 días	- 243,01 días	23,93 horas	24,62 horas
Período de traslación	0,24 años	0,61 años	1 año	1,88 años
Satélites	0	0	1	2

- **Los planetas exteriores o gigantes.** Son los más alejados del Sol. Son Júpiter, Saturno, Urano y Neptuno. Son mucho mayores que los planetas de tipo terrestre, y poseen enormes atmósferas formadas básicamente de hidrógeno y helio, que envuelven un núcleo sólido. Su densidad es muy baja: 1,5 g/cm^3. Tienen un gran número de satélites y presentan anillos, es decir, grupos de partículas de polvo, rocas y hielo, que giran alrededor de los planetas.

  Hasta 1977 se creyó que sólo Saturno poseía un sistema de anillos concéntricos, pero posteriormente se ha observado su presencia en el resto de los planetas exteriores.

	JÚPITER	SATURNO	URANO	NEPTUNO
Distancia al Sol (ua)	5,20	9,55	19,21	30,11
Diámetro ecuatorial (km)	142 984	120 536	51 118	49 528
Masa (kg)	$1{,}90 \cdot 10^{27}$	$5{,}69 \cdot 10^{26}$	$8{,}70 \cdot 10^{25}$	$1{,}03 \cdot 10^{26}$
Gravedad superficial (G)	2,64	1,15	1,17	1,18
Temperatura media (°C)	– 150	– 180	– 210	– 220
Período de rotación	9925 horas	10 656 horas	– 17,24 horas	16,11 horas
Período de traslación	11,85 años	29,42 años	83,75 años	163,72 años
Satélites	16	18	17	8

5.3. *Los satélites*

Son astros que se formaron a partir de la concentración de las partículas procedentes de la nebulosa inicial y quedaron alrededor de los planetas. Además de tener un movimiento de rotación sobre sí mismos, presentan un movimiento de traslación alrededor de su planeta y, junto con él, en torno al Sol.

De los 40 satélites detectados, el más estudiado es la Luna, situada a 384 400 km de distancia. En su superficie se observan diversas formaciones: los **mares,** grandes llanuras formadas por rocas oscuras; las **tierras,** zonas de rocas más claras y más antiguas, y los **cráteres,** debidos a numerosos impactos de fragmentos rocosos procedentes del Sistema Solar.

+

En la actualidad, los satélites de los planetas exteriores se han convertido en uno de los lugares que mayor interés despierta entre los astrónomos. Mundos como Europa, con un enorme océano cubierto de hielo, presentan ambientes en los que tal vez pudiera haberse desarrollado la vida.

LOS MAYORES SATÉLITES DEL SISTEMA SOLAR

Planeta	Satélite	Diámetro (km)	Distancia al planeta (en miles de km)	Período orbital
Tierra	Luna	3476	384,4	27,32 días
Júpiter	Ganímedes	5262	1070	7145 días
	Calisto	4800	1883	16 689 días
Saturno	Rea	1530	527	4518 días
	Tetis	5150	1222	15 945 días
Urano	Titania	1580	436	8706 días
Neptuno	Tritón	2706	355	– 5877 días

Ceres es el mayor de los asteroides del sistema solar, con 933 km de diámetro. Presenta una forma casi esférica. Cuando fue descubierto en 1801 por Giuseppe Piazzi fue considerado un nuevo planeta, y así se mantuvo hasta 1852.

5.4. Asteroides, cometas y meteoritos

Los planetas y los satélites no son los únicos astros que forman el Sistema Solar.

Los *asteroides* y los *cometas* también pertenecen a este sistema planetario, al igual que las rocas de diversos tamaños, procedentes de su fragmentación, denominadas *meteoritos*.

Los asteroides

Son, en general, más pequeños que los planetas y los satélites. Los de menor tamaño son irregulares y los mayores tienen forma esférica.

Estos astros proceden de planetesimales que no llegaron a integrarse en ningún planeta y quedaron girando alrededor del Sol.

Entre las órbitas de Marte y Júpiter se encuentra el cinturón principal de asteroides, mientras que, compartiendo la órbita con Júpiter y controlados por su gravedad, existen los llamados asteroides Troyanos. Probablemente Fobos y Deimos, los satélites de Marte, sean dos asteroides capturados por el planeta.

Los cometas

Tienen el mismo origen que los asteroides y su forma característica los convierte en astros muy populares. Están compuestos por una mezcla congelada de agua, amoníaco y dióxido de carbono, junto con partículas sólidas de polvo.

Entre las órbitas de Saturno y Neptuno se halla un conjunto de cometas denominados Centauros, el mayor de los cuales es Quirón. Aún más exterior es el cinturón de Kuiper, que parece ser la fuente de la mayor parte de los cometas que aparecen periódicamente en el cielo, como el Halley o el Hale-Bopp.

Plutón, que hasta 2006 había sido considerado el noveno planeta del sistema solar, es un objeto más de la multitud que forma este cinturón.

Los cometas describen una órbita muy elíptica alrededor del Sol, de modo que, cuando pasan muy cerca de él, el calor hace que parte de sus componentes se convierta en gas y se desprendan partículas de polvo. En este punto de su recorrido se pueden observar en los cometas: el núcleo, la cabellera y la cola.

Cuando el cometa se aleja del Sol, se va enfriando y se convierte de nuevo en un cuerpo sólido.

Los meteoritos

Son, en general, fragmentos procedentes de asteroides o cometas que caen sobre la superficie de los planetas o de otros astros. Pueden tener el tamaño de un grano de arena, o bien, un diámetro de varios centenares de kilómetros.

Cuando entran en contacto con la atmósfera terrestre, muchos de ellos se ponen incandescentes debido a la fricción y dejan un rastro luminoso. Éste es el origen de las llamadas estrellas fugaces. Normalmente se desintegran antes de llegar a la superficie terrestre, pero si son de gran tamaño pueden conservar parte de su masa y caer sobre la superficie de nuestro planeta. En este caso se denominan *bólidos*.

También se consideran meteoritos algunos fragmentos de la Luna y de Marte que se han localizado en la Tierra, y que se originaron por el impacto de asteroides contra estos astros.

+

En 1996 un equipo de científicos anunció que había hallado indicios de actividad biológica fosilizada en el meteorito ALH84001, recogido en la Antártida y proveniente de Marte.

¿Y PLUTÓN?

En 1930, el astrónomo Clyde Tombaugh descubrió más allá de la órbita de Neptuno un nuevo objeto al que se identificó como el 'planeta X', y que posteriormente fue bautizado con el nombre de Plutón. El nuevo objeto cruzaba en su órbita la de Neptuno, y fue considerado durante 60 años como un planeta extraño en los confines de nuestro Sistema Solar: menor que la Luna, sólido, con una gran excentricidad...

Cuando se descubrió el cinturón de Kuiper, que incorporaba gran cantidad de objetos, empezó a plantearse la posibilidad de que Plutón no fuera más que uno de sus componentes, al igual que Ceres había pasado en 1852 a ser un elemento del cinturón de asteroides.

El descubrimiento en 2003 de Eris, un objeto del cinturón de Kuiper mayor que Plutón, llevó finalmente a la exclusión de este último del selecto club de los planetas, reduciéndolo a ocho, e incorporando a Plutón al grupo de los componentes del cinturón, los cuales están rodeados de objetos con tamaños semejantes al suyo.

ACTIVIDADES

19. ¿Se convertirá el Sol al final de su vida en un agujero negro? ¿Por qué?
20. ¿Que diferencia existe entre un planeta y un satélite? ¿Y entre un planeta y un asteroide? ¿Y entre un asteroide y un cometa?
21. ¿Qué son los planetesimales? ¿Cuál es el origen de éstos?
22. Explica las diferencias entre los planetas interiores y los exteriores.
23. Ganímedes es más grande que Mercurio. ¿Tendría que ser calificado como planeta? ¿Por qué?
24. ¿Cuál es el planeta que presenta los días más largos? ¿Y el que los tiene más cortos?
25. ¿En qué se diferencia un asteroide de un meteorito?

La atmósfera terrestre se originó a partir de los gases liberados por la actividad volcánica.

6. Origen y evolución de la Tierra

A medida que los protoplanetas chocaban entre ellos en el proceso de formación de la Tierra primitiva, la energía liberada por los impactos, así como la desintegración de los elementos radioactivos, incrementaron la temperatura, provocando que los materiales que formaban el planeta se fundieran. De esta manera se produjo una segregación durante la cual los componentes más pesados se desplazaron hacia el centro por efecto de la gravedad, mientras que los más ligeros quedaron en las capas exteriores.

Poco a poco, estos materiales fueron enfriándose y solidificándose, produciendo una delgada corteza. La dinámica interna del planeta condujo a una serie de episodios volcánicos continuados, durante los cuales se liberaron al exterior los gases generados por el magma, tales como CO_2, SO_2, compuestos de nitrógeno y vapor de agua. Este último se condensó en la atmósfera, originando una capa nubosa de gran potencia, que precipitó en forma de lluvia sobre la superficie y originó los océanos.

En la Tierra primitiva, con una atmósfera reductora, pobre en oxígeno, podrían haberse formado una gran variedad de compuestos orgánicos. Algunos científicos afirman que, sólo con la energía que se libera actualmente en la Tierra en forma de relámpagos, las reacciones en la atmósfera habrían podido formar en 100 000 años materia orgánica suficiente como para cubrir la superficie de la Tierra con un espesor de un metro. Éstas moléculas, arrastradas por la lluvia, se acumularían en los océanos, que quedarían convertidos en una suspensión de materia orgánica: algo parecido a un caldo, el caldo primigenio.

+

Existen otras versiones que pretenden explicar la formación de la materia orgánica en la Tierra previa a la aparición de los seres vivos. Según la panspermia, la materia orgánica llegó por el impacto de un cometa. Se ha confirmado la presencia de compuestos complejos de carbono en los núcleos de los cometas y en la materia interestelar.

FORMACIÓN DE LA GLICINA

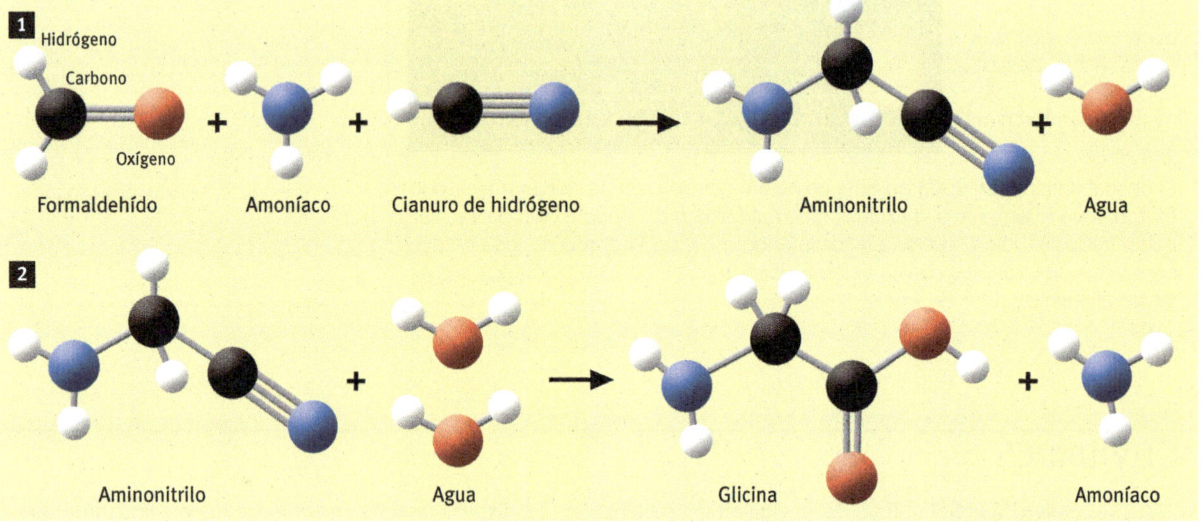

Las reacciones químicas entre los gases de la atmósfera primitiva habrían originado las primeras moléculas orgánicas entre las que habría aminoácidos como la glicina.

A partir de este punto se iniciaría un proceso de evolución química, a través del cual las moléculas orgánicas se irían perfeccionando, creando un sistema de autorreplicación basado en los ácidos nucleicos (ADN y ARN). De este modo, las moléculas orgánicas crearían un entorno propio en el que se aislarían gracias a capas de sustancias grasas o lípidos, que delimitarían los primeros protoorganismos.

Estos primeros organismos, semejantes a las bacterias actuales, se alimentarían de la materia orgánica presente en el agua de los océanos hasta que ésta dejó de ser abundante. Entonces, obligados a buscar nuevas estrategias de obtención de nutrientes, un grupo de seres vivos descubriría una reacción que habría de cambiar radicalmente el aspecto del planeta: la fotosíntesis.

La captación de CO_2 y la liberación a la atmósfera de O_2 por parte de los organismos autótrofos tendrá una serie de consecuencias que marcarán la evolución del planeta y de sus pobladores.

El descenso de la concentración atmosférica de CO_2 provoca	El incremento de la concentración atmosférica de O_2 produce
— Disminución del efecto invernadero — Bajada de las temperaturas en la superficie del planeta — Futuras glaciaciones a escala local y planetaria	— Cambio de la química planetaria: reacciones de oxidación — Extinción de los organismos anaerobios o regresión a hábitats marginales — Posibilidad de metabolismos aerobios, más rentables energéticamente, y con ellos posibilidad del paso a la pluricelularidad — Formación de la capa de ozono y protección de la superficie de los rayos ultravioletas; se hace posible la colonización de la tierra fuera de los océanos

Si observamos las atmósferas de los otros dos planetas interiores que disponen de ella, Venus y Marte, observaremos que se basan fundamentalmente en CO_2 y N_2, tal y como se supone que era la primitiva atmósfera terrestre. La Tierra es el único planeta que conocemos con un porcentaje de oxígeno atmosférico cercano al 20%, producto de la actividad de los seres vivos.

6.1. *Métodos de datación y eras geológicas*

La datación de los acontecimientos geológicos que se han sucedido en el planeta Tierra se ha realizado de diversas maneras. En primer lugar cabe diferenciar entre la datación relativa y la datación absoluta.

La **datación relativa** nos informa de qué materiales son más antiguos que otros, pero no nos dice cuáles son sus edades. Por ejemplo, podemos saber que un fósil A situado en un estrato horizontal situado por encima de otro es más moderno que otro fósil B contenido en el estrato inferior, pero no obtenemos información de cuál es su edad.

Para realizar una **datación absoluta** se utilizan actualmente métodos de radiocronología. Estos métodos se basan en la capacidad de desintegración de los isótopos radiactivos de diferentes elementos. Los elementos radiactivos o padres se desintegran y dan lugar a los elementos radiogénicos. Así, en una determinada roca, a partir del momento en que se forma y a medida que pasa el tiempo, aumenta el porcentaje de los elementos radiogénicos y disminuye el de los radiactivos. Si conocemos la tasa de desintegración del elemento radiactivo en cuestión y los porcentajes en los que aparece en la roca junto con sus elementos hijos, podremos inferir cuál fue la época de formación de dicha roca.

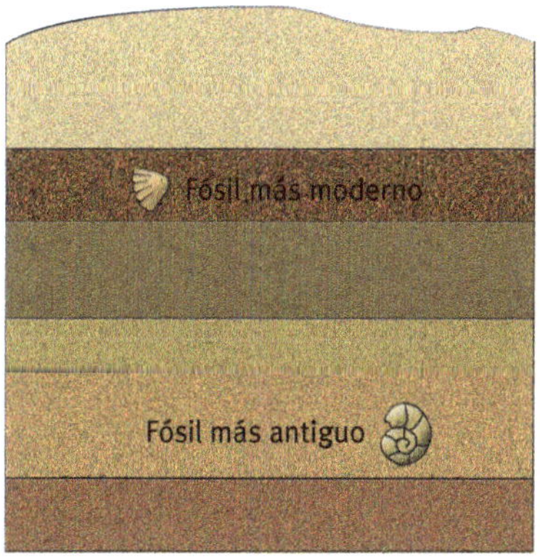

Datación relativa

Isótopo padre	Productos hijos estables	Vida media (millones de años)
U^{238}	Pb^{236} y He	4,5
U^{235}	Pb^{207} y He	0,71
Th^{232}	Pb^{208} y He	14
Rb^{84}	Sr^{87}	51
K^{40}	Ar^{40}, Ca^{40}	1,3

Cuando una roca ígnea se solidifica se forman minerales que atrapan isótopos radiactivos, los cuales iniciarán su desintegración hacia los elementos hijos estables. Los isótopos más utilizados en datación geológica son los de la tabla adjunta. La datación de las rocas nos permite establecer una escala de tiempo geológico sobre la cual situar los acontecimientos y etapas por los cuales ha pasado el planeta. Las eras y períodos en los que se divide la historia de la Tierra no tienen la misma duración, ya que esta escala se estableció en función de las formas de vida fósiles presentes en las diferentes rocas, estableciéndose posteriormente la datación absoluta de las mismas.

C

LA DATACIÓN POR CARBONO-14

Uno de los métodos de determinación de la edad de las sustancias orgánicas por radiocronología es el método del carbono-14. Éste es un isótopo radiactivo del carbono que se origina en las capas altas de la atmósfera y que es incorporado a lo largo de la vida por plantas y animales, manteniéndose una proporción constante entre el carbono-14 y el carbono-12 ordinario.

A partir de la muerte del organismo, la cantidad de carbono-14 empieza a disminuir porque se detiene su incorporación a la vez que va desintegrándose dando lugar a carbono-12. Conociendo la relación existente entre carbono-14 y carbono-12 en una muestra de materia orgánica se puede conocer de forma muy precisa la fecha de la muerte del organismo. El carbono-14 tiene una vida media muy breve, (5730 años) y por ello solo es aplicable a muestras de menos de 40 000 años de antigüedad.

ERAS GEOLÓGICAS DE LA HISTORIA DE LA TIERRA

Era	Período		Época	Inicio hace (m. a.)
CENOZOICO	CUATERNARIO		HOLOCENO	0,01
			PLEISTOCENO	1,8
	TERCIARIO	NEÓGENO	PLIOCENO	5
			MIOCENO	22,5
		PALEÓGENO	OLIGOCENO	37,5
			EOCENO	54,5
			PALEOCENO	65
MESOZOICO	CRETÁCICO		SUPERIOR	100
			INFERIOR	141
	JURÁSICO		MALM	160
			DOGGER	176
			LIAS	195
	TRIÁSICO		SUPERIOR (KEUPER)	212
			MEDIO (MUSCHELKALK)	223
			INFERIOR (BUNTSANDSTEIN)	230
PALEOZOICO	PÉRMICO		SUPERIOR	251
			INFERIOR	280
	CARBONÍFERO		SUPERIOR	325
			INFERIOR	245
	DEVÓNICO		SUPERIOR	360
			MEDIO	370
			INFERIOR	395
	SILÚRICO		SUPERIOR	423
			INFERIOR	435
	ORDOVÍCICO		SUPERIOR	450
			INFERIOR	500
	CÁMBRICO		SUPERIOR	515
			MEDIO	540
			INFERIOR	570
PRECÁMBRICO	ALGÓNQUICO			2650
	ARCAICO			4600

ACTIVIDADES

- **26.** ¿Qué orígenes tienen los componentes de la atmósfera del planeta Tierra? Explica brevemente los procesos que han llevado a su formación.
- **27.** ¿Cómo se originó la materia orgánica en el planeta Tierra?
- **28.** ¿Puede existir vida en un planeta sin oxígeno? Justifica tu respuesta.
- **29.** Comenta el impacto de la aparición de los organismos fotosintéticos sobre el planeta.
- **30.** ¿En qué se diferencia la datación absoluta de la relativa?

31 Lee el siguiente texto acerca de la expansión del universo y responde las cuestiones que se intercalan.

Intentando comprender la expansión del universo

Imagina que estás junto a una carretera y ves venir un automóvil desde lejos. Cuando pasa por delante de ti y se aleja, se produce un cambio en el sonido que percibes. ¿A qué es debido este cambio?

Imagina el automóvil en reposo. Las ondas sonoras que emite presentan unas crestas equidistantes, ya que la frecuencia del sonido es una determinada. En cambio, cuando el automóvil se desplaza, las ondas se apilan ante él mientras que detrás quedan más alejadas. Como consecuencia de ello, la frecuencia del sonido por delante será más alta que por detrás. Así, cuando el coche se acerque el sonido será más agudo, y cuando se aleje, más grave.

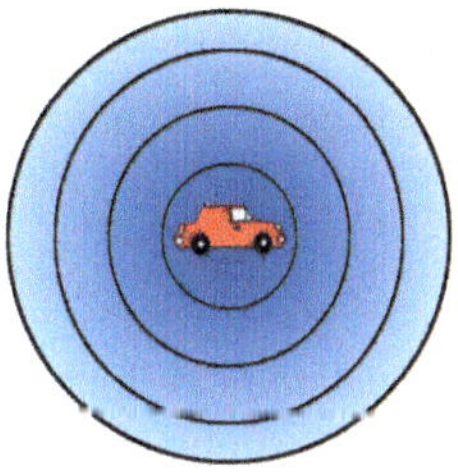

Con las ondas de luz que emiten las galaxias sucede algo parecido. En el caso de la luz visible, las frecuencias más altas corresponden a la luz azul, mientras que las más bajas corresponden a la luz roja. Cuando Hubble examinó la luz emitida por las galaxias descubrió que en todos los casos se halla desplazada hacia el rojo.

— ¿Qué conclusiones pueden extraerse de esta observación?

— Otra de las observaciones de Hubble fue que el desplazamiento hacia el rojo era tanto más intenso cuanto más alejada se hallaba una galaxia. ¿Qué puedes inferir de ello?

— ¿Que sucedería si invirtiéramos el sentido de desplazamiento de las galaxias? ¿Con qué teoría podemos relacionar esta situación?

Esta conclusión, el desplazamiento hacia el rojo de la luz emitida por las galaxias, parecía indicar que nuestra galaxia, la Vía Láctea, se encontraba en el centro del universo, y todas las galaxias se alejaban de ella ordenadamente. Sin embargo, esta posición central privilegiada no parecía demasiado compatible con una visión del universo en la cual habíamos sido desplazados del centro, primero de nuestro sistema solar, después de la galaxia... Por este motivo, el mismo Hubble se encargó de presentar una alternativa a esto. La pregunta que el científico debió formularse podría ser: ¿cómo se puede interpretar que todas las galaxias se alejen de nosotros, más rápidamente cuanto más lejos se encuentran, sin que nos hallemos en el centro del universo?

Para buscar una respuesta a esta pregunta vamos a realizar la siguiente experiencia:

— Toma un globo e hínchalo, pero no demasiado. Señala con un rotulador cinco puntos que llamaremos A, B, C, D y E distribuidos por la superficie, y que representarán cinco galaxias. Con una cinta métrica mide la distancia de cada uno de los puntos al punto A y anótala. Repite la misma operación con otro punto cualquiera, y anota también los resultados. Acaba de hinchar el globo y repite las medidas que tomaste anteriormente, tanto para el punto A como para el otro.

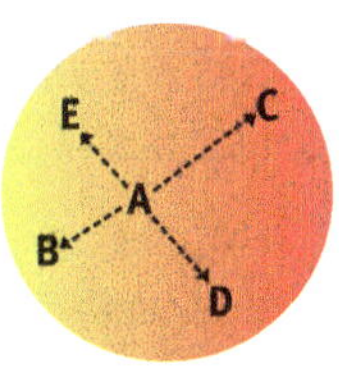

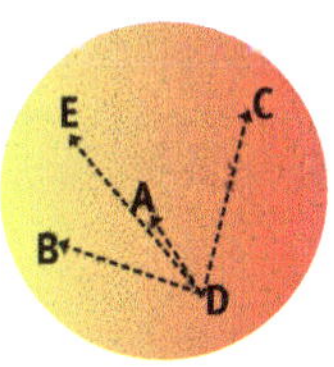

— ¿Cómo han variado las distancias referidas al punto A? ¿Podrías decir que un observador en la galaxia A vería que todas las otras galaxias se alejan de él?

— ¿Qué pasa con las medidas referentes al otro punto?

— Ten en cuenta que las medidas que has tomado se encuentran sobre la superficie del globo, y, por tanto, en dos dimensiones, mientras que el plano sobre el que se hallan los puntos está doblado sobre una tercera dimensión. ¿Qué implica extrapolar estos resultados a nuestro universo?

32 Uno de los principales obstáculos que se tuvo que superar en el proceso de aceptación del modelo heliocéntrico fue que la Tierra, el sistema de referencia firme y estable por antonomasia, pasara a presentar movimientos que, para el observador, no eran evidentes. El movimiento de rotación terrestre puede ser demostrado fácilmente mediante el péndulo de Foucault.

— Busca alguna información sobre este dispositivo. ¿Dónde fue construido el primero? ¿De qué manera demuestra el movimiento de rotación de la Tierra?

— La demostración del movimiento de traslación puede derivarse directamente del paralaje estelar. ¿Qué es el paralaje estelar? Aparte de demostrar el movimiento de traslación, ¿para qué se utiliza en astronomía el paralaje estelar?

33 En julio de 2007 se inauguró en el observatorio del Roque de los Muchachos, en la isla de La Palma, el Gran Telescopio de Canarias, el telescopio reflector mayor del mundo, con un espejo primario de 10,4 m de diámetro.

— Busca información sobre este observatorio y el Gran Telescopio de Canarias. ¿Por qué la isla de La Palma es un lugar privilegiado para instalar un telescopio?

34 Marte ha sido, desde siempre, uno de los planetas que más ha atraído la curiosidad de los seres humanos. Su proximidad a la Tierra, sus semejanzas con nuestro planeta y su color rojo en el firmamento nocturno han provocado un interés que desborda el que producen otros de nuestros vecinos en el Sistema Solar.

— Confecciona un friso cronológico con las diversas misiones que han sido enviadas a Marte y algunos de los resultados y datos obtenidos en ellas. ¿Existe algún proyecto de enviar una misión tripulada al planeta rojo? ¿Cuándo?

35 Una de las unidades de longitud más utilizadas en astronomía es el año luz. Una año luz es la distancia que recorre la luz en el vacío en un intervalo temporal de 365,25 días.

— Si sabemos que la velocidad de la luz en el vacío es de 299 792 458 m/s, ¿a cuántos kilómetros equivale un año luz?

— Sabiendo la distancia en km a la que equivale una ua, averigua cuánto tiempo tarda la luz del Sol en llagar a nuestro planeta.

36 ¿Podemos enviar sondas a otros sistemas solares? Gliese 876 es una estrella de la categoría de las enanas rojas que se halla a 15 años luz de distancia del Sol. Hasta el momento se le han descubierto tres planetas girando a su alrededor: un planeta de 7,5 veces la masa de la Tierra en una órbita de 2 días y dos mucho mayores en órbitas de 30 y 60 días, respectivamente.

— ¿Cuánto tardaría una sonda enviada desde el planeta Tierra en llegar a ese sistema para explorar los nuevos planetas? Busca información sobre la velocidad a la cual viajan las sondas espaciales y realiza los cálculos necesarios.

37 No podemos observar el Sol directamente, ya que dañaría nuestras retinas, pero podemos fabricar una cámara oscura siguiendo las indicaciones del dibujo.

— Partiendo de una caja de zapatos vacía, recorta un rectángulo en uno de los extremos y coloca un papel vegetal translúcido en el hueco resultante. En el lado opuesto haz un orificio, no excesivamente grande, ya que provocaría la desaparición del efecto de cámara oscura, ni muy pequeño, ya que daría una imagen muy débil del Sol. Tapamos la caja y apuntamos con ella al Sol. Al atravesar el orificio, los rayos, proyectarán una imagen del as-

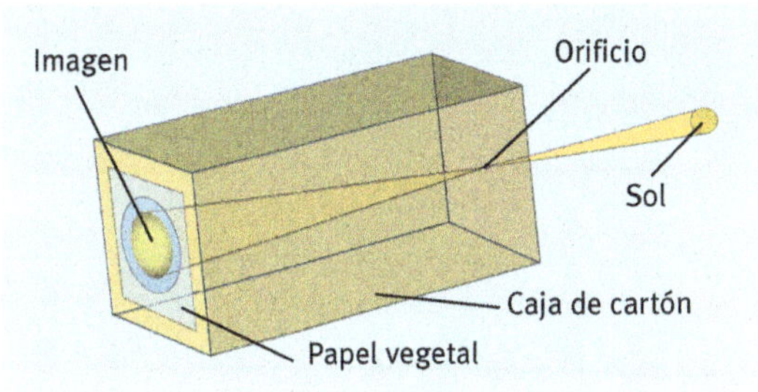

tro sobre la pantalla de papel vegetal. Con una sencilla cámara oscura, Johannes Kepler observó en 1607 las manchas solares.

38 A partir de los datos que hallarás en las tablas de la unidad, calcula la densidad de los ocho planetas del sistema solar y completa el cuadro siguiente.

MERCURIO	VENUS	TIERRA	MARTE

JÚPITER	SATURNO	URANO	NEPTUNO

— Representa los resultados obtenidos en una gráfica. ¿Qué observas en ella? ¿A qué se deben las diferencias que aparecen?

— ¿Podrías utilizar estos datos para establecer una clasificación de los planetas del sistema solar? Justifica la respuesta que has dado y la clasificación resultante.

39 A partir del valor de la gravedad superficial que te suministran las tablas de la unidad, ¿serías capaz de calcular cuál sería tu peso en la superficie de cada uno de los planetas del Sistema Solar? ¿Como crees que afectaría una gravedad mayor que la de la Tierra al aspecto de un planeta y sus habitantes? ¿Y una gravedad menor?

40 La Luna es, en relación al planeta al cual orbita, el satélite mayor de todo el Sistema Solar. Tanto es así que algunos astrónomos han llegado a referirse al sistema Tierra-Luna como a un planeta doble. Con la ayuda de unos prismáticos, intenta localizar algunos de los detalles más relevantes de la geografía lunar. Puedes guiarte por una foto como la adjunta. Una buena idea es comenzar identificando los mares y seguir por los cráteres más importantes. ¿Dónde se produjo el primer alunizaje?

41 Sir Edmund Halley afirmó que los cometas observados en 1531, 1607 y 1682 eran el mismo y predijo su retorno al considerarlo un fenómeno periódico, provocado por el paso del objeto por las proximidades del Sol. Éste se produjo según lo previsto, aunque el astrónomo no pudo verlo: había muerto dieciséis años antes. Desde entonces el cometa fue denominado con su nombre.

— ¿Podrías decir que año sucedió el primer retorno predicho del Halley? ¿Cuál es el período aproximado de este cometa?

— Recopila información sobre las sucesivas apariciones del cometa Halley. ¿Cuál ha sido la última? ¿Para qué fecha se espera nuevamente su visita?

42 Sitúa en la era y el período geológico correspondiente los siguientes acontecimientos:

— Aparición de los primeros seres vivos

— Extinción de los dinosaurios

— Aparición de los primeros seres humanos

— Aparición de los primeros mamíferos

— Primeras plantas con semillas

— Extinción masiva por incremento del oxígeno atmosférico

— Aparición de los primeros insectos

— Origen de los Alpes y el Himalaya

TIC

43 En la página web http://hubblesite.org/gallery/album/ encontrarás una colección de fotografías transmitidas por el telescopio orbital Hubble. Localiza entre ellas:

- Una nebulosa de absorción, una de reflexión y una de emisión.
- Una galaxia espiral, una espiral barrada, una elíptica y una irregular
- Una nebulosa planetaria
- Una supernova

Crea con estas fotografías un documento y comenta sus características.

2

Formación de cristales y rocas

Las rocas presentan una gran variedad de formas, colores y texturas diferentes. En la naturaleza podemos encontrar desde rocas con formas geométricas y cantos angulosos hasta otras de formas pulidas y redondeadas. También podemos encontrar colores muy diversos: existen rocas de color negro intenso, otras que presentan colores vivos y brillantes, e incluso algunas que pueden ser transparentes.

En cuanto a su localización, en algunos casos un determinado tipo de roca puede ser el principal componente de grandes extensiones de terreno de composición homogénea. Sin embargo, en otros casos en una pequeña parcela podemos encontrar rocas de características muy diferentes. ¿Qué fenómenos relacionados con el origen de las rocas pueden explicarnos estas situaciones?

Las rocas son producto de un proceso de cambio continuo debido a la acción de los agentes geológicos internos y externos. La acción de estos agentes, el tipo de material sobre el que actúan y las diferentes condiciones ambientales dan lugar a la gran diversidad de minerales y rocas presentes en la naturaleza.

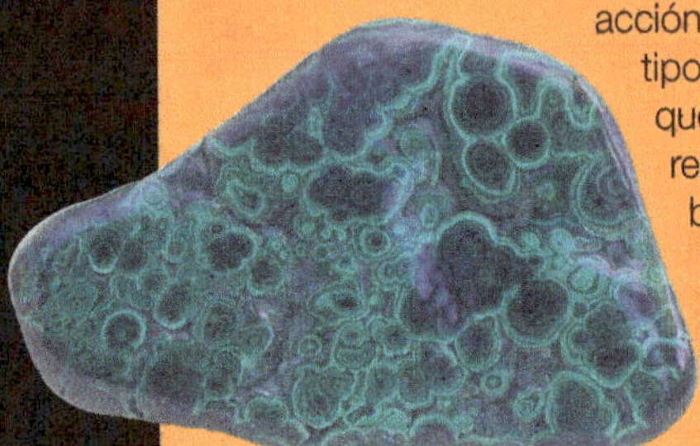

Cuevas de Valporquero (León).

ORGANIZACIÓN DE LOS CONTENIDOS

1. Cristalografía
- 1.1. Elementos de cristalografía
- 1.2. Cristalización
- 1.3. Sistemas cristalinos
- 1.4. Aplicaciones de los cristales

2. Ambientes petrogenéticos

3. El ambiente magmático
- 3.1. Tipos de magmas
- 3.2. Diferenciación magmática
- 3.3. Minerales de origen magmático
- 3.4. Clasificación de las rocas magmáticas
- 3.5. Texturas de las rocas magmáticas

4. El ambiente sedimentario
- 4.1. Ambientes de sedimentación
- 4.2. Litificación
- 4.3. Minerales de origen sedimentario
- 4.4. Clasificación de las rocas sedimentarias
- 4.5. Texturas de las rocas sedimentarias
- 4.6. Fosiles y fosilización

5. El ambiente metamórfico
- 5.1. Factores que intervienen en el metamorfismo
- 5.2. Clases de metamorfismo
- 5.3. Minerales de origen metamórfico
- 5.4. Texturas de las rocas metamórficas

OBJETIVOS

- Conocer los elementos de un cristal y su proceso de formación.
- Diferenciar los distintos ambientes y procesos petrogenéticos.
- Establecer una clasificación de las rocas magmáticas, sedimentarias y metamórficas y explicar sus procesos de formación.
- Conocer los diferentes minerales asociados a cada ambiente petrogenético.

+

La palabra griega *kristal* significa 'hielo'. Y, según cuenta la tradición, la utilizó el naturalista Plinio hace más de 2000 años para designar al cuarzo transparente de los Alpes, que en su época era considerado hielo definitivamente petrificado.

C

Un **mineral** es un material cristalino inorgánico de origen natural que tiene una composición química determinada, mientras que una **roca** es una mezcla consolidada de minerales.

Cristal de cuarzo

1. Cristalografía

En la materia mineral distinguimos entre materia mineral amorfa, sin una estructura interna, y materia mineral cristalina, con una estructura interna característica de cada mineral, también denominada estructura cristalina.

La cristalografía es la ciencia que estudia los sólidos cristalinos.

1.1. *Elementos de cristalografía*

La materia mineral cristalina está constituida por partículas materiales, es decir, átomos, iones o moléculas, que se encuentran unidas entre sí por enlaces químicos.

La disposición en el espacio de estas partículas materiales da lugar a la **celda elemental.** Ésta es un motivo geométrico, o poliedro, que se repite en las tres direcciones del espacio dando lugar a la estructura cristalina propia de cada mineral.

Las partículas materiales se encuentran colocadas en determinadas posiciones fijas del poliedro, como pueden ser los vértices, los centros de las caras y los puntos medios de las aristas.

Como consecuencia de esta disposición, distinguimos diferentes elementos cristalográficos geométricos:

- **Línea de puntos o fila de nudos.** Está formada por las partículas materiales que se encuentran en una misma arista o una diagonal del poliedro.
- **Planos reticulares.** Los planos reticulares están constituidos por todas las partículas materiales que se encuentran en el mismo plano.

  Sabemos, por geometría, que un plano puede definirse mediante dos rectas paralelas, dos rectas secantes o tres puntos no alineados. Por ello, análogamente, un plano reticular puede definirse mediante dos filas de nudos paralelas, dos filas de nudos secantes o tres nudos no alineados.
- **Conjunto reticular.** El conjunto reticular es la representación de la red o estructura cristalina. La repetición de una celda elemental en las tres direcciones del espacio da lugar al conjunto reticular.

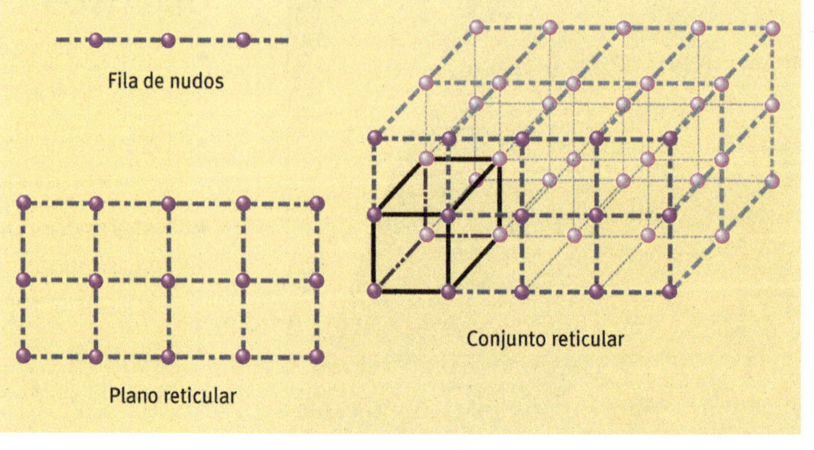

• **Ejes cristalinos.** Estos ejes son las tres filas de nudos que definen la celda elemental según las tres direcciones del espacio, a modo de los tres ejes de un sistema de coordenadas en el espacio. Estos ejes reciben el nombre de anteroposterior, transversal y vertical, y sirven para determinar la posición que ocupan las caras en los cristales.

• **Constantes cristalográficas.** Las longitudes de las aristas *a, b* y *c,* que definen la celda elemental, y los ángulos que forman entre ellas, α, β y γ, se consideran constantes para cada celda elemental y se denominan constantes cristalográficas.

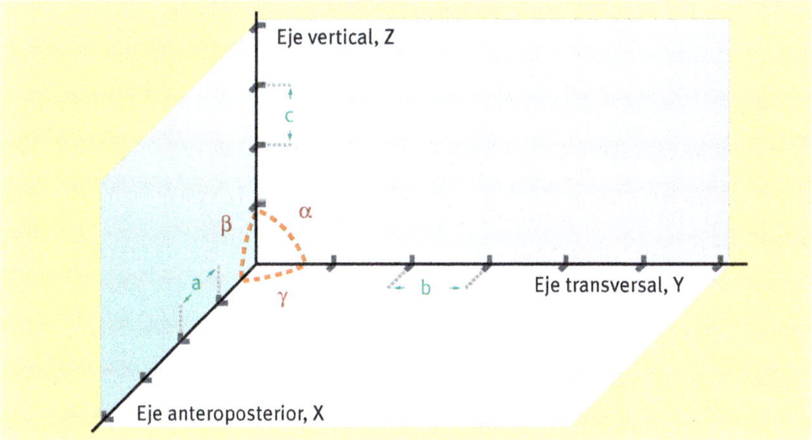

1.2. *Cristalización*

La **cristalización** es el proceso de ordenación de las partículas materiales de un mineral a partir de su estado fluido; esta ordenación tiene como resultado la formación de cristales. Se debe diferenciar este concepto del de **solidificación,** puesto que este último consiste en el paso de un cuerpo de estado líquido a estado sólido por enfriamiento y variación adecuada de la presión. Cuando un cuerpo formado por materia mineral cristalina se solidifica, nos encontramos ante una cristalización.

Algunos minerales pueden formarse por un proceso de **recristalización,** consistente en la formación de una nueva estructura cristalina por la modificación de la estructura cristalina inicial.

Los cristales pueden originarse mediante tres mecanismos distintos:

• A partir de disoluciones, como los cristales de sal al evaporarse el mar.

• A partir de materiales fundidos, como en el enfriamiento de un magma.

• Por disminución de la temperatura en sustancias en estado gaseoso. Es el caso de la formación de los cristales de hielo de la nieve al enfriarse el aire saturado de vapor de agua.

En cualquier caso, la cristalización se produce por la incorporación de partículas materiales a los **núcleos de cristalización.** Son pequeños cristales compuestos por numerosas partículas minerales dispuestas ordenadamente. Estos núcleos de cristalización son capaces de acumular sobre ellos, y siguiendo la misma ordenación, otras partículas disueltas, con el consiguiente crecimiento del cristal.

Los cristales de hielo de la nieve se forman al enfriarse el vapor de agua atmosférico.

Auguste Bravais (1811-1863) fue un físico francés que trabajó en campos tan dispares como la geobotánica, la astronomía, la cristalografía, la meteorología o la hidrografía. En 1845 modificó el trabajo previo de Frankheim, presentando las redes cristalinas que hoy se conocen como redes de Bravais.

1.3. *Sistemas cristalinos*

El crecimiento de cristales a partir de los distintos tipos de celdas elementales da lugar a diferentes estructuras cristalinas o **sistemas cristalinos.**

Se han definido siete sistemas cristalinos elementales: cúbico, tetragonal, hexagonal, romboédrico, rómbico, monoclínico y triclínico. Estos sistemas cristalinos presentan distintos tipos de simetría.

A partir de estos sistemas cristalinos y añadiendo nuevas partículas minerales sin modificar la simetría del sistema, se obtienen otros siete sistemas más complejos. Todos ellos constituyen las denominadas **redes de Bravais.** Con ellas intentó explicar los fenómenos de simetría y anisotropía de los cristales. En el siglo XX su teoría fue plenamente confirmada por las experiencias de difracción de rayos X.

En el recuadro inferior puedes observar las diferentes unidades elementales de las 14 redes de Bravais y ver cómo la presencia de una partícula material centrada en las caras posibilita la formación de una red nueva, al dar un nuevo punto de enlace entre celdas.

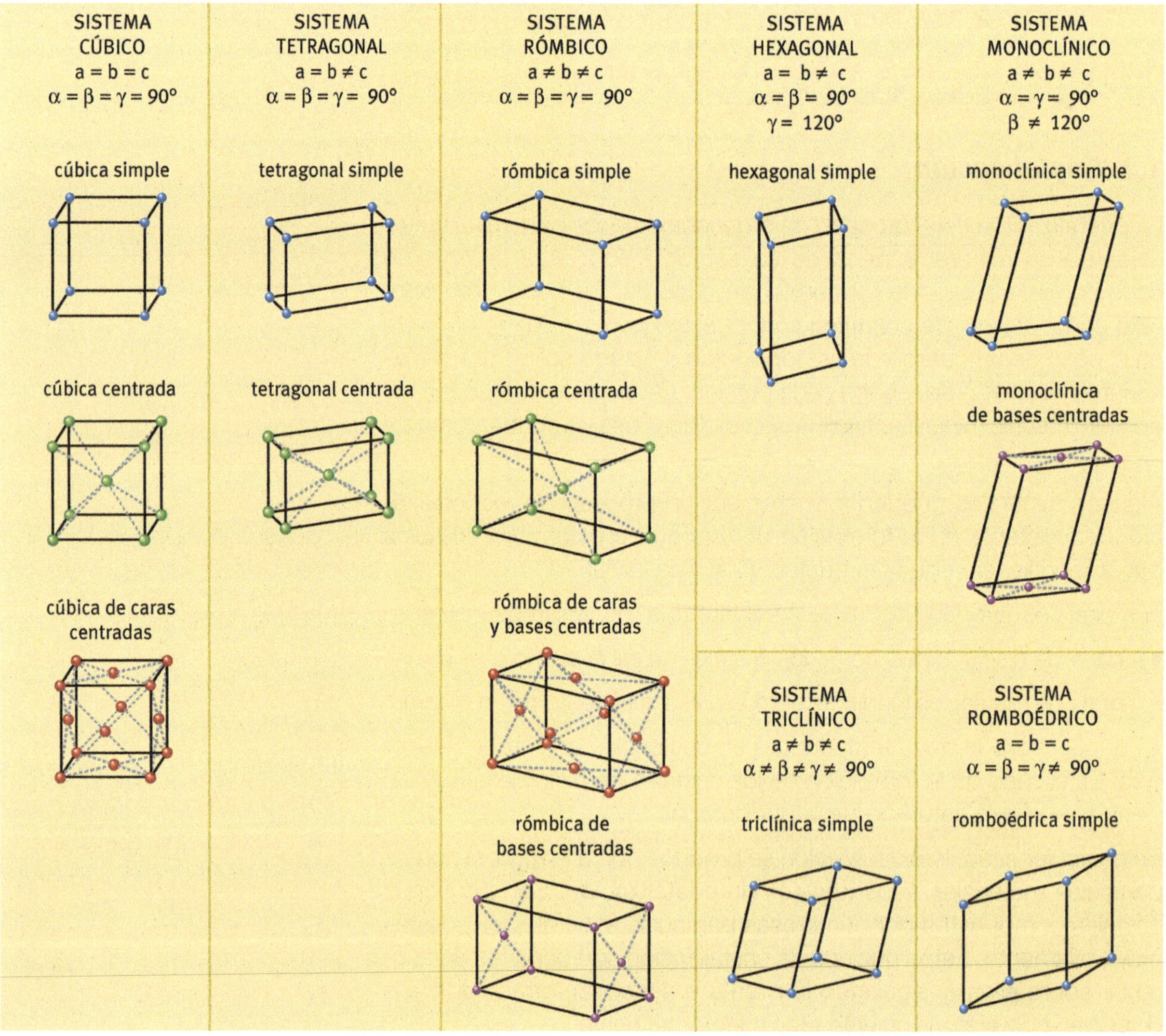

1.4. *Aplicaciones de los cristales*

Desde la Antigüedad, el ser humano ha utilizado los cristales para diferentes actividades. Actualmente, existe multitud de aplicaciones de los cristales, y a medida que la tecnología y la ciencia avanzan, se descubren nuevos usos. Algunos campos en los que se utilizan los cristales son:

- **Industria.** Debido a su composición química, una de las principales aplicaciones de los cristales es su uso para la obtención de determinados minerales. Es el caso de los cristales de calcopirita, de los que se obtiene el cobre, la utilización de los cristales de grafito para la elaboración de minas de lápices.
- **Joyería.** En joyería y decoración se emplea un tipo de cristales denominados *piedras preciosas* que una vez tallados y pulidos adquieren una gran belleza. Son piedras preciosas la esmeralda, el rubí y el zafiro.

El rubí es una piedra preciosa muy utilizada en joyería.

Además, los cristales se pueden utilizar en otros campos, por ejemplo, en la **alimentación,** como los cristales de sal gema, o la **construcción,** como el diamante, que se usa como abrasivo.

2. Ambientes petrogenéticos

Un ambiente petrogenético es un área de la corteza terrestre en la que se dan unas condiciones de temperatura, presión y composición química que hacen posible la formación de un determinado tipo de rocas.

Las rocas se pueden agrupar en tres clases según su origen:

- **Rocas magmáticas o ígneas:** originadas por la consolidación del magma al enfriarse. Reciben el nombre de rocas endógenas.
- **Rocas sedimentarias:** originadas por el depósito o sedimentación de materiales que proceden de la meteorización de rocas preexistentes. También se conocen como rocas exógenas.
- **Rocas metamórficas:** originadas por la transformación de rocas ígneas, sedimentarias u otras rocas metamórficas por la acción de procesos metamórficos protagonizados por variaciones de la presión y la temperatura. Son rocas endógenas.

Así pues, existen tres grandes ambientes de formación de rocas, o ambientes petrogenéticos, que dan origen a cada uno de los tres grupos de rocas definidos. A continuación, estudiaremos estos ambientes.

ACTIVIDADES

1. Define los siguientes conceptos: mineral, roca, materia cristalina, celda elemental, constantes cristalográficas.
2. ¿A partir de qué procesos puede tener lugar la formación de cristales?
3. Busca información sobre cuatro cristales diferentes y sus aplicaciones.
4. Dibuja una celda cristalina con las siguientes constantes cristalográficas: $a \neq b \neq c$; $\alpha = \beta = \gamma = 90°$. ¿A qué sistema cristalino representa?
5. ¿Qué es un ambiente petrogenético? ¿Qué los caracteriza? ¿Cuáles son los diferentes tipos de roca a los que dan lugar?

3. El ambiente magmático

El ambiente magmático, o ígneo, corresponde a las regiones internas de la Tierra, donde las condiciones de presión y temperatura hacen posible la fusión de las rocas. Éstas constituyen una masa de minerales fundidos que se denomina **magma.**

La **lava** es el material magmático emitido a la superficie terrestre en estado fluido.

3.1. *Tipos de magmas*

El magma está compuesto, principalmente, por los elementos químicos más abundantes en la corteza terrestre: silicio (Si), oxígeno (O), calcio (Ca), magnesio (Mg), hierro (Fe), sodio (Na) y potasio (K), además de otros elementos químicos en menor cantidad.

Los átomos de oxígeno y los de silicio se disponen formando tetraedros que presentan cuatro iones de oxígeno en los vértices y un ion de silicio en el interior. Este compuesto, el óxido de silicio (SiO_2), también recibe el nombre de sílice, aunque normalmente lo encontramos en forma de silicatos (SiO_4^{4-}).

Los tetraedros de sílice pueden estar aislados o disponerse formando cadenas según la abundancia de sílice respecto a los otros componentes del magma. Así, distinguimos entre un **magma básico,** que es un magma con poco SiO_2 y, por lo tanto, sin cadenas de sílice, y un **magma ácido,** que es un magma con abundante SiO_2 y con cadenas de sílice.

La presencia de estas cadenas de tetraedros proporciona al magma una mayor viscosidad, por lo que el magma ácido es más viscoso que el magma básico.

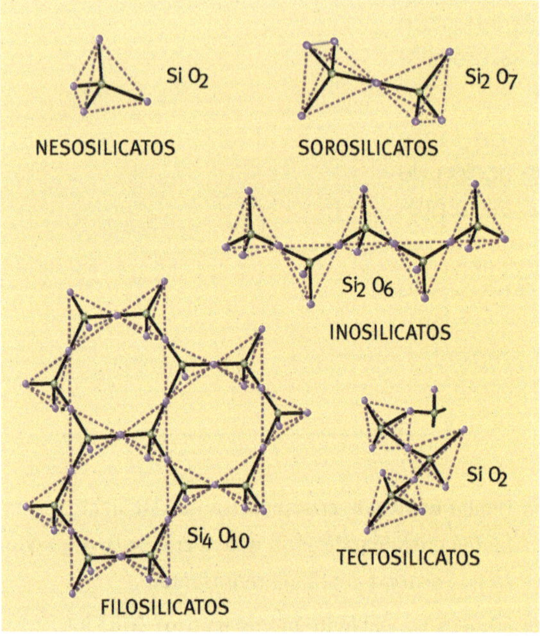

Principales estructuras y grupos de silicatos originados por la unión de la unidad básica SiO_4.

Además, en el magma existen los denominados *componentes minoritarios,* que se encuentran en forma de gases disueltos, como el vapor de agua y el dióxido de carbono. Estos componentes pueden separarse de la masa fundida cuando desciende la temperatura o la presión a la que el magma está sometido.

Asimismo, los magmas pueden originarse de dos formas, por **ascensión** desde el manto de materiales plásticos en fusión o por **anatexia,** o fusión de rocas de la corteza debido a condiciones extremas de presión y temperatura.

Así pues, teniendo en cuenta estos dos criterios, el contenido de sílice y el origen, se considera que las rocas magmáticas se forman a partir de dos tipos de magma:

- El **magma basáltico:** es un magma de naturaleza básica, con escasa sílice y generado en el manto o en las zonas profundas de la corteza terrestre.
- El **magma ácido:** es un magma de naturaleza ácida, con abundante sílice y generado por anatexia o fusión de rocas de la corteza terrestre.

3.2. *Diferenciación magmática*

Como ya hemos visto, existen diferentes tipos de magma según su composición y origen. Estos magmas sufren un proceso de *diferenciación magmática* que da lugar a la formación de las diversas rocas magmáticas. Así pues, la **diferenciación magmática** es el proceso de evolución del magma a diferentes rocas magmáticas a la vez que se enfría y asciende hacia la superficie terrestre.

Cada clase de roca magmática procede de un magma concreto, distinto de los que forman otras rocas y que contiene todos los componentes necesarios para formar la roca. Así, por ejemplo, un magma con abundante sílice formará granitos, que es una roca ácida, y un magma con escasa sílice formará gabros, que es una roca básica.

La diferenciación magmática supone una *cristalización fraccionada* de los minerales, empezando por aquellos que poseen el punto de fusión y la densidad más elevados, como el olivino, y finalizando por los más ligeros y con menor punto de fusión, como los feldespatos.

Se han descrito diversas secuencias de cristalización, estableciendo el orden en que deberían cristalizar los componentes de un magma, que explicarían la formación de distintas rocas a partir de un único magma.

De entre todas las secuencias de cristalización cabe destacar las denominadas **series de Bowen.** Éstas pueden ser *continuas* o *discontinuas.*

- **Serie continua:** todos los minerales pueden coexistir en un mismo espacio y tiempo, como, la serie de los silicatos ricos en calcio y sodio. De esta manera, podemos encontrar la plagioclasa cálcica y el feldespato potásico en una misma área.
- **Serie discontinua:** los minerales ya formados se hacen inestables al disminuir la temperatura o la presión y acaban por reaccionar con el magma residual, dando lugar al siguiente mineral de la serie. Es el caso de los silicatos ricos en hierro y magnesio. Así, el olivino, normalmente, acaba por reaccionar con el magma residual originando piroxeno.

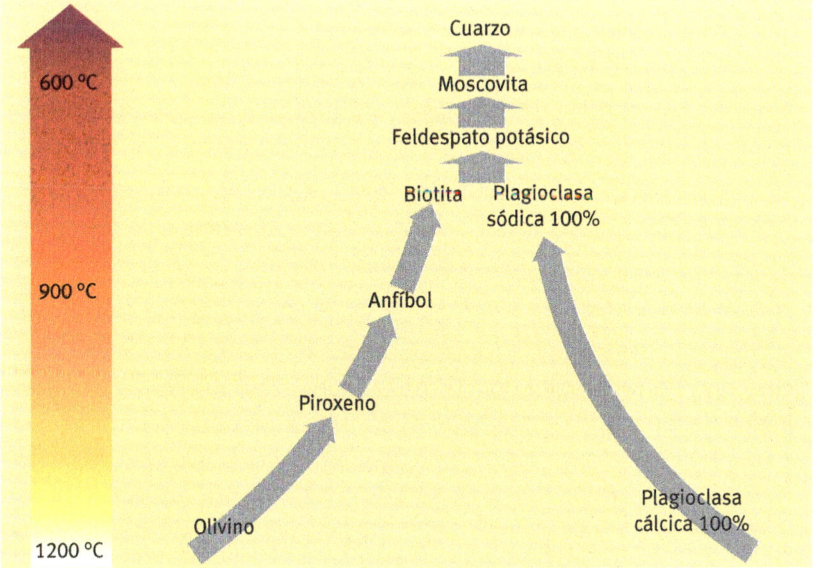

Las series de Bowen muestran la secuencia de cristalización de diferentes minerales de origen magmático según su densidad y punto de fusión. En la imagen, la serie discontinua está a la izquierda y la continua a la derecha.

Tª	Etapa	
1200 °C 700 °C	Ortomagmática	
700 °C 400 °C	Pegmatítica o pneumatolítica	
400 °C 300 °C	Hidrotermal	Hipotermal
300 °C 150 °C		Mesotermal
150 °C 100 °C		Epitermal

Las etapas de consolidación del magma

Crocoíta

Topacio

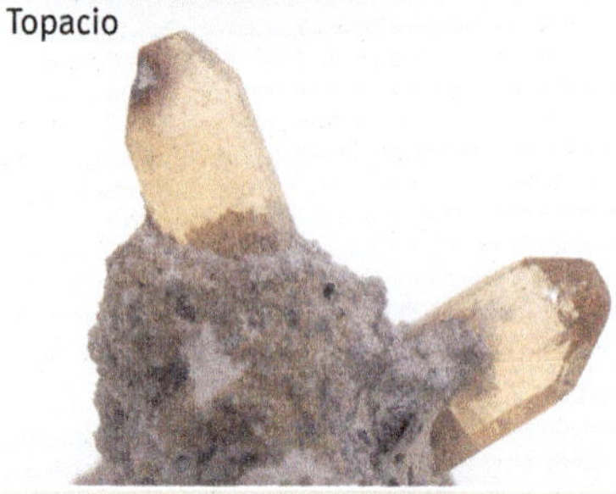

Mispíquel

La crocoíta es un óxido de cromo y plomo que forma cristales prismáticos de color rojo y brillo adamantino. El topacio es una gema propia de yacimientos pegmatíticos. El mispíquel es un mineral que se encuentra en yacimientos hipotermales.

Fases de consolidación del magma

Los procesos de diferenciación magmática están asociados a la evolución de un magma que se enfría y asciende a la superficie consolidándose. Se han definido diferentes etapas de consolidación del magma que llevan asociadas unos determinados yacimientos minerales. Estas etapas son:

- **Etapa ortomagmática**

  Se desarrolla a temperaturas que oscilan entre los 1200 °C y los 700 °C. Se forman minerales denominados pirogenéticos, como la anortita, la magnetita, la pirita...

  En los yacimientos de origen ortomagmático la mena se forma en el seno del magma y puede estar diseminada, como los diamantes dentro de la kimberlita, o bien concentrada, como la cromita dispuesta en capas dentro de la peridotita.

- **Etapa pegmatítica o pneumatolítica**

  Se desarrolla a temperaturas entre los 700 °C y los 400 °C. Cristalizan, sobre todo, el cuarzo y la ortosa, que son los principales componentes de las rocas denominadas pegmatitas. El fluido resultante de la formación de las pegmatitas es rico en elementos volátiles o gases, como el vapor de agua o el dióxido de carbono, que en contacto con otras rocas pueden originar rocas pneumatolíticas.

  Los yacimientos pegmatíticos están asociados a la presencia de pegmatitas y a la formación de cristales grandes y bien formados, entre los que podemos encontrar diversas gemas, como, por ejemplo, el topacio. En los yacimientos pneumatolíticos la mena está más o menos diseminada en el seno de la roca.

- **Etapa hidrotermal**

  La etapa hidrotermal, o de aguas calientes, se desarrolla a temperaturas entre los 400 °C y los 100 °C. En esta etapa el magma es una fase acuosa con numerosas sustancias en disolución e incluso cierta cantidad de sílice.

  Frecuentemente, los yacimientos asociados a la etapa hidrotermal se dividen en tres grupos:

  En los *yacimientos hipotermales,* formados a temperaturas altas, entre los 400 °C y los 300 °C, se encuentra, por ejemplo, el mispíquel.

  En los *yacimientos mesotermales,* formados a temperaturas medias, entre los 300 °C y los 150 °C, se encuentra, por ejemplo, la calcopirita.

  En los *yacimientos epitermales,* formados a temperaturas relativamente bajas, inferiores a 150 °C, encontramos, por ejemplo, el oropimente.

Dado que en un magma los primeros minerales que se separan son los que tienen bajo contenido en sílice, se supone que el magma evoluciona hacia composiciones más ricas en sílice, por lo que, las rocas formadas en las primeras fases son más básicas que las que lo hacen en las últimas. También, se incrementa la proporción de volátiles, que se concentran en los fluidos residuales.

3.3. *Minerales de origen magmático*

Denominamos rocas a los agregados naturales de distintos minerales.

En una roca magmática clasificamos los minerales que la constituyen en función de diferentes factores.

- La **coloración.** Nos permite distinguir entre *minerales leucócratos,* es decir, que son blancos o claros, como el cuarzo; y *minerales melanócratos,* negros u oscuros, como la biotita.
- La **composición química.** Diferenciamos entre *minerales félsicos,* es decir, ricos en feldespato y sílice; y *minerales máficos,* ricos en magnesio y hierro.
- La **abundancia de los minerales** en la roca. De este modo, encontramos los denominados *minerales esenciales,* que caracterizan la roca, como la ortosa en el granito; los *minerales accesorios,* que son frecuentes pero están presentes en pequeña cantidad y permiten distinguir diferentes variedades en cada especie de roca, como el granito con granates; y los *minerales accidentales,* que son muy poco frecuentes en la roca, como, por ejemplo, los anfíboles en los granitos.
- El **origen del mineral.** Distinguimos entre *minerales primarios,* que son aquellos que se han formado durante la cristalización del magma, como por ejemplo, la biotita, y *minerales secundarios,* aquellos que provienen de la alteración de los minerales primarios de la roca, como, por ejemplo, la clorita, que procede de la alteración química de la biotita.

3.4. *Clasificación de las rocas magmáticas*

Para clasificar las rocas magmáticas, es necesario más de un criterio. Veamos a continuación los que más se utilizan.

Índice de color

Este criterio hace referencia a la presencia de minerales félsicos o claros, o de minerales máficos u oscuros.

Composición química

Según su contenido en sílice (SiO_2), las rocas magmáticas se clasifican en ácidas, intermedias, básicas y ultrabásicas.

Profundidad de cristalización

Según la profundidad a la que ha cristalizado el magma, podemos distinguir las siguientes rocas magmáticas:

Las **rocas extrusivas,** también llamadas volcánicas, son aquellas que cristalizan en la superficie de la Tierra o a escasos metros de profundidad.

Las **rocas intrusivas** son aquellas que cristalizan en el interior de la Tierra. Según la profundidad a la que cristalizan, dentro de este último grupo distinguimos entre:

- Las *rocas plutónicas,* cuando la cristalización se produce a profundidades superiores a 1 km.
- Las *rocas hipoabisales,* o *filonianas,* cuando cristalizan a profundidades inferiores a 1 km. Estas rocas se consolidan al inyectarse el magma en grietas, fracturas o fallas.

En el siguiente cuadro puedes observar la clasificación de las rocas magmáticas según su composición química y su profundidad de cristalización. Esta última viene indicada por la temperatura.

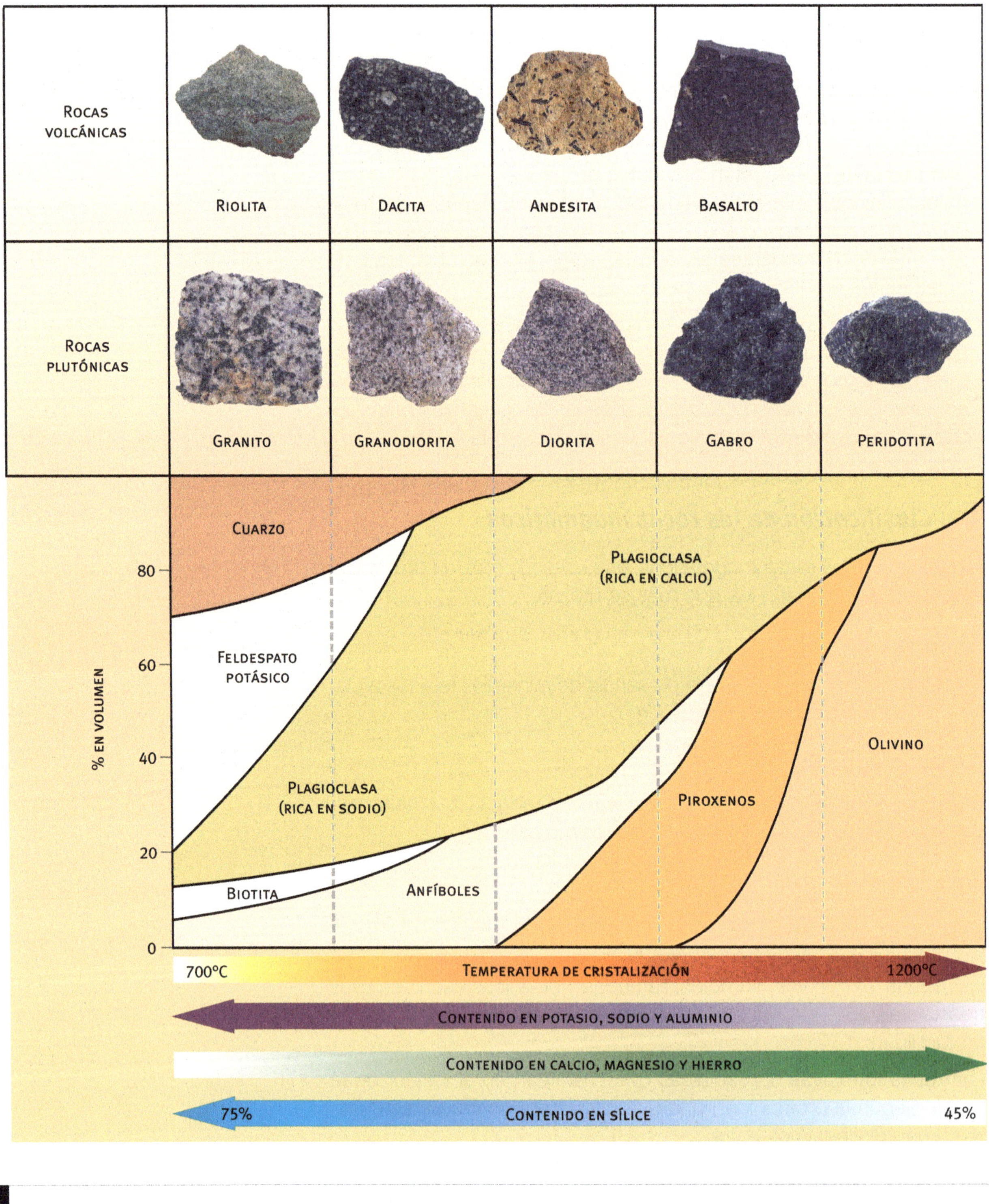

Clasificación de las rocas magmáticas en función de su composición, temperatura y profundidad de cristalización. Observa que la misma composición origina un tipo diferente de roca dependiendo de si la formación se da en profundidad o no.

3.5. *Textura de las rocas magmáticas*

La textura está definida por una serie de características, como las relaciones geométricas, la forma o el tamaño que presentan los minerales componentes. En las rocas magmáticas predominan las texturas siguientes:

- **Textura granular:** los minerales integrantes de la roca han formado cristales más o menos homogéneos y del mismo tamaño, en contacto unos con otros. Esta textura es consecuencia de un enfriamiento lento del magma que ha permitido un buen desarrollo de los cristales.
- **Textura porfídica:** se distinguen unos cristales más o menos desarrollados, denominados *fenocristales,* y una masa de apariencia homogénea que los rodea, llamada *matriz.* A veces, esta matriz, observada al microscopio, presenta unos cristales diminutos llamados *microlitos,* aunque generalmente es amorfa. Esta textura nos indica que se han sucedido dos fases de cristalización. La primera, de enfriamiento lento, permite el crecimiento de los cristales que dan lugar a unos fenocristales. La segunda fase, con un rápido enfriamiento, da lugar a los microlitos que forman la matriz.
- **Textura vítrea:** está presente en rocas constituidas por una masa mineral amorfa o criptocristalina. Estas rocas son consecuencia de un rápido enfriamiento del magma.
- **Textura fluidal:** los cristales están ordenados como si siguieran la corriente de un río, indicando que, antes de su solidificación, el magma discurría sobre el terreno formando corrientes de lava.
- **Textura espumosa:** esta textura muy porosa es consecuencia de los huecos dejados por las muchas burbujas procedentes de las emanaciones de gases que contenía el magma que formó la roca. Estas burbujas han quedado atrapadas por el súbito enfriamiento de la corriente de lava al entrar en contacto con la atmósfera.

La textura también es un criterio útil para clasificar las rocas magmáticas. Aunque éstas son muy variadas, a grandes rasgos nos permiten determinar su origen. Podemos establecer las siguientes relaciones:

- La textura granular, asociada a las rocas plutónicas.
- La textura porfídica, que predomina en las rocas hipoabisales.
- La textura fluidal y la espumosa, características de las rocas volcánicas.

El lamprófido es una roca con textura porfídica.

La riolita es una roca con textura fluidal.

ACTIVIDADES

6. ¿Cómo se clasifican los magmas en función de su contenido en silicio? ¿Podemos relacionar de alguna manera esta composición con su origen?
7. ¿En qué consiste la diferenciación magmática?
8. Comenta las diferencias entre las etapas ortomagmática, pegmatítica y hidrotermal de la consolidación de un magma.
9. Indica tres criterios que nos permitan establecer una clasificación de las rocas magmáticas, y las categorías que resultan de cada uno de ellos.
10. ¿Que roca se ha formado por un proceso de enfriamiento del magma más lento, una con textura granular o una con textura espumosa? Justifica tu respuesta.

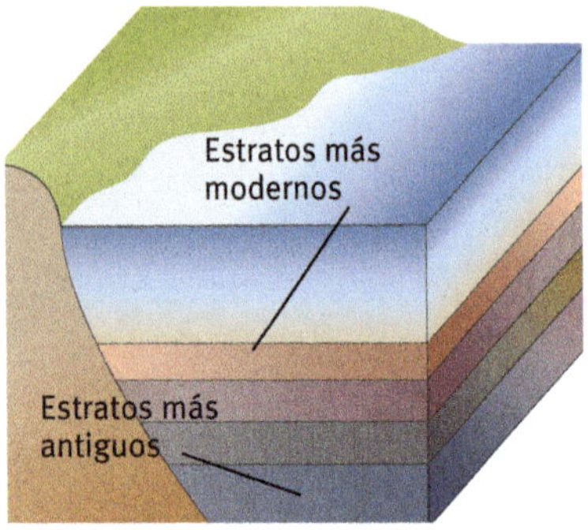

4. El ambiente sedimentario

El ambiente sedimentario se corresponde con las zonas superiores de la corteza terrestre. Estas zonas de la litosfera pueden estar en contacto bien con la hidrosfera o bien con la atmósfera. Así, mientras que las rocas magmáticas se originan en el interior de la Tierra, los procesos de formación de rocas sedimentarias se desarrollan en la zona más superficial.

La característica más remarcable de las rocas sedimentarias es su disposición en capas, denominada **estratificación.** Los sedimentos, al depositarse, constituyen capas o estratos superpuestos horizontalmente y separados por superficies paralelas de discontinuidad.

Cada estrato representa un período sedimentario y cada superficie o plano de discontinuidad, una interrupción en la sedimentación o un cambio de la naturaleza del sedimento.

El estudio de los diferentes estratos tiene un importante valor geológico, puesto que nos permite interpretar la evolución del lugar a lo largo del tiempo. Por lo general, los estratos inferiores son más antiguos que los superiores; a partir de este precepto, podemos analizar las etapas por las que ha pasado un determinado ambiente.

C

El **principio de superposición de Steno** establece que los estratos inferiores son más antiguos que los superiores.

4.1. *Ambientes de sedimentación*

Se consideran ambientes de sedimentación aquellas extensiones de la superficie terrestre aptas para detener el transporte de partículas, provocar su sedimentación y albergar su litificación. Estos ambientes de sedimentación pueden ser *continentales, mixtos* o *marinos.*

Cada uno de ellos se caracteriza por ciertas condiciones físicas, químicas y biológicas, que darán como resultado un tipo de rocas diferenciado. El estudio de estas rocas nos aportará pistas sobre el ambiente en que se formaron.

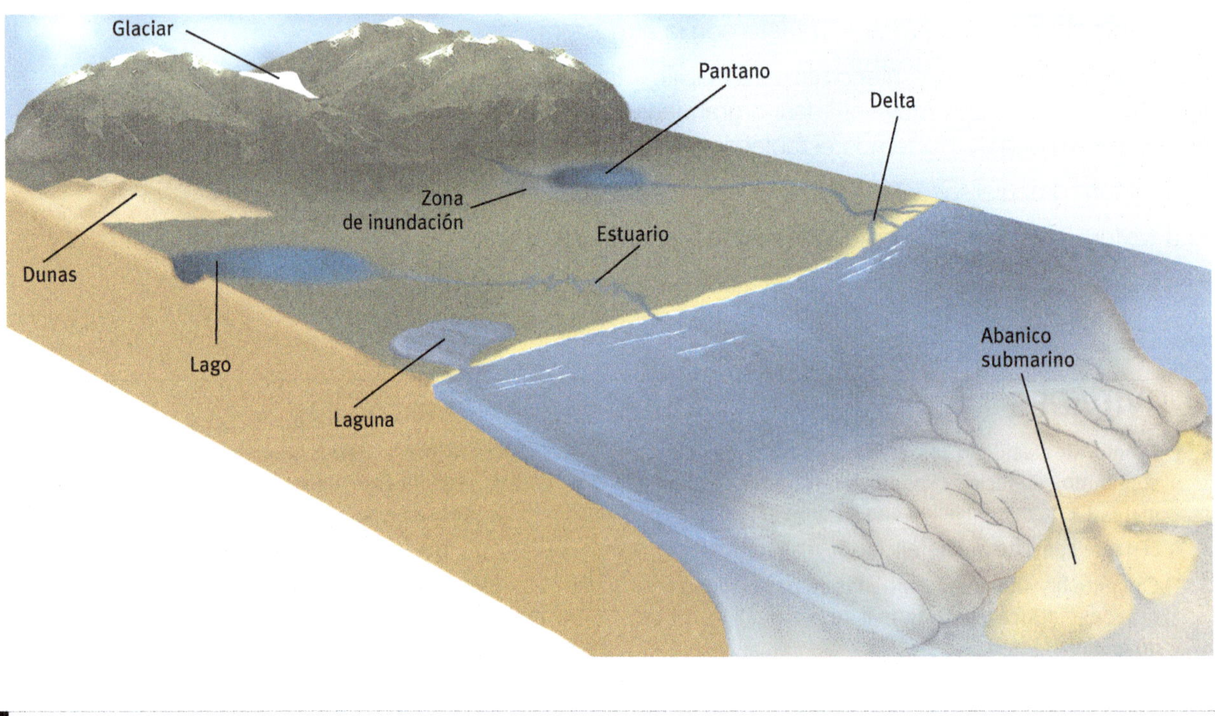

Representación de diversos ambientes de sedimentación: continentales, mixtos y marinos. Las diferentes condiciones que se dan en cada uno de ellos determinarán el tipo de roca originado.

4.2. *Litificación*

Los materiales que forman las rocas sedimentarias proceden de rocas preexistentes, erosionadas por los agentes geológicos externos. Los materiales o detritos producidos por la erosión son depositados en el fondo del mar, de los lagos y de los ríos, o bien en la superficie terrestre.

Al principio, los materiales que han de originar las rocas sedimentarias están sueltos. La **litificación** es el conjunto de transformaciones que éstos experimentan hasta formar una roca compacta. Los pasos que tienen lugar durante el proceso de litificación son los siguientes:

- **Compactación:** es la reducción del espacio existente entre las partículas y la expulsión de parte del agua contenida en el sedimento. Estos dos procesos son consecuencia de la presión ejercida por las capas superiores de sedimentos.
- **Cementación:** consiste en la unión de las partículas del sedimento gracias a la precipitación de los minerales disueltos en el agua. Por los espacios que todavía quedan después de la compactación discurren disoluciones cargadas de sustancias que precipitan en dichos espacios actuando como cemento.
- **Diagénesis:** se trata de un cambio estructural y químico que experimentan las partículas del sedimento al variar las condiciones de presión y temperatura respecto a las condiciones iniciales. De este modo, se pueden llegar a formar nuevos minerales.
- **Mesomatismo:** consiste en la sustitución de un mineral del sedimento por otro de distinta composición. Es el caso de la mineralización de fósiles que se silicifican.

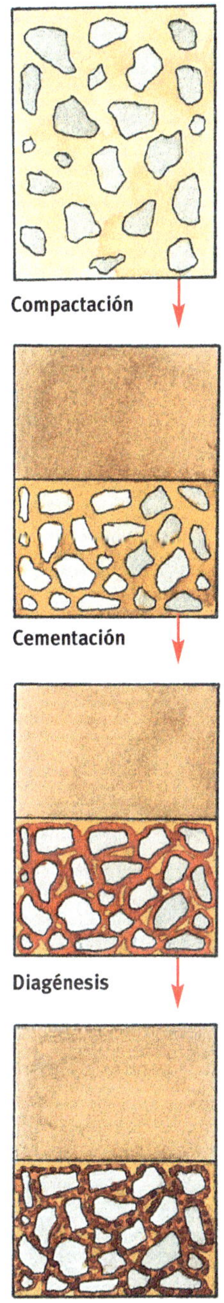

4.3. *Minerales de origen sedimentario*

El principal criterio de clasificación de los minerales que forman las rocas sedimentarias es su origen, según el cual distinguimos de mineral:

- **Alotígenos:** cuando el mineral ya estaba formado antes del proceso de sedimentación.
- **Autígenos:** cuando el mineral se forma a lo largo de las diversas fases sedimentarias, desde la fuente de aporte de sedimentos hasta la litificación.

Así, por ejemplo, en un conglomerado los minerales que forman los clastos acostumbran a ser alotígenos, mientras que la matriz, o cemento que los une, suele estar formada por minerales autígenos.

No obstante, en las rocas sedimentarias podemos encontrar ejemplos en los que coexisten fracciones de un mismo mineral de origen alotígeno y de origen autígeno. Así, en el caso de un determinado conglomerado pueden aparecer granos de cuarzo provenientes de fragmentos de rocas preexistentes y que han sido transportados hasta la cuenca sedimentaria (alotígenos) junto con cuarzo de grano muy fino en la matriz que los envuelve, que ha precipitado a partir de sílice coloidal (autígeno).

4.4. *Clasificación de las rocas sedimentarias*

Atendiendo al origen de los componentes de las rocas sedimentarias, éstas pueden clasificarse en tres grandes grupos: *rocas de origen detrítico, rocas de origen físico-químico* y *rocas de origen orgánico.*

Rocas de origen detrítico

Las rocas sedimentarias de origen detrítico resultan de la sedimentación de materiales de grosores diversos por acción, principalmente, de la gravedad. Por este motivo, las rocas sedimentarias de origen detrítico son llamadas también de origen mecánico.

Los materiales componentes de las rocas detríticas son fragmentos más o menos angulosos, denominados **clastos,** que son transportados por los agentes geológicos externos a mayor o menor distancia de la fuente de aporte.

■ Conglomerado

Este transporte provoca que los clastos choquen unos con otros, se partan en trozos, se redondeen, se pulimenten y se vayan transformando sucesivamente en cantos rodados, gravas, arenas... Cuando se somete a los procesos de litificación, esta diversidad de materiales acaba por originar rocas compactas.

Estos dos criterios, el tamaño de los fragmentos y la redondez de sus cantos, nos permiten establecer una clasificación de las rocas detríticas.

Tamaño de los fragmentos	Elementos constitutivos	Morfología de los elementos	Rocas	Granulometría
> 200 mm	Bloques	Variable	Conglomerado	Grosera
200-20 mm	Fragmentos Cantos	Angulosa Redondeada	Brecha Pudinga	
20-2 mm	Gravas	Variable	Arcosa	Media
2-0,063 mm	Arenas	Variable	Arenisca	
0,063-0,004 mm	Limos	Angulosa	Limolita	Fina
0,004 mm	Arcillas	Angulosa	Perlita	

Rocas de origen físico-químico

Podemos distinguir dos grupos de procesos: los físicos, como la evaporación, y los químicos, como la precipitación, que dan lugar respectivamente a dos tipos distintos de rocas: las *evaporitas* y las *calizas.*

- **Evaporitas:** se forman por evaporación de aguas saladas, marinas o continentales, en zonas cálidas y secas. Se supone que el proceso es similar al que tiene lugar en las salinas.

  Estas rocas se originan por precipitación de materiales disueltos en el agua cuando ésta se evapora. Las más importantes son la anhidrita, el yeso y la sal gema.

Así, en el caso de un golfo marino semiobturado por una barra litoral que sólo recibe aporte de agua marina durante las mareas altas y en donde el ambiente es seco y cálido, las aguas encerradas se irán evaporando y, las sales disueltas irán depositándose en el fondo.

La secuencia en la cual se depositen las sales vendrá determinada por su solubilidad: primero precipitarán las menos solubles y finalmente, a medida que se incrementa la salinidad, las más solubles. En las primeras fases precipitarán el yeso y la anhidrita, después la sal gema y en las etapas finales las sales de potasio y magnesio.

- **Calizas:** todas las formaciones rocosas en las que el anión predominante es el carbonato $(CO_3)^{2-}$. Este carbonato puede tener tres orígenes: precipitado a partir de disoluciones que lo contienen, acumulado a partir de los restos de seres vivos con exoesqueletos o conchas calcáreas y a partir de restos de rocas carbonatadas preexistentes.

Las **calizas químicas** son rocas carbonatadas de origen químico obtenidas de la precipitación del carbonato de calcio. Destacan las *calizas pisolíticas,* formadas por granos de carbonato cálcico del tamaño de un guisante; las *calizas litográficas,* con grano muy fino y presencia habitual de fósiles, y el *travertino calizo,* constituido por delgados depósitos de estructura esponjosa surgidos junto a manantiales o sobre plantas por evaporación de aguas con carbonato cálcico.

Las **calizas orgánicas** son rocas formadas por materiales carbonatados de origen orgánico. En este grupo están las *lumaquelas*, unas rocas constituidas por conglomerados de conchas de lamelibranquios, un tipo de moluscos.

Las **calizas clásticas** son rocas carbonatadas formadas por fragmentos de rocas calizas más antiguas. Se denominan *cantos calizos* si aparecen sueltas, o *conglomerados calizos* si aparecen cementados. Se clasifican según los criterios de las rocas detríticas.

Rocas de origen orgánico

Estas rocas resultan de la acumulación de restos de tejidos animales y vegetales. Cabe destacar dos grupos de rocas: *silíceas* y *carbonosas.*

- **Silíceas:** se trata de fangos que contienen restos de esqueletos silíceos. Entre estas rocas distinguimos: las diatomitas, con restos de algas diatomeas; las radiolarias, con cápsulas de protozoos radiolarios; y las espongiolitas, con restos de esponjas silíceas.

- **Carbonosas:** son rocas formadas a partir de la transformación de restos con abundancia de carbono orgánico. Destaca el **carbón** originado por la acumulación y compactación de restos vegetales en ambientes pantanosos o deltaicos. Según su riqueza en carbono, distinguimos entre turba, lignito, hulla y antracita.

Dentro de este grupo encontramos, también, los **hidrocarburos** que provienen de la descomposición de la materia orgánica, principalmente plancton. De entre los hidrocarburos destacan el petróleo y el gas natural.

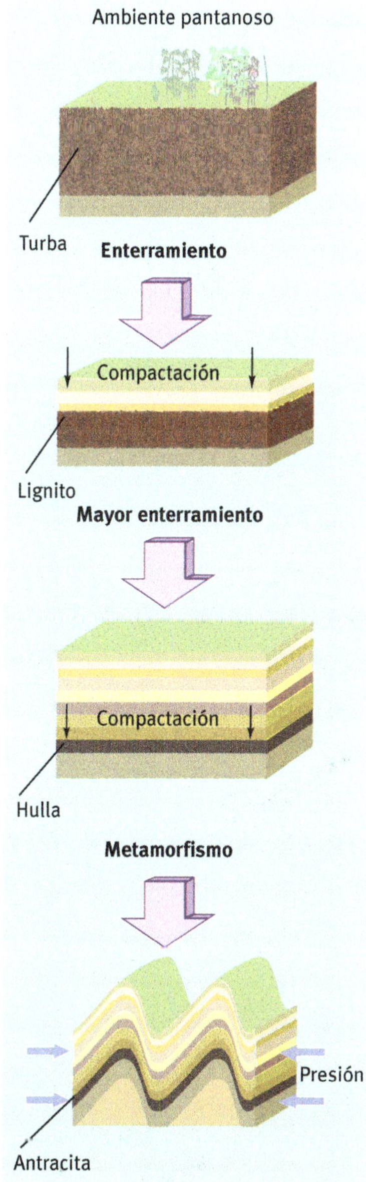

Formación de carbón a partir del enterramiento de restos de materia orgánica en un ambiente pantanoso

4.5. *Texturas de las rocas sedimentarias*

No es posible establecer una clasificación de las rocas sedimentarias en función de su textura, ya que no se puede asociar una textura a un único tipo de rocas y, de igual manera, se pueden encontrar distintas texturas en rocas sedimentarias de un mismo tipo.

La textura adquiere cierta relevancia en la identificación de las rocas sedimentarias detríticas. Algunos aspectos texturales que deben tenerse en cuenta para describir este grupo de rocas son los siguientes:

- **Tamaño del grano:** es importante observar si existen diferentes tamaños del grano en una misma roca. De este modo, podemos distinguir diversos elementos mezclados, como las arenas y los limos; a su vez dentro de cada una de estas categorías podemos comprobar si el grano es constante o bien varía, como en el caso de que existan distintos tipos de arena.
- **Morfología de los granos:** puede variar de muy angulosa a muy redondeada y, a su vez, presentar baja o alta esfericidad; es decir, la morfología de los granos indicará si el tiempo de transporte ha sido menor o mayor.
- **Carácter de *clast-supported* o *matrix-supported:*** indica si en una roca sedimentaria predominan los clastos o la matriz, respectivamente.

4.6. *Fósiles y fosilización*

Trilobites, fósil característico del Cámbrico

Las rocas sedimentarias pueden contener restos de organismos o de su actividad que han quedado conservados durante el proceso de formación de la roca. Estos restos se conocen con el nombre de fósiles, y nos ayudan a interpretar el pasado geológico. Su estudio nos permite conocer cómo eran los seres vivos, ahora extinguidos, que poblaban el planeta en otras épocas. A partir de este conocimiento podemos saber cuáles eran las condiciones ambientales que se daban en el pasado en un área determinada y también podemos datar una roca.

Vamos a conocer que procesos tienen lugar desde la muerte de un organismo hasta su transformación en un fósil observable en la superficie terrestre.

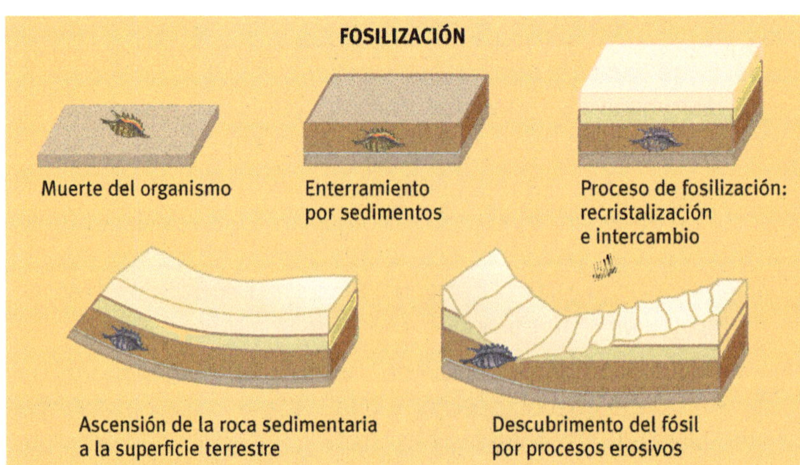

El proceso de fosilización consiste en la sustitución de la materia orgánica del organismo muerto por materia mineral. Para que pueda llevarse a cabo se han de dar ciertas circunstancias, ya que de lo contrario la acción de los organismos carroñeros y descomponedores haría desaparecer los restos. Para que un organismo fosilice tiene que quedar cubierto de forma casi inmediata por sedimentos y preservarlo de la actuación de los agentes que provocarían su desaparición. Es mucho más frecuente en medios acuáticos que en medios terrestres, por lo que la fosilización de los organismos acuáticos es más corriente que la de los terrestres. Otro factor determinante para la fosilización es que su cuerpo presente partes duras: los organismos blandos, como medusas o gusanos, fosilizan raras veces, mientras que aquellos que poseen caparazones, huesos, conchas... constituyen la mayor parte del registro fósil.

Cuando el sedimento ha cubierto el cadáver, la materia orgánica suele sufrir un proceso de putrefacción debido a la actividad microbiana. Las partes duras, en cambio, pasarán por una serie de procesos de intercambio y sustitución de minerales con el sedimento en el que están englobados. Según naturaleza, se pueden distinguir diversos tipos de fosilización:

C

Al margen de la fosilización, otras formas de conservación de restos animales son la inclusión en ámbar, una resina fósil, o la congelación, como ha sucedido con los mamuts siberianos.

Carbonatación	Es el más común. Las partes duras del organismo quedan substituidas por carbonato cálcico en forma de calcita. Las conchas, corales... estan formados fundamentalmente por calcita.
Silicificación	Los restos del organismo quedan substituidos por sílice en forma de calcedonia. Suele darse en restos de foraminíferos, equinodermos y troncos fósiles.
Piritización	Cuando los gases de azufre producidos por la descomposición de la materia orgánica reaccionan con las sales de hierro del agua del mar, dan lugar a sulfuros de hierro como la pirita, que reemplazan los componentes del fósil.
Fosfatación	Es el proceso más común en la fosilización de vertebrados, dado el elevado contenido en fosfatos de los huesos. También interviene el carbonato cálcico.
Carbonificación	Frecuente en vegetales. Los restos vegetales enterrados en ambientes sin oxígeno inician un proceso de conversión en turba. Ésta, con el paso del tiempo y sometida a presiones elevadas incrementa su porcentaje de carbono al perder componentes líquidos y gaseosos.

Algunas veces las rocas sedimentarias no contienen restos de los seres vivos, pero sí de su actividad, como huellas, madrigueras, coprolitos... que nos aportan información sobre su presencia y sus hábitos.

ACTIVIDADES

- **11.** ¿Por qué en una serie los estratos inferiores son más antiguos que los superiores? ¿Puede suceder lo contrario?
- **12.** Haz un esquema de los diferentes ambientes de sedimentación, clasificándolos en continentales, mixtos y marinos.
- **13.** ¿Qué és la litificación? ¿Cuáles son los pasos que se suceden a lo largo de la formación de la roca sedimentaria?
- **14.** ¿En qué consiste el metasomatismo?
- **15.** Realiza un esquema de la clasificación de las rocas sedimentarias según su origen, indicando las diferentes subcategorías.
- **16.** ¿Qué es un fósil? ¿Qué tipo de información nos aporta cuando aparece en una roca? Indica las diferentes etapas en la formación de un fósil.
- **17.** Comenta la frase: 'El registro fósil es un registro incompleto'.

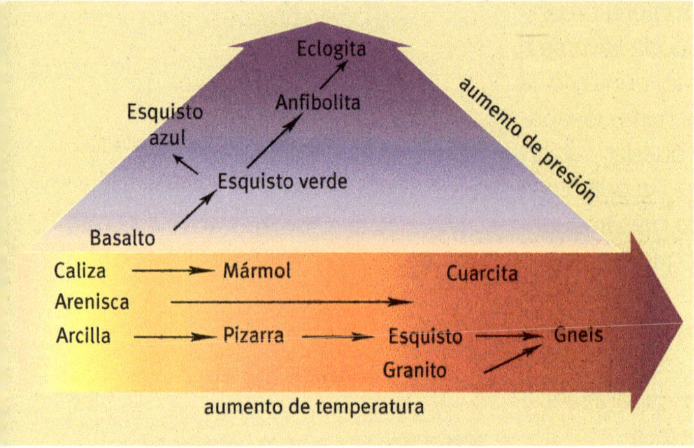

Formación de rocas metamórficas en función del incremento de presión y temperatura

5. El ambiente metamórfico

Por metamorfismo entendemos el conjunto de transformaciones sufridas por una roca preexistente (magmática, sedimentaria o ya metamorfizada anteriormente), debido a variaciones en las condiciones físico-químicas de formación, principalmente presión y temperatura.

El estudio del metamorfismo comprende los pasos intermedios por los que pasa una roca metamórfica y la caracterización de su estadio final con el fin de conocer la evolución metamórfica de una zona y su relación con el proceso que la ha creado.

Estas transformaciones pueden efectuarse por **metamorfismo isoquímico,** es decir, mediante reacciones en estado sólido entre los bordes de los granos minerales sin que haya un cambio de la composición química global de la roca, o por **metasomatismo,** con aportes o migraciones de elementos con el consiguiente cambio químico. Estas transformaciones metamórficas se dan siempre sin que la roca primitiva se funda. Como resultado, la estructura inicial de la roca queda alterada y se forman nuevos minerales por transformación de los antiguos.

5.1. *Factores que intervienen en el metamorfismo*

Hemos visto que los factores que condicionan el metamorfismo de una roca son fundamentalmente la presión y la temperatura. Sin embargo, también hay que considerar la acción de otras sustancias sobre la roca y el tiempo que dura el proceso metamórfico.

Incremento de la temperatura

El incremento de la temperatura puede deberse a varios motivos. Por un lado, se sabe que en la corteza terrestre se produce un aumento de temperatura de al menos 10 °C por cada kilómetro de profundidad. Además, el ascenso de masa magmática hacia la corteza afecta térmicamente a las rocas próximas a los materiales fundidos. Por último, las fricciones que se producen entre los dos bloques de una falla o entre dos placas tectónicas desprenden calor. Estas variaciones de temperatura afectan a las rocas. En general, es necesario un aumento de temperatura para que dos compuestos químicos reaccionen entre sí: el incremento de la temperatura facilita las reacciones químicas.

De igual forma, un aumento de temperatura provoca un incremento de la vibración de las partículas de las redes cristalinas. Así, éstas pueden adecuarse a las nuevas condiciones en que se halla la roca.

Incremento de la presión

Igual que sucede con la temperatura, el incremento de presión puede tener diferentes causas. La **presión litostática,** o presión vertical realizada por una columna de materiales de la corteza terrestre sobre un punto, es una de las causas principales del aumento de la presión en las rocas.

Otra causa del incremento de presión son las fuerzas orogénicas que ejercen unas presiones enormes en los puntos sobre los que actúan. También cabe señalar que los gases que se escapan de las masas magmáticas ejercen una presión considerable sobre las rocas más próximas.

Como respuesta a las variaciones de presión, las rocas pueden modificar su estructura «aplanando» los minerales. Dichos minerales tienden entonces a orientarse o situarse en planos perpendiculares a la dirección en la que se ejerce la presión y adoptan formas aciculares.

Otra respuesta al incremento de presión es que se produzcan cambios profundos en la red cristalina, originándose nuevas ordenaciones y, por tanto, nuevos minerales.

ROCA PRIMARIA
(Magmática, sedimentaria o metamórfica)

+

Incremento de presión y /o temperatura

=

Roca metamórfica

Metasomatismo

En ocasiones, puede suceder que una roca entre en contacto con sustancias químicas nuevas en su entorno, bien porque se han segregado de un magma en fase de consolidación, bien porque han sido expulsadas de otras rocas sometidas a grandes presiones.

Según la naturaleza química de estas sustancias y de la roca que las recibe, pueden producirse reacciones químicas que generan nuevos minerales pero que conservan las estructuras originales. Este proceso se conoce con el nombre de **metasomatismo.**

Tiempo

El tiempo que dura el metamorfismo influye de forma muy importante en los cambios mineralógicos y estructurales de las rocas.

Así, por ejemplo, una elevada presión pero de corta duración puede provocar una profunda transformación mecánica y una escasa alteración química; es decir, se transforma la textura de la roca pero los minerales siguen siendo los mismos. Por el contrario, una presión no muy intensa pero continuada modificará las redes cristalinas y generará nuevas especies minerales.

C

La clasificación de las rocas metamórficas es muy compleja puesto que debe considerar diversos parámetros, por ello, la clasificación más completa es aquella que considera la clase de metamorfismo y el tipo de roca primaria.

5.2. Clases de metamorfismo

Según las condiciones de presión y temperatura que deben soportar las rocas, se distinguen las siguientes clases de metamorfismo.

Pirometamorfismo

Se debe, principalmente, a las elevadas temperaturas a que se ven sometidas las rocas situadas en la superficie de la Tierra o a escasa profundidad cuando entran en contacto con las coladas de lava volcánica.

Metamorfismo de contacto o termometamorfismo

Se desarrolla en la zona de contacto de una intrusión de magma y debido principalmente a una elevada temperatura y a una cierta presión, como sucede, por ejemplo, en las chimeneas volcánicas.

El metamorfismo de contacto puede ser isoquímico, si no se dan variaciones de elementos, o metasomático, si se producen aportes de gases que se escapan de la intrusión magmática.

+

El termometamorfismo genera unos depósitos, denominados **skarns**, donde se suelen encontrar minerales tales como la magnetita, la scheelita...

C

En muchos casos las observaciones de campo, es decir, la observación a macroescala permite asociar una roca a un determinado tipo de metamorfismo.

Metamorfismo regional

Afecta a amplias regiones que se corresponden con zonas de actividad orogénica sometidas a grandes presiones.

Este tipo de metamorfismo incluye también el llamado *metamorfismo de hundimiento,* consecuencia del sobrepeso que se produce sobre las capas inferiores debido a las espesas capas de sedimentos superiores.

En el metamorfismo regional los dos factores físico-químicos que intervienen son, fundamentalmente, la presión y la temperatura, que aumentan con la profundidad.

Así pues, en función de la profundidad, distinguimos tres zonas distintas del metamorfismo regional:

- **Epizona:** es la zona más superficial en la que encontramos minerales cuya composición es rica en radicales -OH, como, por ejemplo, el talco.
- **Mesozona:** es la zona intermedia; en ella se encuentran minerales como las micas.
- **Catazona:** es la zona más profunda, donde, en general, los minerales carecen de radicales -OH, como, por ejemplo, los granates.

Más allá de la catazona la temperatura es tan elevada que se produce una fusión parcial de la roca. Éste es prácticamente el ámbito de la *anatexia,* o fusión de rocas de la corteza por condiciones extremas de temperatura o presión. Aquí se encuentra un tipo de rocas metamórficas muy características, las *migmatitas,* en las que coexisten restos de partes sólidas con productos del magma recristalizado.

Metamorfismo profundo

El factor que más influye es la presión litostática, característica de zonas profundas de la corteza de entre 10 y 30 km.

Dinamometamorfismo

Está caracterizado por elevadísimas presiones y bajas temperaturas. Es típico de rocas sometidas a esfuerzos orogénicos a escasa profundidad.

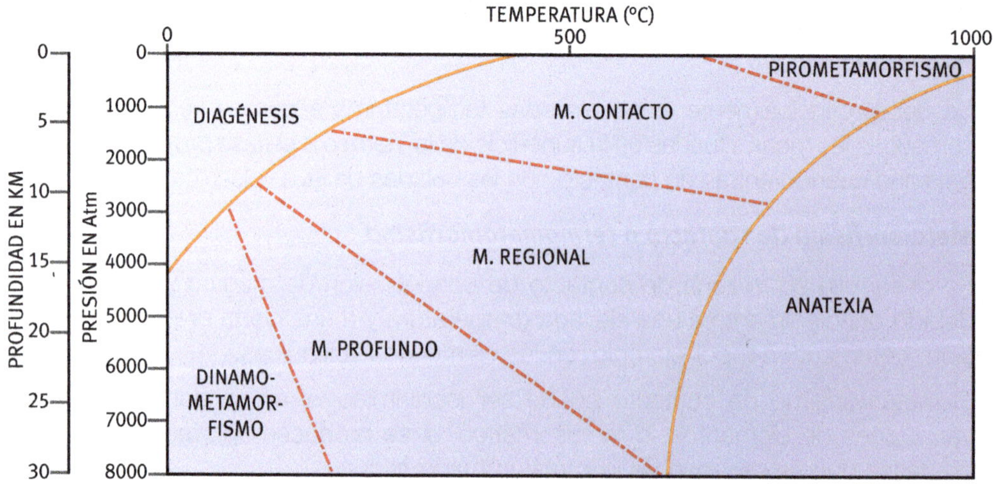

El diagrama recoge las presiones y las temperaturas que caracterizan a cada una de las clases de metamorfismo que acabamos de estudiar.

5.3. *Minerales de origen metamórfico*

Cualquier mineral de la corteza terrestre puede verse afectado por variaciones de presión o temperatura y por la presencia de diversos compuestos químicos. Por lo tanto, la variedad de minerales y rocas que han sido originadas por metamorfismo es muy grande.

De cara al estudio o caracterización mineralógica del metamorfismo, es muy importante tener en cuenta lo que se denomina **campo de estabilidad** o **de equilibrio** de un mineral. Éste es el rango de presión y temperatura para el que el mineral permanece estable.

Así, existen minerales con campos de estabilidad muy amplios que admiten grandes variaciones de presión y temperatura, y cuya presencia en rocas metamórficas poco nos indica sobre la formación de éstas; son los denominados minerales resistentes a los procesos metamórficos. Estos minerales aportan información sobre la mineralogía de las rocas antes del proceso de metamorfismo. Es el caso del cuarzo y la calcita, minerales estables en condiciones muy diversas de metamorfismo.

Sin embargo, existen otros minerales con un campo de estabilidad más reducido y que, sometidos a variaciones de presión y temperatura, se transforman en otros; son los minerales típicamente metamórficos. Es el caso de la andalucita, la distena...

Andalucita

Un importante criterio de clasificación de los minerales es la afinidad geoquímica. Teniendo en cuenta esta afinidad, se trata de establecer las asociaciones de minerales más frecuentes con el fin de caracterizar las condiciones físico-químicas de formación en cada caso.

Así, la asociación de varios minerales metamórficos nos permite conocer más exactamente las condiciones del proceso metamórfico que los ha originado. Una asociación de minerales que caracterizan unas condiciones concretas de metamorfismo se denomina **facies metamórfica.**

La siguiente imagen muestra las principales facies metamórficas e indica para qué valores de presión y temperatura aparecen y cuáles son sus campos de estabilidad.

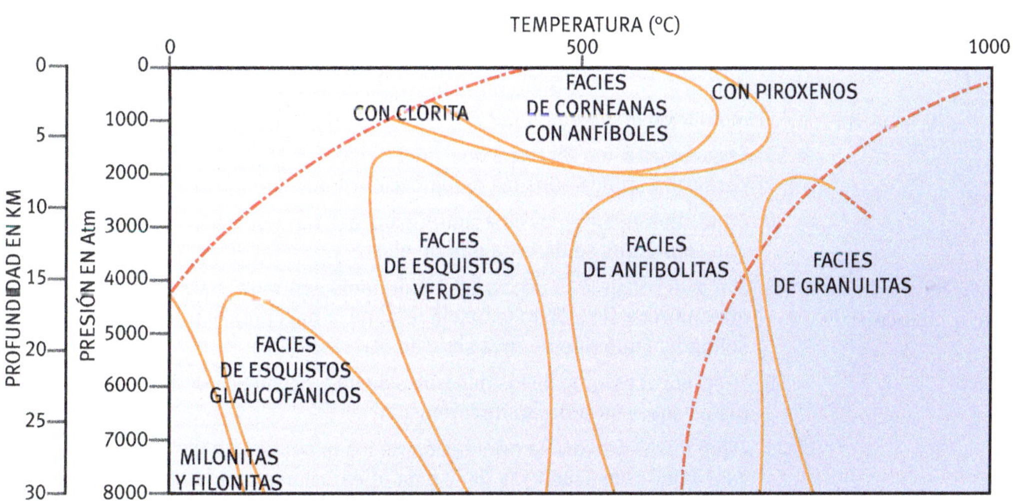

C

Textura granoblástica

Aproximadamente todos los granos minerales son de igual dimensión.

Textura porfiroblástica

Se distinguen unos granos minerales bien desarrollados envueltos por una matriz de grano fino.

Textura poiquiloblástica

Es un caso particular de la textura anterior, en la que se presentan unas inclusiones de uno o varios minerales.

5.4. *Texturas de las rocas metamórficas*

Ya sabemos que los minerales de las rocas metamórficas crecen en un entorno sólido, por lo que se oponen en su expansión a los minerales adyacentes. Ésta es una diferencia muy significativa respecto a los minerales de las rocas ígneas, que crecen inmersos en un fluido, el magma.

Por otro lado, todos los minerales metamórficos se desarrollan sometidos a elevadas presiones. Estos dos factores proporcionan a las rocas metamórficas unas texturas muy características: la **textura granoblástica**, la **textura porfiroblástica** y la **textura poiquiloblástica**.

En las rocas metamórficas, además, cabe indicar la presencia de estructuras nuevas originadas por el metamorfismo y de unas estructuras relictas, es decir, estructuras que se han heredado de la roca primaria y que han resistido al metamorfismo.

La tabla siguiente recoge las nuevas texturas más relevantes generadas por el metamorfismo.

Texturas foliadas Los minerales presentan una orientación perpendicular a las fuerzas de compresión.	**Pizarrosidad** Resulta de la recristalización de los minerales en pequeños microcristales que se alinean paralelamente, de forma que pueden separarse fácilmente en láminas planas.
	Esquistosidad Es consecuencia de temperaturas y presiones mayores que pueden formar cristales más grandes que dan a la roca un aspecto escamoso.
	Bandeado gnésico Se da a temperaturas y presiones muy elevadas, que provocan que los minerales puedan migrar y separarse, dando un patrón de líneas.
Texturas no foliadas	Suelen darse en rocas formadas por un único mineral. En estos casos se forman nuevos cristales relativamente grandes y las impurezas aparecen como bandas de otros colores.

Microfotografía del mármol. Textura granoblástica

ACTIVIDADES

- **18.** Define el concepto metamorfismo. ¿Qué transformaciones puede acarrear el metamorfismo a una roca?
- **19.** Esquematiza los factores que intervienen en el metamorfismo, indicando para cada uno de ellos las modificaciones que comportan.
- **20.** ¿Qué diferencia el pirometamorfismo del termometamorfismo? Indica en qué ambientes se da cada uno de ellos.
- **21.** ¿En qué consiste la anatexia? ¿Podríamos continuar hablando de una roca metamórfica si se diera la fusión total de la roca preexistente? Una vez consolidada, ¿qué tipo de roca sería en ese caso?
- **22.** ¿Por qué el estudio de los minerales de una roca metamórfica nos puede dar pistas sobre su proceso de formación?
- **23.** ¿Qué factor provoca la orientación de los minerales en las rocas metamórficas? ¿Qué consecuencias tiene esta orientación sobre la roca?

24 Vamos a estudiar las propiedades físicas de los minerales siguiendo las pautas indicadas a continuación.

Cada mineral presenta una serie de características físicas fácilmente determinables ya sea a simple vista o por pruebas relativamente sencillas. Utilizando muestras minerales del laboratorio del instituto, realizaremos un análisis de las propiedades físicas de algunos de ellos (halita, mica, galena, calcedonia...).

Analiza las siguientes características y haz una ficha para cada uno de ellos:

- Forma externa y simetría: ¿Puedes ver los cristales que lo forman? ¿Puedes identificar el sistema cristalino?
- Exfoliación: ¿Tiene tendencia a romperse dando lugar a superficies aplanadas?
- Peso específico: ¿Puedes calcular la densidad del mineral? Su peso específico será el cociente entre su densidad y la del agua a 4 °C.
- Dureza: ¿En qué grado se resiste a ser rayado? Para asignarle un grado de dureza a un mineral se compara con el patrón de la escala de Mohs. Un mineral rayará a todos los de número inferior y será rayado por todos los de número superior.
- Superficies de fractura: ¿Cómo son las superficies resultantes cuando se rompe? (regular, irregular, astillosa, curvada, ganchuda...).
- Brillo: ¿Cuál es el aspecto de su superficie cuando refleja la luz? (metálico, adamantino, vítreo, resinoso, sedoso...).
- Color: ¿Presenta algún color o por el contrario es incoloro?
- Raya: ¿Cómo es la raya que queda en su superficie cuando lo rascamos con una placa de loza blanca no vidriada? ¿El polvo producido tiene algún color característico?

Supón que encuentras en el campo un mineral de aspecto vítreo y quieres saber si es un diamante. ¿Qué prueba sencilla podrías realizar para confirmarlo?

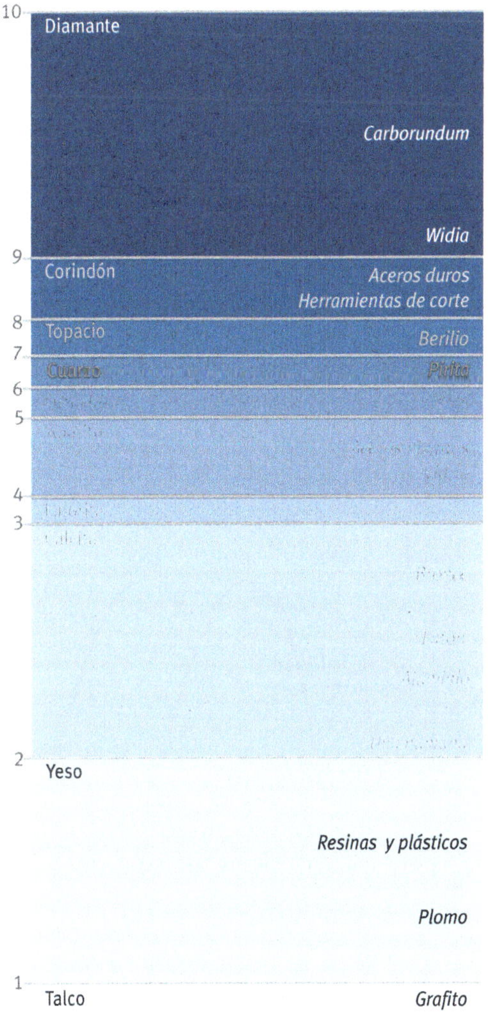

Escala de Mohs. Equivalencia de la escala de Mohs con diversos objetos y materiales.

Fractura astillosa. Distena

Fractura concoidea. Obsidiana

Fractura regular. Halita

25 Para identificar de ciertas rocas y minerales es necesario realizar determinadas pruebas químicas. Una de las más habituales es la siguiente:

— Procura obtener muestras de caliza y dolomía. La caliza está formada fundamentalmente por calcita ($CaCO_3$), mientras que el componente fundamental de la dolomía es la dolomita ($CaMg(CO_3)_2$). Ambas rocas pueden ser diferenciadas fácilmente mediante una prueba química sencilla.

— Deja caer una gota de ácido clorhídrico concentrado sobre cada una de las muestras. ¿Qué observas?

Si sabemos que el ácido clorhídrico reacciona con el carbonato de calcio produciendo ácido carbónico, que se descompone en CO_2 y agua, provocando efervescencia, ¿cuál de las dos muestras corresponde a la caliza y cuál a la dolomía?

26 Observa la siguiente gráfica y responde las preguntas que tienes a continuación.

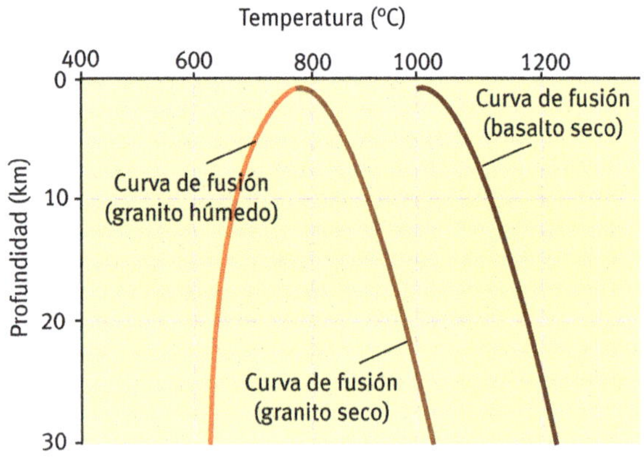

— ¿Cuál de las dos rocas (basalto y granito) es una roca ácida y cuál una roca básica? Comenta cuál de las dos (en seco) se fundirá más fácilmente e intenta razonar el porqué en función de su composición.

— Justifica a partir de la gráfica la afirmación de que los magmas basálticos se generan en zonas profundas de la corteza terrestre.

— ¿Qué influencia tiene la presencia de volátiles (agua) sobre la temperatura de fusión de las rocas? ¿Cómo afecta a la influencia de la presión? Justifica tus respuestas a partir de la gráfica.

27 Examina muestras de las siguientes rocas :

Rocas magmáticas: granito, diorita, gabro, basalto, obsidiana, pumita.

Rocas sedimentarias: conglomerado, brecha, arenisca, caliza, antracita.

Rocas metamórficas: esquisto, pizarra, mármol, gneis.

— Confecciona una ficha para cada una de ellas en la que hagas constar su nombre, clasificación, origen, textura, componentes...

28 Examina la ilustración siguiente y responde a las preguntas.

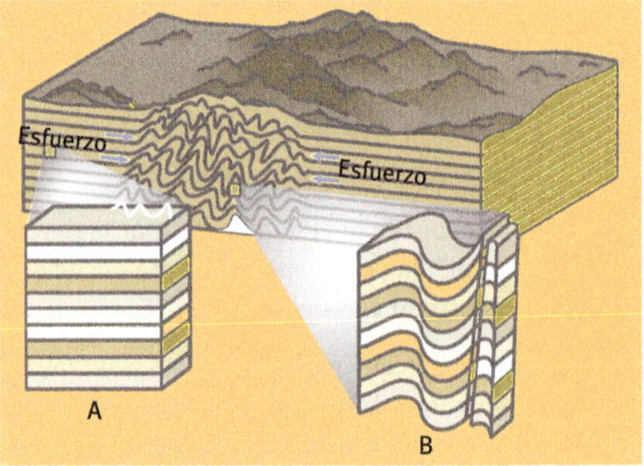

— En la zona A, ¿qué tipo de roca encontramos? ¿Qué elementos te permiten realizar esta afirmación? ¿Y en la zona B? ¿Por qué?

— Si originariamente ambas zonas tenían la misma composición a nivel de rocas, ¿qué proceso ha dado lugar a la diferenciación actual? ¿Cuál ha sido el factor principal? ¿Qué efectos habrá tenido sobre la textura de la roca y sus componentes?

29 Observa la ilustración correspondiente a un corte geológico del Gran Cañón del Colorado en Estados Unidos, y responde a las preguntas siguientes.

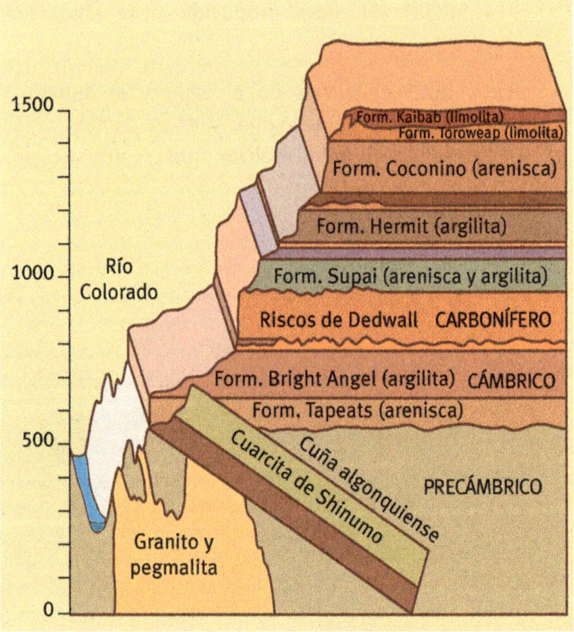

— ¿Qué tipo de rocas es el componente fundamental a partir de los 500 m de altitud? ¿Qué te permite afirmarlo?

— Ordena de la más antigua a la más moderna las siguientes formaciones: Kaibab, Hermits, Tapeats, Supai, Coconino. Justifica tu respuesta.

— ¿En cuál de las formaciones podrías hallar fósiles de trilobites? ¿Y de anfibios? ¿Por qué?

TIC

30 En la página http://www.sciencecourseware.org/VirtualDating/ hallarás explicaciones y ejercicios (en inglés) sobre técnicas de datación.

— ¿Qué es la vida media de un elemento radiactivo?

— Si un isótopo tiene una vida media de 100 millones de años, ¿cuál será la edad de una muestra que contiene un 15% de isótopos originales? Básate en la gráfica interactiva de la web.

— Explica el proceso a seguir para datar una muestra de roca por el método del Rb-Sr.

— ¿Para qué materiales es adecuada la datación con C14? ¿Permite datar helechos del Carbonífero? ¿Por qué?

31 La luz polarizada y el estudio de las rocas: el microscopio petrográfico.

La luz normal se transmite mediante ondas que oscilan en múltiples planos, pero si la hacemos atravesar determinados medios, llamados polarizadores, únicamente consigue atravesarlos la luz que oscila en un plano determinado: a partir de aquel punto todas las ondas de luz oscilarán siguiendo el mismo plano.

La luz que cumple con esta característica se conoce como luz polarizada. Si interponemos nuevamente un polarizador a un haz de luz polarizada, ésta únicamente conseguirá atravesarlo si coincide su plano de polarización con el del nuevo filtro.

Uno de los instrumentos fundamentales de los geólogos para la identificación de los minerales es el microscopio petrográfico. Este tipo de microscopio dispone de un polarizador o prisma de Nicol que permite polarizar la luz con la que examinamos una fina lámina de roca, y otro polarizador, el analizador, que determina el ángulo que ha sido desviada la luz al atravesar la muestra.

El ángulo desviado vendrá determinado por la estructura cristalina de los diferentes minerales, permitiendo una determinación de los mismos.

Examina diversas muestras de rocas magmáticas y metamórficas con el microscopio petrográfico.

Muestra de gabro al microscopio petrográfico

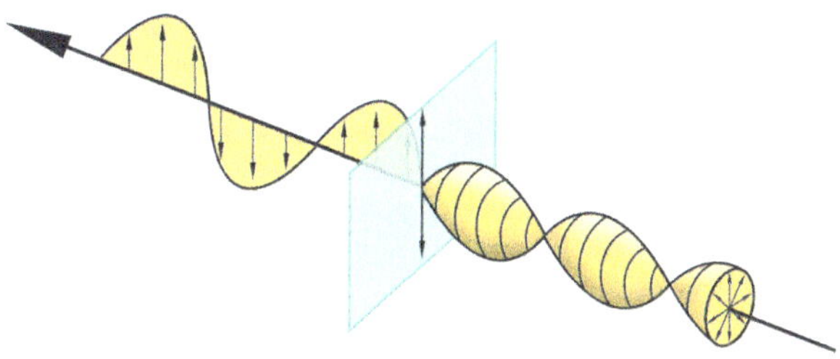

Acción de un polarizador sobre un haz de luz.

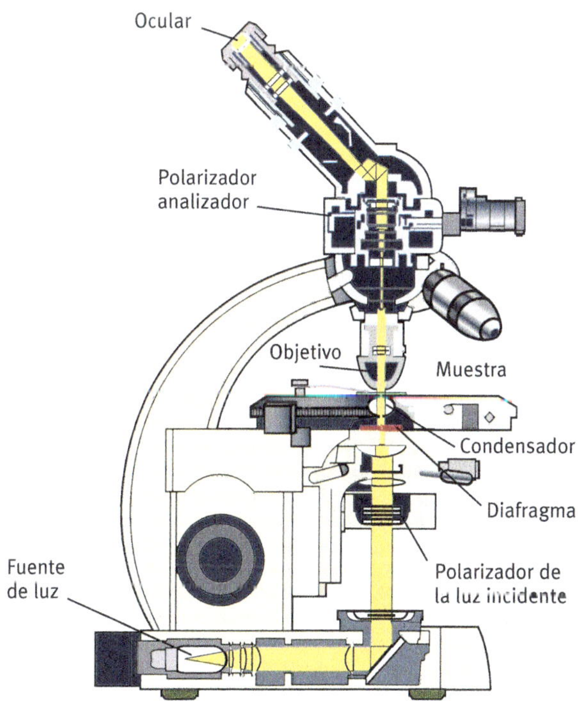

Microscopio petrográfico

3

La actividad geológica interna

El interior de la Tierra es un medio difícil de estudiar por su inaccesibilidad y por las condiciones que imperan de elevadas temperaturas y presiones.

A pesar de estas dificultades, existen metodologías que nos permiten conocer numerosos aspectos de este medio.

El análisis de las ondas sísmicas, el estudio de la actividad volcánica y la interpretación de numerosos aspectos del relieve nos informan de la existencia en el interior de la Tierra de diferentes capas de materiales, fluidos y sólidos.

El comportamiento de estas capas del interior de la Tierra nos explica muchos de los fenómenos geológicos que suceden en la superficie del planeta, tanto a pequeña como a gran escala.

El conocimiento de todos estos fenómenos nos permite valorar el riesgo que representa la realización de determinadas intervenciones en el medio ambiente, así como la prevención de catástrofes relacionadas con las actividades humanas.

Volcán Semeru en erupción. (Java, Indonesia)

ORGANIZACIÓN DE LOS CONTENIDOS

OBJETIVOS

- Relacionar la actividad que tiene lugar en los límites de las placas litosféricas con el desplazamiento que experimentan los continentes.
- Comprender, a partir de datos obtenidos por la sismología, la estructura interna de la Tierra.
- Diferenciar los pliegues de las fallas e identificar las partes de ambos tipos de deformaciones.
- Mostrar interés por la identificación en el entorno de formaciones geológicas relacionadas con la actividad interna de la Tierra.

El sondeo más profundo realizado hasta la actualidad es el Kola Superdeep Borehole, situado en la península de Kola, en Escandinavia. Se inició en 1970 y se dio por finalizado en 1994; alcanzó los 12 261 m.

1. Geofísica

La **geofísica** es la ciencia que, utilizando los métodos de la física, estudia la totalidad de la Tierra, desde el centro hasta la atmósfera superior.

En la unidad anterior hemos visto que la Tierra está formada por diferentes materiales, y sabemos que en su interior tienen lugar distintos fenómenos. Para estudiar la composición de la Tierra y su actividad interna se utilizan diversos métodos que dividimos en: **métodos directos** y **métodos indirectos.**

Los métodos directos se basan en el estudio de los materiales y son:

- **Estudio de las rocas de la superficie.** Se basa en la observación directa de las rocas. Si se conoce su composición, localización, deformación..., se puede obtener información del interior de la Tierra, ya sea porque las rocas estudiadas se han formado en su interior o porque la actividad de nuestro planeta las ha modificado.
- **Estudio de los meteoritos.** Los meteoritos son fragmentos de roca procedentes del espacio que caen sobre la superficie terrestre. Su origen es el mismo que el de la Tierra y, por lo tanto, su análisis químico proporciona información sobre cuál puede ser la composición interna de nuestro planeta.
- **Realización de sondeos.** Los sondeos son excavaciones de 10 a 15 centímetros de diámetro y de varios kilómetros de profundidad. El sondeo más profundo hasta la actualidad alcanzó más de 12 km y se realizó en la península de Kola (Rusia) en el año 1984. Este método permite conocer la composición y estructura de los materiales a diferentes profundidades.

La limitación impuesta por los métodos directos obliga a los geólogos a recurrir a los métodos indirectos para estudiar la corteza terrestre e investigar las otras capas del planeta.

A continuación describiremos algunos de los métodos indirectos que se utilizan para investigar el interior del planeta: la *gravimetría,* el *geomagnetismo,* la *geotermia* y la *sismología.*

1.1. *Gravimetría*

La gravimetría estudia el campo gravitatorio de la Tierra y sus anomalías.

La ley de gravitación universal formulada por Newton en 1686 establece que *la fuerza de atracción que existe entre dos cuerpos cualesquiera es directamente proporcional al producto de sus masas e inversamente proporcional al cuadrado de la distancia que separa sus centros.*

$$F = G \frac{m_1 \cdot m_2}{d^2}$$

Luego, si M es la masa de la Tierra y R su radio, la fuerza con que la Tierra atraería a un objeto de masa m situado sobre su superficie es:

$$F = G \frac{M \cdot m}{R^2}$$

Dividiendo ambas partes de la igualdad por la masa *m* del objeto, obtenemos la aceleración *(g)* con la que la Tierra atrae al objeto. Sabemos que:

$$F = m \cdot a \qquad \text{luego: } \frac{F}{m} = G\frac{M}{R^2} = g$$

Si sabemos que la densidad se define como: $d = \dfrac{M}{V}$ y consideramos que la Tierra es una esfera, tenemos que:

$$d = \frac{M}{(\frac{4}{3}\pi R^3)}, \text{ de donde: } M = \frac{4}{3}(d\pi R^3)$$

Sustituyendo este valor en el correspondiente de la expresión de la aceleración de la gravedad, se cumple que:

$$g = \frac{4}{3}(Gd\pi R)$$

Fíjate en que *g* depende únicamente del radio de la Tierra *R* y de la densidad *d*.

Si en el cálculo teórico de *g* introducimos todas las correcciones necesarias en función de la latitud y la altitud, dicho valor sólo depende de la densidad del material. Sin embargo, al efectuar la experiencia de determinar el valor de la gravedad en áreas montañosas para observar cómo la masa montañosa lo afecta, se obtienen valores menores de los esperados.

Este hecho se explicaría mediante la presencia de materiales ligeros debajo de las montañas que compensarían el material adicional que suponen dichas montañas. Estos estudios de gravimetría dieron lugar a diferentes hipótesis de la estructura interior de la Tierra.

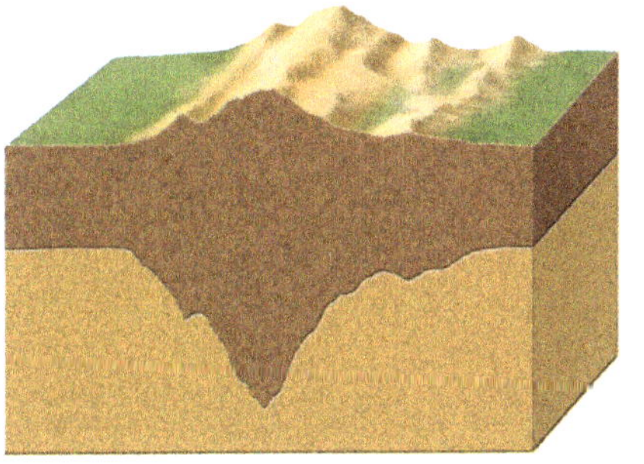

Unos bloques de madera de diferentes grosores flotan sobre el agua alcanzando diferentes alturas. En la litosfera terrestre sucede lo mismo: las placas litosféricas de mayor grosor (que coinciden con los sistemas montañosos) se hunden más en el manto que las zonas de la litosfera de menor grosor (que coinciden con la litosfera terrestre de zonas bajas o con la oceánica).

Isostasia

Al admitir que las montañas poseen raíces incrustadas en materiales más densos, se acepta que el material pueda desplazarse en profundidad, estableciéndose un equilibrio hidrostático, o **isostasia,** en un nivel inferior. Éste se denomina de compensación y se correspondería con el límite entre el material superior (ligero y rígido), llamado *litosfera,* y el material inferior (denso y plástico), llamado *astenosfera.*

Así pues, consideramos la isostasia como la condición de equilibrio que regula la posición de los materiales de la litosfera. Este equilibrio no es un hecho estático, sino que fenómenos como la formación o fusión de casquetes glaciares, o la erosión del relieve y su sedimentación en las cuencas alteran su nivel. El principio de isostasia fue formulado por vez primera por Airy y Pratt en 1850.

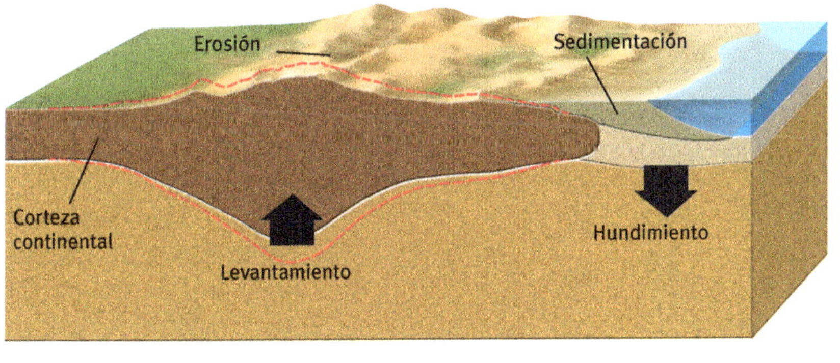

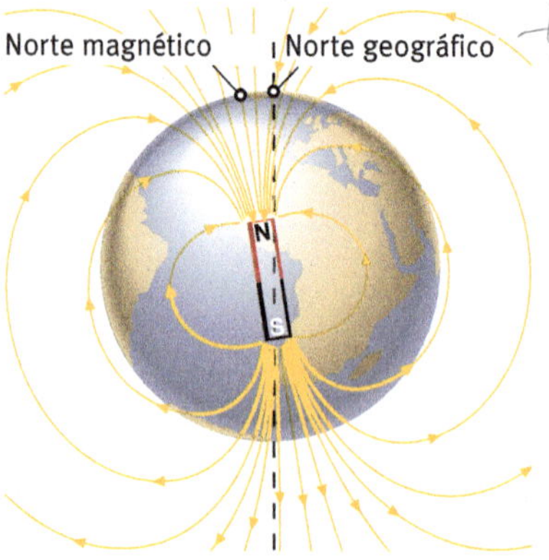

1.2. *Geomagnetismo*

El estudio de los efectos producidos por el campo magnético terrestre, o geomagnetismo, es otro método indirecto utilizado para investigar el interior del planeta. El campo magnético terrestre es similar al que produciría un dipolo magnético situado en el centro de la Tierra. Fíjate, en la imagen de la izquierda, que el eje del dipolo forma un cierto ángulo respecto al eje de rotación de la Tierra, con el polo norte magnético en el hemisferio Norte y el polo sur magnético en el hemisferio Sur.

En el campo magnético terrestre se detectan diferencias entre los valores medidos en el campo con un magnetómetro y los valores calculados teóricamente. Estas anomalías son consecuencia, principalmente, de la presencia de minerales o rocas con propiedades magnéticas. De entre los minerales con propiedades magnéticas cabe destacar la magnetita y, de entre las rocas, los gabros y basaltos, con abundantes silicatos ferromagnesianos.

C

El campo magnético de la Tierra se define en cada punto de la superficie terrestre mediante tres magnitudes: *declinación* (D), *inclinación* (I) e *intensidad* (F).

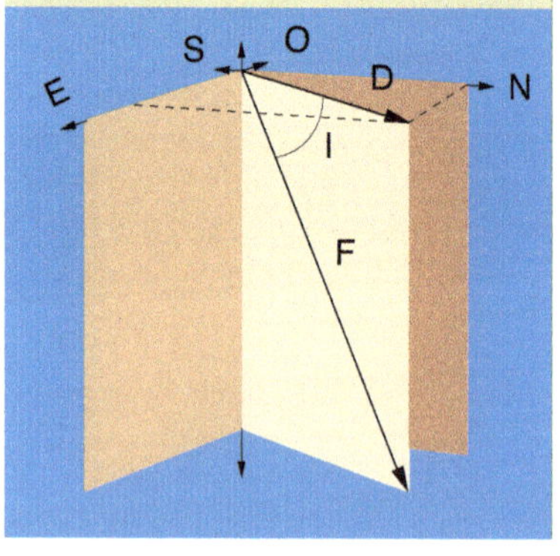

Diversas investigaciones han comprobado que el campo magnético terrestre varía a lo largo del tiempo, con un período del orden de siglos. Este fenómeno recibe el nombre de **variación secular.** Aunque el registro de variaciones obtenido directamente es muy corto, poseemos un registro de las variaciones del campo magnético en tiempos pasados, recogido y preservado en las rocas.

Al efectuar estudios de magnetismo, se descubrió la existencia en los cinco continentes de rocas magnetizadas en sentido diferente al actual campo magnético, algunas con variaciones de hasta 180°. Estos últimos casos son consecuencia de las denominadas **inversiones magnéticas,** o cambios de polaridad del campo magnético terrestre.

Sobre las rocas volcánicas que solidifican en superficie actúa, en el proceso de su formación, el campo magnético terrestre. Por ello, los minerales ferromagnesianos de la roca en formación adoptan la dirección del campo magnético, como lo haría la aguja imantada de una brújula. Así pues, al solidificar la roca, dichos minerales quedan orientados indicando la dirección y el sentido del campo magnético en ese momento.

En los últimos 3,5 millones de años ha habido por lo menos 10 inversiones de polaridad en el campo magnético de la Tierra. Las épocas en las cuales la polaridad coincide con la actual se las denomina épocas normales, mientras que a las de polaridad opuesta, épocas invertidas. A su vez, en cada época pueden aparecer periodos relativamente breves de cambio de polaridad, llamados eventos. El gráfico siguiente muestra las inversiones de la polaridad magnética de la Tierra en los últimos 4 millones de años.

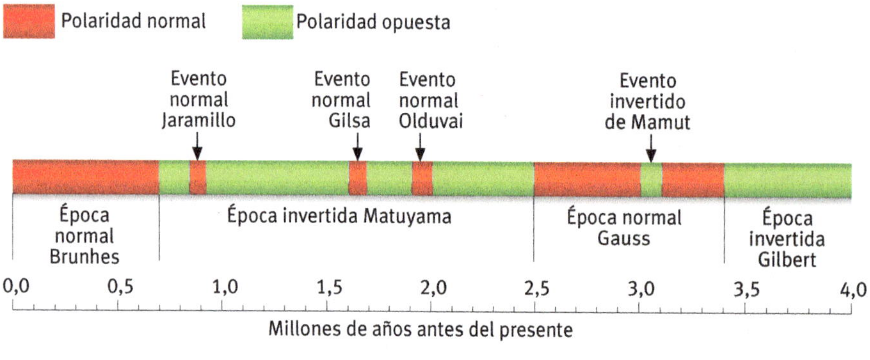

1.3. *Geotermia*

La geotermia es la parte de la geofísica que estudia el calor y las temperaturas internas en la Tierra. Se tiene conocimiento del calor interno de la Tierra por los volcanes y, además, por el desprendimiento de energía que se registra en las áreas tectónicamente activas. Sin embargo, la mayor parte del flujo calórico tiene lugar por conducción, y ha podido ser detectado mediante sondeos.

El ascenso de la temperatura con relación a la profundidad se denomina **gradiente térmico**. Por convenio se acepta que el valor medio del aumento del gradiente es de 1 °C por cada 33 m de profundidad.

Este ascenso únicamente es aceptable para los primeros 5 o 6 km de profundidad, y a partir de esta profundidad se supone una elevación de la temperatura mucho menor. En caso contrario, obtendríamos unas temperaturas en el centro de la Tierra incompatibles con el estado en que se supone se encuentra la materia del núcleo.

Además, cabe destacar que este aumento del gradiente geotérmico para los primeros kilómetros tampoco se distribuye uniformemente, puesto que depende de dos factores, fundamentalmente, la conductividad de las rocas y el propio flujo de calor.

Se han observado y delimitado áreas, tanto en el ámbito oceánico como en el continental, en las que el flujo de calor es significativamente diferente. Estas zonas reciben el nombre de **áreas geotérmicas**. Así, en el ámbito oceánico se ha detectado un mayor flujo de calor en las zonas de las dorsales oceánicas, que disminuye a medida que nos alejamos de ellas. En cuanto al ámbito continental, las áreas orogénicamente activas son las que poseen un flujo calorífico mayor que el resto.

C

En algunos casos, encontramos zonas de la parte superior del manto que presentan un flujo de calor hacia un área central de una placa litosférica.

Esta columna de magma ascendente, o **pluma térmica**, puede llegar a la superficie de la litosfera y producir una surgencia magmática, o **punto caliente**.

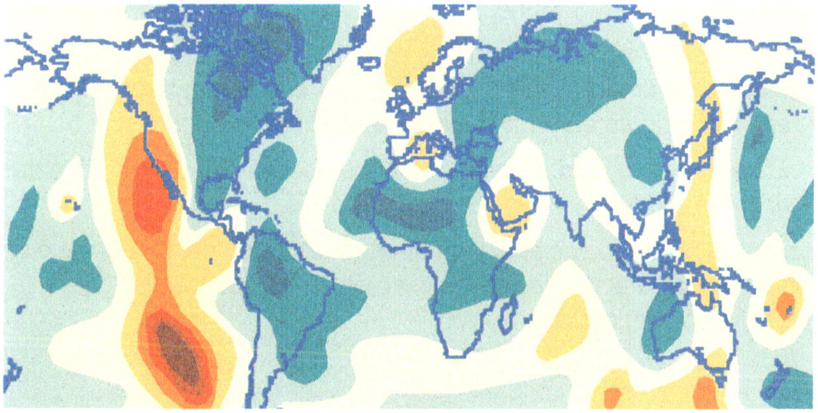

El calor de la zona está representado mediante una gradación de colores: el rojo representa las zonas más cálidas, y el verde, las más frías.

En general, se puede considerar que el flujo calorífico es significativamente superior al valor medio en las zonas activas de la corteza terrestre. Una excepción son las fosas abisales, que, a pesar de ser áreas de gran actividad, poseen un flujo calorífico muy bajo.

Se supone que el origen del calor interno de la Tierra es, fundamentalmente, consecuencia de procesos radiactivos. Luego la fuente de energía que provoca fenómenos como el vulcanismo o la deformación de la corteza terrestre es el calor producido por la desintegración de isótopos radiactivos.

A

Podemos ejemplificar el proceso de una **erupción volcánica** con una bebida con gas.

En estas bebidas, la presión del interior de las botellas es elevada y, por ello, una gran parte del dióxido de carbono se mantiene disuelto, mientras que el resto se encuentra en forma de diminutas burbujas. Si agitamos la botella, la presión aumenta y todo el gas se disuelve.

Al abrir la botella, la presión disminuye bruscamente y se forman grandes burbujas de gas que arrastran el líquido hacia el exterior.

Actividad volcánica

La actividad volcánica es una manifestación de la energía interna de la Tierra.

Esta actividad se produce mediante la emisión rápida del magma, procedente del interior de la Tierra, a la superficie.

El magma se encuentra sometido a elevadas presiones y, debido a ello, contiene disueltas grandes cantidades de vapor de agua, dióxido de carbono y otros gases. A medida que se acerca a la superficie, la presión disminuye y los gases forman grandes burbujas que lo arrastran hacia el exterior a través de grietas o fracturas. Esta emisión del magma se denomina **erupción volcánica.**

Los materiales emitidos en una erupción volcánica son:

- **Gases.** El más abundante es el vapor de agua y, en menor cantidad, se expulsa dióxido de carbono, dióxido de azufre, nitrógeno, hidrógeno, etc.
- **Lava.** Es el magma en estado fluido, que discurre por la superficie terrestre a mayor o menor velocidad según su viscosidad.
- **Productos piroclásticos.** Son fragmentos de roca arrancados del conducto por el que asciende el magma o fragmentos de lava que se han solidificado al contactar con el aire. Estos fragmentos se clasifican, de mayor a menor tamaño, en: bombas, lapilli y cenizas.

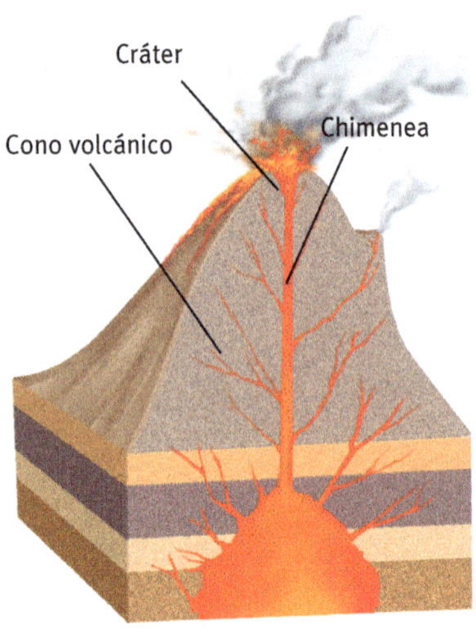

Las erupciones volcánicas pueden ser de distinto tipo:

- **Erupciones explosivas.** Son las más violentas; en ellas se liberan grandes cantidades de material piroclástico.
- **Erupciones efusivas.** Se producen emisiones continuas de lava muy fluida sin explosiones.
- **Erupciones extrusivas.** La lava es tan viscosa que se acumula a la salida originando un tapón en forma de cúpula. La lava que sigue ascendiendo se abre paso por otras grietas del cono.
- **Erupciones mixtas.** La lava es menos viscosa que en el caso anterior y se producen explosiones moderadas.

La forma del cono volcánico, es decir, la elevación del terreno que se produce por la acumulación de los materiales expulsados, está relacionada con el tipo de erupción.

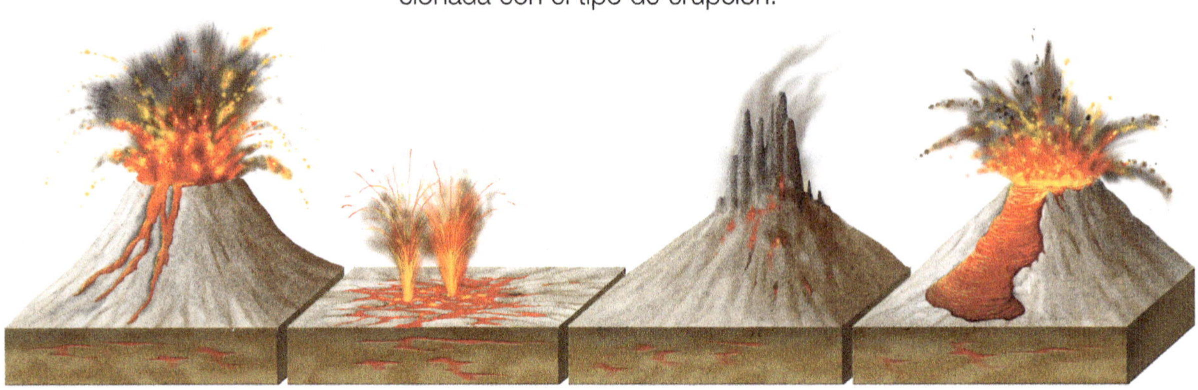

Erupción explosiva — Erupción efusiva — Erupción extrusiva — Erupción mixta

1.4. *Sismología*

Los seísmos o terremotos son vibraciones del terreno debidas a la fricción o fractura de bloques de la litosfera o bien originadas como consecuencia de la actividad magmática.

Los diversos procesos que se producen en el interior de la Tierra provocan la acumulación de tensiones en algunas zonas. Cuando las tensiones son excesivas, las rocas se rompen y se libera una gran cantidad de energía. Esta energía se transmite en todas las direcciones en forma de ondas sísmicas; su velocidad de propagación depende de las propiedades físicas del medio que atraviesan.

La zona del interior de la Tierra donde se produce la fractura es el **foco sísmico** o **hipocentro.**

El **epicentro** es el punto de la superficie terrestre que se encuentra sobre la vertical del hipocentro; en él se registra la máxima intensidad.

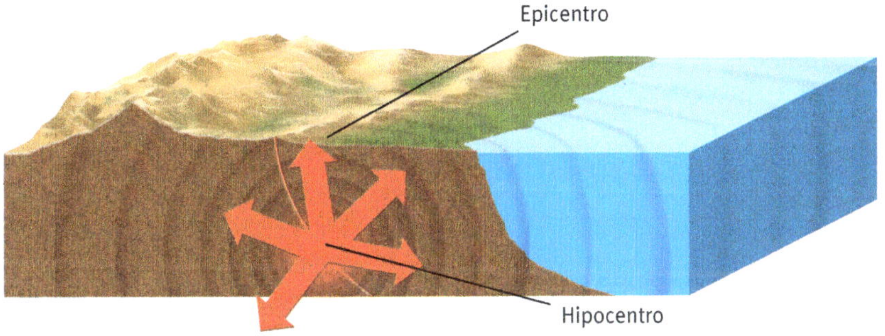

A

Según la profundidad del hipocentro, los terremotos pueden clasificarse en:

- Terremotos de foco superficial, de 0 a 70 km.
- Terremotos de foco intermedio, de 70 a 300 km.
- Terremotos de foco profundo, de más de 300 km.

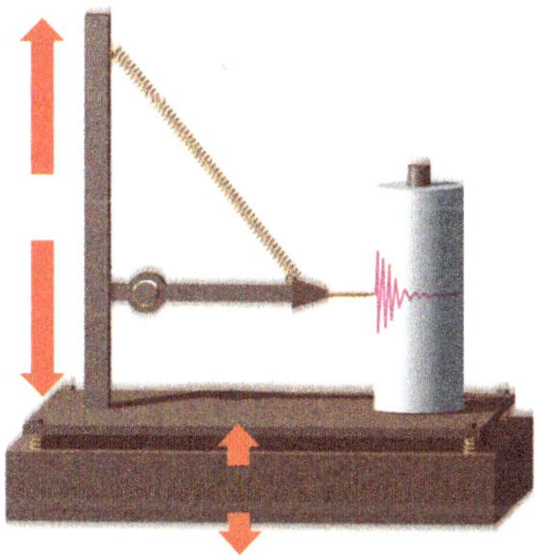

Sismógrafo vertical

Los **sismógrafos** son los aparatos que registran las ondas sísmicas producidas durante un terremoto. Los sismógrafos más sencillos son de dos tipos: los que detectan los movimientos horizontales y los que detectan los movimientos verticales. Ambos tienen como fundamento una esfera muy pesada, que está unida a una pluma, y un rodillo, sujeto a la base, que oscila cuando se produce una vibración. Por el rodillo pasa una tira de papel sobre la que la pluma escribe constantemente.

El registro de la pluma sobre el papel constituye el **sismograma.** En caso de inactividad sísmica, las posiciones relativas del rodillo y la esfera se mantienen y la pluma traza una línea prácticamente recta. Cuando se produce un seísmo, la pluma dibuja una línea quebrada con picos que reflejan los diferentes tipos de ondas sísmicas.

- **Ondas P** o **longitudinales.** Las partículas vibran paralelamente a la dirección de la onda. Atraviesan cualquier tipo de material, a más velocidad cuanto más rígido sea. Son las primeras en detectarse.
- **Ondas S** o **transversales.** Las partículas vibran perpendicularmente a la dirección de propagación. No se propagan en los medios fluidos, y su velocidad aumenta a medida que el material es más rígido.
- **Ondas superficiales.** Se propagan por la superficie y son la causa de los desperfectos. Éstas pueden ser de dos tipos:
  - **Ondas L** o **Love.** Provocan movimientos laterales de las partículas.
  - **Ondas R** o **Rayleigh.** Originan movimientos en forma de olas en las partículas.

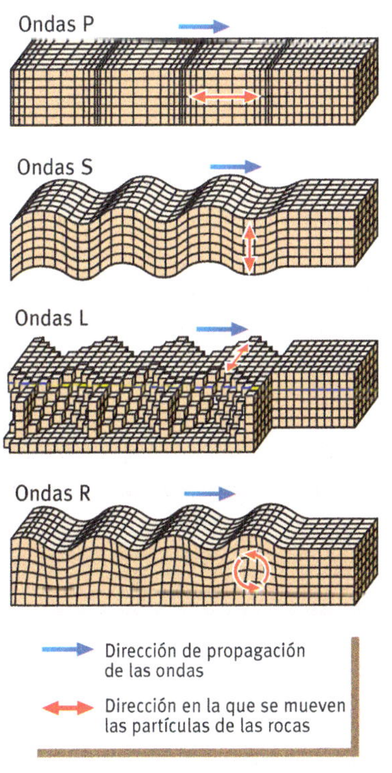

Consideremos el siguiente ejemplo: se produce un terremoto de hipocentro H y sus ondas son registradas por los sismógrafos de cuatro observatorios situados en distintos puntos de la superficie del planeta.

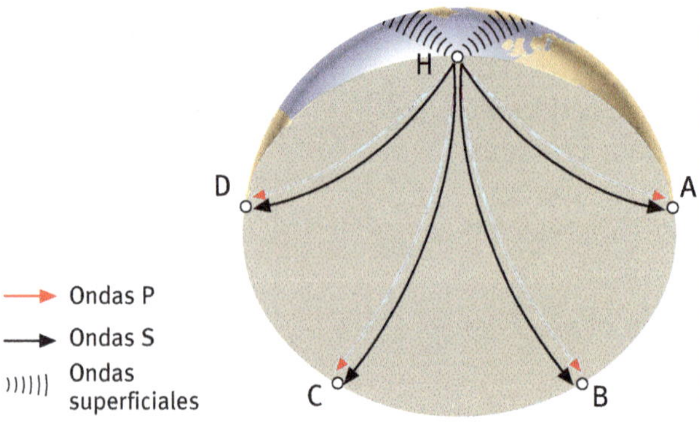

Si la Tierra fuese homogénea, la velocidad de propagación de las ondas sería constante, ya que no atravesarían materiales de distinta composición y estructura.

Las únicas diferencias entre los registros de los diferentes observatorios se deberían a que las ondas llegan en momentos distintos, a causa de las diferencias entre las distancias que separan el hipocentro de cada observatorio.

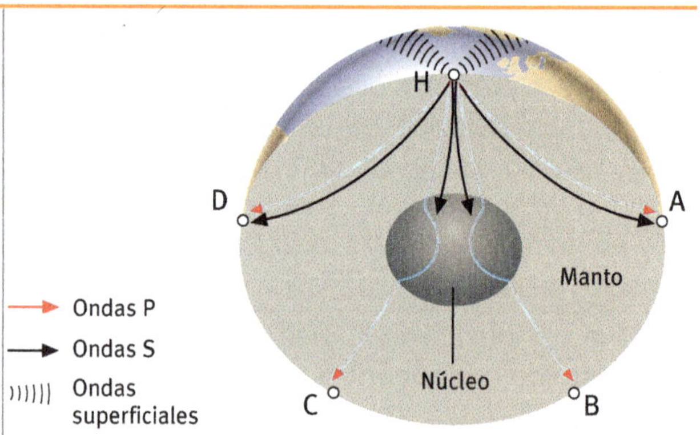

En realidad, los observatorios A y D reciben las ondas P y S, mientras que B y C sólo reciben las ondas P. La desaparición de las ondas S nos indica la presencia de una capa fluida.

La velocidad de propagación de las ondas no es constante y, por tanto, la distancia de los observatorios al hipocentro no es el único factor determinante de las diferencias en el registro sísmico.

Corte transversal de la Tierra

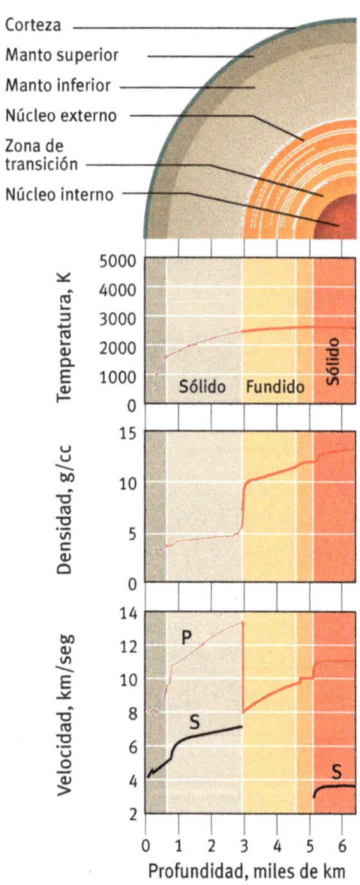

Con los sismogramas procedentes de varios observatorios se trazan las gráficas sísmicas, es decir, gráficas que relacionan la velocidad de propagación de las ondas P y S con la profundidad de la Tierra.

A partir del estudio de estas gráficas, hemos podido comprobar que nuestro planeta no es homogéneo, sino que presenta **capas** de distinta composición y estructura, y también unas **discontinuidades,** es decir, zonas que separan materiales distintos.

Observa la gráfica lateral, que nos proporciona la siguiente información:

- De 0 km a 70 km: la velocidad de las ondas P y S se mantiene constante. Esta zona corresponde a la primera capa, que tiene un grosor máximo de 70 km.
- 70 km: se observa una discontinuidad muy marcada.
- 70 km-2900 km: la velocidad aumenta mucho hasta los 700 km, a partir de este valor, se incrementa con más lentitud. El cambio de velocidad indica una discontinuidad; por tanto, la segunda capa se divide en dos partes.
- 2900 km: hay una discontinuidad importante y las ondas S dejan de propagarse.
- 2900 km-6371 km: la velocidad de las ondas P aumenta hasta los 5100 km y a partir de este valor es constante. El cambio de velocidad se debe a una discontinuidad en la capa más interna de la Tierra.

Observa en el dibujo de la izquierda las capas en las que se divide el interior de la Tierra y las distintas discontinuidades que las separan.

La estratificación en capas de diferente composición se produjo en las primeras fases de formación del planeta, cuando éste pasaba por un estado de fusión, probablemente como consecuencia del hundimiento de los materiales más pesados a medida que los menos densos ascendían. A causa de esta diferenciación podemos identificar tres regiones principales en el interior de la Tierra:

- La **corteza**, capa muy fina, de entre 3 y 70 km de espesor, compuesta fundamentalmente por oxígeno, silicio, aluminio y hierro.
- El **manto**, una capa de roca sólida muy rica en sílicio, magnesio, oxígeno y hierro, que alcanza los 2900 km de profundidad.
- El **núcleo**, una esfera compuesta por hierro y níquel con un radio de 3486 km.

La temperatura y la densidad condicionan el carácter plástico o rígido de los materiales que constituyen las capas terrestres. Así, en función de este comportamiento se distinguen nuevas capas en el interior de la Tierra.

- **Litosfera.** Comprende la corteza oceánica y continental, y los primeros 100 km del manto superior. Es sólida y tiene un comportamiento rígido.
- **Astenosfera.** Corresponde a la porción intermedia del manto superior. Es sólida, pero contiene una parte de materiales fundidos que le confieren un carácter plástico.
- **Mesosfera.** Comprende una parte del manto superior y todo el manto inferior. La mesosfera es rígida.
- **Endosfera.** Corresponde a todo el núcleo. El núcleo interno es sólido y tiene un comportamiento rígido mientras que el externo es fluido y tiene un comportamiento plástico.

Estudios recientes indican la ausencia de la capa fluida que representa la astenosfera bajo bastantes zonas de la corteza terrestre, por lo cual aquella no representaría realmente un nivel universal. Se trataría más bien de zonas en las cuales hallaríamos material semifundido proveniente de penachos o plumas térmicas que alcanzan la zona directamente desde la interfase entre núcleo y manto. Esto implicaría que la litosfera reposa sobre el manto sólido.

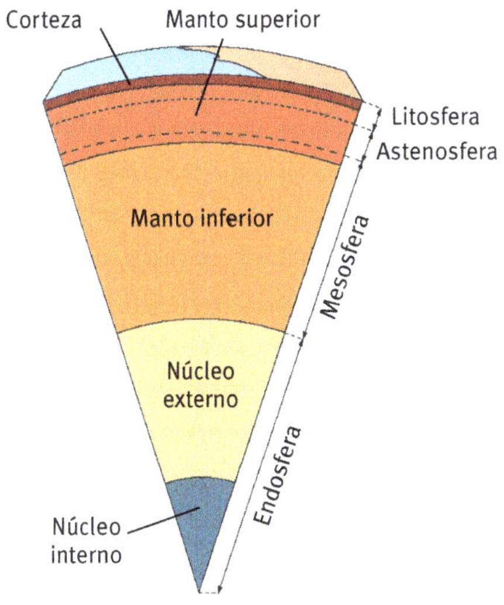

C

Andrija Mohorovicic (1857-1936), científico croata, fue profesor de geofísica y astronomía de la Universidad de Zagreb. Postuló que la Tierra estaba formada por capas concéntricas situadas alrededor de un núcleo. En su honor, la discontinuidad que separa la corteza terrestre del manto recibe el nombre de **discontinuidad de Mohorovicic** o Moho.

ACTIVIDADES

1. ¿Cómo podemos conocer la estructura y composición del interior de la Tierra? Enuncia cinco métodos de estudio del interior terrestre y clasifícalos en directos e indirectos.
2. Basándote en la isostasia, ¿podrías justificar por qué la corteza terrestre tiene un grosor de tan sólo 3 km en los fondos oceánicos, mientras que bajo el Himalaya alcanza los 70 km?
3. ¿A qué nos referimos cuando hablamos de un periodo de polaridad normal? ¿Y cuándo nos referimos a uno de polaridad invertida?
4. ¿Cuál se cree que es la fuente del calor interno de nuestro planeta? ¿Es homogéneo el flujo geotérmico en toda la superficie de la Tierra? ¿De qué factores depende?
5. ¿Qué tipos de erupciones volcánicas conoces? ¿A qué son debidas estas tipologías?
6. ¿Qué es el epicentro de un terremoto? ¿Y el hipocentro?
7. Explica cuáles son las tipologías de ondas sísmicas. ¿En qué se diferencian?
8. ¿Por qué las ondas sísmicas nos permiten estudiar el interior de la Tierra?

2. Tectónica global

Aunque hasta mediados del siglo XX los científicos no aceptaron la idea de que los continentes se desplazan, ya en el primer cuarto del siglo XVII el filósofo inglés F. Bacon puso de manifiesto la coincidencia entre la costa occidental de África y la costa oriental de Sudamérica.

Todas las modificaciones que tienen lugar en la superficie terrestre y que hemos estudiado encuentran su explicación en la teoría de la **tectónica de placas,** que se desarrolló a finales de los años sesenta, gracias a los avances en el conocimiento de la *deriva continental,* las *corrientes de convección* y la *expansión y subducción de la litosfera.*

La tectónica de placas, o tectónica global, propone que la litosfera está formada por una serie de placas en continuo movimiento, debido a la actividad de la astenosfera.

Para los detractores de la presencia de la astenosfera, el movimiento de las placas se daría gracias a un movimiento de convección global del manto terrestre, que se comportaría de manera plástica aun hallándose en estado sólido.

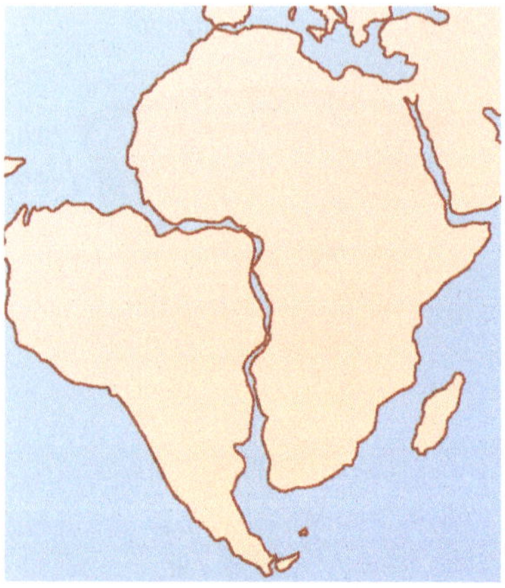

Pruebas geográficas: encaje entre las costas de África y Sudamérica

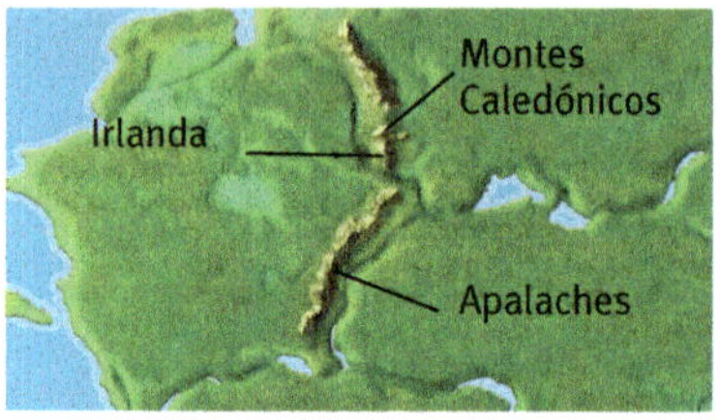

Pruebas geológicas: continuidad de cordilleras antiguas

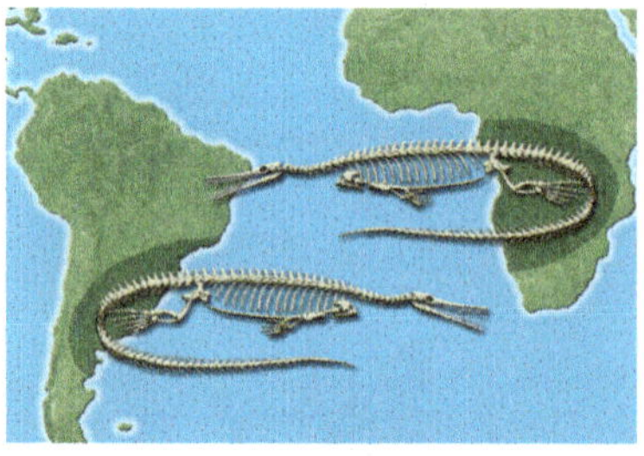

Pruebas paleontológicas: distribución de fósiles de *Mesosaurus*

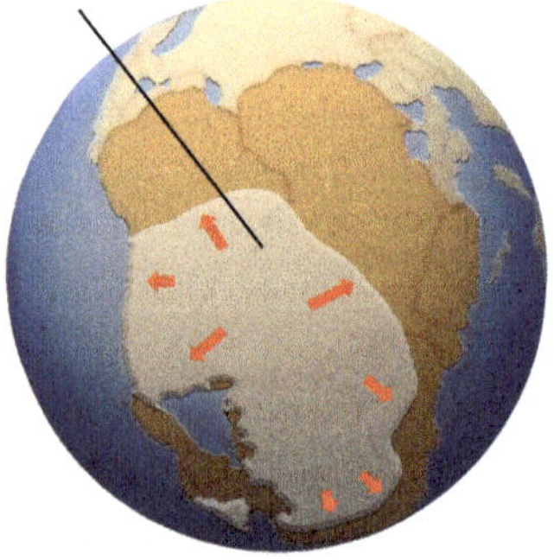

Pruebas paleoclimáticas: señales de desplazamiento del casquete de hielo

2.1. *Deriva continental*

El precursor de la teoría de la tectónica de placas fue el geofísico y meteorólogo alemán A. L. Wegener, que hizo públicas sus hipótesis por primera vez en una conferencia en el año 1912. En ellas defendía la idea de que los continentes habían estado unidos en el pasado formando un continente único y que se fueron desplazando los unos respecto a los otros por un movimiento que denominaba **deriva continental.** Para defender esta teoría, Wegener aportó numerosas pruebas.

- **Pruebas geográficas.** Los perfiles de algunos continentes encajan como las piezas de un rompecabezas, a pesar de que en la actualidad se encuentren muy separados. Éste es el caso de los continentes situados a ambos lados del océano Atlántico.
- **Pruebas geológicas.** Las formaciones geológicas, como pliegues, fallas, cordilleras, etc., que se interrumpen bruscamente en la línea de costa de algunos continentes tienen continuidad en otros continentes de los que están separados. Por ejemplo, una antigua cadena montañosa que tiene continuidad en Norteamérica y Europa.
- **Pruebas paleontológicas.** Algunos fósiles de plantas y de animales que no existen en la actualidad son idénticos a ambos lados del Atlántico. *Mesosaurus* es un reptil fósil que se ha localizado en África y América del Sur.
- **Pruebas paleoclimáticas.** El estudio de los restos fósiles y de las rocas sedimentarias de una zona permite reconstruir el clima del pasado. Esta reconstrucción indica la existencia de antiguas unidades climáticas en áreas actualmente separadas. Por ejemplo, existen indicios de que hace millones de años, grandes superficies de África, Sudamérica y Australia estuvieron cubiertas de hielo.

A continuación exponemos qué movimientos se produjeron en las placas a partir del Pérmico. El continuo movimiento de las placas nos ha permitido deducir que los procesos de acercamiento y separación de los continentes han sucedido más de una vez desde que se formó la Tierra.

HACE 225 MILLONES DE AÑOS

Durante el Pérmico, toda la Tierra se encontraba unida en un único continente, al que se ha dado el nombre de **Pangea.** El océano que rodeaba este continente recibía el nombre de **Pantalasa.**

HACE 180 MILLONES DE AÑOS

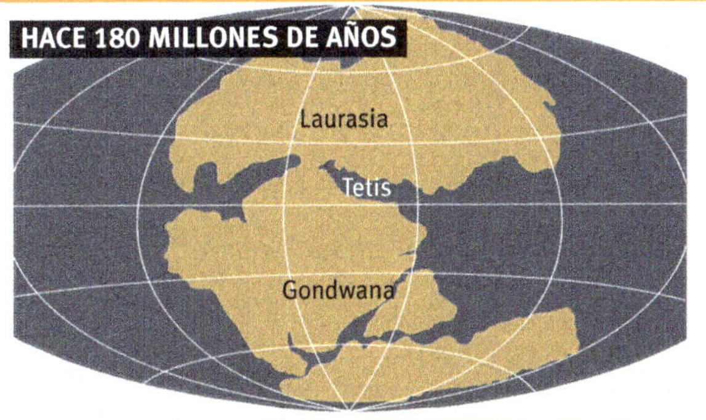

A finales del Triásico, Pangea se rompió en dos continentes: **Gondwana** al Sur y **Laurasia** al Norte; entre ambos quedó el mar de **Tetis**. Gondwana incluiría las actuales América del Sur, África, Australia, la Antártida y la India; y Laurasia, Europa, Asia y América del Norte.

HACE 135 MILLONES DE AÑOS

A finales del Jurásico, Gondwana empezó a fracturarse, y se separaron la Antártida y Australia hacia el Sur y la India hacia el Norte. Laurasia se desplazó hacia el Norte.

HACE 65 MILLONES DE AÑOS

A finales del Cretácico, Gondwana y Laurasia se acabaron de fragmentar en varios bloques, que se desplazaron en diferentes direcciones y sentidos. Se abrió el océano Atlántico.

EN LA ACTUALIDAD

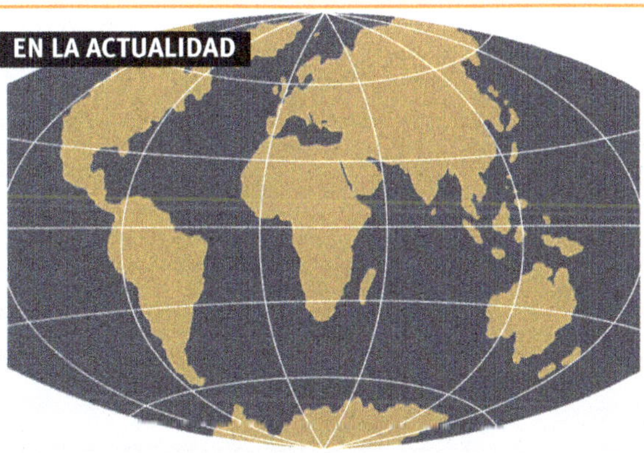

Durante los últimos 65 millones de años se han unido los continentes de América del Norte y América del Sur; la India se ha incorporado al continente asiático y la Antártida y Australia se han separado.

EN EL FUTURO

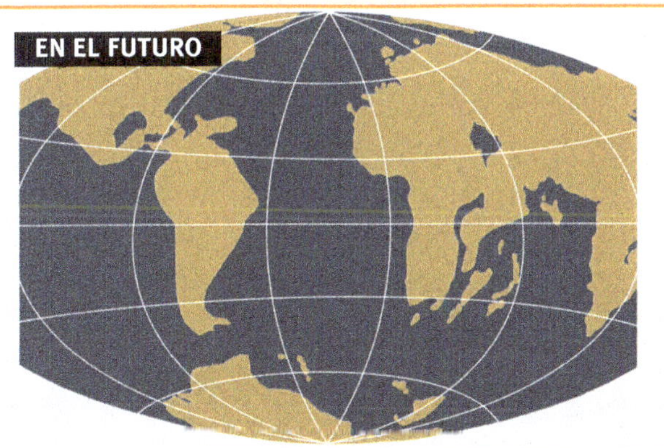

Los continentes siguen su evolución y dentro de 100 millones de años el aspecto de la superficie terrestre será diferente. África se habrá fracturado por el Rift y el océano Atlántico será más ancho.

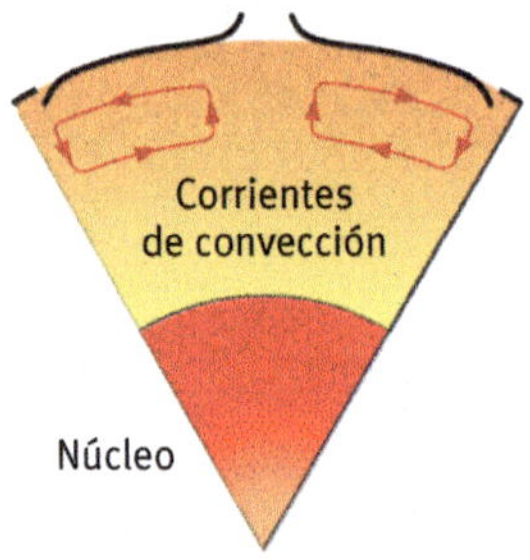

Desplazamiento de las placas causado por las células convectivas de la zona superior del manto (astenosfera)

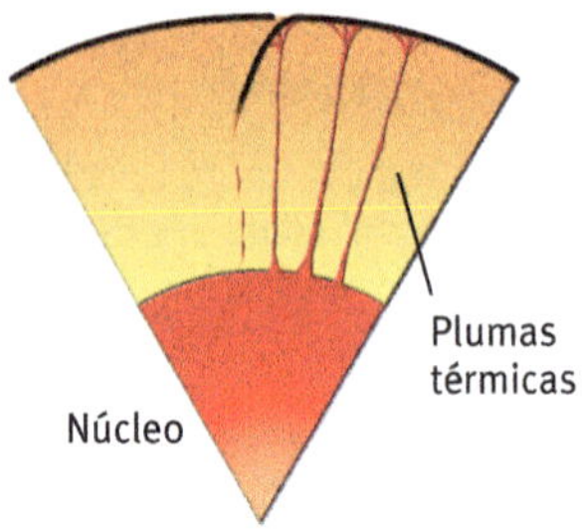

Desplazamiento de las placas debido al tirón provocado por la subducción de la placa litosférica

2.2. *Corrientes de convección*

Una vez aceptada la hipótesis de que los continentes se desplazan, es necesario conocer cuál es el motor que genera dicho desplazamiento. Así, en 1929 el geólogo inglés A. Holmes presentó su teoría de que los materiales del manto que se encuentran en un estado fluido, debido a las elevadas temperaturas, producen un flujo en forma de corrientes de convección. Los materiales cálidos ascienden, se enfrían y descienden, originando el movimiento de los continentes.

Este sistema convectivo se planteó en un principio de tal manera que las placas litosféricas eran arrastradas pasivamente por el flujo de la astenosfera, fluida. En la actualidad se ha postulado un modelo en el cual la placa litosferica se hunde en uno de sus márgenes hasta el núcleo, donde el material se calienta, funde y asciende nuevamente. La placa sería arrastrada, por tanto, por el hundimiento de uno de sus extremos en el manto. Los penachos o plumas térmicas que ascienden desde la interfase manto-núcleo constituirían un segundo flujo convectivo, que podría perforar la litosfera y crear puntos calientes que provocarían la rotura de los continentes.

Las tensiones y los movimientos de la litosfera dan lugar a una gran actividad volcánica y sísmica.

Observa en el siguiente mapa la distribución de los volcanes activos en la actualidad y los epicentros de los terremotos recientes.

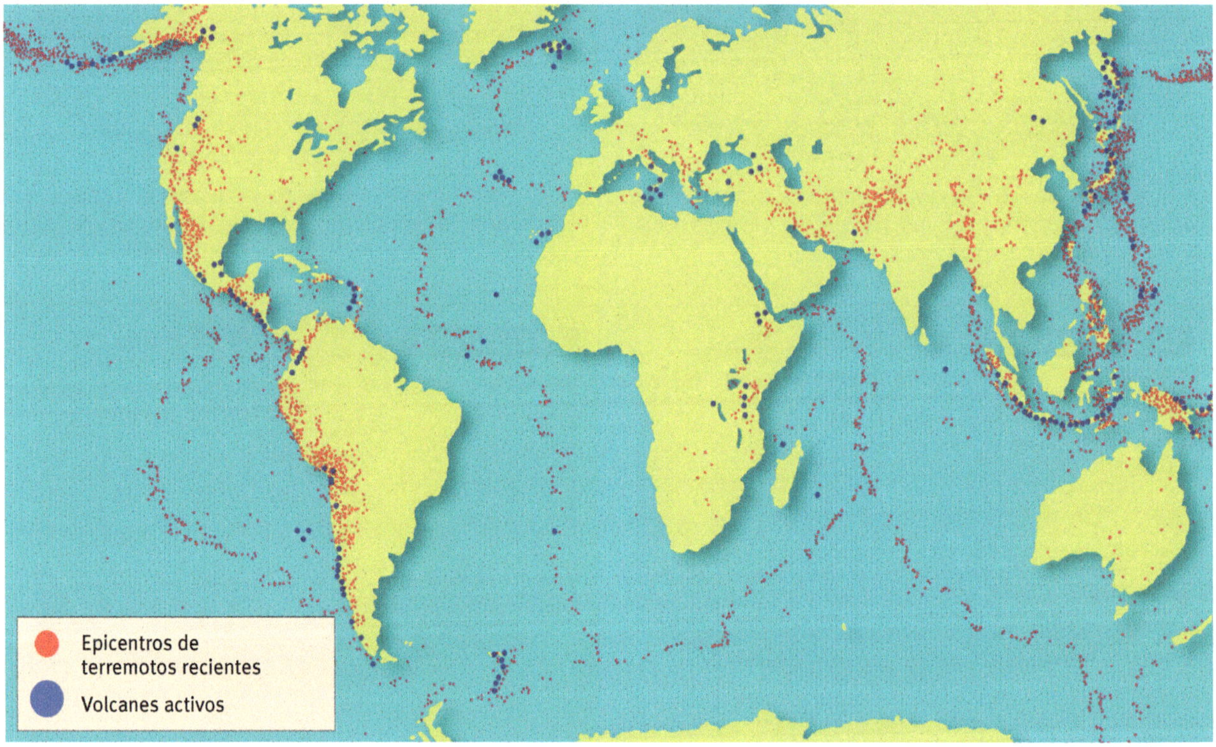

Los volcanes y terremotos no se producen con la misma probabilidad en todos los puntos de la Tierra. Las zonas donde se localizan, las **zonas activas,** son lugares en los que la litosfera está fracturada.

2.3. *Expansión y subducción de la litosfera*

De las teorías anteriores surgieron otras dos: la *expansión de los fondos oceánicos* y la *subducción de la corteza* en las fosas oceánicas.

Expansión de los fondos oceánicos

Esta teoría considera las grandes cordilleras submarinas o dorsales oceánicas como grietas de la corteza por las que asciende el magma hacia la superficie. Una vez consolidado, el magma es empujado a un lado y a otro de la dorsal por nuevo magma procedente del interior, provocando así que el fondo oceánico se expanda. En el caso del modelo mixto de circulación del manto, la dorsal se separaría a causa de las tensiones divergentes producidas por el hundimiento del margen opuesto de la placa.

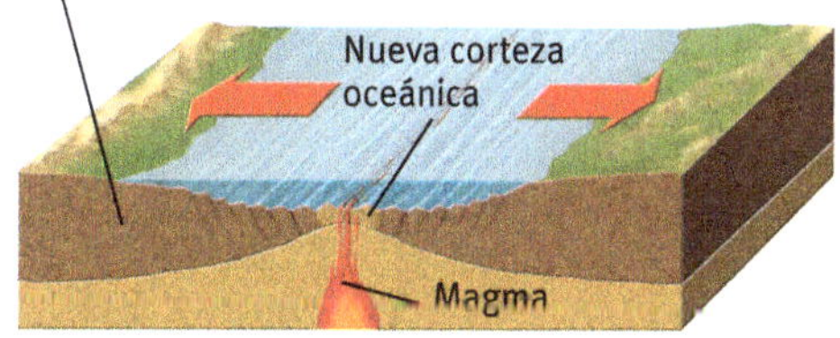

Expansión del fondo oceánico

La presencia, a ambos lados de la dorsal, de una serie, completamente simétrica, de bandas de rocas que muestran un cambio de polaridad magnética es la prueba definitiva que corrobora la expansión del fondo oceánico. Dichas bandas magnéticas de rocas están formadas por basaltos expulsados por la dorsal. Al solidificar, los minerales magnéticos (férricos y ferromagnesianos) que forman parte de los basaltos indican la posición del eje magnético en ese preciso momento.

Subducción de la corteza

Actualmente se admite que el volumen de la Tierra ha permanecido constante desde que se consolidó su corteza hace 3500 millones de años y que por lo tanto su superficie tampoco ha variado. Si esto es así, parece lógico pensar que si en las dorsales se produce la expansión de la corteza oceánica, deben existir otros lugares en los que la corteza se destruya. La teoría del hundimiento o subducción propone que una parte de la corteza se hunde en el manto, siguiendo unos planos inclinados, hasta quedar completamente asimilada por éste a unos 700 km de profundidad. Las zonas donde se da este proceso se denominan **zona de Benioff.** Los lugares susceptibles de producirse dicho fenómeno parecen ser las fosas oceánicas. Algunas de las observaciones que parecen corroborar dicha suposición son:

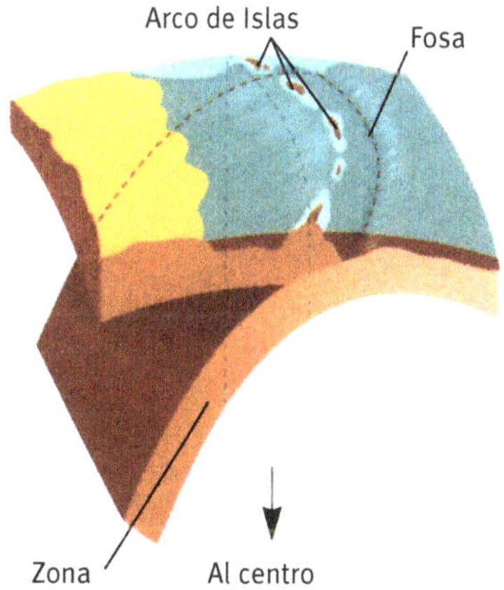

- En las fosas oceánicas existen disminuciones del campo gravitatorio, lo que significa que se producen pérdidas de masa.
- La zona de Benioff es un plano inclinado por el que se desliza friccionando una placa litosférica por debajo de otra, provocando seísmos y fusión de material por rozamiento. Este hecho explicaría los abundantes seísmos y la presencia de volcanes cerca de las fosas.
- Las prospecciones sismológicas confirman dicho hundimiento al poner de manifiesto que las ondas S no disminuyen su velocidad entre los 100 y los 400 km por debajo de las fosas oceánicas, que sería lo esperado por la sismología. Esto nos permite deducir que en ese lugar la astenosfera está atravesada por una capa de material rígido.

En el modelo mixto de circulación del manto, la placa que padece la subducción lo hace hasta el límite entre el manto y el núcleo exterior, el llamado nivel D.

C

La teoría de la tectónica de placas revolucionó las ideas fijistas, según las cuales la estructura de la Tierra no había experimentado cambios desde su formación.

Esta teoría es equivalente en importancia a la teoría de la evolución de las especies propuesta por Darwin.

2.4. *Tectónica de placas*

El estudio sísmico efectuado por B. Gutenberg en 1959 confirmó las hipótesis anteriores. Gutenberg postuló la existencia de una capa rígida, la litosfera, situada sobre una masa a alta temperatura y en estado plástico, la astenosfera.

Los datos aportados recientemente por la tomografía sísmica ponen en duda la existencia de esa capa, aunque como hemos visto, el fenómeno del desplazamiento de las placas puede ser explicado por otros mecanismos.

La teoría de placas supone que la litosfera es frágil y rígida, y como consecuencia de ello, al ser sometida a las tensiones se fractura. Así, la litosfera está dividida en fragmentos que se denominan **placas litosféricas** y que encajan entre sí a modo de las piezas de un puzzle.

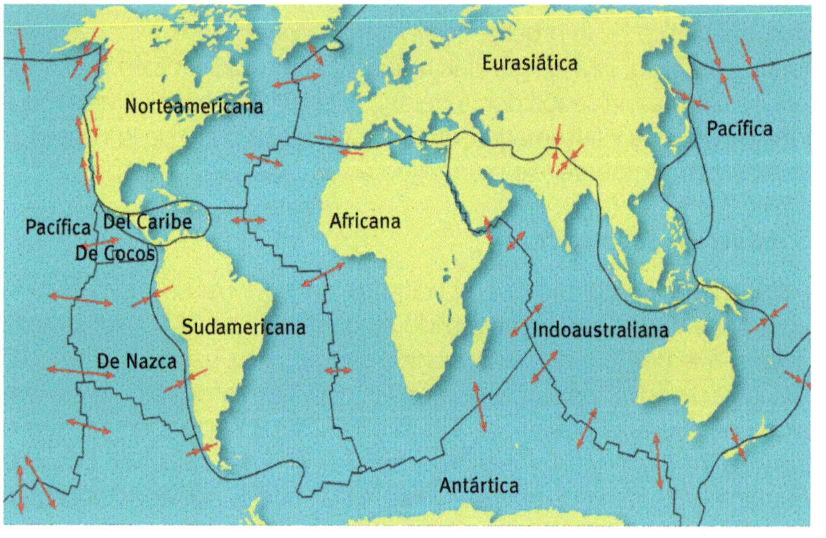

El hecho de que unas placas se desplacen respecto a otras hace que los límites o bordes entre ellas sean áreas geológicamente muy activas. Distinguimos tres clases de límites o bordes: *divergentes, convergentes* y *transformantes.*

Límites divergentes

Son límites en los que las dos placas se separan; corresponden a una zona en la que el magma asciende.

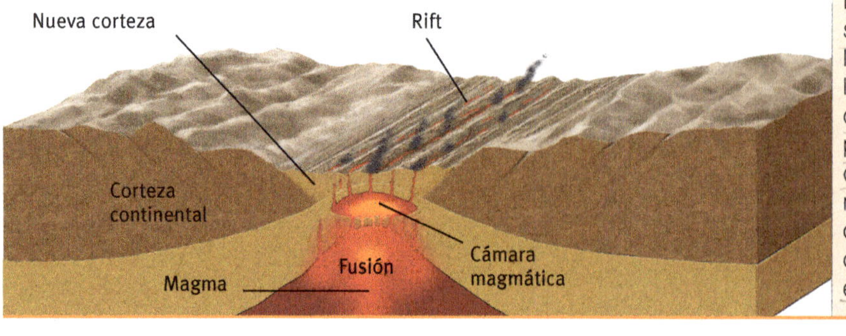

En la zona en la cual dos placas se separan se origina una enorme grieta, llamada **rift**, cuya anchura oscila entre los 25 y los 50 km. Allí, las fracturas originadas son rellenadas por magma proveniente de la zona inferior, hecho que genera cordilleras submarinas llamadas **dorsales oceánicas**. Este proceso produce la expansión del fondo oceánico, a un ritmo que puede oscilar entre 2 y 20 cm al año.

Límites convergentes

Son límites en los que las placas se aproximan y chocan. Corresponden a zonas donde el magma desciende. Existen diversos tipos de límites convergentes.

Convergencia entre dos placas oceánicas.

Convergencia de una placa oceánica y una continental.

Convergencia de dos placas continentales.

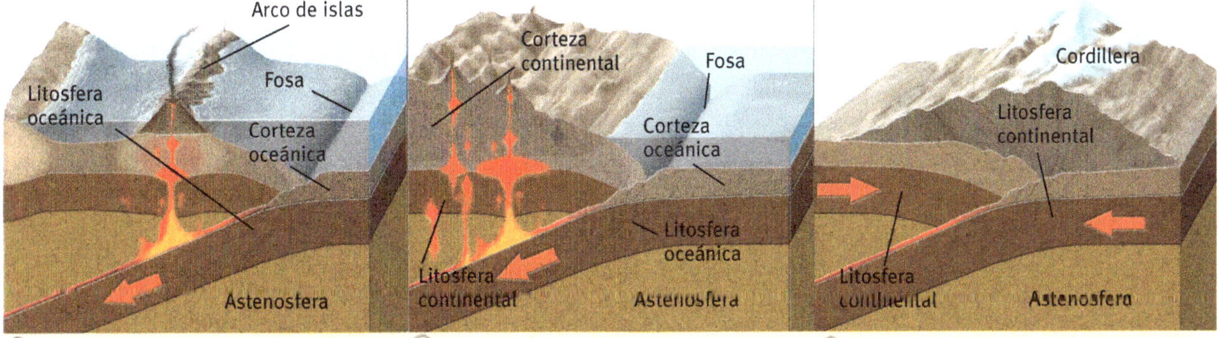

La placa más densa se hunde bajo la otra. El magma producido por la fusión de la placa asciende, provocando erupciones submarinas que darán origen a **arcos insulares**. En estas zonas de subducción son muy frecuentes los terremotos.

La placa oceánica, más densa, subduce bajo la continental, originando una fosa. La compresión en el margen continental provoca la formación de una **cordillera perioceánica**, con gran actividad volcánica y sísmica.

Cuando una placa oceánica subduce bajo una continental puede dar a largo plazo el choque entre dos continentes. Dada su baja densidad, la corteza continental no se hunde, y las fuerzas de compresión provocan la formación de cordilleras en el interior del continente.

En los procesos de subducción la litosfera se destruye, pero en el caso del choque de dos placas continentales, aumenta su grosor.

Límites laterales o transformantes

Son límites en los que las placas se desplazan lateralmente. El roce entre las placas hace que las rocas se rompan y se produzcan grandes fallas. El desplazamiento de los bloques origina una gran actividad sísmica.

2.5. *El ciclo de Wilson*

Todos los procesos expuestos anteriormente dan lugar a un ciclo de fusión y ruptura de placas, llamado **ciclo de Wilson**. El ciclo comienza con la fragmentación de un gran escudo continental por acción de un punto caliente, la formación de un rift y su conversión en un océano. El desplazamiento de las placas continentales originadas por la fragmentación y la subducción de las placas oceánicas conducirá finalmente a una colisión que originará un nuevo supercontinente. Llegados a este punto, se inicia una nueva fase de rotura por efecto de un nuevo punto caliente.

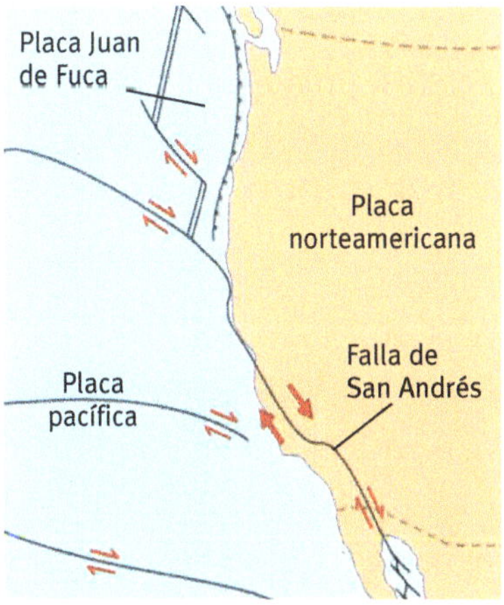

La falla de San Andrés es un ejemplo del límite lateral entre placas.

ACTIVIDADES

9. Enuncia las pruebas que respaldan la deriva continental. ¿Quién fue el primero en postularla?

10. ¿Cómo se explica el movimiento de las placas litosféricas? ¿En qué difieren las dos hipótesis presentadas en el tema?

11. Haz un esquema de los diferentes límites de placa, indicando sus características.

12. ¿En qué consiste el ciclo de Wilson? ¿Cómo se llama el supercontinente que constituía la superficie del planeta hace 225 millones de años?

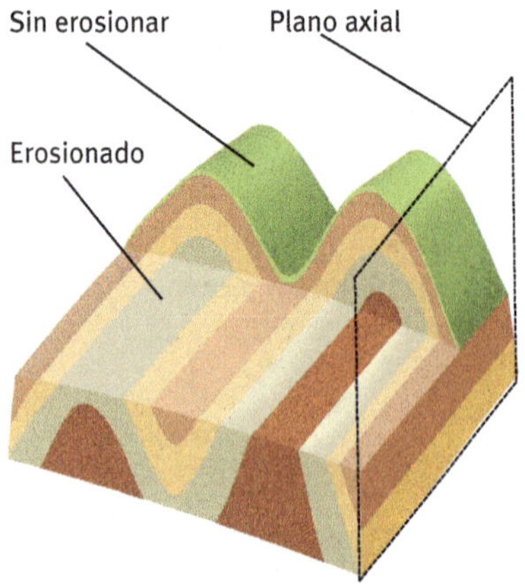

En los estudios de campo, los pliegues se suelen hallar erosionados, presentando en superficie estructuras más o menos paralelas, como muestra la ilustración.

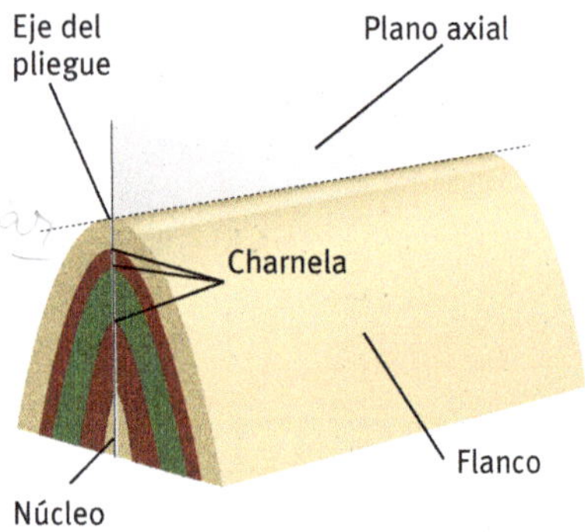

3. Deformaciones

La actividad interna de la Tierra no siempre se manifiesta de forma brusca e inmediata como en el caso de los volcanes o terremotos. También da origen a procesos continuos y muy lentos, como compresiones de las rocas que provocan su deformación.

Estas deformaciones dan lugar a pliegues cuando no se supera la capacidad plástica de las rocas, es decir, la capacidad de deformarse sin romperse. Cuando se supera el límite de plasticidad y las rocas se rompen, originan las fracturas.

En las rocas sedimentarias, tanto los pliegues como las fracturas son muy evidentes, ya que se altera la disposición horizontal que habían adoptado los estratos durante el proceso de sedimentación. La inclinación que adquieren los estratos respecto al plano horizontal a causa de este desplazamiento se conoce como **buzamiento**.

3.1. *Pliegues*

Son ondulaciones de los estratos rocosos. En ellos se distinguen:

— **Charnela:** línea que une los puntos de máxima curvatura de los distintos estratos.

— **Flancos:** zonas situadas a ambos lados de la charnela.

— **Plano axial:** plano que divide el pliegue en dos partes, con inclinaciones opuestas.

— **Núcleo:** zona de máxima curvatura del pliegue.

Los pliegues pueden ser:

Pliegue en forma de bóveda. Pliegue en forma de cubeta.

Generalmente, los pliegues aparecen asociados, de modo que se alternan pliegues anticlinales con pliegues sinclinales. Los pliegues también pueden clasificarse según la inclinación del plano axial. Así, se distinguen pliegues *rectos*, *inclinados* y *tumbados*.

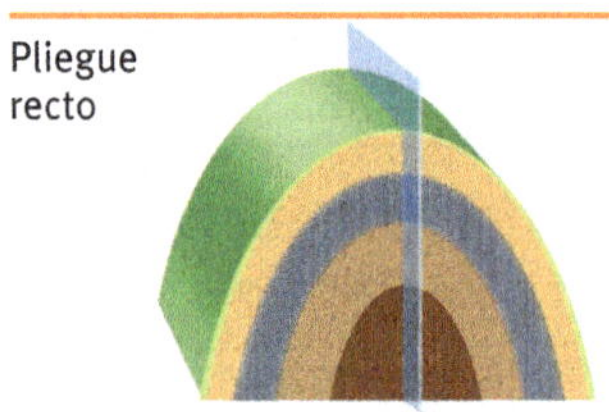

El plano axial es vertical.

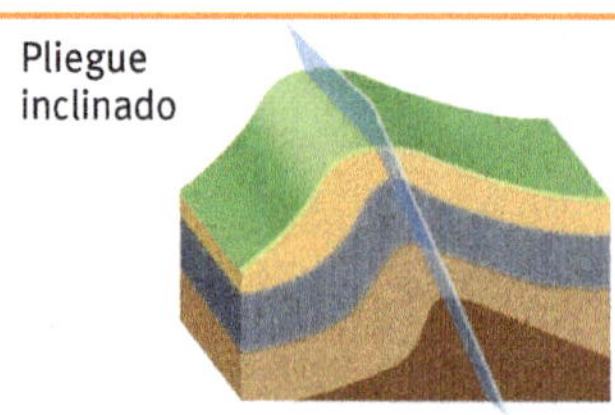

El plano axial está inclinado.

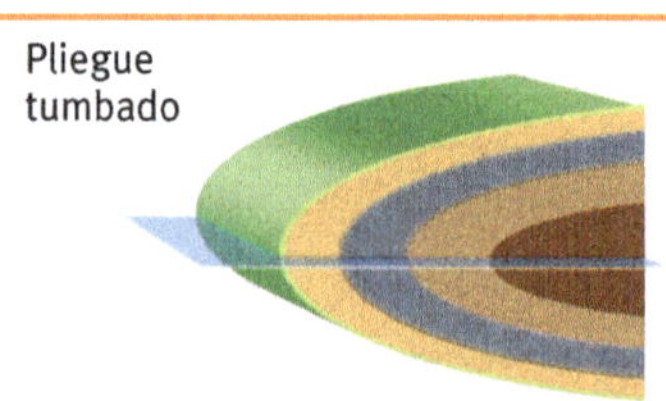

El plano axial es horizontal.

3.2. *Fracturas*

Las fracturas son roturas en las rocas que dan lugar a bloques separados de éstas.

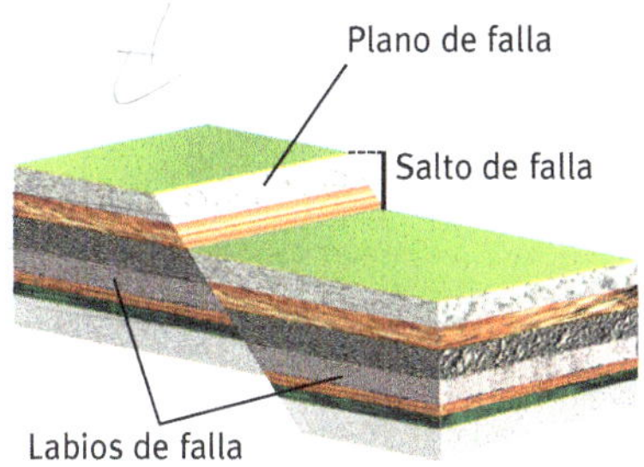

- Las **fallas** son un tipo de fractura en la que los bloques se desplazan uno respecto a otro. Los elementos de una falla son:
  - — **Plano de falla:** superficie de fractura.
  - — **Labios de falla:** los dos bloques desplazados.
  - — **Salto de falla:** desnivel que existe entre ambos labios.

Las fallas se clasifican según la inclinación del plano de falla y el movimiento de los bloques.

Falla normal

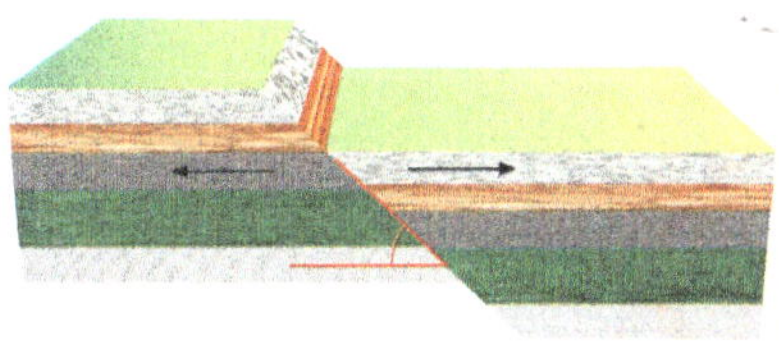

Se produce un desplazamiento vertical de un bloque respecto a otro porque tienden a separarse. Está causada por un esfuerzo de **distensión**.

Falla inversa

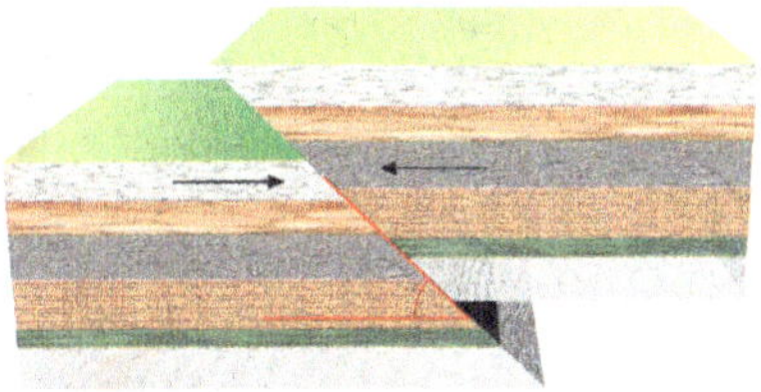

Se produce un desplazamiento vertical de un bloque respecto a otro porque tienden a acercarse. Está causada por un esfuerzo de **compresión**.

Falla horizontal o de desgarre

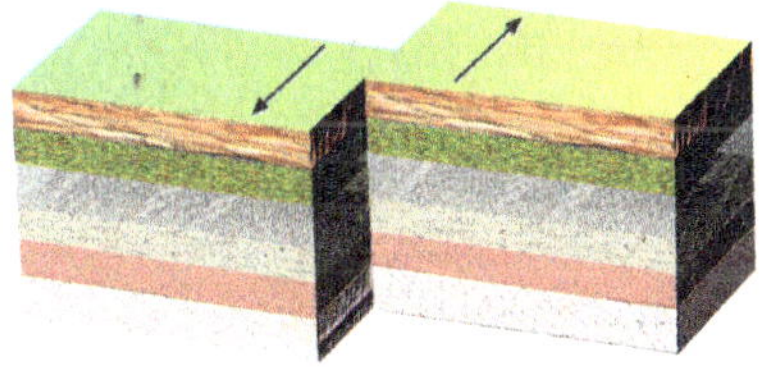

Se produce un desplazamiento horizontal de un bloque respecto a otro.

Al igual que en el caso de los pliegues, las fallas suelen estar asociadas, configurando las *fosas tectónicas* y los *macizos tectónicos*.

Fosa tectónica

Sistema de fallas con el bloque central hundido.

Macizo tectónico

Sistema de fallas con el bloque central elevado.

- Las **diaclasas** también son fracturas, pero en ellas no se produce el desplazamiento de los bloques.

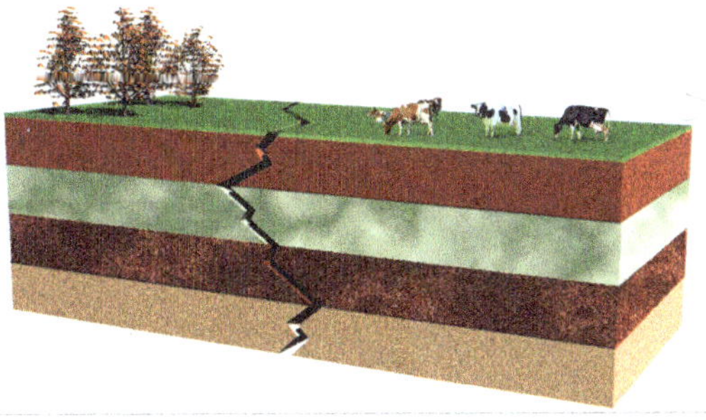

Las deformaciones de las rocas son procesos muy frecuentes en la superficie terrestre, pero su identificación en la naturaleza y su posterior estudio requieren cierta habilidad y práctica que las descripciones teóricas no nos proporcionan.

Por ello, mostramos a continuación el procedimiento que se sigue para llevar a cabo la identificación y el estudio de un pliegue y una falla.

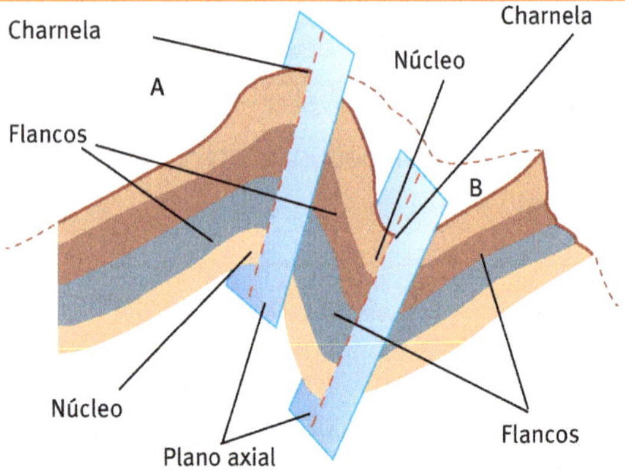

- Las ondulaciones de los estratos de las rocas nos indican que se trata de un **pliegue**.
- En la parte A del esquema, los estratos presentan una curvatura hacia abajo y en la parte B la curvatura es hacia arriba. Por tanto, se trata de la asociación de un pliegue **anticlinal** con un pliegue **sinclinal**.
- Los planos axiales de ambos pliegues están inclinados, por tanto, se trata de un pliegue anticlinal **inclinado** y un pliegue sinclinal **inclinado**.

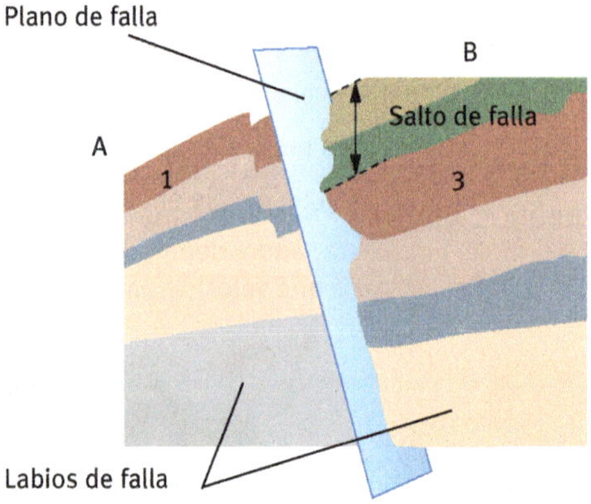

- En las rocas se ha producido una **fractura**. Los estratos de los bloques resultantes no están alineados a ambos lados de la fractura, sino que se han desplazado; por tanto, se trata de una **falla**.
- El estrato 1 del labio A coincide con el estrato 3 del labio B, lo que nos indica un desplazamiento vertical de los labios.
- El plano de falla se dirige hacia el labio hundido, por tanto, es una **falla normal**.

ACTIVIDADES

- **13.** ¿Qué determina que un esfuerzo aplicado sobre una roca dé lugar a un pliegue o a una falla?
- **14.** ¿Qué es el plano axial? ¿Y la charnela?
- **15.** Dibuja un anticlinal y un sinclinal con sus partes.
- **16.** ¿Qué tipo de falla obtendremos si se somete una roca a una distensión? ¿Y si las fuerzas son de compresión?
- **17.** ¿Cuál es la diferencia entre una falla y una diaclasa?

18 Observa el siguiente perfil geológico entre los continentes africano y americano. En la zona inferior se muestra el perfil del fondo oceánico sobre la línea marcada.

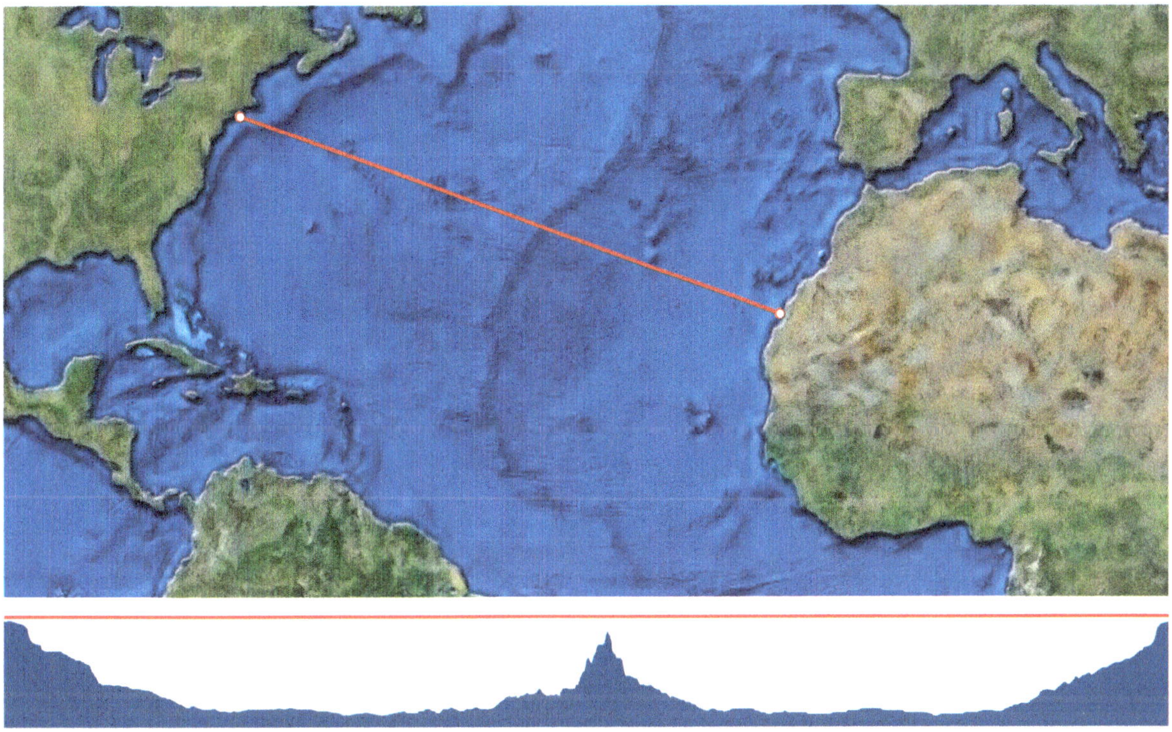

— ¿Qué estructura se observa en la zona central? ¿A qué tipo de límite de placas corresponde?

— ¿Podrías indicar cuál será el desplazamiento de las placas a ambos lados?

— ¿Qué movimiento relativo presentan la placa norteamericana y la africana entre ellas?

— ¿Puedes observar la línea de encaje entre las costas atlánticas de ambos continentes?

— Si ambas líneas de costa, ahora separadas unos 6300 km, se hallaban unidas en la Pangea hace 225 millones de años, ¿puedes calcular a qué velocidad media se ha producido la expansión del océano Atlántico?

La imagen siguiente representa la edad de la corteza oceánica en la cuenca del océano Atlántico y del océano Pacífico.

— ¿Mediante qué técnica crees que pueden haberse obtenido estos datos?

— ¿Cuál de los dos océanos se expande más rápidamente? Justifica tu respuesta.

— ¿Qué ha sucedido con las zonas de placa más antigua de la placa de Nazca? ¿Cómo se llama este fenómeno? ¿Qué consecuencias ha tenido este proceso?

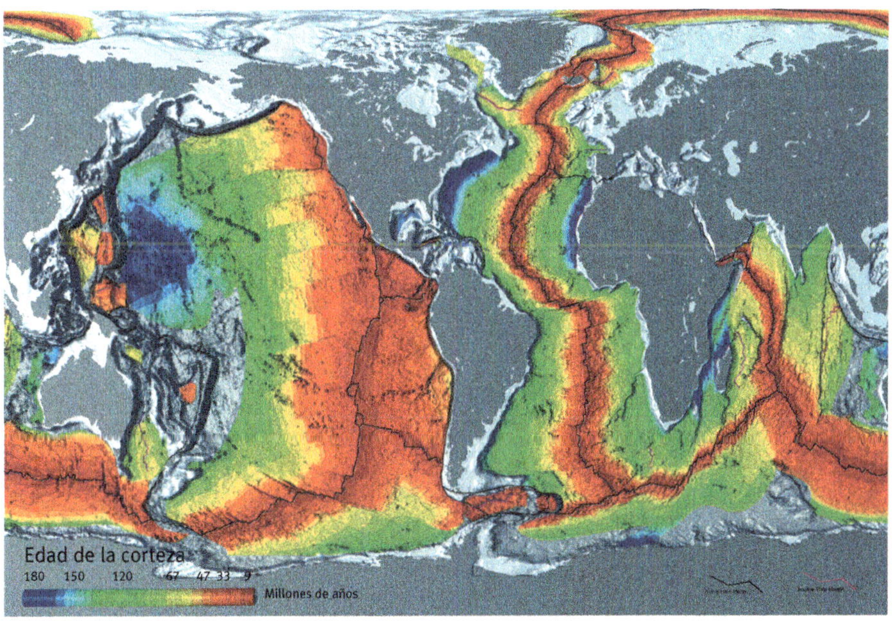

19 Observa el siguiente sismograma:

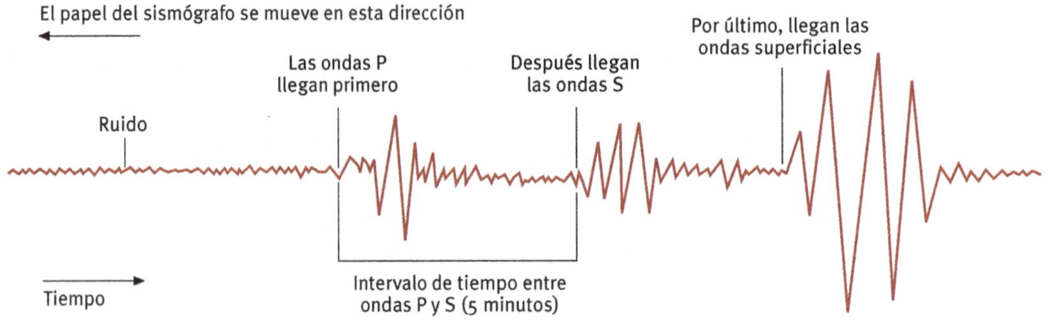

— ¿Cuáles son las ondas sísmicas que viajan más rápido? ¿Cuáles son las últimas en llegar?

— ¿Qué características tiene cada una de ellas?

20 Conociendo el intervalo temporal que se da entre la llegada de las ondas P y las ondas S se puede conocer la distancia al epicentro del terremoto a partir de la siguiente gráfica.

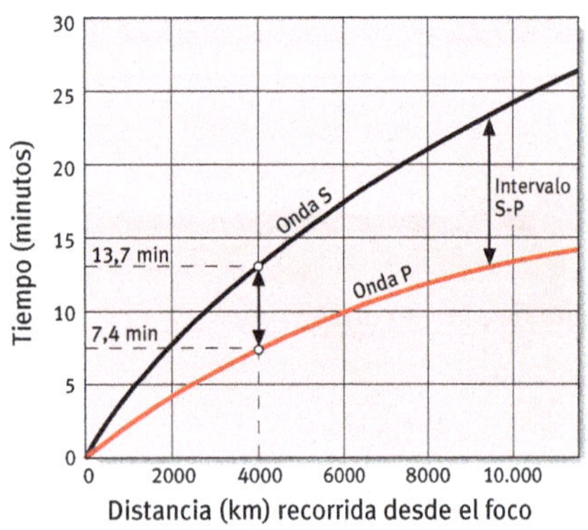

— Calcula la distancia entre el epicentro del terremoto y el lugar en el cual ha sido registrado el sismograma del ejercicio anterior.

21 Relaciona los siguientes conceptos con los límites de placa correspondientes:

	Divergente	Convergente	Transformante
Fosa oceánica			
Rift			
Subducción			
Desplazamiento lateral de las placas			
Arcos insulares			
Dorsales oceánicas			
Formación de cordilleras			
Expansión del fondo oceánico			

22 En el planisferio siguiente aparecen las distribuciones de los fósiles de *Cynognathus*, *Lystrosaurus*, *Mesosaurus* y *Glossopteris*.

— ¿Puedes utilizar los datos que se presentan como una prueba de la deriva continental? Justifica tu respuesta y dibuja las posiciones relativas de los continentes que permitirían esta distribución de los fósiles.

23 Hace 200 millones de años, la totalidad de la Antártida, la India, Madagascar, la mitad meridional del continente africano, el sur de Australia y la zona oriental de Argentina estaban cubiertos por un casquete glaciar. Busca información sobre la climatología actual de estas zonas.

— ¿Qué factores crees que han influido en los cambios que puedes observar? ¿Qué influencia habrá tenido este cambio sobre los seres vivos que habitaban estas zonas?

24 El siguiente mapa presenta la distribución de terremotos, la posición de un arco volcánico y la situación de una fosa.

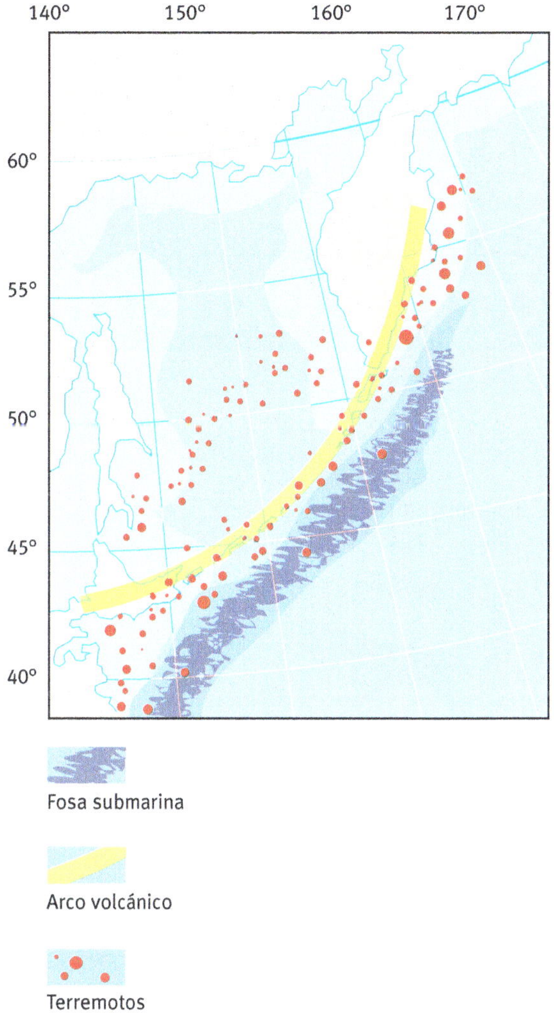

— Interpreta qué fenómeno tiene lugar en la litosfera de esta zona. Justifica tu respuesta y dibuja cómo podría ser el perfil de los límites de placa asociados a esta situación.

TIC

25 Puedes profundizar en tus conocimientos sobre terremotos en la página web de Terremoto virtual. En ella puedes llevar acabo simulaciones de cálculo de la posición del epicentro de un seismo o calcular su magnitud.

p://www.sciencecourseware.org/eec/Earthquake_es/

— ¿Qué mide la magnitud de un terremoto? ¿Qué es un nomograma Richter'?

26 ¿Qué factor determinará el comportamiento de una roca cuando ésta sea sometida a un esfuerzo de compresión? ¿En qué caso se originará un pliegue? ¿Y una falla?

27 En la actualidad, la separación de la parte oriental de África del resto de la placa africana genera una serie de tensiones que se manifiestan en la formación del valle del Rift. ¿Qué tipo de fallas crees que se encuentran asociadas a este proceso?

— Busca información sobre el valle del Rift. ¿Qué sucederá en el futuro en esta zona del planeta? Justifica tu respuesta.

28 ¿De qué manera se puede diferenciar sobre el terreno un sinclinal de un anticlinal si han sido erosionados? Haz un dibujo de lo que observaríamos en la superficie en uno y otro caso.

29 Identifica las deformaciones de las rocas presentadas en las fotografías A y B, y realiza un dibujo esquemático en el cual muestres las diferentes partes con sus nombres. Trata de identificar los siguientes elementos:

charnela/ flanco/ labio elevado/ labio hundido/ plano axial/ plano de falla / salto de falla/ núcleo

4

La actividad geológica externa

La acción geológica realizada por los agentes geológicos externos como el agua, el viento o los seres vivos es, seguramente, la más fácil de reconocer en nuestro territorio.

La actividad geológica externa da lugar a muchos de los paisajes a los que estamos acostumbrados. Los valles y terrazas fluviales, las dunas costeras o los circos glaciares son consecuencia de estos procesos geológicos, aunque el tiempo que han necesitado para su formación difiere en cada caso.

Esta actividad geológica genera, además, los procesos indispensables para completar el ciclo de las rocas, haciendo posible la erosión, el transporte y la sedimentación de los distintos tipos de roca.

Sin duda, el ser humano es actualmente uno de los mayores agentes geológicos que actúan sobre la superficie del planeta. Sus efectos sobre el territorio son sobradamente conocidos por todos nosotros, y los límites y las interacciones de esta intervención sobre los procesos geológicos naturales es una de las cuestiones que más se plantean y estudian a día de hoy los científicos.

Arco natural (Biarritz, Francia).

ORGANIZACIÓN DE LOS CONTENIDOS

OBJETIVOS

- Diferenciar los procesos geológicos externos e identificar los agentes que los producen.
- Comprender la relación entre el clima y los agentes y procesos geológicos externos.
- Comprender las etapas de formación del suelo y su composición.
- Mostrar interés por conocer la relación existente entre los procesos geológicos internos y los externos.
- Ser consciente de la necesidad de conservar y proteger el suelo como recurso imprescindible para la vida en la Tierra.

El **relieve** es el conjunto de formas de la superficie terrestre (montañas, colinas, llanos, deltas, valles...) que caracterizan un paisaje.

Los procesos geológicos internos, como la actividad volcánica o la formación de pliegues y fallas, son procesos que, generalmente, provocan la creación de relieve.

Como veremos en esta unidad, los procesos geológicos externos suelen tener un efecto contrario, ya que actúan sobre la superficie terrestre suavizando las formas del relieve.

1. Procesos geológicos externos

Los procesos geológicos externos son las modificaciones del relieve producidas por agentes geológicos como el agua, el viento, etc., los cuales actúan sobre la superficie terrestre y desgastan las rocas.

Se distinguen los siguientes procesos geológicos externos:

- **Meteorización:** fragmentación de una roca o transformación química de los minerales que la componen.
- **Erosión:** desgaste de las rocas. Su intensidad depende de si previamente ha tenido lugar un proceso de meteorización.
- **Transporte:** traslado de los materiales resultantes de la meteorización o de la erosión.
- **Sedimentación:** depósito de los materiales transportados.

1.1. *Meteorización*

La acción de la temperatura, del agua y de los seres vivos sobre las rocas produce su fragmentación o su transformación. La intensidad y la rapidez de este proceso dependen de la composición de las rocas, del clima de la zona y del tiempo de exposición de las rocas a estas condiciones.

La meteorización puede ser:

Meteorización física o mecánica

Es la rotura de la roca en pequeños fragmentos sin alterar su composición química. Se distinguen diversos tipos.

Termoclastismo

Los cambios bruscos de temperatura que se producen diariamente en algunas zonas provocan sucesivas dilataciones y contracciones en las rocas, que debilitan su estructura hasta fragmentarlas.

Este proceso es frecuente en zonas con clima desértico, donde las variaciones de temperatura entre el día y la noche pueden superar los 30 °C.

Desierto de piedras (Marruecos).

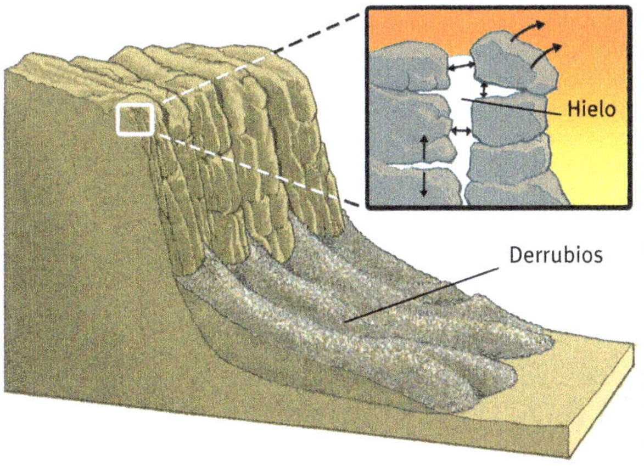

Crioclastismo o gelifracción

Cuando se hiela el agua que hay en las fisuras o los poros de las rocas, ésta aumenta su volumen y provoca fuertes presiones que crean grietas en las rocas o ensanchan las existentes.

Este fenómeno se produce en climas fríos.

Haloclastismo

Cuando la concentración de sales del agua de las fisuras es muy elevada, la disolución se satura y se forman cristales de sal. Estos cristales, como en el caso de la gelifracción, originan fracturas y agrandan las existentes.

Este proceso es frecuente en zonas cercanas al mar y con temperaturas elevadas.

Efecto de descompresión

Se presenta en las rocas que se forman en el interior de la Tierra bajo unas condiciones de temperatura y de presión superiores a las de la superficie terrestre. A medida que estas rocas afloran, la temperatura y la presión disminuyen y las rocas experimentan una descompresión.

Como consecuencia de este proceso, se originan grietas en las rocas o se desprenden fragmentos en forma de láminas.

Las raíces de los árboles y las galerías que excavan algunos animales también contribuyen a la formación de grietas en las rocas o al ensanchamiento de las existentes.

Meteorización química

Es la transformación química o la disolución de los minerales de las rocas, lo que facilita, en general, la acción de otros procesos geológicos externos, como la erosión y el transporte.

El agua es la causa de la mayoría de estos procesos y, por ello, la meteorización química es característica de climas húmedos. A continuación, mostramos los principales procesos de meteorización química.

- **Hidrólisis.** Es una reacción entre el agua y ciertos minerales, principalmente silicatos, que altera la red cristalina de éstos. Así, las micas y los feldespatos (ortosa) dan lugar, por hidrólisis, a nuevos minerales, como los óxidos de hierro y de aluminio o la caolinita.

$$2AlKSi_3O_8 + 2H_2O \rightarrow Al_2Si_2O_5(OH)_4 + K_2O + 4SiO_2$$

Ortosa — Caolinita — Potasa — Sílice

La meteorización química del feldespato implica la disgregación de las rocas graníticas, dando lugar al *lehm*.

La hidrólisis de los feldespatos es muy frecuente en las rocas graníticas, las cuales se transforman en un compuesto arenoso denominado *arenisca* o *lehm*.

- **Disolución.** El agua disuelve las rocas que contienen minerales formados por cloruros, como la halita, la silvina, la carnalita, etc.

  Estas sales disueltas pueden volver a precipitar si cambian las condiciones ambientales.

  La **carbonatación** es un caso particular de disolución. El agua reacciona con el dióxido de carbono atmosférico y forma el ácido carbónico, que es capaz de disolver, principalmente, las rocas constituidas por calcita (carbonato de calcio), como las calizas.

$$CO_2 + H_2O \rightarrow H_2CO_3 \rightarrow HCO^-_3 + H^+$$

Dióxido de carbono — Ácido carbónico — Ion bicarbonato

$$CaCO_3 + H_2CO_3 \rightarrow Ca(HCO_3)_2 + H_2O$$

Calcita — Bicarbonato cálcico

El bicarbonato cálcico resultante puede precipitar de nuevo.

La carbonatación de las rocas calizas es la responsable del *modelado cárstico*.

- **Hidratación.** Las moléculas de agua se insertan en las estructuras cristalinas de los minerales aumentando su volumen. Así, cuando la anhidrita se hidrata, se transforma en yeso.

$$CaSO_4 + 2H_2O \rightarrow CaSO_4 \cdot 2H_2O$$

Anhidrita — Yeso

Los óxidos de hierro dan coloración rojiza a los suelos de los que forman parte.

- **Oxidación.** El oxígeno atmosférico que se encuentra disuelto en el agua oxida los minerales formados por hierro y magnesio. Estos minerales oxidados son insolubles y precipitan.

  La oxidación del piroxeno de hierro II forma la limonita, un mineral que proporciona una coloración ocre-rojiza a la superficie de las rocas.

$$4FeSiO_3 + O_2 + 2H_2O \rightarrow 4FeO\ (OH) + 4SiO_2$$

Piroxeno de hierro II — Limonita

Como ocurre en la meteorización física, en la química también intervienen los seres vivos. La mayoría de las plantas y los animales segregan ácidos orgánicos que reaccionan con los minerales de las rocas.

1.2. *Erosión*

Al igual que la meteorización, la erosión depende del tiempo de exposición a los agentes geológicos, de la estructura y de la composición de las rocas. Además, es mucho más intensa si previamente ha tenido lugar un proceso de meteorización.

Este desgaste se produce de dos formas:

- Las corrientes de agua y de viento desprenden partículas de los materiales que encuentran a su paso.

  Cuanto mayores son la velocidad y la fuerza de las corrientes, mayor es la erosión que provocan.

- A su vez, las partículas sólidas que transportan el agua o el viento impactan contra la superficie de las rocas, lo que contribuye también a su desgaste.

  Este tipo de erosión se denomina **abrasión.**

1.3. *Transporte*

El agua y el viento son los medios de transporte de las partículas resultantes de la meteorización o de la erosión.

Se distinguen dos formas de transporte:

- **Químico.** Las partículas se transportan disueltas como iones en el agua.
- **Físico.** El agua o el viento transportan las partículas sólidas por arrastre, rodadura, saltación, en suspensión...

Las partículas de arena transportadas por el viento desgastaron la zona media de la esfinge de Gizeh, en Egipto. La parte inferior, que había quedado cubierta por la arena, no presenta marcas tan evidentes.

C

El transporte físico

Arrastre y rodadura

Las partículas más grandes y pesadas son arrastradas o se transportan rodando por el sustrato.

Saltación

Cuando impactan contra el sustrato, algunas partículas rebotan y vuelven a caer repetidamente.

En suspensión

Las partículas más pequeñas se mantienen sin depositarse sobre el sustrato, debido a las turbulencias o los remolinos.

El **modelado cárstico** es un paisaje que se origina por carbonatación de las rocas calizas. Si las formaciones son superficiales, reciben el nombre de **exocársticas,** y, si son internas, **endocársticas**.

Las principales formas exocársticas son:

- **Dolinas**. Cavidades de diversos tamaños en forma de embudo, que se encuentran aisladas o en grupos.
- **Lapiaz**. Superficies rocosas con numerosas canalizaciones y surcos.

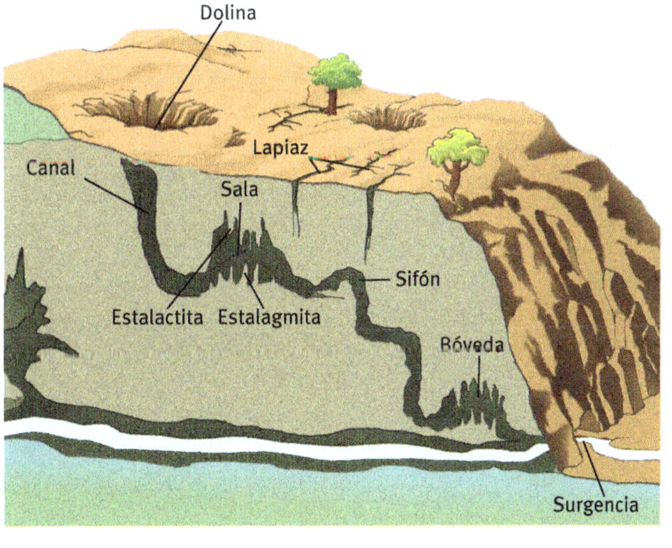

Las **cuevas** son las formaciones endocársticas más importantes. Son cavidades de grandes dimensiones formadas a partir de grietas de las rocas.

En ellas, generalmente se forman *estalactitas* y *estalagmitas*, que son estructuras producidas por la precipitación del bicarbonato cálcico.

Las estalactitas cuelgan del techo de las cuevas, mientras que las estalagmitas se forman en el suelo.

Las partículas transportadas por el río sedimentan al entrar éste en contacto con el agua del mar.

1.4. *Sedimentación*

Cuando disminuye la capacidad de transporte del agua y del viento, las partículas sólidas transportadas se depositan por efecto de la gravedad. Este proceso tiene lugar en áreas deprimidas como los cauces de los ríos, los lagos, los mares, etc. La sedimentación puede ser:

- **Sedimentación repentina.** Todos los sedimentos se amontonan a la vez y quedan mezclados los grandes con los pequeños.
- **Sedimentación gradual.** Los sedimentos más grandes quedan depositados en primer lugar y después lo hacen los más pequeños.

Si el transporte es por disolución, las partículas disueltas precipitan por evaporación del agua o por saturación de la disolución.

ACTIVIDADES

1. ¿Qué se entiende por proceso geológico externo? ¿Cuáles son estos procesos?
2. Comenta las diferencias existentes entre meteorización física y química. ¿Qué factores actúan en cada caso?

C

El clima es el conjunto de las condiciones atmosféricas que caracterizan una región, como la temperatura, la humedad, las precipitaciones...

La determinación de los climas se basa en el registro y tratamiento estadístico de los datos recopilados durante largos períodos de tiempo en una misma área.

Las características de un clima quedan reflejadas en los gráficos llamados climogramas, en los cuales se representan la media de las precipitaciones y temperaturas de cada mes.

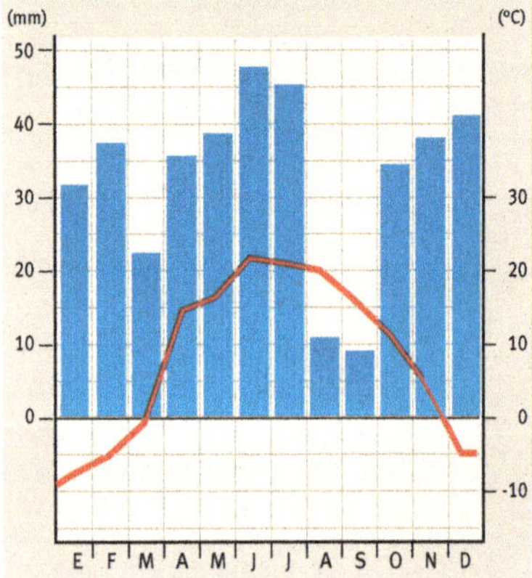

2. Agentes geológicos

Los agentes responsables de los procesos de meteorización, erosión, transporte y sedimentación que hemos descrito son, principalmente, el agua, el viento y los seres vivos. Estos factores, denominados **agentes geológicos externos,** están condicionados básicamente por el **clima.**

Las condiciones atmosféricas que determinan el clima dependen de la radiación solar. Observa en la siguiente tabla cómo incide la radiación solar sobre la superficie de la Tierra.

TOTAL DE ENERGIA INCIDENTE	**100 %**
ENERGÍA PERDIDA O ABSORBIDA POR LA ATMÓSFERA	
Absorción por la atmósfera, incluidas las nubes	19 %
Pérdidas por radiación hacia el espacio	9 %
Pérdidas por reflexión en las nubes y el suelo	25 %
ENERGÍA TOTAL PERDIDA	**53 %**
ENERGIA RECIBIDA EN EL SUELO	
Por penetración directa	41 %
Por radiación difusa	6 %
ENERGÍA TOTAL RECIBIDA EN EL SUELO	**47 %**

A su vez, también puede cuantificarse y desglosarse cómo es devuelta a la atmósfera la energía absorbida por el suelo:

ENERGÍA DEVUELTA A LA ATMÓSFERA	
Por radiación de onda larga	14 %
Por evaporación y condensación	23 %
Por conducción	10 %
ENERGIA TOTAL DEVUELTA POR EL SUELO	**47 %**

La energía necesaria para que los agentes geológicos actúen procede del Sol.

Así, la mayor parte de la energía procedente del Sol recibida en la superficie terrestre se irradia a la atmósfera y determina la **temperatura** del aire.

El aire que está en contacto con la superficie se calienta y asciende; en esta zona se origina un núcleo de bajas presiones. El aire frío de las capas altas desciende y provoca la formación de un núcleo de altas presiones. La diferencia entre el núcleo de altas presiones y el de bajas presiones origina una circulación de aire entre ellos, que recibe el nombre de **viento.**

Por otra parte, cuando la temperatura es elevada, aumenta la capacidad de la atmósfera de contener vapor de agua. Cuando una masa de aire caliente y húmedo se enfría, el vapor de agua se condensa a consecuencia de la pérdida de esta capacidad, y se producen las precipitaciones.

Si conocemos el clima de una zona, podemos deducir los agentes geológicos y los procesos a través de los cuales modelarán el relieve.

C

La diferencia de temperatura entre el mar y la tierra provoca la formación de la **brisa**. Sobre la costa (zona de mayor temperatura) se observa una zona de bajas presiones en superficie provocada por el ascenso del aire caliente. Este ascenso provoca el desplazamiento de masas de aire marinas, más frías, que ocupan su lugar. En las horas nocturnas, el enfriamiento más rápido de la costa invierte el proceso.

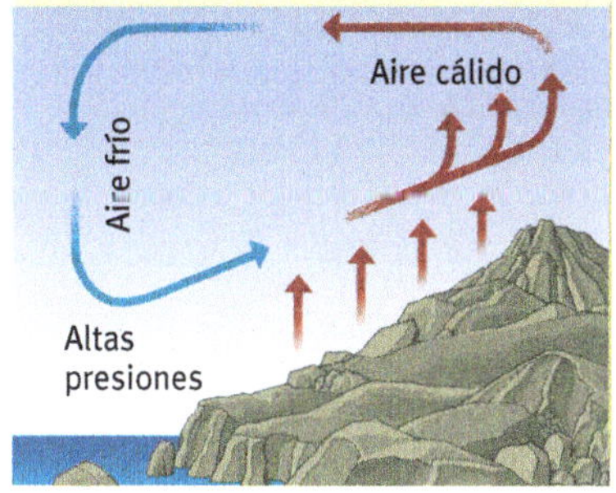

CLIMA	PROCESOS Y AGENTES GEOLÓGICOS
Clima frío	Acción del hielo. Crioclastismo. Bajo nivel de meteorización química a causa de la temperatura. Erosión por ventiscas.
Clima seco	Oscilación de temperaturas. Salinidad del agua. Termoclastismo y haloclastismo. Meteorización química poco efectiva por falta de humedad. Erosión por viento y agua de lluvia que circula por la superficie.
Clima templado	Acción de todos los agentes geológicos a nivel moderado. Meteorización química más importante que la física. Erosión por el agua de los ríos.
Clima tropical	Los agentes dominantes son el agua y los seres vivos. Importante meteorización química. Erosión por ríos y torrentes.

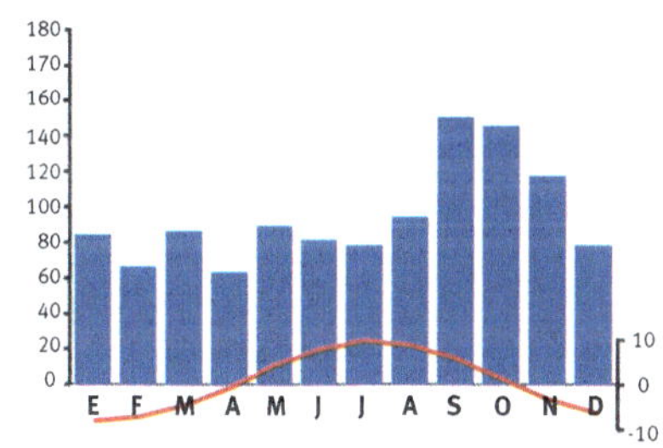

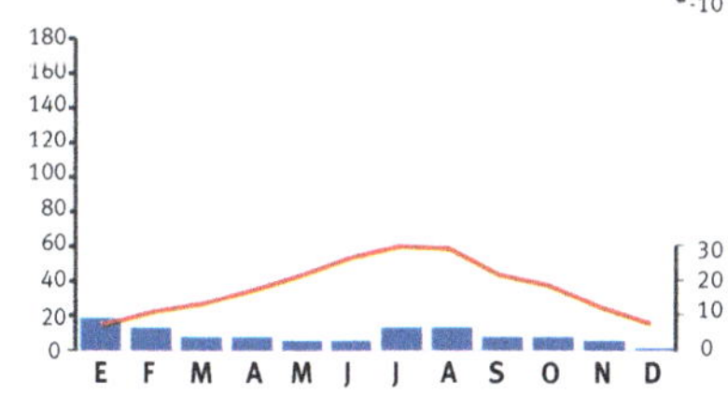

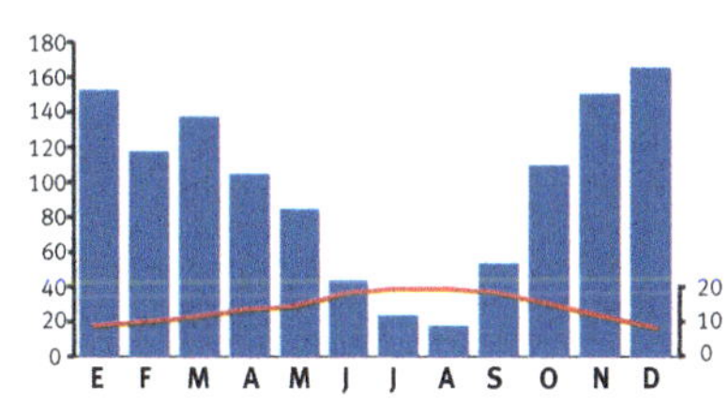

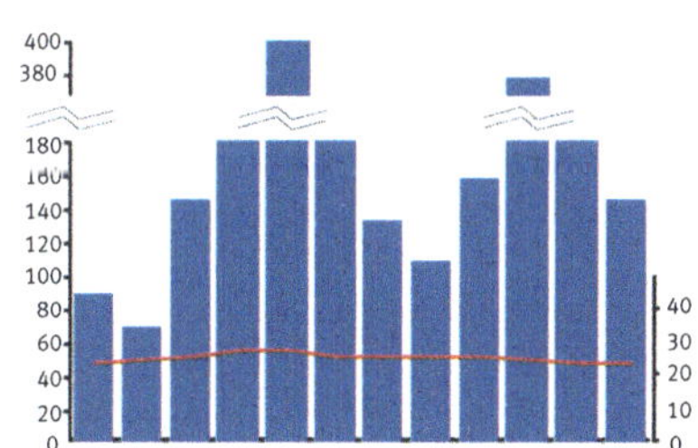

ACTIVIDADES

3. A partir de los climogramas que aparecen en esta página, describe las características fundamentales de cada una de las categorías de clima en lo referente a temperatura y precipitaciones.

4. Construye el climograma que representa los siguientes datos (El Cairo), indica a que tipología de clima corresponde y cuáles serán los principales agentes y procesos geológicos que actuarán en la zona. Justifica tu respuesta.

	E	F	M	A	M	J	J	A	S	O	N	D
T (°C)	13,3	14,7	17,5	21,1	25,0	27,5	28,3	28,3	26,1	24,2	20,0	15,0
P (mm)	5	5	5	2	2	2	0	0	2	2	2	5

DISTRIBUCIÓN DEL AGUA EN EL PLANETA		
Depósito	**Volumen (en millones de km³)**	**Porcentaje (%)**
Océanos y mares	1370	97,25
Casquetes y glaciares	29	2,05
Agua subterránea	9,5	0,68
Lagos	0,125	0,01
Humedad del suelo	0,065	0,005
Atmósfera	0,013	0,001
Ríos	0,0017	0,0001
Biomasa	0,0006	0,00004

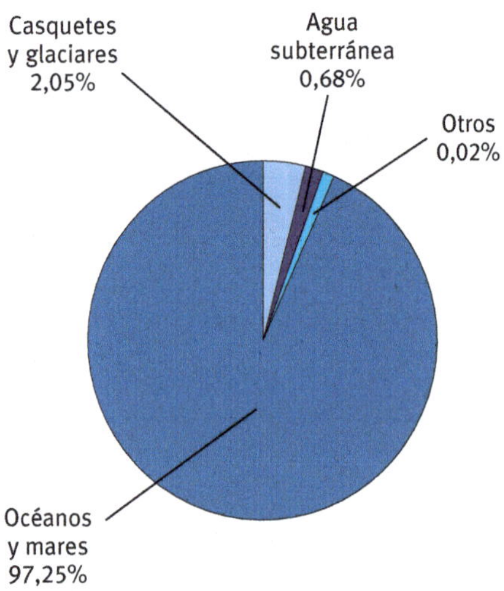

Distribución del agua en el planeta

3. El agua

El agua es la sustancia más abundante en la superficie de la Tierra y constituye la hidrosfera. El 97,2 % del agua se encuentra en los océanos y los mares, y sólo el 2,8 % corresponde a las aguas continentales o dulces. Las aguas continentales están constituidas por:

- El agua que discurre por los *ríos* y los *torrentes*.
- El agua acumulada en *lagos*, *lagunas* y *embalses*.
- Las *aguas subterráneas* presentes en el interior de la corteza terrestre.
- El *hielo* de los *glaciares* y la *nieve* de las altas montañas.

Por otro lado, una pequeña proporción de agua se encuentra en forma de vapor en la atmósfera y contenida en los seres vivos.

El agua no permanece inmóvil, sino que está sometida a un continuo movimiento durante el cual, además, se producen cambios de un estado a otro. Ello origina el denominado **ciclo del agua.**

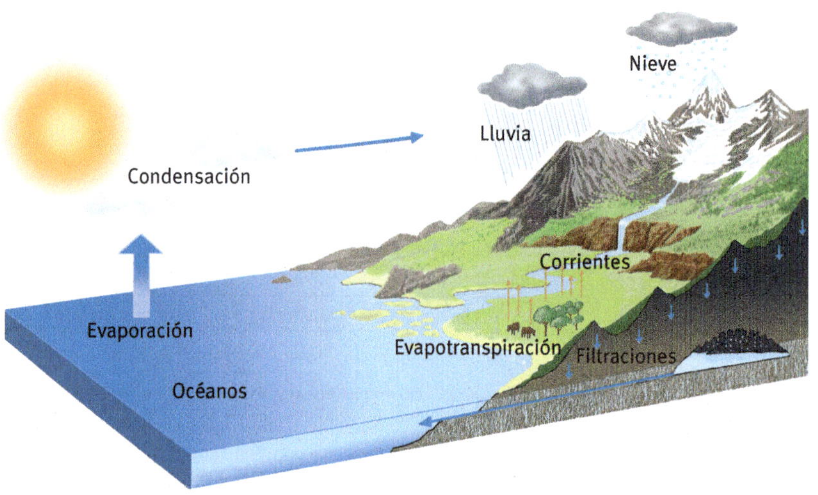

— El agua de los océanos y de los continentes se evapora cuando el Sol la calienta, y se incorpora a la atmósfera en forma de vapor de agua, formando las nubes.

— Las nubes, transportadas por el viento, llegan a zonas más frías de la atmósfera, donde el vapor de agua se condensa y da lugar a precipitaciones líquidas o sólidas.

— Parte del agua de las precipitaciones discurre por la superficie terrestre en forma de:

  - **Ríos y torrentes.** Circulan por un cauce fijo.
  - **Aguas de escorrentía.** Circulan sin un cauce fijo por las zonas de máxima pendiente.
  - **Aguas subterráneas.** Circulan por capas de distinta profundidad de la corteza terrestre debido a la infiltración del agua.

— El agua que circula por la superficie terrestre, en un momento u otro, vuelve al **mar,** con lo que comienza de nuevo el ciclo.

Una parte del agua de la superficie terrestre es absorbida por los seres vivos. Éstos, a su vez, devuelven un gran porcentaje de ésta a la atmósfera mediante la transpiración y la respiración.

A continuación, vamos a describir las modificaciones que el agua que circula por la superficie origina en el paisaje.

3.1. *Los ríos*

Son corrientes de agua que discurren de forma permanente y continua por un cauce fijo. Nacen en las sierras altas y fluyen pendiente abajo hasta desembocar en un río mayor, un lago o el mar.

El agua de los ríos puede proceder de la lluvia, de los torrentes, de las aguas subterráneas, del deshielo y de otros ríos de menor tamaño.

En general, el **caudal,** es decir, el volumen de agua que transporta el río, varía según las estaciones del año.

Si representamos en una gráfica la relación entre las diferentes altitudes del curso del río, desde su nacimiento hasta su desembocadura, y la distancia recorrida, se observa un progresivo descenso de la **pendiente.**

Valle fluvial. Río Verdon (Francia).

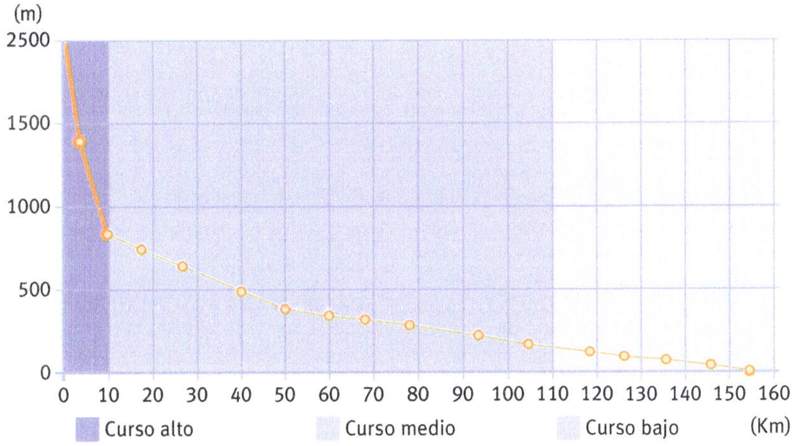

Esta representación gráfica se denomina **perfil longitudinal** de un río, y en ella se distinguen:

- El **curso alto,** que es la parte inicial del recorrido, caracterizado por la fuerte pendiente.
- El **curso medio,** la zona media del recorrido, de pendiente suave.
- El **curso bajo,** o tramo final, en el que no hay prácticamente pendiente.

En estas tres partes del río se producen simultáneamente los procesos de erosión, transporte y sedimentación. Pero en cada una de ellas predomina un proceso u otro, según sean el caudal o la pendiente.

La formación del relieve más característica de los ríos son los **valles fluviales,** los cuales tienen una forma de V característica, como muestra la fotografía de la derecha. Esta forma es más o menos acusada según la zona del recorrido del río.

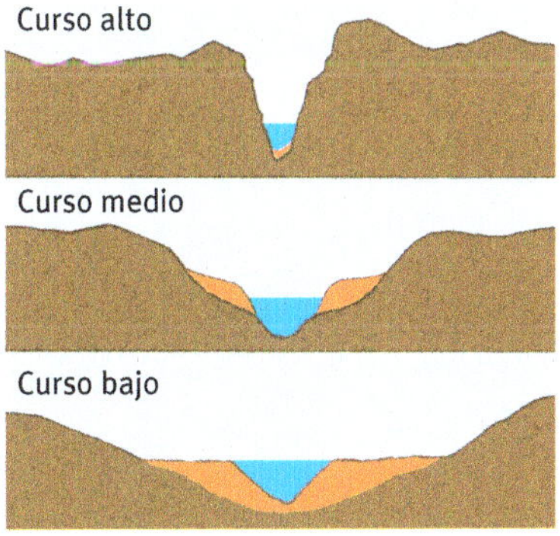

Sección de un río en el curso alto, el curso medio y el curso bajo

PARTES	CARACTERÍSTICAS Y PROCESOS	RELIEVE
Curso alto  Río Guadalquivir en la Sierra de Cazorla (Jaén)	La pendiente es muy acusada; el agua circula a gran velocidad y con mucha fuerza. Predomina la **erosión** de las rocas. Los fragmentos resultantes son redondeados por efectos de la meteorización y de la abrasión, y son transportados hacia el curso medio; cuando su peso es excesivo, sedimentan.	• El agua debe superar desniveles considerables y da lugar a *cascadas* y *rápidos*. • La erosión del agua origina profundos *valles fluviales* muy estrechos, también llamados *gargantas, cañones, desfiladeros,* etc. • Se forman las *marmitas de gigante*, cavidades cilíndricas en el lecho del río, debidas a la erosión producida por el movimiento giratorio de los fragmentos rocosos que transporta el agua.
Curso medio  Meandros del Guadalquivir en Almodóvar del Río (Córdoba)	La pendiente es más suave y el agua fluye a menor velocidad y con menos fuerza. Predomina el **transporte**, ya sea químico o físico.	• Al disminuir la intensidad de la erosión, el *valle* se ensancha. • Aparecen los *meandros*, curvas del cauce del río debidas a la circulación del agua por las zonas en que los materiales ofrecen menos resistencia a su paso. La velocidad es máxima en la parte cóncava y mínima en la convexa, lo que produce erosión en la primera y sedimentación en la segunda. • También se forman las *terrazas aluviales*, depósitos escalonados de aluviones a ambos lados del cauce del río. Son zonas interesantes para el cultivo.
Curso bajo  El Guadalquivir en Doñana (Huelva)	Corresponde a terrenos sin apenas pendiente, por tanto la velocidad del agua disminuye. Cerca de la desembocadura se mezclan el agua del río y la del mar. Predomina la **sedimentación** de los materiales que el río arrastra.	• El *valle* es muy amplio. • Se forman también *meandros* y grandes *terrazas aluviales*. • Cuando el río desemboca en zonas donde las mareas y las corrientes son suaves, se forman los *deltas*, depósitos de sedimentos de forma triangular que se adentran en el mar. • Si la desembocadura tiene lugar en zonas de mareas y corrientes intensas se forman los *estuarios*, donde los sedimentos son arrastrados mar adentro y el río se abre al mar en una amplia desembocadura.

3.2. *Los torrentes*

Son corrientes de agua que discurren de forma temporal por un cauce fijo. El agua procede de las lluvias intensas o del deshielo y, por ello, se localizan en zonas de montaña con fuertes pendientes.

Una diferencia básica entre un río y un torrente es que la longitud del cauce de este último suele ser mucho menor. En los torrentes se distinguen tres partes. En cada una de ellas predomina uno de los procesos geológicos, según sean la pendiente del terreno y el caudal.

- **Cuenca de recepción:** tiene forma de embudo y en ella se recoge el agua procedente de las lluvias o el deshielo. El agua circula a gran velocidad debido a la fuerte pendiente y la erosión es muy intensa. Se originan barrancos, y son frecuentes los deslizamientos y corrimientos de tierras. Es frecuente la formación de cascadas.
- **Canal de desagüe:** es la zona intermedia, donde el agua fluye a gran velocidad. Predomina el transporte, aunque la erosión también es importante, originando cavidades en el lecho.
- **Cono de deyección:** es la zona final, que desemboca en un río. La pendiente disminuye y con ello la velocidad y la fuerza del agua. Predomina la sedimentación, y los materiales se depositan en forma de abanico.

Los torrentes son corrientes temporales de agua que circulan por zonas de gran pendiente.

3.3. *Las aguas de escorrentía*

Las aguas de escorrentía son corrientes de agua temporales sin un cauce fijo. Fluyen formando una película que recorre la superficie del terreno hasta alcanzar la base de las montañas, el cauce de un río o el de un torrente. Se originan en zonas con pendientes muy acusadas, cuando el aporte de agua, ya sea de las lluvias o del deshielo, es superior a la capacidad de absorción del terreno.

La acción de estas aguas es básicamente erosiva y depende de la cantidad de agua, de la pendiente, de la presencia de vegetación y, sobre todo, de la naturaleza de las rocas. Las rocas que se erosionan más fácilmente son las arcillas y los conglomerados. Las partículas transportadas sedimentan en la base de las montañas, o bien, pasan a formar parte de los materiales que transportan los ríos y los torrentes.

Las aguas de escorrentía dan lugar a formaciones características, como:

- **Cárcavas y barrancos:** surcos de profundidad variada que aparecen en las laderas de las montañas.
- **Ramblas:** depresiones donde confluye el agua que desciende por las cárcavas y barrancos. En períodos de lluvias intensas se convierten en avenidas de agua.
- **Pirámides de tierra o chimeneas de las hadas:** estructuras verticales producidas por la erosión diferencial de un terreno. En ocasiones, los materiales menos erosionables aparecen en forma de sombrero sobre los más alterables.
- **Badlands:** grandes extensiones de terreno en zonas áridas con numerosos barrancos de paredes verticales.

Cárcavas en las Bardenas Reales

Chimeneas de las hadas en Capadocia

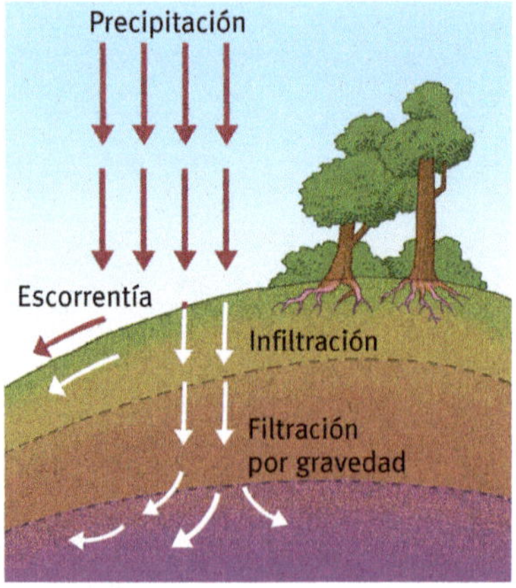

El agua de infiltración tiene un papel fundamental en el ciclo hidrológico.

En el paisaje podemos identificar terrenos por los que discurren ríos, torrentes o aguas de escorrentía. Hay que fijarse en:

- Presencia de un cauce.
- Formaciones del relieve (meandros, desfiladeros...).
- Forma de los fragmentos arrastrados por el agua.
- La observación de la vegetación proporciona información útil sobre la frecuencia y la cantidad de precipitaciones y, por tanto, sobre la posibilidad de que los cursos de agua sean permanentes o temporales.

3.4. *Las aguas subterráneas*

Son aquellas que se encuentran en los poros de las rocas, o bien, en las grietas o las fisuras. En el primer caso, el agua circula con dificultad y en el segundo, se forman verdaderos ríos subterráneos.

El agua procede de la **infiltración,** es decir, del paso de las aguas superficiales (precipitaciones, ríos, torrentes, etc.) hacia el interior del terreno debido a la fuerza de la gravedad.

La cantidad de aguas subterráneas que se pueden encontrar en un terreno depende de los siguientes factores:

- **Permeabilidad de las rocas.** Es la capacidad que tienen las rocas de permitir el paso del agua o del aire a través de ellas. Esta propiedad está relacionada con la **porosidad,** es decir, la cantidad de espacios vacíos, o poros, que tiene una roca. Los poros están comunicados entre sí y con la superficie, y se llenan de aire y de agua procedentes del medio. Cuanto mayores son y más comunicación existe entre ellos, más permeables son las rocas; es el caso de la caliza. También puede darse el caso de rocas no porosas que son permeables, debido a que tienen fracturas.
- **Evapotranspiración.** Este término se refiere a la pérdida de agua por evaporación del agua del suelo y por la transpiración de las plantas.
- **Pendiente del terreno.** Si ésta es muy acusada, el agua fluye a más velocidad y se infiltra en menor cantidad.
- **Vegetación.** Las raíces de las plantas aumentan el tamaño de los espacios que hay entre las rocas, lo que favorece la infiltración del agua. También contribuye a ello la existencia de una capa abundante de vegetación, que frena las aguas de escorrentía.

Los vegetales ceden por el proceso de transpiración gran cantidad de agua a la atmósfera. Se ha calculado que una planta de trigo transpira alrededor de 204 litros de agua durante los meses más calurosos. Una hectárea plantada de trigo cedería a la atmósfera algo más de 3 000 000 de litros de agua en ese período.

Las aguas subterráneas discurren hacia el interior a través de las rocas permeables y porosas. Cuando estas aguas llegan a una roca impermeable, se acumulan formando depósitos de agua dulce, denominados **acuíferos.**

Veamos algunos conceptos relacionados con los acuíferos.

Un acuífero presenta las siguientes partes:

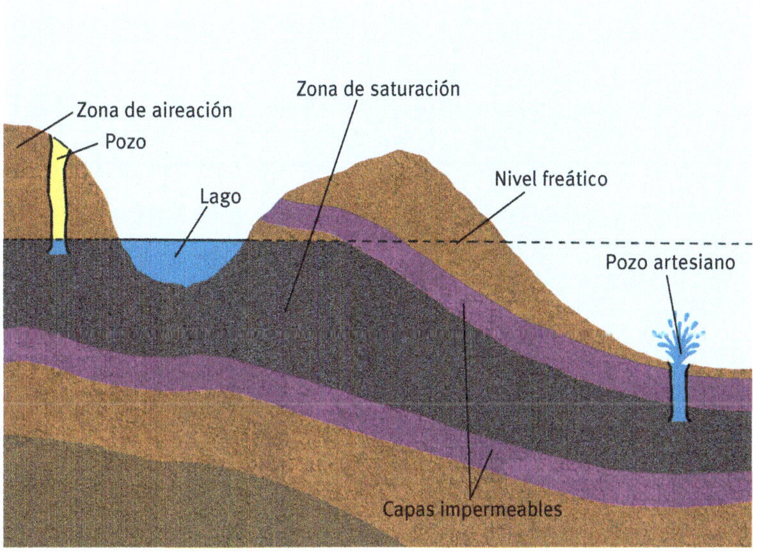

- **Zona de aireación.** Es la parte más cercana a la superficie. En ella, los poros de las rocas no están saturados, porque es una zona afectada por los procesos de evapotranspiración y porque en ella se encuentran las raíces de las plantas.
- **Nivel freático.** Es el límite o superficie de separación entre la zona de aireación y la de saturación. La profundidad a la que se encuentra este nivel varía en función de las precipitaciones y de las extracciones de agua de la zona de saturación.
- **Zona de saturación.** En ella los poros de las rocas están saturados de agua. Por debajo de esta zona se encuentran las rocas impermeables.

El agua presente en los acuíferos puede emerger a la superficie en forma de:

- **Fuentes.** Son flujos de agua subterránea que emergen en cualquier punto de la superficie, el suelo, los ríos, los océanos, etc. También se producen en los valles cuando el nivel freático intersecta con la superficie.
- **Pozos.** Son construcciones hechas por el ser humano para la extracción del agua de los acuíferos. Los pozos artesianos son un caso particular en el cual el agua sale al exterior por la presión debida a que el acuífero se halla entre dos capas impermeables que alcanzan alturas superiores a la de la perforación.

Las aguas de los acuíferos constituyen un recurso hídrico muy importante, ya que se utilizan para la agricultura o para el abastecimiento de las poblaciones. Su excesiva explotación o su contaminación, causada por las actividades agrícolas y ganaderas, ponen en peligro su futuro. Por tanto, es necesario conservarlos y explotarlos racionalmente.

Las aguas subterráneas son responsables de diversos procesos de meteorización, erosión, transporte y sedimentación en el interior de la corteza terrestre.

Estas aguas disuelven las rocas solubles, principalmente las rocas calizas, mediante reacciones de carbonatación, y dan lugar a las formas **endocársticas,** que hemos tratado anteriormente.

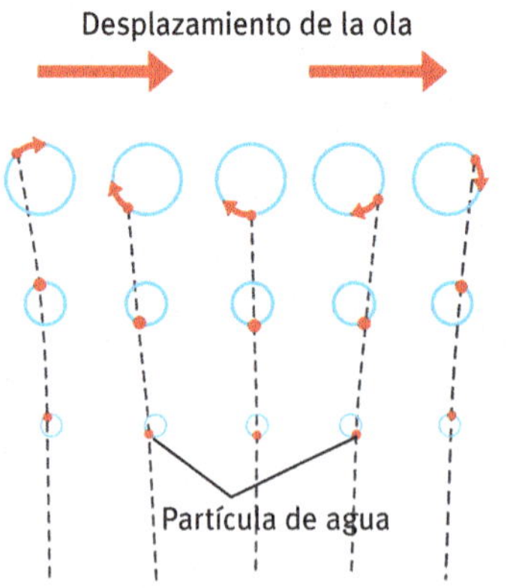

Movimiento de las partículas en una ola

3.5. *Los océanos y los mares*

Son grandes acumulaciones de agua y ocupan las dos terceras partes de la superficie de nuestro planeta.

Los océanos y los mares presentan movimientos del agua debidos a las diferencias de temperatura y de salinidad, a la acción del viento y a la atracción gravitatoria del Sol y la Luna. Distinguimos tres tipos de movimientos:

- **Olas.** Son movimientos ondulatorios de las partículas de agua producidos por el viento. Las partículas de agua se mueven continuamente y describen una órbita circular.

  Observa la figura de la derecha, en ella se representan diversas partículas contiguas en la superficie del agua y la órbita que describe cada una. Cada partícula se encuentra en una posición distinta de esta trayectoria, por lo que en la superficie del agua se forma una ondulación.

  Todas las partículas empiezan y terminan el movimiento en el mismo punto, de modo que la masa de agua no se traslada.

  Cuando este movimiento tiene lugar en zonas próximas a la costa y por tanto, poco profundas, el movimiento de las partículas se ve alterado por el rozamiento con el fondo.

  Así, la longitud de la onda disminuye, mientras aumenta su altura, hasta que la parte superior de la ola cae hacia delante, originando lo que habitualmente se conoce como el **romper de las olas.**

Corrientes marinas superficiales

- **Corrientes.** En este caso se desplazan masas de agua tanto a nivel superficial como en zonas profundas. Las corrientes superficiales dependen principalmente de los vientos, mientras que las corrientes profundas se deben a las diferencias existentes de temperatura y salinidad.
- **Mareas.** Son movimientos periódicos de ascenso y de descenso del nivel del mar debido a la atracción gravitatoria de la Luna y del Sol sobre la Tierra. El movimiento de ascenso determina un avance de la masa de agua hacia el continente; el máximo nivel que alcanza se denomina *pleamar*. El movimiento de descenso implica un retroceso, y el nivel más bajo que se alcanza se llama *bajamar*.

Todos estos movimientos son los causantes de los procesos geológicos que tienen lugar en la **línea de costa,** es decir, en la zona de contacto entre el mar y la tierra.

Las olas son las principales responsables de la erosión, mientras que las corrientes son la causa fundamental del transporte de los materiales. Las mareas favorecen la acción de las olas y de las corrientes, ya que modifican temporalmente la línea de costa.

Las olas son el principal elemento erosivo de la línea de costa, y actúan principalmente mediante la presión del impacto y la abrasión provocada por los fragmentos de roca en suspensión.

Esta acción erosiva da como resultado la formación de **acantilados** litorales. Al retroceder éstos por la pérdida de sustentación en su base, dejan tras de sí una plataforma de abrasión. Otras formas producidas por la acción erosiva sobre la línea de costa son los **arcos** y las **chimeneas,** constituidas al desmoronarse el techo de un arco.

Otro factor que aporta elementos a la fisonomía costera es la deposición de sedimentos por parte de las corrientes litorales. Las **flechas** son barras de arena que se prolongan desde tierra en la zona de la desembocadura de una bahía, sin llegar a cerrarla por completo. En caso de que esto suceda hablamos de una **barra de bahía.** Los **tómbolos** son cúmulos de arena que unen una isla con la costa.

Las barras de bahía son producidas por el depósito de sedimentos, separando la bahía del mar abierto. Playa en Córcega.

La línea de costa experimenta una evolución a lo largo del tiempo como consecuencia de la relación existente entre el aporte de sedimentos y la erosión que producen las olas. La actividad humana puede alterar estos factores, provocando un incremento de la erosión o su atenuación, aumentando o disminuyendo el depósito de sedimentos, etc.

Las costas se pueden clasificar en cinco categorías según su origen, los cambios que han experimentado respecto del nivel del mar y su desarrollo posterior.

- **Costas de inmersión:** se dan como consecuencia del hundimiento de la corteza terrestre o por el aumento del nivel del mar. En este tipo de costas se pueden presentar estructuras tales como rías, fiordos, acantilados...
- **Costas de emersión:** son fruto de la elevación de la corteza terrestre o del descenso del nivel del mar. Los antiguos fondos oceánicos quedan al descubierto, originando llanuras costeras, con extensas playas de pendiente suave.
- **Costas neutras:** se han formado por aporte de materiales al agua, como, por ejemplo, costas en delta, costas volcánicas originadas por coladas de lava o costas de arrecifes coralinos.
- **Costas de falla:** se originan al producirse una falla en la cual se hunde el bloque contiguo al mar, quedando el plano de falla como un acantilado.
- **Costas compuestas:** consisten en una combinación de los tipos anteriores.

Las playas de arena extensas y de pendiente suave se presentan en costas de emersión. Playa en Fuerteventura.

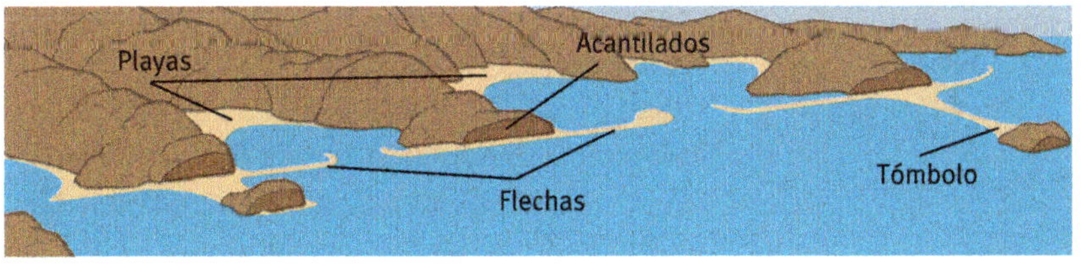

Formaciones más frecuentes en una costa de inmersión

C

LAS GLACIACIONES

A lo largo de la historia de la Tierra se han sucedido episodios geoclimáticos de enfriamiento conocidos como glaciaciones.

En estos períodos, la nieve acumulada en invierno no llega a fundirse completamente durante el verano debido a las bajas temperaturas, y se congela. Esto conduce a un crecimiento de los glaciares y de los casquetes polares, que llegan a alcanzar las latitudes medias. Este aumento de la superficie helada aumenta el reflejo de la radiación solar, potenciando el enfriamiento planetario.

Este proceso se asocia a un descenso del nivel del mar debido a la inmovilización de grandes cantidades de agua en forma de hielo. A lo largo del pleistoceno se han sucedido cuatro períodos glaciales (Günz, Mindel, Riss y Würm), aunque en la historia del planeta se cuentan un mínimo de doce.

3.6. *Los glaciares*

Son grandes masas de hielo situadas en las cimas de las altas montañas y en los polos. Este hielo procede de la acumulación de la nieve que cae, año tras año, durante el invierno y que no se funde completamente en la primavera. Según su situación geográfica, existen dos tipos de glaciares:

- **Casquetes polares o *inlandsis*.** Se encuentran en las zonas polares, uno en Groenlandia (polo norte) y el otro en la Antártida (polo sur). El hielo se acumula sobre grandes extensiones de terreno y puede alcanzar entre 2000 y 4000 metros de profundidad.
- **Glaciares de valle o glaciares alpinos.** Se encuentran en zonas de alta montaña y tienen un curso largo y estrecho.

Los glaciares se deslizan muy lentamente por sus laderas, como si se tratase de un río de hielo, y erosionan el lecho. Los materiales arrancados son transportados a zonas de menor altitud, donde sedimentan.

En los glaciares alpinos se distinguen tres partes en las que, como en el caso de los ríos y los torrentes, predomina un proceso geológico y, por tanto, un modelado del relieve.

La formación del relieve más característica de los glaciares son los **valles glaciares,** en forma de U. Esta forma será más o menos acusada según la parte del glaciar en que se encuentre.

Un glaciar se compone de:

- El **circo glaciar.** Es la zona de acumulación del hielo y la nieve. El peso de la masa de nieve de la zona superior compacta las capas inferiores, convirtiéndolas en hielo. Éste inicia su desplazamiento lentamente hacia las zonas más bajas, provocando procesos de abrasión al rozar contra las rocas. Este proceso erosivo originará amplios valles de fondo plano y paredes verticales en forma de U, los valles glaciares.
- La **lengua del glaciar**. Al desplazarse, el hielo supera las prominencias rocosas que obstaculizan su camino a lo largo de la pendiente. Una vez lo ha logrado, su velocidad aumenta, lo cual provoca grandes tensiones, que agrietan su superficie. En esta zona se dan procesos de erosión y transporte. Los materiales transportados se denominan morrenas, y están formadas por fragmentos angulosos de rocas de tamaños variables.

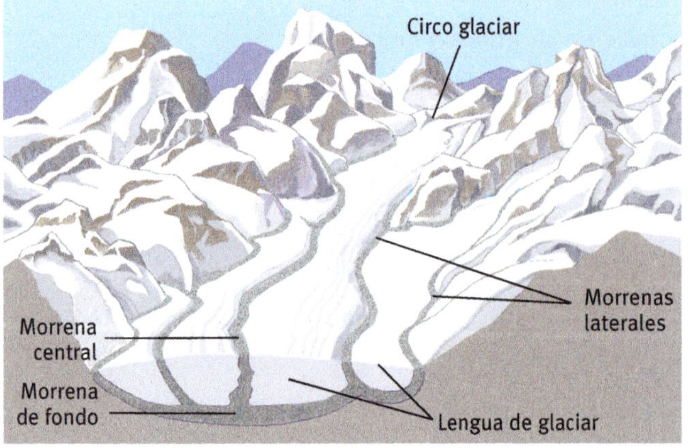

- El **frente del glaciar**. La parte inferior del glaciar es aquella en la que se da una pérdida de hielo por fusión, porque las temperaturas son más altas. La pendiente también es menor, y se dan procesos de sedimentación. Los sedimentos del glaciar se acumulan de modo caótico, ya que no existe estratificación o distribución por tamaños. El agua del deshielo puede originar un río o un lago.

4. El viento

El viento es el movimiento del aire producido por las diferencias de presión atmosférica.

Las características de las zonas con clima seco son favorables para la acción del viento, ya que la humedad baja y la escasa cobertura vegetal facilitan la erosión.

La **erosión eólica,** es decir, la producida por el viento es de dos tipos.

- **Deflación.** Es el barrido y transporte de las partículas que se encuentran sobre las rocas. La intensidad de este proceso depende de la fuerza del viento.
- **Abrasión eólica.** Consiste en el desgaste producido por el impacto de las partículas que transporta el viento en suspensión contra la superficie de las rocas. La intensidad de la abrasión está en función de la cantidad de partículas que lleve el viento.

El **transporte** implica pequeñas partículas que pueden viajar a kilómetros de distancia por suspensión, arrastre o saltación.

Cuando la velocidad del viento disminuye o las partículas chocan con algún obstáculo, tiene lugar la **sedimentación**. Ésta puede dar origen, entre otras formaciones, a las **dunas**, acumulaciones de arena móviles de diferentes formas y tamaños, o a **loess**, formados por la acumulación de limos, partículas de menor tamaño que la arena, que tras ser transportadas se depositan en las zonas húmedas.

El viento es el agente geológico más importante en los desiertos y es la causa de que en ellos se distingan dos zonas: una de dunas, que se denomina **erg,** y una zona de piedras o **reg.**

Dunas en el Sáhara (Marruecos)

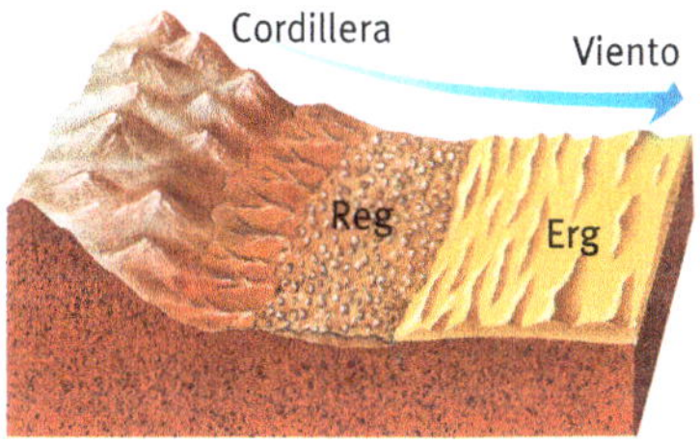

El reg y el erg son fruto de la erosión de una cordillera por el viento

ACTIVIDADES

5. ¿De dónde procede la energía que mueve el ciclo del agua? Esquematiza este ciclo, indicando cada uno de los compartimentos por los cuales circula el agua.
6. Define los siguientes conceptos: caudal, perfil longitudinal, curso alto, medio y bajo, meandro, terraza aluvial, delta y estuario.
7. ¿Cuáles son las diferencias fundamentales entre un río y un torrente?
8. ¿Qué factores afectan a la capacidad erosiva de las aguas de escorrentía? Justifica la influencia de cada una de ellas.
9. ¿Qué es el nivel freático? ¿Qué dos zonas delimita?
10. ¿Cuáles son los movimientos del agua de los mares y océanos? ¿A qué se debe cada uno de ellos?
11. Explica las diferencias entre una costa de inmersión y una de emersión, comentando los trazos característicos de cada una de ellas.
12. Comenta las características principales de la erosión debida a la acción de los glaciares.
13. ¿Por qué el viento es un agente erosivo especialmente importante en las zonas de clima seco?
14. Define los siguientes conceptos: deflación, abrasión eólica, duna, loess, reg, erg.

La materia orgánica acumulada en el suelo aportará al descomponerse gran cantidad de componentes orgánicos a éste.

5. El suelo

Como resultado de la meteorización de las rocas de la superficie terrestre se forma el **suelo,** una capa delgada que cubre algunas zonas de la superficie. En ella se desarrollan las raíces de las plantas y, por ello, es una zona imprescindible para la vida. En el desarrollo del suelo a partir de una roca madre desnuda intervienen fundamentalmente cinco factores que se denominan formadores del suelo.

- La **roca madre original.** Constituye la base de la mayoría de suelos. El proceso de formación se inicia por una meteorización de la roca madre por acción de los agentes geológicos externos, que la fracturan, favoreciendo la alteración química de los minerales que la forman. Estos procesos darán lugar a los compuestos que determinarán una parte de las características del suelo resultante.
- El **relieve.** La circulación del agua por la superficie del terreno será tanto más rápida cuanto más acusada sea la pendiente, predominando la erosión y dificultando la acumulación. Así, las zonas en pendiente presentarán suelos más delgados que las áreas llanas. Por otra parte, las pendientes débiles favorecen la penetración del agua de la lluvia, con lo cual se potencian los procesos de alteración química.
- El **tiempo.** Los procesos físicos, químicos y biológicos que llevan a la formación del suelo actúan a lo largo del tiempo, provocando cambios graduales en su estructura y composición. A partir de una roca madre se inicia el proceso de meteorización física y química, que lleva a una disgregación y alteración del material. La colonización de estos fragmentos por organismos vivos llevará a un incremento de la alteración por la acción de las sustancias químicas que segregan y por el efecto mecánico de sus raíces. Además, el aporte de materia orgánica procedente de los seres vivos que mueren o las hojas caídas incorporará nuevos compuestos al suelo. Con el tiempo, la lluvia arrastrará sustancias desde la parte superior del suelo hacia el interior, generando una nueva zona intermedia. Cuando sobre un suelo han actuado todos los procesos durante suficiente tiempo, éste alcanza una etapa en la cual su estructura o perfil no variará significativamente. Hablamos de suelo maduro.
- El **clima.** Es el elemento más importante en la formación del suelo. Entre los diversos factores que agrupa, la humedad, la temperatura y el viento son los que tendrán una mayor influencia.

  La **lluvia** aporta al suelo el agua necesaria para que se lleven a cabo las reacciones químicas y es imprescindible para la actividad de los seres vivos. También es la responsable del arrastre de materiales hacia la zona intermedia o, en casos de lluvias excesivas, de la eliminación del suelo de sustancias solubles o coloides. En los climas secos la evaporación provoca el ascenso hacia la superficie del agua de las capas inferiores, que arrastra con ella las sales disueltas. Este proceso da lugar a la formación de costras de carbonato cálcico o yeso.

  Las **temperaturas** altas favorecen las reacciones químicas y la actividad bacteriana, por lo cual los suelos de climas cálidos y húmedos están muy alterados químicamente; por este motivo, apenas presentan en su superficie materia orgánica en descomposición por la rapidez con la cual se dan estos procesos. En cambio, en los climas continentales fríos, la materia orgánica en descomposición cubre el suelo del bosque.

El **viento** incrementa la evaporación y puede eliminar materiales de la superficie si no existe la protección de una cubierta vegetal.

- Los **seres vivos**. Son factores muy influyentes en la formación del suelo. Por una parte actúa la microflora, compuesta por bacterias, hongos, líquenes... que intervienen en los procesos de descomposición química, tanto de la roca madre como de la materia orgánica acumulada. Los árboles, arbustos y hierbas, que constituyen la macroflora, son a la vez condicionados por el tipo de suelo y participan en su formación. Los vegetales absorben componentes de las capas inferiores (potasio, calcio, magnesio) incorporándolos a su organismo y depositándolos posteriormente en la superficie cuando se descomponen, lo que mantiene la fertilidad del suelo. La macroflora realiza el principal aporte de materia orgánica, que al oxidarse dará lugar al humus.

En un suelo maduro se pueden distinguir las siguientes capas u horizontes:

- **Horizonte A:** Es la capa superior, habitualmente de color oscuro, rica en materia orgánica. En la zona más superficial o horizonte A00 se acumulan hojas caídas y residuos orgánicos.
- **Horizonte E:** Se encuentra entre el A y el B; está formado por partículas minerales de color claro. Es una zona de tránsito de materiales arrastrados por el agua hacia el horizonte B.
- **Horizonte B:** Es la zona de acumulación de materiales (coloides) arrastrados por la lluvia desde el horizonte A. En él se pueden encontrar también fragmentos provenientes del horizonte C.
- **Horizonte C:** Está constituido por el material de la roca madre meteorizado.
- **Horizonte D:** Es la roca madre inalterada.

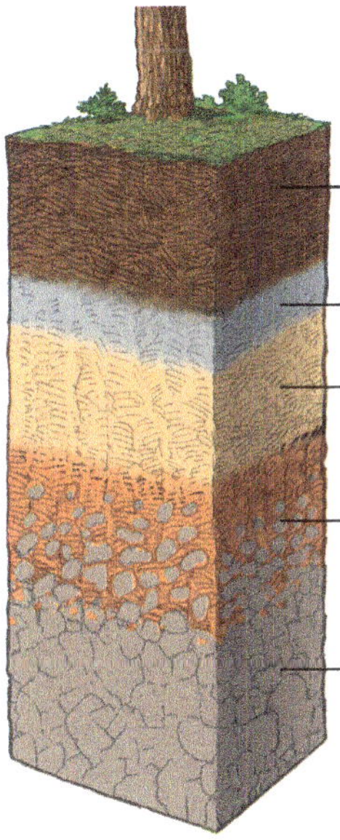

Horizontes en un suelo maduro

Los horizontes A, E y B constituyen el *solum*, el suelo propiamente dicho, mientras que el C representa el subsuelo.

A veces, los suelos no se forman a partir de la roca que constituye el horizonte D, sino que se originan por sedimentación de materiales procedentes de la erosión de rocas de otros lugares, como por ejemplo el loess o los suelos aluviales.

En el suelo se encuentran aproximadamente:

- Un 50% de materiales sólidos, sobre todo fragmentos de rocas y minerales de tamaño variable y restos procedentes de seres vivos (raíces muertas, hojas, excrementos de animales...).
- Un 25% de materiales líquidos en forma de agua con diversos tipos de sustancias disueltas.
- Un 25% de materiales gaseosos, en concreto aire procedente de la atmósfera que está en contacto con el suelo, y productos del metabolismo respiratorio de los seres vivos que lo habitan.

Como hemos visto,el proceso de formación del suelo es muy lento, especialmente en aquellos terrenos en los cuales la escasez de lluvias retarda los procesos químicos que lo generan, como en la zona mediterránea. La actividad humana provoca en muchos casos pérdidas de suelo o favorece la erosión de los mismos, de manera que esta disminución no puede ser compensada por los procesos naturales de generación de suelo; esta actividad, por tanto, puede conducir a la desertización.

Algunas de las acciones humanas que acentúan la erosión del suelo son:

- La **tala indiscriminada** de árboles y la **minería a cielo abierto**: son dos tipos de actividad que eliminan la cobertura vegetal, con lo cual se incrementa la erosión causada por las aguas de escorrentía o el viento. Es especialmente grave en zonas con fuerte pendiente, donde puede llegar a producir corrimientos de tierra.
- El **uso incorrecto del suelo y el agua**. En zonas áridas, el riego con agua con sales conduce a la salinización del suelo, haciendo imposible a medio plazo el crecimiento de la vegetación. Ciertas técnicas de cultivo intensivo también favorecen la desaparición del suelo.
- El **sobrepastoreo** provoca la pérdida de la cobertura vegetal y deja desprotegido el suelo.
- La **compactación del suelo** a causa de la circulación provoca un endurecimiento e impermeabilización del mismo, que dificulta el crecimiento de la vegetación y facilita la erosión.

La tala indiscriminada priva al suelo de la protección de la cobertura vegetal, favoreciendo su erosión.

ACTIVIDADES

15. ¿Puede una misma roca madre dar lugar a dos tipos de suelo diferentes? Justifica tu respuesta.

16. ¿Dónde será más fácil la generación de suelo, en la ladera de una montaña o en un valle? ¿Por qué?

17. ¿Qué papel juegan los seres vivos en la formación del suelo?

18. ¿Todos los suelos presentan todos los horizontes indicados? ¿Por qué?

19. Un cultivo en terrazas en la ladera de una montaña, ¿ayuda a conservar el suelo o lo dificulta?

20. ¿En qué zona será más frágil el suelo, en un bosque atlántico o en un encinar mediterráneo? Justifica tu respuesta.

6. El ciclo geológico

En esta unidad hemos descrito la actividad modificadora de la superficie de nuestro planeta, que tiene lugar por acción de los procesos geológicos externos.

Éstos actúan de manera conjunta con los internos, descritos en la unidad anterior, provocando desplazamientos y transformaciones de los materiales. Dichos desplazamientos y transformaciones se repiten a lo largo del tiempo de una forma cíclica y constituyen el **ciclo geológico.** Este ciclo fue formulado por primera vez por James Hutton en el s. XVIII.

El ciclo se inicia cuando los magmas procedentes del interior de la corteza terrestre emergen generando las rocas volcánicas. El enfriamiento de éstos en profundidad daría lugar, como hemos visto, a las rocas plutónicas.

Cuando estas rocas de origen magmático afloran a la superficie terrestre quedan expuestas a la acción de los agentes geológicos externos que provocan su meteorización, erosión y transporte, depositándose los fragmentos en las cuencas sedimentarias. Los procesos de compactación y diagénesis los acaban convirtiendo en rocas sedimentarias, las cuales a su vez pueden experimentar nuevamente el mismo proceso.

Las rocas sedimentarias y las magmáticas pueden ser sometidas a determinadas condiciones de presión o temperatura, que las transformaran en rocas metamórficas que en condiciones extremas pueden llegar a fundirse y originar nuevamente magmas.

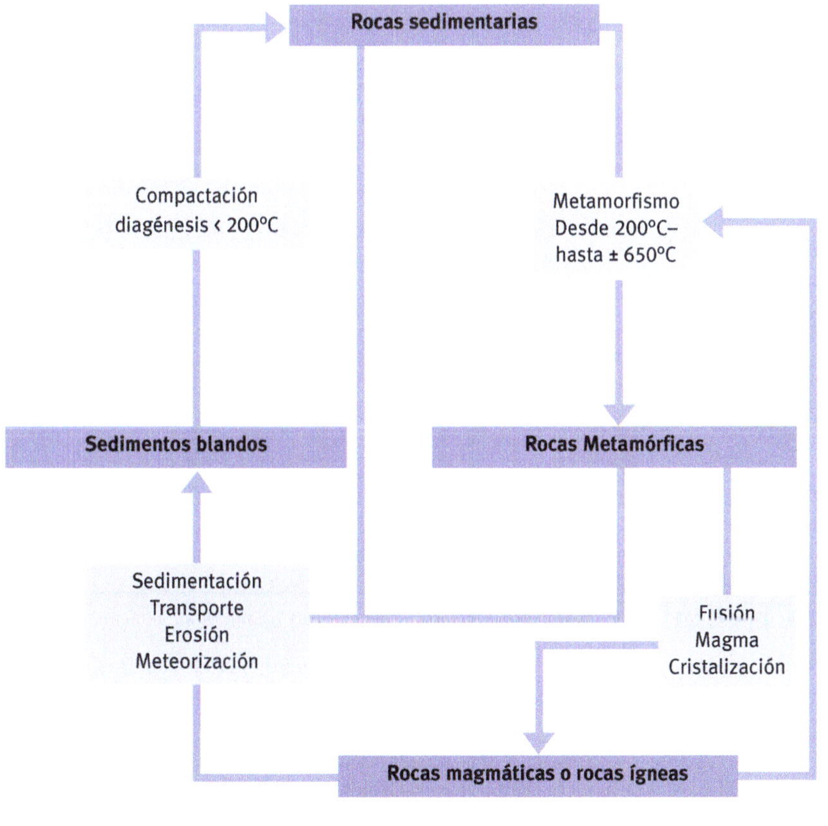

+

James Hutton (1726-1797), científico británico considerado como el padre de la geología moderna, defendió la teoría del plutonismo, en contra de las teorías neptunistas en boga en el momento, que consideraban que las rocas de la superficie terrestre habían precipitado a partir de una gigantesca inundación. Hutton, en cambio, aseguraba que éstas eran producto de la solidificación del magma interior de la tierra.

Igualmente, Hutton fue un defensor del actualismo, teoría según la cual la historia geológica pasada puede interpretarse a partir de los procesos observables en la actualidad, sin necesidad de recurrir a grandes episodios catastróficos, tal y como era propio en la época.

En 1785 publicó su teoría en la cual enunciaba una serie de ciclos en la historia de la tierra que implicaban la formación de rocas, su posterior erosión y depósito en cuencas marinas, y su posterior ascenso a la superficie, iniciándose nuevamente el proceso.

Los procesos geológicos provocan dos efectos en la litosfera.

Por un lado, a partir de los procesos geológicos internos se pueden producir importantes acumulaciones de materiales en unas zonas concretas de la litosfera. En ellas se observa un aumento del espesor de esta capa.

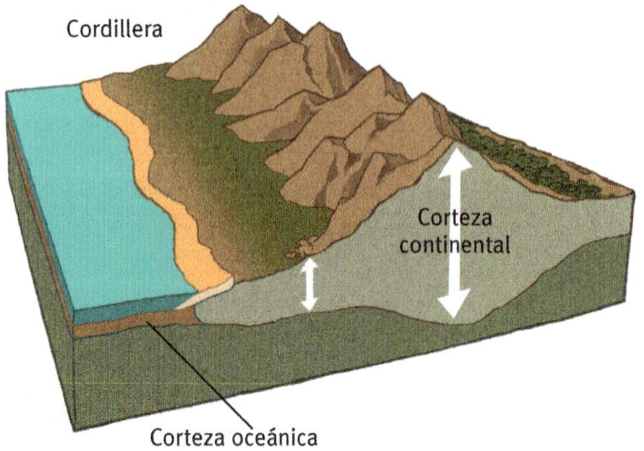

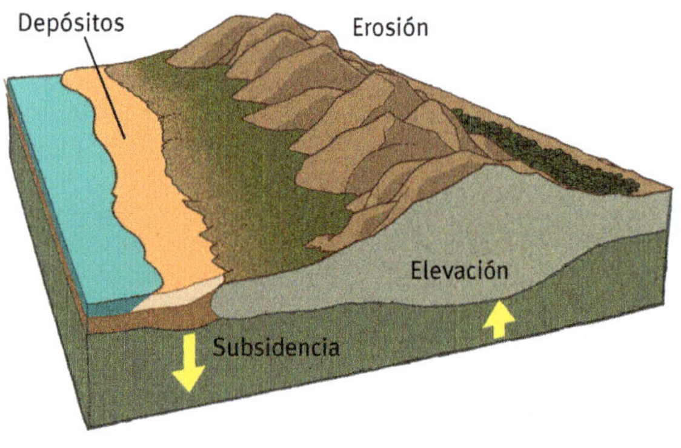

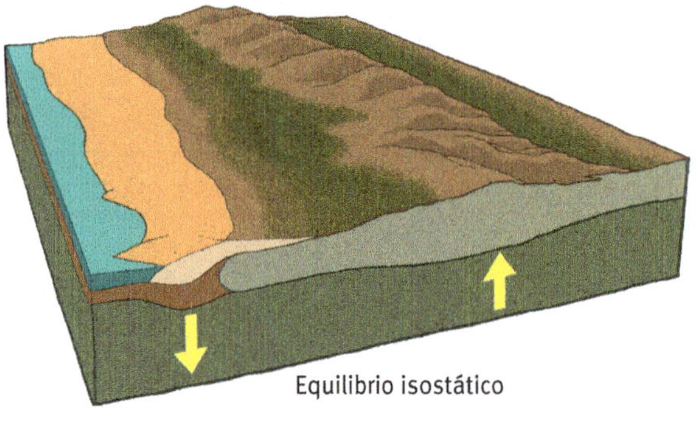

Por otro lado, los procesos externos desgastan las rocas que se encuentran en la superficie, lo que provoca una disminución del espesor de la litosfera.

Los procesos orogénicos que conducen a la formación de cordilleras, como los plegamientos, provocan un incremento del grosor de la litosfera y por lo tanto del peso que debe soportar localmente la capa superior del manto.

Como consecuencia de este aumento de peso la litosfera tiende a hundirse de forma gradual y espontánea hasta alcanzar el nivel de equilibrio. A este proceso se le denomina **subsidencia**.

La acción continuada de los procesos geológicos de meteorización, erosión y transporte a lo largo de los años produce una disminución del grosor de la litosfera en determinadas zonas, que quedan liberadas de peso, produciéndose un movimiento ascendente hasta un nuevo nivel de equilibrio.

De igual manera, las cuencas en las cuales se acumulan los sedimentos procedentes de la erosión incrementan su potencia, padeciendo un proceso de subsidencia que las hace hundirse hasta alcanzar el **equilibrio isostático.**

Este proceso también se da en zonas donde se acumulan grandes cantidades de hielo, las cuales padecen en primer lugar un proceso de subsidencia a causa del peso del casquete acumulado y una elevación posterior al retirarse éste. En Europa, la península de Escandinavia se halla aún en proceso de elevación como consecuencia de la retirada del casquete de hielo producido durante la última glaciación.

ACTIVIDADES

- **21.** Sitúa en un esquema del ciclo geológico las siguientes rocas: basalto, conglomerado, mármol, granito, pizarra, arenisca, caliza, gneis. Explica los procesos que relacionan los diferentes tipos de roca.
- **22.** ¿Qué procesos provocan el desplazamiento vertical de la corteza terrestre? ¿Qué es la subsidencia? ¿A qué es debida?

ACTIVIDADES DE SÍNTESIS

23 Según los últimos resultados obtenidos por los robots y las naves en órbita alrededor de Marte, el planeta habría tenido en el pasado unas condiciones marcadas por la presencia de agua líquida. El estudio de determinadas estructuras geológicas nos permite establecer paralelismos entre los agentes causantes de ellas en nuestro planeta y los que las produjeron en el planeta rojo.

A continuación te presentamos una colección de parejas de fotografías de la Tierra y de Marte. Identifica las estructuras que aparecen en ellas y comenta, para cada una de ellas, las causas implicadas en su formación.

A

B

C
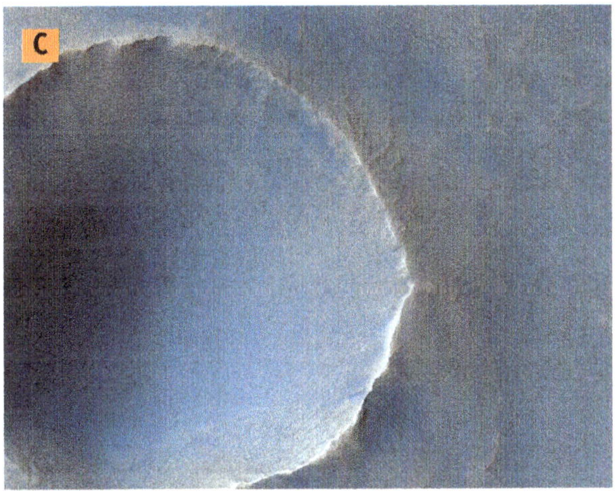

D
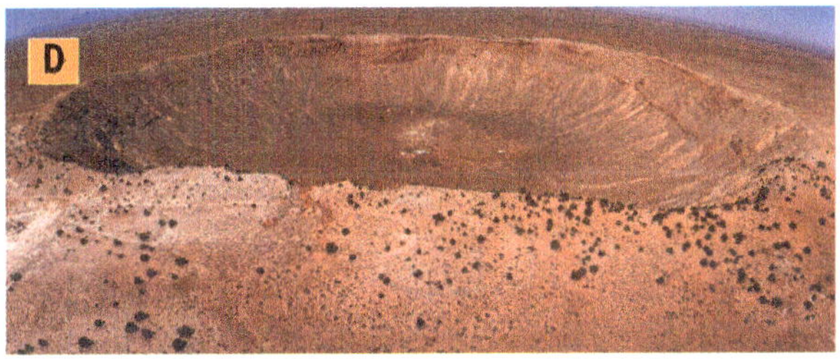

E

F

24 ¿Qué diferencia hay entre meteorización y erosión? ¿Cómo se relacionan ambos procesos entre sí? Pon un ejemplo.

25 Completa una tabla como la siguiente sobre las causas y efectos de los diferentes tipos de meteorización:

		CAUSAS	RESULTADO
Meteorización química	Hidrólisis		
	Disolución		
	Hidratación		
	Oxidación		
Meteorización física	Termoclastismo		
	Crioclastismo		
	Haloclastismo		
	Descompresión		

26 A partir de los datos que hallarás a continuación, construye los climogramas de las tres estaciones siguientes y señala cuáles serán los principales procesos y agentes geológicos externos que actuarán en la formación del relieve en cada una de ellas. Sitúalas en un mapa.

		E	F	M	A	M	J	J	A	S	O	N	D
Sevilla	**T (°C)**	10	11,5	14	16,5	20	24,5	27,5	28	24,5	19,5	14,5	11
	P (mm)	55	73	84	58	33	23	2	2	28	66	94	71
Georgetown	**T (°C)**	26	26	26,5	27	26,5	26,5	26,5	27	27,5	27,5	27	26,5
	P (mm)	203	114	175	139	289	302	254	175	81	78	155	287
Upernivik	**T (°C)**	–21,5	–23	–21	–14	–3,5	2	5	5	1	–4	–9,5	–16,5
	P (mm)	10	12	17	15	15	15	23	28	28	28	28	15

	SEVILLA			GEORGETOWN			UPERNIVIK		
Importancia:	**Alta**	**Media**	**Baja**	**Alta**	**Media**	**Baja**	**Alta**	**Media**	**Baja**
Crioclastismo									
Haloclastismo									
Meteorización química									
Acción de los seres vivos									
Viento									
Ríos y torrentes									

27 A partir de los datos siguientes, dibuja el perfil longitudinal del río Duero. Señala en el perfil el curso alto, medio y bajo del río, e indica las características de cada uno de ellos.

Longitud desde el nacimiento (km)	0	10	30	50	100	150	200	300	400	500	600	650	700	897
Altura sobre el nivel del mar (m)	2150	1250	1100	1050	950	850	800	700	650	600	250	100	50	0

28 Identifica las siguientes formaciones del relieve, y el agente geológico responsable de su formación.

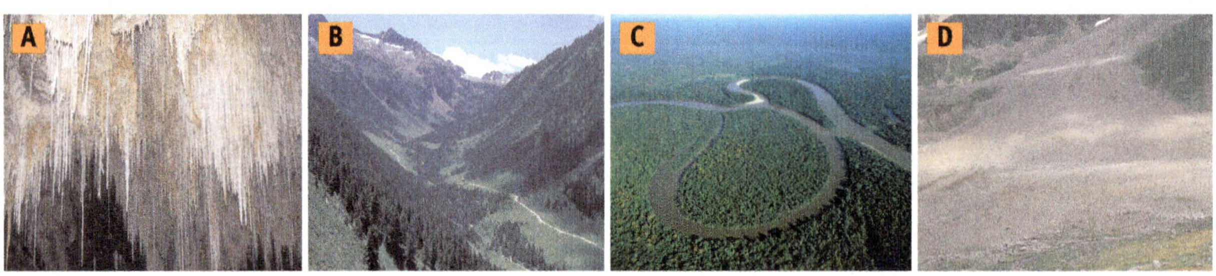

29 Un delta se forma por el depósito de los sedimentos transportados por un río al entrar en contacto con un lago, el mar o el océano. En los dibujos siguientes se muestra la evolución de un delta a lo largo del tiempo.

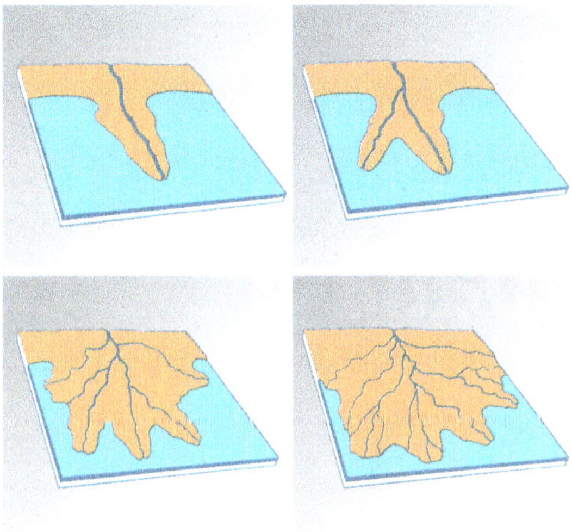

— ¿Qué factores crees que determinarán el ritmo de crecimiento del delta? ¿Cómo crees que afectará a este proceso la construcción de una presa en el curso medio del río? Justifica tu respuesta.

30 Dos granjas vecinas excavan un pozo cada una de ellas, ambos de la misma profundidad. Mientras que el pozo de una de ellas produce agua en abundancia, el otro no. Describe una situación en la cual pudiera darse este caso, razonando los motivos.

31 Identifica los modelos de costa representados por las fotografías y comenta sus características.

32 ¿Cómo se puede interpretar la presencia de valles en U en Europa en zonas de latitudes medias, alejadas de las zonas con glaciares?

33 ¿Podemos considerar que el suelo es un ente dinámico? ¿A qué influencias responde?

— Explica a grandes trazos el proceso de evolución de un suelo a partir de una roca madre.

34 Identifica en la siguiente foto los horizontes del suelo, y haz un esquema en tu cuaderno indicando cada uno de ellos.

— ¿A qué es debida la diferencia de color entre el horizonte A y el B?

— Compara la medida de las partículas que conforman los horizontes A y B. ¿A qué es debida esta diferencia? Comenta el origen de cada uno de estos horizontes.

— ¿Dirías que éste es un suelo maduro o inmaduro? ¿Por qué? ¿Qué caracteriza a los suelos maduros?

35 Relaciona los siguientes procesos de degradación del suelo con sus consecuencias

• Salinización	• Pérdida de profundidad del suelo
• Erosión	• Disminución de la infiltración del agua de la lluvia
• Compactación	• Dificultad de la absorción de agua por las plantas a causa de la elevada concentración iónica
	• Problemas para la penetración de las raíces
	• Incremento de la cantidad de sedimentos arrastrados por las aguas de escorrentía

TIC

36 En *https://www.mapya.es/app/Condicional/Documentos/Ficha%20suelos.pdf* hallarás la Ficha de conservación y protección del suelo publicada por el Ministerio de Agricultura, Pesca y Alimentación, en la cual se presentan una serie de medidas para reducir el impacto de la actividad agrícola y ganadera sobre el mismo.

Una vez la hayas leído, enuncia tres acciones que ayuden a conservar el suelo agrícola, y justifica su efectividad en función de lo que has aprendido al respecto de la erosión. ¿Qué otros beneficios se obtienen de la realización de éstas prácticas? ¿Por qué?

5

La geología de España

Nuestro territorio presenta una gran variedad de formaciones geológicas diferentes fruto de la acción de los agentes geológicos tanto internos como externos.

La tectónica de placas ha originado grandes accidentes geográficos como los Pirineos o las islas Canarias, mientras que otros agentes, como el agua de los ríos o el hielo, han realizado transformaciones de carácter regional o local sobre el relieve preexistente.

El análisis de las características geológicas de un territorio mediante la observación de imágenes de satélite, los sistemas de información geográfica (SIG) o bien las observaciones directas sobre el terreno, permiten la elaboración de mapas geológicos que nos muestran toda esta información.

Toda este trabajo resulta de gran importancia para temas tan diversos como la planificación de los usos del territorio, la prevención de catástrofes naturales o el conocimento de los recursos minerales.

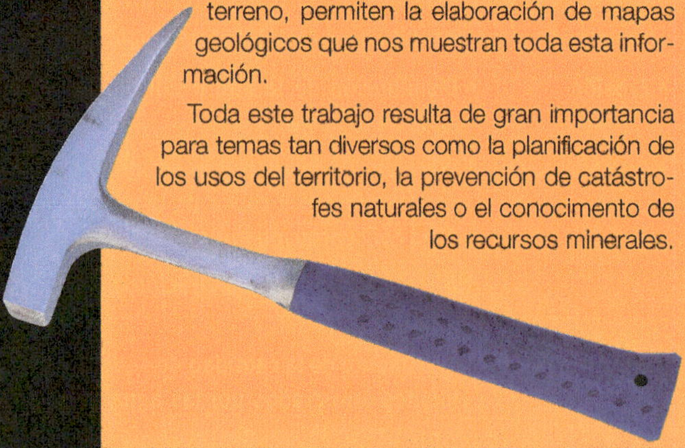

Las Médulas (León)

ORGANIZACIÓN DE LOS CONTENIDOS

1. La situación de la Península Ibérica y de las islas Canarias

2. Origen y evolución de nuestro territorio

2.1. La placa ibérica

2.2. Las islas Canarias

3. Unidades geológicas del territorio español

3.1. El basamento hercínico

3.2. Las cordilleras alpinas

3.3. Las cuencas alpinas

3.4. Zonas de actividad volcánica reciente

4. Métodos de estudio en geología

4.1. Teledetección

4.2. Sistemas de información geográfica

4.3. Trabajo de campo

4.4. Mapa geológico

4.5. Corte geológico

5. Aplicaciones de la geología

5.1. Investigación y prospección de recursos naturales

5.2. Evaluación de riesgos geológicos e impactos medioambientales.

OBJETIVOS

- Comprender, según la teoría de la tectónica de placas, el origen de la Península Ibérica y los fenómenos volcánicos que dieron lugar a las islas Canarias.
- Conocer la constitución litológica de los terrenos que forman la Península y los archipiélagos balear y canario.
- Valorar la importancia de los recursos geológicos de España y la necesidad de explotarlos racionalmente.
- Conocer y practicar los principales métodos de estudio de la geología y sus aplicaciones.

1. La situación de la Península Ibérica y de las islas Canarias

En las unidades anteriores, hemos descrito la dinámica interna y externa de la Tierra. Ahora, vamos a centrarnos en el estudio de nuestro país: veremos cuál es su origen, cómo se ha formado el relieve y qué tipo de rocas lo constituyen.

Nuestro relieve se encuentra repartido entre dos placas tectónicas, ambas situadas en el hemisferio norte:

- La Península Ibérica y las islas Baleares son la parte emergente de la **placa ibérica,** la cual se separó de otras placas mayores, la placa eurasiática y la africana, hace 200 millones de años, aproximadamente.
- El archipiélago canario está situado en la **placa africana.**

En la siguiente imagen, puedes apreciar la situación y los límites de ambas placas.

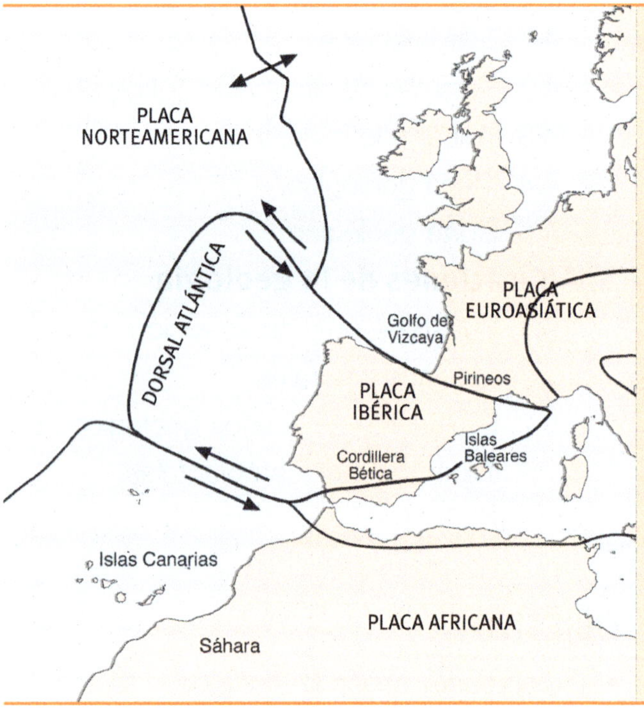

Placa ibérica

- Al Norte, limita con la placa eurasiática por los Pirineos y el golfo de Vizcaya.
- Al Este, limita con la placa de Alborán. En su extremo oriental se encuentra una llanura abisal, perteneciente a la corteza oceánica del fondo del mar Mediterráneo. Entre la Península y esta zona, emerge el archipiélago balear.
- Al Sur, limita con la placa de Alborán mediante la cordillera Bética, y con la placa africana por medio de una falla lateral.
- La dorsal atlántica, que recorre el fondo de este océano, constituye el límite occidental con la placa norteamericana.

Islas Canarias

Están situadas al noroeste de la placa africana. La actividad de esta placa ha dado lugar a diversos fenómenos de vulcanismo, que son el origen de las islas. El archipiélago está situado a 100 km de la costa africana, frente al Sáhara, y está formado por siete islas principales y seis islotes.

El estado actual de nuestro territorio es el resultado de una constante transformación de la corteza terrestre que ha durado millones de años.

A continuación, vamos a ver cuáles han sido los procesos que se han ido sucediendo desde la formación de la placa ibérica y las Canarias hasta llegar a nuestros días.

ACTIVIDADES

■ **1.** Identifica los bordes de placa que limitan o afectan al bloque ibérico en el mapa que acompaña la tabla anterior y explica los procesos que se asocian a cada uno de ellos.

2. Origen y evolución de nuestro territorio

La deriva de continentes es la causa de la distribución de las tierras que actualmente forman la superficie de nuestro planeta, y de los océanos y mares que las rodean. Vamos a analizar cómo ha afectado este fenómeno al territorio español.

2.1. *La placa ibérica*

Las etapas de acercamiento y separación de la placa ibérica respecto a las que la rodean se han sucedido a lo largo de las distintas eras geológicas, y con ello han originado la formación de la Península, con sus unidades de relieve, y de las aguas que la circundan.

Del mismo modo que los movimientos de distensión producen una separación entre las placas y la abertura de los mares, los movimientos de compresión dan lugar a diferentes relieves, como las cordilleras. Este proceso de creación de relieve recibe el nombre de **orogenia.**

A lo largo de la historia geológica se han producido tres orogenias, que han afectado a la placa ibérica:

— **Orogenia caledoniana,** que tuvo lugar durante el Paleozoico, en los períodos Ordovícico y Silúrico. Apenas quedan restos de esta orogenia en el planeta y no hay rastros en la Península Ibérica, debido a la acción de la erosión.

— **Orogenia herciniana,** que se produjo a finales del Paleozoico, en el período Carbonífero. Se manifiesta en las montañas más antiguas y, por tanto, más erosionadas.

— **Orogenia alpina,** que actuó durante el Cenozoico, en los períodos Terciario y Cuaternario. Corresponden a este proceso las montañas más modernas.

La reconstrucción histórica de los procesos que han originado nuestro territorio puede resumirse en varias etapas de compresión y de distensión.

Como ya hemos visto, desde su formación se han sucedido episodios de convergencia de placas continentales que por colisión han originado supercontinentes. Estas grandes masas emergidas padecen posteriormente procesos de rotura y disgregación, que acaban culminando en un nuevo proceso de convergencia: es el **ciclo de Wilson**.

La placa ibérica también se ha visto implicada en estos desplazamientos, lo que ha conllevado no sólo su cambio de posición, sino elevaciones, deformaciones y transgresiones marinas, así como un cambio de latitud, desde las posiciones cercanas al polo Sur que ocupaba a finales de la era Paleozoica hasta su situación actual en las latitudes medias. De la misma manera, el desplazamiento de los grandes continentes y su interacción con las corrientes marinas ha provocado numerosos cambios climáticos, entre los que se incluyen glaciaciones y períodos áridos, que han hecho que el territorio que hoy en día habitamos haya tenido bien poco en común en el pasado con aquello que actualmente nos rodea.

TIEMPO GEOLÓGICO

Hace... millones de años	Era	Período
	Cenozoico	Cuaternario*
65		Terciario*
	Mesozoico	Cretácico
		Jurásico
250		Triásico
	Paleozoico	Pérmico
		Carbonífero**
		Devónico
		Silúrico***
		Ordovícico***
550		Cámbrico
	Precámbrico	

* Orogenia alpina
** Orogenia herciniana
*** Orogenia caledoniana

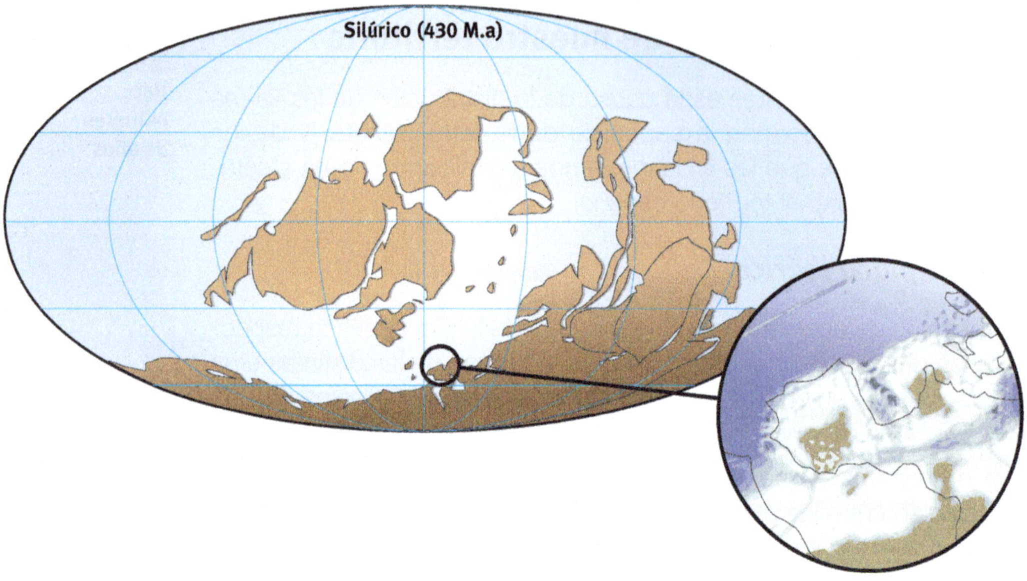

Posición de la placa ibérica en el Silúrico. Una gran parte de ella forma parte del fondo oceánico.

Durante el Silúrico lo que en el futuro sería la Península Ibérica constituía una parte del fondo del océano. A finales del período Carbonífero, hace 280 millones de años, se inició un nuevo proceso de convergencia que llevó a todas las tierras emergidas a colisionar dando lugar a la Pangea. Ligado a estos acontecimientos tuvo lugar la orogenia herciniana, que elevó una gran cadena montañosa entre las actuales Norteamérica y Europa, el macizo hercínico o Varisca. El hecho de tratarse de una cadena de gran antigüedad y, por tanto, haber estado sometida a un largo proceso erosivo, ha hecho que hoy en día únicamente se puedan apreciar sus basamentos más profundos. Éstos afloran no únicamente en la Península Ibérica, sino también en Inglaterra, Francia, centro de Europa, norte de África y Norteamérica.

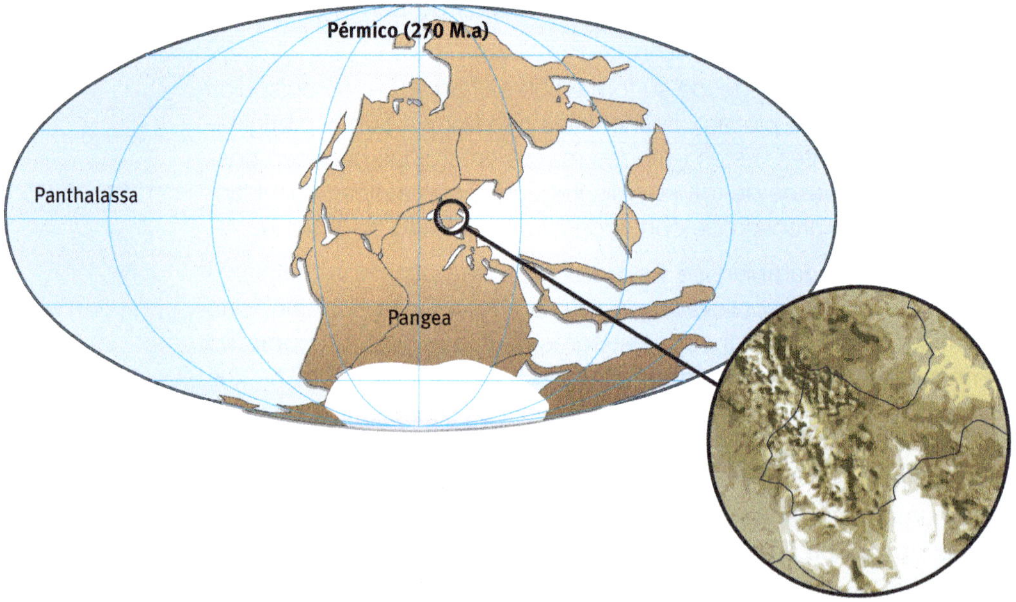

Posición de la placa ibérica en el Pérmico formando parte del macizo hercínico

En los inicios del Pérmico los procesos erosivos que actuaron sobre el macizo hercínico provocaron la acumulación de sedimentos en cuencas de tipo continental. Ya a mediados del Mesozoico, el bloque ibérico se vio afectado por episodios de fractura, provocados por el desplazamiento de las placas al iniciarse la disgregación de la Pangea. Esto provocó el hundimiento de grandes bloques de la litosfera, lo cual permitió la entrada del agua oceánica desde el Este, inundándose gran parte de Europa y dando origen al mar de Tetis.

También en el Mesozoico, a finales del Triásico, se inició la separación de América y Eurasia con la apertura del Atlántico, lo cual generó la aparición de una nueva cuenca oceánica situada al Oeste del bloque ibérico. Durante este proceso la placa ibérica se vio sometida a un giro contrario a las agujas del reloj que la desgajó de la placa eurasiática, originando el golfo de Vizcaya.

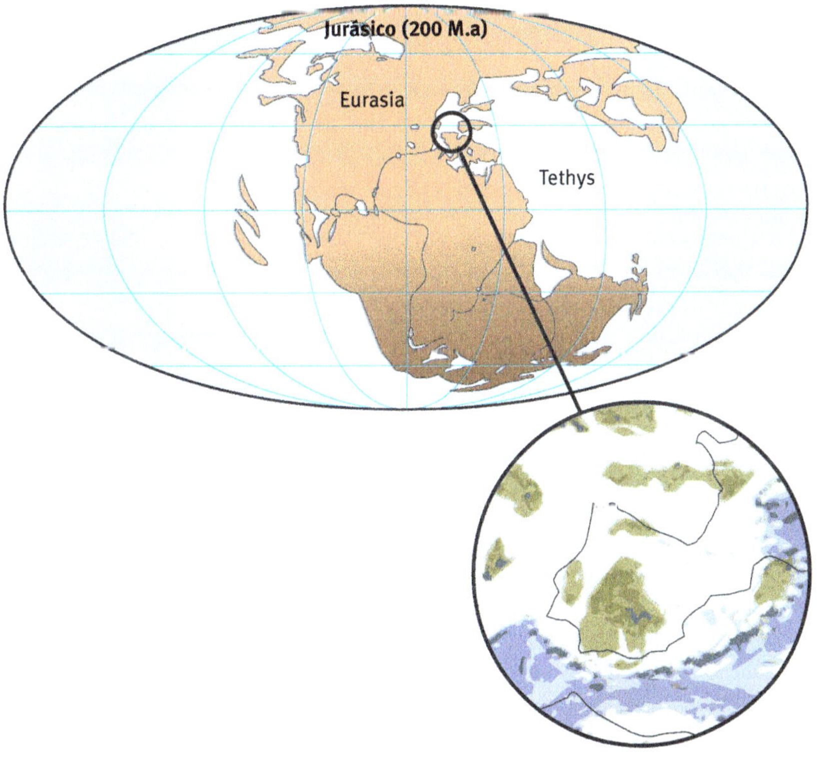

Posición de la placa ibérica en el Jurásico

Debido a estos procesos, en períodos como el Jurásico el bloque quedó convertido en una isla, emergiendo únicamente algunas zonas del oeste peninsular y quedando en el noreste y sudoeste grandes cuencas sedimentarias marinas.

Hace unos 65 millones de años se inició una nueva etapa compresiva en la cual la placa africana se desplazó hacia el Norte, colisionando con la placa eurasiática. Esto dio lugar al levantamiento de las cuencas sedimentarias de las zonas noreste y sudoeste, y su plegamiento originó nuevas cadenas montañosas: los Pirineos, la cordillera Ibérica y la cordillera Bética. Este proceso se incluye en el fenómeno global de la orogenia alpina, que originó también los Alpes, el Rift o el Atlas.

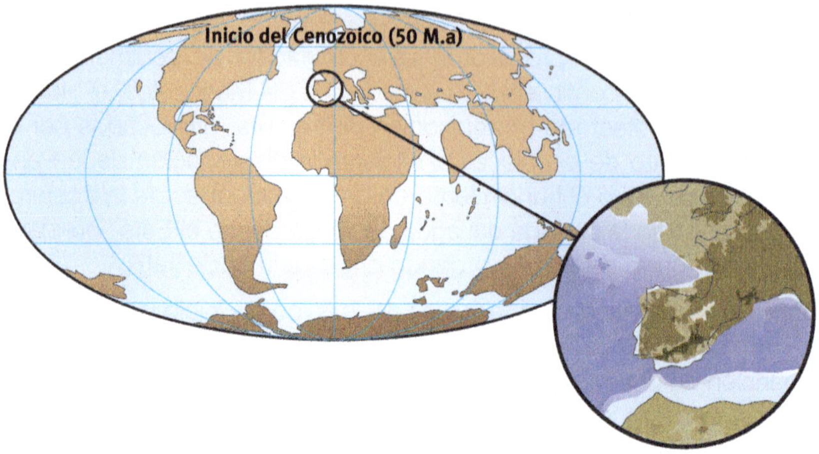

Posición de la placa ibérica al inicio del Cenozoico

Como consecuencia del empuje de África sobre la placa eurasiática, el estrecho de Gibraltar se cerró hace 6,5 millones de años. El mar Mediterráneo llegó a secarse al quedar aislado del océano Atlántico dando lugar a grandes extensiones de depósitos salinos. Esta situación perduró hasta la reapertura del estrecho hace 5 millones de años, con la consecuente entrada de agua oceánica.

El proceso de compresión de la placa africana sobre la euroasiática aún continúa actualmente, lo cual se manifiesta en una cierta actividad sísmica a nivel de los Pirineos y la cordillera Bética.

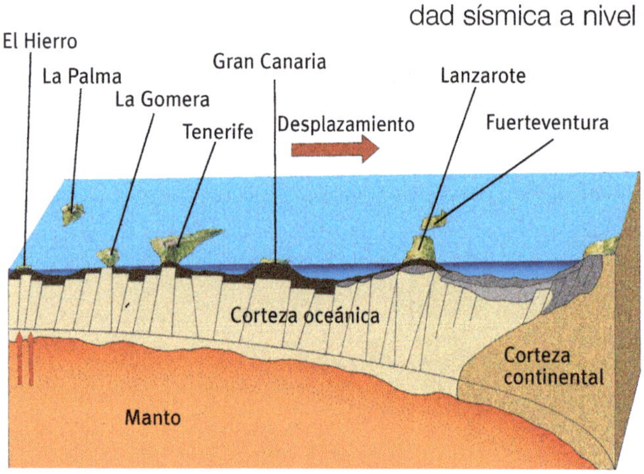

2.2. *Las islas Canarias*

En el Oligoceno, hace unos 30 millones de años, se dieron los primeros episodios volcánicos en la corteza oceánica frente a las costas africanas que dieron origen a las islas Canarias. Estos episodios, inicialmente submarinos, emergieron finalmente manteniendo su actividad hasta la actualidad. El origen más aceptado en la actualidad para este vulcanismo es el levantamiento de bloques del fondo oceánico como respuesta a la compresión resultante de la expansión del fondo oceánico atlántico sobre el continente africano.

ACTIVIDADES

- **2. ¿Qué se entiende por orogenia? Nombra y sitúa temporalmente las orogenias que han afectado a la Península Ibérica.**
- **3. ¿Qué posición ocupaba la placa ibérica en la época de la Pangea? ¿Qué es el mar de Tetis? ¿Con qué mar actual se puede relacionar?**
- **4. ¿Qué cordilleras origina en la Península la orogenia alpina? ¿Qué otras cordilleras aparecen en la misma época?**
- **5. Explica las diferencias que existen entre el origen de las islas Canarias y el de las islas del Japón.**

3. Relieve del territorio español

Como hemos visto, el relieve actual de la Península Ibérica es el fruto de su historia geológica, la cual se refleja tanto en las estructuras como en la composición de las rocas. Para su estudio dividiremos la Península en cuatro grandes unidades geológicas:

- El **basamento hercínico,** originado en la era Paleozoica como consecuencia de la orogenia hercínica, que constituirá los terrenos más antiguos.
- Las **cordilleras alpinas**, generadas a inicios del Terciario por la orogenia alpina.
- Las grandes **cuencas alpinas**, asociadas a la formación de las cordilleras anteriores.
- Las **zonas de actividad volcánica** reciente, que incluye las islas Canarias y las regiones volcánicas peninsulares.

■ Relieve de la Península Ibérica

■ Unidades geológicas de la Península Ibérica

3.1. *El basamento hercínico*

Se trata del afloramiento de lo que había sido la cadena hercínica originada a consecuencia de la colisión entre las placas de Norteamérica y Eurasia a finales del Carbonífero. En la Península Ibérica a los restos de esta cadena hercínica se les da el nombre de macizo hespérico y están representados por elevaciones poco accidentadas y montañas de altitud media, como consecuencia de la prolongada exposición a los procesos erosivos. La Meseta central y el macizo Gallego son los ejemplos más representativos.

Está formado fundamentalmente por rocas sedimentarias paleozoicas o precámbricas, metamorfizadas en mayor o menor grado, junto con rocas plutónicas. Las características de estas rocas hacen que la zona de la Península en la que aparecen, fundamentalmente la zona Oeste, sea denominada la España silícea.

■ El macizo Gallego forma parte del basamento hercínico peninsular, originado en el Paleozoico.

3.2. *Las cordilleras alpinas*

El Aneto, situado en la zona axial, es el pico más alto de los Pirineos, con 3404 m.

La orogenia alpina dio lugar fundamentalmente a la formación de tres grandes cordilleras en la Península:

Los Pirineos

Se originan por el choque entre la placa ibérica y la europea, dándose un ligero proceso de subducción de la primera. Se extienden, además de en la zona del istmo que une la Península a Europa, hacia el Oeste hasta la plataforma cantábrica, y hasta el Este hacia los Alpes. Podemos diferenciar varias unidades estructurales, especialmente en la zona del Pirineo central y oriental, como son el Pirineo axial, la zona surpirenaica, la zona norpirenaica y las zonas del antepaís plegadas.

La zona axial constituye el núcleo de la cordillera, y es de origen hercínico, formado por rocas metamórficas y granito. Las zonas norpirenaica y surpirenaica, también conocidas como Prepirineo, están constituidas principalmente por rocas sedimentarias de origen marino, muy deformadas, entre las que abundan rocas carbonatadas y de tipo detrítico. El antepaís se compone de materiales postorogénicos, deformados débilmente, que descansan sobre el zócalo hercínico.

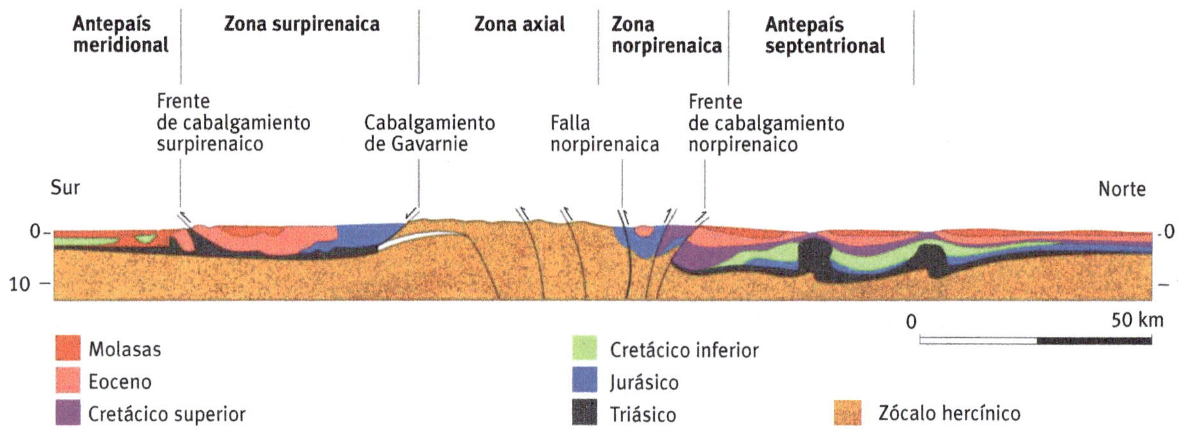

Corte geológico de los Pirineos en su sector central. Principales unidades estructurales.

La cordillera Bética

El Mulhacén (3478 m), en las Cordilleras Béticas, se originó por el choque entre la placa Ibérica y la de Alborán.

Cuando la zona Pirenaica no pudo absorber una mayor deformación, los plegamientos se trasladaron hacia el Sur, hacia zonas más débiles, provocando la aparición de una nueva cordillera que actuará como frontera entre las placas eurasiática y africana. La cordillera Bética se extienden desde el estrecho de Gibraltar hasta las islas Baleares, las cuales representan sus estribaciones más orientales, y al otro lado del Mediterráneo, por el Norte de Marruecos, Túnez y Argelia. Son el resultado de la interacción entre la placa ibérica y la africana, que aprisionan entre ambas a la microplaca de Alborán. En esta cordillera se halla el pico más alto de la Península, el Mulhacén, con 3478 m, y aún presenta una cierta actividad sísmica. La litología es fundamentalmente de roca caliza y detrítica, con afloramientos silíceos en algunas zonas centrales.

La cordillera Ibérica

A diferencia de las anteriores, la cordillera Ibérica no se forma en el límite de una placa, sino como consecuencia de los esfuerzos transmitidos hacia su interior. Esto provoca la fractura del basamento, dando fallas que al desplazarse deforman los sedimentos superiores. Los restos de la base hercínica afloran en determinadas zonas, como el Moncayo o el sistema Central, mientras que el resto está constituido por rocas sedimentarias del Mesozoico.

La transmisión de fuerzas desde los márgenes de placa en la época de la orogenia alpina también reactivará otras fallas, originando relieves sobre el basamento hercínico como Sierra Morena, los montes de Toledo, el sistema Central o la cordillera Cantábrica.

Estas zonas, en las cuales las deformaciones hacen aflorar las rocas sedimentarias del Mesozoico, fundamentalmente calcáreas, constituyen la denominada España caliza.

El Moncayo es la cumbre más alta del sistema Ibérico (2314 m).

3.3. *Las cuencas alpinas*

La gran deformación sufrida durante la era Terciaria por la placa ibérica llevó a la constitución no únicamente de cordilleras montañosas, sino también de cuencas sedimentarias.

En el interior de la Península la reactivación provocada por las tensiones transmitidas desde el borde de las placas dio como resultado el hundimiento de amplios bloques de falla. La consecuencia fue la aparición de las dos grandes cuencas del Duero y el Tajo. Estas cuencas, originariamente unidas, se independizaron la una de la otra al producirse la elevación del sistema Central en el Eoceno.

Las cuencas del Ebro y del Guadalquivir se asocian a la formación de las cadenas montañosas que las delimitan. La cuenca del Ebro se halla ligada a la formación de los Pirineos y el sistema Ibérico, mientras que la del Guadalquivir lo está con la cordillera Bética.

El relleno de estas cuencas se dio a lo largo de la era Terciaria con sedimentos de ambientes continentales provenientes de los ríos. Son los terrenos de formación más reciente y constituyen la España arcillosa.

La cuenca del Guadalquivir se originó a partir de la formación de la cordillera Bética durante la orogenia alpina.

C

En la Península existen restos de vulcanismo de intraplaca en las comarcas de la Garrotxa (Girona) y el Campo de Calatrava (Ciudad Real), y de margen de placa en Cabo de Gata (Almería).

El cono volcánico del Teide, en Tenerife, es el pico más alto del Estado español, con 3718 m.

3.4. *Zonas de actividad volcánica reciente*

La mayor zona volcánica del estado español la constituyen las islas Canarias, que cuentan con un registro de actividad volcánica durante los últimos 30 millones de años, y que se mantiene activo en la actualidad.

Las primeras etapas de formación de las islas Canarias datan del Cretácico, cuando tras la separación de las placas africana y americana se inició un proceso de sedimentación en la nueva cuenca resultante. Posteriormente una serie de episodios de vulcanismo submarino provocaron el apilamiento de lavas que generó el complejo basal de las islas. A finales del Mioceno la elevación de bloques a causa de la presión de la litosfera oceánica sobre el continente africano permitió la construcción de estructuras volcánicas sobre ellos, dando lugar a las series basálticas antiguas. Tras una pausa, se inició una nueva serie eruptiva que originó las series volcánicas modernas.

Los materiales emitidos por los volcanes modifican el paisaje en poco tiempo, como ocurrió con la formación del campo de volcanes de Timanfaya, en Lanzarote. A lo largo de una etapa que duró sólo seis años (1730-1736), se originó un conjunto de formaciones volcánicas que se extiende por una superficie de 5200 hectáreas.

La última erupción que ha tenido lugar en las islas fue la del Teneguía, en La Palma, entre octubre y noviembre de 1971, en la cual el volcán lanzó 40 millones de metros cúbicos de materiales, que fluyeron hacia el mar y formaron una plataforma costera que incrementó la superficie de la isla.

Los procesos erosivos que tienen lugar sobre estas rocas en los períodos de ausencia de erupciones dan como resultado la acumulación de materiales, principalmente en las playas, que generan las turbiditas.

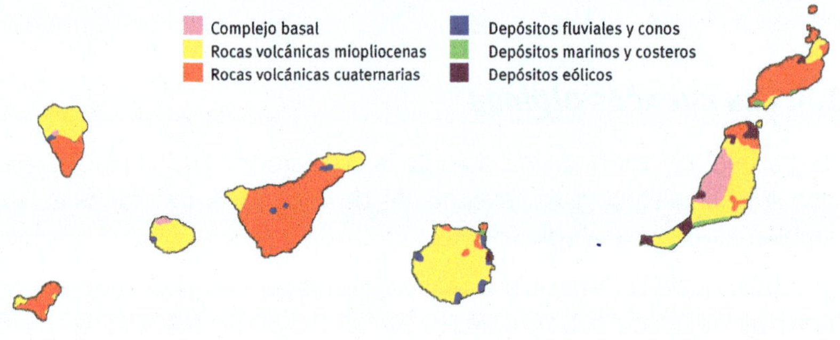

Mapa geológico de Canarias. Unidades geológicas

ACTIVIDADES

6. ¿A qué es debido que las estribaciones de las cordilleras del noroeste peninsular sean menos elevadas que las del sudeste?
7. ¿Qué tipo de rocas constituyen mayoritariamente el macizo hespérico? ¿Se pueden localizar en otras partes de la Península? ¿En cuáles?
8. ¿A qué es debido que una gran parte del sistema Ibérico esté formado por calizas de origen marino?
9. ¿Cuáles son las unidades estructurales que configurán los Pirineos? ¿Cuál es la constitución litológica de cada una de ellas?
10. ¿Qué territorios constituyen la llamada España arcillosa? ¿Cuál es su historia geológica?
11. ¿En qué época se inició el vulcanismo que originó las islas Canarias? ¿De qué etapas consta? ¿Cuándo fue la última erupción?

4. Métodos de estudio en geología

A lo largo de la unidad, hemos descrito los distintos aspectos geológicos que caracterizan nuestro territorio. Para poder describir con detalle las características geológicas de una zona se sigue una serie de procedimientos que implican el uso de un conjunto de herramientas, de las cuales trataremos a continuación.

4.1. *Teledetección*

Entendemos por teledetección, o percepción remota, la medida de la magnitud de una propiedad o característica de un objeto que no se efectúa mediante un contacto directo con éste.

- La **fotografía aérea** se considera el primer método de teledetección, aunque anteriormente ya se había iniciado el desarrollo de este campo a partir de observaciones efectuadas desde globos aerostáticos.

  Este método consiste, básicamente, en conseguir una serie de fotografías según la dirección marcada por el vuelo del avión, de manera que se solapen ligeramente. Este solapamiento hará posible su interpretación en tres dimensiones mediante el uso del estereoscopio.

- En la década de los años sesenta, el uso de satélites que contaban con sensores de múltiples espectros acoplados a computadoras reemplazó a los métodos anteriores. Son los llamados **satélites de observación.**

  La radiación solar que llega a la esfera terrestre es reflejada con distintas modificaciones según la estructura y composición de los diversos elementos de la superficie terrestre. Esta radiación es capturada por los sensores de los satélites, que retransmiten esta información a estaciones terrestres para su posterior procesamiento y análisis.

Como consecuencia del uso de ordenadores, la ciencia de la teledetección hizo grandes avances al reproducir las imágenes obtenidas en formato digital, es decir, como series de números con las que un ordenador puede operar.

La **imagen digital** debe imaginarse como una red de múltiples celdas que contienen un número que indica el valor de su brillo.

Estas celdas se denominan **píxeles.** Generalmente, las celdas con números bajos indican áreas más oscuras, escasamente reflectantes, y, por el contrario, las celdas con números altos indican áreas más claras, altamente reflectantes.

La importancia de las imágenes digitales radica en el hecho de que, al estar constituidas por números, son susceptibles de ser modificadas mediante un ordenador para destacar o resaltar aquello que más interese, como, por ejemplo, los límites de un territorio, las llanuras de inundación de un río...

Además del análisis e interpretación de estructuras y formaciones geológicas, la teledetección tiene múltiples aplicaciones en el estudio y gestión del medio. Algunos ejemplos de estas aplicaciones son los siguientes:

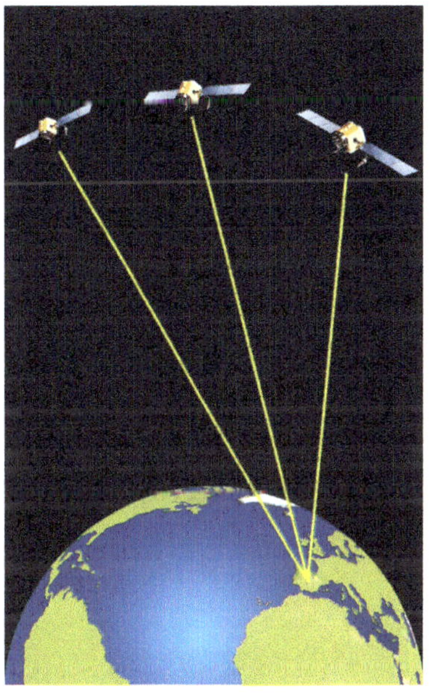

Para calcular la posición, el GPS mide el tiempo que tarda en llegar la señal emitida por tres satélites al punto de la Tierra en el que se halla.

C

El **GPS**, o sistema de posicionamiento global, es uno de los denominados sistemas de navegación por satélite.

La red de satélites de que consta el sistema GPS está formada por 24 satélites que giran alrededor de la Tierra en seis órbitas distintas. Este complejo recibe el nombre de Constelación Operativa del GPS.

La señal del sistema de posicionamiento global procedente de estos satélites proporciona información sobre el tiempo, la posición y la velocidad del receptor.

— Cálculo de la temperatura del agua del mar y caracterización de las grandes corrientes marinas.

— Estudio y seguimiento de grandes masas de fitoplancton marino, así como de los bancos de peces que las depredan.

— Análisis del flujo de sedimentos arrastrados por los ríos hacia el mar, que puede ayudar a prever y evaluar el impacto ambiental de fenómenos como, por ejemplo, las riadas, o actuaciones como la construcción de un pantano o el desvío de un río.

— Detección y localización de los focos de los incendios, que pueden ayudar a deducir el origen intencionado o no del fuego según la época y el lugar en que éste se produzca.

— Análisis de la producción y evolución de las superficies forestales.

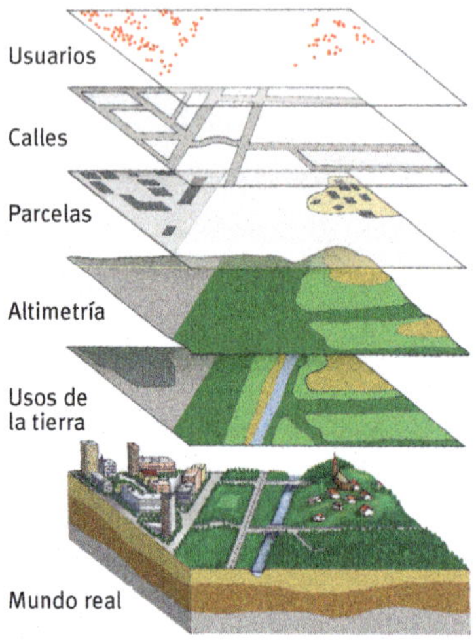

■ Un sistema SIG utiliza toda una serie de informaciones gráficas y no gráficas distribuidas en capas relacionadas con un mapa digitalizado. Esto permite establecer relaciones, generar nueva información y construir modelos y simulaciones. Ejemplo de capas de información en un mapa digital.

4.2. *Sistemas de información geográfica*

Un sistema de información geográfica (SIG) es un programa informático que consiste en una base de datos que almacena información que puede relacionarse sobre un mapa digital con localizaciones concretas. El acceso a esta información puede hacerse a partir de diversas capas temáticas, que, al superponerse, facilitan relaciones entre los datos que de otra forma serían difíciles de detectar.

Las funciones que puede desempeñar un SIG son:

— Obtención, almacenamiento y manipulación de la información

— Análisis y generación de modelos

— Presentación de la información

Un SIG puede ayudar a resolver problemas que van desde la localización de un determinado lugar, al análisis de pautas de evolución espacial o temporal de determinadas características, o al establecimiento de modelos para ciertos parámetros que permitan generar simulaciones o predicciones.

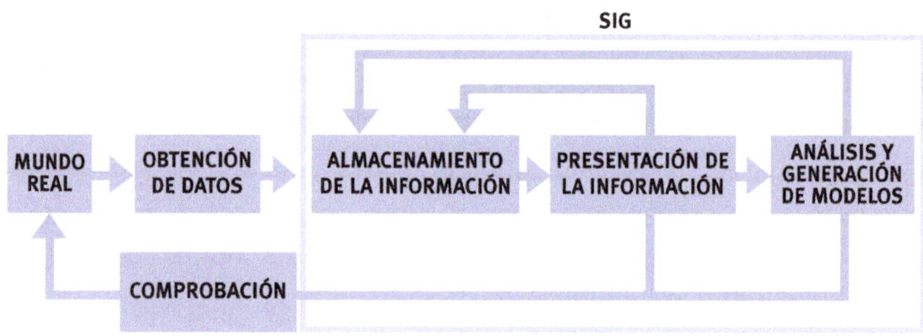

Otro sistema relacionado con las utilidades de los SIG es el GPS o Sistema de posicionamiento global, que permite la localización de cualquier objeto sobre la superficie del planeta con una precisión inferior al metro. El sistema GPS se halla bajo control del gobierno de los Estados Unidos; por este motivo, en la actualidad, la Unión Europea está desarrollando su propio sistema de posicionamiento global, llamado Galileo, que se prevé que esté operativo para usos comerciales en 2011.

4.3. *Trabajo de campo*

En las visitas al campo, el geólogo estudiar sobre el terreno la composición y la disposición de los diversos materiales del área analizada.

Esta fase del estudio es muy importante puesto que permite obtener una información de primera mano, es decir, sin apenas distorsiones, y que debe ser lo suficientemente exhaustiva y detallada como para que, una vez se haya finalizado el reconocimiento, no sea necesario volver a la zona.

A continuación veremos cómo debe prepararse una visita al campo para que sea lo más provechosa posible:

— Delimitar la zona que se va a estudiar y recabar el máximo de información posible a partir de la bibliografía ya existente.

— Preparar el material necesario:

- Mapa de la zona y brújula.
- Martillo de geólogo.
- Lápiz y cuaderno.
- Lupa.
- Bolsas de plástico.
- Rotuladores permanentes.
- Cámara fotográfica.
- Cinta métrica.

— Aplicar los conocimientos adquiridos para reconocer y distinguir las estructuras, a macroescala, mesoescala y microescala, así como los criterios para reconocer e identificar los diversos materiales.

— Recoger y seleccionar aquellas muestras que no lleguemos a identificar con los criterios de campo y necesiten un estudio posterior.

— Referenciar cualquier incidencia que suceda, tanto su localización en el mapa como su explicación en el cuaderno.

Una vez finalizado el trabajo de campo, el geólogo ha de ser capaz de deducir e interpretar las características estructurales y litológicas de la zona estudiada. A partir de este trabajo de campo se puede confeccionar una cartografía geológica que sintetice correctamente toda la información obtenida.

Posteriormente, los materiales que no hayan sido identificados en el campo se analizarán en el laboratorio mediante el microscopio petrográfico o aplicando otras técnicas, para determinar de qué clase de mineral se trata.

ACTIVIDADES

12. ¿En qué consiste la teledetección?

13. ¿De qué manera se lleva a cabo la localización por medio de un GPS? ¿Qué otros sistemas de localización por satélite conoces?

14. ¿Qué ventajas tiene la imagen digital sobre la analógica? ¿Qué es un píxel?

15. ¿Qué diferencia un sistema de información geográfica de un programa de realización de mapas?

16. ¿Qué elementos componen un SIG? ¿Qué procesos se pueden llevar a cabo con él?

17. Realiza un inventario del material necesario para realizar un trabajo de campo en geología.

La fotografía por satélite permite estudiar y controlar la evolución de la deforestación en amplias zonas de nuestro planeta.

+

Casiano del Prado (Santiago de Compostela, 1797 - Madrid, 1866). Geólogo e ingeniero español, fue director de las minas de Riotinto y de Almadén, las cuales contribuyó a mejorar mediante una serie de estudios geológicos. En 1849 entra a formar parte de la comisión encargada de la elaboración de la Carta Geológica de Madrid y la Carta General del Reino como responsable de la geología y paleontología. Destacó por sus estudios de campo, su participación en la elaboración de mapas geológicos y la publicación de obras y estudios geológicos. Su trabajo fue reconocido internacionalmente, entrando a formar parte de la Geological Society inglesa y la Societé Geologique de France.

4.4. *Mapa geológico*

Los mapas geológicos representan la distribución sobre la superficie del terreno de los distintos tipos de rocas que lo constituyen, así como los fenómenos geológicos que han afectado a dichos terrenos. Estos fenómenos se hacen más evidentes cuando afectan a rocas sedimentarias, dispuestas en estratos, que pueden deformarse o erosionarse. Los estratos inferiores son, normalmente, más antiguos que los superiores, tal como hemos visto al hablar del **principio de superposición** de Steno.

Los mapas van acompañados de una leyenda que explica el significado de los colores y los símbolos empleados, según un código que se utiliza internacionalmente. La información que contiene un mapa geológico es la siguiente:

— La **constitución litológica** del terreno. A cada tipo de terreno le corresponde un color. En lugar de colores, se puede utilizar una trama.

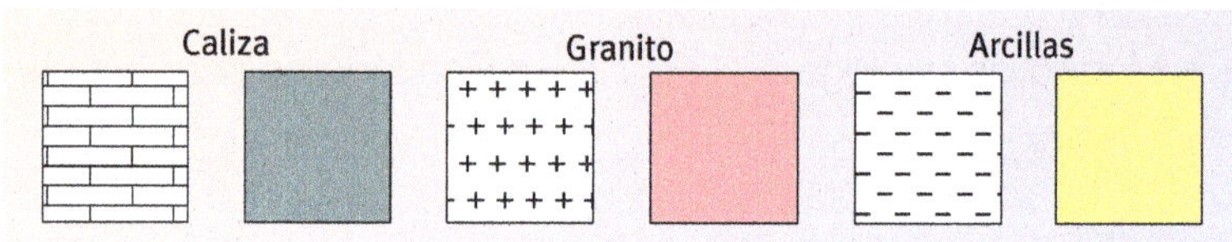

Además del color, a cada unidad geológica se le asigna un conjunto de letras. La primera representa la edad de la unidad geológica; así, K representa cretácico, T terciario, Q cuaternario... Las minúsculas representan el nombre de la unidad, si lo tiene, o su composición si no lo tiene.

— Los **fenómenos tectónicos** (fallas, pliegue) y los **fenómenos de erosión y sedimentación** (terrazas aluviales, conos de deyección).

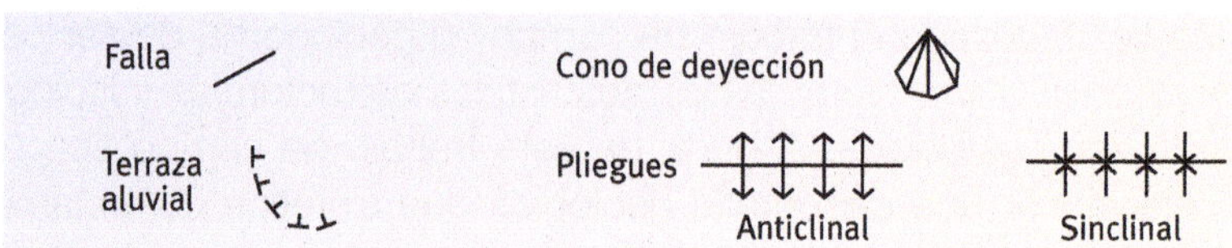

En el caso de las fallas, la línea en el mapa indica el contacto de falla, es decir, la zona en la cual se ponen en contacto materiales que de forma normal no lo estarían. En cuanto a los pliegues, la línea sigue la cresta del mismo, mientras que las flechas indican el sentido de la inclinación de los estratos que lo conforman.

— El **tipo de contacto** entre las distintas rocas, que puede ser:

Concordante

Discordante

- **Concordante.** Se da cuando los estratos tienen la misma inclinación. Es signo de que los estratos son paralelos y de que no se ha producido una erosión importante.
- **Discordante.** El ángulo de inclinación cambia de un estrato a otro; por tanto, los estratos no son paralelos. Se produce cuando hay cambios como una erosión del terreno y la posterior sedimentación de rocas, un deslizamiento de tierras o un afloramiento de rocas plutónicas.

— El **buzamiento.** Puede haber fuerzas que plieguen los estratos, y éstos se inclinan. Esta inclinación también se refleja en el mapa geológico.

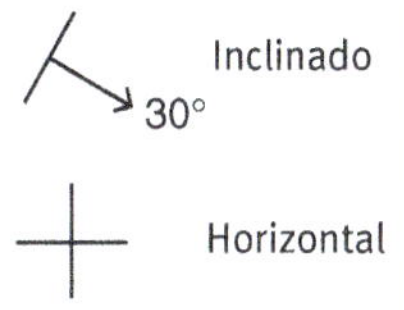

El símbolo indica el sentido de la inclinación; o buzamiento; y los grados, el ángulo respecto a la horizontal del terreno.

Podemos igualmente hallarnos ante rocas que, como el granito, no presenten organización en estratos. En tales casos no se indica nada. Si los estratos se hallan perfectamente horizontales, los representaremos mediante una cruz con ambos lados de la misma longitud, mientras que si se hallan en posición vertical, lo haremos con una cruz con uno de los brazos más corto, indicando el más largo la dirección de los estratos.

A continuación, vamos a interpretar un mapa geológico.

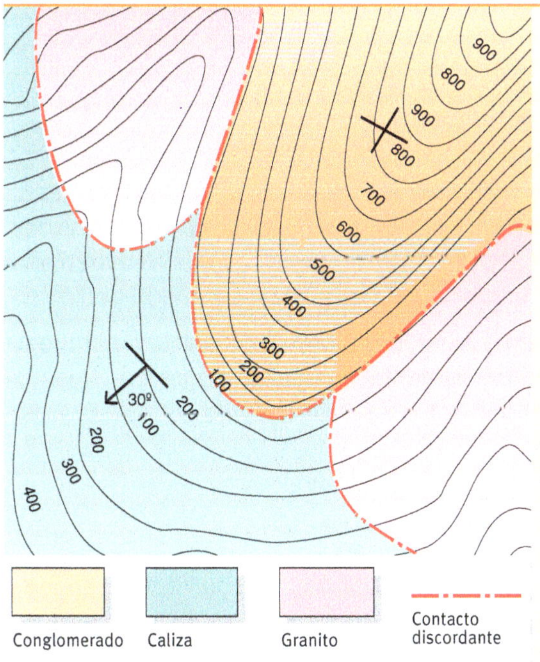

Distinguimos dos grupos de rocas:

- Magmática plutónica: granito.
- Sedimentarias: calizas y conglomerados.

Disposición de las rocas:

- El granito no forma estratos, ya que es una roca plutónica.
- Los estratos de caliza presentan un buzamiento de 30°.
- Los conglomerados forman estratos horizontales.

Contacto entre las rocas:

- Todos los contactos son discordantes.

La interpretación es la siguiente:

— En esta zona se depositaron sedimentos de roca caliza.
— Luego actuaron fuerzas orogénicas que los plegaron.
— Se produjo una intrusión de magma y se formó granito.
— Después, hubo una etapa de erosión.
— Luego, tuvo lugar una sedimentación de materiales y la formación de conglomerados.
— Se erosiona el terreno hasta el relieve actual.

ACTIVIDADES

18. Interpreta el siguiente mapa geológico. Indica los diferentes procesos que han tenido lugar y el orden en que se han dado.

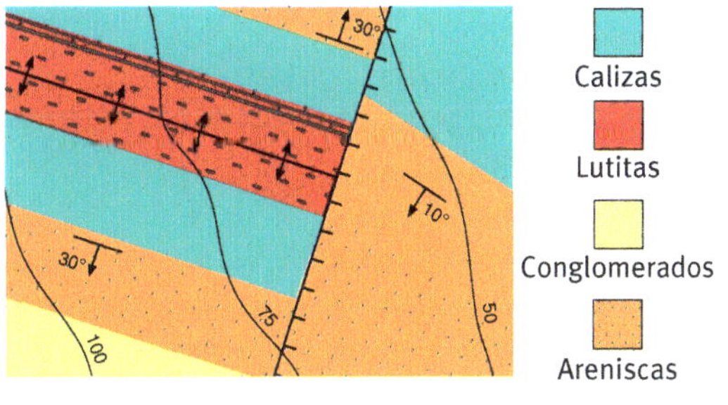

4.5. *Corte geológico*

Como ya sabes, a partir de los mapas topográficos se pueden realizar representaciones del relieve en sentido vertical, perpendicular a la superficie de la Tierra, que ponen de manifiesto las diferentes altitudes del relieve a lo largo de una línea de corte imaginaria. Recordemos cómo se dibuja un perfil topográfico.

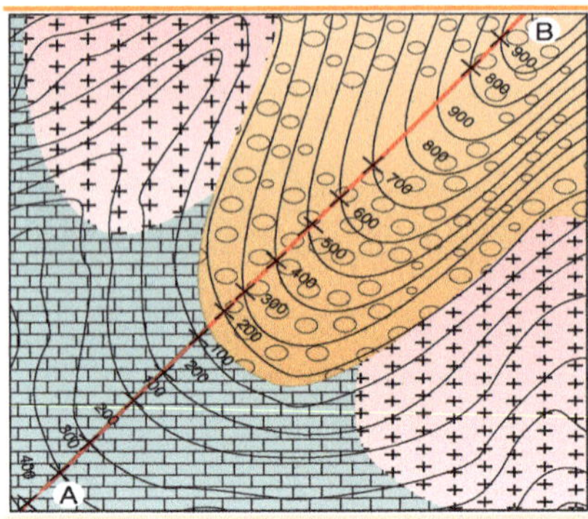

Trazamos sobre el mapa una línea recta que corte las curvas de nivel por el lugar que interese. Se marcan los puntos de intersección con las curvas de nivel.

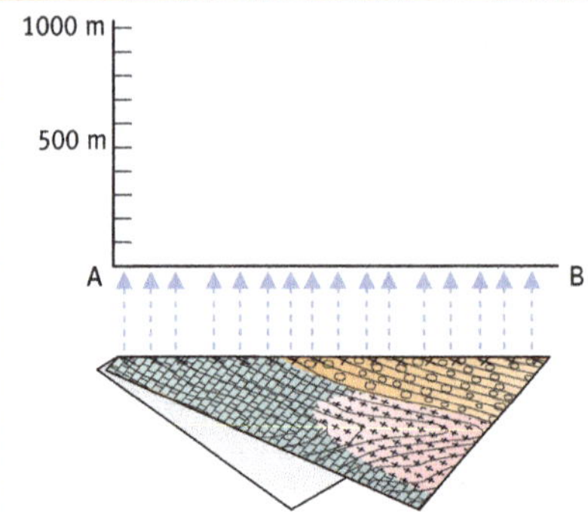

En un papel milimetrado se trazan dos ejes de coordenadas que representan la distancia horizontal del mapa (abscisas) y la altura del relieve (ordenadas). Los puntos se trasladan sobre el eje de abscisas.

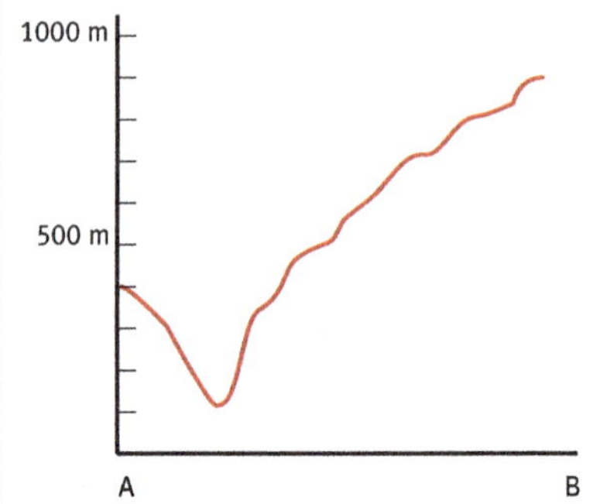

Se representan estos puntos en la gráfica, a la altura que les corresponda. Al unir los puntos, queda dibujado el perfil topográfico.

El **corte geológico** proporciona la visión vertical de los distintos accidentes geológicos, las unidades litológicas y, también, la disposición en profundidad de los materiales que las forman. Es una herramienta muy útil para completar la información que ofrece el mapa geológico.

El corte geológico se dibuja a partir del mapa geológico y, para realizarlo, es preciso trazar previamente el perfil topográfico. Veamos cómo se dibuja un corte geológico.

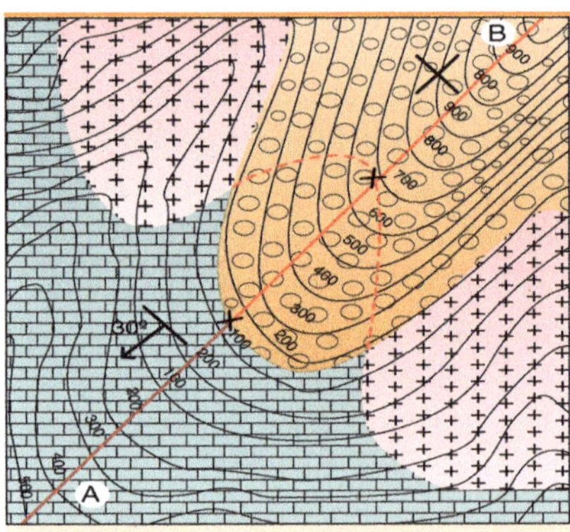

Trazamos sobre el mapa geológico una línea recta, por el lugar por el que se pretende hacer el corte. Se marcan los puntos de intersección de la línea con el lugar de contacto de las distintas rocas.

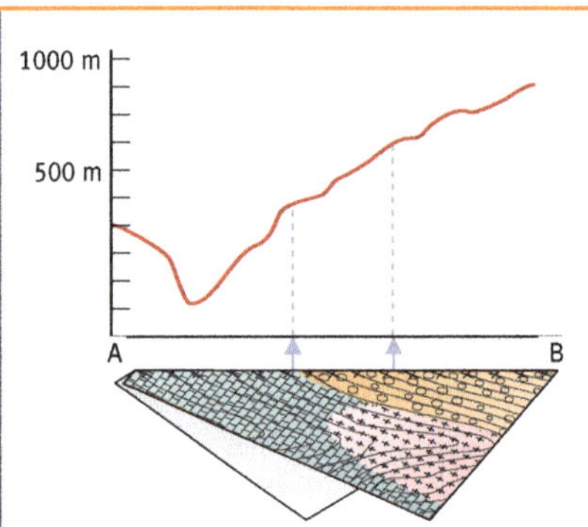

Los puntos de intersección se proyectan sobre el perfil topográfico, de forma similar a como se hace para proyectar las curvas de nivel.

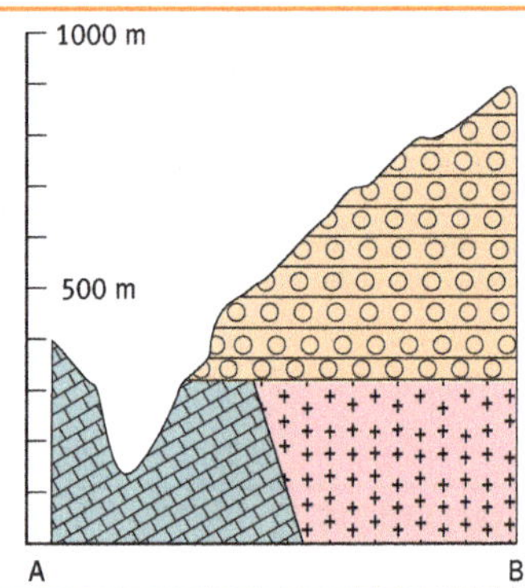

Con un transportador de ángulos, se dibujan los estratos con la inclinación indicada por el buzamiento. Se pintan con los colores o tramas correspondientes, que se indican en la leyenda del mapa. Se interpreta el corte, en el que pueden distinguirse las fases señaladas en el mapa geológico anterior.

Vamos ahora a interpretar un corte geológico.

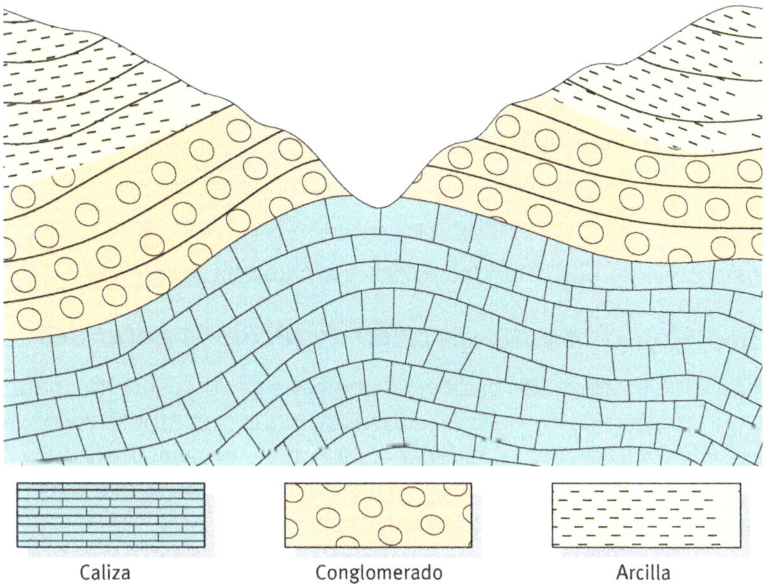

Distinguimos:

- Tres estratos: roca caliza, conglomerados y arcilla.
- Los estratos están inclinados.

La interpretación del corte es la reconstrucción histórica de los fenómenos que han tenido lugar en esta zona. Podemos deducir que hubo:

— Tres etapas de sedimentación distintas. Por el principio de superposición de Steno, el orden de antigüedad de los estratos es: roca caliza, conglomerados y arcilla.

— Sobre estos estratos han actuado fuerzas orogénicas que los han plegado, originando un pliegue anticlinal.

— Los terrenos se han erosionado posteriormente hasta formar el relieve actual, debido a la actividad de un río.

ACTIVIDADES

19. Realiza el corte geológico indicado en el siguiente mapa e interpreta los procesos que han tenido lugar.

— Indica la secuencia en que se depositaron los estratos.

— Razona si se trata de un pliegue anticlinal o sinclinal.

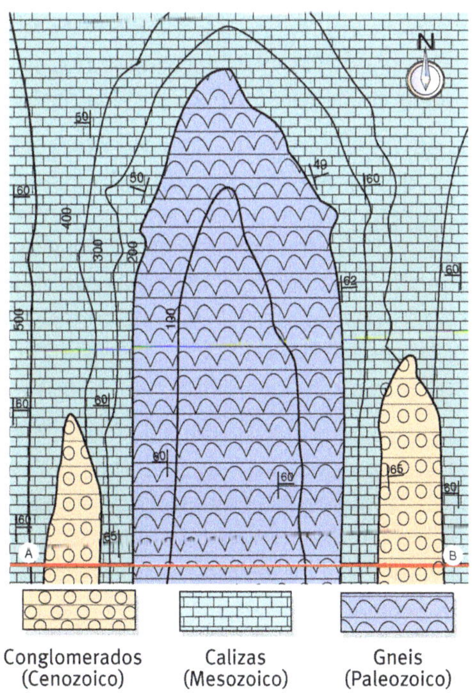

5. Aplicaciones de la geología

La Geología, además de proporcionarnos conocimiento sobre la estructura y historia del planeta, tiene un gran número de aplicaciones que afectan a nuestra vida diaria. El trabajo de los geólogos se centra en los siguientes campos:

- Investigación y prospección de recursos naturales
- Evaluación de riesgos geológicos e impactos medioambientales
- Estudios geológicos para obra civil y control de sondeos
- Cartografía geológica del territorio

Vamos a conocer con más detalle las dos primeras aplicaciones.

5.1 *Investigación y prospección de recursos naturales*

La civilización moderna se basa en el consumo de una serie de materiales que extraemos de la corteza terrestre y que nos proporcionan la energía y la materia para abastecer la industria y el consumo particular. Los yacimientos a partir de los cuales pueden ser explotados estos recursos son limitados y se hallan situados en zonas concretas del planeta, por lo cual resulta de gran importancia su localización y estudio.

Ciertos recursos, que llamamos **recursos renovables,** se recuperan de forma relativamente rápida, como por ejemplo la madera de los bosques. Otros, en cambio, tienen unos períodos de recuperación tan largos que desde un punto de vista práctico, se dispone de ellos de forma limitada. Estos últimos son los **recursos no renovables,** como el carbón, el petróleo o los metales. Desde el punto de vista de su uso, distinguimos entre recursos energéticos y recursos minerales.

Recursos energéticos

La energía de la cual depende nuestra sociedad todavía se genera en su mayor parte a partir de recursos geológicos no renovables de origen fósil: el carbón, el petróleo y el gas. Estas tres fuentes de energía se originaron a partir de la transformación de la materia orgánica.

El **carbón** tiene su origen en la acumulación de restos vegetales en ambientes pobres en oxígeno de zonas pantanosas. Posteriormente, los procesos de transformación a los que se ve sometido este depósito puede dar lugar a diferentes tipos de rocas ricas en carbono, como el lignito o la antracita. Más del 70% del consumo de carbón se destina en la actualidad a la producción de energía eléctrica en centrales térmicas. En la Península Ibérica los yacimientos de carbón más importantes se localizan en las cuencas asturianas de hulla y antracita.

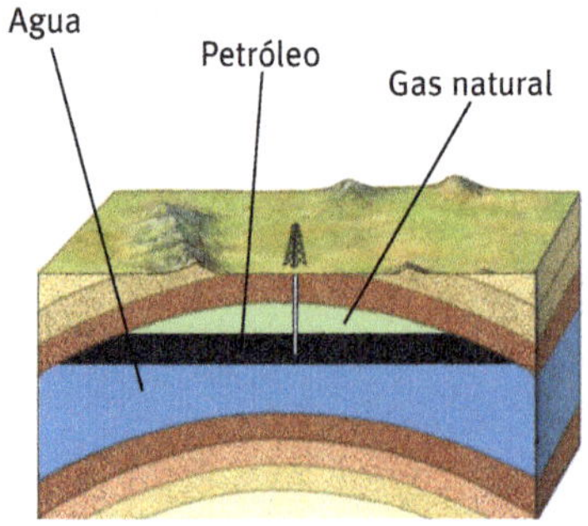

Las trampas petrolíferas son disposiciones de los estratos que permiten la acumulación del petróleo y el gas.

La formación del **petróleo** y el **gas natural** se asocia a cuencas oceánicas en las cuales se produce una sedimentación de abundantes restos de seres vivos, que quedan enterrados y sufren una transformación en ausencia de oxígeno. A medida que estos productos se van generando, suelen migrar a través de las capas permeables en las que se acumulan. Si la disposición de los estratos es la adecuada, el petróleo puede acumularse hasta cantidades que hagan rentable su explotación en las llamadas trampas petrolíferas. En la Península Ibérica existen yacimientos petrolíferos frente a las costas de Tarragona y Vizcaya; y yacimientos de gas en Vizcaya.

Otro recurso no renovable de carácter energético es el **uranio**, utilizado como combustible en las centrales nucleares. En la península podemos hallar depósitos de uranio en Salamanca y Badajoz.

Recursos minerales

Llamamos recursos minerales al conjunto de minerales que tienen utilidad en la industria y de los cuales se puede disponer para su aprovechamiento. Estos recursos incluyen los yacimientos que ya se hallan en explotación, los que no son explotables en la actualidad, por agotamiento o imposibilidad técnica, e incluso aquellos que, a pesar de no haber sido descubiertos, se supone su existencia. Se llama mena a los minerales que pueden ser explotados obteniendo un beneficio económico.

Distinguimos entre minerales metálicos y no metálicos:

- La mayor parte de depósitos de **minerales metálicos**, como el oro, el cobre la plata, se han originado por procesos ígneos, y están relacionados con la segregación magmática. Otras menas metálicas, como la calcopirita (cobre), la galena (plomo) o la magnetita (hierro) se hallan asociadas a zonas de metamorfismo de contacto. En otros casos, los procesos de meteorización pueden dar lugar a la formación de acumulaciones minerales que en la roca originaria se hallaban dispersos, dando lugar al enriquecimiento secundario. La bauxita, la principal mena de aluminio, se origina así. Son muy importantes las expotaciones de Andalucía, con minas de hierro, cobre y plomo en Huelva, o plata en Córdoba. En Galicia existen yacimientos de cobre y en Murcia de plomo.
- Entre los recursos **minerales no metálicos** podemos contar los materiales de construcción y los minerales industriales.

  Uno de los recursos minerales más utilizados como **materiales de construcción** son los áridos (arena, roca triturada y grava), el yeso para recubrimientos, la arcilla para ladrillos y tejas y el cemento, fabricado a base de roca caliza y lutitas. La combinación del cemento con los áridos da lugar al hormigón. Otro elemento importante entre los materiales de construcción son las rocas ornamentales, como el granito, el mármol, la pizarra o ciertos tipos de roca calcárea, utilizadas directamente en el recubrimiento de superficies.

  Las explotaciones de recursos destinados a la construcción suelen ser de tipo local, dado su bajo precio y el encarecimiento que representa su transporte. Existen importantes canteras de granito y pizarra en Galicia y Cantabria, caolín en Guadalajara, yeso y arcilla en la Comunidad Valenciana y de rocas volcánicas en Canarias.

  Los **minerales industriales** se utilizan para la obtención de productos químicos o fertilizantes, compuestos de nitrato, fosfato y potasio. Éste último se obtiene sobre todo de rocas salinas evaporíticas, como la potasa, de la cual hay yacimientos en explotación en Cataluña. Otras explotaciones de recursos no metálicos en la Península son las minas de volframio en Salamanca o las salinas en Alicante.

La pirita es una de las menas de hierro más comunes.

5.2 *Evaluación de riesgos geológicos e impactos medioambientales*

Se considera **riesgo natural** a una condición de la naturaleza que implica una amenaza para la salud, el bienestar o la seguridad de un grupo de personas, actividades o economía de una comunidad. Se evalúan dos componentes: la probabilidad de que sucedan y los costes que producen en caso de suceder. Ciertos riesgos, como la pérdida de suelo, la desertización o las inundaciones, pueden verse acentuados por la acción humana al alterar el equilibrio natural. A este tipo de riesgos se le conoce como **riesgos inducidos**.

A pesar de que muchos de los fenómenos que provocan riesgos para el ser humano no pueden ser controlados, sí se puede intentar minimizar sus efectos. Para ello hay que actuar sobre cuatro aspectos:

PREDICCIÓN DEL RIESGO	PREVISIÓN DEL RIESGO	PREVENCIÓN DEL RIESGO	ORDENACIÓN DEL TERRITORIO
Consiste en anticipar, basándose en el conocimiento del sistema que lo provoca y de los indicios, el lugar y el momento en que se desencadenará un determinado fenómeno.	A partir del registro histórico de las manifestaciones de un fenómeno y del conocimiento de sus causas, puede determinarse cuáles son las zonas en las que tiene más probabilidad de manifestarse, y las consecuencias que puede tener.	Consiste en la adopción de medidas que prevengan la aparición del fenómeno causante, o bien medidas de alarma y control para minimizar sus efectos, como los planes de evacuación ante una erupción.	Se trata de la planificación y adecuación de los usos del suelo de un territorio con el objeto de minimizar las condiciones de riesgo.

Finalmente, los riesgos naturales pueden clasificarse en dos grandes categorías: los de carácter **meteorológico,** relacionados con el clima, y los **geotécnicos,** relacionados con la liberación de energía interna.

TIPO DE RIESGO		MEDIDAS DE GESTIÓN
Geotécnico	Seísmos y tsunamis	Reconocimiento de las zonas de riesgo, predicción mediante precursores (pequeños seísmos, gravimetría...), construcción de edificios sismorresistentes...
	Erupciones volcánicas	Predicción por precursores (seísmos, fumarolas, ruidos...), ordenación territorial, planes de evacuación...
Meteorológico	Sequía	Correcta gestión de los recursos hídricos.
	Inundaciones	Predicción de lluvias torrenciales, uso correcto del suelo, control urbanístico, obras hidráulicas (presas, diques...), planes de evacuación.
	Tormentas tropicales, huracanes	Predicción de fenómenos meteorológicos, planes de evacuación...
	Desprendimientos y deslizamientos de tierra	Estudio de los materiales, la tipología de los suelos y los planos de debilidad; descartar la urbanización, promover la cimentación y drenaje de taludes. Estos fenómenos pueden ser desencadenados tanto por causas ligadas a la meteorología como a la geotectónica (seísmos).
	Erosión de las playas	Construcción de espigones o puertos, restauración de las playas por aporte de materiales, posibilitar la regeneración natural.
	Erosión del suelo	Uso racional del suelo: evitar agricultura intensiva, sobrepastoreo y talas indiscriminadas; control de incendios forestales.

20 En terrenos del Cretácico con una edad de 120 millones de años se han hallado en la Rioja restos fósiles y huellas de dinosaurios (icnitas), entre los cuales se cuentan *Ouranosaurus*, *Iguanodon*, *Deynonichus*, *Psitacosaurus* o *Hypselosaurus*.

— Busca información sobre la ubicación de la placa Ibérica en el Cretácico. ¿Cuáles eran las condiciones ambientales en la Península Ibérica en aquella época? Localiza en un mapa de aquella época el lugar que ocuparía La Rioja y especifica en qué difiere su localización en el globo de la actual (latitud, posición respecto a los océanos y mares, orientación...).

— Se cree que la zona en la cual se han hallado las huellas constituía un amplio delta que desembocaba sobre el mar de Tetis, que penetraba a través del valle del Ebro. La composición litológica de la zona en el Cretácico inferior presenta rocas provenientes de depósitos aluviales, detríticas y calizas lacustres, con influencia marina. ¿Cómo podrías justificar la afirmación anterior referente al ambiente a partir de la composición litológica?

— De esta misma época se han datado arrecifes de coral fósiles en la zona de las Cordilleras Béticas. Si sabemos que los corales, en la actualidad, requieren para su crecimiento una temperatura entre los 20 y los 28 °C, ¿qué podemos inferir sobre las condiciones climáticas de la Península en el Cretácico?

Éste es un mapa geológico de la zona en la que se hallan las icnitas.

8. Materiales aluviales y detríticos.
7. Terciario postectónico: Rocas sedimentarias continentales de sistemas aluviales y lacustres.
6. Terciario pretectónico y sintectónico: Rocas sedimentarias continentales de sistemas aluviales y lacustres.
5. MEGACICLO 4: Rocas sedimentarias continentales en la parte inferior, pasando a marinas en el resto.
3-4. MEGACICLO 3: Depósitos aluviales y carbonatos lacustres con influencia marina ocasional.
4a. Rocas calcáreas con frecuentes icnitas de dinosaurio.
2. MEGACICLO 2 : Depósitos de plataformas marinas carbonatadas.
1. MEGACICLO 1 : Inicio con materiales continentales, seguidos de depósitos marinos someros y final con depósitos de costa.

— El mapa presenta una zona de tránsito entre dos unidades bien diferenciadas: la cuenca del Ebro y la Cordillera Ibérica. Señala cada una de las zonas y comenta sus diferencias a nivel de composición litológica. Resume el proceso de formación de estas estructuras.

— Localiza en el mapa: un pliegue anticlinal, un pliegue sinclinal y una falla.

— ¿Cuál crees que es la causa de la abundancia de pliegues y fracturas en esta zona?

Si examinas la dirección de las fallas y los pliegues, ¿qué detectas? ¿A qué puede ser debido? ¿Puedes determinar la dirección y el sentido de los esfuerzos a los que ha sido sometido el terreno?

— Además de en Enciso, existen otras localidades en La Rioja en las cuales han aparecido icnitas de dinosaurio del Cretácico. Puedes obtener más información sobre la composición geológica del terreno en el SIG del gobierno de la Rioja en http://www.iderioja.larioja.org/cartografia/index.php?map=RIOJA_C04&&&lang=es

ACTIVIDADES DE SÍNTESIS

S

21 El mapa que aparece a continuación muestra la distribución de los epicentros de terremotos producidos en la Península Ibérica durante los últimos decenios.

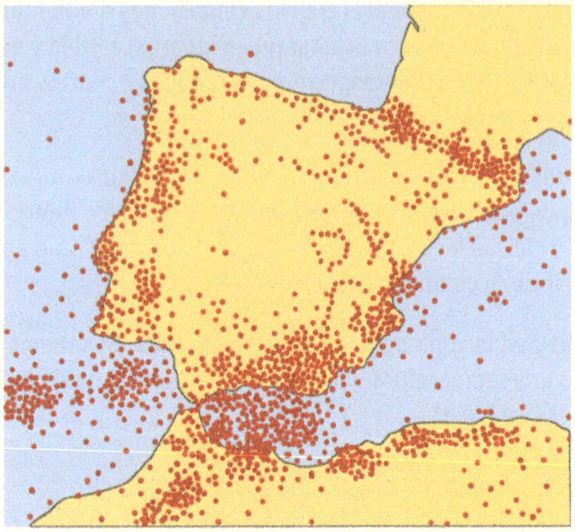

— ¿A qué estructuras asociarías esta distribución? Comenta cómo la interrelación entre estas estructuras puede ser la causante de la actividad sísmica.

— ¿Existe algún tipo de vulcanismo asociado a margen de placa en la Península Ibérica?

— Dado que en las zonas en las que se distribuyen estos sismos podemos localizar ciudades, ¿podríamos considerar que existe un riesgo natural? ¿Por qué? ¿Qué acciones se podrían llevar a cabo para gestionar dicho riesgo?

22 Basándote en los datos que aporta el siguiente mapa geológico del fondo oceánico frente a las costas del norte de España, responde a las siguientes preguntas.

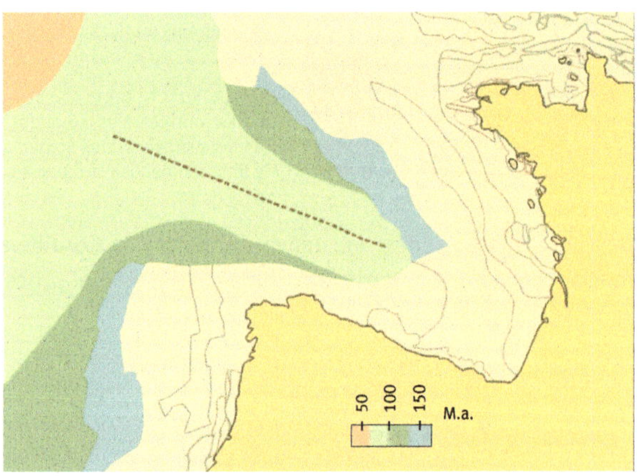

— ¿En qué período geológico se produce la apertura del Golfo de Vizcaya? Justifica tu respuesta.

— ¿A qué estructura presente en el mapa se asocia este proceso?

23 Los Pirineos y la cordillera Bética fueron originadas por la misma orogenia. Pero ¿fue simultánea su formación?

— Justifica tu respuesta y explica con qué orogenia se relacionan y la época geológica en que se produce ésta, así como los principales relieves que generan.

TIC

24 En *http://www.mapasdelapalma.es/* tienes acceso a un sistema de información geográfica (SIG) que presenta datos sobre la geología de la isla de La Palma. Utilízalo para responder a las siguientes preguntas:

— ¿En qué zona de la isla se hallan mayoritariamente los conos volcánicos?

— ¿En qué zona se hallan los diques?

— ¿Cuánto mide la isla en su parte más ancha? ¿Y cuál es la longitud máxima de la isla?

— ¿En qué municipio se halla el Parque Nacional?

— ¿En qué zona las precipitaciones son mayores?

— ¿Qué municipio tiene la temperatura media más alta?

25 La actividad humana provoca cambios en la superficie del planeta y en su atmósfera. La mayor parte de los procesos que hemos examinado funcionan gracias a un frágil equilibrio entre diversos factores:

- La circulación atmosférica y la de los océanos, son consecuencia, entre otros motivos de la temperatura del planeta.

• La formación del suelo y su erosión, el depósito de sedimentos en costas y deltas, la riqueza de los acuíferos, la cobertura boscosa... se mantienen en sus niveles a causa de una determinada correlación de variables.

El ser humano ha alcanzado la capacidad de modificar de forma apreciable estos factores, con lo cual su intervención puede llevar asociadas consecuencias no deseadas.

En *http://www.greenpeace.org/espana/* hallarás información sobre las campañas que esta organización está llevando a cabo en el estado español sobre algunos de estos impactos de la actividad humana.

Por grupos, reunid información sobre alguno de los siguientes temas:

— El calentamiento global, sus causas y sus consecuencias

— La sequía y la gestión adecuada del agua

— La salinización de los acuíferos litorales y el turismo

— La erosión del suelo

— La deforestación en la Península y los incendios forestales

— La erosión de las costas en el litoral del estado español

— La minería a cielo abierto

Preparad una exposición ante la clase en la que se expongan las raíces del problema, su situación actual y las posibles acciones para corregirlo.

27 Muchas ciudades del litoral mediterráneo están construidas sobre los cauces de antiguas ramblas. Este hecho provoca que en épocas de lluvias torrenciales se den con frecuencia inundaciones. ¿Qué acciones se pueden llevar a cabo para evitar o minimizar los efectos de estos episodios?

28 Según el principio de superposición, indica cuál es el orden de los estratos en este corte geológico.

— En el corte aparece una falla. Sitúala e indica cuál es el terreno que se ha levantado y cuál el que se ha hundido.

— El esfuerzo al cual se ha sometido el terreno ¿ha sido compresivo o distensivo? Argumenta tu afirmación a partir de los fenómenos que observas en el corte.

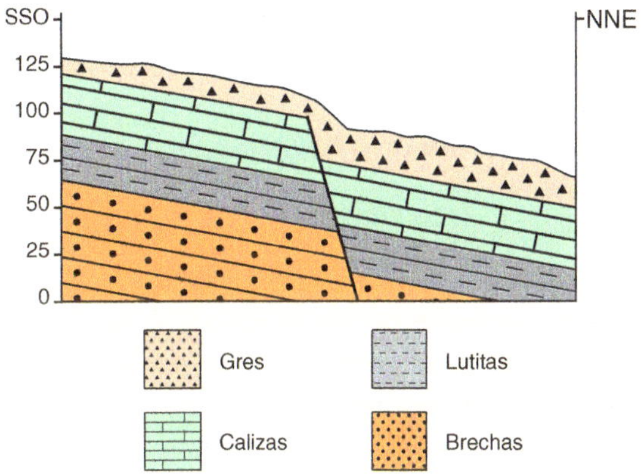

26 Fotocopia el mapa adjunto y realiza el corte geológico indicado AB. Una vez realizado el corte, responde a las siguientes preguntas.

— ¿Qué estructura se observa en el corte geológico?

— ¿Cuáles son los materiales más antiguos? ¿Cuáles se depositaron primero?

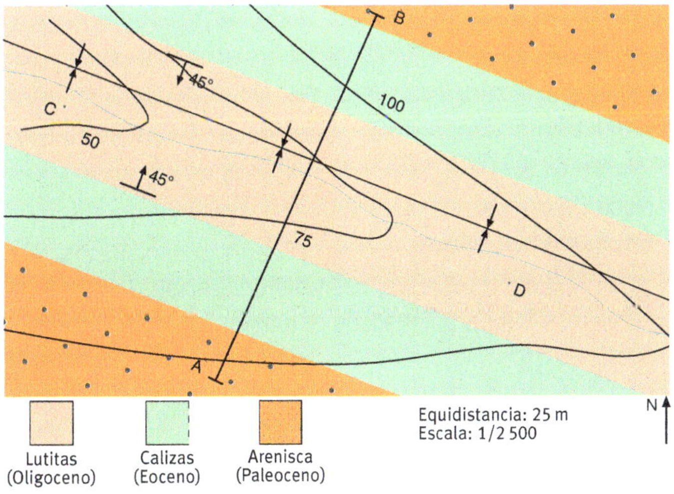

PRÁCTICA 1 Clasificación de rocas

OBJETIVOS

- Utilizar una clave dicotómica para identificar muestras de rocas.
- Reconocer y analizar las distintas características de las rocas.

El uso de las claves dicotómicas es muy habitual en las ciencias, en particular es muy útil para la identificación de rocas u organismos. Vamos a utilizar la clave dicotómica adjunta para clasificar las muestras de la colección de rocas de que dispongamos en el aula.

Una clave dicotómica está compuesta por diferentes parejas de características que se excluyen mutuamente. Debes utilizar la clave empezando por observar qué característica de la primera pareja cumple tu muestra; una vez identificada la característica que corresponde a la muestra analizada se indica la siguiente pareja de características que debes analizar.

Debes repetir esta operación hasta que la característica identificada te indique una roca concreta en vez de otro par de características.

1. Roca totalmente formada por cristales visibles a simple vista 2
   No se observan las características anteriores 8
2. Se observa direccionalidad en los elementos de la roca, los minerales están entremezclados al azar. . 3
   No se observan las características anteriores 4
3. Roca formada por granos gruesos, de más de 2 mm, con foliación irregular .... **GNEIS**
   No se observan las características anteriores .... **ESQUISTO**
4. Roca que da efervescencia con HCl, no raya al vidrio .... **MÁRMOL**
   No se observan las características anteriores 5
5. Roca de color oscuro .... **GABRO**
   Roca de color claro 6
6. Cristales pequeños, de 1-2 mm .... **APLITA**
   Cristales más grandes 7
7. Cristales mayores de 1 cm .... **PEGMATITA**
   Cristales menores de 1 cm .... **GRANITO**
8. Roca formada por algunos cristales plenamente formados inmersos en una matriz amorfa 9
   Roca con numerosas pequeñas cavidades 14
9. Roca muy porosa 10
   No se observan las características anteriores 12
10. Roca que reacciona con HCl dando efervescencia, puede presentar restos vegetales........ **TRAVERTINO**
    No se observan las características anteriores 11
11. Roca de color claro .... **PUMITA**
    Roca de color oscuro........ **PIROCLASTO**
12. Roca que presenta bandas de diferente coloración .... **RIOLITA**
    No se observan las características anteriores 13
13. Roca de color claro .... **PÓRFIDO**
    Roca de color oscuro .... **BASALTO**

14. Roca compuesta de partículas identificables a simple vista 15
No se observan las características anteriores 18

15. Partículas menores de 2 mm .... **ARENISCA**
Partículas mayores de 2 mm 16

16. La mayoría de los elementos mayores (clastos) sin redondear .... **BRECHA**
La mayoría de los elementos mayores (clastos) redondeados 17

17. Los clastos presentan estructuras concéntricas,
de orígen biológico .... **CALIZA NUMMULÍTICA**
Los clastos no presentan estructuras concéntricas .... **CONGLOMERADO**

18. Roca que reacciona con HCl produciendo efervescencia 19
Roca que no reacciona con HCl 20

19. Roca pulverulenta, con olor a tierra mojada .... **LUTITA**
No se observan las características anteriores .... **CALIZA**

20. Roca brillante, textura laminar muy marcada .... **PIZARRA**
No se observan las características anteriores 21

21. Roca oscura 22
Roca clara .... **CUARCITA**

22. Roca de textura vítrea .... **OBSIDIANA**
No se observan las características anteriores 23

23. Roca de estructura leñosa 24
No se observan las características anteriores 25

24. Roca con restos vegetales visibles .... **TURBA**
No se observan las características anteriores .... **LIGNITO**

25. Roca mate, mancha los dedos .... **HULLA**
No se observan las características anteriores ... **ANTRACITA**

Observa a continuación a qué ambiente petrogenético pertenece cada una de las rocas que aparecen en la clave dicotómica.

Magmáticas	Gabro, granito, aplita, pegmatita, pórfido, basalto, riolita, pumita, piroclasto y obsidiana.
Sedimentarias	Brecha, conglomerado, arenisca, lutita, travertino, caliza nummulítica, caliza, turba, lignito, hulla y antracita.
Metamórficas	Mármol, gneis, esquisto, pizarra y cuarcita.

CUESTIONES

1. Identifica las muestras que te ha proporcionado tu profesor o profesora y clasifícalas según su ambiente petrogénico de origen.
2. Indica qué rocas de la clave dicotómica son rocas sedimentarias de origen orgánico.
3. Cita tres rocas magmáticas de la clave dicotómica que sean extrusivas. Razona tu respuesta.

6

Las moléculas de la vida

El silicio es uno de los elementos más abundantes de la litosfera; no obstante en la biosfera es un componente minoritario.

La estructura y las funciones características de los seres vivos son posibles gracias a combinaciones moleculares de un número limitado de elementos. Las características de cada una de estas moléculas están íntimamente ligadas a su estructura.

A lo largo de esta unidad conoceremos la relación entre la estructura y la función de las moléculas de la vida.

■ Molécula de dextrosa al microscopio de luz polarizada

ORGANIZACIÓN DE LOS CONTENIDOS

1. El agua

2. Los compuestos de carbono

3. Los glúcidos

3.1. Monosacáridos

3.2. Disacáridos

3.3. Polisacáridos

4. Los lípidos

4.1. Lípidos complejos

4.2. Lípidos sencillos

5. Las proteínas

6. Los ácidos nucleicos

7. El origen de la vida y su evolución

OBJETIVOS

- Conocer las diferencias entre la composición de la materia viva y la materia inerte.
- Comprender la relación entre las características químicas del agua y su importancia para los seres vivos.
- Identificar las características y las propiedades de los diferentes tipos de compuestos de carbono.
- Conocer las teorías más destacadas sobre el origen de la vida y el proceso que condujo a la aparición de los precursores de las células.

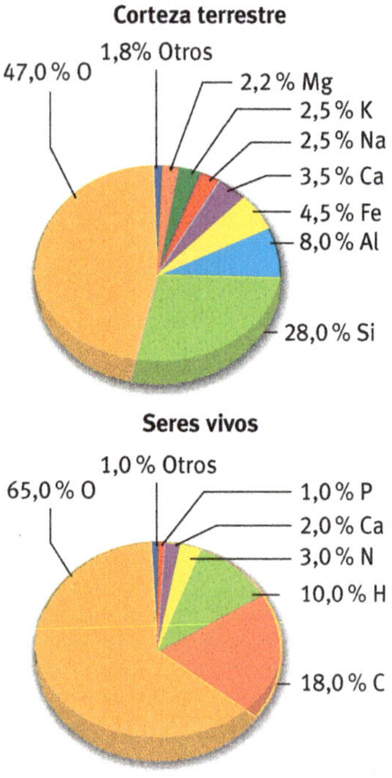

Diagramas comparativos de la composición química de la corteza terrestre y los seres vivos

Contenido hídrico en algunos seres vivos	
Animales	**%**
Feto humano (3 meses)	94
Hombre adulto	63
Medusa	95
Insecto	72
Vegetales y hongos	**%**
Algas	98
Setas	91
Liquen	55

Contenido hídrico en algunos órganos	
Animales	**%**
Cerebro	86
Sangre	79
Hueso	22
Vegetales	**%**
Patata	78
Garbanzo	11
Grano de arroz	10

En la Tierra hay más de 90 elementos químicos naturales. Estos elementos como tales, o bien combinados, componen toda la materia. Los datos que aparecen en el diagrama lateral sobre la composición de la corteza terrestre y de los seres vivos nos permiten observar que la proporción en que se encuentran los distintos **elementos** es diferente.

En la corteza terrestre predominan el silicio y el oxígeno. Éste es también muy abundante en los seres vivos; sin embargo, el silicio supone una proporción mínima que queda incluida en el porcentaje de *Otros.*

Los principales elementos que constituyen el cuerpo de los seres vivos (oxígeno, carbono, hidrógeno, nitrógeno, fósforo y azufre), los elementos biogénicos o bioelementos, presentan ciertas características:

- Pueden enlazar con más de un átomo (excepto el hidrógeno) y constituir moléculas de gran tamaño o **macromoléculas.**
- Son elementos muy ligeros que forman enlaces covalentes y constituyen moléculas muy estables. Su ligereza ha facilitado la adaptación de los seres vivos al campo gravitatorio terrestre.

Las biomoléculas se encuentran formando parte de la compleja estructura de los seres vivos. Existen dos grupos:

- **Biomoléculas inorgánicas:** se pueden encontrar también en la materia mineral. Incluyen el agua (la más abundante) y las sales minerales.
- **Biomoléculas orgánicas:** exclusivas de los seres vivos. Incluyen los glúcidos, los lípidos, las proteínas y los ácidos nucleicos.

1. El agua

Es imprescindible para el desarrollo de la vida. El estudio de su composición, de algunas de sus propiedades físicas y químicas y de su influencia en los seres vivos nos permite comprender el porqué.

Composición. Cada molécula de agua contiene 2 átomos de hidrógeno y uno de oxígeno, que se unen mediante **enlaces covalentes,** compartiendo electrones para alcanzar su configuración más estable.

Aunque en la molécula el número de capas negativas y positivas es igual y, por tanto, es neutra, las cargas se encuentran repartidas de una forma ligeramente desigual. Así, las negativas se concentran cerca de los átomos de oxígeno, mientras que las positivas lo hacen cerca de los de hidrógeno. Por esto se dice que la molécula de agua presenta **polaridad,** ya que en ella se distinguen dos zonas con cargas de distinto signo, por lo que consideramos las moléculas de agua como **dipolos.**

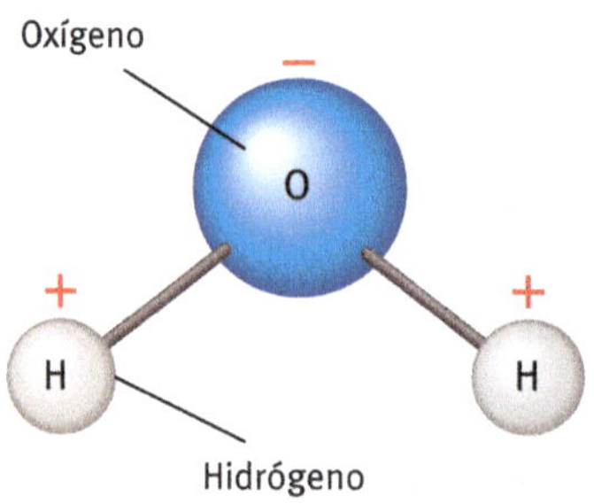

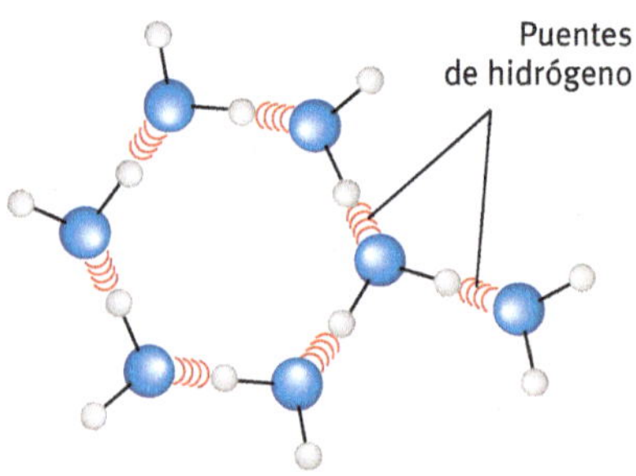

De este modo, entre una molécula de agua y las que la rodean se establecen fuerzas de atracción que tienden a unir cargas de signo opuesto. Estas uniones reciben el nombre de **puentes de hidrógeno.**

Propiedades. La polaridad de la molécula de agua y la existencia de los puentes de hidrógeno explican dos propiedades que hacen del agua una sustancia imprescindible para los seres vivos: su gran capacidad disolvente y su elevado calor específico.

- El agua tiene una gran **capacidad disolvente.**

  Cuando entra en contacto con una **sustancia iónica,** como el cloruro de sodio, el agua es capaz de separar los iones de esta sustancia y rodearlos con moléculas que se orientan en la posición correspondiente a la carga de cada ion. De esta forma se disuelve en el agua una gran cantidad de sustancias.

  También es capaz de disolver sustancias que, aunque no son iónicas, tienen **polaridad** en su molécula, como glúcidos, aminoácidos y otros compuestos que trataremos más a fondo en esta unidad.

  Además, existe una gran variedad de sustancias que contienen en su molécula **grupos hidrofóbicos** (partes de la molécula insolubles en agua) y **grupos hidrofílicos** (partes solubles en agua). Estas sustancias también se dispersan en el agua, como los glicerofosfolípidos y otros componentes de las células que se disponen formando pequeñas esferas, con la parte hidrofílica de sus moléculas en dirección al agua y el resto hacia el interior de las esferas.

- El agua tiene elevado **calor específico,** ya que los puentes de hidrógeno limitan, en parte, la movilidad de las moléculas y, por tanto, se retarda el incremento de la agitación térmica frente al calor. Así, el agua debe absorber más calor que otras sustancias para aumentar su temperatura; y también es mayor la cantidad de calor que el agua debe desprender para que su temperatura descienda.

Influencia en los seres vivos. Estas dos propiedades suponen para los seres vivos ventajas importantes:

- En el agua que constituye la mayor parte del citoplasma y el medio extracelular se disuelven numerosas sustancias. De este modo las sustancias son **transportadas** y participan en las **reacciones metabólicas** que tienen lugar en los organismos.

- El agua tiene un **efecto moderador de la temperatura** de los seres vivos. Debido a su elevado calor específico, el contenido en agua de los seres vivos amortigua las variaciones de la temperatura corporal, que podrían producirse como consecuencia de los cambios bruscos de la temperatura ambiental. Esto favorece el mantenimiento, dentro de unos límites, de una temperatura más estable.

ACTIVIDADES

1. Nombra dos elementos químicos y di cuál es su función específica en los seres vivos.
2. Explica por qué las sustancias polares se disuelven en el agua mientras que las no polares no lo hacen.

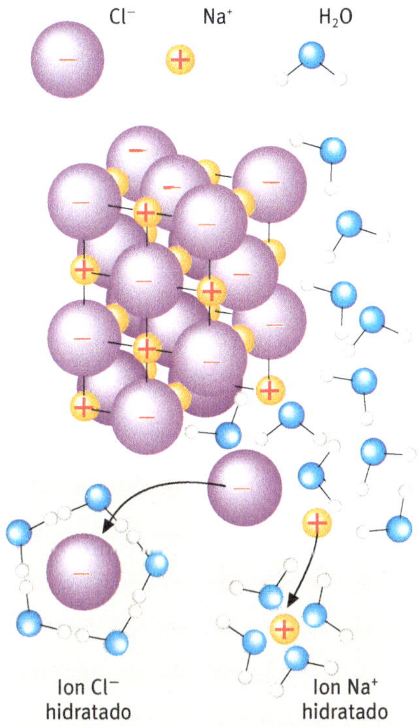

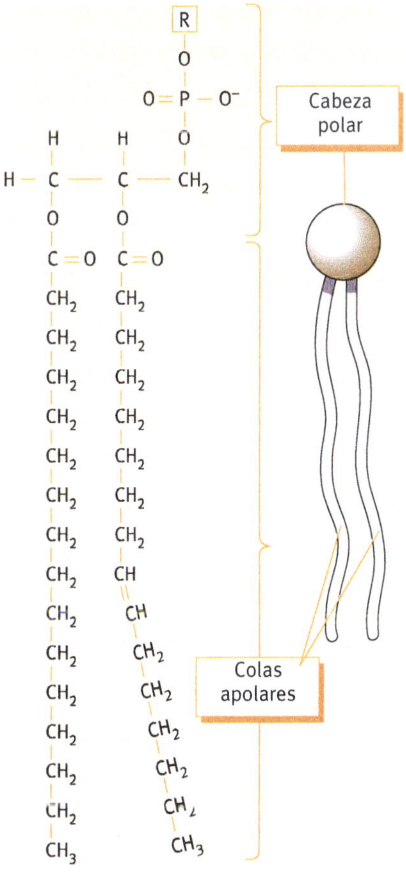

Representación de una molécula de glicerofosfolípido con la cabeza polar o hidrofílica y las colas apolares o hidrofóbicas

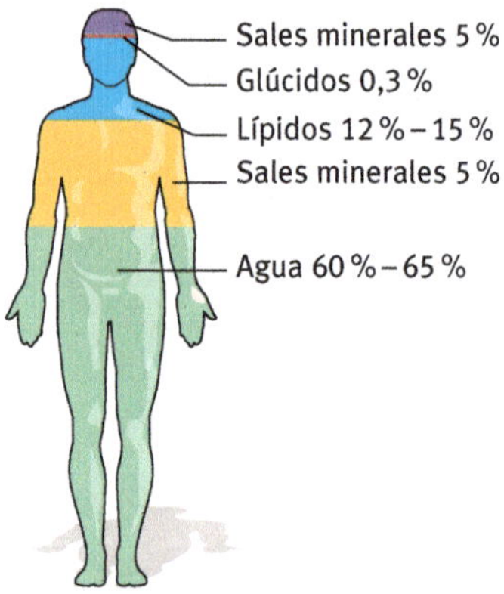

Porcentaje de biomoléculas orgánicas e inorgánicas del cuerpo humano

2. Los compuestos de carbono

Entre los elementos químicos que intervienen en la composición de la materia viva, el carbono es el segundo más abundante. Por sus características físico-químicas, el átomo de carbono puede formar cuatro enlaces covalentes con otros tantos átomos. Si enlaza consigo mismo, forma cadenas en las que los enlaces entre los átomos de carbono pueden ser sencillos, dobles e incluso triples.

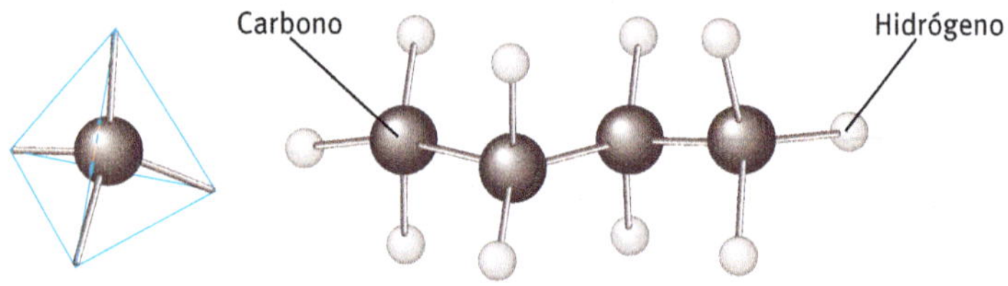

A lo largo de estas cadenas, se pueden establecer enlaces con distintos átomos o **grupos funcionales** de átomos. Algunos de los grupos funcionales más frecuentes son los siguientes.

—OH	Hidroxilo	—CO—	Cetona	$-NH_2$	Amina
—CHO	Aldehído	—COOH	Ácido carboxílico	$-CONH_2$	Amida

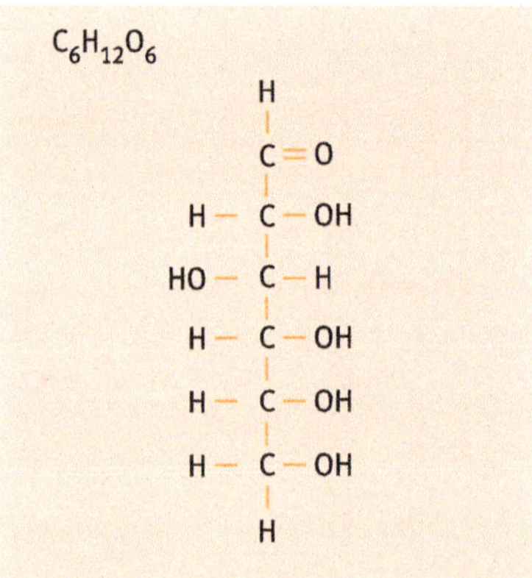

Glucosa: fórmula molecular y fórmula desarrollada

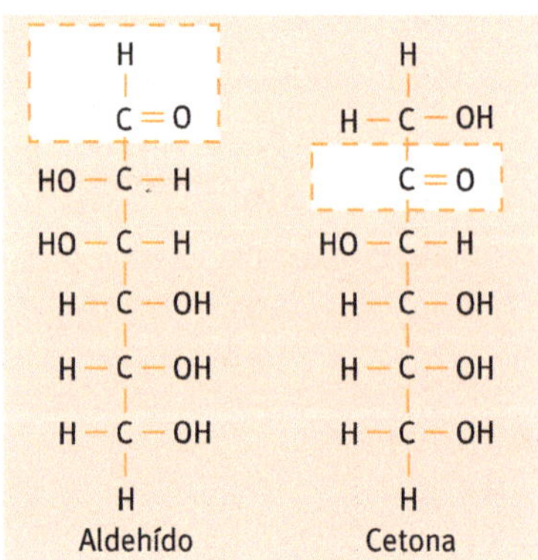

La combinación de distintos grupos funcionales a lo largo de las cadenas de carbono origina el grupo de **compuestos de carbono,** componentes fundamentales de la materia viva que pueden formar largas cadenas carbonadas lineales, ramificadas, en forma de anillo, etc. Las moléculas de estos compuestos se representan mediante dos tipos de fórmulas. Las **fórmulas moleculares** expresan la composición elemental y la proporción de cada elemento en la molécula. Con la **fórmula desarrollada,** además, es posible observar la forma en que se unen los átomos. Los compuestos de carbono o biomoléculas orgánicas presentan funciones muy diversas: pueden ser componentes estructurales de los seres vivos, proporcionar energía, controlar reacciones metabólicas...

Existen cuatro tipos fundamentales de biomoléculas orgánicas: *glúcidos, lípidos, proteínas* y *ácidos nucleicos.*

3. Los glúcidos

Son biomoléculas orgánicas formadas por carbono, oxígeno e hidrógeno. Constituyen un grupo de sustancias muy extenso y variado. Químicamente son polihidroxialdehídos o polihidroxicetonas (un grupo aldehído o un grupo cetona y numerosos grupos hidroxilo) o bien compuestos formados por la unión de moléculas de este tipo.

Se suelen distinguir tres grandes grupos de glúcidos: los *monosacáridos*, los *disacáridos* y los *polisacáridos*.

3.1. *Monosacáridos*

Son los más sencillos. Constan de una cadena que tiene entre 3 y 7 átomos de carbono. Según el número de carbonos que presentan, se clasifican en triosas, tetrosas, pentosas, hexosas y heptosas. Además, los monosacáridos con un grupo aldehído se denominan **aldosas** y los que presentan un grupo cetona, **cetosas**. Así, un monosacárido con 4 átomos de carbono y un grupo aldehído es una aldotetrosa; si presenta un grupo cetona y 5 átomos de carbono es una cetopentosa.

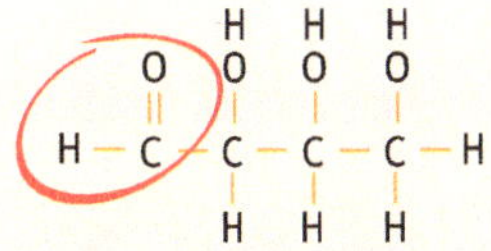

Aldotetrosa

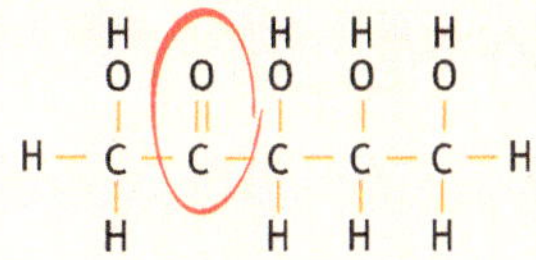

Cetopentosa

Características. En general, los monosacáridos son dulces, tienen color blanco y son solubles en agua.

Función. Son las sustancias que las células utilizan para obtener energía. Cuando tiene lugar una reacción química, se produce siempre la ruptura de unos enlaces y la formación de otros nuevos. Si los enlaces que se rompen poseen más energía que los que se forman, esta reacción *desprende energía*, es una **reacción exergónica.** Si, por el contrario, los nuevos enlaces son más energéticos que los de las moléculas iniciales, esta reacción *necesita un aporte de energía*, es una **reacción endergónica**.

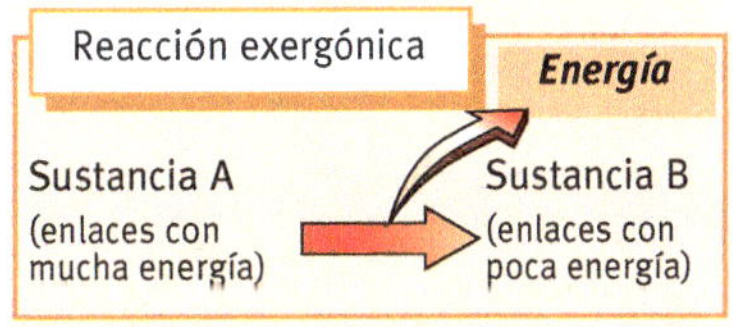

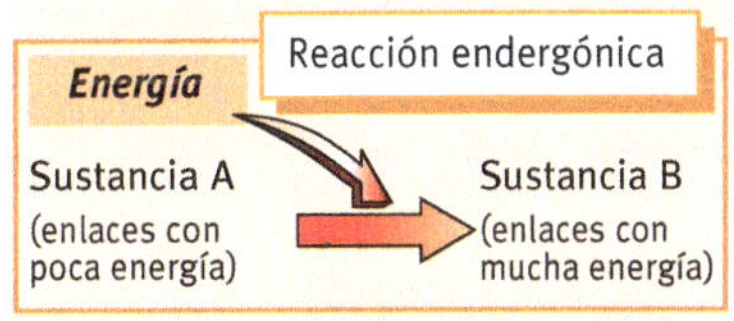

Los monosacáridos reaccionan con el oxígeno y como resultado de ello, se obtienen moléculas con enlaces de menor energía que los del monosacárido. Esta reacción, por tanto, produce energía que la célula aprovecha para llevar a cabo otras reacciones que la necesitan.

En los seres vivos, el monosacárido más utilizado como fuente de energía es la **glucosa,** que se encuentra en la fruta y la miel; cada gramo de glucosa que reacciona con el oxígeno proporciona unas 4 kcal (aproximadamente 16 720 julios). Otros monosacáridos muy abundantes en la naturaleza son la **fructosa,** también presente en la fruta, y la **galactosa** de la leche. En disolución, estos monosacáridos adoptan una estructura cíclica que se representa mediante polígonos.

Glucosa

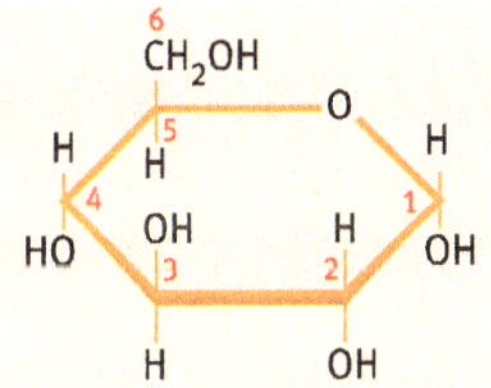

Fructosa

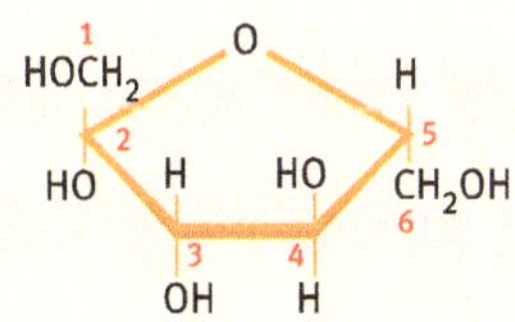

Galactosa

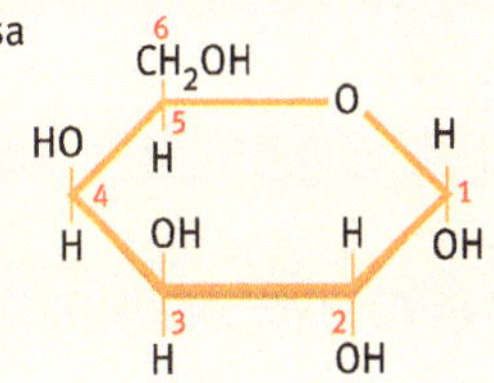

■ Los vértices de los polígonos representan la posición de los átomos de C. El carbono 1 es el carbono que presenta el grupo aldehído o bien, el más cercano al grupo cetona.

3.2. *Disacáridos*

Están formados por la unión de dos monosacáridos. En la mayoría de los casos, la unión se realiza entre el carbono de un grupo hidroxilo de un monosacárido y el carbono del grupo aldehído o cetona del otro. Al producirse el enlace, se desprende una molécula de agua. Este tipo de **enlace** se denomina **O-glucosídico.**

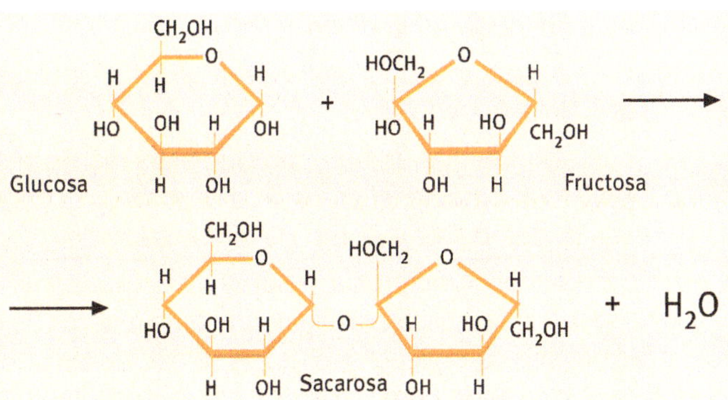

C

El **enlace O-glucosídico** recibe este nombre debido al oxígeno (O) que queda en el punto donde se produce la unión entre las moléculas.

La reacción inversa (la obtención de monosacáridos libres) es la **hidrólisis;** necesita incorporar una molécula de agua.

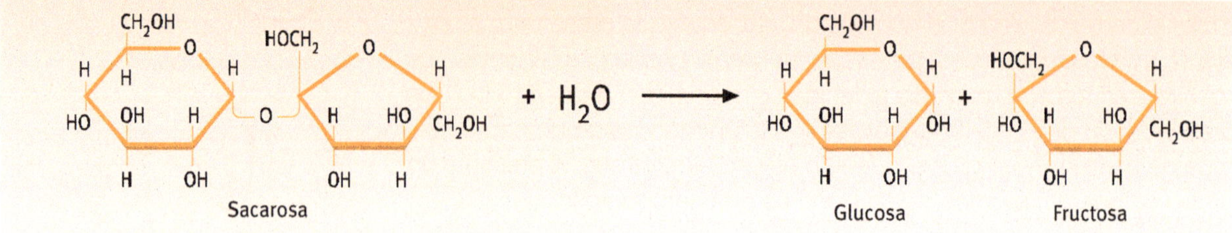

C

La **intolerancia a la lactosa** es la incapacidad para digerir la lactosa: se presenta cuando el intestino delgado no produce suficiente enzima lactasa.

Cuando las personas con intolerancia a la lactosa comen o beben productos lácteos, pueden presentar síntomas como distensión abdominal, exceso de gases intestinales, náuseas, diarrea... Los síntomas se alivian generalmente con la supresión de los productos lácteos en la dieta, en cuyo caso se deben incorporar en la alimentación otras fuentes de calcio.

Características. Los disacáridos tienen, en general, sabor dulce, son cristalizables y solubles en agua.

Función. Están relacionados con el aporte energético a las células, pues su hidrólisis produce monosacáridos que pueden utilizarse para la obtención de energía.

Un ejemplo de disacáridos es la **lactosa,** que se encuentra exclusivamente en la leche y está formada por la unión de una molécula de **galactosa** y una de **glucosa.** Otro disacárido es la **sacarosa,** que se extrae de la remolacha y de la caña de azúcar para obtener el azúcar de mesa; está formada por la unión de una molécula de **fructosa** y una de **glucosa.** La **maltosa** es un disacárido formado por la unión de **dos** moléculas de **glucosa,** y se obtiene de la hidrólisis del almidón, un polisacárido de reserva.

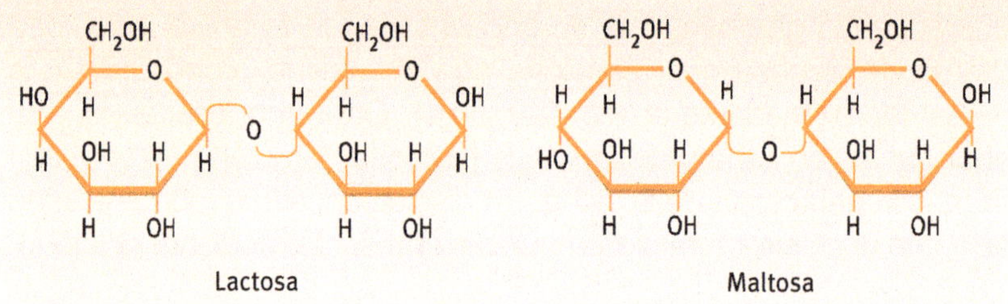

C

Cuando una sustancia está formada por la unión de varias moléculas menores recibe el nombre de **polímero**.

Cada una de las pequeñas moléculas que se han unido recibe el nombre de **monómero**.

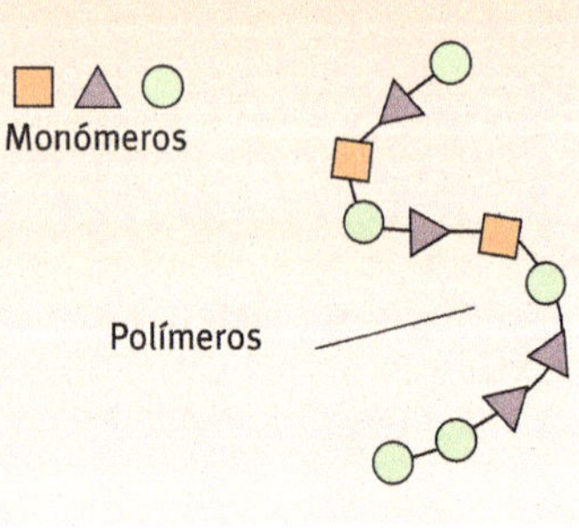

3.3. *Polisacáridos*

Pertenece a este grupo la mayor parte de los glúcidos que se encuentran en la naturaleza. Se trata de compuestos que contienen gran número de monosacáridos, unidos entre ellos mediante enlaces O-glucosídicos, formando cadenas lineales o ramificadas. La unidad que se repite más a menudo es la *glucosa*.

Características. Los polisacáridos, en general, no son dulces, no cristalizan y no son solubles en agua.

Función. Los más abundantes actúan como sustancias de *reserva de energía,* o bien tienen características *estructurales* como constituyentes de diversas partes del organismo de los seres vivos.

Entre los polisacáridos con **función de reserva** vamos a referirnos al *almidón* y al *glucógeno*. Estas sustancias actúan como reserva de energía, porque al hidrolizarse se obtienen los monómeros que las constituyen, los cuales se utilizan en los procesos de obtención de energía.

El **almidón** es la sustancia de reserva en los vegetales. Es muy abundante en las semillas y en los tubérculos. Forma grandes cadenas ramificadas constituidas por la unión de miles de moléculas de glucosa.

El **glucógeno** desempeña en los animales la misma función que el almidón en los vegetales. Es especialmente abundante en el hígado, en el que puede constituir el 10 % de su peso, y también en los músculos.

El glucógeno se encuentra en forma de moléculas ramificadas de gran tamaño similares a las de almidón. La única diferencia entre el glucógeno y el almidón es la frecuencia en que se ramifica la cadena: en el almidón las ramificaciones están separadas por un número de moléculas de glucosa que oscila entre 24 y 30; en el glucógeno este número va de 8 a 12.

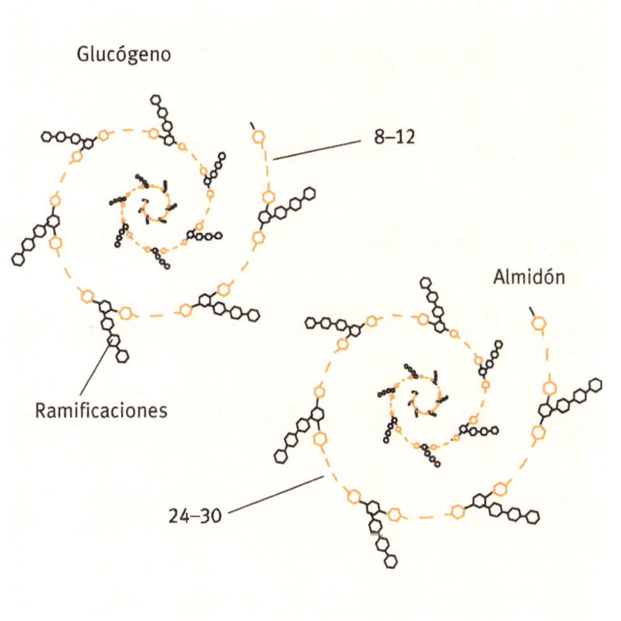

Los polisacáridos con función **estructural** constituyen partes esenciales del cuerpo de los seres vivos. Es el caso de la **celulosa,** que forma la pared celular de las células de los vegetales, y también de la **quitina,** el principal componente del exoesqueleto de los insectos, de los crustáceos y de la pared que recubre las células de los hongos. Las subunidades que lo constituyen son moléculas de N-acetilglucosamina.

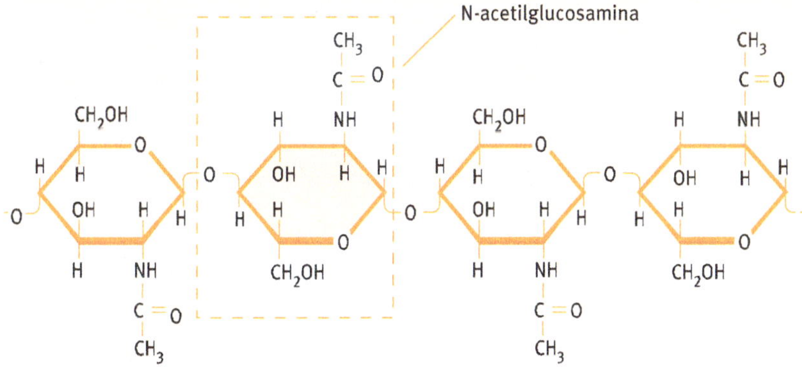

Fragmento de polímero de quitina

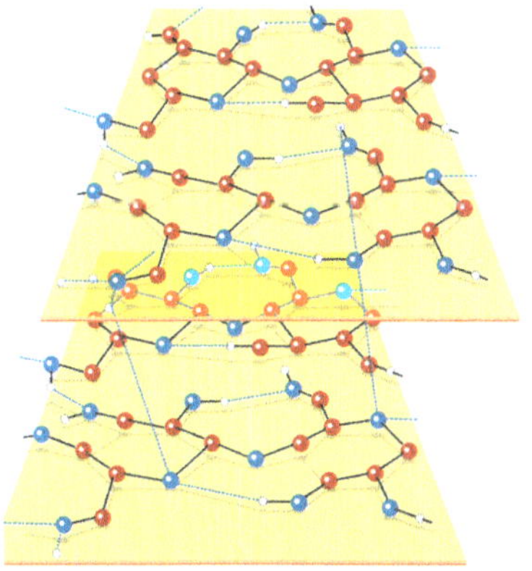

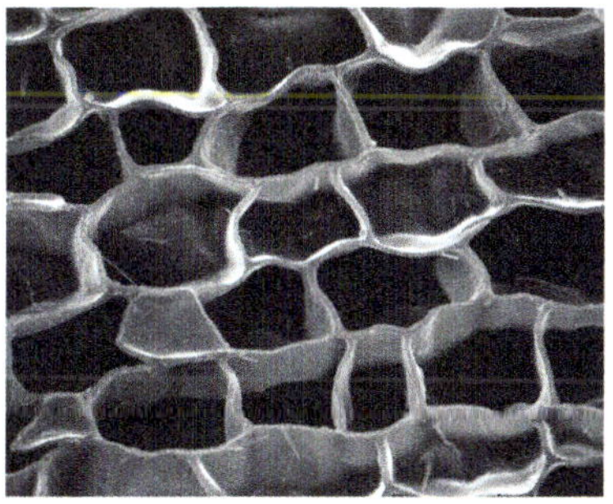

Estructuras de celulosa. Paredes celulares de células vegetales vistas al microscopio electrónico de barrido. 800x.

ACTIVIDADES

3. Compara las estructuras y las funciones del almidón, el glucógeno, la celulosa y la quitina.
4. Explica cuál es la estructura de un polisacárido. Haz un esquema para explicar el enlace O-glicosídico entre dos monosacáridos
5. La fórmula general de los monosacáridos es $(CH_2O)_n$, donde n es el número de átomos de carbono. Escribe la fórmula molecular de una triosa, una tetrosa, una pentosa y una hexosa.
6. Confecciona un esquema para clasificar a los glúcidos en el que se refleje su composición, características, función y ejemplos de cada grupo.

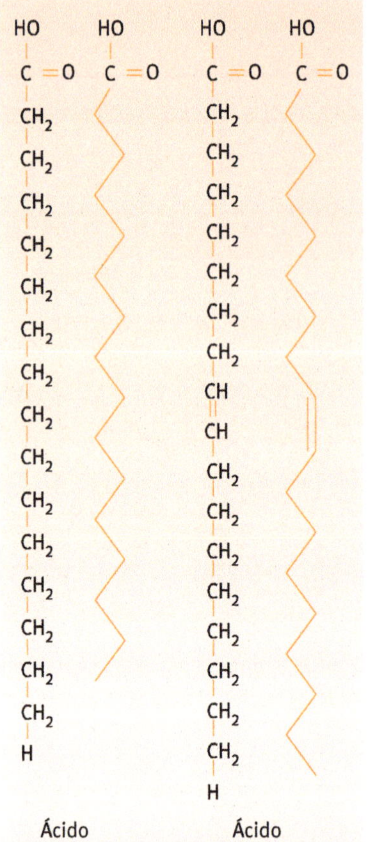

■ Moléculas de ácidos grasos; junto a la fórmula desarrollada, la representación esquemática habitual de este tipo de moléculas.

C

Los ácidos grasos son moléculas **anfipáticas**, presentan una zona hidrófila con afinidad por el agua, formado por el grupo carboxilo y una zona hidrófoba constituida por la cadena hidrocarbonada que es repelida por el agua.

+

Los ácidos grasos insaturados que tienen un único doble enlace se denominan **monoinsaturados**. Los que tienen más de uno son **poliinsaturados**.

Los ácidos grasos insaturados están presentes en los alimentos de origen vegetal y en el pescado, y son más adecuados para seguir una dieta sana que los saturados, que predominan en la carne.

4. Los lípidos

Son un grupo de biomoléculas orgánicas formadas por carbono, oxígeno e hidrógeno y, en ocasiones, otros elementos: fósforo, nitrógeno, azufre... Constituyen un grupo muy extenso de biomoléculas; no son solubles en agua sino en alcohol, acetona...

Se distinguen dos tipos de lípidos, los *lípidos complejos* y los *lípidos sencillos.*

4.1. *Lípidos complejos*

Son los lípidos que contienen *ácidos grasos,* sustancias formadas por una larga cadena de átomos de carbono con un grupo carboxilo en un extremo. Un ácido graso es **saturado** cuando todos los enlaces entre los carbonos son **sencillos.** En cambio, es **insaturado** cuando tiene algún **doble** enlace entre los carbonos. El *ácido palmítico* es saturado y el *oleico* es insaturado.

En general, los ácidos grasos tienen una cadena con un número par de átomos de carbono que oscila entre 14 y 22, aunque lo más frecuente es que tengan entre 16 y 18 átomos de carbono.

Clasificación y función. Se distinguen distintos tipos de lípidos complejos: los *triacilgliceroles,* los *glicerofosfolípidos*, los *esfingolípidos* y las *ceras.*

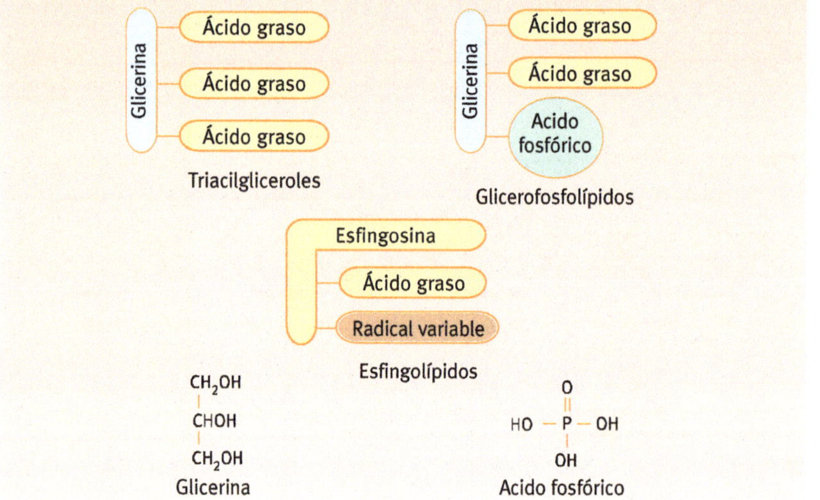

- **Triacilgliceroles.** Son compuestos en los que intervienen una molécula de *glicerina* y tres *ácidos grasos*. Cada uno de los tres grupos hidroxilos de la glicerina se halla unido a un ácido graso mediante **enlaces éster.** La esterificación es la reacción mediante la cual se unen el carbono de un grupo hidroxilo con el carbono de un grupo carboxilo con la pérdida de una molécula de agua.

  Estos compuestos constituyen el grupo de lípidos más abundante. Su misión es actuar como sustancias de **reserva energética,** ya que los ácidos grasos proporcionan mucha más energía que los monosacáridos, aproximadamente 9 kcal/g (algo más de 37 600 J/g).

Los triacilgliceroles que se presentan en estado sólido a temperatura ambiente se conocen con el nombre de *grasas,* mientras que los que lo hacen en estado líquido son *aceites.*

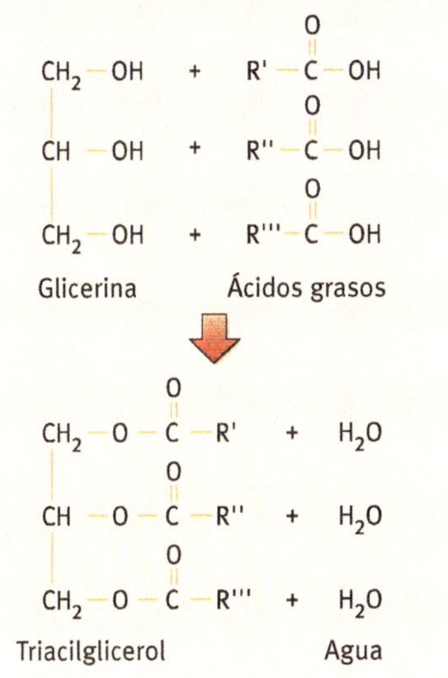

- **Glicerofosfolípidos.** Se diferencian de los triacilgliceroles en que uno de los grupos hidroxilo de la *glicerina* se encuentra unido a una molécula de *ácido fosfórico.* La única parte de la molécula que es soluble en agua es el ácido fosfórico.

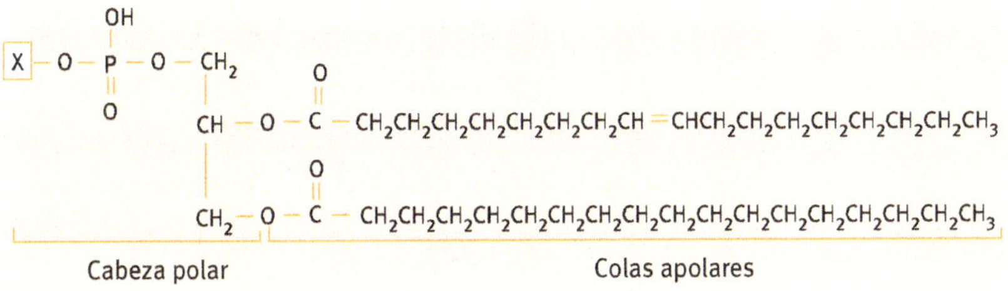

Por esta razón, al estar en contacto con el agua, las moléculas de glicerofosfolípidos se disponen con sus extremos polares hacia fuera, dejando en la parte interna los restos apolares. Son componentes fundamentales de las membranas celulares.

- **Esfingolípidos.** Están formados por una molécula de *esfingosina* (o un derivado de ella), un *ácido graso* y un *radical de cabeza polar.* Su principal función biológica es formar parte de las membranas celulares. Son especialmente abundantes en las células del tejido nervioso.
- **Ceras.** Son sustancias resultantes de la esterificación de *ácidos grasos de cadena larga* con *monoalcoholes* también *de cadena larga.* Impermeabilizan diversas partes y órganos de los vegetales (las hojas, los frutos) y de los animales (la piel, el pelo, las plumas...).

4.2. *Lípidos sencillos*

No contienen ácidos grasos. Tienen composición variada y funciones diversas.

- Muchas **sustancias aromáticas** de los vegetales son lípidos sencillos, es el caso del *mentol*, el *alcanfor* y el *geraniol.*
- Entre los lípidos sencillos también se encuentran sustancias muy importantes para la regulación de los procesos metabólicos: las **vitaminas,** como la *A*, la *D,* la *E* y la *K*, o las **hormonas** sexuales masculinas y femeninas, andrógenos y estrógenos, respectivamente.
- El **colesterol** pertenece también al grupo de lípidos sencillos; se caracteriza por tener un grupo alcohol y forma parte de las membranas celulares. Se fabrica en el hígado a partir de ácidos grasos. También se ingiere ya que forman parte de los tejidos grasos de los animales.

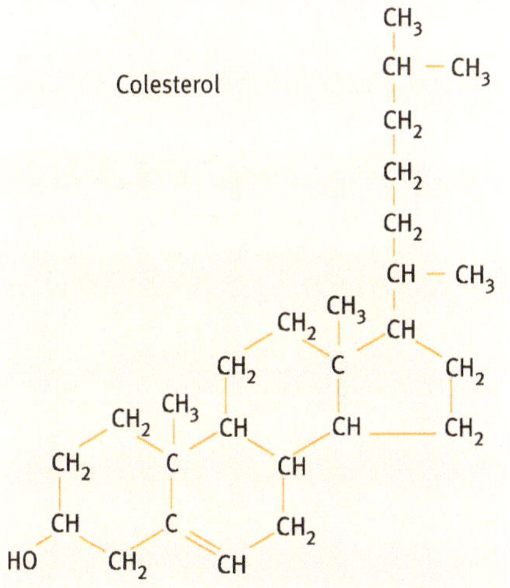

ACTIVIDADES

7. Compara las estructuras y las propiedades físicas de los triacilgliceroles y los glicerofosfolipidos.

8. Explica por qué los triacilgliceroles no pueden ser componentes significativos de las bicapas lípidicas.

+

El término **proteína** se suele utilizar para indicar estructuras tridimensionales de *miles* de aminoácidos.

Polipéptido indica compuestos de *muchos* aminoácidos.

Por esta razón, a nivel práctico, los términos *polipéptido* y *proteína* son intercambiables.

5. Las proteínas

Las proteínas son las biomoléculas orgánicas más abundantes en las células. Todas las proteínas contienen carbono, oxígeno, hidrógeno y nitrógeno; además, la mayoría contiene azufre y, algunas, fósforo, hierro, cinc y cobre.

Composición. Las proteínas son grandes moléculas formadas por la unión de subunidades más pequeñas llamadas **aminoácidos.**

Existen 20 aminoácidos diferentes y todos tienen una estructura básica idéntica: un grupo *amino*, un grupo *carboxilo* y un carbono central unido a un *radical* que varía de un aminoácido a otro.

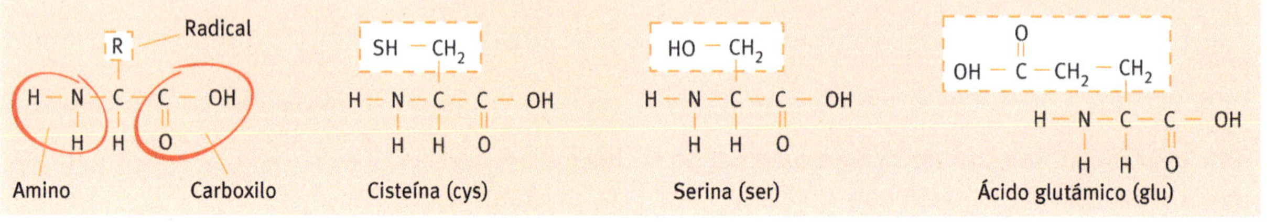

El enlace se produce entre el grupo carboxilo de un aminoácido y el amino del siguiente; esta unión libera una molécula de agua. Este enlace es covalente y se denomina **enlace peptídico.** Debido a ello, las moléculas formadas se pueden denominar también **polipéptidos**.

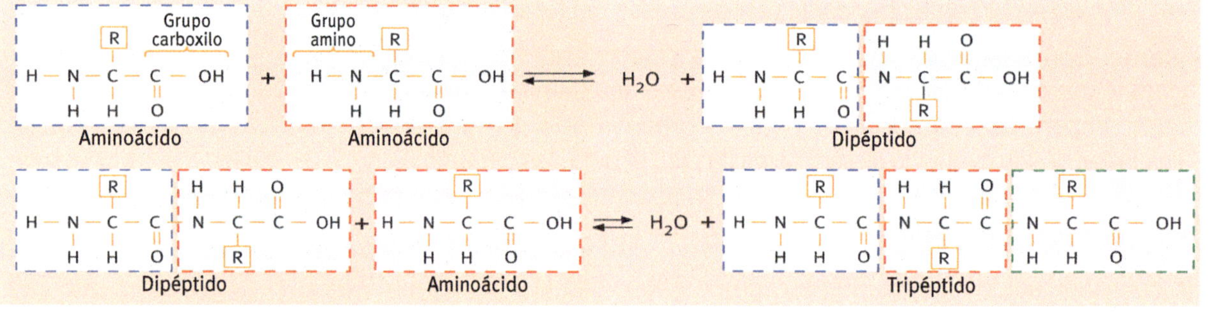

C

En los mamíferos, los **aminoácidos esenciales** son aquellos que no pueden ser sintetizados por las células y han de formar parte, necesariamente de la dieta.

Características. Las proteínas forman soluciones coloidales que pueden precipitar en coágulos, al añadir sustancias ácidas o básicas, o cuando se calientan; así sucede con la *albúmina* del huevo. Algunas pueden cristalizar, como el *citocromo,* que transporta electrones en las reacciones que se producen durante la respiración celular.

Clasificación. Las proteínas se pueden clasificar en dos grandes grupos: *proteínas simples* y *proteínas conjugadas.*

- Las **proteínas simples** u **holoproteínas** están formadas exclusivamente por cadenas de polipéptidos; por tanto, su hidrólisis produce únicamente aminoácidos. Entre las holoproteínas más conocidas están las del grupo de las albúminas.
- Las **proteínas conjugadas** o **heteroproteínas** están formadas por cadenas de péptidos unidas a otro tipo de compuestos que reciben el nombre de grupo prostético. Si el grupo prostético es un glúcido, la heteroproteína se denomina glucoproteína; si es una sustancia lipídica recibe el nombre de lipoproteína.

+

Muchas macromoléculas producen **soluciones coloidales**, es decir, se dispersan en un medio acuoso en forma de partículas de tamaño comprendido entre 1 nm y 1 μm.

En la organización en el espacio de una proteína podemos distinguir cuatro niveles o estructuras que son sucesivamente más complejos. Estas estructuras son:

- **La estructura primaria.** Es la secuencia de aminoácidos que se suceden en la cadena, uno a continuación de otro. Esta estructura determina la estructura tridimensional de la proteína. Se representa de la siguiente manera:

  Ala–Cys–Leu–Val–Lys–Ser

- **La estructura secundaria.** Se da cuando la cadena de aminoácidos se pliega sobre sí misma, se establecen puentes de hidrógeno en diferentes partes de la molécula y ésta adquiere una estructura tridimensional. Existen dos tipos de estructura secundaria: la *hélice* α y la *conformación* β.

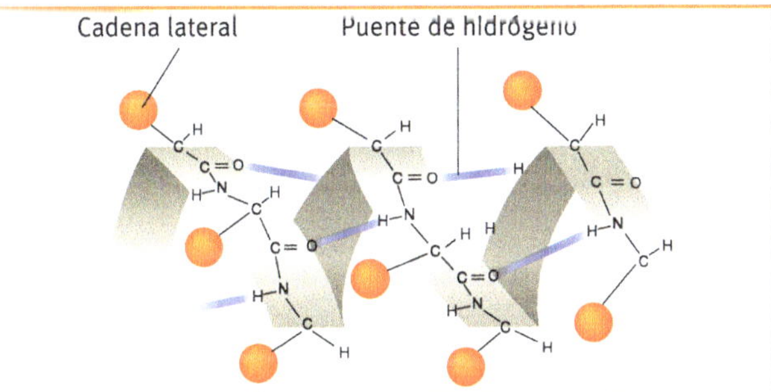

La **hélice** α es una estructura helicoidal característica de las proteínas que forman estructuras resistentes. Un ejemplo es la queratina, proteína que se encuentra en el cabello, las uñas y las plumas.

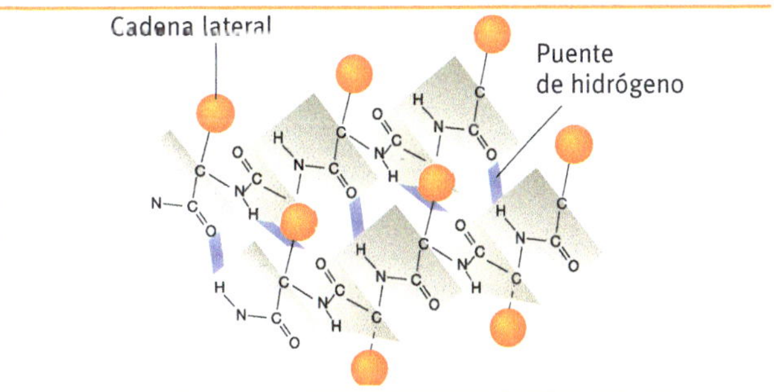

La **conformación** β es una estructura plana que se pliega en forma de zigzag, y es característica de las proteínas que forman filamentos suaves y flexibles. Un ejemplo es la fibroína, que se encuentra en la seda.

- La **estructura terciaria.** Se da en aquellas proteínas en que la estructura secundaria se pliega sobre sí misma. Los enlaces más importantes que mantienen la estructura terciaria son los puentes de hidrógeno entre cadenas laterales y los puentes disulfuro entre aquellas zonas de la proteína en las que existen átomos de azufre. Un ejemplo es la mioglobina en los músculos de los vertebrados

- La **estructura cuaternaria.** Está constituida por varias cadenas polipeptídicas que se unen mediante enlaces no covalentes, para formar una gran proteína. Un ejemplo es la hemoglobina que contiene los eritrocitos de la sangre.

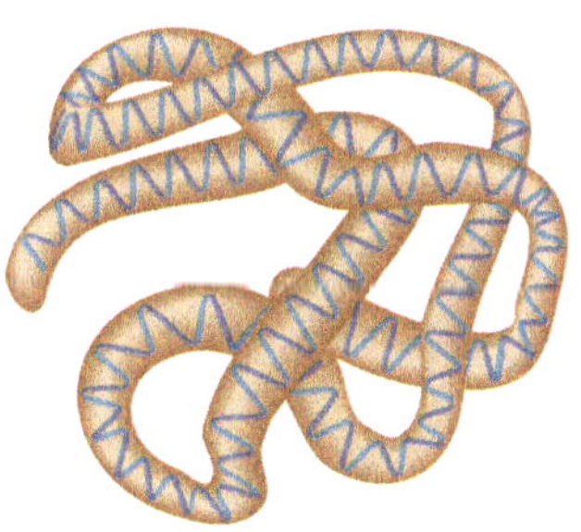

■ Estructura terciaria

■ Estructura cuaternaria

C

Los enzimas son un grupo de proteínas cuya función es controlar la velocidad de las reacciones químicas que tienen lugar en los seres vivos; es decir, que tienen función **catalizadora.**

La presencia de un enzima determinado puede aumentar o retardar la velocidad de una reacción química. Algunos son imprescindibles para que tengan lugar ciertas transformaciones.

Por ejemplo, la **glucógeno-sintasa** es un enzima que hace posible la síntesis de glucógeno a partir de glucosas.

La **glucógeno-fosforilasa** provoca la reacción contraria y hace que el glucógeno se descomponga en glucosa.

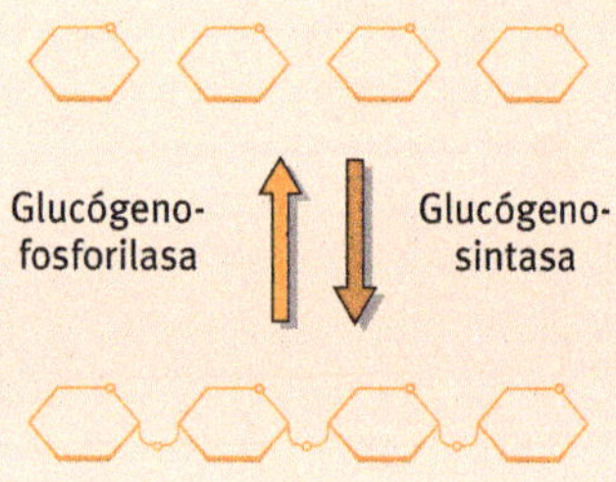

Cuando las proteínas son sometidas a la acción del calor o a valores de pH extremos, pierden su configuración tridimensional y, por tanto, sus propiedades físicas y sus funciones biológicas. Este proceso se conoce con el nombre de **desnaturalización** de la proteína.

Las posibilidades de combinación en cuanto al número y tipo de aminoácidos que se unen en las cadenas son muy numerosas. De ahí la gran variedad de funciones que desempeñan las proteínas.

FUNCIONES	EJEMPLOS
Estructural	Colágeno que forma los huesos y los tendones; queratina del pelo, las uñas y las plumas.
De reserva	Ovoalbúmina de la clara de huevo; caseína de la leche.
De regulación	Hormonas como la del crecimiento o la insulina.
De control metabólico	Enzimas como la glucógeno-sintasa.
Defensiva	Anticuerpos para combatir las infecciones.
Transportadora	Hemoglobina de la sangre.
Contráctil	Miosina de los músculos.

Cada proteína tiene una secuencia de aminoácidos concreta. De ella depende que la molécula se pliegue correctamente, es decir, que adquiera su correcta **conformación.**

Cualquier error en la posición de los aminoácidos puede provocar que la proteína no se pliegue correctamente y, por tanto, que no tenga la estructura tridimensional que le permite realizar su función. Esto puede alterar el funcionamiento de todo el organismo. Por este motivo, el análisis de la secuencia de aminoácidos puede ayudar en el desarrollo de pruebas diagnósticas y terapias eficaces. Así por ejemplo, el cambio de un aminoácido por otro en la molécula de hemoglobina provoca la anemia falciforme.

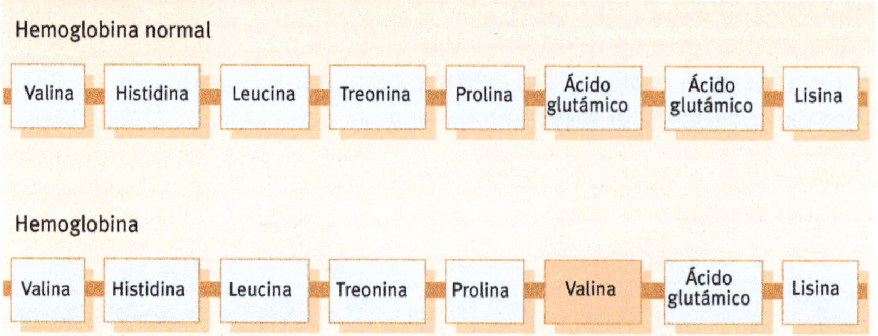

En la **anemia falciforme** los glóbulos rojos están deformados porque el aminoácido cambiado respecto a la hemoglobina normal hace que la molécula se pliegue de manera incorrecta. Los glóbulos rojos son más frágiles y se rompen con facilidad, lo que provoca la anemia.

6. Los ácidos nucleicos

Al igual que los polisacáridos, los lípidos y las proteínas, los ácidos nucleicos son moléculas de gran tamaño formadas por la unión de subunidades menores, es decir, son polímeros.

Composición. Los monómeros de los ácidos nucleicos se denominan **nucleótidos.** En la estructura de un nucleótido intervienen los siguientes compuestos:

Base nitrogenada

Pentosa

Nucleósido

Ácido fosfórico

NUCLEÓTIDO

DOBLE CADENA DE ADN

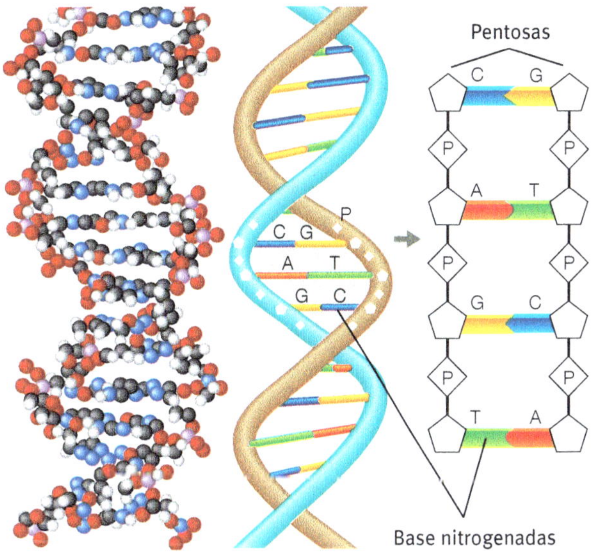

Clasificación. Existen dos tipos de ácidos nucleicos, el **ácido desoxirribonucleico** o **ADN** y el **ácido ribonucleico** o **ARN.** Se diferencian en la pentosa y en el tipo de nucleótidos. Existen cuatro nucleótidos componentes del ADN y reciben el nombre de desoxirribonucleótidos.

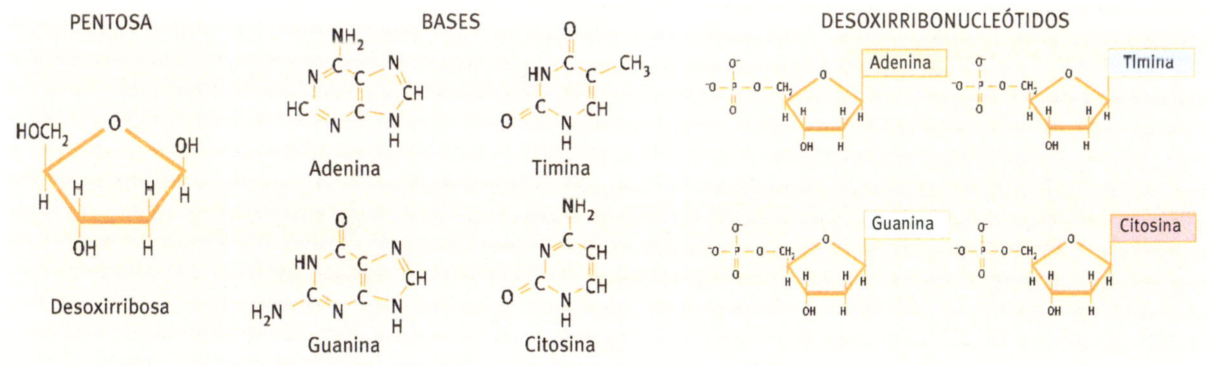

En el caso del ARN existen cuatro **ribonucleótidos.**

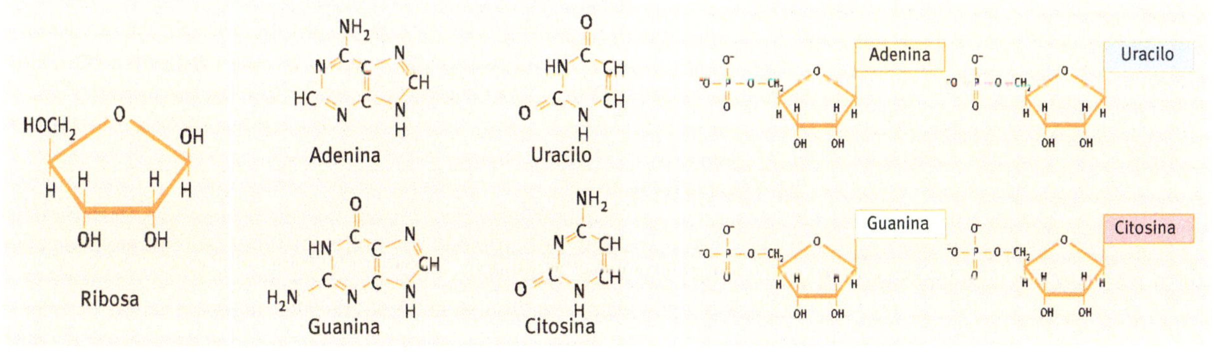

Los nucleótidos se unen uno tras otro mediante **enlaces fosfodiéster,** constituyendo largas cadenas que, en ocasiones, pueden ser dobles y arrolladas en espiral.

Función. El ADN forma los cromosomas de los seres vivos y contiene la información necesaria para la síntesis de todas las proteínas que ha de sintetizar la célula. A continuación, descubriremos de modo resumido la función del ADN.

En las células eucariotas, el ADN no sale nunca del núcleo y, sin embargo, la síntesis de proteínas tiene lugar en el citoplasma celular. Por tanto, se produce una transferencia de información desde el núcleo hasta el citoplasma interviene el ARN mediante dos procesos:

— **Transcripción:** el ARN se sintetiza en el núcleo de la célula como pequeñas réplicas del ADN y sale al citoplasma.

— **Traducción:** en el citoplasma el ARN se asocia a un ribosoma y en este orgánulo se produce la unión de los aminoácidos.

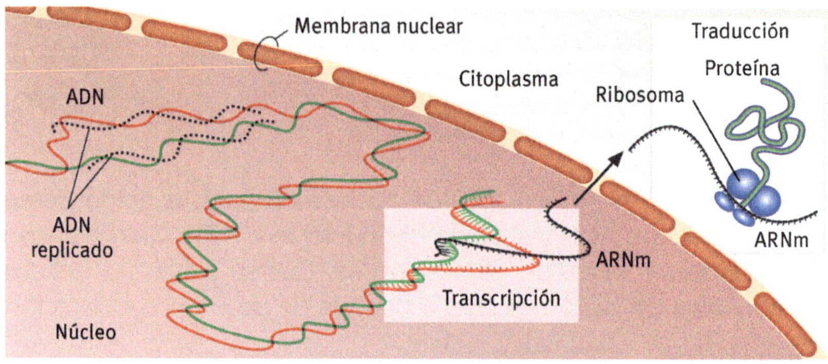

La unión se realiza interpretando el **código genético,** según el cual cada grupo de tres bases nitrogenadas (tripletes) del ARN determina la unión de un aminoácido. A continuación puedes ver algunos ejemplos de tripletes y el aminoácido que les corresponde.

G	C	A	Alanina (Ala)	G	G	A	Glicina (Gly)
A	G	A	Arginina (Arg)	C	A	U	Histidina (His)
U	G	U	Cisteína (Cys)	U	U	A	Leucina (Leu)
A	U	G	Metionina (Met)	U	U	U	Fenilalanina (Phe)

Así, la molécula de ARN se va «leyendo» y se va colocando un aminoácido tras otro en la cadena polipeptídica en formación, tal como indica la secuencia de nucleótidos del ARN. Los ácidos nucleicos determinan la posición de cada aminoácido en todas las proteínas.

+

El español **Severo Ochoa** (1905-1993) fue el científico que descifró el código genético. Por sus trabajos recibió el premio Nobel en 1959.

La información que contiene el ADN es de vital importancia para la vida de un individuo, pero también para los miembros de una especie en conjunto. La transmisión de esta información de una generación a la siguiente se garantiza durante la reproducción, ya que pasa a los descendientes una copia de cada cromosoma.

ACTIVIDADES

9. ¿De qué depende la función de una proteína? Describe brevemente las funciones que se les atribuyen.

10. Responde las preguntas siguientes sobre esta cadena de ácido nucleico:

................AAGGCCTTAGACG.............

— ¿Es un fragmento de ADN o de ARN? Justifica la respuesta.

— Escribe la cadena complementaria.

7. El origen de la vida y su evolución

Las biomoléculas que acabamos de describir se encuentran en todos los seres vivos. La morfología, la organización y las funciones de un ser vivo dependen del ensamblaje y la actividad de glúcidos, lípidos, proteínas y ácidos nucleicos. Debido a este principio universal, decimos que los seres vivos presentan **unidad química.**

Al ser uniforme la composición química de todos los seres vivos, es lógico pensar que tienen un **origen común,** es decir, que todos ellos proceden de un grupo de organismos de estructura muy sencilla, a partir de los cuales han evolucionado las distintas formas de vida.

El origen de la Tierra se remonta a 4600 millones de años. La evidencia fósil de vida más temprana conocida tiene unos 3500 millones de años, como prueban los estromatolitos de Warrawoona, en Australia. Se trata de estructuras fósiles formadas probablemente por microorganismos semejantes a las cianobacterias y por carbonato cálcico, un producto de su actividad metabólica.

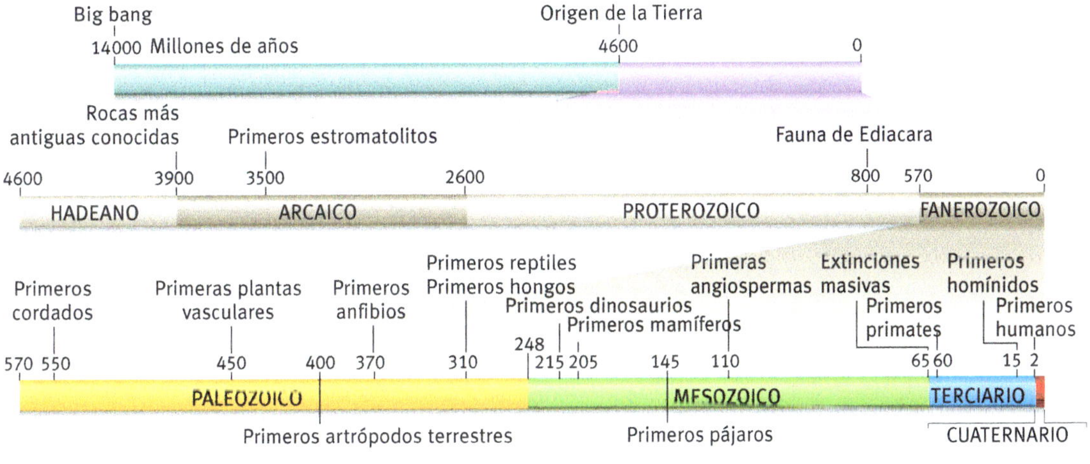

Es imposible descubrir exactamente cómo surgió la vida, ya que los estudios paleontológicos y de laboratorio únicamente proporcionan indicios sobre el origen de la vida. Al existir una falta de información directa, los científicos proponen muchas teorías, a veces opuestas. La tarea del científico, en este caso, consiste en demostrar que lo que se argumenta en su teoría pudo ser posible.

El punto de partida es la consideración de que, para la formación y actividad de una célula, es preciso:

- El ensamblaje de las biomoléculas necesarias para constituir las membranas y otros orgánulos celulares (lípidos y proteínas).
- La presencia de sustancias que proporcionen la energía necesaria para las actividades celulares (glúcidos).
- La existencia de moléculas que almacenen información y puedan autocopiarse para transmitirla de una generación a la siguiente (ácidos nucleicos).

Por tanto, la primera gran cuestión que se ha de resolver es: ¿cómo se originaron las biomoléculas?

Los estromatolitos son formaciones sedimentarias de tipo calcáreo que se originan por la actividad de seres unicelulares. Existen estromatolitos fósiles con más de 3500 millones de años de antigüedad.

Los científicos suponen que todas se formaron a partir de los elementos químicos y de las moléculas inorgánicas que existían en la Tierra en esas fases tempranas de su existencia: la **era prebiótica**.

El escenario en que hizo su aparición la vida es motivo de controversia, pero parece seguro que, una vez que se formó la superficie sólida del planeta, se produjeron numerosas e intensas erupciones volcánicas que emitían a la atmósfera gases diversos y simples, sobre todo vapor de agua, nitrógeno y dióxido de carbono, así como cantidades menores de amoniaco, metano y sulfuro de hidrogeno.

La mayoría de los científicos acepta que así se formó una atmósfera reductora, es decir, con muy poco oxígeno libre. Cuando la temperatura fue inferior al punto de ebullición del agua, el vapor se condensó y formó grandes océanos, que disolvieron parte de las sustancias que formaban la atmósfera, generando una sopa prebiótica. Éste es el punto de partida de la primera teoría sobre el origen de la vida. Se debe al científico británico J. B. S. Haldane y al ruso A. I. Oparin, que dieron a conocer sus ideas en 1929 y 1934, respectivamente.

Condiciones ambientales de la Tierra primitiva. La atmósfera estaba compuesta, principalmente, por gases fruto de la elevada actividad volcánica.

Su teoría propone que las sustancias que componían la atmósfera y los océanos eran los ingredientes necesarios para la síntesis de las primeras moléculas orgánicas simples relacionadas con la vida. Para que esto fuera posible debían darse dos condiciones:

- La **ausencia de oxígeno libre.** Este gas, tan necesario para la mayoría de las formas de vida actuales, tendría efectos destructores en las fases iniciales de la vida, ya que oxidaría cualquier inicio de molécula orgánica, destruyéndola.
- Un **aporte de energía.** Si las biomoléculas se produjeron a partir de los compuestos inorgánicos que formaban la atmósfera y los océanos primitivos, se necesitaba energía, ya que las uniones entre los átomos de carbono tienen más energía de enlace que los de las sustancias iniciales.

  La ausencia de oxígeno supone la no existencia de capa de ozono; por ello, la superficie de la Tierra recibía gran cantidad de energía en forma de radiaciones ultravioletas del Sol, además de descargas eléctricas en forma de relámpagos, calor procedente de la desintegración de elementos radiactivos, viento, etc.

Así pues, según Haldane y Oparin, las reacciones químicas que produjeron las primeras moléculas orgánicas fueron:

$$\text{Compuestos inorgánicos sencillos } (CH_4 + NH_3 + H_2S + ...) \xrightarrow{E} \text{Monosacáridos, ácidos grasos, aminoácidos...}$$

Estas ideas no prosperaron entre los científicos de la época, porque les recordaban demasiado a la teoría de la generación espontánea, superada desde hacía años. Pero, en 1953, H. C. Urey y S. L. Miller, basándose en las teorías de Haldane y Oparin, realizaron un experimento muy importante en la historia de la biología: simularon experimentalmente que el proceso supuesto por estos científicos era posible. Para ello idearon el siguiente montaje, mediante el que intentaban reproducir las condiciones de la Tierra en estas fases iniciales.

+

Durante siglos, se creyó que de la materia inerte podían surgir seres vivos. Esta teoría, denominada de la **generación espontánea**, sostenía, por ejemplo, que la carne se convertía en gusanos o que el brillo del rocío de la mañana engendraba luciérnagas. Louis Pasteur, en 1874, descartó definitivamente estas creencias.

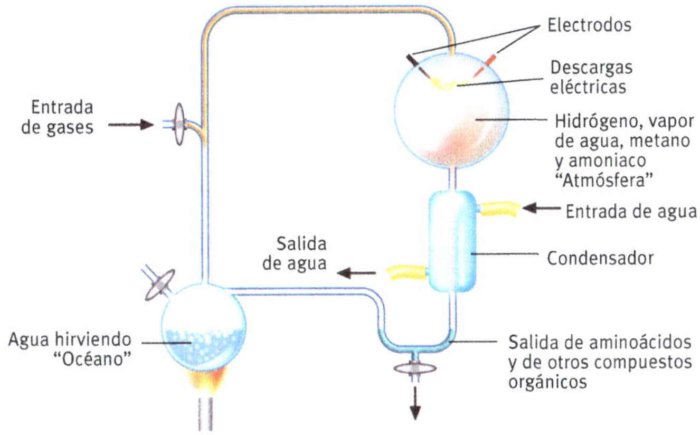

Tal como muestra el dibujo, aplicaron descargas eléctricas (que simulaban los rayos) a una mezcla de gases que constituían los compuestos fundamentales de la atmósfera primitiva: metano, amoniaco, vapor de agua e hidrógeno. Esta mezcla de gases estaba contenida en un recipiente cerrado a 80 °C.

El montaje se mantuvo en funcionamiento durante una semana. Transcurrido este tiempo analizaron los diferentes compartimentos y encontraron diferentes compuestos orgánicos solubles en agua: varios aminoácidos y otros compuestos relacionados con el metabolismo de los seres vivos. En experimentos posteriores se comprobó la síntesis del uracilo y la citosina.

De ahí concluyeron que si las condiciones de la Tierra eran realmente las que ellos habían conseguido reproducir en sus experiencias, la síntesis de compuestos orgánicos a partir de elementos y compuestos inorgánicos sencillos era posible.

Durante el siglo XX se realizaron muchísimos experimentos para originar componentes fundamentales de la materia viva. Uno de los más remarcables fue el realizado por Joan Oró en 1960. A partir de cianuro de hidrógeno (HCN), constituyente principal de las nubes interestelares y de los núcleos de los cometas, se produce la síntesis en el laboratorio de adenina, compuesto orgánico de gran importancia en la transmisión de la información genética.

Además de los interrogantes sobre cómo era la Tierra hace 4600 millones de años, existen muchas otras cuestiones por resolver.

¿Cómo se forman polímeros a partir de monómeros?

- La formación de polímeros debía superar una dificultad: si las reacciones de síntesis de polímeros tuvieron lugar en un medio acuoso, al ser las reacciones reversibles, la presencia de agua favorecía la hidrólisis en lugar de la síntesis.

  Una posible solución a esta dificultad es que existieran mareas que, al retirarse, dejaran charcas arcillosas aisladas. En ellas se produciría evaporación del agua y se debieron de **concentrar los monómeros,** que vieron de esta forma facilitada su polimerización. Así, S. W. Fox demostró que mezclas secas de aminoácidos puros se polimerizan espontáneamente, a temperaturas de unos 130 °C.

C

Stanley L. Miller (1930-2007) químico y biólogo estadounidense conocido por sus estudios sobre el origen de la vida. Sus estudios iniciaron una división nueva de la biología: la exobiología.

La exobiología es la rama de las ciencias biológicas que se ocupa del estudio de la existencia de vida en otros lugares del universo aparte de la Tierra. Relacionada con la astronomía, la cosmología, la biología, la astrofísica, etc., busca responder cuestiones fundamentales: ¿Qué es la vida? ¿Cómo evoluciona? ¿Hay vida en otros lugares del universo?

Miller fue profesor de Química en la Universidad de California a partir de 1958. Posteriormente fue director de un grupo de investigación del centro especializado en exobiología de la NASA. Obtuvo numerosos reconocimientos a su trabajo, entre ellos la Medalla Oparin.

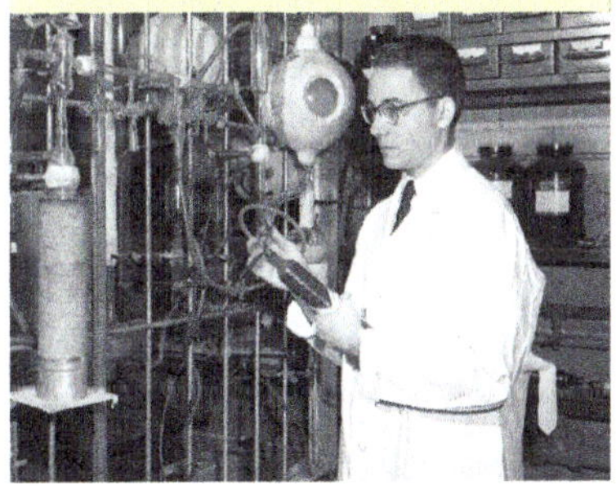

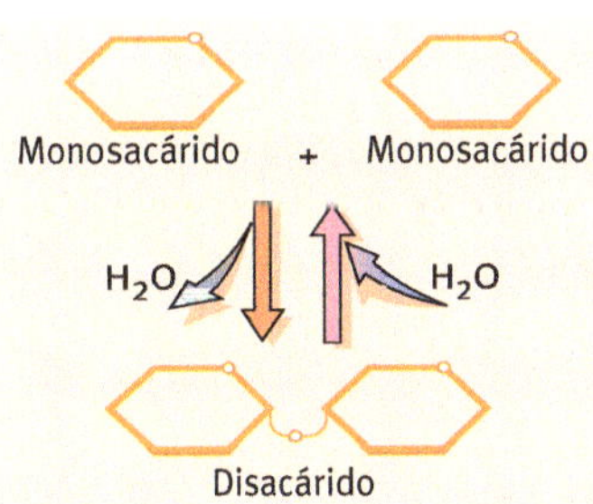

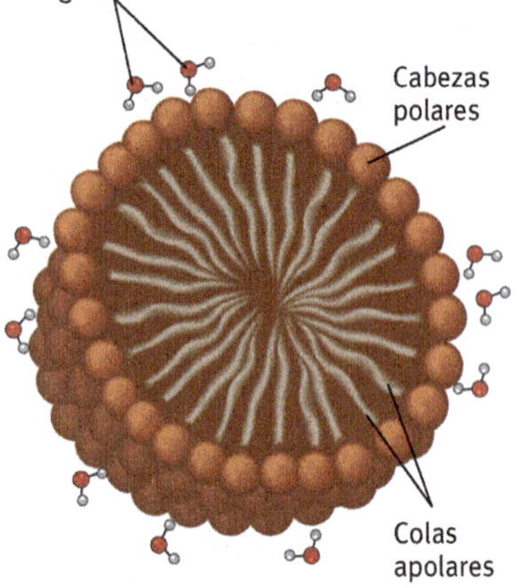

En las micelas las cabezas polares de sus componentes se disponen hacia fuera y las colas apolares hacia dentro.

¿Cómo sucede un aislamiento entre un medio externo y uno interno?

- El **aislamiento** del medio exterior proporciona cierta protección frente a los efectos adversos del entorno, permitiendo que en el medio interno se mantengan concentraciones elevadas de componentes que de otra manera se difundirían y dispersarían. Las sustancias, cuando se hallan en mayor concentración pueden reaccionar más rápido y aumentar su eficacia en la producción de polímeros y otros tipos de reacciones químicas.

  En sus experiencias, Oparin y Fox observaron que cuando se mezclan con agua, algunos polímeros forman unas pequeñas gotitas (de hasta 500 micras de diámetro), denominadas **coacervatos** o **micelas.** Estas microesferas sirven para imaginar el comportamiento de gotitas similares que pudieron haberse producido.
- Actualmente las teorías se dirigen a la búsqueda de moléculas que combinan en su estructura una parte hidrofóbica y una parte hidrofílica. Estas moléculas en contacto con el agua forman **vesículas** membranosas (sacos llenos de liquido) que crecen y se dividen.

¿Cómo se originó la capacidad de reproducción?

- Este punto es aún confuso. Si las vesículas contuvieran alguna molécula con **capacidad informativa,** estaríamos ante el precursor de una célula. La capacidad reproductiva debió de desarrollarse **paralelamente** a la capacidad de **sintetizar sustancias** y la de **obtener energía** para la construcción de las propias estructuras.

El flujo de información genética tal como funciona hoy día es el paso de ADN a ARN y de éste a proteínas. Y aunque se necesita ADN para producir proteínas también necesitamos para transcribir, traducir, replicar, etc. Por tanto, ¿qué macromolécula fue primero: las proteínas o el ADN?

Según diversas hipótesis, el ARN surgió antes que el ADN y las proteínas en las primeras etapas de la evolución. El ARN controlaría la traducción de proteínas, y el ADN sería posterior. Así se intenta explicar de qué modo se obtuvieron las que se consideran las primeras células: seres similares a las células procariotas.

Sin embargo, estas teorías sólo son respuestas provisionales. Con toda seguridad, el desarrollo de nuevas líneas de investigación puede modificar cualquiera de los planteamientos anteriores y posiblemente nos acerquemos, cada vez más, al conocimiento del origen de la vida.

Más tarde aparecieron las primeras células eucariotas. Según parece, este proceso fue mucho más complicado y requirió unos 2000 millones de años. Por tanto, es posible que a partir de un organismo ancestral la vida celular haya evolucionado en tres grandes dominios, **Bacteria** y **Archaea**, que incluyen microorganismos unicelulares procariotas (sin núcleo), y **Eucarya**, que incluye seres unicelulares y pluricelulares eucariotas (con núcleo).

La aparición de seres pluricelulares y la diferenciación en especies es un proceso mucho mejor conocido, ya que los fósiles y los conocimientos sobre genética permiten reproducir la historia de la vida hasta hoy.

11 Los investigadores han estudiado los componentes esenciales de los seres vivos, encontrando un conjunto de características comunes. Lee el siguiente artículo sobre las características que se consideran esenciales en el desarrollo y evolución de los seres vivos y responde las cuestiones que tienes a continuación:

Los cuatro ingredientes de la vida

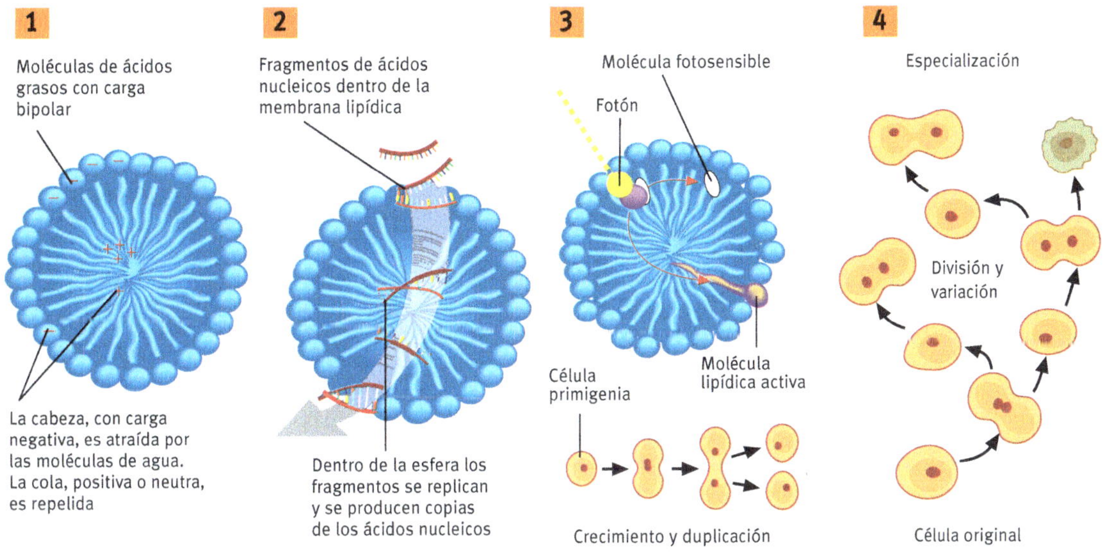

1. Un contenedor

La primera forma de vida –o su precursor– tuvo que aislarse del exterior en una burbuja de ácidos grasos. Éstos tienden a formar vesículas de forma espontánea debido a que son moléculas bipolares, con una cabeza hidrófila –amante del agua– y una cola hidrófoba.

2. La herencia

Debía poseer información genética, es decir, instrucciones sobre cómo operar y reproducirse, y contar con algún modo de replicarse y ejecutar las instrucciones. El interior de las burbujas lipídicas facilitaba la integridad y la movilidad del ADN o del ARN primordial.

3. El metabolismo

Su primer alimento pudieron ser precursores de ácidos grasos con moléculas fotosensibles unidas a su cabeza, lo que los hacía solubles. Así penetraban en la burbuja; una vez dentro, la luz rompía el capuchón fotosensible para dejar al descubierto su cabeza con carga negativa.

4. La evolución

Las copias del material genético que hacía de sí mismo no podían ser completamente idénticas. Debía permitir una cierta variabilidad aleatoria de la información para que pudiera ponerse en marcha la denominada evolución darwiniana.

Texto e imágenes adaptados de E. Coperías. Milagro en el mar. Muy Interesante. 2005

- Observa los dibujos y resume la idea principal del texto que acompaña a cada imagen. A continuación, ordena cronológicamente los siguientes procesos: especialización, replicación, metabolismo, formación de burbujas lipídicas.
- ¿Qué condiciones iniciales previas deberían existir para que se formasen las vesículas? Justifica la respuesta.
- ¿Por qué las vesículas no podían presentar barreras que las aislasen totalmente de su entorno?
- Observa la figura 3. ¿Qué fuente de energía utilizaban las primeras formas de vida? ¿Cómo obtienen en la actualidad los seres vivos la energía para realizar sus procesos metabólicos?
- A partir de los conocimientos que se han expuesto en esta unidad, reflexiona por qué el ADN ha sustituido al ARN como material genético en la mayoría de los seres vivos.
- Lee atentamente el texto de la figura 4. ¿Por qué los sistemas que se replican deben sufrir cambios?
- Realiza una búsqueda de información de la teoría endosimbiótica postulada por Lynn Margulis que explica la aparición de la célula eucariota y haz un breve resumen de la idea principal.

12 Por qué el agua forma gotas casi esféricas en la superficie de un coche recién encerado?

13 ¿Crees que podría existir la vida sin agua? Justifica tu respuesta.

14 Observa el gráfico adjunto. Muestra la evolución de la concentración de la glucosa en diversos momentos después de haber ingerido un alimento.

A partir de la información que suministra el gráfico:

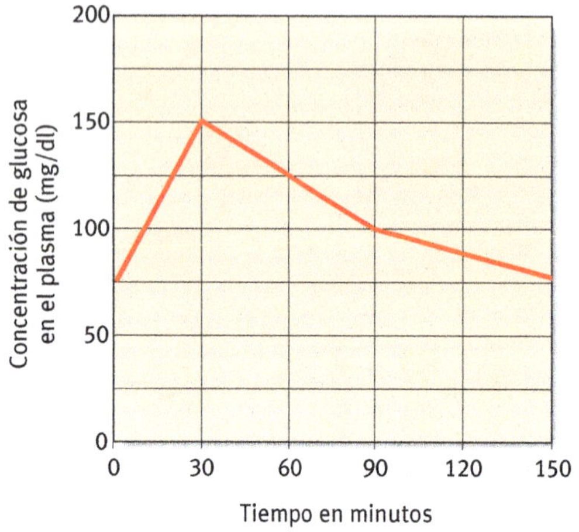

— Completa la tabla siguiente:

Tiempo	minuto 0	minuto 30	minuto 90	minuto 150
Concentación de glucosa				

— Justifica el aumento y descenso de la glucosa en el plasma después de la ingestión del alimento.

15 La celulosa está presente en las dietas de adelgazamiento. No obstante, las vacas ganan mucho peso cuando pastan. ¿Cómo justificarías estos hechos?

16 Observa la siguiente reacción:

$$CH_3-(CH_2)_{14}-\overset{\overset{\displaystyle O}{||}}{C}-OH + NaOH$$

ácido palmítico

$$CH_3-(CH_2)_{14}-\overset{\overset{\displaystyle O}{||}}{C}-O\,Na + H_2O$$

palmitato sódico (jabón)

Contesta las siguientes preguntas:

— ¿Qué tipo de biomolécula orgánica es el ácido palmítico?

— Explica qué átomos se intercambian en la reacción.

— ¿Por qué el jabón se considera una molécula amfipática? Justifica la respuesta.

— Busca información sobre la elaboración del jabón artesanal. Describe los ingredientes que se utilizaban, y el proceso a seguir.

17 ¿Es correcta esta afirmación "la ingesta de grasas produce un gran incremento de peso"? Justifica la respuesta.

18 Identifica la fuente más probable (CH_4, NH_3, H_2O o H_2) de cada átomo del aminoácido alanina generado en el experimento de Urey-Miller.

TIC

19 Busca en Internet la información de cesárea y resuelve las cuestiones:

— Busca la estructura química de los siguientes aminoácidos: cisteína, serina y glutamina.

— Construye el tripéptido formado por los tres aminoácidos.

— Escribe una estructura primaria de una proteína formada por diez aminoácidos combinando solamente estos tres aminoácidos.

20 ¿Por qué los enzimas pierden su actividad cuando se les somete a altas temperaturas?

21 Completa el siguiente cuadro:

Tipo de estructura	Posibilidades	Enlaces implicados
	casi infinitas	
estructura secundaria	helice α	
		puentes disulfuro
		enlaces no covalente

22 Los virus pueden tener como material genético: ADN de cadena sencilla o doble, o ARN de cadena sencilla o doble.

En la siguiente tabla se indica la composición de bases nitrogenadas de cuatro virus:

	Adenina	Guanina	Citosina	Timina	Uracilo
Virus 1	30,9	19,9	19,8	29,4	----
Virus 2	23,3	21,1	19,8	35,8	----
Virus 3	23,1	20,2	29,1	----	34,6
Virus 4	30,8	18,6	18,6	----	31,7

— Indica y justifica qué tipo de ácido nucleico tiene cada uno de los virus

S

23 Observa las moléculas de la figura y contesta las siguientes cuestiones:

A

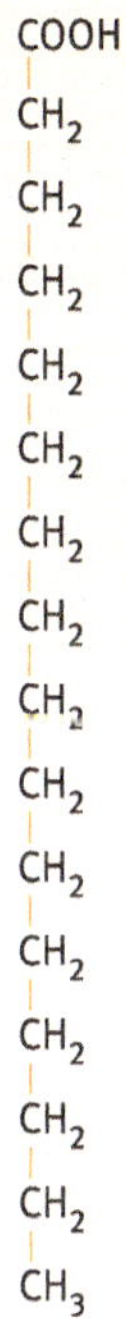

B

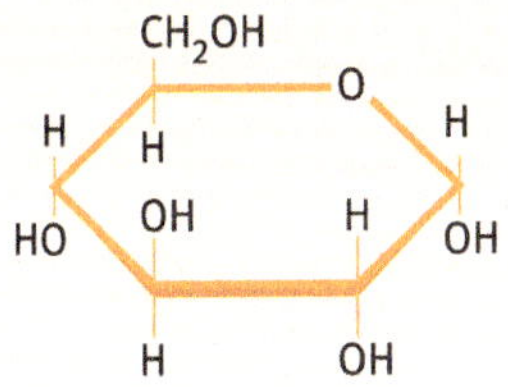

C

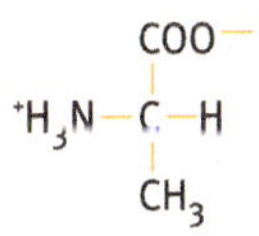

D

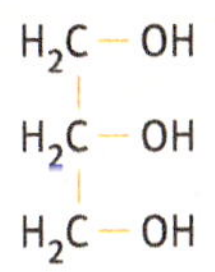

E

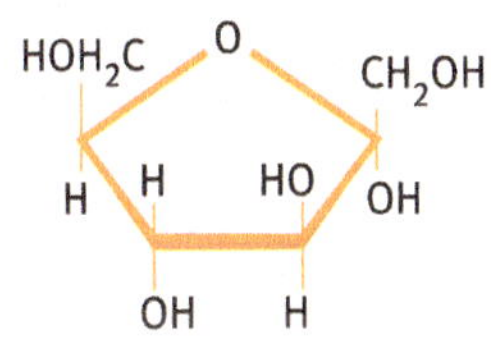

F

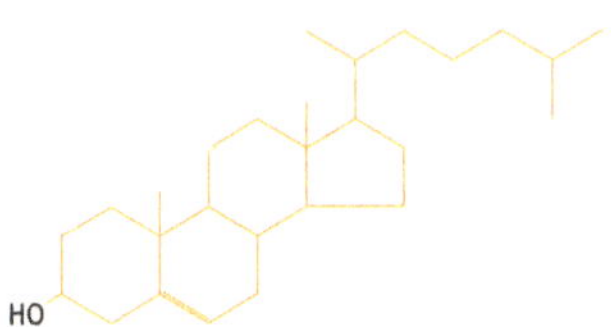

G

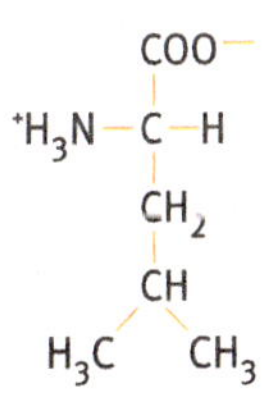

— De las moléculas de la figura, ¿cuáles son monosacáridos?

— Escribe la fórmula resultante de su unión. ¿Qué tipo de biomolécula se forma?

— ¿Cómo se denomina este enlace?

— ¿Cuál o cuáles moléculas de la figura son ácidos grasos?

— ¿Con cuál de las moléculas de la figura se combinan los ácidos grasos para almacenar energía en la célula? Nombra este compuesto y escribe su formula general.

— ¿Qué tienen en común las moléculas A y F? Explícalo.

— ¿Cuál o cuáles moléculas de la figura son aminoácidos?

— Escribe la fórmula resultante de su unión. Indica cómo se denomina este enlace y nombra el compuesto resultante.

24 Si comparamos las estructuras primarias de la insulina de diversos mamíferos, observamos que existe una secuencia pequeña que las diferencia. Observa las estructuras primarias de las insulina de buey y de cordero.

Insulina de buey: ala-ser-val

Insulina de cordero: ala-gly-val

— Explica qué significa “estructura primaria de la insulina”. Explica la estructura general de sus monómeros.

— ¿Qué relación existe entre las secuencias de la insulina y del ADN?

25 Explica la importancia del experimento que realizaron Miller y Urey. Justifica si este experimento demuestra la aparición de la primera célula a partir de moléculas orgánicas.

TIC

26 Abre la dirección de Internet: *recursos.cnice.mec.es/biosfera/alumno/2bachillerato/biomol/contenidos.htm*

Empleando la información que te ofrecen las ventanas, contesta las actividades siguientes:

— Indica qué procedimientos se utilizan para extraer las biomoléculas de los seres vivos.

— Busca el contenido que hace referencia al agua: describe la unión entre dos moléculas de agua e indica qué tipo de enlace se forma.

— Indica 4 propiedades del agua y relaciónalas con las funciones de los seres vivos.

— Observa la animación de la síntesis de la maltosa y de la lactosa, y escribe las reacciones de síntesis que aparecen.

— Indica en qué grupo de biomoléculas se encuentra el colesterol y qué otros compuestos pertenecen a este grupo.

— Observa y explica qué le ocurre al aminoácido alanina cuando cambia el pH del medio.

— Busca en algún diccionario especializado la definición de renaturalización.

— Accede a la página de enzimas. Observa la animación y explica cómo interviene un enzima en una reacción.

— El ADN presenta cuatro niveles de estructuración diferentes. Descríbelos brevemente.

— En la célula aparecen cuatro tipos de ARN. Confecciona una tabla en la cual aparezca el tipo, la función y la localización.

7

Células y virus

Todos los seres vivos están formados por células. En los organismos unicelulares se trata de una única célula que realiza todas las funciones. En el caso de los organismos pluricelulares se trata de un conjunto más o menos numeroso de células coordinadas ente sí.

La célula es la unidad estructural y funcional de la vida y su estudio es primordial para comprender el funcionamiento de cualquier ser vivo. No obstante, su conocimiento no se consolidó hasta mediados del siglo XIX gracias sobretodo a la utilización del microscopio óptico como método de observación.

Por otro lado, los virus no presentan estructuras celulares; ello junto con su incapacidad para sobrevivir fuera de una célula, ha provocado que, desde su descubrimiento, se pusiera en entredicho su clasificación como ser vivo.

Paramecio al microscopio electrónico de barrido. 335 x

ORGANIZACIÓN DE LOS CONTENIDOS

1. La observación al microscopio

2. Las membranas celulares

2.1. Estudio de la membrana plasmática

3. La célula procariota: las bacterias

4. La célula eucariota

4.1. Componentes

5. La función de nutrición

5.1. La fotosíntesis

5.2. La respiración celular

5.3. La nutrición en las bacterias

6. La función de relación

7. La función de reproducción

7.1. La división celular en las bacterias

7.2. La división celular en la célula eucariota

8. Los virus

9. Las enfermedades infecciosas

OBJETIVOS

- Comprender los principios de la teoría celular y relacionarlos con la morfología y las funciones de las células.
- Entender las características y el funcionamiento del microscopio óptico y las diferencias entre éste y el electrónico.
- Conocer las estructuras de las células procariotas y eucariotas y los procesos que tienen lugar en ellas.
- Comprender las principales características de la fotosíntesis y la respiración celular.
- Analizar y comprender las diferencias entre células y virus.

El ojo humano no puede percibir objetos de diámetro inferior a 0,1 mm.

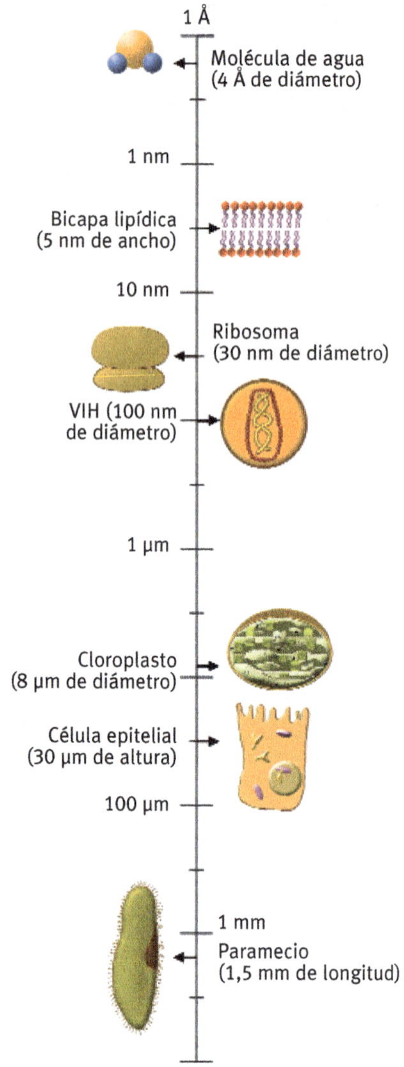

Tamaños de diferentes estructuras y tipos de células

En 1838 Matthias J. Schleiden (botánico) y en 1839 su colega Theodor Schwann (zoólogo) enuncian dos de los principios de la **teoría celular**:

- Todos los seres vivos están compuestos por células.
- La célula es la unidad estructural de la vida.

En 1858 Rudolf Virchow (patólogo) postula el tercer principio de la teoría celular:

- Las células provienen por división de una célula preexistente.

El tamaño medio de las células procariotas oscila entre 1 y 5 µm de longitud y el de las células eucariotas entre 10 y 30 µm. A causa del tamaño de las células, la biología celular depende de los instrumentos de observación y de la tecnología para avanzar en el conocimiento de la morfología y la fisiología celulares.

1. La observación al microscopio

Los microscopios son instrumentos que aumentan el tamaño de la imagen de los objetos que se observan a través de ellos.

Los primeros microscopios posibilitaron, en el siglo XVII, el descubrimiento de las células. En la actualidad existen diversos tipos de microscopios.

Microscopio óptico. Presenta distintos tipos de lentes que constituyen la **parte óptica;** las lentes producen la refracción de la luz y forman imágenes de mayor tamaño que el objeto que se observa.

La **parte mecánica** permite el funcionamiento y ajuste de los elementos de la parte óptica.

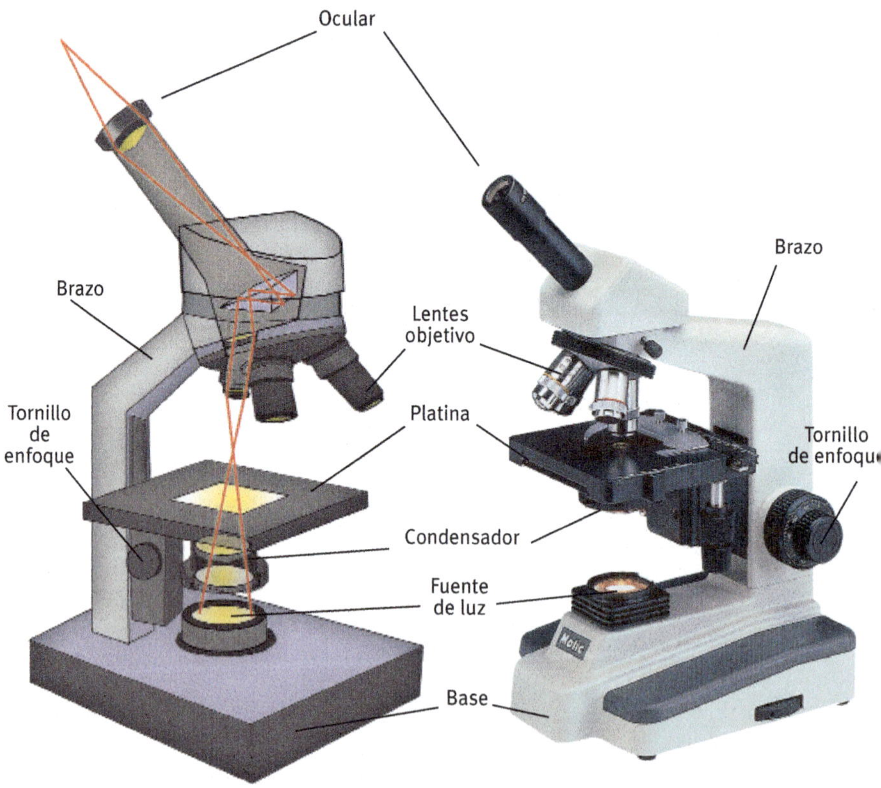

El microscopio necesita luz para funcionar; por ello suele llevar una bombilla incorporada.

— La luz entra por la parte inferior. La cantidad de luz se regula con el **diafragma,** de modo que se ilumina la muestra con la intensidad más adecuada.

— El **condensador** concentra los rayos de luz sobre la muestra.

— La **muestra** está montada sobre un portaobjetos y, normalmente, está fijada y teñida.

— La luz atraviesa la muestra y sigue su recorrido en sentido ascendente hasta el **objetivo,** cuya lente refracta la luz y produce un aumento en la imagen de la muestra.

— La lente del **ocular** produce una segunda refracción y un segundo aumento del tamaño de la imagen.

— Finalmente, en los ojos del observador se forma la imagen definitiva, que es **invertida** y **mayor** que la muestra.

En los microscopios se valoran dos características:

- El **aumento** o número de veces que incrementa el tamaño de la imagen de la muestra. En los microscopios ópticos se calcula multiplicando los aumentos indicados en el ocular por los del objetivo.
- El **poder de resolución,** que es la capacidad para distinguir separadamente dos puntos que están muy próximos.

Un microscopio óptico puede proporcionar 1500 aumentos y 0,2 µm de poder de resolución.

Para la observación al microscopio óptico, debe hacerse una **preparación,** una muestra de lo que se desea observar, modificada con técnicas diversas para facilitar su observación.

Las preparaciones se llevan a cabo sobre un portaobjetos siguiendo, en general, estos pasos:

— **Obtención de la muestra.** Para que pueda ser atravesada por la luz, la muestra debe ser muy delgada.

Si las células son parte de un tejido, se efectúan *cortes muy finos* que se depositan sobre el portaobjetos; si se trata de células incluidas en un medio líquido, se realiza una *extensión* de este líquido sobre el portaobjetos.

— **Fijación de las células.** Hace que los componentes celulares se estabilicen preservando su apariencia estructural. Se suele hacer con calor, o aplicando alcohol o formaldehído. También favorece la absorción del colorante que se aplica a continuación.

— **Tinción.** El colorante **reacciona** y se une a los componentes celulares. Se suelen combinar distintos colorantes **específicos** para diferentes orgánulos. Ello permite identificarlos fácilmente al observar.

La preparación se suele cubrir con un cubreobjetos. Si se quiere conservar la preparación por un tiempo prolongado, se aplican productos que impidan su putrefacción, y se sella el cubreobjetos.

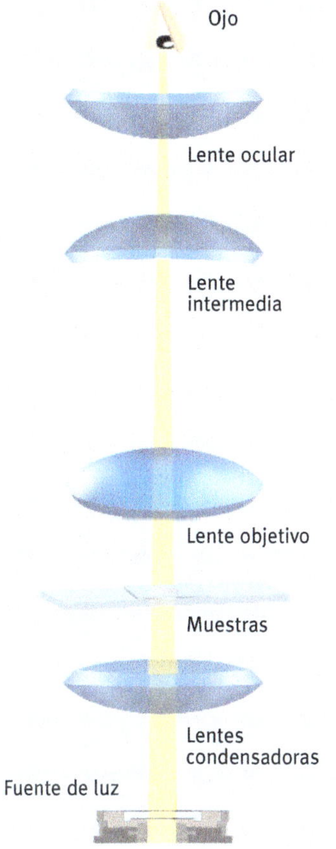

Recorrido de la luz a través de los elementos ópticos del microscopio óptico

+

También es frecuente realizar preparaciones mediante la técnica del **frotis** o **extensión**. Se usa para observar sangre y se basa en la extensión de una gota de sangre sobre un portaobjetos utilizando otro portaobjetos. A continuación se fija y se tiñe y tampoco en este caso se utiliza cubreobjetos.

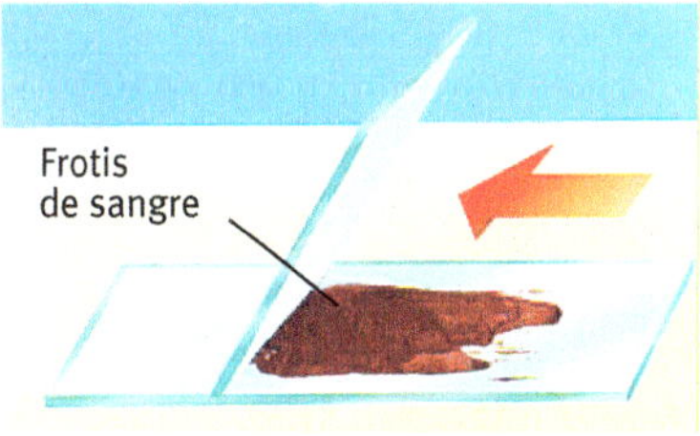

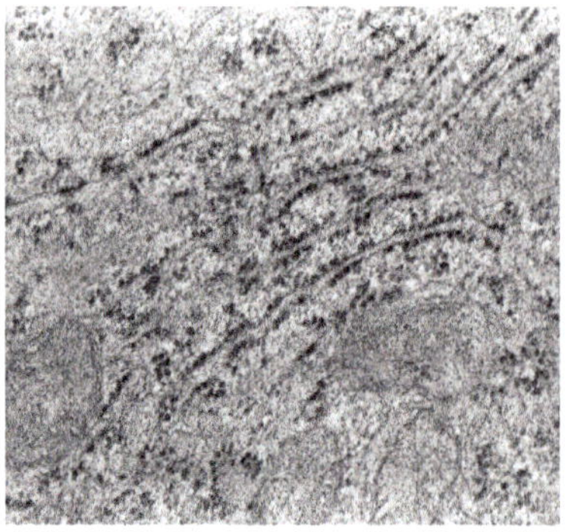

Micrografía del retículo endoplasmático rugoso de células del epidídimo. Microscopio electrónico de transmisión a 72.000 x.

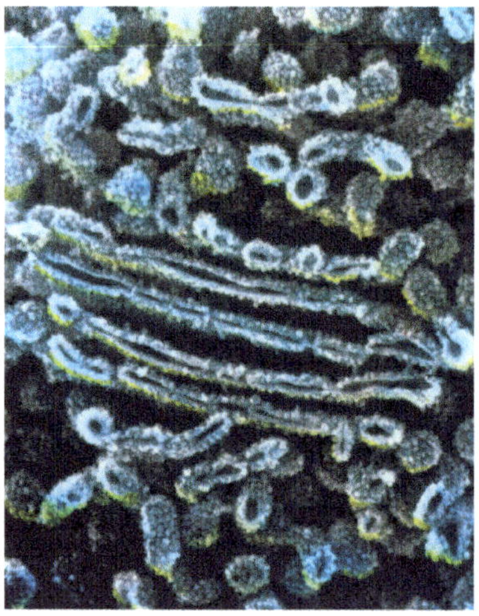

Micrografía del retículo endoplasmático rugoso realizada con un microscopio electrónico de barrido.

Microscopio electrónico. Existen diversos tipos de microscopios electrónicos. Los principales son los microscopios electrónicos de transmisión y los microscopios electrónicos de barrido.

En el microscopio electrónico **de transmisión** se sustituye la luz por haces de electrones. Ello permite un poder de resolución de unos 1,5 nm y hasta 500 000 aumentos. Así, la capacidad para distinguir detalles es mucho mayor que en el microscopio óptico.

— Los electrones se producen en el **cañón de electrones** y se aumenta su velocidad mediante un **acelerador.**

— El haz de electrones entra en una serie de lentes electromagnéticas que generan campos electromagnéticos.

— La muestra ultrafina ha recibido un tratamiento especial con fijadores basado en glutaldehído y tetraóxido de cobre. Se coloca sobre un portaobjetos especial que consiste en una rejilla de cobre y se tiñe con soluciones de metales pesados para aumentar la dispersión de los electrones y mejorar el contraste.

— Los electrones llegan a la muestra, la atraviesan y son desviados de su trayectoria original por los átomos con los que interfieren.

La desviación de los electrones depende de las características de cada parte de la muestra. La imagen se observa en una pantalla fluorescente a partir de las desviaciones de los electrones: cuanto **más desviados** salen de su trayectoria original, **más oscura** es la imagen.

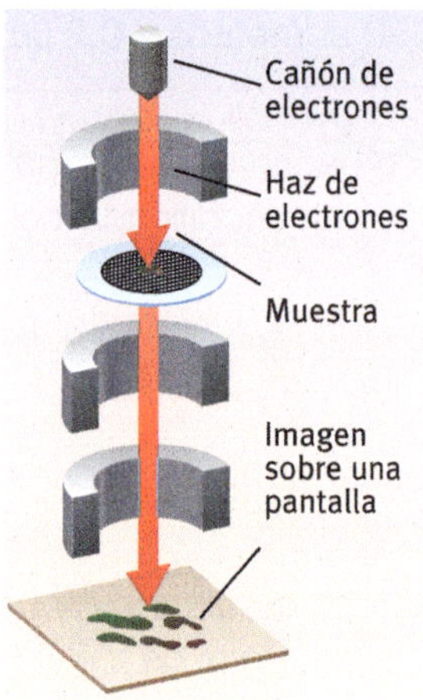

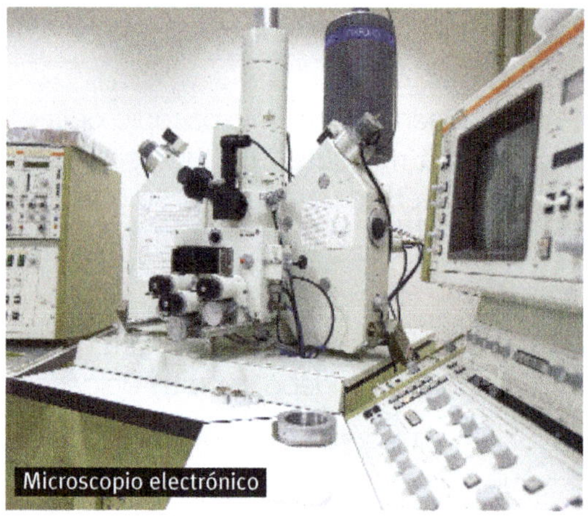

El microscopio electrónico **de barrido** tiene una aplicación diferente. En este caso, el haz de electrones se concentra en un punto que recorre toda la muestra, pero sin atravesarla. Los electrones salen rebotados y forman una **imagen de la superficie** de la muestra. Esto permite una observación tridimensional de las muestras: las zonas hundidas aparecen oscuras y las que sobresalen se ven claras.

Las imágenes que proporcionan los microscopios electrónicos son en blanco y negro, pero normalmente se colorean para distinguir los detalles más destacados. Durante el siglo XX, los avances en las técnicas de observación y el extraordinario desarrollo de las investigaciones bioquímicas han permitido conocer, con bastante detalle, cómo son las células y cómo desarrollan sus actividades.

2. Las membranas celulares

La membrana celular es una estructura común a todas las células. La membrana hace posible que las células sean unidades estructurales y funcionales, ya que las separa del entorno. Su aparición fue un paso decisivo en la desarrollo de las primeras formas de vida.

Todas las membranas biológicas, ya sea la membrana plasmática, que envuelve a la célula y limita el espacio interior o las membranas que delimitan la mayoría de estructuras internas de la célula, presentan una estructura general común. Por este motivo, algunos autores denominan a las membranas biológicas membranas unitarias.

2.1. *Estudio de la membrana plasmática*

La membrana plasmática forma una capa que envuelve a la célula y limita el espacio interior, donde se localizan los diferentes organulos celulares. Vamos a conocer su estructura y sus funciones.

Estructura de la membrana: presenta un espesor de entre 5 y 10 nm. Al ser tan fina, no se aprecia cuando se examina una preparación celular al microscopio óptico. Por este motivo, la membrana plasmática no pudo estudiarse hasta finales de 1950, cuando progresaron las técnicas de preparación de tejidos para su observación con el microscopio electrónico.

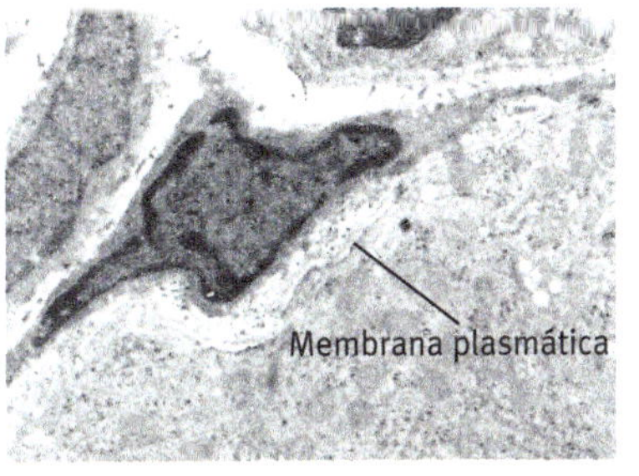

Membrana plasmática de una célula del intestino delgado vista con el microscopio de transmisión a 21.000 aumentos

El microscopio electrónico permite distinguir en la membrana plasmática **tres capas**: oscura, clara y oscura. Las colas hidrofóbicas de la bicapa lipídica corresponden a la parte clara que se observa al microscopio electrónico, mientras que las cabezas hidrofílicas son las líneas oscuras.

Según el modelo que propusieron J. Singer y G. Nicolson en 1972, denominado el modelo de **mosaico fluido**, la composición química de las membranas consta de compuestos lipídicos y compuestos proteicos. Este modelo, actualmente en vigencia, establece que la membrana presenta una estructura cambiante, en la que sus componentes se desplazan de forma similar a como sucede en el seno de un fluido.

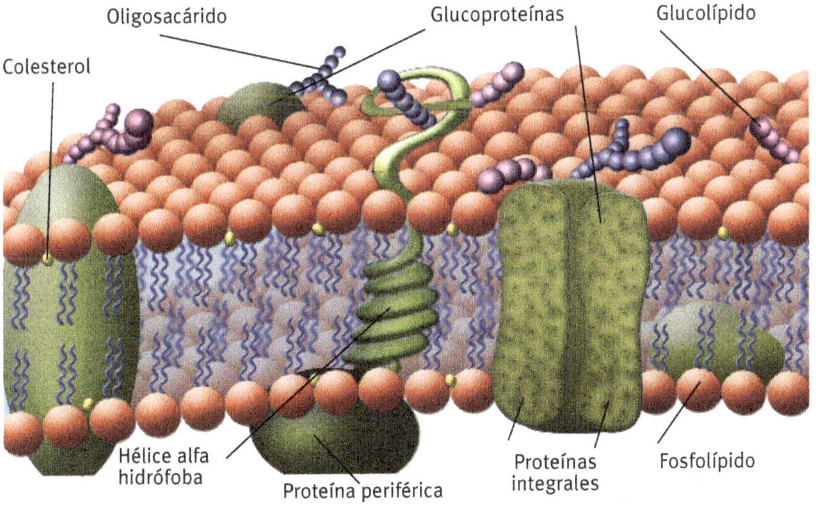

Dibujo de una membrana plasmática

COMPOSICIÓN.

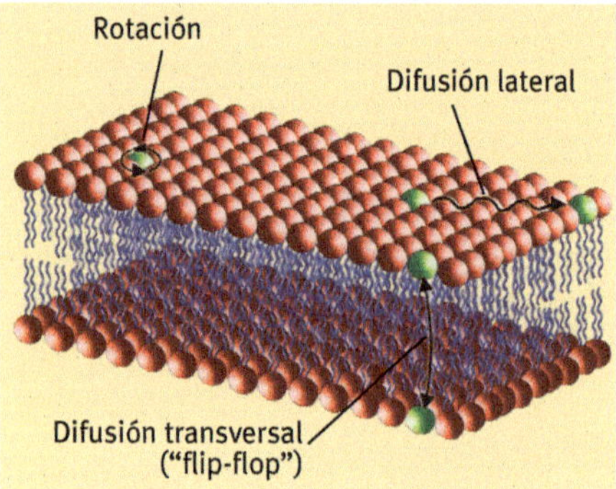

Movimientos de las moléculas de fosfolípidos en la membrana

- **Compuestos lipídicos.** Constituyen la estructura básica de la membrana. Se trata principalmente de **fosfolípidos** que se disponen con sus colas, o restos hidrofóbicos, enfrentadas y las cabezas, o restos hidrofílicos, expuestas al medio exterior y al interior. Forman la denominada **bicapa lipídica.**

  En menor abundancia se encuentran los **glucolípidos** (constituidos por glúcidos unidos de forma covalente a los extremos polares de los lípidos), y el colesterol. Los glucolípidos parecen intervenir en actividades como el crecimiento celular y las interacciones entre células.

  El **colesterol**, en ciertas células animales, puede constituir hasta el 50% de los lípidos de la membrana. Es más pequeño que otros lípidos de la membrana e interfiere en los movimientos de los fosfolípidos.

  La bicapa lipidica presenta fluidez porque sus componentes, como los fosfolípidos, pueden desplazarse libremente. Los lípidos pueden realizar desplazamientos de difusión lateral, de difusión transversal (flip-flop) y de rotación. La fluidez de la membrana se incrementa conforme se eleva la temperatura y también en bicapas con una mayor cantidad de lípidos insaturados.

- **Compuestos proteínicos.** Entre los lípidos se insertan las proteínas de membrana. Existen diversos tipos, entre ellas se encuentran:

  Proteínas integrales o transmembranosas que penetran en toda la bicapa lipidica y sobresalen por ambos extremos.

  Proteínas periféricas que se localizan fuera de la bicapa lipídica, en el lado citoplasmático o en la superficie extracelular.

  Las proteínas de membrana desarrollan funciones diversas:

  — Participan en el **transporte** de sustancias de un lado a otro de la membrana.

  — Son enzimas y **controlan** reacciones bioquímicas.

  — Actúan como marcadores que identifican a las células e intervienen en el **reconocimiento** celular por parte de hormonas u otras sustancias.

  Algunas proteínas no ocupan posiciones fijas, sino que pueden **desplazarse** a lo largo de la membrana, principalmente por movimientos de rotación y de difusión lateral. Otras dejan entre ellas espacios que actúan como **poros,** a través de los cuales se establece comunicación entre el exterior y el interior de la célula.

  También aparecen en la superficie extracelular de la membrana las glucoproteínas. En las glucoproteínas, los glúcidos, principalmente oligosacáridos, se unen mediante enlaces covalentes a las proteínas.

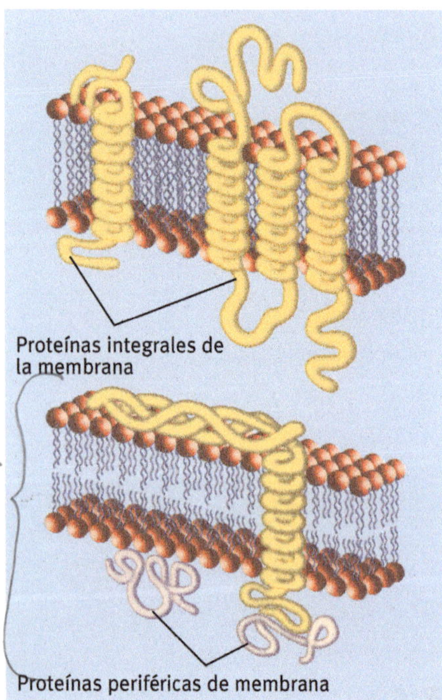

Funciones de la membrana. Son básicas para la actividad celular.

- **Protege** las células y **mantiene** en el interior de éstas las **condiciones necesarias** para el desarrollo de las funciones vitales.
- Permite la **comunicación intercelular**, proporcionando mecanismos de contacto, comunicación y adhesión entre células adyacentes.
- **Regula el intercambio** de sustancias entre medio interno y externo.

El paso de sustancias a través de la membrana es un proceso **selectivo** y **controlado.** Este control es fundamental para el correcto funcionamiento de la actividad celular.

El **transporte de sustancias** se puede producir por distintos mecanismos.

Transporte

- Moléculas pequeñas e iones
  - Transporte pasivo
    - Difusión simple
    - Difusión facilitada
    - Ósmosis
  - Transporte activo
- Moléculas grandes (macromoléculas)
  - Exocitosis
  - Endocitosis
    - Pinocitosis
    - Fagocitosis

El **transporte pasivo** no requiere gasto de energía y se puede hacer mediante:

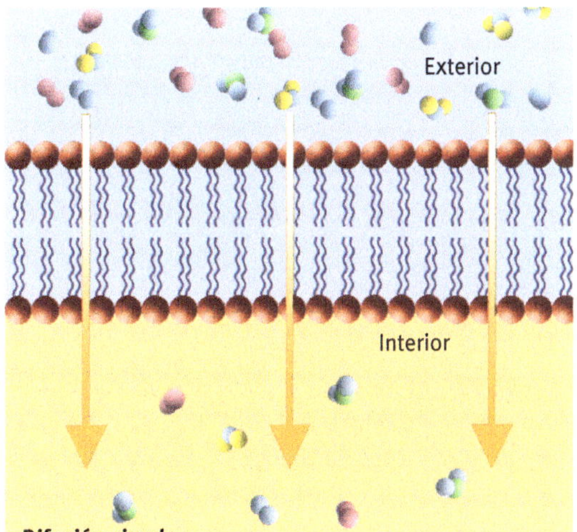

Difusión simple

- **Difusión simple.** Cuando una membrana separa dos disoluciones de diferente concentración, el **soluto** atraviesa la membrana a favor de **gradiente de concentración;** es decir, de la zona más concentrada a la menos concentrada. Al final del proceso se igualan las concentraciones a ambos lados de la membrana.

  La velocidad de penetración de una molécula varía ampliamente entre diferentes sustancias. En general, cuanto menor es su tamaño (menor peso molecular) más rápidamente atraviesan la membrana.

  El **dióxido de carbono,** el **oxígeno** y el **agua** atraviesan la membrana de este modo, puesto que:

  — El dióxido de carbono y el oxígeno son solubles en lípidos y se mueven con facilidad por la **bicapa lipídica.**

  — La molécula de agua, a pesar de ser polar, pasa por las **aberturas** que se crean en la membrana a causa del movimiento espontáneo de los lípidos.

- **Ósmosis o difusión del agua a través de las membranas.** Cuando una membrana separa dos disoluciones de diferente concentración, pero la naturaleza del soluto le impide atravesar la membrana, es el **agua** la que se desplaza a través de la membrana para igualar las concentraciones.

  La disolución con menor concentración se denomina **hipotónica,** y la de mayor concentración, **hipertónica.**

  Cuando el paso del agua iguala las dos concentraciones, ambas disoluciones son **isotónicas.**

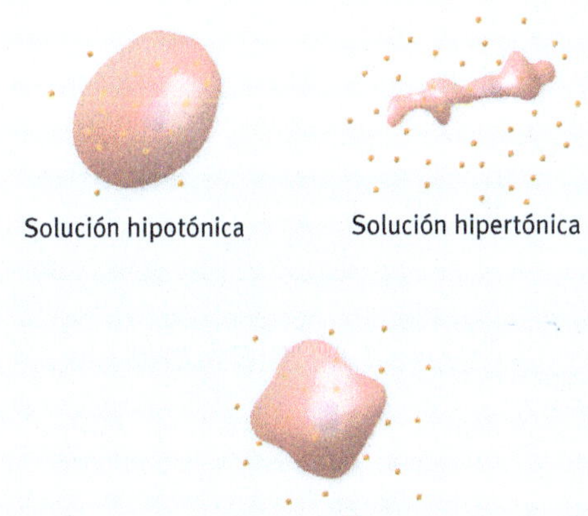

Ósmosis

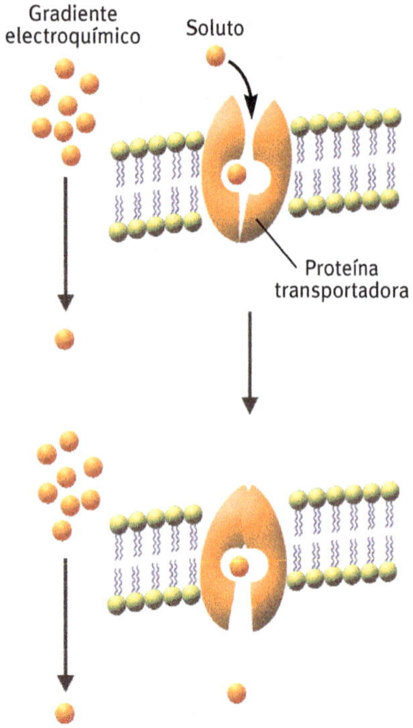

Difusión facilitada a favor del gradiente de concentración

En función de las características del medio externo, la ósmosis puede representar un inconveniente para algunas células.

Por ejemplo, los organismos unicelulares que viven en las aguas dulces se encuentran rodeados de un medio hipotónico y, por ósmosis, el agua tiende a entrar en la célula.

Un aumento en el contenido celular de agua diluye excesivamente las sustancias del interior de la célula y las funciones no se desarrollan adecuadamente. Si continúa entrando agua, la célula podría estallar, fenómeno que recibe el nombre de **lisis.** Por esta razón, este tipo de organismos poseen mecanismos para expulsar agua al exterior mediante un transporte que requiere gasto de energía.

- **Difusión facilitada.** Es un transporte a favor de **gradiente de concentración** y se lleva a cabo mediante **proteínas transportadoras** que atraviesan la bicapa lipídica. Las proteínas que intervienen en el transporte cambian su configuración para poder realizarlo. De este modo atraviesan la membrana sustancias como glucosa, aminoácidos, nucleótidos...

El **transporte activo** tiene lugar cuando las proteínas transportan sustancias **en contra de su gradiente de concentración** y, por tanto, requiere un gasto energético en forma de consumo de ATP (adenosín trifosfato).

Esta situación se da con frecuencia en los organismos. Por ejemplo, las células del hígado almacenan glucosa. En su interior, habitualmente, la concentración de glucosa es más elevada que en el medio exterior. Por transporte activo, las células del hígado siguen captando glucosa para almacenarla.

Además, la mayoría de las células mantiene concentraciones internas de iones sodio (Na^+) e iones potasio (K^+) diferentes del medio exterior:

- El Na^+ exterior puede estar hasta 14 veces más concentrado que el Na^+ interior (la concentración intracelular es de 5 –15 mM, y la concentración extracelular de 145 mM).
- El K^+ dentro de la célula está entre 10 y 30 veces más concentrado que en el exterior (la concentración intracelular es de 140 mM y la concentración extracelular de 5 mM).

Estas diferencias son necesarias para que tengan lugar diversas reacciones bioquímicas, y se mantienen gracias a un mecanismo de transporte activo denominado **bomba de sodio-potasio.**

Estas proteínas de transporte, que utilizan la energía liberada por la hidrólisis del ATP, se denominan ATPasas de transporte.

Aunque el proceso no está del todo aclarado, sabemos que tiene lugar gracias a los cambios en la conformación que presentan las proteínas transportadoras.

- Tienen un centro de unión para K^+ por la parte interior de la membrana; es decir, la que está en contacto con el citoplasma.
- Tienen otro centro de unión para Na^+ por la parte de la membrana que está en contacto con el exterior.

+

El **adenosín trifosfato** o **ATP** es una molécula con enlaces ricos en energía. Cuando los enlaces se rompen, esta energía se libera y se utiliza en las reacciones endergónicas; el ATP se transforma en **ADP** o **adenosín difosfato** que puede pasar de nuevo a ATP, mediante la energía desprendida en las reacciones exergónicas.

El cambio de conformación de las proteínas hace posible el transporte de los iones en contra de su gradiente de concentración.

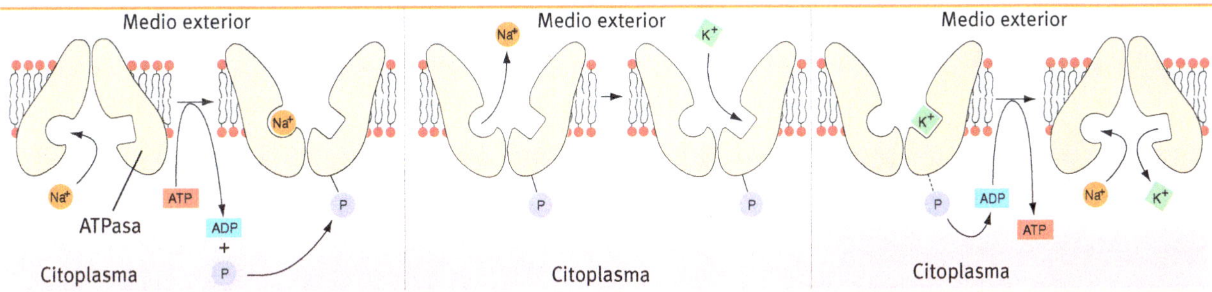

Se incorpora un Na^+ y se produce la hidrólisis de una molécula de ATP; la unión del P provoca el cambio en la conformación de la proteína.

El cambio de conformación libera el Na^+ en el exterior de la célula y la proteína queda en condiciones de aceptar K^+.

La entrada del K^+ produce la liberación del P, así se vuelve a la conformación inicial, se libera el K^+ y la proteína puede aceptar Na^+.

Este proceso tiene lugar de manera continua y, como consecuencia de ello, se mantienen las diferencias de concentración.

Independientemente de los mecanismos descritos hasta ahora, existe otro tipo de transporte, que tiene lugar en el caso de las **macromoléculas** y de **partículas** tales como enzimas, ácidos nucleicos...

- Las sustancias producidas por la célula y que deben salir de ella se transportan por el citoplasma celular, encerradas en vesículas limitadas por una membrana.

  Cuando la vesícula llega a la membrana plasmática, se fusionan ambas membranas y la sustancia del interior de la vesícula es vertida al exterior de la célula. Este proceso recibe el nombre de **exocitosis.**

- Por el proceso inverso, o **endocitosis,** las sustancias del exterior pueden penetrar en el interior: la membrana se invagina y produce una vesícula que se cierra sobre sí misma; ésta se desplaza hacia el interior de la célula. Existen dos tipos de endocitosis:

  — La **fagocitosis** consiste en la incorporación de partículas sólidas de gran tamaño o de restos celulares.

  — La **pinocitosis** consiste en la incorporación de pequeñas cantidades de liquido.

Las membranas plasmáticas están presentes en todas las células. La presencia de membranas que aíslan el material genético de las células del resto del citoplasma determina la existencia de dos tipos de células: las células *procariotas* y las células *eucariotas.*

+

Por cada molécula de ATP que se utiliza en una bomba de sodio-potasio, se expulsan 3 Na^+ y se captan 2 K^+.

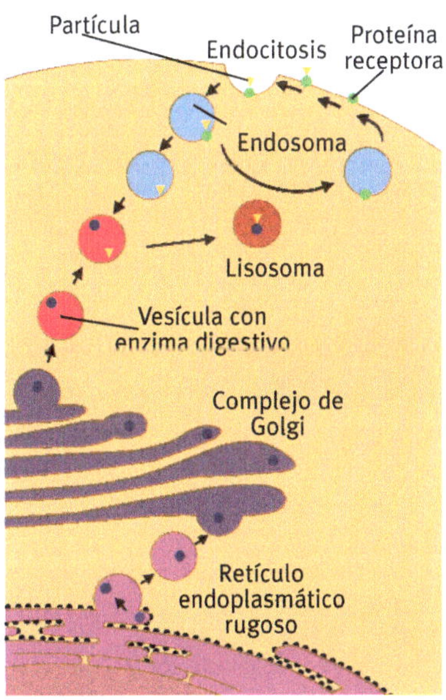

Relación entre el retículo endoplasmático, el aparato de Golgi, los lisosomas y los procesos de endocitosis. Los endosomas recogen partículas que serán digeridas por enzimas en los lisosomas.

ACTIVIDADES

1. Indica la principal ventaja de los microscopios electrónicos sobre los microscopios ópticos. ¿Qué tipo de microscopio utilizarías para ver las estructuras tridimensionales de una célula?

2. ¿Cuál es la finalidad de la tinción de las muestras en la microscopia óptica?

3. Explica porqué la glucosa no atraviesa con facilidad la membrana celular. ¿Cómo la puede atravesar?

3. La célula procariota: las bacterias

Las células **procariotas** son células sin núcleo, es decir, que el espacio de la célula en que se encuentra el material genético no está limitado por membranas. Las bacterias son el ejemplo más común de seres procariotas y presentan las siguientes **características.**

- Son seres unicelulares que presentan tamaños que oscilan entre 0,1 y 0,2 µm de anchura a más de 50 µm de diámetro y entre 1 µm y 500 µm de longitud.
- La mayoría son células de forma esférica, cilíndrica o espiral y reciben el nombre de *cocos, bacilos* y *espirilos* respectivamente. Otras tienen forma de coma y se denominan *vibrios*.
- Existen numerosos grupos de bacterias, adaptados a una gran variedad de hábitats (aguas termales, fondos marinos, el cuerpo humano, el suelo...).

En la célula bacteriana se distinguen los siguientes **componentes.**

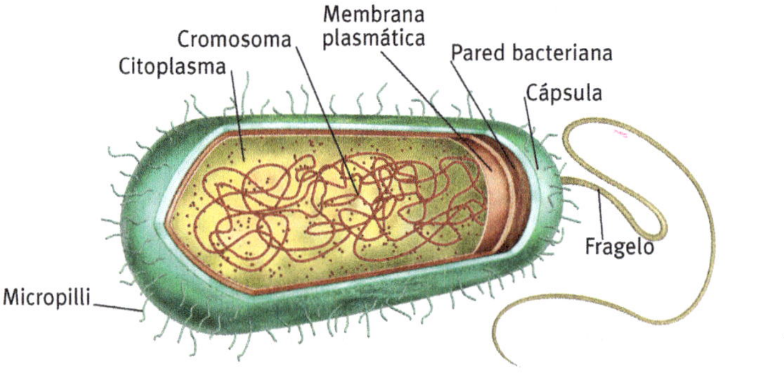

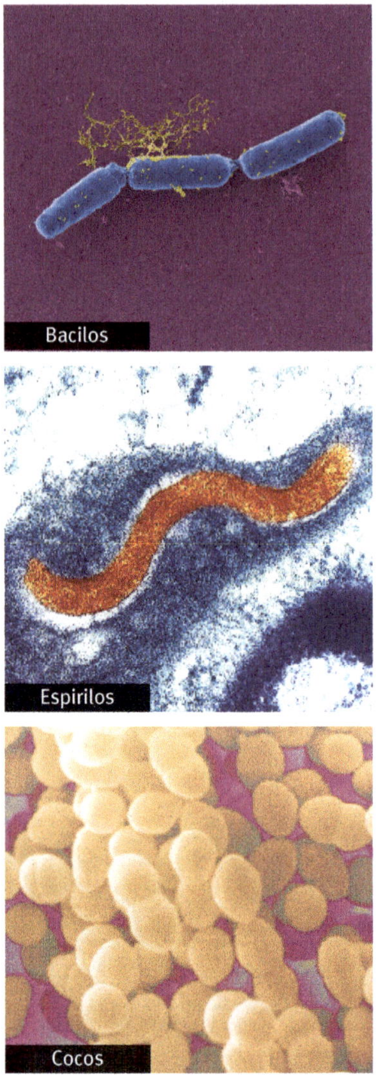

Formas celulares representativas de cada tipo de morfología celular de los procariotas

Pared bacteriana. Es exterior a la membrana plasmática. Está compuesta por sustancias diversas, de las que la más abundante es un peptidoglicano denominado **mureína.** Ésta es un polímero formado por cadenas en las que alternan como monómeros la *N-acetilglucosamina* y el *ácido N-acetilmurámico,* dos derivados de azucares, unidos por enlaces O-glucosídicos, y un pequeño grupo de aminoácidos.

La estructura básica del peptidoglicano está constituida por una malla, en la que las cadenas de derivados de azúcares se conectan entre sí, por puentes peptídicos, a través de los aminoácidos. Desempeña dos funciones básicas:

- Da **forma** y **consistencia** a la célula.
- Mantiene diferencias de **concentración** entre el interior de la célula y el exterior.

La acción de ciertos antibióticos como la penicilina se basa en impedir las uniones laterales de las moléculas de peptidoglicano, de modo que la pared se debilita y la célula muere por lisis.

Las bacterias pueden presentar dos modelos diferentes de pared, que se observan al realizar la tinción de Gram, ideada por Hans Christian Gram (1853-1938). Este tipo de tinción distingue a las bacterias en dos grandes grupos: las **Gram positivas** (si quedan teñidas) y las **Gram negativas** (las que no quedan teñidas).

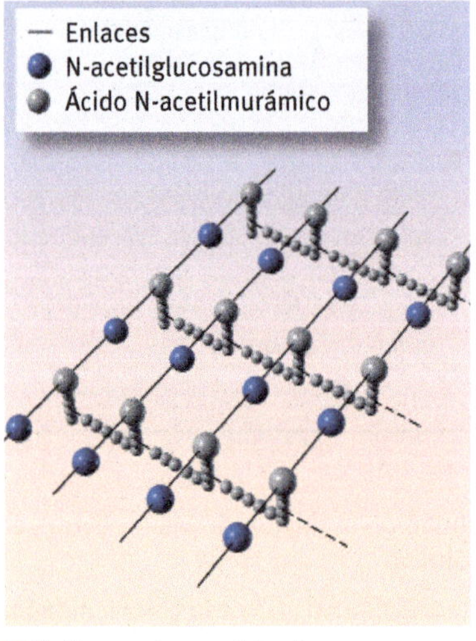

Polímero de peptidoglicano

Las bacterias Gram positivas presentan una pared más ancha (10 a 80 nm) formada hasta en un 90% por peptidoglicano; también suelen estar presentes en pequeñas cantidades otra clase de componentes, como los ácidos teicoicos (polisacáridos ácidos).

Las bacterias Gram negativas tienen una pared más delgada (menos de 10 nm), compuesta por varias capas y más compleja que las Gram positivas. Está constituida por una capa de peptidoglicano, alrededor del 10% de la pared, y por otra membrana más externa compuesta principalmente de lipopolisacáridos que confiere toxicidad para los animales. Entre las Gram negativas patógenas para el hombre y otros mamíferos se encuentran especies de los géneros *Salmonella* y *Escherichia.*

Como resultado de la tinción de Gram se observa:

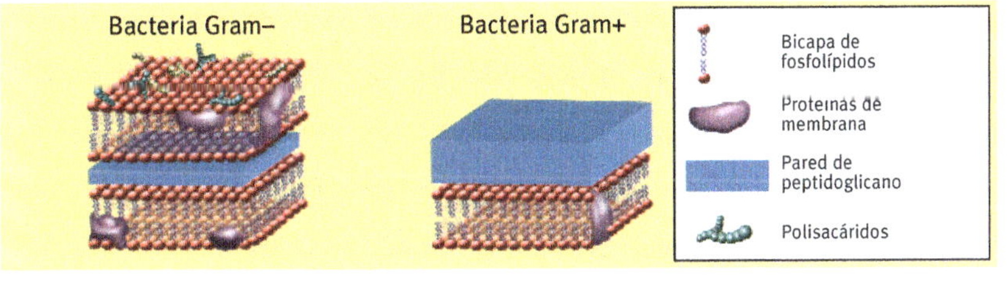

Algunas bacterias tienen, además, una **cápsula** externa compuesta por diversos polisacáridos, que les permite fijarse al sustrato donde viven (el tubo digestivo del hombre, la piel externa de los peces...). Las bacterias encapsuladas a menudo son muy resistentes a la acción de las células fagocitarias del sistema inmunitario, como ocurre con *Streptococcus pnumoniae*.

Membrana plasmática. Presenta las características estructurales de las membranas, que hemos expuesto anteriormente. Debido a la simplicidad de la célula bacteriana, la membrana es fundamental, porque aloja las sustancias que hacen posible el desarrollo de muchas actividades. Por ello, se repliega hacia el interior y forma los **mesosomas,** que permiten aumentar la superficie de la membrana.

Entre las numerosas actividades fundamentales para la célula bacteriana que tienen lugar en la membrana plasmática, destacamos:

- **Transporte** de sustancias.
- **Síntesis** de la propia membrana y de la pared.
- Obtención de **energía.**
- **Captación de energía luminosa,** en las bacterias fotosintéticas.
- Control de la **distribución del material celular** a las células hijas durante la reproducción.

Citoplasma. Es la sustancia que ocupa el interior de la célula, en el que se encuentran incluidos otros componentes celulares (cromosoma, ribosomas...). Contiene diversas sustancias de reserva en disolución:

- **Almidón, glucogéno y ácido ß-hidroxibutírico:** son polímeros en forma de gránulos que actúan como reserva de carbono y energía.
- **Polifosfato y azufre:** constituyen grandes reservas en forma de gránulos de fosfato inorgánico y azufre elemental.

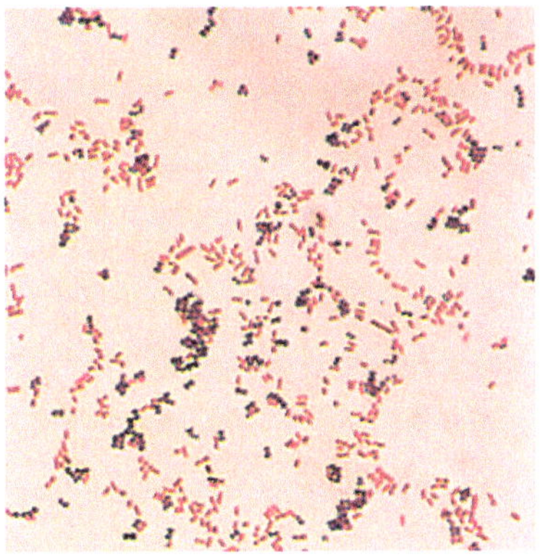

Bacterias teñidas con la tinción Gram. Las bacterias Gram positivas quedan teñidas de morado y las Gram negativas de rosa.

C

Principales pasos en la tinción de Gram:

- El frotis se fija con calor. Se tiñen las celulas con cristal violeta, adquiriendo un color azul-violeta.
- Se añade una solución de lugol.
- Se decolora con alcohol.
- Las celulas Gram positivas mantienen el color azul-violeta y las Gram negativa se decoloran.
- Se tiñen con un colorante de contraste, la safanina.
- Las celulas Gram negativas adquieren un color rosa o rojo.

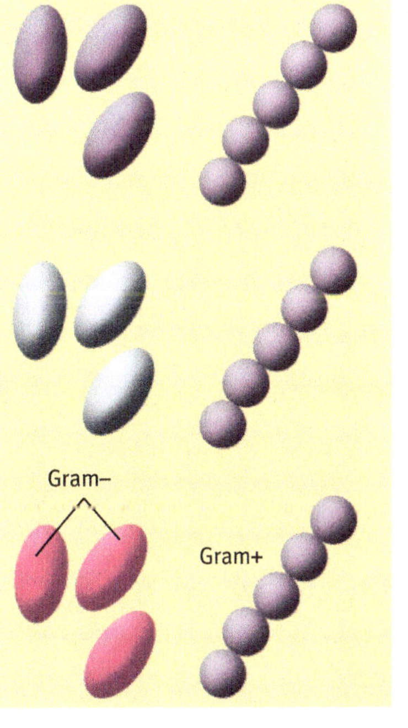

C

La mayoría de las bacterias contienen unas moléculas pequeñas de ADN circular extracromosómico, denominadas **plásmidos**. Los plásmidos no son esenciales para las bacterias pero suelen contener genes que les confieren propiedades metabólicas especiales según el ambiente en que se encuentran.

Según el hábitat de los diferentes grupos de bacterias, en el citoplasma se pueden encontrar:

- **Vacuolas de gas** que favorecen la flotabilidad de las bacterias que viven en el medio acuático. Tienen aspecto cilíndrico y están rodeadas de una capa proteica de 2 nm de espesor.
- **Carboxisomas,** estructuras en forma de poliedro que contienen los enzimas necesarios para la fijación del dióxido de carbono atmosférico.

En una célula bacteriana se encuentran además:

Cromosoma o **genóforo.** Constituido por una doble cadena de ADN cerrada sobre sí misma y replegada; ocupa una zona del citoplasma que se ha denominado la *región nuclear bacteriana* o *nucleoide*. Esta zona no se encuentra limitada por ningún tipo de membrana. El cromosoma controla las funciones de la célula y transmite las características propias de cada especie de una generación a la siguiente.

Ribosomas. Contienen RNA y su función es sintetizar las proteínas de la célula.

Flagelo. Está presente en numerosos grupos de bacterias. Se compone de subunidades, constituidas por una proteína denominada **flagelina,** que se acoplan helicoidalmente dejando un espacio central hueco.

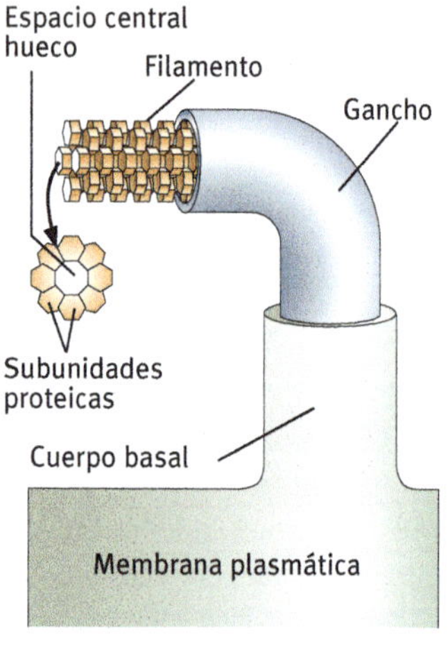

El flagelo hace posible el desplazamiento de las células y consta de:

- **Filamento** externo.
- **Gancho** o curvatura del filamento.
- **Cuerpo basal,** que es el mecanismo de anclaje del flagelo a la célula y responsable del movimiento rotatorio del flagelo.

Según la posición y el número de los flagelos, distinguimos los diferentes tipos de disposición flagelar:

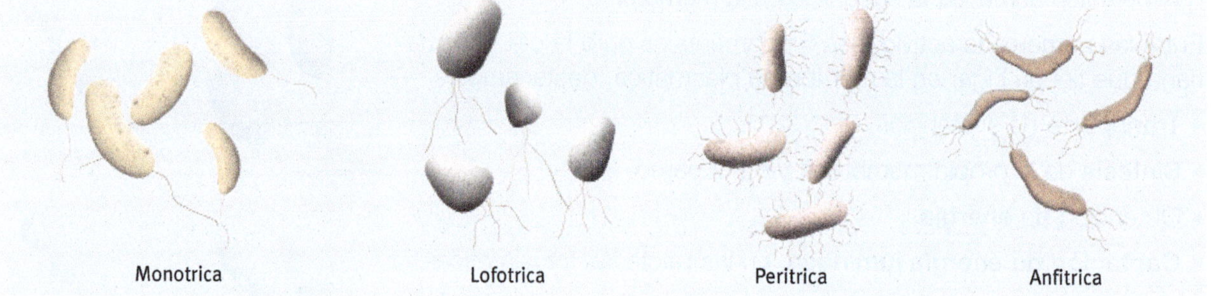

+

Muchas bacterias presentan unos apéndices filamentosos que se denominan **fimbria** o **pili**, también proteicos. Sus funciones son variadas, pero destaca su importancia en la adherencia a superficies y, en el caso de las patógenas, sirve para el reconocimiento de las células que éstas infectan.

ACTIVIDADES

4. Explica en qué se diferencian los modelos de pared de las bacterias Gram positivas de las bacterias Gram negativas.

5. ¿Qué tipos de inclusiones citoplasmáticas presentan las bacterias?

— ¿Cuál es la función de las vacuolas de gas? ¿Cómo es la estructura que permite retener el gas en su interior?

4. La célula eucariota

El paso a la vida eucariota. Un hecho decisivo en el proceso evolutivo fue la transición de formas de vida sencillas como las bacterias, a células más complejas como las que integran el cuerpo de los animales y vegetales. Estas últimas son células **eucariotas** y se caracterizan por:

- Tener una zona limitada por membranas, el **núcleo,** donde se alojan los cromosomas.
- Poseer orgánulos (mitocondrias y cloroplastos en la célula vegetal y mitocondrias en la animal) que contienen **ribosomas** y **ADN.**

Esta estructura de mitocondrias y cloroplastos recuerda mucho a la de las bacterias primitivas. Esta observación constituye la base de la **teoria endosimbiótica**, según la cual los cloroplastos y las mitocondrias evolucionaron a partir de seres procariotas primitivos que se hicieron **simbiontes;** es decir, se asociaron a otras células mayores para obtener beneficio mutuo.

El hecho de que mitocondrias y cloroplastos estén rodeados de una **doble membrana** refuerza esta teoría. Los científicos creen que si la incorporación se hubiera producido por **endocitosis,** las células pequeñas habrían quedado rodeadas de su membrana y de la porción de membrana plasmática de la célula mayor que, a modo de vacuola, las englobaría.

Como en toda relación de simbiosis, los dos componentes de esta asociación debían obtener beneficios:

- Las células mayores recibían **energía** proporcionada por la fotosíntesis o la degradación de biomoléculas que llevaban a cabo los seres procariotas primitivos.
- Las células procariotas incorporadas obtenían **nutrientes,** captados por la célula mayor, además de protección y estabilidad.

En la actualidad, algunas células eucariotas contienen en su interior bacterias y algas unicelulares. Así podemos suponer que esta simbiosis es posible y puede ser duradera.

4.1. *Componentes*

Las células eucariotas presentan **núcleo.** Además, tienen las siguientes características:

- Suelen ser de mayor tamaño que las procariotas, ya que pueden medir entre 10 y 100 μm.
- Tienen membranas internas que crean distintos compartimentos especializados en funciones diferentes. Son los **orgánulos celulares.**
- En el interior de la célula se localiza el **hialoplasma,** el líquido en el que están inmersos el núcleo y el resto de los orgánulos celulares. La parte correspondiente al núcleo recibe el nombre de *nucleoplasma* y el resto es el *citoplasma.*

Existen dos modelos de célula eucariota: la **animal** y la **vegetal.**

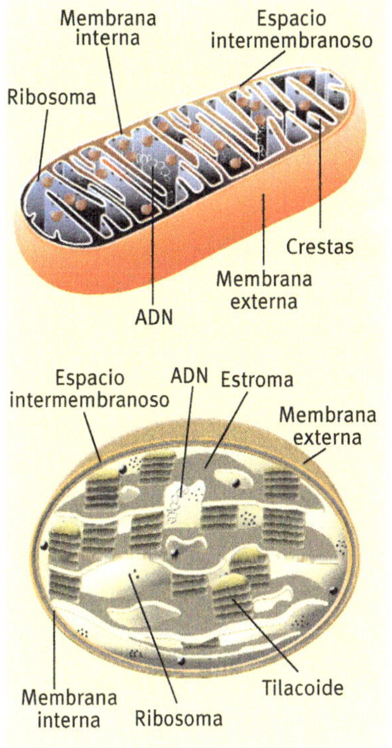

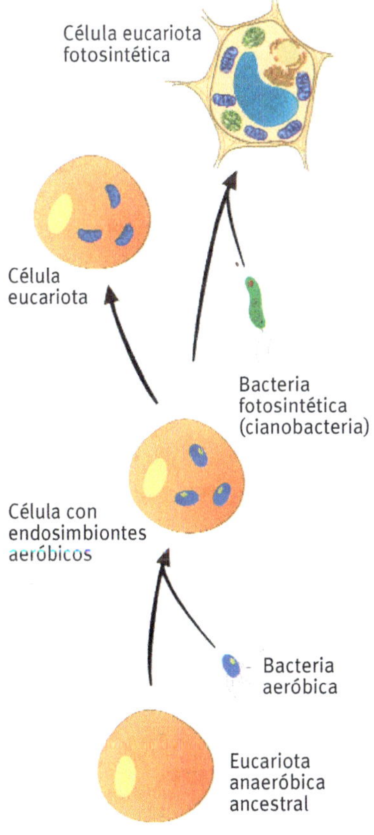

Representación del origen evolutivo de las mitocondrias y de los cloroplastos

Célula animal

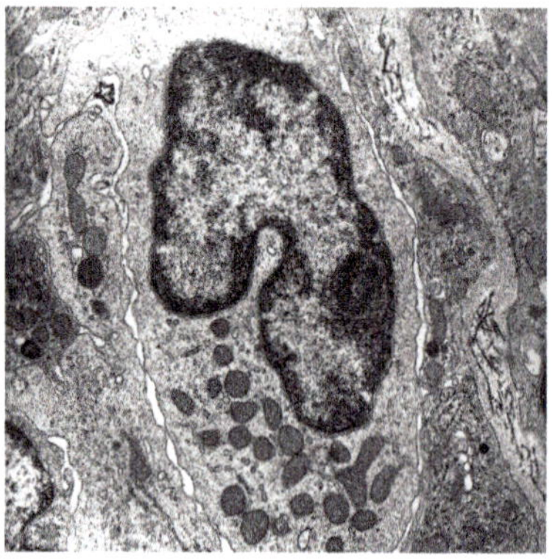

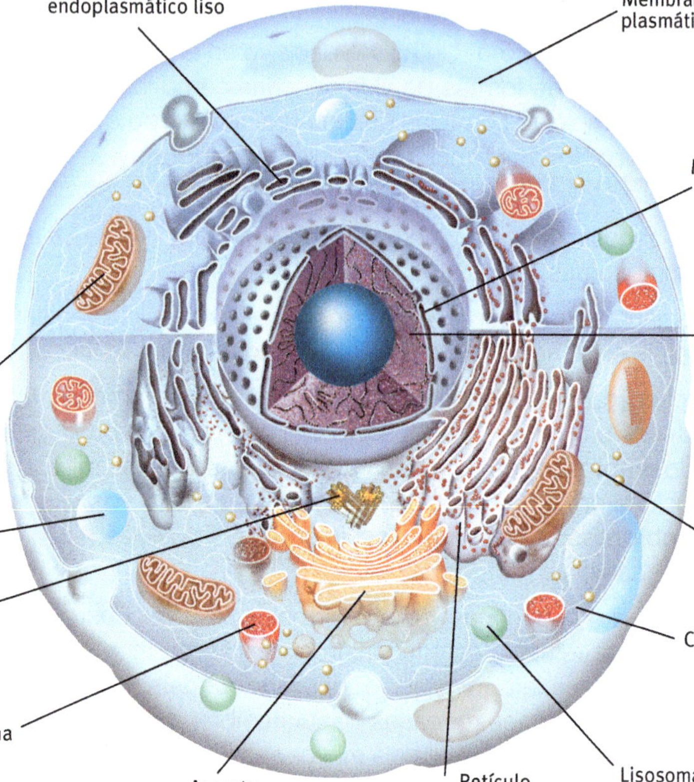

Célula vegetal

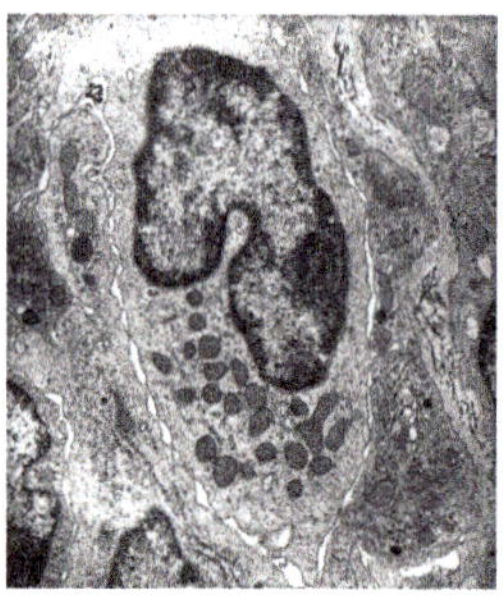

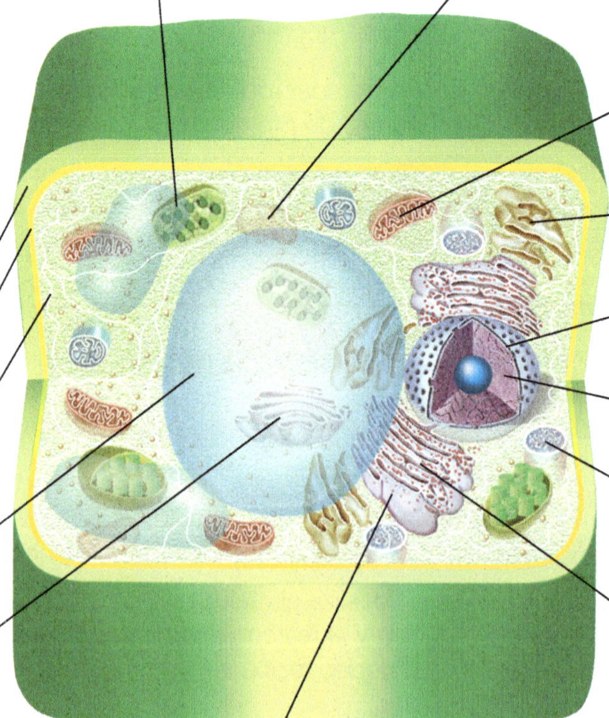

	ESQUEMA	CARACTERÍSTICAS
Membrana plasmática	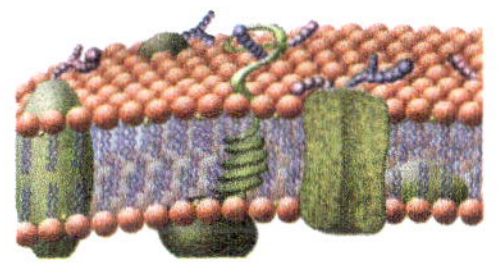	Es una doble capa de lípidos con proteínas y glúcidos insertados. Delimita la célula y a su través tienen lugar los intercambios de sustancias.
Núcleo	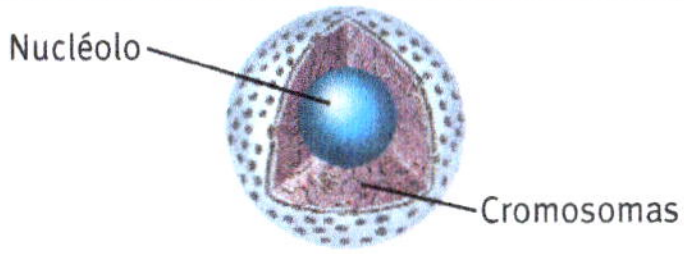	Delimitado por la **envoltura nuclear**; a través de poros tienen lugar intercambios entre el nucleoplasma y el citoplasma. Contiene los cromosomas. También se encuentra en él el **nucléolo**, que está relacionado con la síntesis de RNA, a partir del ADN cromosómico.
Citoesqueleto		Es un conjunto de filamentos de composición proteica, y longitud y grosor variables. Dan forma a la célula y son los responsables del movimiento celular.
Ribosomas		Constan de dos subunidades de diferente tamaño, formadas por RNA y proteínas. Se encuentran dispersos por el citoplasma o asociados a las membranas del retículo endoplasmático. Sintetizan las proteínas.
Retículo endoplasmático		Es un conjunto de sáculos y túbulos aplanados. Existe una continuidad entre el retículo endoplasmático y la envoltura nuclear. Sintetiza y transporta lípidos y proteínas de membrana. Se distinguen el liso y el rugoso, ya que este último tiene ribosomas adosados.
Aparato de Golgi	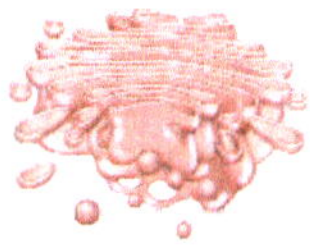	Es un conjunto de vesículas y sáculos aplanados y superpuestos. Transforma, empaqueta y selecciona macromoléculas para su transporte a otros orgánulos o al exterior de la célula.
Mitocondrias	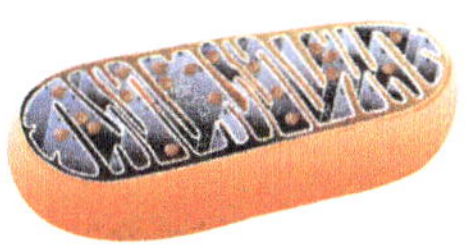	Tienen doble membrana; en la más interna presentan unos repliegues llamados **crestas;** el líquido interior es la **matriz.** En la matriz mitocondrial se encuentran ADN y ribosomas. Producen energía utilizable para la célula a partir de sustancias con enlaces ricos en energía.
Lisosomas y peroxisomas	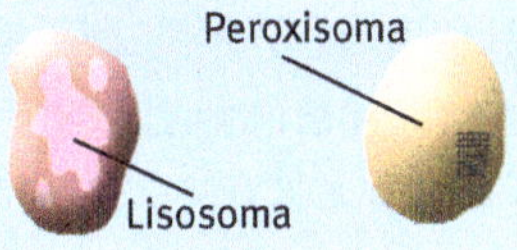	Son vesículas que contienen enzimas. En los lisosomas se hallan enzimas hidrolíticos y en los peroxisomas, oxidativos. Intervienen en los procesos de degradación de sustancias.
Centríolos		Son exclusivos de las células animales; en general, se presentan dos centríolos en posición perpendicular uno respecto a otro. Intervienen en el movimiento de los cromosomas durante la reproducción celular.
Vacuolas		Son vesículas cuya función primordial es el almacenaje de sustancias diversas (nutritivas, toxicas...). Suelen ser mayores en las células vegetales que en las células animales.
Pared celular		Es exclusiva de las células vegetales. Es rígida y rodea la membrana plasmática. Está compuesta de fibras de celulosa y una matriz de proteínas, otros polisacáridos, agua y sales minerales. Sirve de exoesqueleto a la célula, confiriéndole forma y protección.
Cloroplastos		Se encuentran únicamente en las células vegetales. Están limitados por una doble membrana. En su interior, se encuentra un líquido denominado **estroma** y unos sacos aplanados, los **tilacoides**, que se agrupan y forman los **grana**. Contienen los pigmentos para la fotosíntesis.

Las células desarrollan las funciones vitales características de la vida: la *nutrición,* la *relación* y la *reproducción.*

5. La función de nutrición

Consiste en la captación de materia para crecer, reponer las partes de la célula que envejecen y disponer de materias primas para las distintas actividades celulares. También se obtiene energía, que se consume en estas actividades. Todos los procesos tienen lugar mediante reacciones bioquímicas.

El conjunto de reacciones que se producen en el interior de las células de los seres vivos recibe el nombre de **metabolismo.** Son reacciones totalmente interrelacionadas y catalizadas por enzimas específicos. El conjunto de reacciones interrelacionadas constituye una **vía** o **ruta metabólica.**

Distinguimos dos grupos de reacciones metabólicas:

- *Metabolismo de síntesis* o *anabolismo.*
- *Metabolismo de degradación* o *catabolismo.*

El **metabolismo de síntesis** o **anabolismo** es el conjunto de reacciones bioquímicas que tienen por objeto la obtención de moléculas complejas y ricas en energía, tales como glúcidos, ácidos grasos..., a partir de moléculas simples. Estas reacciones consumen energía, que se incorpora a las moléculas sintetizadas como energía química de enlace. Son, por tanto, reacciones **endergónicas.**

Reacción endergónica

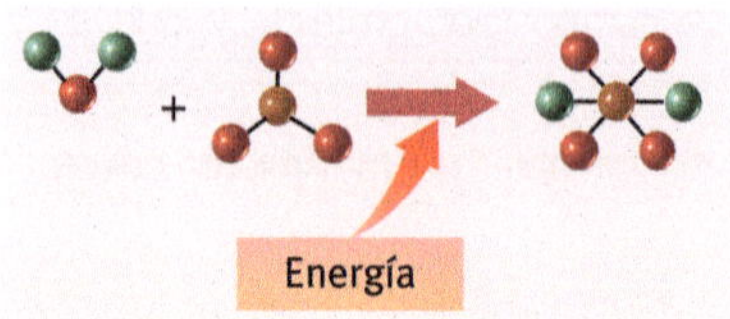

Reacción exergónica

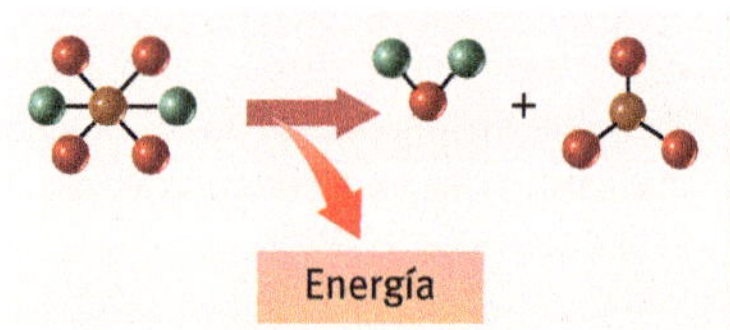

El **metabolismo de degradación** o **catabolismo** agrupa todas las transformaciones bioquímicas que las células realizan a partir de moléculas complejas energéticamente ricas. Así se produce energía química disponible y se obtienen productos finales más simples. Estas reacciones producen energía y son, por tanto, **exergónicas.**

Una parte de la energía obtenida en los procesos catabólicos se utiliza para la realización de los anabólicos, de modo que las reacciones se encadenan.

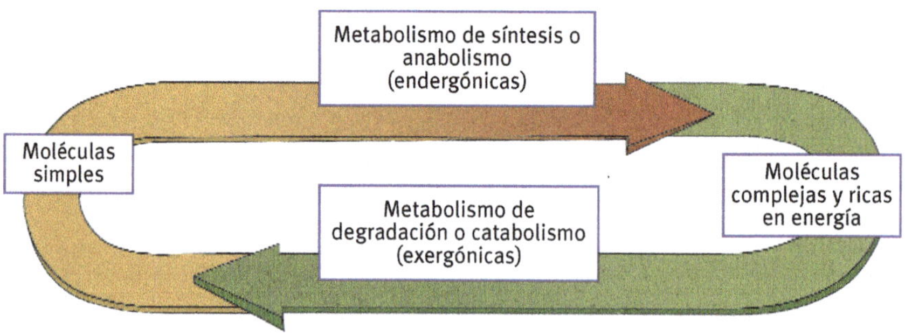

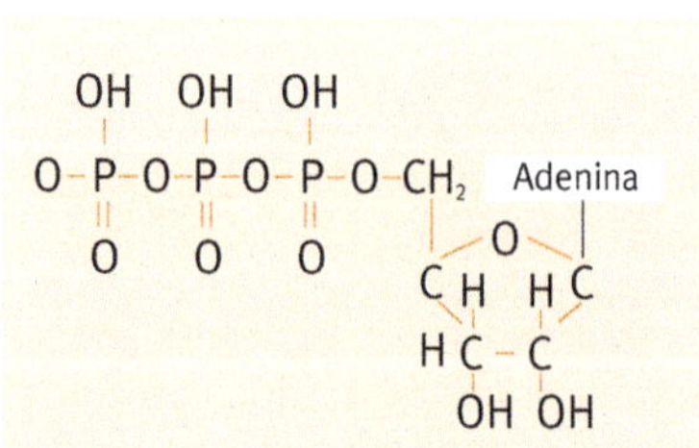

■ Molécula de ATP

La energía que se produce o se consume en el metabolismo, se transporta y almacena mediante los enlaces moleculares ricos en energía del ATP. Al romperse dicho enlace, se libera energía que puede utilizarse para las diversas actividades vitales.

La materia y la energía que proporciona la nutrición ponen en marcha todas las reacciones metabólicas. El proceso se inicia con la entrada de nutrientes del exterior.

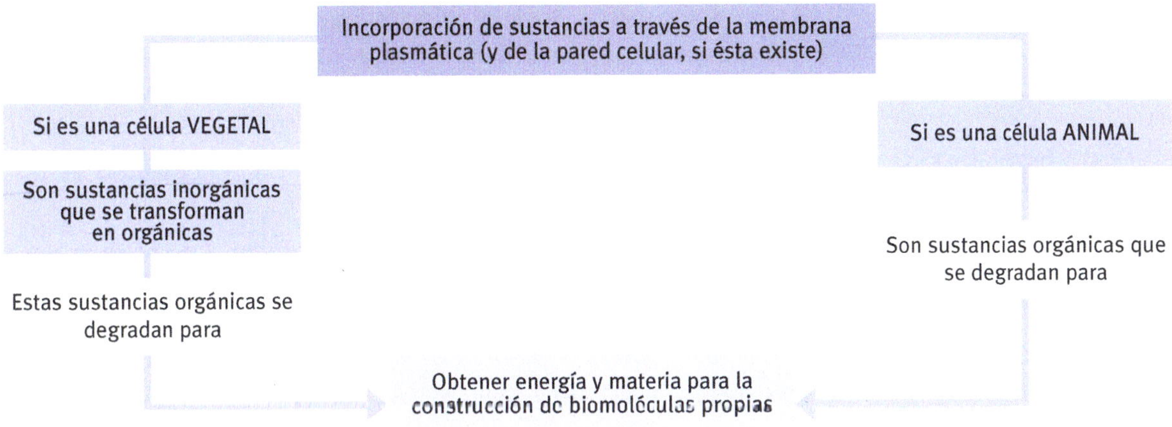

La célula eucariota incorpora carbono y energía para construir sus biomoléculas orgánicas. Según la procedencia del carbono y la energía distinguimos los siguientes tipos de nutrición:

- La célula vegetal se caracteriza por presentar una **nutrición autótrofa,** ya que sintetiza biomoléculas orgánicas a partir de materia inorgánica; su fuente de carbono para la formación de sus biomoléculas es el dióxido de carbono del medio. Los vegetales también tienen **nutrición fotótrofa,** ya que la energía necesaria para la biosíntesis procede de la energía lumínica transformada durante la fotosíntesis en energía química.
- La célula animal presenta una **nutrición heterótrofa,** ya que obtiene el carbono de biomoléculas orgánicas procedentes de otros organismos. La célula animal también tiene **nutrición quimiótrofa,** ya que al no poder aprovechar la energía lumínica, depende de las biomoléculas orgánicas que toma del exterior y de la energía química que éstas contienen.

ACTIVIDADES

6. Completa la tabla.

TIPO DE NUTRICIÓN	SE UTILIZA MATERIA...	SE UTILIZA ENERGÍA...
Fotoautótrofa	Inorgánica	....
Fotoheterótrofa	....	....
Quimioautótrofa	....	....
Quimioheterótrofa	....	....

7. ¿Qué pruebas respaldan la teoría de la endosimbiosis?
8. ¿Cuáles son las principales diferencias morfológicas entre una célula eucariota animal y una célula eucariota vegetal?
9. ¿En qué se diferencian los organismos fotoheterótrofos y fotoautótrofos? ¿Qué similitudes presentan?

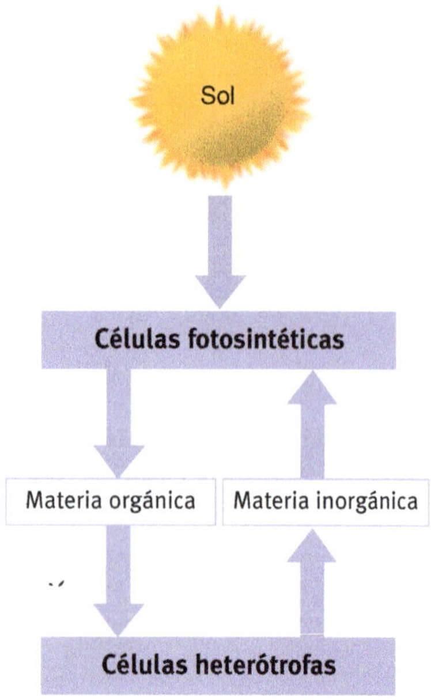

En último término, la energía del Sol es la fuente de energía para todos los seres vivos.

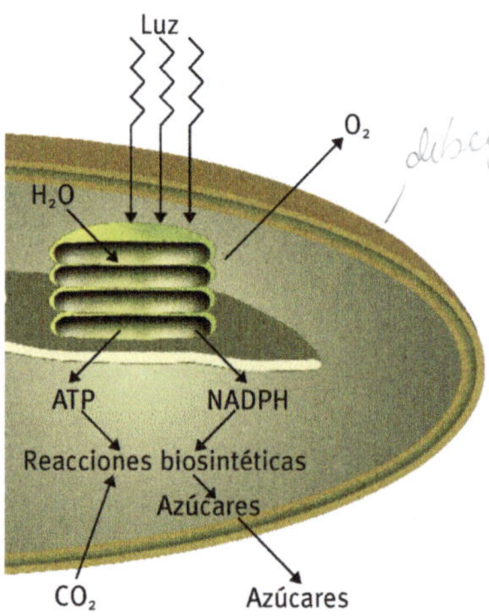

5.1. *La fotosíntesis*

Los organismos fotoautótrofos obtienen mediante la fotosíntesis la materia que será utilizada en procesos posteriores. Aunque los vegetales no son los únicos seres fotosintéticos en el ecosistema del planeta, la importancia cuantitativa de la fotosíntesis vegetal es enorme. Su objetivo consiste en obtener moléculas orgánicas (glúcidos) a partir de moléculas inorgánicas (CO_2 y H_2O). Para realizar la fotosíntesis se necesita:

- **Luz.** La más eficaz es la de longitud de onda correspondiente a nuestro espectro visible.
- **Pigmentos.** El más abundante es la **clorofila.** Las moléculas de pigmentos están agrupadas en las membranas de los tilacoides y forman los *fotosistemas.* Cuando la luz incide sobre los fotosistemas, se produce el desprendimiento de electrones activados.
- Moléculas **transportadoras** de electrones que están relacionadas con los pigmentos y pueden ser proteínas o bien compuestos lipídicos. Estas moléculas captan los electrones desprendidos y los transportan hasta el **aceptor** final, una sustancia denominada $NADP^+$ (nicotinamina adenina dinucleótido fosfato).
- Un **espacio cerrado** para que los electrones activados pasen de una molécula a otra sin dispersarse. El **cloroplasto** es este espacio cerrado; contiene la clorofila, las moléculas transportadoras y las aceptoras; así la eficacia en la transferencia de electrones es máxima.

En conclusión, en las células vegetales se dan las circunstancias que permiten la fotosíntesis, cuya fórmula global es la siguiente:

$$\text{Energía lumínica} + CO_2 + H_2O \longrightarrow O_2 + \text{Glúcidos}$$

En el desarrollo de la fotosíntesis se distinguen dos fases:

Fase dependiente de la luz

- Es imprescindible la presencia de **luz.**
- Se produce en la **membrana de los tilacoides.**

De modo simplificado, las reacciones fotoquímicas que tienen lugar son:

— Se capta la energía lumínica, que se invierte en:

- **Activar** la clorofila para que se desprendan electrones.
- **Romper** moléculas de agua.

$$H_2O \rightarrow 2\,H^+ + 1/2\,O_2 + 2\,e^-$$

Los electrones (e^-) pasan a la clorofila para reponer los que se han desprendido, y también a las moléculas transportadoras.

— Cuando llegan al $NADP^+$ junto a los protones (H^+) forman NADPH.

$$NADP^+ + 2\,H^+ + 2\,e^- \longrightarrow NADPH + H^+$$

— Paralelamente al proceso anterior se sintetiza ATP; este proceso se activa mediante la circulación de los protones (H^+).

$$ADP + P \rightarrow ATP$$

Estas moléculas contienen en sus **enlaces** la **energía química** procedente de los electrones activados.

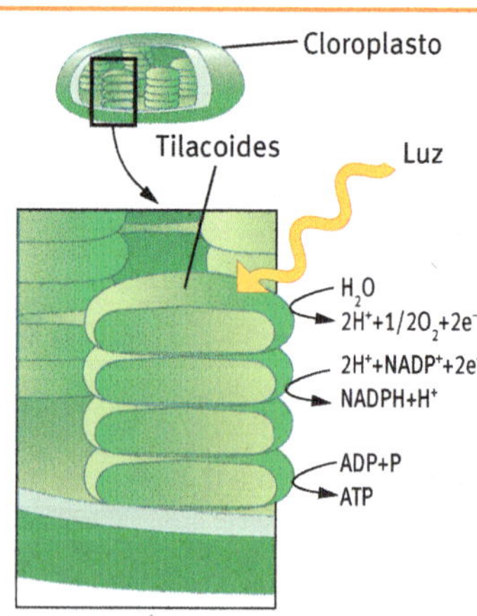

Fase independiente de la luz

- No requiere la presencia de luz.
- Tiene lugar en el **estroma del cloroplasto.**

Esta fase también se llama **fijación del carbono.** En ella, el CO_2 atmosférico se incorpora, mediante un conjunto de reacciones biosintéticas, a pequeños compuestos de carbono para formar glucosa.

— A partir de un compuesto de 5 átomos de carbono, denominado **ribulosa-1, 5-bisfosfato** y una molécula de CO_2, se obtienen dos moléculas de **3-fosfoglicerato**.

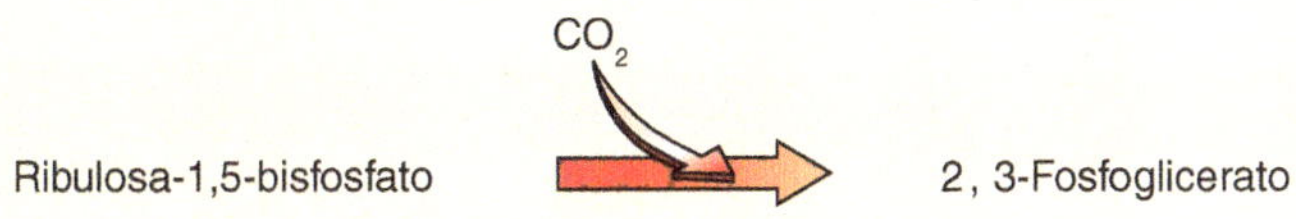

— A partir de los compuestos de tres carbonos que se obtienen, se encadenan diversas reacciones químicas en las que se gastan **NADPH** y **ATP,** obtenidos en la fase dependiente de la luz, para obtener una molécula de glucosa.

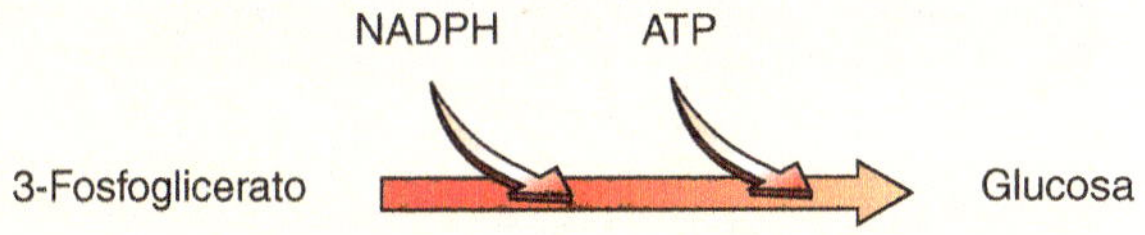

— Esta molécula de glucosa se polimeriza y origina almidón.

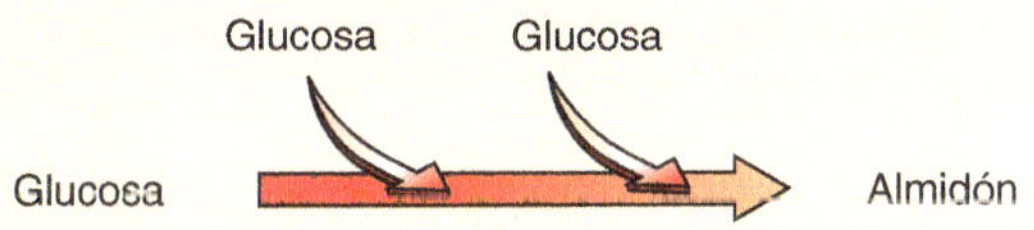

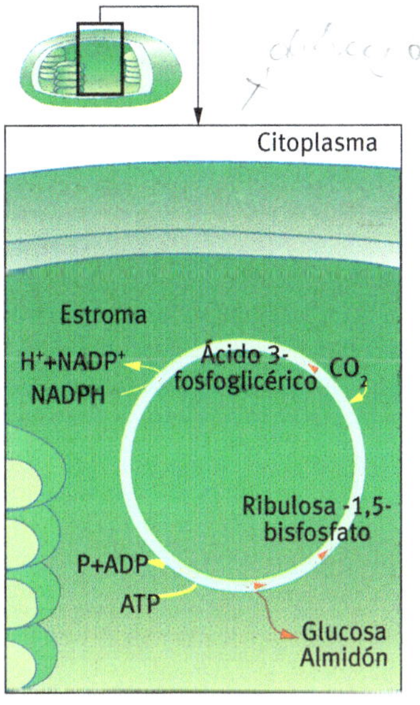

Los glúcidos obtenidos de la fijación del carbono se utilizan, además, en la síntesis de otras biomoléculas: aminoácidos, lípidos y nucleótidos.

5.2. *La respiración celular*

En las células vegetales, el proceso de **respiración celular** tiene lugar a partir de la glucosa obtenida en la fotosíntesis. En las células animales, se lleva a cabo a partir de la glucosa ingerida en los alimentos.

En ambos tipos de células, el proceso se desarrolla con las mismas características.

El objetivo de la respiración es la obtención de ATP y NADH (nicotinamina adenina dinucleótido), que pueden trasladarse por la célula proporcionando energía a las diferentes actividades celulares.

La respiración necesita:

- Monómeros de las grandes biomoléculas, principalmente **glucosa.**
- Moléculas transportadoras de electrones, principalmente proteínas, que constituyen una cadena por donde circulan los electrones: la *cadena respiratoria.* La sustancia que finalmente acepta los electrones es el **oxígeno.**
- Un espacio cerrado ya que, también en este caso, se produce transferencia de electrones. Este espacio es la **mitocondria.**

En suma, tanto en las células vegetales como en las animales, tiene lugar la respiración celular, cuya ecuación química global es:

glucosa

$$C_6H_{12}O_6 + O_2 \rightarrow CO_2 + H_2O + ATP$$

De modo simplificado, las reacciones que tienen lugar durante la respiración son las siguientes:

En el **citoplasma**:

- Mediante la ruta de la **glucólisis**, la glucosa se convierte en piruvato, obteniéndose dos moléculas de ATP

En la **mitocondria**:

- El piruvato obtenido en la glucólisis penetra en la mitocondria y se transforma en Acetil-CoA y CO_2.
- El Acetil-CoA entra en una ruta cíclica: el **ciclo de Krebs**, generándose ATP, NADH, CO_2 y H_2O.

El médico **Hans Adolf Krebs** (Hildesheim, Alemania, 1900 - Oxford, 1981) identificó en colaboración con el bioquímico Kurt Henseleit, el conjunto de reacciones químicas conocidas posteriormente con el nombre de "ciclo de la urea".

En 1935 fue nombrado profesor y director del Departamento de Bioquímica en Sheffield (Inglaterra), y comenzó sus trabajos sobre el metabolismo celular, fundamentalmente en la transformación de los nutrientes en energía.

El grupo de Krebs descubrió la acción catalizadora del citrato, que condujo al conocimiento definitivo del llamado "ciclo del ácido citríco o ciclo de Krebs". Albert Lipmann, con su estudio sobre el coenzima A, completó el conocimiento de dicho ciclo, y ambos recibieron en 1953 el Premio Nobel de Medicina.

Este ciclo es la vía fundamental para la degradación de la mayoría de los compuestos orgánicos.

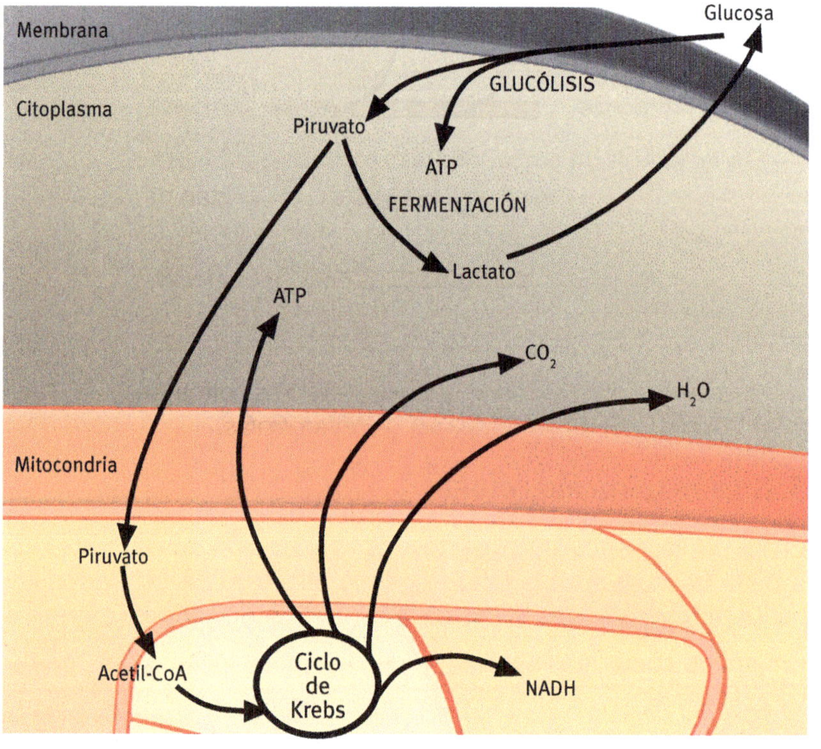

La degradación con oxígeno recibe el nombre de **aeróbica** y produce el máximo rendimiento energético. En algunos casos la degradación se produce sin oxígeno, es decir, de manera **anaeróbica.** Es el caso de la reacción que tiene lugar en las células musculares cuando se ven sometidas a un esfuerzo extremo que provoca una demanda de oxígeno superior a la cantidad que proporciona el aparato circulatorio. Por ello se produce una degradación sin oxígeno o **fermentación.** El piruvato no entra en la mitocondria y, en el citoplasma, se transforma en lactato. El rendimiento energético de esta reacción es menor que si tuviera lugar mediante oxígeno, ya que el lactato todavía contiene energía química en sus enlaces.

5.3. *La nutrición en las bacterias*

La diversidad en las formas de vida de las bacterias hace que en los diferentes grupos se encuentren representados distintos tipos de nutrición, que dependen de la fuente de energía y la fuente de materia que utilizan.

Fuente de energía \ Fuente de materia	MATERIA INORGÁNICA (CO_2, H_2S, S, H_2, NH_3)	MATERIA ORGÁNICA (GLÚCIDOS)
LUZ	Nutrición fotoautótrofa Ejemplo: *Chlorobium*	Nutrición fotoheterótrofa Ejemplo: *Rhodospirillum*
COMPUESTOS QUÍMICOS (GLÚCIDOS)	Nutrición quimioautótrofa Ejemplo: *Nitrobacter*	Nutrición quimioheterótrofa Ejemplo: *Escherichia*

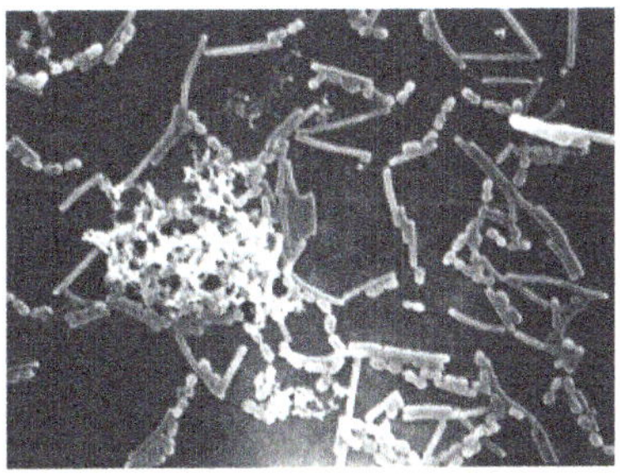

Micrografías de diferentes especies de *Lactobacillus* responsables de las fermentaciones lácticas. Microscopio de barrido, 3 000 x.

Bacterias fotosintéticas. Poseen pigmentos fotosintéticos, denominados **bacterioclorofilas,** para captar la **energía lumínica.** La fotosíntesis se desarrolla de modo similar a la que tiene lugar en el cloroplasto de las células vegetales, aunque en lugar de agua utilizan compuestos como H_2S y por ello, no se libera oxígeno.

Si utilizan materia **inorgánica** para obtener compuestos de carbono, son bacterias *fotoautótrofas*. Algunos grupos utilizan materia **orgánica;** en este caso, son bacterias *fotoheterótrofas.*

Bacterias quimiosintéticas. Utilizan **energía química** procedente de la oxidación de moléculas inorgánicas sencillas.

Para obtener compuestos de carbono, algunos grupos utilizan materia **inorgánica,** son bacterias *quimioautótrofas;* otros emplean materia **orgánica** y son, por tanto, *quimioheterótrofas.*

En este último caso, para obtener ATP se puede utilizar la materia orgánica de dos modos:

- Por degradación **aeróbica,** al igual que hemos visto en las células eucariotas. A partir del ácido pirúvico, se obtiene acetil-coenzima A y finalmente CO_2.
- Por degradación **anaeróbica** por medio de **fermentaciones,** o bien degradando totalmente la materia orgánica y utilizando nitritos o azufre como aceptores de electrones al final del proceso.

+

Las bacterioclorofilas absorben energía de las radiaciones luminosas de longitud de onda correspondiente a la zona roja del espectro.

ACTIVIDADES

10. Explica brevemente de qué manera el proceso de la fotosíntesis es inverso al proceso de la respiración celular.

11. ¿Cuál es la fuente de carbono de las bacterias quimioheterótrofas? ¿En qué se diferencian de las fotoautótrofas?

12. Explica si son verdaderas o falsas las siguientes afirmaciones:

    — La fotosíntesis la realizan las células vegetales; en cambio la respiración solamente la realizan las células animales.

    — Sin el ciclo de Krebs no es posible obtener energía de la glucosa.

13. ¿Sería posible la realización de la fase independiente de la fotosíntesis, sin realizar previamente la fase dependiente? Razona la respuesta.

6. La función de relación

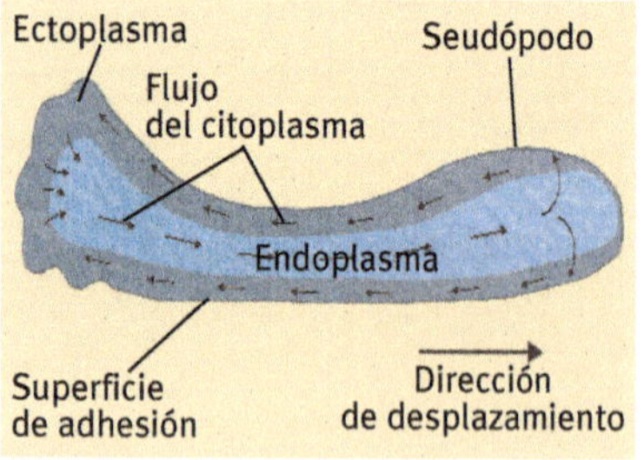

Movimiento ameboide de progresión mediante seudópodos

Consiste en captar las condiciones del medio extracelular y elaborar las respuestas más indicadas para sobrevivir en cada caso. Para ello, las células deben presentar **sensibilidad** respecto a ciertos **estímulos** (luz, sustancias químicas, temperatura, contacto...).

- Las células sensibles a la **luz** tienen orgánulos especializados en captarla, como las **manchas oculares** en las algas unicelulares.
- Si son sensibles a determinadas **sustancias químicas,** las proteínas situadas en la parte exterior de las membranas plasmáticas reaccionan a las fluctuaciones que experimenta el medio extracelular y la reacción modifica las condiciones del interior de la célula.

Las respuestas son las reacciones de la célula ante los estímulos. Las más evidentes son los movimientos de desplazamiento o **tactismos,** los cuales son positivos, si la célula se aproxima hacia la fuente del estímulo, y negativos en caso contrario.

En las células pueden darse movimientos internos mediante **corrientes citoplasmáticas** debidas a cambios de viscosidad del hialoplasma. Responden a estímulos que provocan el movimiento de los orgánulos.

Las células también pueden desplazarse:

- Mediante **seudópodos.** Éstos se producen en células libres, como las amebas y los macrófagos. Son prolongaciones del citoplasma que arrastran a la célula y la desplazan. El movimiento que se produce es **ameboide** y está relacionado con la presencia de agrupaciones de filamentos de actina, una proteína que se puede contraer en el interior de los seudópodos.
- Mediante **cilios y flagelos.** Éstos son estructuras tubulares, rodeadas de membrana plasmática, que se prolongan fuera de la célula. Este tipo de movimiento se denomina **vibrátil.**
  - — Los cilios son pequeños (0,2 micras de diámetro y 2-10 micras de longitud) y numerosos.
  - — Los flagelos son mayores (0,2 micras de diámetro y 200 micras de longitud). Se presentan en número reducido: uno o muy pocos.

Las células con pared exterior rígida (bacterias, algas unicelulares) suelen desplazarse mediante flagelos. Los flagelos de las células eucariotas tienen una estructura diferente a la de los flagelos bacterianos. El flagelo eucariota presenta partes diferenciadas:

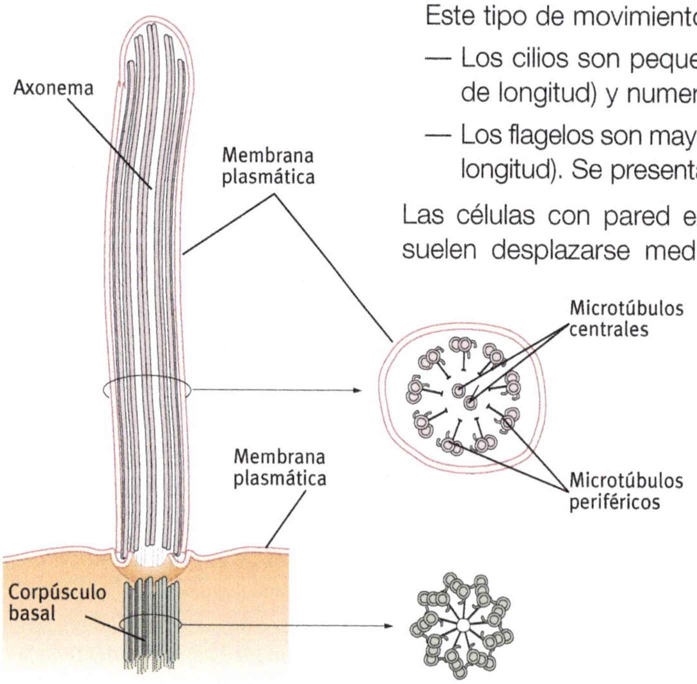

- El **axonema,** con dos microtúbulos centrales rodeados por la membrana plasmática y 9 pares de microtúbulos periféricos.
- El **corpúsculo basal,** que es un cilindro situado en la base del flagelo; consta de 9 tripletes de tubos periféricos y carece del central.

La estructura de los cilios es semejante a la de los flagelos; sin embargo, se mueven de modo diferente:

- Los cilios baten **bruscamente,** de modo similar a un látigo, y después recuperan su posición original. Con su movimiento, por ejemplo, las células del epitelio traqueal expulsan sustancias nocivas del tracto respiratorio de los vertebrados.
- Los flagelos tienen un **suave movimiento** ondulante que se origina en la base y se prolonga hasta el extremo. Los flagelos, por ejemplo, permiten el desplazamiento de los espermatozoides en el medio.

Ante condiciones ambientales muy desfavorables (calor, escasez de humedad, congelación, presencia de productos tóxicos), algunas células, como las bacterianas, producen **esporas.** Son estructuras muy resistentes que engloban una parte del citoplasma y el cromosoma. Las esporas pueden resistir mucho tiempo; cuando las condiciones mejoran, la espora *germina,* es decir, absorbe agua, activa su metabolismo y la célula se reproduce.

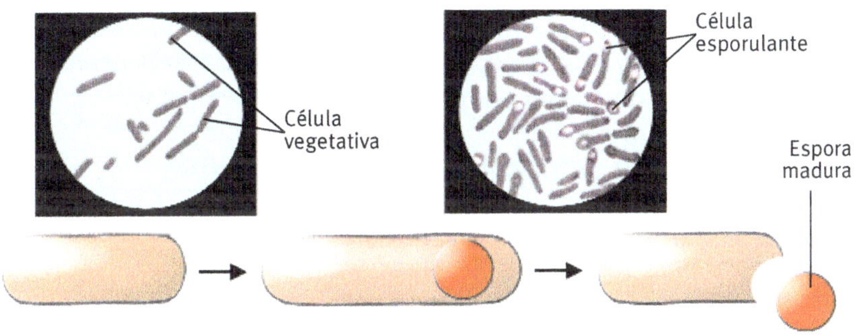

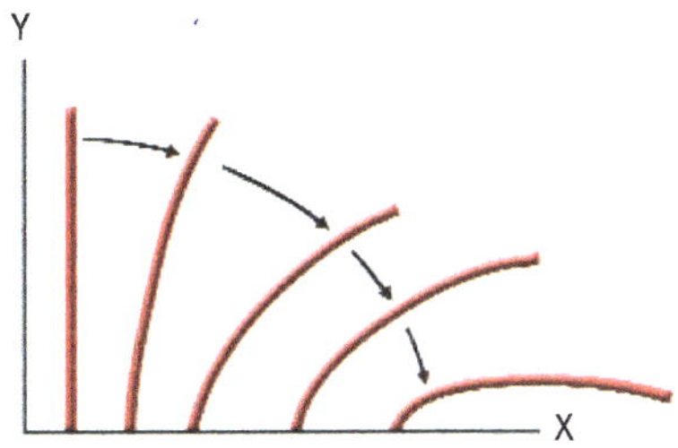

Secuencia del golpe de barrido de un cilio

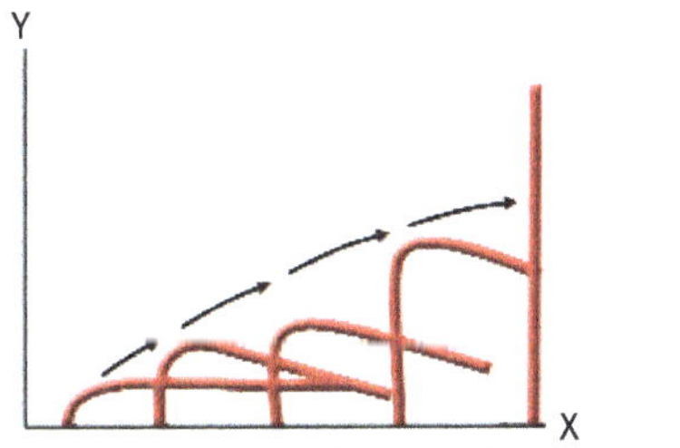

Golpe de recuperación de un cilio

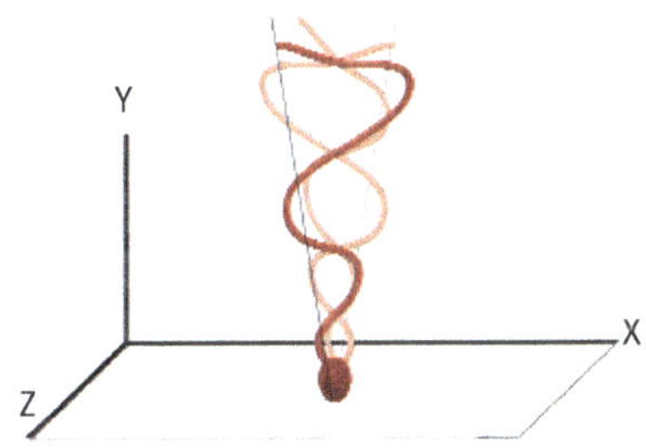

Movimiento del flagelo

Representación en ejes de coordenadas del movimiento de cilios y flagelos

7. La función de reproducción: la división celular

Uno de los principios fundamentales de la teoría celular es que toda célula procede de otra célula. Por un proceso de **división celular,** a partir de una célula madre se obtienen dos células hijas con sus mismas características morfológicas y fisiológicas. Aunque los tres modelos celulares que acabamos de estudiar presentan peculiaridades en su proceso de división, el objetivo que se pretende es común: conseguir la duplicación de la célula, de modo que las dos células hijas reciban la **dotación cromosómica** característica de cada especie y los **orgánulos necesarios** para desarrollar su actividad.

7.1. *La división celular en las bacterias*

En la división celular intervienen los mesosomas. El proceso se desarrolla del modo siguiente:

— El cromosoma se duplica:

- La doble hélice de ADN **se abre** y se separan las dos cadenas sencillas. A medida que se va abriendo, a partir de cada cadena, **se sintetiza** una cadena complementaria.
- Se obtienen **dos réplicas** iguales del cromosoma.

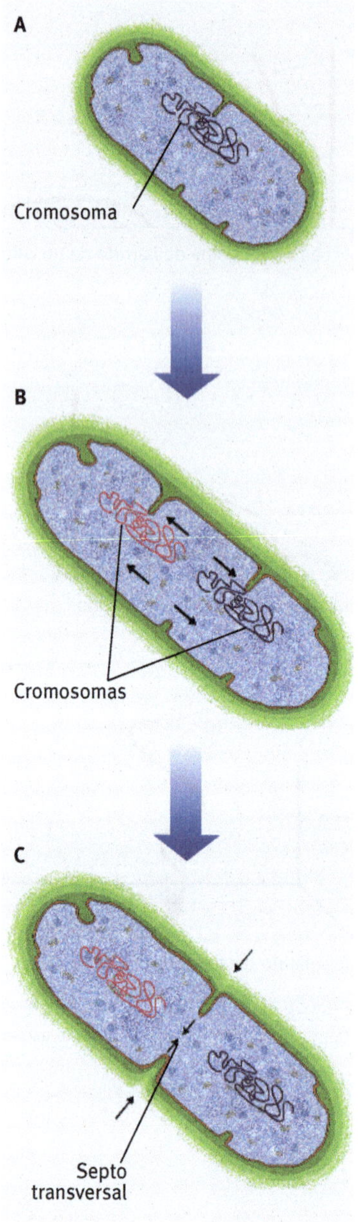

- Cada nuevo cromosoma queda sujeto a un mesosoma. Este **anclaje** permite que se separen y que cada uno quede situado en una de las células hijas.
- Se duplican los orgánulos celulares y la célula **aumenta** de tamaño.
- Se forman unas invaginaciones de la membrana, que crecen hacia el interior. Este **tabique transversal** o septo separa dos zonas del citoplasma.
- En la zona de la separación, se segregan las sustancias que formarán la **pared bacteriana** y las dos células hijas quedan separadas. Las células hijas formadas son genéticamente idénticas a la célula madre. Así, por sucesivas divisiones se pueden llegar a formar colonias de células iguales denominadas **clones**.

Las mutaciones y su rápida división celular originan la elevada diversidad de las bacterias, que posibilita una gran adaptabilidad a los cambios ambientales.

En condiciones favorables y sin limitaciones de nutrientes, las células bacterianas se dividen con una frecuencia muy elevada: una única célula original puede dar lugar, en sólo 14 horas, a una estirpe de descendientes de 250 000 bacterias. Es el caso de la bacteria *Escherichia coli*, que puede completar su división en unos 20 minutos.

7.2. *La división celular en la célula eucariota*

La célula eucariota presenta núcleo y, por ello, la división celular sigue un proceso más complejo que en la célula procariota. Tomando como referencia la célula animal, se pueden distinguir varias etapas.

Interfase

Es el tiempo que transcurre entre el final de una división celular y el principio de la siguiente. Es la etapa más larga, y hay una gran actividad bioquímica. Se caracteriza por el desarrollo de las funciones propias de la célula. El material genético del núcleo celular esta formado principalmente por ADN y proteínas, y se denomina **cromatina**. Los cromosomas no son visibles al microscopio, porque están **desespiralizados,** debido a la actividad metabólica en la célula.

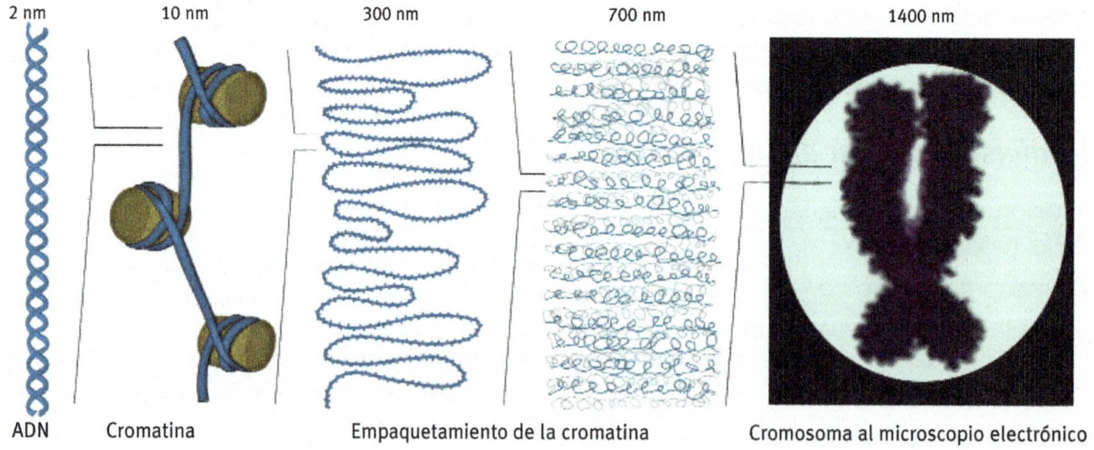

- Al aproximarse el momento de la división, se **duplican** los cromosomas.
  — Las dobles hélices de cada cromosoma **se abren** y **se copian.**
  — Se obtiene **una réplica** de cada cromosoma que permanece unida al cromosoma original por una zona llamada **centrómero.** Las dos copias de cada cromosoma reciben el nombre de **cromátidas.**
- Los cromosomas se **condensan** y son **perfectamente visibles.**
- Al mismo tiempo se **duplican** los diferentes orgánulos; cabe destacar la duplicación de los **centríolos,** por la que se obtienen dos parejas. La célula **aumenta** de tamaño.

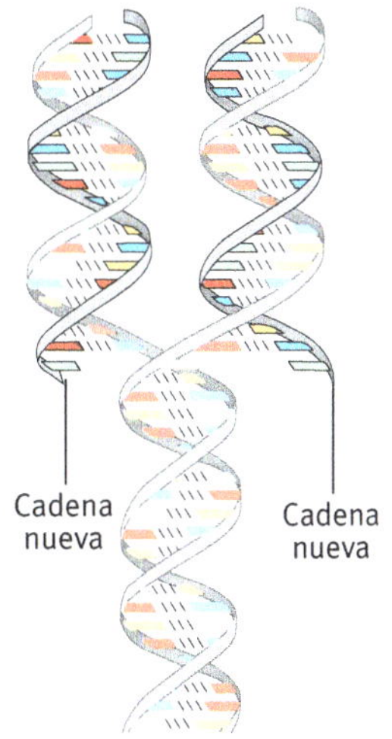

A continuación, se inicia la **división celular:** la **mitosis** es el proceso de reparto de los cromosomas duplicados a dos células hijas idénticas a la célula madre, y la **citocinesis** es el reparto del citoplasma y los orgánulos. En la mitosis distinguimos:

Profase *(dura un 40% del tiempo total de la mitosis)*

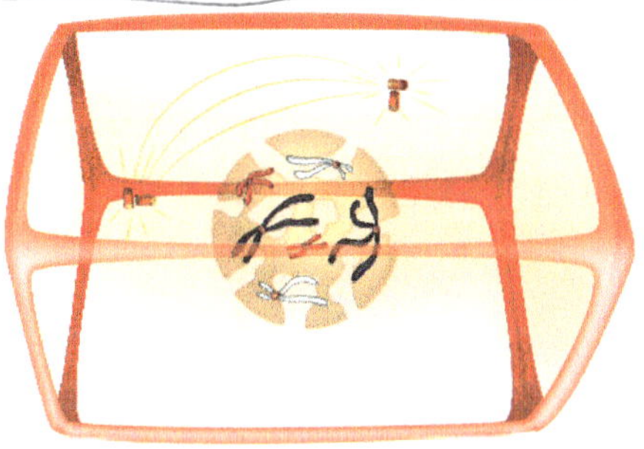

- La célula aparece más esférica y el citoplasma más viscoso. La envoltura nuclear se **dispersa** y **desaparece** el nucléolo.
- Cada pareja de centríolos se desplaza hacia uno de los polos de la célula. Entre los centríolos van apareciendo microtúbulos que se disponen formando el inicio del **huso mitótico.**

Metafase *(dura un 20 % de la mitosis)*

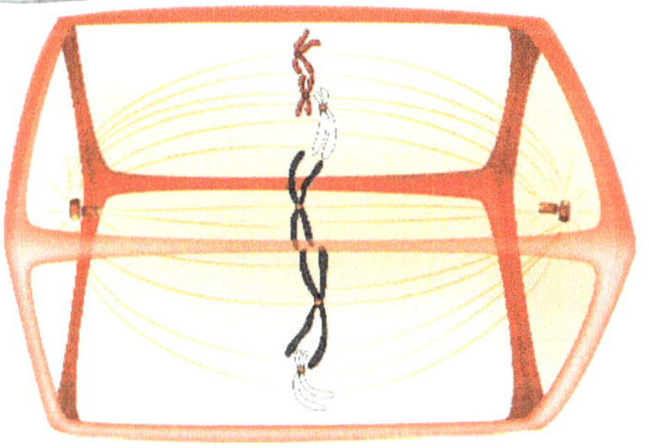

- En los centrómeros que unen las dos cromátidas se encuentran unos discos proteicos, los **cinetocoros**. A ellos se unen unas fibras del huso llamadas **fibras cinetocóricas**. Las cromátidas unidas a las fibras cinetocóricas se sitúan en la zona modia formando la **placa ecuatorial**.
- El resto de las fibras del huso se dispone de polo a polo de la célula; son las **fibras polares**.

Anafase *(dura el 10 % de la mitosis)*

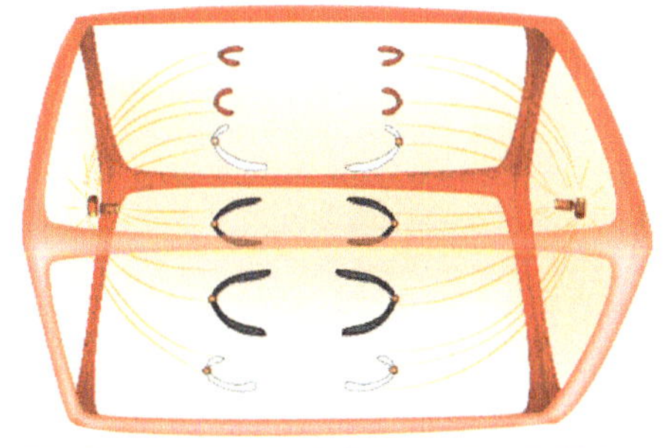

- Los **cinetocoros se dividen**; las dos **cromátidas hermanas se separan** y cada una se desplaza hacia un polo de la célula, aparentemente a causa de la tracción de las fibras cinetocóricas.
- Aparecen dos grupos de cromátidas que se mueven **simultáneamente** hacia los **polos** de la célula.

Telofase *(dura el 30% de la mitosis)*

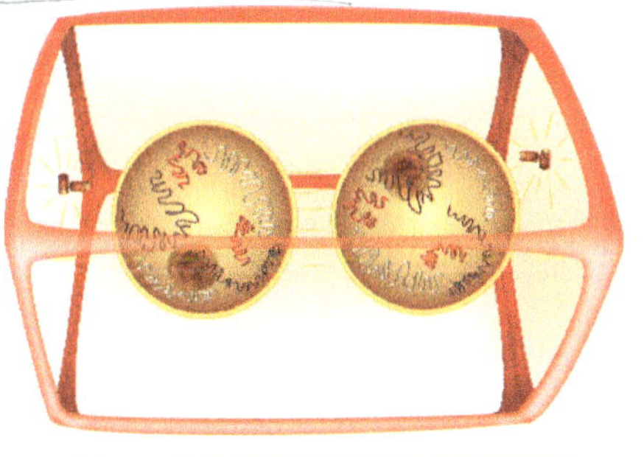

- En cada polo de la célula se sitúan dos grupos iguales de cromátidas. Las fibras **se dispersan.**
- Se inicia la formación de las **envolturas nucleares.**
- La imagen del interior del núcleo se va haciendo difusa por la **desespiralización** de los cromosomas y aparece un **nucléolo** en cada núcleo.

Durante la telofase, en las células animales se inicia la **citocinesis** o división del citoplasma.

- Aparecen unas **invaginaciones** de la membrana en la zona media de la célula, que progresan hacia el interior. Parece ser que este movimiento está provocado por fibras de la proteína **actina** que actúan como una lazada que constriñe a la célula.
- La conexión entre las dos células va estrechándose y, finalmente, se **separan.**

Las células hijas son menores que la célula madre, pero durante la interfase, que se inicia a continuación, alcanzan el tamaño y todas las características que tenía la célula madre.

En algunas células, como algas y hongos, a la duplicación de los núcleos o mitosis no le sigue inmediatamente la división del citoplasma o citocinesis. En estos casos se constituyen células con más de un núcleo.

En la **célula vegetal,** al no presentar centriolos y tener además una pared celular rígida, el proceso presenta algunas variaciones:

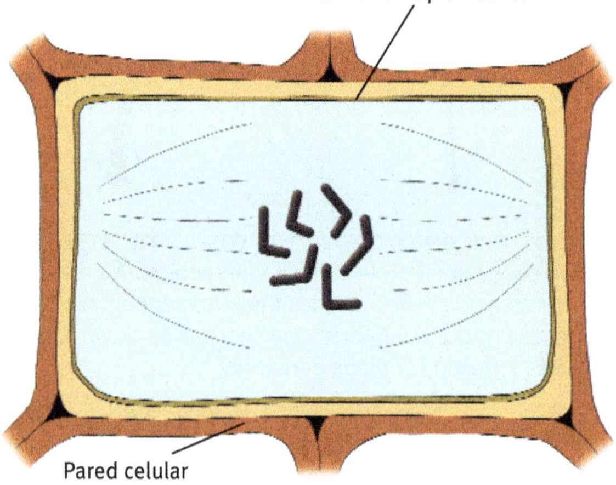

Durante la separación de los dos grupos de cromosomas, se observa la presencia de **fibras** interpuestas entre ellos. De modo similar a las fibras del huso mitótico en la célula animal, controlan el **movimiento** de las cromátidas.

Guiadas por estas fibras, las cromátidas se separan y se disponen en dos grupos que formarán parte de los núcleos de las células hijas.

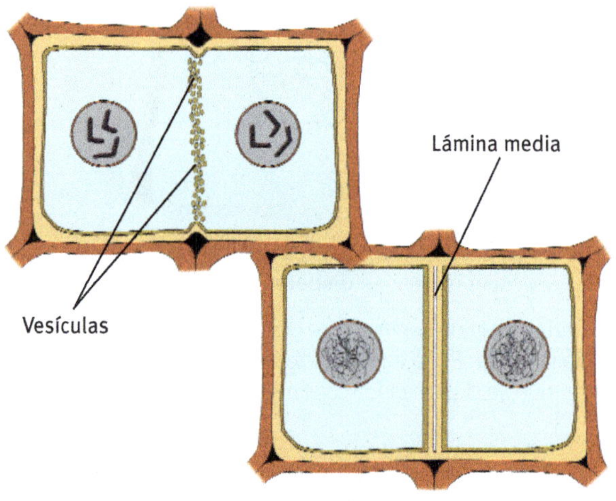

Durante la citocinesis aparecen **vesículas** procedentes del complejo de Golgi que contienen **polisacáridos**, situadas en la zona media de la célula.

Se fusionan entre ellas y, posteriormente, con la membrana plasmática. Entre las dos células hijas se forma un depósito de polisacáridos, como la **pectina**, denominado **lámina media**. Con la adición de celulosa, segregada por cada célula hija, se formará la pared celular.

ACTIVIDADES

- **14.** ¿Por qué motivo se produce en las células eucariotas el proceso de la mitosis?
- **15.** Explica que mecanismo produce el movimiento de los cromosomas durante la anafase.
- **16.** Contrasta los fenómenos que ocurren durante la citocinesis en las células animales y vegetales.

8. Los virus

Los virus no tienen estructura celular. Por ello son muy diferentes de cuanto hemos estudiado hasta el momento; sólo podemos encontrar los siguientes paralelismos:

- Están compuestos por biomoléculas.
- Los ácidos nucleicos actúan como moléculas informativas.

Por sí solos no pueden llevar a cabo ninguna función vital. Su supervivencia se basa únicamente en infectar una célula y utilizar los componentes de ésta para reproducirse. Por esta razón, se consideran **parásitos intracelulares** obligados de las células.

Los virus que infectan bacterias se denominan **bacteriófagos;** los que infectan células animales o vegetales reciben el nombre de **virus animales** o **virus vegetales.**

Aunque existen distintos tipos, con aspecto diferente, todos ellos tienen la misma composición:

- Una **cápsida** de proteínas que constituye la cubierta externa del virus.
- Una molécula de **ácido nucleico** en el interior, que puede ser ADN o ARN de cadena sencilla o doble. Permite la **síntesis de las proteínas** de la cápsida y de otras necesarias para infectar las células y reproducirse.

El **ciclo vital** de los virus se divide en dos estados: uno *extracelular* y otro *intracelular*.

En **estado extracelular**, la partícula vírica denominada **virión** es metabólicamente inactiva.

En **estado intracelular** tiene lugar la replicación del virus: se producen nuevas copias del genoma del virus y se sintetizan los componentes de su cubierta proteica. Este proceso se denomina infección.

Los virus, una vez han infectado la célula, pueden desarrollar dos modalidades de ciclo vital: un **ciclo lítico** o un **ciclo lisogénico.**

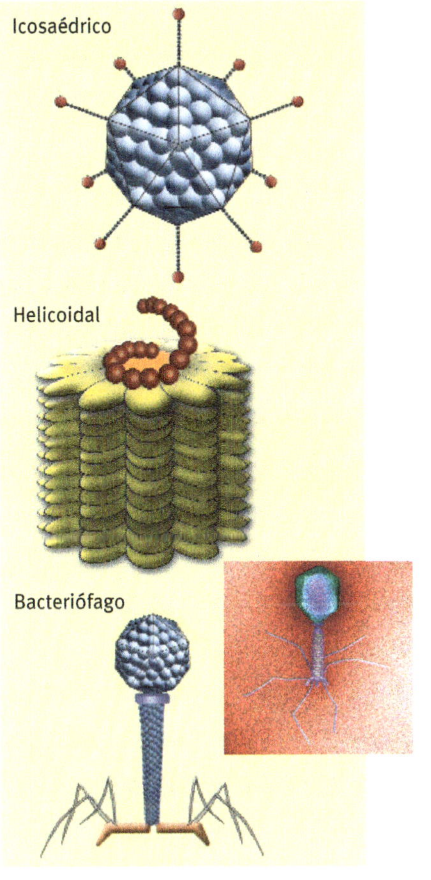

Tipos de virus según su forma. En la parte superior, micrografía de un bacteriófago.

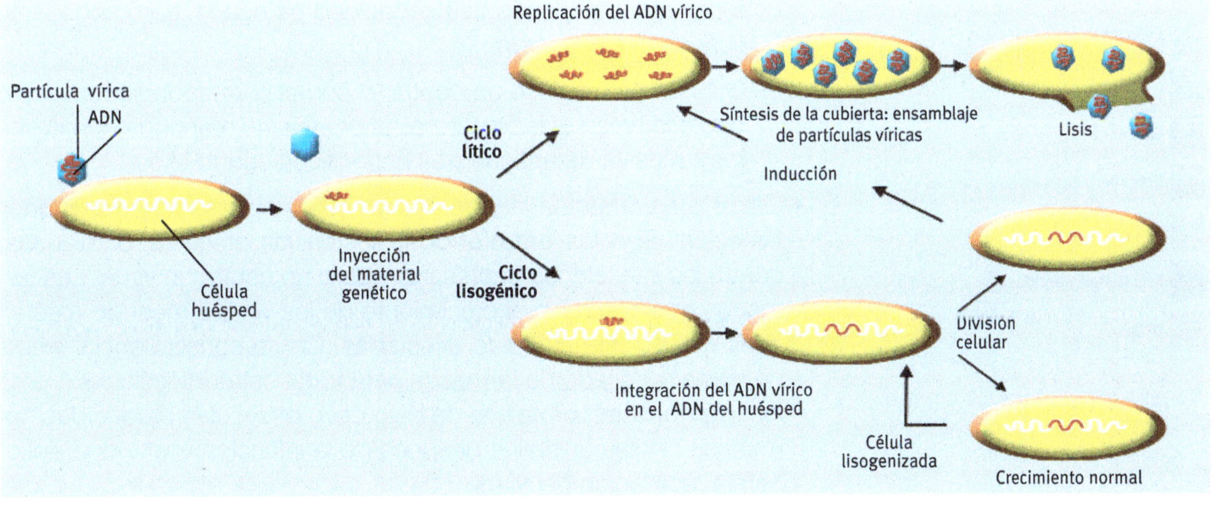

CICLO LÍTICO	CICLO LISOGÉNICO
• Si el virus contiene ADN, **se incorpora** a los cromosomas de la célula. Si es RNA, se transforma en ADN y se incorpora al cromosoma.	• El ADN vírico se **incorpora** en el celular y permanece **inactivo** durante un tiempo indeterminado, que depende del propio virus y de la célula.
• Desde el cromosoma **dirige** la actividad celular y produce copias de sí mismo sintetizando ácidos nucleicos y cápsidas proteicas.	• Cuando la célula se divide, sus **descendientes** llevan **integrada** en sus cromosomas una copia del ADN del virus. Éste se encuentra en estado de provirus o **profago**.
• Las dos piezas se ensamblan y se constituyen los nuevos virus, que producen la **lisis** de la célula para salir al exterior e infectar nuevas células.	• Una alteración de las condiciones ambientales puede provocar la activación del ciclo lítico. Se formarán nuevos virus que producen la lisis de la célula

Enfermedades bacterianas	Enfermedades víricas
Tétanos	Resfriado
Tos ferina	Gripe
Cólera	Sida
Difteria	Sarampión

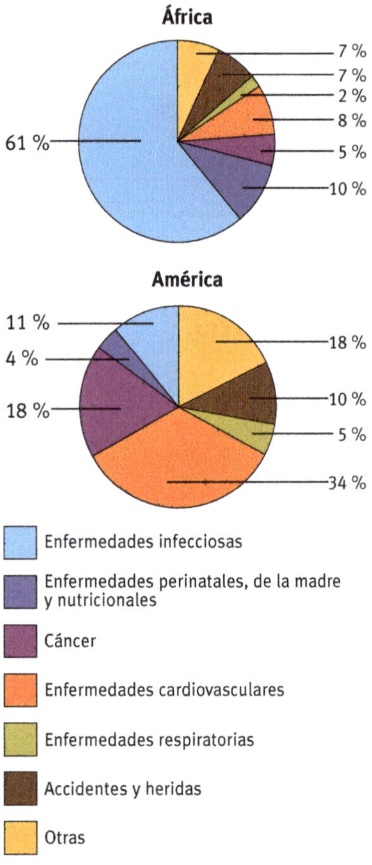

Principales causas de muerte en América y África. Las causas varían notablemente según el grado de desarrollo de los países.

9. Las enfermedades infecciosas

Los virus, numerosas especies bacterianas y otros tipos de microorganismos producen trastornos a los seres vivos a los que infectan. Provocan **enfermedades infecciosas.** Nuestro organismo tiene mecanismos de defensa contra este tipo de enfermedades.

- Poseemos varios tipos de **barreras externas** que impiden la entrada de los microorganismos: la piel, las lágrimas, la saliva, las mucosas de la boca y las vías respiratorias son algunos ejemplos.
- El **sistema inmunitario** combate los microorganismos que consiguen superar las barreras externas. Lo hace mediante:
  - — La producción de células que fagocitan a los agentes patógenos; por ello, se denominan **fagocitos.**
  - — La estimulación de células denominadas **linfocitos** que refuerzan la acción de los fagocitos y, además, sintetizan **anticuerpos** específicos contra el microorganismo invasor.

Pero nuestros sistemas de defensa no son siempre efectivos; en estos casos deben aplicarse dos tipos de medidas:

- Unas que protejan del contagio de la enfermedad; las más efectivas son las **vacunas:**
  - — Se pone en contacto el organismo con partículas víricas o bacterias que han perdido su capacidad patógena, pero son identificadas por el cuerpo.
  - — Cuando se produce un contacto con el correspondiente virus o bacteria en su estado natural, el organismo responde con la síntesis masiva de anticuerpos específicos contra ellos.
- Otras que combatan la enfermedad cuando ya se ha producido la infección. Son los **antibióticos,** sustancias efectivas para luchar contra las infecciones bacterianas, pero no contra un virus. Los fármacos que controlan el crecimiento de los virus a menudo afectan a las células que han sido infectadas. Como consecuencia, existe un número limitado de fármacos para tratar estas infecciones, como los **interferones**, proteínas de bajo peso molecular, producidas por muchas células animales después de la infección viral y que detienen la replicación del virus.

17 Lee el siguiente artículo sobre los radicales libres y su relación con las enfermedades degenerativas y responde las preguntas que tienes a continuación:

ESTRÉS OXIDATIVO

Los radicales libres son moléculas extremadamente inestables que en su estructura atómica presentan un electrón sin emparejar, confiriéndoles una gran capacidad para formar enlaces con otro átomo o molécula, para conseguir su estabilidad electroquímica. Dentro de este concepto encontramos los ROS (Radical Oxigen Species) que se forman como productos del metabolismo normal del oxigeno.

La mitocondria constituye la principal fuente de radicales libres: al nivel de la cadena de transporte de electrones se genera un gradiente eléctrico, a través de la membrana mitocondrial interna, que aporta energía suficiente para formar ATP. El oxígeno actúa como aceptor final de electrones y en este proceso se forman diversas moléculas con diferente grado de oxidación que pueden originar radicales libres (óxido nítrico, radical hidroxilo, peróxido de hidrógeno...).

Nuestro organismo ha desarrollado un elaborado mecanismo de defensa antioxidante que protege los tejidos del daño oxidativo, mediante enzimas.

En los procesos de estrés oxidativo la producción de ROS es muy alta y sobrepasa las defensas antioxidantes de nuestro sistema, produciendo alteraciones moleculares que afectan todo tipo de moléculas biológicas: los ROS reaccionan químicamente con lípidos, proteínas, carbohidratos y ácidos nucleicos.

La oxidación de las proteínas se encuentra muy involucrada en diversas enfermedades. El daño proteico que se da en situaciones de estrés oxidativo puede ser fruto de una oxidación directa del ROS sobre las proteínas. El estrés oxidativo se ha relacionado, con la muerte celular, debido al envejecimiento, y con diversas enfermedades como la aterosclerosis, cáncer, cataratas, artritis, diabetes, enfermedades neurodegenerativas... Actualmente existe una gran tendencia a relacionar el estrés oxidativo con enfermedades como el Alzheimer o el Parkinson. Está documentado que el daño oxidativo interviene en procesos de agregación de proteínas, desregulación del calcio, mal funcionamiento mitocondrial, inflamación crónica, alteración de la función antioxidante... hechos, entre otros, presentes en estas enfermedades.

El cerebro posee ciertas características que lo hacen especialmente vulnerable al estrés oxidativo: consume un importante porcentaje de oxígeno, es un órgano rico en ácidos grasos poliinsaturados (su estructura de dobles enlaces aumenta la probabilidad de ser oxidados), presenta cierta tendencia a acumular iones metálicos (los desequilibrios en la cantidad de hierro y cobre pueden producir consecuencias neurológicas severas), y tiene una baja cantidad de sistemas antioxidantes. Por tanto, una reducción de la expresión de proteínas antioxidantes, conduce al estrés oxidativo, acumulando proteínas en las células cerebrales y acelerando la neurodegeneración.

Enfermedades neurológicas como las de Alzheimer, Parkinson, Huntington o enfermedades priónicas, están asociadas a la agregación de proteínas. No esta claro, pero, si la agregación de proteínas es consecuencia de esta situación o bien, el hecho de que se estructuren en forma de agregados, provoca una desregulación celular que afecta a la mitocondria, aumentando la formación de ROS y favoreciendo la oxidación de las proteínas. (...)

Anna Martínez Casals
Adaptado y traducido de OMNIS CEL·LULA (junio 2007)

- ¿Qué diferencia existe entre el oxígeno molecular y los radicales libres de oxigeno?
- ¿Qué orgánulo interviene en la formación de radicales libres?
- Indica qué transporte de los estudiados interviene en la formación de radicales libres.
- ¿Qué es el estrés oxidativo? Explica su toxicidad para el organismo.
- Indica tres enfermedades que estén relacionadas con el estrés oxidativo.
- Busca información y elabora un informe de la enfermedad del Alzheimer o de la enfermedad del Parkinson: características generales de la enfermedad, diagnóstico y pronóstico.
- ¿Por qué las anteriores enfermedades se las considera enfermedades neurodegenerativas?
- Indica tres motivos que expliquen la vulnerabilidad del cerebro al estrés oxidativo.
- Discute si la alimentación puede influir en el estrés oxidativo.

18 ¿Qué estructura celular representa la figura? Identifica sus diferentes elementos.

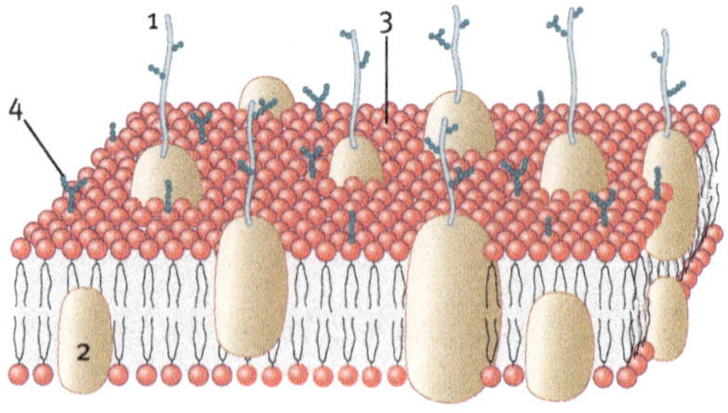

19 Indica y razona qué mecanismo de transporte utilizará la célula para introducir las partículas o sustancias siguientes:

— Oxígeno
— Glucosa
— Gotas de líquidos extracelulares
— Partículas sólidas de gran tamaño

20 Supón que existen las siguientes concentraciones dentro y fuera de una célula de la retina:

	CONCENTRACIÓN (mM)	
	Dentro	**Fuera**
K^+	100	5
Na^+	10	140
Cl	10	100

- Explica qué tipo de mecanismo de transporte utilizaría la célula para: a) introducir K^+; b) introducir Na^+; c) extraer Cl^-

- Explica qué tipo de célula presentaría una cantidad de mitocondrias mayor, una célula muscular de la piernas de un atleta o una célula vegetal de la epidermis de la cebolla. Justifica la respuesta.

21 Completa cada una de las frases, indicando el orgánulo que le corresponde.

El ciclo de Krebs se produce en ....

La fase lumínica de la fotosíntesis se produce en .... y la fase oscura en ....

La .... es exclusiva de las células vegetales. Es una estructura rígida que rodea la ....

Los enzimas hidrolíticos se encuentran en ....

22 Observa la siguiente imagen de una célula.

- Indica si se trata de una célula eucariota animal o vegetal. Justifica la respuesta.

- Identifica cada una de las estructuras numeradas.

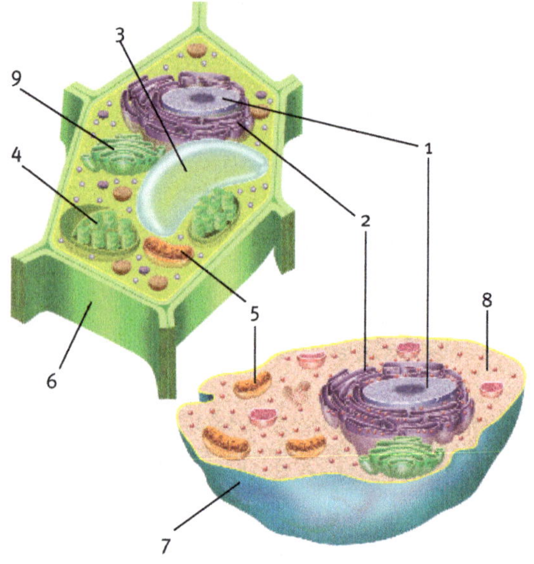

23 En términos generales. ¿En qué se diferencian dentro de la fotosíntesis las reacciones independientes de la luz de las que dependen de la luz?

- ¿Cuáles son los principales productos de ambos tipos de reacciones?

24 ¿Cuáles son los productos finales de la degradación de la glucosa por vía aerobia y por vía anaerobia?

- Razona cuál de las dos vías es más rentable energéticamente.

25 Contrasta los movimientos que describe un cilio y un flagelo.

26 Explica mediante qué mecanismo una bacteria percibe la dirección en que se encuentra una sustancia y se dirige hacia ella.

27 Observa las diferentes etapas de la mitosis:

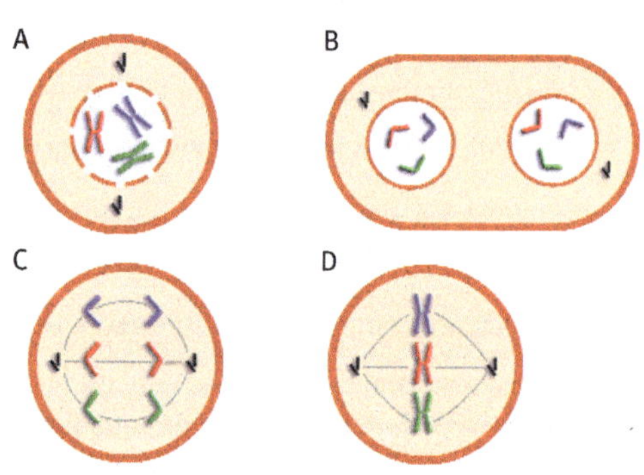

- Indica qué etapa representa cada una de las figuras. Justifica la respuesta.

28 ¿Qué características de la estructura y función de un virus indicarías para argüir que los virus son organismos vivos?

29 Da una definición de profago. Describe en qué etapa del proceso de replicación de un virus se encuentra.

30 ¿Puede suponer alguna ventaja para el virus tener una fase extracelular metabólicamente no activa?

31 ¿Por qué hay tan pocos fármacos antivíricos eficaces?

32 Actualmente se sugiere la existencia en una parte de la población de un desarrollo rápido de resistencias especificas a microorganismos que producen enfermedades. Busca información sobre este tema y explica por qué crees que aparece la resistencia a los antibióticos.

— ¿Qué medidas deberían tomarse para evitar el desarrollo de resistencias a los antibióticos?

33 Observa el diagrama de la página 180 sobre las causas de muerte en diferentes grupos de países.

- Compara la mortalidad producida por las enfermedades infecciosas en los países desarrollados y en los países en desarrollo.
- Describe los principales sistemas de defensa que son utilizados para combatir las enfermedades infecciosas.

S

34 La célula eucariota presenta una estructura muy compleja. Completa la siguiente tabla que relaciona orgánulos celulares y estructuras celulares con la función que realizan en la célula:

Orgánulo celular	Estructura	Función
Aparato de Golgi		
	Contiene la cromatina	
		Almacenar substancias diversas en las células vegetales
	Dos subunidades formadas por ARN y proteínas	

a) Los espermatozoides requieren oxigeno para obtener la energía necesaria para su movimiento.

- Indica qué orgánulo celular utilizará para obtener la energía.
- Haz un dibujo de la ultraestructura de este orgánulo, señalando sus partes principales.
- Explica dos semejanzas que puede tener este orgánulo con las bacterias actuales.

b) ¿Las células procariotas son, en general, más grandes o más pequeñas que las eucariotas? Indica cuáles son las características de las células procariotas y de las células eucariotas en los aspectos siguientes:

- localización del ADN
- presencia y localización de ribosomas
- metabolismo aeróbico o anaeróbico
- tipo de organización (unicelular o pluricelular)

c) Si se colocan células sanguíneas en una solución de concentración mucho más baja que el suero fisiológico, ¿qué modificaciones de forma y volumen observaríamos? Dibújalo y explica a qué son debidos estos cambios.

TIC

35 Consulta la siguiente web www.elpais.com/todo-sobre/tema/gripe/aviar/121/ en la que encontrarás amplia información sobre diversos aspectos de una de las pandemias de esta década, la gripe aviar.

— Dirígete a *Animaciones/Gráficos* de esta página, abre la animación: *Virus de la gripe aviar* y contesta las siguientes preguntas:

a) ¿Cómo se denomina el virus de la gripe aviar? ¿A qué animales infecta?

b) Explica los pasos de infección del virus en una célula sana.

c) ¿En qué país se inicio la epidemia de la gripe aviar?¿A que países se extendió la epidemia en el 2004?

d) Explica a qué países se continuó dispersando la enfermedad a lo largo del año 2005. ¿Cuál fue el motivo de esa dispersión?

— Visita el gráfico: Esquema de la propagación de la gripe aviar.

e) Explica como se puede propagar la enfermedad desde las aves hasta los seres humanos.

— Accede a la animación: La expansión del virus H5N1.

f) Realiza un esquema que indique cronológicamente los humanos infectados por la propagación del virus desde el año 2003 hasta el año 2006.

— Dirígete a Noticias:

g) Indica las dos últimas noticias que aparecen sobre la gripe aviar. Explícalas brevemente.

8

Los seres pluricelulares

Un organismo pluricelular está formado por un conjunto de células, todas ellas con la misma información genética. Sin embargo, si comparamos células nerviosas con células epiteliales de un mismo organismo podemos apreciar que son muy diferentes, tanto estructural como funcionalmente. Ello es debido a que las células de un mismo individuo se especializan en actividades y funciones concretas, agrupándose las que son similares entre sí.

De hecho, los tumores son conjuntos de células que se dividen contínuamente sin especializarse y que, por tanto, son incapaces de llevar a cabo su función.

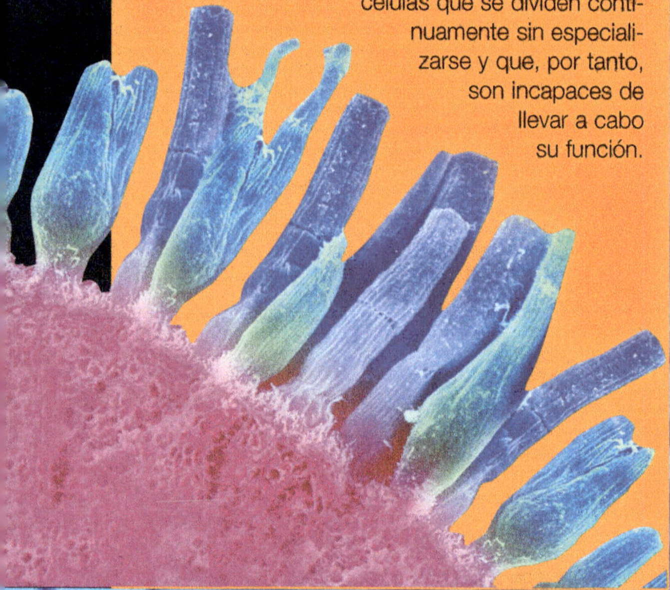

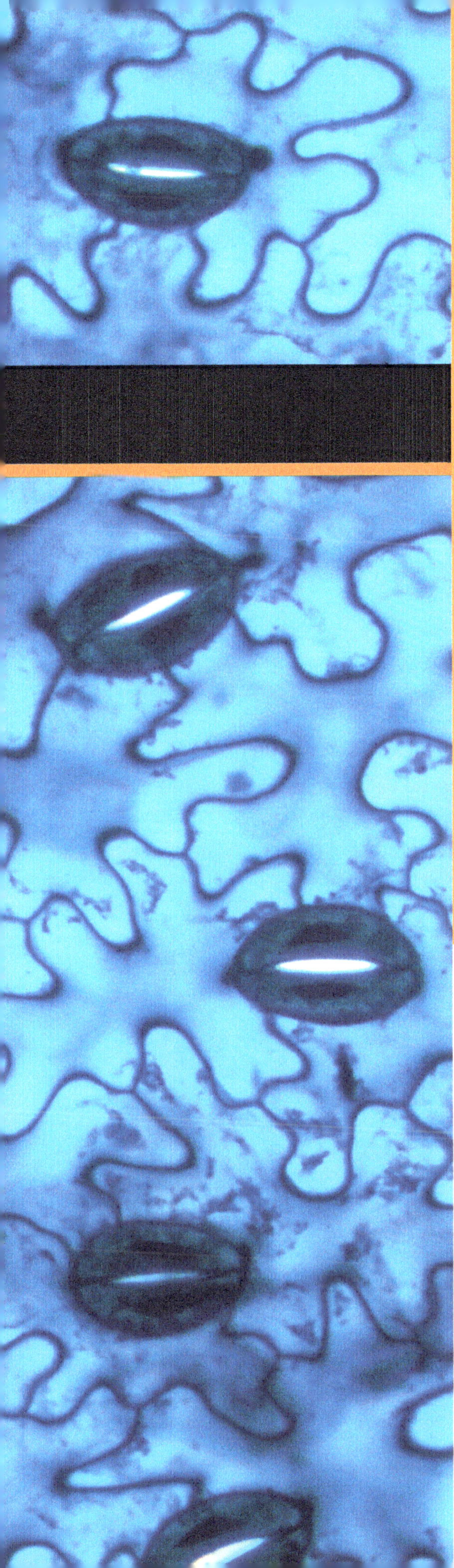

ORGANIZACIÓN DE LOS CONTENIDOS

1. La organización pluricelular
- 1.1. El medio interno
- 1.2. Presencia de distintos tipos de células

2. El desarrollo embrionario y la diferenciación celular
- 2.1. Desarrollo embrionario en las angiospermas
- 2.2. Desarrollo embrionario en los animales

3. Tejidos vegetales
- 3.1. Meristemos
- 3.2. Tejidos conductores
- 3.3. Tejidos protectores
- 3.4. Parénquimas
- 3.5. Tejidos de sostén

4. Tejidos animales
- 4.1. Tejido epitelial
- 4.2. Tejidos conectivos
- 4.3. Tejido muscular
- 4.4. Tejido nervioso

5. Nomenclatura y clasificación de los seres vivos

OBJETIVOS

- Conocer las características generales de la organización pluricelular.
- Identificar los procesos más significativos del desarrollo embrionario en vegetales y animales.
- Reconocer las características generales de los diferentes tejidos vegetales y animales y relacionarlas con su función.
- Comprender los conceptos de género y especie.
- Conocer una de las posibles clasificaciones de los seres vivos y valorar su utilidad.

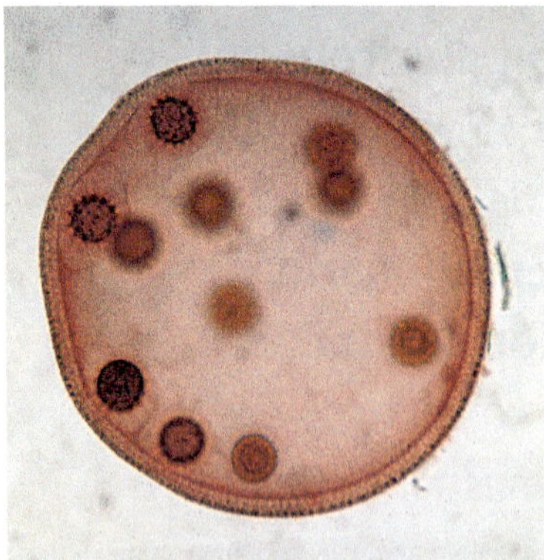

Volvox es un alga colonial de forma esférica. Presenta una organización pluricelular muy sencilla: un conjunto de células superficiales que realizan las funciones de nutrición y relación, y las células encargadas de la reproducción, las oosporas, que se encuentran en el interior de la esfera.

1. La organización pluricelular

Las células eucariotas tienen orgánulos que actúan como «compartimentos» dentro de la célula, en los cuales se desarrollan funciones específicas. Las sustancias que deben participar en cada función se concentran en estos «compartimentos», lo que incrementa la eficacia de los distintos procesos.

Para desarrollar funciones más variadas y específicas, debería aumentarse la cantidad y el tipo de «compartimentos» y con ello el volumen del organismo. La manera más eficaz de conseguirlo sin que se vea afectada la actividad celular es mediante la presencia de muchas células que trabajan coordinadamente, a la vez que desarrollan funciones diversas, constituyendo un individuo.

Así se forman los seres pluricelulares, que presentan ciertas ventajas respecto a los unicelulares:

- Mayor **independencia** del medio exterior.

  Los seres unicelulares se ven muy afectados por cualquier cambio que sucede en su entorno.

  En los seres pluricelulares, la presencia de muchas células hace necesaria la existencia de un medio interno, que presenta, como ventaja frente a los anteriores, características más constantes que el medio exterior.

- Posibilidad de desarrollar **funciones más complejas.**

  En los seres unicelulares, su célula realiza las funciones vitales propias de ésta y que son las mismas que desarrolla el individuo.

  En los seres pluricelulares cada célula, de manera individual, desarrolla sus funciones vitales; pero, al mismo tiempo, se produce una especialización, de modo que diferentes tipos de células llevan a cabo funciones específicas dentro del individuo.

En un organismo pluricelular complejo, como el ser humano, podemos observar diferentes niveles de organización celular:

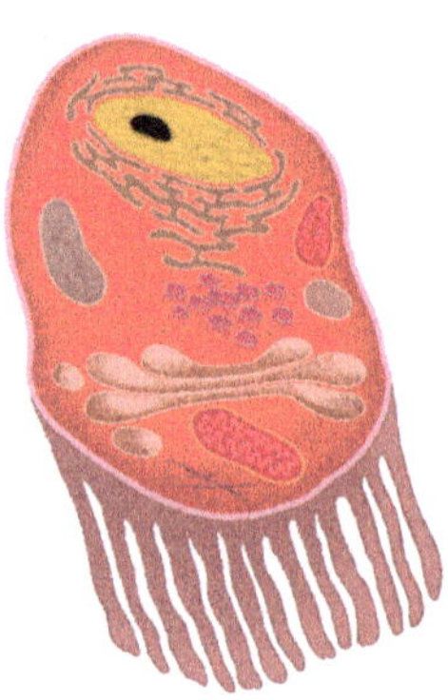

Nivel celular: célula epitelial

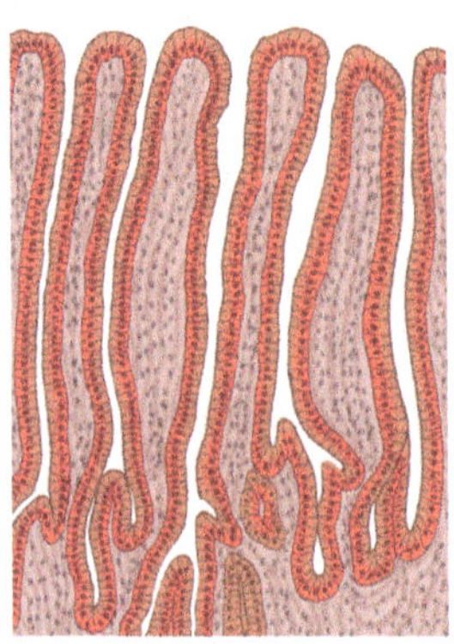

Nivel tisular: tejido epitelial

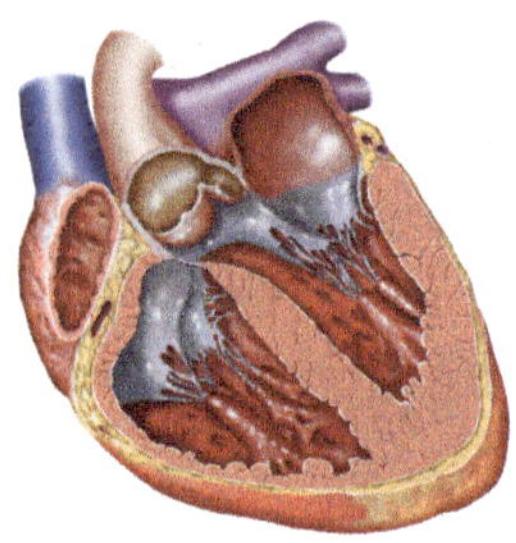

Nivel de órgano: corazón

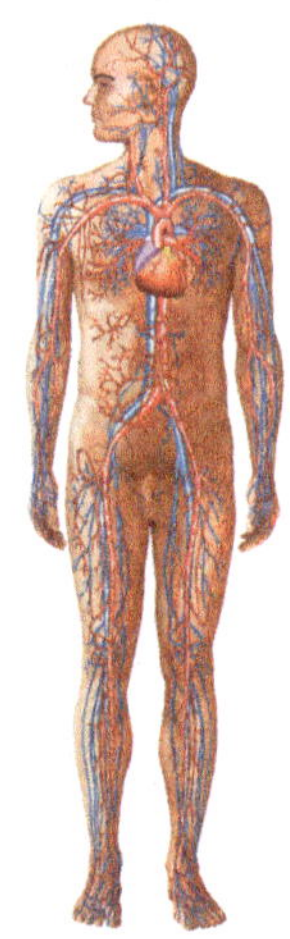

Nivel de aparato o sistema: sistema circulatorio

La **organización pluricelular estricta** es propia de una gran diversidad de organismos y se caracteriza por la presencia de:

- Un **medio interno** distinto del exterior.
- **Distintos tipos de células** que, a lo largo del proceso evolutivo, adquieren funciones especializadas.

1.1. *El medio interno*

Es el conjunto de líquidos y fluidos que rodea las células de un organismo pluricelular. Las células realizan todos los intercambios que necesitan en él.

En las plantas superiores, el medio interno está formado por la **savia bruta** y la **savia elaborada,** que circulan por la planta transportando agua y sales minerales, y los productos de la fotosíntesis, respectivamente.

En los animales acuáticos que tienen organización simple, el medio interno es, a menudo, agua del exterior del cuerpo. En los más complejos y de mayor tamaño, así como en los animales terrestres, el medio interno mantiene una composición distinta de la del medio exterior. Como componentes del medio interno distinguimos:

- El **líquido circulante** que comunica entre sí a todas las células, como la hemolinfa, la sangre y la linfa.
- El **líquido intersticial** en el que viven inmersas las células.

En los invertebrados, el líquido circulante recibe el nombre de *hemolinfa* y tiene un componente líquido, con pigmentos respiratorios disueltos.

En los vertebrados, el líquido circulante constituye un sistema doble. Hay un sistema circulatorio que transporta *sangre* y un sistema complementario de conducción de la *linfa.*

La **sangre** está compuesta por *plasma,* un líquido formado por agua y diversas sustancias disueltas (glúcidos, lípidos, péptidos, sodio, potasio, calcio). En él se encuentran inmersas las células sanguíneas: glóbulos rojos, glóbulos blancos y plaquetas.

La **linfa** es un líquido de composición similar al plasma que contiene sustancias y células especializadas en la defensa del organismo, como los linfocitos.

El líquido intersticial, también de composición similar al plasma, rodea las células de la mayoría de los tejidos, con las cuales intercambia sustancias. Posteriormente se reabsorbe y vuelve a la sangre y a la linfa.

El sistema circulatorio sanguíneo y el linfático permiten:

- Llevar hasta las células las sustancias que éstas deben utilizar.
- Recoger de las células las sustancias que producen, para su excreción o para su transporte a otras células del organismo; esto último establece comunicación entre distintas células.

En todos los seres pluricelulares, el medio interno se mantiene en unas condiciones físicas y químicas relativamente constantes. El mantenimiento del equilibrio del medio interno recibe el nombre de **homeostasis.**

C

El **plasma** es la parte líquida de la sangre y contiene tres tipos de sustancias disueltas: **proteínas** (albúminas, fibrinógeno y globulinas); **sales inorgánicas** (cloruros, bicarbonatos, sodio, potasio, calcio, magnesio, etc.), y **otras moléculas** transportadas por la sangre; éstas pueden proceder de la digestión, como la glucosa o los aminoácidos, o bien del metabolismo celular, como la urea, el ácido úrico o el CO_2.

El **suero** se diferencia del plasma porque no tiene fibrinógeno, una proteína que al transformarse en fibrina es la responsable de la coagulación de la sangre. Así, cuando la sangre se coagula aparece un líquido amarillento (el suero) alrededor del coágulo. Éste está constituido por una red de fibrina que atrapa las células sanguíneas y tapona las heridas.

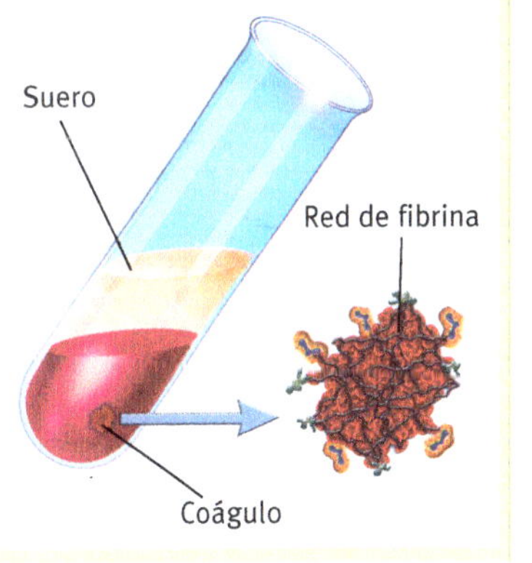

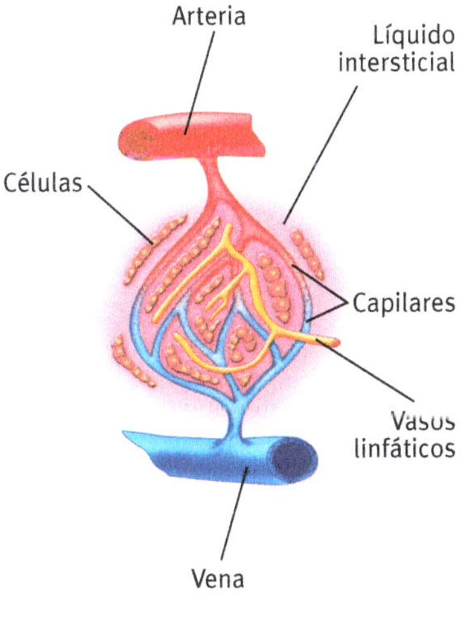

Los mecanismos que permiten la homeostasis son diversos: desde la presencia de cubiertas protectoras, como la cutícula de los vegetales o la piel de los vertebrados, hasta los más sofisticados mecanismos bioquímicos. Así, la regulación de los niveles de glucosa en sangre en el ser humano **(glucemia)** viene determinada por la secreción de insulina, una hormona que estimula la entrada de glucosa a las células, o bien, por la secreción de glucagón, que degrada el glucógeno de reserva celular, permitiendo la salida de glucosa a la sangre.

La insulina y el glucagón son hormonas antagónicas, es decir, que ejercen funciones opuestas; esto es necesario para mantener los niveles de glucosa en sangre estables (80-120 mg/100 ml). Ambas hormonas son sintetizadas en el páncreas, en los islotes de Langerhans, donde se encuentran las células α y β.

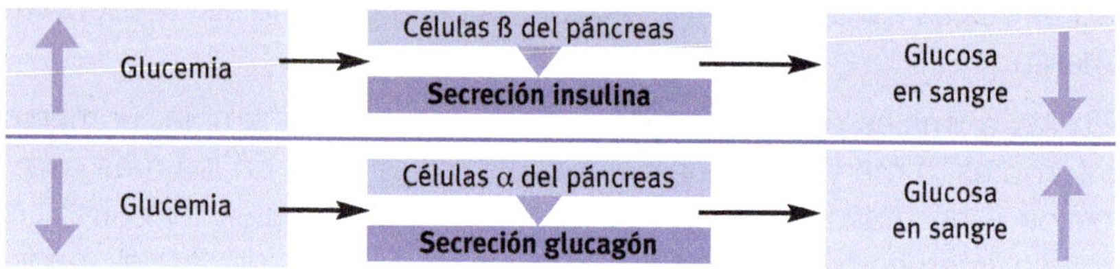

1.2. *Presencia de distintos tipos de células*

La morfología de las diferentes células que pueden formar parte de un mismo organismo es muy variada y está relacionada con la función que desempeñan. Así, los espermatozoides tienen una estructura adaptada al movimiento; la morfología de las células ciliadas ofrece la máxima superficie de intercambio posible; los adipocitos han simplificado su estructura para contener una vacuola de reserva de gran volumen, y la morfología de las neuronas está adaptada a tener numerosas conexiones con otras células.

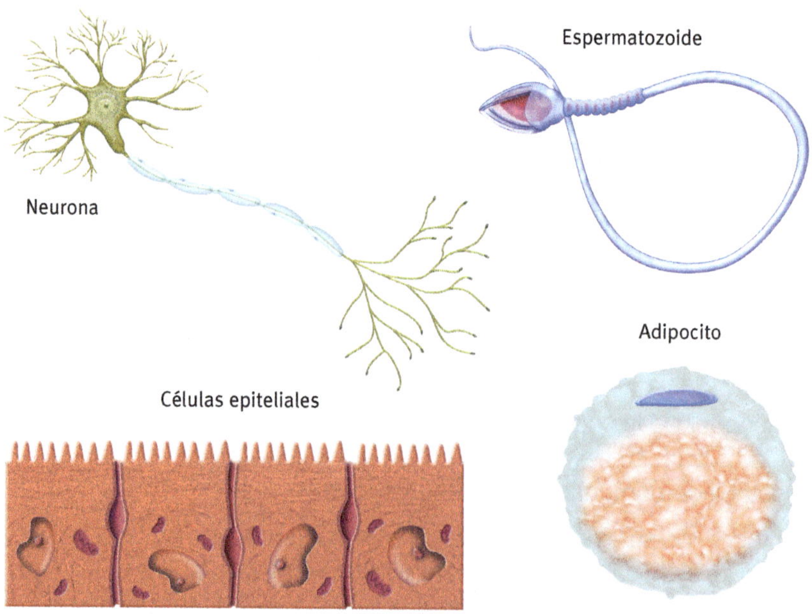

Diversidad de morfologías celulares

2. El desarrollo embrionario y la diferenciación celular

Todas las células de un organismo pluricelular derivan, por mitosis, de una sola célula original.

El **desarrollo embrionario** es el proceso por el cual, a partir de una sola célula inicial, se constituye un organismo pluricelular completo. Durante este desarrollo se producen la **multiplicación** y la **diferenciación celular,** es decir, la aparición de distintos tipos de células que presentan diferencias morfológicas y de funcionamiento y que irán constituyendo los distintos tipos de tejidos, órganos y aparatos o sistemas.

El proceso es muy diferente según se trate de vegetales o de animales. Vamos a presentar un ejemplo de cada uno.

2.1. *Desarrollo embrionario en las angiospermas*

Los espermatófitos son las plantas más numerosas de la Tierra y se clasifican en gimnospermas y angiospermas. En las plantas con flor o angiospermas, después de la fecundación, se forma la semilla. La **semilla** está formada por el embrión, encargado de originar la nueva planta; el endosperma, que es el tejido que nutre el embrión durante el proceso de la germinación, y por unas capas protectoras de todo el conjunto, normalmente endurecidas, llamadas testa.

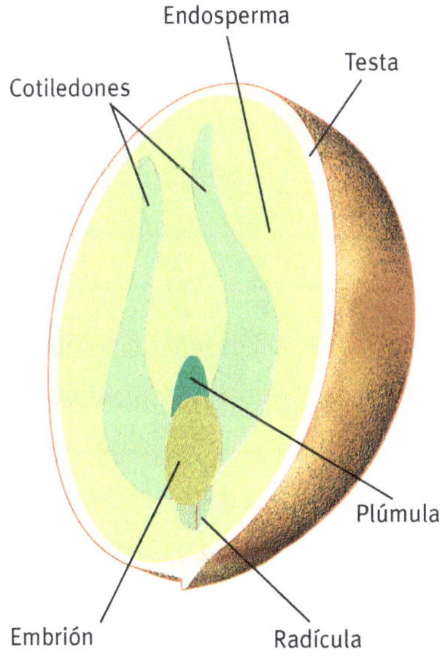

En el caso de las plantas dicotiledóneas, además de la radícula y la plúmula, en el embrión también pueden distinguirse los cotiledones. Éstos son unas hojas modificadas, denominadas primordiales, que sirven para alimentar a la plántula una vez ha germinado hasta que ésta dispone de verdaderas hojas que puedan realizar la fotosíntesis.

En las primeras fases del desarrollo del embrión se produce la diferenciación en éste de tres tipos de tejidos.

- **Meristemos primarios,** que son tejidos especializados en el crecimiento de la planta. Posteriormente, se diferenciarán en otros tipos de tejidos.
- **Tejidos vasculares,** con la misión de conducir la savia.
- **Tejidos epidérmicos** que forman la cubierta externa y protectora de la planta.

Estos tejidos constituyen las partes de un embrión; durante la germinación, a partir del embrión se desarrolla la planta adulta.

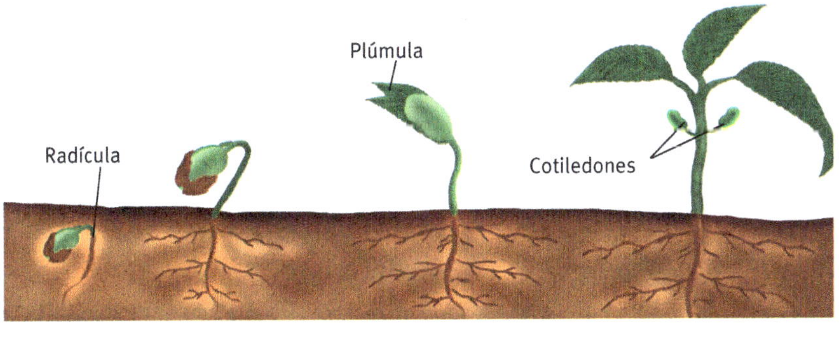

2.2. *Desarrollo embrionario en los animales*

Es un proceso que se desarrolla de forma muy diversa, según el grupo de animales de que se trate. A pesar de ello, en la mayoría de los casos podemos distinguir dos etapas comunes.

— **Segmentación.** El zigoto que resulta de la fecundación se divide por mitosis y da lugar a una esfera que contiene 2, 4, 8... células, según progresan las sucesivas divisiones del zigoto.

Se forma una esfera maciza de células denominada **mórula.** Hasta ese momento, las células, por lo general, no están diferenciadas y cada una de ellas, denominada **blastómero,** puede dar lugar a un nuevo individuo si se separa de las demás.

A continuación, las células se desplazan y dejan en el centro de la esfera una cavidad, el **blastocele.** Empiezan a especializarse y pierden, por tanto, la capacidad de generar un individuo cada una. Es el estadio de **blástula.**

— **Gastrulación.** Se produce una invaginación hacia el interior del blastocele que da lugar a la aparición de dos capas de células denominadas **ectodermo,** la exterior, y **endodermo,** la interior. En este estadio el embrión se denomina **gástrula** y la cavidad interna, **arquenterón.** Por la zona donde se produce la invaginación queda el **blastoporo,** que comunica el exterior con el arquenterón.

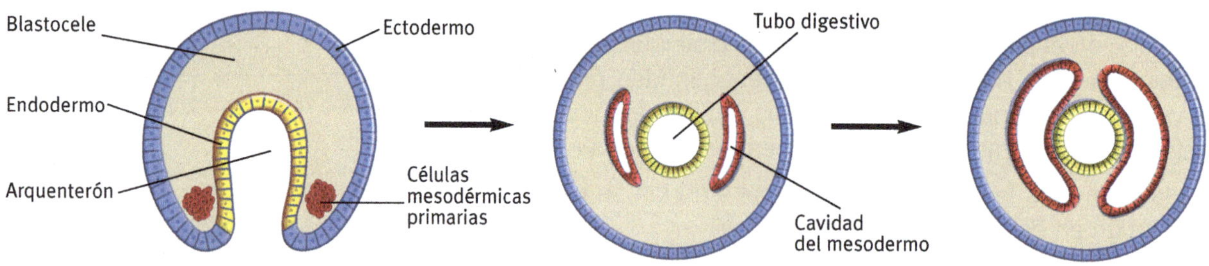

Gastrulación. Desarrollo de un animal triblástico.

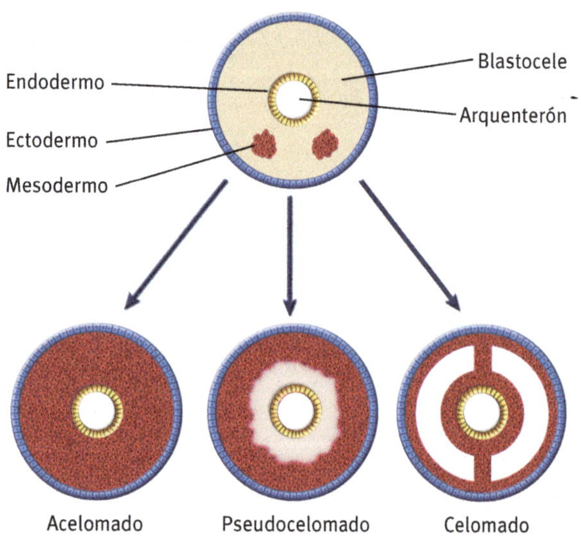

A partir de aquí, el desarrollo embrionario se produce según las características de cada grupo animal:

- **Animales diblásticos:** provienen de una gástrula diblástica en la que no se desarrollan las células mesodérmicas primarias. Todas sus células provienen del ectodermo y del endodermo. El arquénteron actúa como cavidad digestiva. Las esponjas y los cnidarios son animales de este grupo.
- **Animales triblásticos:** la gástrula diblástica continúa su desarrollo y genera, a partir de las células del endodermo, unas células primarias que dan lugar al mesodermo, una tercera capa situada entre el ectodermo y el endodermo. Si entre mesodermo y endodermo no queda espacio vacío, se habla de animales **acelomados,** como los platelmintos. Si existe este espacio, se denomina *pseudoceloma* y son animales **pseudocelomados.**

Si la cavidad está totalmente rodeada por el mesodermo, se denomina *celoma* y se trata de animales **celomados,** como los anélidos, los artrópodos y los vertebrados.

Después de la gastrulación, las células desarrollan plenamente su capacidad de diferenciación, dando lugar a los distintos partes y órganos de cada individuo. En el cuadro siguiente se muestra la capa embrionaria de la que derivan diversas estructuras de los vertebrados.

CAPA EMBRIONARIA	PARTES QUE DERIVAN
Ectodermo	Epidermis de la piel. Órganos anexos a la piel (uñas, pelo...). Encéfalo y sistema nervioso.
Endodermo	Revestimiento del tubo digestivo y del respiratorio. Glándulas como el hígado, el páncreas...
Mesodermo	Dermis de la piel. Revestimiento de cavidades internas. Órganos excretores y reproductores. Órganos circulatorios. Músculos y esqueleto.

ACTIVIDADES

1. Un solo embrión, ¿puede generar dos o más individuos? En caso afirmativo, explica qué estadio del desarrollo embrionario está implicado. Razona también si los individuos que provienen del mismo embrión tienen o no la misma información genética.
2. La información genética del zigoto, ¿es la misma que la que se encuentra en una célula de la mórula? Razona la respuesta.
3. ¿Qué diferencias hay entre el arquénteron y el celoma?

3. Tejidos vegetales

En el reino vegetal se distinguen dos grandes modelos estructurales:

- Los **talófitos**, constituidos por un talo, es decir una masa de células indiferenciadas en la que no se distinguen tejidos. Poseen un órgano de fijación, el rizoide; un órgano de sostén, el cauloide y un filoide que tiene una estructura laminar fotosintética, que recuerda a las hojas. Las algas, los musgos y las hepáticas son talófitos.
- Los **cormófitos**, constituidos por un cormo, una estructura en la que las células están agrupadas en tejidos y constituyen diferentes órganos, como la raíz, el tallo y las hojas. Los pteridófitos y los espermatófitos son cormófitos. Los espermatófitos, o plantas con semilla, se dividen a su vez en angiospermas y gimnospermas.

A continuación vamos a describir las principales características de los tejidos vegetales de las plantas cormófitas.

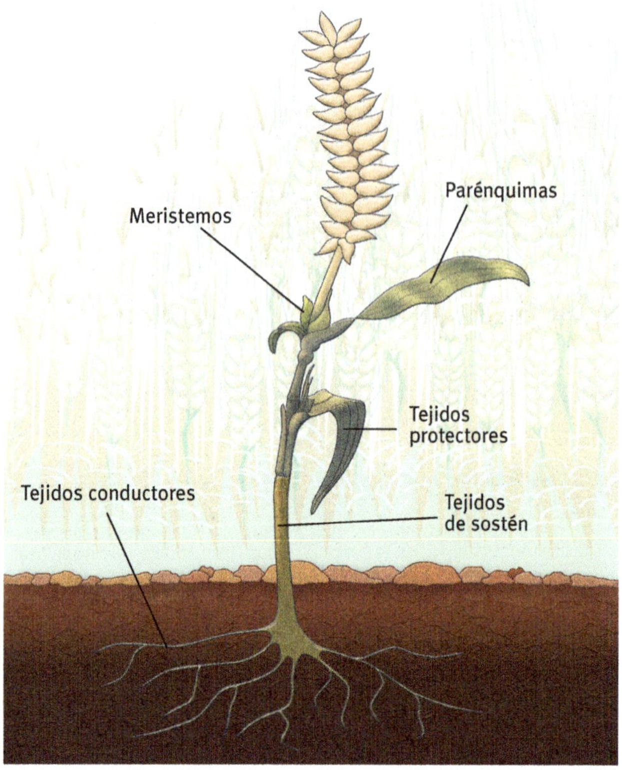

LOS TEJIDOS VEGETALES	
Meristemos	
	Embrionario
	Primario
	Secundario
	Cambium
	Felógeno
Tejidos conductores	
	Xilema o leño
	Floema o líber
Tejidos protectores	
	Epidérmico
	Suberoso
	Cutículas y pelos
Parénquimas	
	Clorífico
	De reserva
Tejidos de sostén	
	Colénquima
	Esclerénquima

3.1. *Meristemos*

Se caracterizan por estar poco diferenciados y por su gran capacidad de división. Son los responsables de la multiplicación celular.

Sus células presentan paredes celulares finas, con pocas vacuolas, de pequeño tamaño y núcleos grandes.

Se distinguen los siguientes tipos de meristemos:

- **Embrionario.** Forma el embrión en la fase de semilla y puede permanecer en reposo durante mucho tiempo.
- **Primario o apical.** Está especializado en el crecimiento en longitud de la planta.

  Se localiza en los extremos de la raíz (cofia) y el tallo (yemas terminales y axilares).
- **Secundario.** Produce el crecimiento en grosor de la planta.

  Se localiza en toda la planta (partes gruesas del tallo y raíz) y está formado por células que, en determinadas épocas del año, recuperan la capacidad de dividirse, para diferenciarse en tejidos concretos. Se distinguen:

  — El **cámbium,** que se encuentra intercalado entre los tejidos que conducen la savia y los regenera periódicamente.

  — El **felógeno,** que está bajo la epidermis y produce entre ésta y el propio tejido una nueva capa de células que forman un tejido protector, denominado *suberoso.* Produce el crecimiento en grosor de tallos y raíces.

En los troncos de los árboles, el cámbium produce cada primavera nuevas células que constituyen el xilema y el floema. Las células del año anterior se secarán para formar parte de la corteza, generándose así los anillos de crecimiento del árbol.

3.2. *Tejidos conductores*

Son los tejidos que se ocupan de la función de transporte de la savia. Sus células tienen forma alargada. Los tabiques de separación entre ellas están ausentes, o bien están dispuestos de manera oblicua para favorecer la circulación a través de ellos. Pueden ser de dos tipos:

- **Xilema o leño.** Transporta **savia bruta** en sentido ascendente (desde la raíz hasta las hojas) y procede de la división y especialización de las células del cámbium, que proliferan hacia la parte interior de la planta. Está formado por células muertas situadas una sobre otra sin tabique de separación y, por tanto, constituyen estructuras tubulares. Las células que forman estructuras tubulares se llaman traqueidas y tráqueas. Se trata de células muertas recubiertas de lignina; la lignina no recubre uniformemente los vasos en su totalidad.
- **Floema o líber.** Transporta la **savia elaborada** que resulta de la fotosíntesis, tanto en sentido ascendente como descendente, y procede de la especialización de las células del cámbium que se dividen hacia el exterior. Sus células se disponen del mismo modo que las del xilema, pero están vivas, con las paredes celulares de separación entre ellas perforadas por numerosos orificios. Se denominan células cribosas o tubos cribosos y son alargadas.

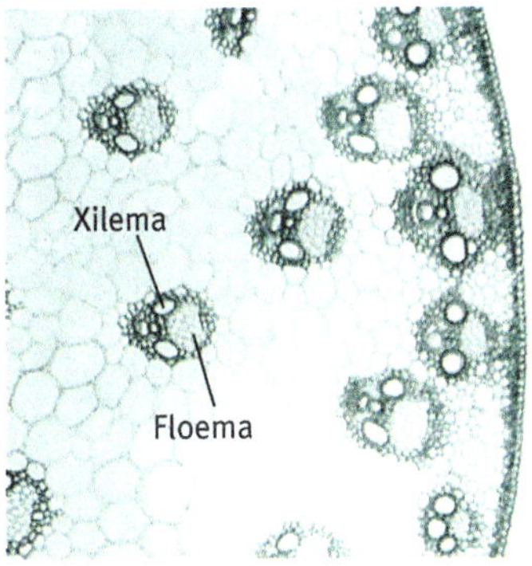

Corte transversal de un tallo de planta monocotiledónea visto al microscopio óptico. La tinción permite distinguir los diferentes tejidos. Los conductores se agrupan en pequeñas zonas distribuidas por el interior del tallo.

3.3. *Tejidos protectores*

Recubren la planta, protegiéndola de la acción de agentes externos y evitando la pérdida de agua. Las características de sus células varían según el tipo de tejido protector de que se trate:

- **Epidérmico.** Forma la epidermis y está constituido por una sola capa de células vivas sin clorofila adosadas unas a otras.

  La epidermis se encuentra perforada regularmente por unos orificios llamados **estomas,** que tienen capacidad para regular su grado de apertura según las condiciones ambientales.

  Este tejido puede presentar algunas **modificaciones:**

  — **Cutícula:** es una capa formada por sustancias céreas, principalmente *cutina*, que impregna la pared de las células que está en contacto con la atmósfera.

  — **Pelos o tricomas:** son prolongaciones de la epidermis, formadas por una o varias células. Pueden ser absorbentes, como los de las raíces, o secretores, por ejemplo, de sustancias urticantes como los de las hojas, los tallos, etc.

- **Tejido suberoso o súber.** Se origina a partir de la especialización de las células del felógeno y sustituye a la epidermis para reforzar las partes de la planta que crecen en grosor.

  La pared de sus células está impregnada de *suberina* (de naturaleza lipídica) totalmente impermeable. Este aislamiento hace que sean células muertas.

  El súber presenta unos orificios denominados **lenticelas,** por donde los tejidos adyacentes intercambian gases.

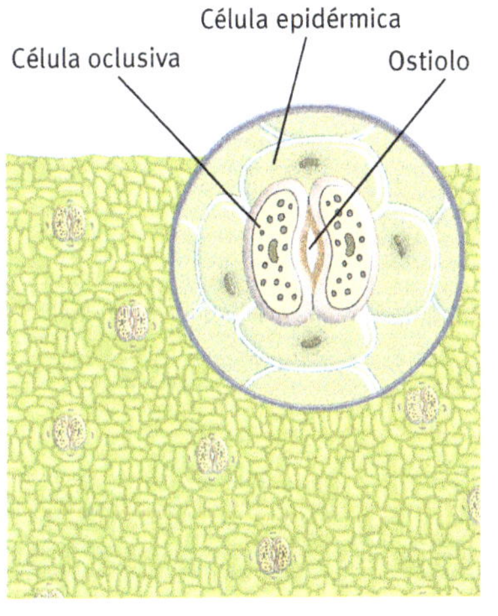

Los estomas regulan el intercambio gaseoso mediante la obertura o cierre de su ostiolo gracias a la presencia de dos células oclusivas.

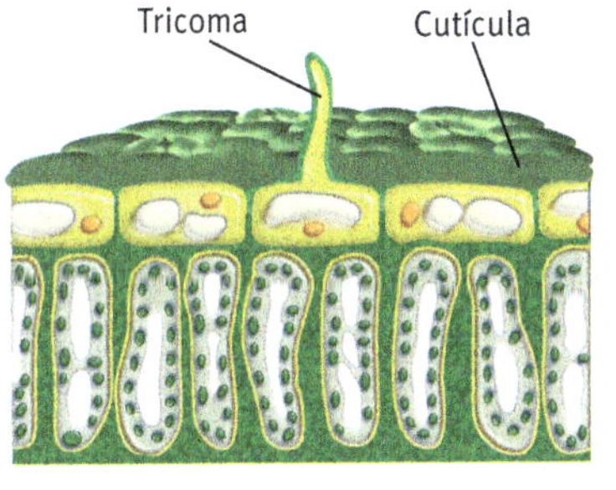

Corte del tejido epidérmico de una hoja

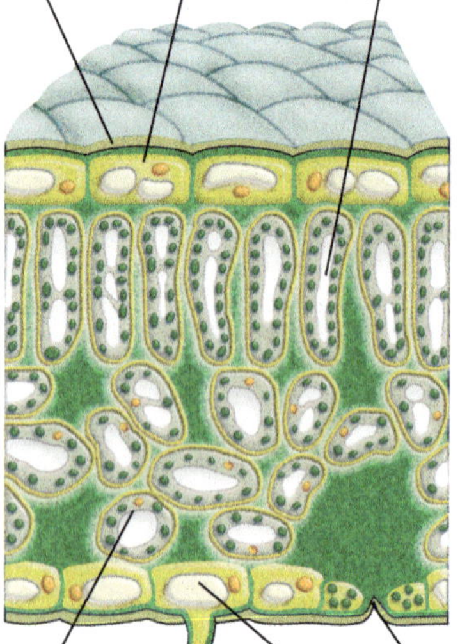

Corte transversal de una hoja. La zona del haz está compuesta por parénquima clorofílico en empalizada y es rica en cloroplastos. El envés está compuesto por parénquima clorofílico lagunar, con abundantes espacios intercelulares donde encontramos los estomas que permiten el intercambio gaseoso.

+

En determinados tejidos se localizan células o grupos de células con actividad **secretora**.

En la epidermis se pueden intercalar células que segregan esencias, como sucede en la planta del romero o la menta.

En otros casos existen tubos y bolsas que producen y segregan otras sustancias, como la resina de los pinos o el látex de las higueras.

3.4. *Parénquimas*

Son tejidos que dan cuerpo a las distintas partes de la planta, a la vez que se especializan en funciones diversas. Sus células son vivas, grandes, con numerosas vacuolas y cloroplastos. La pared celular está bien desarrollada. Entre cada célula y las contiguas existen uniones denominadas **plasmodesmos,** por donde se establece el transporte de sustancias entre las células.

Según la función, los parénquimas se clasifican en:

- **Clorofílico.** Es el principal tejido fotosintético; por ello, sus células contienen abundantes cloroplastos. Se localiza en las partes verdes: hojas y tallos herbáceos. Se distinguen dos capas:
  - — La capa más próxima a la epidermis está formada por el *parénquima clorofílico en empalizada;* tiene células prismáticas y muy juntas, para un aprovechamiento más eficaz de la luz.
  - — Debajo de éste se encuentra el *parénquima clorofílico lagunar,* formado por células más redondeadas que dejan huecos entre unas y otras. Por ellos circulan gases como CO_2, O_2 y vapor de agua.
- **De reserva.** Sus células tienen grandes vacuolas que acumulan productos elaborados por la planta (almidón, disacáridos, ácidos...). Se localizan en ciertos órganos específicos como los tubérculos.

3.5. *Tejidos de sostén*

Confieren a la planta solidez y consistencia. Están formados por células de paredes gruesas y adosadas estrechamente. Existen dos tipos:

- **Colénquima.** Lo constituyen células vivas que han engrosado los ángulos de sus paredes celulares con depósitos de celulosa. Esto le confiere gran resistencia y, a la vez, flexibilidad. Se localiza en las partes de la planta que están en crecimiento.
- **Esclerénquima.** Está formado por células muertas que han sufrido un proceso de lignificación. Se localiza en la cáscara de algunos frutos, como las nueces, o en fibras flexibles, como el lino y el cáñamo.

ACTIVIDADES

4. Relaciona los siguientes conceptos:

1. Produce el crecimiento en grosor de tallo y raíces	a) Xilema
2. Proceden de las células del cámbium que se dividen hacia la parte exterior	b) Floema
3. Se encuentra perforada por unos orificios que permiten el intercambio gaseoso	c) felógeno
4. Tejido formado por células vivas con gran cantidad de vacuolas que almacenan almidón	d) Epidermis
5. Prodecen de las células del cámbium que se dividen hacia la parte interior.	e) Parénquima

5. Explica qué se entiende por intercambio gaseoso. Explica por dónde se realiza y como se regula.

4. Tejidos animales

Tejido epitelial	De revestimiento Glandular
Tejidos conectivos	Conjuntivo Adiposo Cartilaginoso Óseo
Tejido muscular	Estriado Cardíaco Liso
Tejido nervioso	Neuronas Células gliales

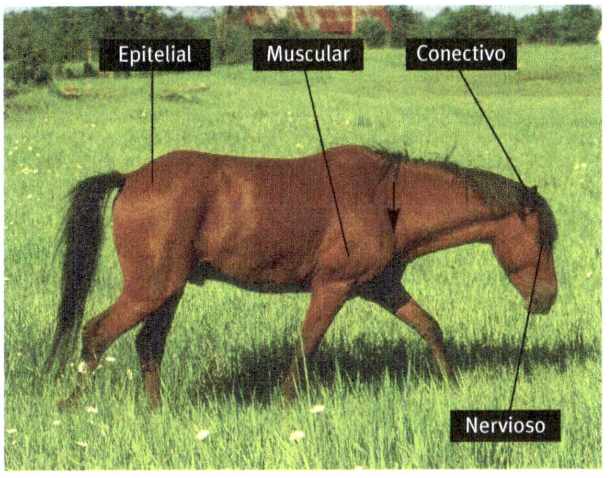

A continuación, vamos a describir las características de los principales tejidos que forman los animales, tomando como ejemplo a los vertebrados. Observa algunos órganos donde se localizan estos tejidos.

4.1. *Tejido epitelial*

Agrupa diversos tipos de tejidos que pueden tener función protectora, secretora o de absorción. El tejido epitelial recubre la superficie del cuerpo y las cavidades y conductos del organismo. Sus células son geométricas, más o menos regulares y están unidas íntimamente por desmosomas, hecho que provoca que el espacio intercelular sea muy pequeño. Por debajo del epitelio encontramos una **membrana basal**, formada por una red de fibras que conecta con el tejido conectivo.

> **C**
>
> La **sangre** es un tejido formado por diversos tipos de células rodeadas de un medio líquido. Estas características lo diferencian del resto de los tejidos.

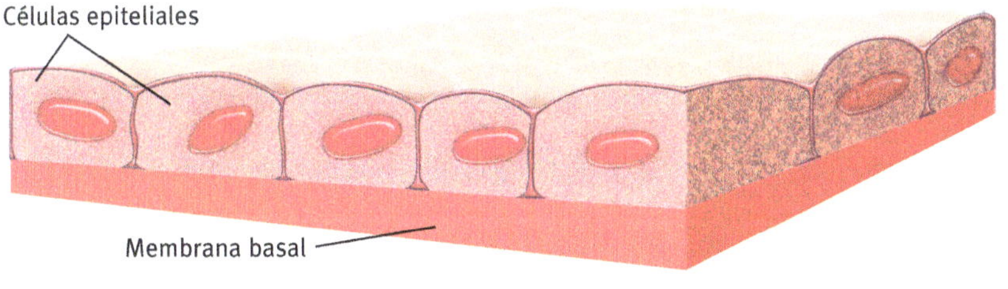

Existen dos grandes grupos de epitelios: el *epitelio de revestimiento,* que tiene una función protectora y de absorción, y el *epitelio glandular,* cuya función es la secreción de sustancias.

- **Epitelio de revestimiento.** Está especializado en la protección de las estructuras que se encuentran por debajo de él. Las células del epitelio de revestimiento pueden constituir una o varias capas. Según esto podemos distinguir:
  - **Epitelio simple:** formado por una sola capa de células.
  - **Epitelio estratificado:** formado por más de una capa de células.
  - **Epitelio pseudoestratificado:** Las células tienen una distribución irregular y al microscopio da la impresión de que se trata de un epitelio estratificado.

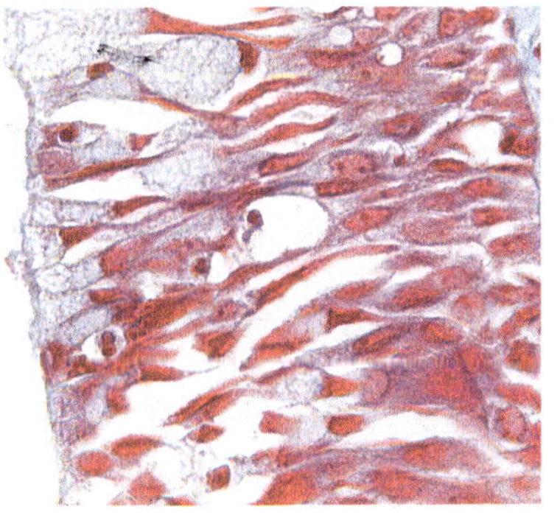

Epitelio olfativo visto al microscopio óptico (500x)

Las células, además, pueden presentar formas muy diversas: pueden ser planas, cúbicas o cilíndricas. Así teniendo en cuenta la forma de las células y el número de capas que lo constituyen, podemos distinguir los siguientes tipos de epitelio de revestimiento:

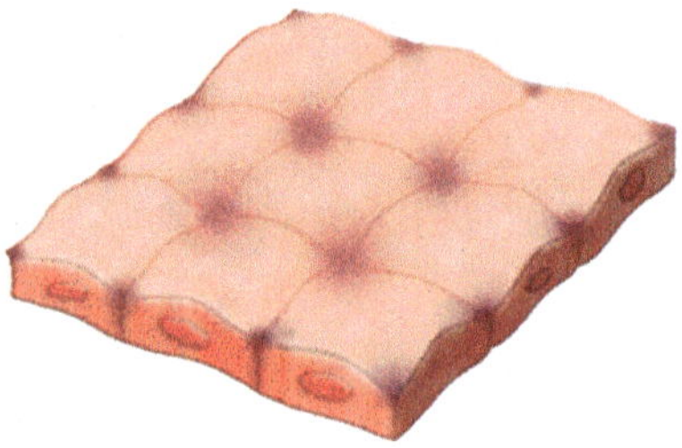

Epitelio plano simple: Tapiza la pared interna de los vasos sanguíneos y de órganos como el corazón. En este caso, la capa recibe el nombre de **endotelio.**

También tapiza cavidades internas procedentes del celoma, como la peritoneal y la pleural. En este caso se denomina **mesotelio.**

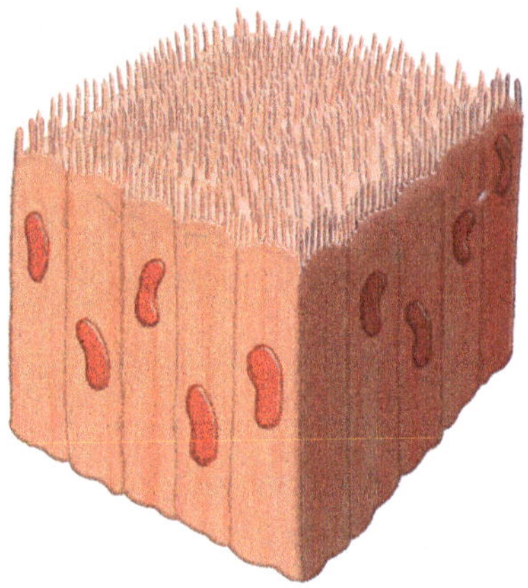

Epitelio cilíndrico: Es el epitelio que presentan las microvellosidades intestinales; permite aumentar la superficie de absorción.

Epitelio cúbico: Tapiza las glándulas salivares, el riñón, el páncreas y tambien las células pigmentadas de la retina.

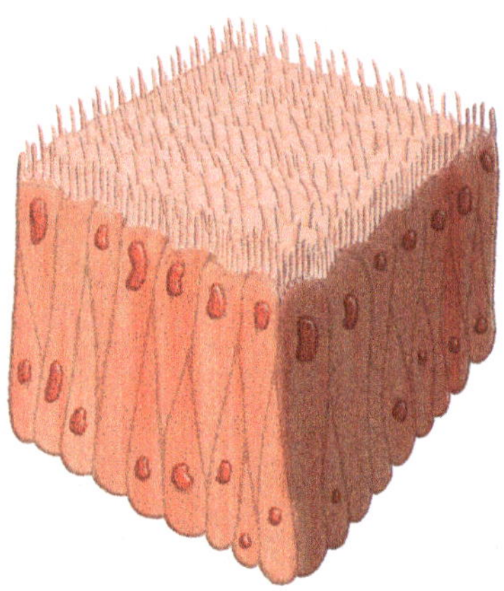

Epitelio cilíndrico pseudoestratificado: La disposición de los núcleos produce una imagen similar a la que se observa cuando existen varias capas. Presenta cilios, por lo que se denomina también *epitelio vibrátil*. Recubre la mayor parte de las vías respiratorias.

Epitelio plano estratificado: Recubre la superficie externa de la mayoría de los vertebrados, formando la epidermis de la piel.

La parte superior de la epidermis es una capa córnea; cuando ésta falta, la piel recibe el nombre de **mucosa.**

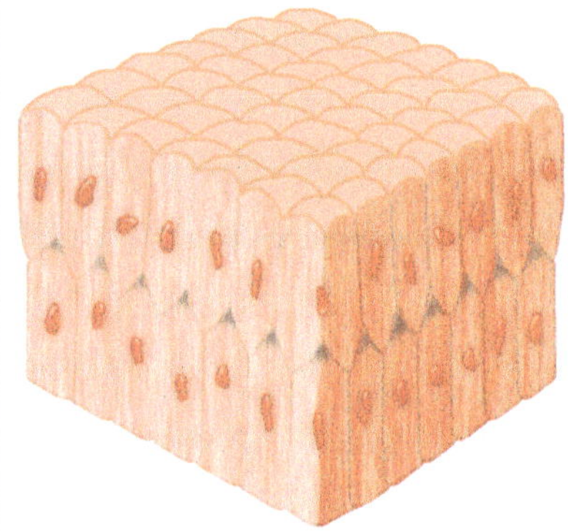

Epitelio cilíndrico estratificado: Recubre algunas zonas del aparato digestivo en los vertebrados inferiores; en los superiores está muy poco extendido, normalmente en zonas de transición entre epitelio pavimentoso pluriestratificado y cilíndrico pseudoestratificado.

- **Epitelio glandular.** Constituye las **glándulas,** órganos especializados en la fabricación y secreción de sustancias.

  Según su estructura, se distinguen dos tipos de glándulas:

  — **Unicelulares,** como las glándulas **caliciformes,** que se encuentran en la pared del intestino delgado y segregan una sustancia, denominada *mucus,* que facilita el tránsito a lo largo del intestino y lo protege de la acción de los enzimas digestivos.

  — **Pluricelulares,** como la glándula **tiroides,** que produce diversas hormonas, entre ellas la *tiroxina.*

  Según dónde vierten sus productos, existen tres tipos de glándulas:

  — **Exocrinas:** vierten las sustancias que producen exclusivamente al exterior del organismo, como las **glándulas lacrimales.**

  — **Endocrinas:** segregan sustancias exclusivamente al medio interno, como la **hipófisis** que produce, la *hormona del crecimiento.*

  — **Mixtas:** segregan dos tipos de sustancias diferentes, unas al exterior y otras al medio interno. Es el caso del **páncreas,** que segrega *jugo pancreático* al interior del tubo digestivo, una zona considerada exterior al cuerpo, y también segrega hormonas como *insulina* y *glucagón* a la sangre, para el control de la glucemia.

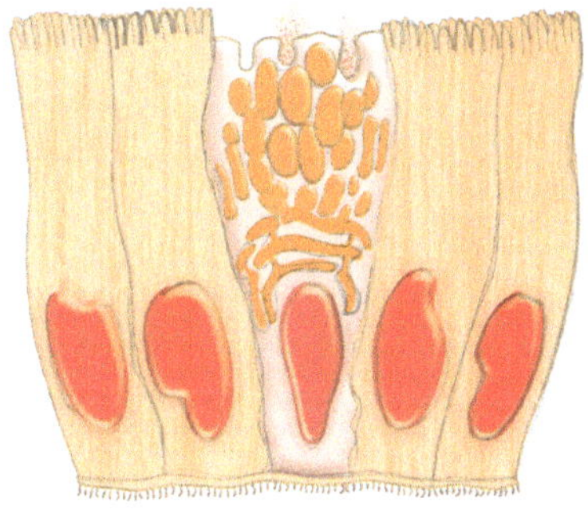

Glándula caliciforme

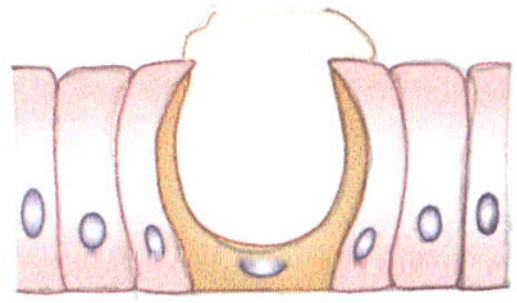

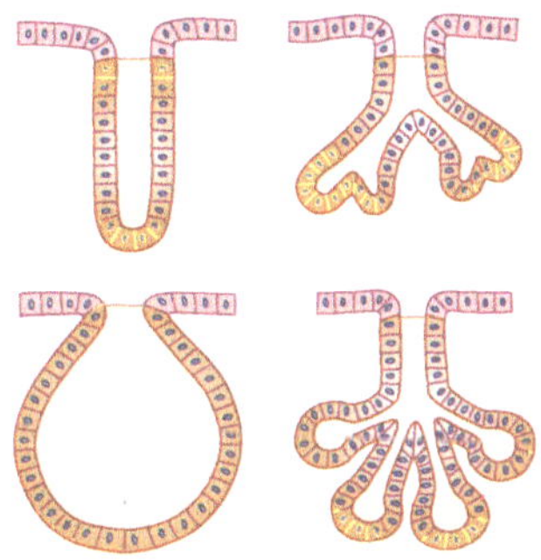

Distintos tipos de glándulas pluricelulares

4.2. *Tejidos conectivos*

Son un grupo de tejidos con funciones diversas: unen entre sí distintos órganos; llenan huecos y espacios entre los diferentes tejidos, y también constituyen elementos de soporte y de movimiento. Los tejidos conjuntivo, adiposo, cartilaginoso y óseo son variedades del tejido conectivo. Todos se caracterizan por tener gran cantidad de sustancia intercelular y pocas células.

Vamos a conocer los componentes del **tejido conectivo básico**:

CÉLULAS		Fibroblastos	Constituyen las fibras que forman la sustancia intercelular. Tienen forma alargada o estrellada
		Macrófagos	Son células de gran tamaño, fagocitarias y con capacidad migratoria.
		Mastocitos	Células redondeadas que contienen gran cantidad de heparina y otras sustancias que intervienen en la coagulación de la sangre.
MATRIZ	Fibras	Colágeno	Fibras consistentes y elásticas, en forma de haces. Se llaman fibras blancas. Por cocción generan gelatina.
		Elastina	Proteína fibrosa que puede estirarse y recuperar su forma original. Se llaman fibras amarillas. No se disuelven por cocción.
		Reticulina	Tienen la misma composición que las anteriores, pero se disponen como una red.
	Sustancias no fibrosas	Mucopolisacáridos	Compuestos formados por una parte proteínica y una parte glucídica.
		Sales	Diversas como fosfatos o carbonatos.
		Agua	

Vamos a conocer las características de los principales tipos de tejido conectivo:

- **Tejido conjuntivo:** Tiene la función de unir órganos y tejidos. Según la cantidad de fibras que lo constituyen puede ser laxo (pocas fibras) o fibroso (muchas fibras). El laxo recubre, principalmente, la superficie de los órganos y rellena los huecos entre ellos. El fibroso forma estructuras muy resistentes que por su función son sometidas a fuertes tensiones, como los tendones.
- **Tejido adiposo:** Es un tejido formado por unas células llamadas adipocitos, caracterizadas por poseer una gran vacuola llena de lípidos, que obliga al nucleo celular a desplazarse hacia un extremo de la célula. Este tejido constituye la principal reserva de energía de muchos organismos, aunque tambien lo encontramos realizando una función protectora bajo la piel, alrededor de los riñones, del corazón...

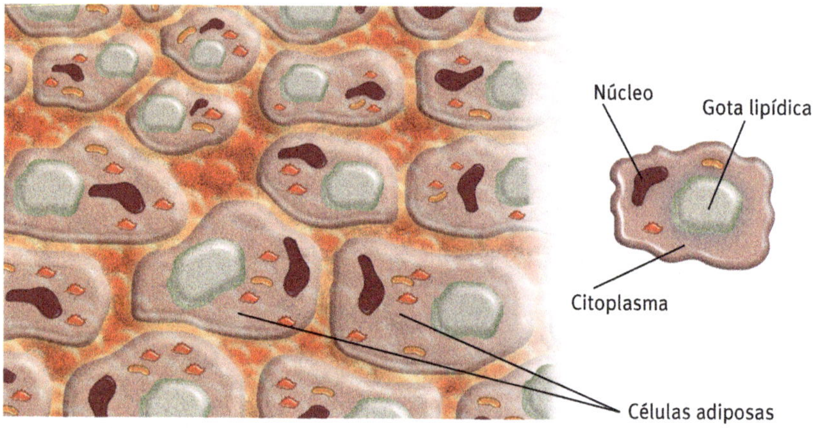

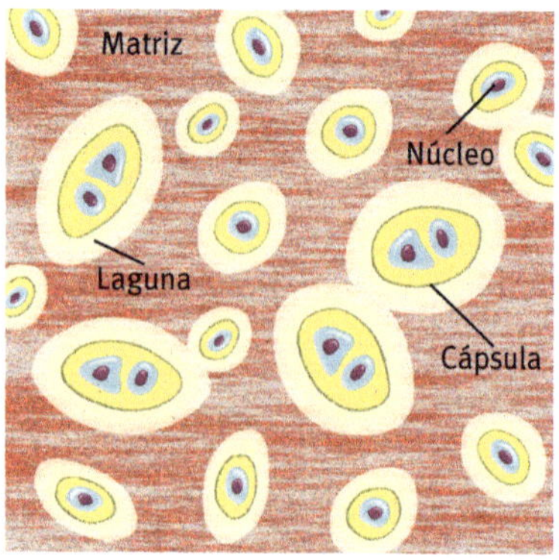

■ Condrocitos. Tejido cartilaginoso.

- **Tejido cartilaginoso:** Está formado por unas células llamadas condrocitos. En las primeras épocas de la vida es el único componente de las estructuras esqueléticas; al avanzar el desarrollo, es reemplazado por tejido óseo. Recubre el interior de las articulaciones y evita su desgaste. También se encuentra en el pabellón auditivo y el extremo de la nariz.
- **Tejido óseo:** Las células de este tejido se llaman osteocitos. Este tejido forma los huesos y, por tanto, participa en el movimiento y en la protección de diversos órganos. Si es denso y sin huecos, se denomina **compacto;** si presenta huecos, es **esponjoso.** Muchos huesos presentan ambos tipos de tejido.

■ Osteocitos. Tejido óseo.

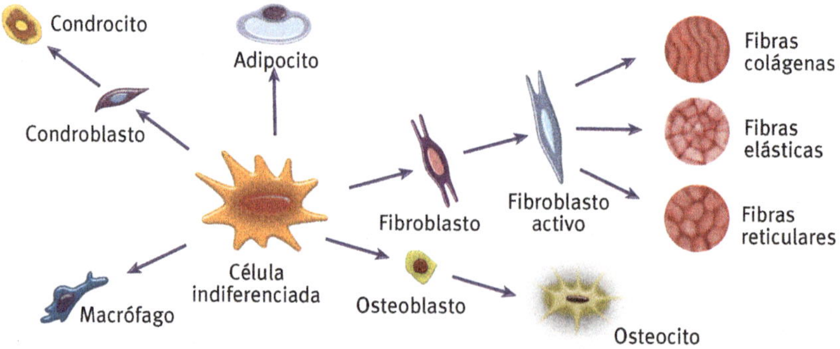

4.3. *Tejido muscular*

Está formado por células denominadas **miocitos.** En su interior contiene las *miofibrillas,* formadas por **actina** y **miosina,** dos proteínas con capacidad contráctil. El tejido muscular puede ser:

- **Estriado.** Células grandes que han fusionado sus membranas celulares y presentan numerosos núcleos. Las miofibrillas están ordenadas y esto les da un aspecto estriado. La contracción es rápida y voluntaria (músculos que participan en la marcha).
- **Cardíaco.** Células estriadas mononucleadas, de contracción rápida e involuntaria. Se encuentran en el corazón y mantienen el latido cardíaco de manera constante a lo largo de toda la vida.
- **Liso.** Células pequeñas alargadas con un solo núcleo. Las miofibrillas se encuentran desordenadas y no se observan estrías. La contracción es lenta e involuntaria (movimientos del tubo digestivo).

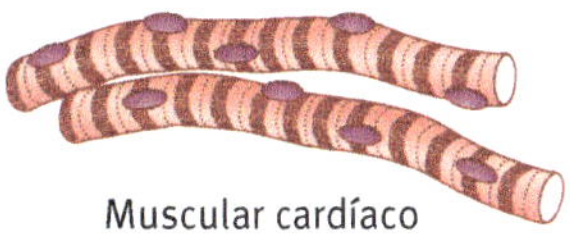
Muscular cardíaco

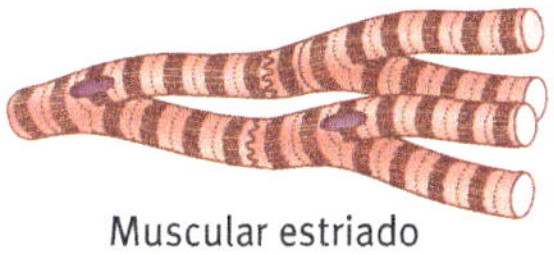
Muscular estriado

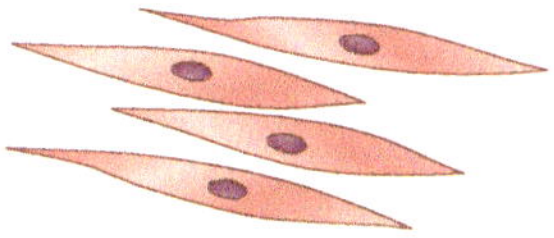
Muscular liso

4.4. *Tejido nervioso*

Está formado por dos tipos de células: *neuronas* y *células gliales.*

- Las **neuronas** son células muy especializadas que se caracterizan por su capacidad para generar y transmitir **impulsos nerviosos,** pequeñas corrientes eléctricas entre las neuronas. Constan de:
  - — Un **cuerpo neuronal,** con el núcleo y gran parte del citoplasma.
  - — Varias **dendritas,** prolongaciones de escasa longitud que rodean el cuerpo neuronal.
  - — Un **axón o cilindro eje,** que es una larga prolongación que parte del cuerpo neuronal.

  La transmisión del impulso nervioso siempre se efectúa desde la dendrita al axón. La conexión entre dos neuronas tiene lugar sin contacto físico, a través del espacio que las separa, la **sinapsis.**
- Las **células gliales** se interponen entre las neuronas y pueden ser:
  - — **Astrocitos.** Tienen aspecto estrellado y transportan sustancias nutritivas desde la sangre a las neuronas. También actúan como soporte, enlazando entre sí los distintos componentes del tejido.
  - — **Microglia.** Tienen aspecto espinoso y son móviles; fagocitan sustancias de desecho y desempeñan una función defensiva.
  - — **Células de Schwann.** Contienen *mielina,* una sustancia de gran importancia en la transmisión del impulso nervioso. Forman pequeñas envolturas en diferentes zonas a lo largo del axón.

Los **estímulos** son variaciones del ambiente externo o del interior del cuerpo. Son transmitidos a los centros nerviosos que elaboran respuestas que serán enviadas a los órganos efectores. Éstos llevan a cabo la respuesta.

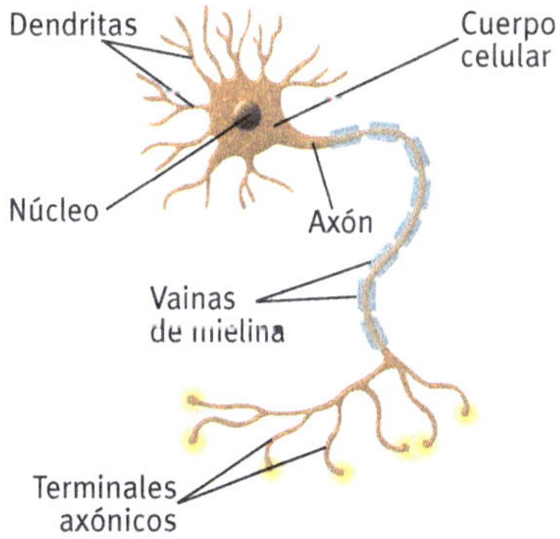

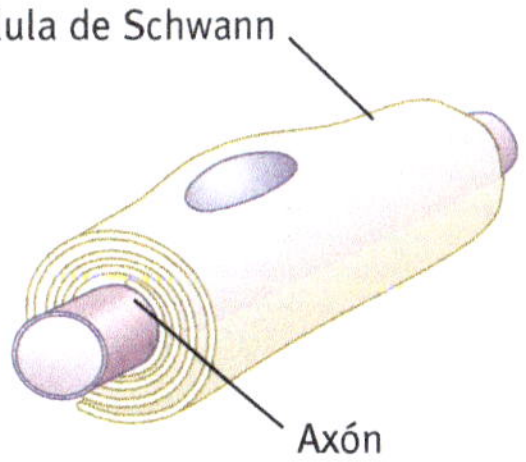

ACTIVIDADES

6. Explica la función de los siguientes tejidos y células: adipocitos, tejido óseo, tejido glandular, fibroblastos, mastocitos, microglia.
7. Contesta las siguientes preguntas que se refieren a características del tejido muscular:
   a) Cómo se llaman las células que lo constituyen?
   b) Qué proteínas tienen estas células que permiten la contracción muscular?
   c) Qué quiere decir que el tejido muscular liso tiene una contracción involuntaria?

5. Nomenclatura y clasificación de los seres vivos

Si algo caracteriza a la vida es la gran diversidad de formas en que se presenta. Los distintos tipos de organización celular y la especialización en tejidos, órganos y aparatos o sistemas dan como resultado la existencia de individuos con características diversas entre sí.

De esta gran diversidad, surgen dos necesidades:

- Utilizar nomenclaturas de uso internacional, que permitan identificar con claridad a los seres vivos.
- Clasificarlos para facilitar su estudio, es decir, agruparlos según tengan características similares entre ellos y distintas de las de otros grupos.

La especie es un concepto básico tanto para la nomenclatura como para la clasificación de los seres vivos. Una **especie** es un grupo de individuos que se reproducen entre sí, originando seres fértiles y similares a sus progenitores.

C

Una **especie endémica** es aquella que es originaria y exclusiva de un área geográfica.

En España existen numerosas especies endémicas, como, por ejemplo, el pinsapo y el cedro canario entre los vegetales y el lince ibérico y el samarugo entre los animales.

Se cree que existen más de 5 millones de especies diferentes y no se sabe con certeza si en realidad son muchas más. Algunos autores hablan de hasta 50 millones.

El nombre de una especie se compone siempre de dos palabras en lengua latina, es un sistema de **nomenclatura binomial.**

- La primera palabra es el nombre del *género* o **nombre genérico.** Corresponde a diversas especies con características similares. La primera letra del nombre genérico se escribe siempre en mayúscula.
- La segunda palabra es la que se ha asignado a una especie, dentro de su género; es el **epíteto específico.** La primera letra se escribe siempre en minúscula.

Canis lupus

Canis latrans

Canis familiaris

C

Los nombres científicos de los seres vivos se escriben siempre en letra cursiva o bien subrayados.

En el ejemplo anterior se observa que todos los animales representados pertenecen al género *Canis.*

Las distintas especies se nombran añadiendo al nombre genérico anterior el epíteto específico característico de cada una: *Canis lupus,* el lobo; *Canis latrans,* el coyote, y *Canis familiaris* para el perro.

Es muy importante destacar dos aspectos:

- El nombre genérico tiene sentido en sí mismo, ya que identifica a un conjunto de especies de características similares ese género. Así, se puede hablar de *Ciconia* como el género al que pertenecen todas las especies de cigüeñas; el género *Pinus* incluye las diferentes especies de pinos, y el género *Homo* identifica a las distintas especies correspondientes al ser humano actual y sus antepasados directos.
- La palabra que concreta la especie no tiene sentido en sí misma. Ello se debe a que en ocasiones simplemente describe el color, la forma o el lugar donde suele habitar un ser vivo. Por ello, hay individuos muy diferentes con un mismo epíteto específico. Existe una especie de cigüeña que se denomina *Ciconia nigra*, mientras *Pinus nigra* es una especie de pino. El epíteto *nigra* describe simplemente el color del plumaje en un caso y el de la corteza del árbol en el otro.

A lo largo de la historia se han establecido numerosas clasificaciones de los seres vivos. En el siglo XVIII, Linneo desarrolló el sistema de nomenclatura binomial y estableció que los seres vivos se podrían agrupar de forma jerarquizada. Cada categoría recibe el nombre de **taxón**. Las especies se agrupan en géneros, los géneros en familias, las familias en órdenes, los órdenes en clases, las clases en filums y los filums en reinos. Observa los taxones correspondientes al ser humano y al ratón común.

Estas dos especies filogenéticamente lejanas comparten, sin embargo, el epíteto 'melanogaster' en su nombre científico. En este caso, 'melanogaster', hace referencia únicamente a la coloración oscura de ambos animales.

Categorías taxonómicas	FILUM	CLASE	ORDEN	FAMILIA	GÉNERO	ESPECIE
TAXONES del ratón común	Cordados	Mamíferos	Roedores	Múridos	*Mus*	*Mus musculus*
TAXONES del ser humano	Cordados	Mamíferos	Primates	Homínidos	*Homo*	*Homo sapiens*

A continuación, se describen algunas características que permiten diferenciar los cinco reinos de seres vivos.

REINO	CÉLULAS	NUTRICIÓN	ORGANIZACIÓN	PARED CELULAR
Móneras	Procariota	Autótrofa o heterótrofa	Unicelular	Con mureína
Protoctistas	Eucariota	Autótrofa o heterótrofa	Unicelular o pluricelular poco especializada	Presente en ciertos grupos y de composición variable
Hongos	Eucariota	Heterótrofa	Unicelular o pluricelular poco especializada	De quitina
Plantas	Eucariota	Autótrofa	Pluricelular	De celulosa
Animales	Eucariota	Heterótrofa	Pluricelular	No tienen

C

Carl von Linneo (1707-1778) fue un naturalista sueco que desarrolló las principales categorías en que se organizan los seres vivos y el sistema de nomenclatura binomial.

Estudió medicina, aunque mostraba una verdadera afición por la botánica. En 1735 publicó su *Systema naturae* (Sistema natural), donde se propone de forma esquemática un sistema de clasificación taxonómica para los reinos animal, vegetal y mineral. El sistema de clasificación que propuso para el reino vegetal continúa vigente hoy en dia.

En las siguientes unidades de este libro se describen las características de distintos grupos de seres vivos. Por ello, presentamos la siguiente clasificación basada en los 5 reinos y ordenada según la escala evolutiva.

Esta clasificación, aunque contiene aspectos que están en discusión entre los especialistas, es la que en su conjunto es más aceptada por la mayor parte de la comunidad científica.

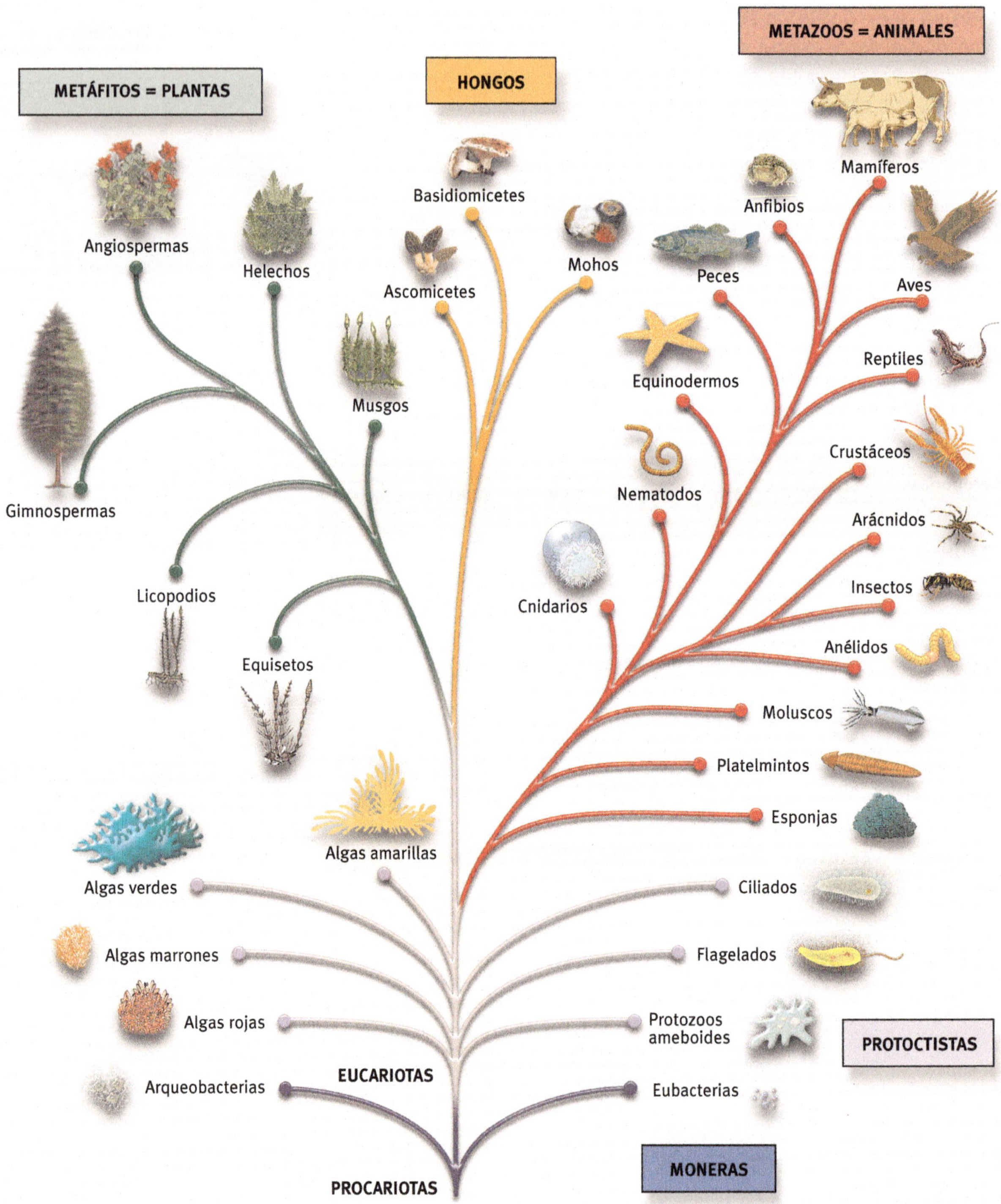

9 Lee el siguiente texto sobre el uso de la genética en el estudio de la evolución y responde las preguntas que tienes a continuación.

CARL WOESE Y LAS NUEVAS PERSPECTIVAS EN LA EVOLUCIÓN

Los científicos, cuando descubren especies o géneros nuevos se sienten orgullosos, pero no llegan a superar al orgullo que siente Carl Woese, biólogo molecular, por descubrir un dominio entero de vida, las archeas.

Fue en 1977 cuando salió a la luz una publicación junto con su colega Ralph Wolff, en la que se mostraba que un grupo de microbios poco conocidos, llamados arqueobacterias, estaban de hecho mucho mas cercanos evolutivamente con el género eucaria que con el resto de las bacterias. Este descubrimiento, y las herramientas de biología molecular que lo hicieron posible, han revolucionado nuestra visión de la historia evolutiva.

Una de las herramientas más utilizadas en biologia molecular es la comparación de los ácidos nucleicos entre diferentes especies. Todas las formas de vida utilizan el mismo código genético: el ADN y el ARN. Estas cadenas difieren entre especies. Los científicos utilizan el grado de similitud entre cadenas de ácidos nucleicos para establecer la clasificación de los seres vivos, es decir quieren establecer una taxonomía molecular.

Tras una gran labor experimental, Woese se centró en un tipo de ARN, el "ARN mitocondrial 16s". Esta secuencia de código genético se encuentra en el genoma de todos los seres vivos. Es una secuencia perfectamente conservada, es decir, ha evolucionado muy lentamente, por lo que puede ser utilizada para rastrear los cambios evolutivos sucedidos a lo largo de períodos de tiempo muy largos.

Las diferencias entre los ARN mitocondrial 16s de las diferentes especies se puede visualizar gráficamente en el llamado "árbol universal de la vida" o "árbol filogenético molecular de la vida". En este dibujo, la distancia entre dos especies cualesquiera, trazada a lo largo de las líneas que las conectan, es proporcional a las diferencias entre su ARN mitocondrial. Las especies con secuencias prácticamente idénticas están presumiblemente relacionadas y son representadas en el gráfico unas cerca de las otras.

Cuando se emplea este método con plantas y animales, estos trazos en el "árbol de la vida" son muy similares a los de los árboles evolutivos deducidos de la anatomía estructural. Pero la gran sorpresa llegó cuando se aplicó esta técnica al mundo microbiológico.

Se vio que las arqueobacterias ocupan un espacio muy grande, diferente del que ocupan las eucariotas y las bacterias. Basándose en estos descubrimientos, Woese propuso en 1990 la división, aceptada hoy, de las formas de vida en un trabajo titulado "Hacia un sistema natural de organismos: Propuesta de los dominios Archaea, Bacteria y Eucaria". A pesar de su nombre, el dominio archaea no es más antiguo que el de la bacteria. La mayoría de las especies de este dominio evitan el oxígeno (son anaeróbicos) y buscan las altas temperaturas (son termófilos).

Texto adaptado de http://astrobiologia.astroseti.org/articulo.php?num=197

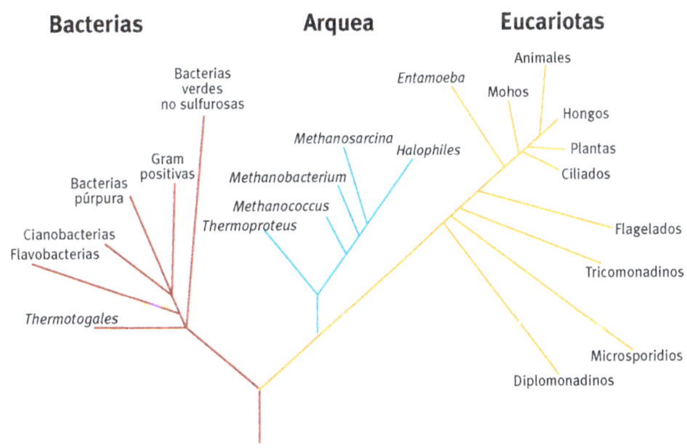

- Explica qué ideas principales se extraen de la lectura del texto y qué dominios establece Woese para clasificar los seres vivos.
- ¿Qué cambio de concepción sobre los reinos o dominios supuso el descubrimiento de Woese?
- Busca información sobre los objetos de estudio de la filogenia y la taxonomía, y explica sus diferencias.
- ¿Quién es Carl Woese? Explica brevemente su trayectoria y sus descubrimientos científicos. *http://es.wikipedia.org/wiki/Carl_Woese*
- Woese trabajó con un tipo de ARN mitocondrial llamado por su velocidad de centrifugación 16s. Este tipo de ARN se llama así porque es exclusivo de las mitocondrias, y no existe fuera de estos orgánulos. Pero, ¿sabrías explicar qué es el ARN? ¿Qué funciones realiza? ¿Qué tipos de ARN conoces a parte del que se nombra en el texto? Para hallar información sobre estas cuestiones puedes visitar esta web: *http://www.um.es/molecula/anuclo3.htm*
- ¿Qué son las arqueobacterias? ¿Qué las diferencia de las eubacterias? Cita algunos ejemplos de diferentes hábitats colonizados por arqueobacterias. En esta dirección web puedes encontrar la información que necesitas. *http://www.biologia.edu.ar/bacterias/arqueobacterias.htm*

10 Di qué células predominan en los siguientes tejidos:

a) cartilaginoso
b) nervioso
c) conjuntivo
d) adiposo

11 ¿Qué quiere decir que una célula está especializada?

12 ¿Qué diferencias hay entre...

a) el suero y el plasma?

b) la sangre y la linfa?

c) células beta y alfa del páncreas?

13 Dibuja una semilla y señala sus partes.

14 ¿Qué es la gastrulación? Explica en qué se basa el proceso.

15 Asocia los siguientes tejidos, sistemas o órganos con la capa embrionaria que los ha originado:

1. Sistema nervioso
2. Aparato excretor
3. Aparato digestivo
4. Tejido epitelial
5. Aparato respiratorio
6. Tejido conjuntivo

a. Endodermo
b. Mesodermo
c. Ectodermo

16 ¿Qué gases circulan a través del parénquima clorofílico?¿Cuál es la función del parénquima clorofílico?

17 Dibuja un estoma y explica qué función lleva a cabo?

18 Este dibujo representa un corte vertical del extremo terminal de una raíz. Sitúa floema, xilema, pelos absorbentes, zona de crecimiento y cofia.

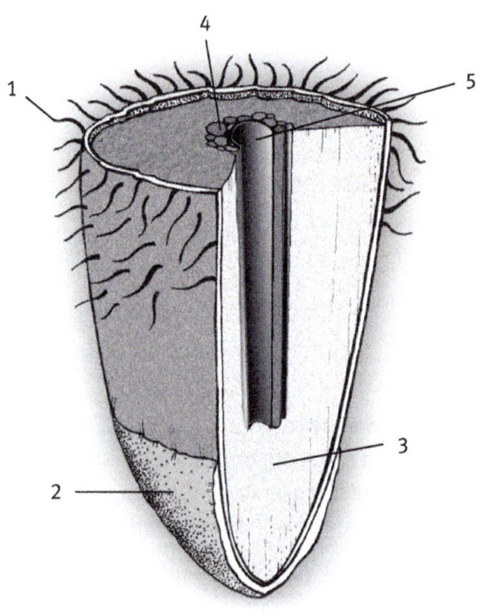

19 Este esquema representa un tejido conjuntivo ideal. Identifica sus componentes:

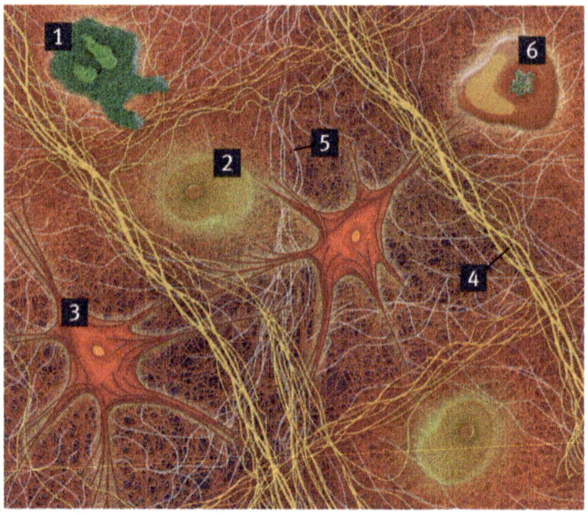

20 Se ha descubierto que una sustancia tiene como efecto la inhibición de la formación de fibroblastos. ¿Qué efectos podría presentar una persona que ingiere de forma habitual esta sustancia?

21 ¿Por qué aparece gelatina al cocer carne, como por ejemplo en un cocido?

22 ¿Qué tipo de células del tejido conectivo se encargan de eliminar a través de la fagocitosis las bacterias y las células muertas?

23 Indica cómo se llaman estos tipos de células:

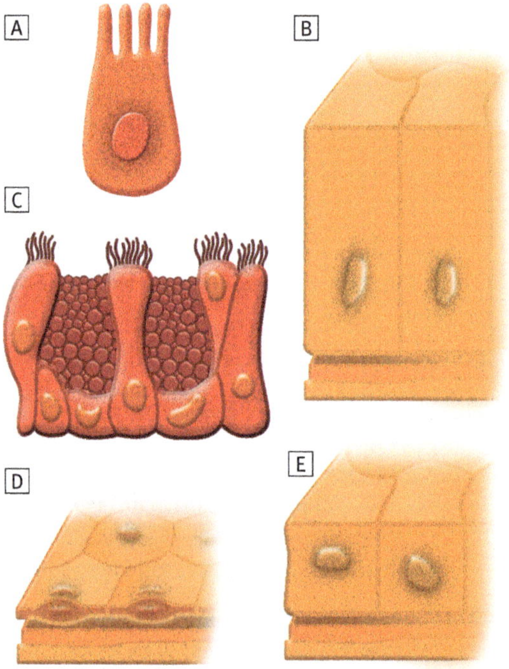

24 Este dibujo representa un tejido que se ha separado de la estructura sobre la que descansa. ¿Podrías decir qué son 1 y 2?

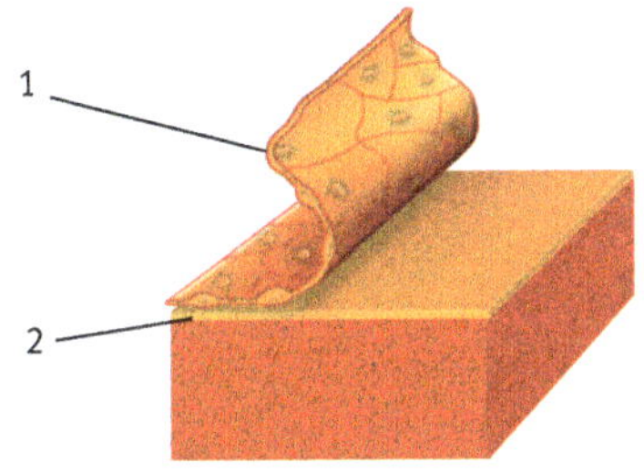

25 Completa el siguiente esquema, con la información sobre el desarrollo embrionario de los animales.

Desarrollo embrionario

Se distinguen dos etapas comunes

....

Aparecen dos capas de células

.... y ....

Si no se diferencia ninguna otra capa, los animales son

....

Si se diferencia otra capa, ésta se denomina

....

y los animales son

....

26 Estos esquemas representan los dos tipos principales de glándulas: endocrinas y exocrinas. Asigna a cada dibujo el tipo de glándula representada y explica el porqué de tu decisión.

— ¿Cómo funciona una glándula exocrina?

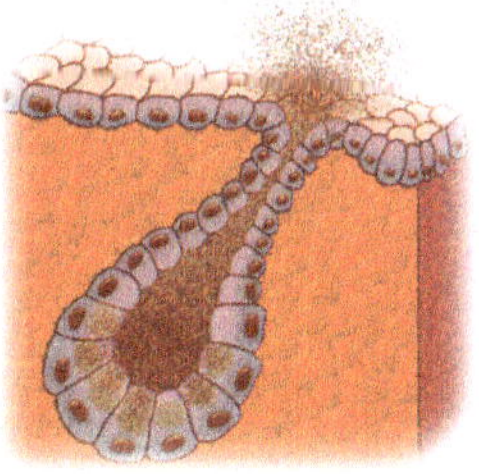

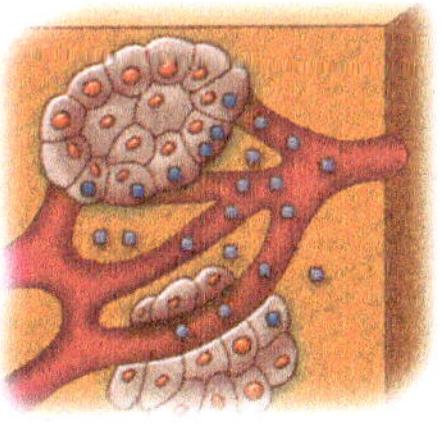

27 Identifica qué tipo de tejido conectivo representa la siguiente imagen microscópica y señala los constituyentes de este tipo de tejido.

Nota: los grumos rojizos son grasa teñida con sudan III. La imagen está tomada a 100 x.

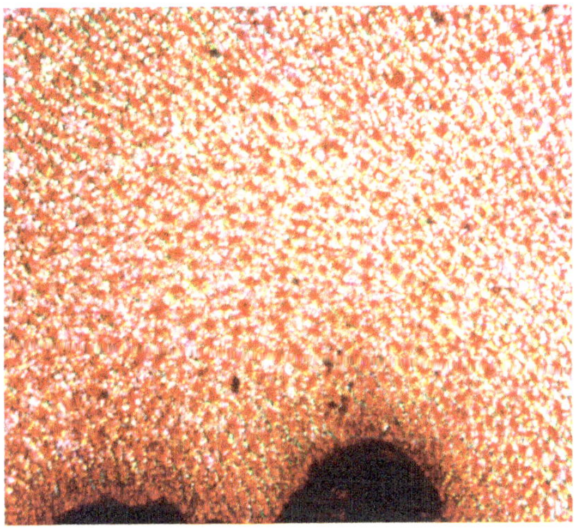

28 Busca información y clasifica los siguientes organismos según el reino al cual pertenecen: feófitos, espermatófitos, poríferos, esporozoos, sarcodinos, dinoflagelados, platelmintos, ascomicetes, briófitos, arqueobacterias.

TIC

29 Visita la web http://recursos.cnice.mec.es/biologia/

Elige biología y geología de 1° Bachillerato, la unidad 2: formas de organización de los seres vivos y contesta estas preguntas:

a) ¿Qué es la teoria de la endosimbiosis? ¿Quién la propuso?

b) ¿Cuál es la ventaja de la pluricelularidad?

c) Explica qué son y cuáles son los tejidos parenquimáticos que se describen en la web.

d) ¿Cómo se realiza la absorción de agua por parte de la raíz de la planta? En tu explicación haz referencia al término ósmosis.

e) Explica qué son las células madre, las células pluripotenciales y las totipotenciales.

f) Resume el proceso de clonación para obtener células de diferentes tejidos.

TIC

30 En la dirección electrónica

http://fq.cebollada.net/cts/wq/intro.htm

tienes una webquest titulada "el asesino en casa: la diabetes". Visítala y a través de sus páginas guiadas realiza una presentación de formato libre de 15 minutos sobre la diabetes.

PRÁCTICA 2 Mitosis en células de raíz de cebolla

OBJETIVOS

- Realizar preparaciones para observar células eucariotas mediante el microscopio óptico.
- Reconocer el proceso de la mitosis y distinguir sus diferentes fases.

Las células del tejido meristemático del extremo apical de la raíz de la cebolla se dividen constantemente, facilitando la observación de células en diferentes fases del proceso de reproducción celular conocido como mitosis.

Material

- Microscopio
- Papel de filtro
- Portaobjetos
- Cubreobjetos
- Lanceta
- Aguja enmangada
- Pinzas de madera
- Palillos
- Tijeras
- Mechero
- Vaso de precipitados
- Vidrio de reloj
- Orceína A
- Orceína B
- Bulbos de cebolla

Proceso

— Antes de realizar la práctica se debe llenar un vaso de precipitados con agua y debe colocarse un bulbo de cebolla sujeto con dos o tres palillos. El bulbo debe situarse de manera que la parte inferior esté en contacto con el agua. Transcurridos 3 o 4 días aparecerán numerosas raíces de unos 3 o 4 cm de longitud.

— Cortar con las tijeras de 2 a 3 mm del extremo de las raíces, ya que en esta zona se encuentran las células en división.

— Verter en un vidrio de reloj 2-3 ml de orceína A y depositar en él los trozos de las raíces.

— Sujetar el vidrio de reloj con las pinzas y calentarlo suavemente con el mechero (es importante que se evite la ebullición, retirando si es necesario el vidrio del reloj) hasta que se desprendan vapores tenues.

— Colocar con la aguja enmangada uno de los extremos de las raíces en un portaobjetos, añadir una gota de orceína B y dejar actuar 1 minuto.

— Con la lanceta corta los extremos. Colocar encima el cubreobjetos y, con la ayuda de un papel de filtro, realizar una suave presión en la zona del cubreobjetos para que las raíces queden extendidas. Esta técnica es conocida como "squash". Se ha de evitar que el cubreobjetos resbale, limpiando el exceso de colorante con el papel de filtro.

— Quitar el papel de filtro y observar al microscopio.

CUESTIONES

1. Dibuja las características morfológicas de las células vegetales.
2. Indica el porcentaje de células que se encuentran en mitosis.
3. Dibuja las diferentes fases de la mitosis que puedas identificar con el microscopio.
4. ¿Qué color presentan los cromosomas que observas?

PRÁCTICA 3 Identificación de lípidos

> **OBJETIVOS**
> - Aplicar técnicas sencillas para la identificación de ciertas propiedades de los lípidos.

El Sudan III es un colorante específico de las grasas, cuya disolución es muy inestable.

En la reacción de saponificación los ésteres de ácidos grasos reaccionan en presencia de un agente alcalino (hidróxido sódico o potásico), obteniéndose la sal sódica o potásica del ácido graso (jabón) y la glicerina.

Material

- Aceite de oliva
- Agua
- Sudan III
- Tinta china roja
- Alcohol de 96°
- Agua destilada
- Solución de NaOH al 20%
- Gradilla
- Vaso de precipitados
- Mechero
- Varilla de vidrio
- Tubos de ensayo

Proceso

1. Tinción de una grasa con Sudan III

— Preparar el Sudan III cada vez que se realiza la práctica. Disolver unos 10 mg de Sudan III en polvo en 20 gotas de alcohol de 96° y 6 gotas de agua destilada. El sobrenadante es una disolución saturada de Sudan III.

— Colocar en un tubo de ensayo 2 ml de aceite. Añadir unas gotas de Sudan III.

— Colocar en otro tubo de ensayo 2 ml de aceite. Añadir unas gotas de tinta roja.

— Agitar ambos tubos y dejar reposar.

— Observar y anotar las diferencias que existen en los dos tubos.

2. Reacción de saponificación

— Colocar en un tubo de ensayo 2 ml de aceite y 2 ml de la solución de NaOH al 20%.

— Agitar enérgicamente con una varilla de vidrio. Colocar el tubo al baño María de 20 a 30 minutos.

— Dejar reposar.

— Observar la aparición de tres capas.

CUESTIONES

1. A la vista de los resultados obtenidos, ¿qué observas en el tubo con Sudan III?¿Y en el tubo con tinta roja?
2. Describe las tres capas que aparecen en la reacción de saponificación.
3. ¿A qué corresponde cada una de las tres capas que aparecen en el tubo? ¿Por qué se produce esta disposición de capas?

9

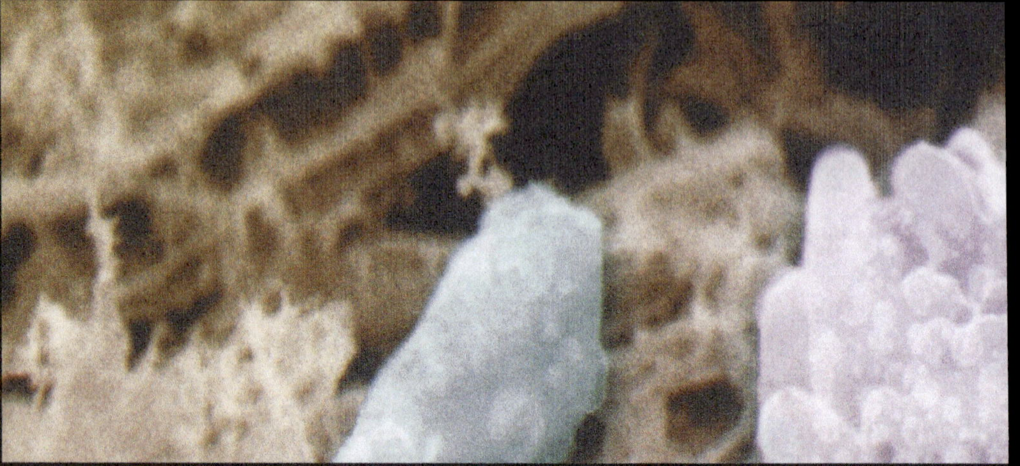

La nutrición: captación de nutrientes

La nutrición es una de las tres funciones vitales de los seres vivos. Las características de los sistemas que presentan los organismos para desarrollar esta función nos permiten diferenciar claramente los seres vivos que pertenecen al reino vegetal de los del reino animal.

En la captación y transformación de los nutrientes es donde se diferencian las dos grandes estrategias propias de cada uno de estos grupos de seres vivos. Así, la fotosíntesis es característica de los vegetales, mientras que la digestión es la estrategia propia de los animales.

En otro orden de cosas, la nutrición constituye una de las principales necesidades del ser humano; en la actualidad, se calcula que los déficits nutritivos constituyen, todavía, una de las principales causas de muerte de una parte de la población. A pesar de ello, se considera que la producción mundial de alimentos en la actualidad sería suficiente para evitar este problema si se pusieran los medios adecuados para distribuirlos y administrar estos alimentos entre toda la población.

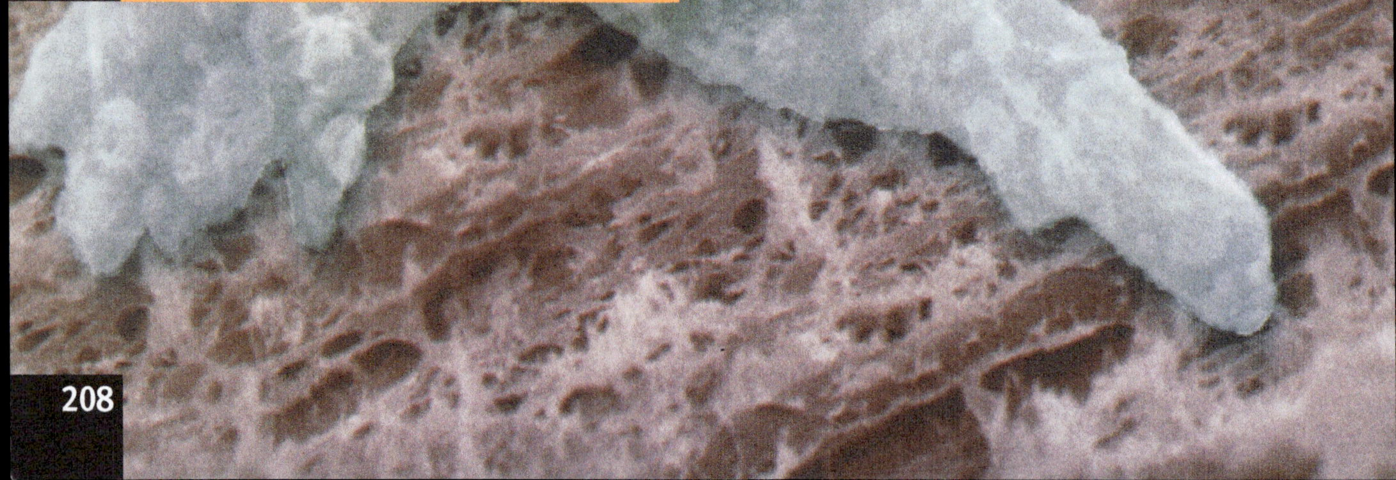

y excreción

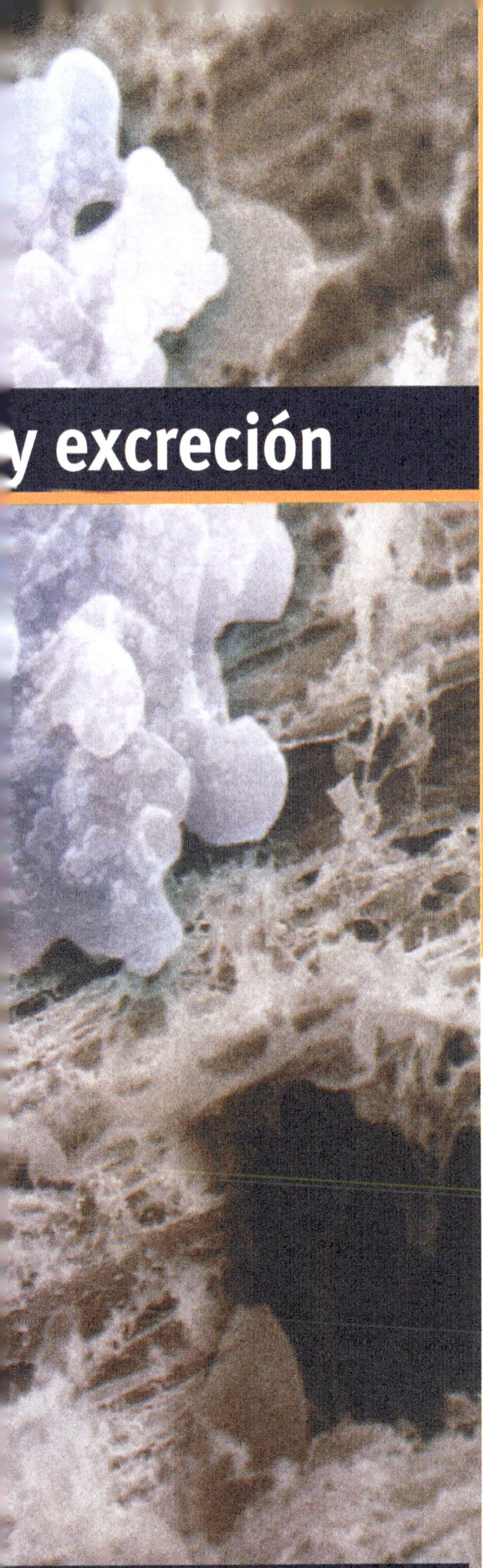

Ameba al microscópio electrónico de barrido.

ORGANIZACIÓN DE LOS CONTENIDOS

1. La función de nutrición

1.1. Ciclo de la materia

1.2. Flujo de energía

1.3. Pirámides de biomasa y de producción

2. La captación y la transfromación de nutrientes en los vegetales

2.1. Transformación de los nutrientes: metabolismo

3. La captación y la transformación de nutrientes en los animales

3.1. Invertebrados

3.2. Vertebrados

3.3. Digestión

3.4. Transformación de los nutrientes: metabolismo.

3.5. Metabolismo del ser humano

4. La excreción en los vegetales

5. La excreción en los animales

5.1. Invertebrados

5.2. Vertebrados

OBJETIVOS

- Comprender las relaciones debidas a la alimentación que se establecen entre los seres vivos, así como los ciclos de la materia y el flujo de energía que tienen lugar en los ecosistemas.
- Conocer el proceso de captación y transformación de nutrientes en los vegetales y los animales, y varios ejemplos significativos desde el punto de vista evolutivo.
- Interpretar las principales reacciones del metabolismo en vegetales y animales.
- Entender el proceso de excreción en los vegetales y los animales.

La célula es la unidad básica *de estructura* de todos los seres vivos. Constituye la unidad *funcional* ya que en ella se desarrollan las tres funciones vitales: nutrición, relación y reproducción.

En los seres pluricelulares aparecen distintos tipos de células, con funciones especializadas. Así, existen tejidos, órganos y aparatos o sistemas que trabajan coordinadamente en el desarrollo de cada función vital.

En esta unidad y la siguiente, describiremos los aspectos fundamentales de la función de nutrición y las adaptaciones que, a lo largo del proceso evolutivo, se han producido en distintos grupos de seres vivos.

1. La función de nutrición

La **nutrición** es el proceso mediante el cual los seres vivos toman nutrientes del exterior y los transforman y utilizan para construir las biomoléculas que componen su organismo y para realizar sus actividades vitales. En el proceso de nutrición se distinguen las siguientes fases:

— **Captación** de los nutrientes del medio.

— **Transformación** de éstos para obtener materia y energía utilizables.

— **Utilización** de la materia y la energía obtenidas en los procesos metabólicos, para el desarrollo de las funciones vitales.

— **Excreción** de las sustancias de desecho que se producen durante las fases anteriores.

Para nutrirse, los seres vivos deben partir de sustancias iniciales, que podemos considerar materias primas, y también de energía.

- La nutrición de los vegetales es **autótrofa:** la fuente de materia es **inorgánica** y la de energía, la **luz.**
- La nutrición de los animales es **heterótrofa.** La materia **orgánica** sintetizada por los seres con nutrición autótrofa es la fuente de materia y de energía.

De este modo, se establece una relación de dependencia entre los seres vivos debida a su nutrición, ya que unos se alimentan de otros. Estas relaciones se suelen representar en forma de cadenas y redes tróficas. En ellas se distinguen diferentes **niveles tróficos.**

- Los seres con nutrición autótrofa se encuentran al inicio de las cadenas y redes tróficas y, por ello, son **productores** de materia y de energía para el resto de los seres vivos. Las *plantas* y *algas* son seres productores.
- Los **consumidores** son los seres vivos que se alimentan de las sustancias elaboradas por los productores. Los *animales* son siempre consumidores. Podemos distinguir diferentes tipos de consumidores:
  — Consumidores **primarios,** que se alimentan de los productores.
  — Consumidores **secundarios** y terciarios, que se alimentan de otros consumidores.
- Algunos seres heterótrofos se alimentan de materia orgánica, procedente de seres que han muerto. Parte de esta materia la incorporan a su organismo y, el resto, la degradan a sustancias inorgánicas sencillas. Estos consumidores especiales son los **descomponedores,** entre los que se encuentran algunos *hongos* y *bacterias.*

Los líquenes son un ejemplo de simbiosis entre un hongo y una alga.

Para aumentar sus posibilidades de supervivencia, los seres vivos han desarrollado diversas estrategias relacionadas con la función de nutrición. Veamos algunos ejemplos.

- La **depredación.** El depredador caza y mata a la presa que le sirve de alimento. En general, los depredadores suelen ser más grandes y robustos que sus presas y se encuentran en menor número. Por ejemplo, el *zorro* es un depredador y el *conejo* su presa.
- El **parasitismo.** El parásito se alimenta a costa de otro organismo al que perjudica. Al contrario de los depredadores, los parásitos no devoran a sus víctimas, son más pequeños y a menudo son mucho más numerosos que sus huéspedes. La *garrapata* parasita al *perro.*
- La **simbiosis.** Es una dependencia entre dos organismos de la que ambos obtienen beneficio. Los *líquenes* son una simbiosis entre un *alga* y un *hongo;* el hongo aprovecha la capacidad fotosintetizadora del alga para nutrirse y las células del alga aprovechan agua y sales minerales captadas por el hongo.
- El **comensalismo.** Es una relación en la que el comensal se aprovecha de restos de alimento que ha conseguido un individuo de otra especie, y que luego ha abandonado. Un ejemplo lo constituye el *buitre* que come parte de la presa cazada por un *león.*

La nutrición de todas las formas de vida del planeta se puede interpretar como un proceso global en equilibrio, pues la materia y la energía van pasando de un ser vivo a otro a través de innumerables cadenas y redes tróficas. Así se constituye, de modo continuo en todos los ecosistemas, un ciclo de la materia y un flujo de energía.

1.1. *Ciclo de la materia*

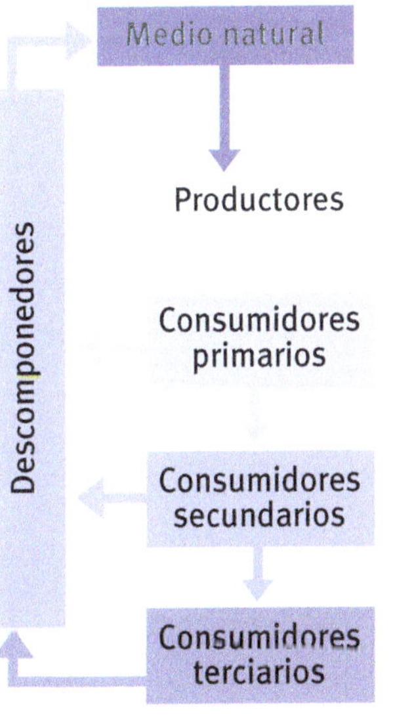

Ciclo de la materia

Mediante la fotosíntesis, una planta incorpora a su organismo materia inorgánica del medio; cuando es comida por un animal herbívoro, esta materia pasa a formar parte del animal. Si éste es devorado por un depredador, la materia queda incorporada al cuerpo de este último.

Con la muerte, la materia de todos los seres vivos llega a los descomponedores, los cuales la transforman en materia inorgánica que vuelve de nuevo al medio físico.

Así se completa el **ciclo de la materia** en todos los ecosistemas y, globalmente, en el gran ecosistema de la biosfera.

Ramón Margalef (1919-2004) está considerado el pionero de la ecología moderna. Realizó trabajos sobre comunidades de algas de agua dulce y se convirtió en un gran limnólogo. Fue el primer catedrático de ecologia de España.

Precursor de la ecología española y reconocido científico a nivel internacional, aplicó la física teórica y modelos matemáticos para desarrollar una ecología teórica, como en la dinámica de las poblaciones.

1.2. *Flujo de energía*

El Sol es la fuente de energía de la que dependen todos los seres vivos del planeta. Esta energía es captada por los pigmentos de los seres fotosintéticos (sobre todo, de los vegetales), que la **absorben** y la transforman en energía química. Ésta recibe el nombre de **energía fijada,** porque queda incorporada a las cadenas y redes tróficas.

Del mismo modo que ocurre en el ciclo de la materia, en cada eslabón de las cadenas y redes se produce una transferencia de energía de un ser vivo a otro. Cada ser vivo utiliza esta energía y, en parte, la disipa durante los procesos en que la consume, como el movimiento o la producción de calor.

Por esta razón, y a diferencia de lo que sucede con la materia, la energía no sigue un proceso cíclico, ya que no puede ser recuperada e incorporada de nuevo a los ecosistemas. En este caso, hablamos de **flujo de energía,** como el tránsito de unos niveles tróficos a otros, la posterior disipación al medio y la pérdida de energía final.

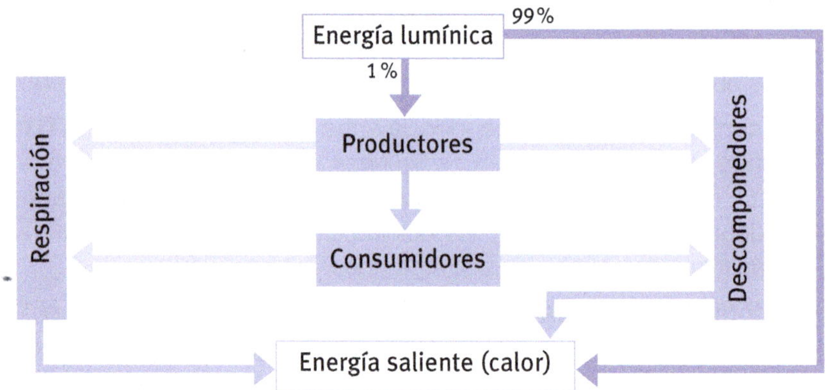

La energía entra en la biosfera en forma de energía lumínica. Tan sólo un 1 % de la energia solar que llega a la superficie terrestre es aprovechada por los productores. La energía se incorpora a la red alimentaria en forma de energía química y sale del ecosistema en forma de energía calorífica.

1.3. *Pirámides de biomasa y de producción*

Las relaciones que acabamos de describir implican a todos los seres vivos del planeta, pero se estudian localizadas en las distintas parcelas en las que se divide la biosfera: los ecosistemas.

- **Biomasa** de un ecosistema es la cantidad de masa que representan todos los seres vivos que habitan en él. Se puede medir en gramos de carbono por metro cuadrado.
- **Producción** de un ecosistema es el aumento de biomasa del ecosistema por unidad de tiempo. Se puede medir en gramos de carbono por metro cuadrado al año.

Estas magnitudes se representan en diagramas denominados pirámides de biomasa y producción, respectivamente. Para elaborarlos, se miden tanto la biomasa como la producción de cada nivel trófico por separado. Cada piso de la pirámide se dibuja con un tamaño proporcional a los valores encontrados para ese nivel trófico en un ecosistema concreto.

En las pirámides de producción el tamaño de los pisos se reduce aproximadamente un 90% en cada nivel, debido a la disipación de la energía que se produce. De un nivel trófico al siguiente tan sólo se transfiere como media un 10% de la energía. Es la llamada **ley del 10%**.

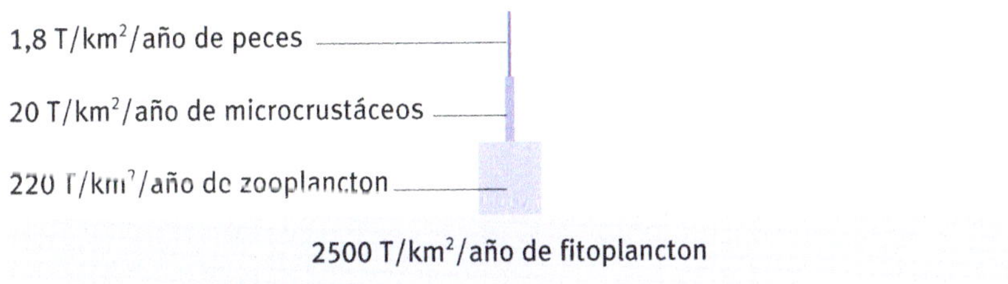

Pirámide de producción (T/km²/año)

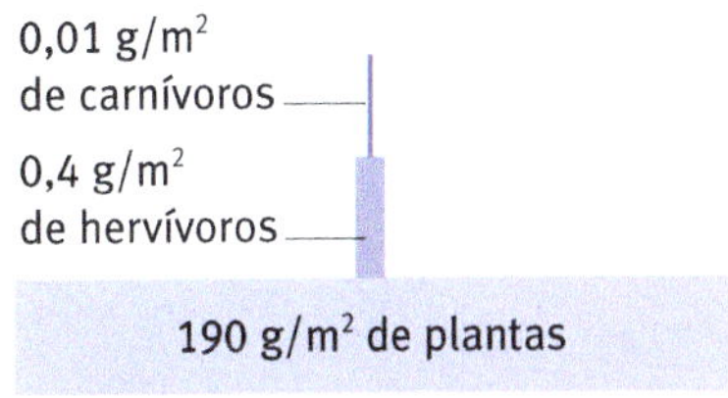

Pirámide de biomasa en un prado (g/m²)

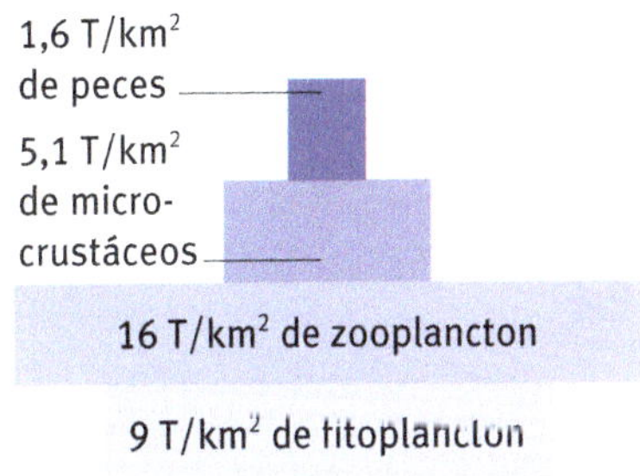

Pirámide de biomasa en el mar (T/km²)

Las pirámides de biomasa, al contrario que las de producción, no tienen porqué tener una forma piramidal.

2. La captación y la transformación de nutrientes en los vegetales

Desde el punto de vista cuantitativo, los vegetales son los productores más importantes. Se encuentran en la base de las pirámides de biomasa y de producción de los ecosistemas y, por tanto, su proceso de nutrición es imprescindible para la vida del resto de los organismos.

Los nutrientes vegetales pueden ser **micronutrientes**, si se necesitan en poca cantidad, calcio, magnesio, azufre, hierro… o **macronutrientes**, si se precisan en grandes cantidades, como agua, dióxido de carbono, oxígeno, nitrógeno, fósforo, potasio… Estos nutrientes se captan de modo diferente, según se trate de plantas talófitas o cormófitas.

- Las **talófitas**, que viven en ambientes acuáticos o húmedos, no tienen tejidos diferenciados. Por tanto, captan los nutrientes a través de toda la superficie de la planta.
- Las **cormófitas**, que viven en ambientes terrestres, poseen tejidos y órganos especializados en la captación de los diferentes nutrientes:
  - — Toman el **oxígeno** y el **dióxido de carbono** del **aire.** Lo hacen a través de los **estomas,** que son las aberturas que se encuentran en las hojas, especialmente en el envés.
  - — Absorben **agua** del **suelo,** a través de la **zona pilífera** de la raíz.

ACTIVIDADES

1. Teniendo en cuenta las definiciones de biomasa y de producción, determina a cuál de los dos conceptos hacen referencia las siguientes frases:

   — En un bosque hay 2700 g/m² de hojarasca.

   — En un bosque se miden 269 g de hojas/m² y año.

2. Suponiendo una cadena trófica lineal, determina la producción de ranas y serpientes en un área pantanosa de 1 km² en la que se producen 200 g/m² de hojas al año.

3. ¿Porqué las plantas talófitas no necesitan tener un sistema de transporte?

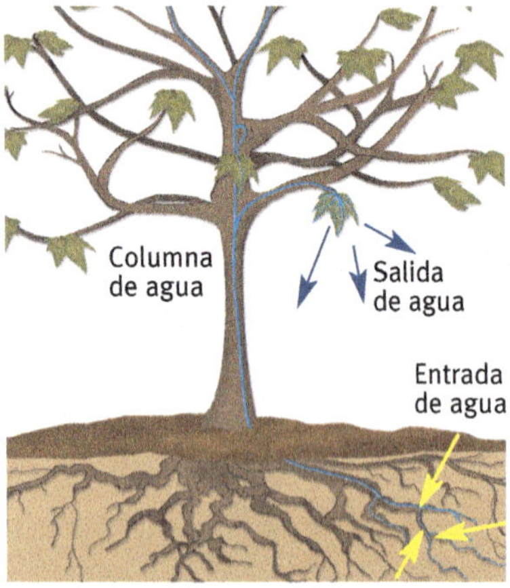

C

Las plantas propias de lugares muy secos presentan algunas **adaptaciones** a la escasez de agua:

- **Captan más agua** desarrollando raíces en profundidad o en superficie.
- **Almacenan agua** en sus tejidos de reserva. Son plantas carnosas o crasas.
- **Gastan menos agua** en la transpiración, reduciendo el número y tamaño de las hojas, engrosando la cutícula protectora y transformando las hojas en espinas.

C

Los nutrientes captados por las raíces (sabia bruta) ascienden a todos los órganos de la planta mediante dos procesos.

- La **presión radical**. El agua entra en la raíz por osmosis. La fuerza de absorción de cada célula se suma y provoca la presión radical, que hace que el agua penetre en la raíz.
- La **transpiración**. Cuando el agua se evapora de las células de las hojas, los solutos se concentran y el agua de las células adyacentes se desplaza hacia las células del mesófilo. Las células adyacentes concentran sus solutos y hacen que el agua de las otras células colindantes vaya hacia ellas. Este fenómeno se repite hasta llegar al xilema, por lo que el agua sube osmóticamente a través de éste.

Está formada por células con prolongaciones que se introducen entre las partículas del suelo. En esta zona de la raíz se inicia el recorrido de los vasos conductores ascendentes.

— Incorporan **micronutrientes** por la raíz, ya que se encuentran disueltos en el agua del suelo.

Existen algunos **factores** que condicionan la captación de nutrientes: la *abertura o el cierre de los estomas* y las *características del suelo.*

a) La **abertura** o el **cierre de los estomas** dependen, a su vez, de:

- La *intensidad luminosa.* Cuanta más luz reciban las hojas, mayor es la abertura de los estomas: la luz estimula la fotosíntesis, y por tanto se incrementa el intercambio de gases.
- La *temperatura.* Al aumentar la temperatura, se agranda la abertura de los estomas: como el agua tiene una elevada capacidad calorífica, el incremento de la transpiración permite eliminar calor para evitar que se eleve la temperatura de las hojas.

  La pérdida de vapor de agua por los estomas de las hojas es el motor que **impulsa la absorción de agua** en las raíces, ya que la salida de agua crea un déficit de presión en los vasos conductores que ascienden desde las raíces. Este déficit de presión se comunica a lo largo del vaso conductor hasta la zona pilífera de la raíz donde, para compensarlo, entra agua desde el suelo.

b) Entre las **características del suelo,** destacamos sus propiedades físicas y químicas y los seres vivos que habitan en él.

- Propiedades físicas.

  — La **textura** es la proporción de arena, limo y arcilla que contiene un suelo.

  — La **estructura** depende del tamaño y la forma de los grumos que configuran las partículas de humus, arcillas, arenas y limos.

  En general, una textura **franca;** es decir, entre 30 y 50 % de arena, 40 y 60 % de limo y 15 y 30 % de arcilla; y una estructura de **granos** de diámetro entre 1 y 5 mm proporcionan características adecuadas al desarrollo de la mayoría de las plantas.

- Propiedades químicas.

  — El **pH** es el logaritmo negativo de la concentración de iones hidronio H_3O^+ ($-\log [H_3O^+]$). En general, los suelos aptos para el crecimiento de las plantas tienen valores de pH entre 5 y 10.

- Presencia de seres vivos.

  — La materia orgánica procedente de los seres vivos que mueren y las sustancias que éstos excretan se convierte en materia inorgánica que **enriquece** el suelo de nutrientes.

  — La presencia de animales excavadores o el crecimiento de las raíces remueve el suelo y contribuye a su **aireación.**

  Son especialmente importantes los microorganismos que intervienen en los procesos relacionados con el nitrógeno.

El nitrógeno del aire es el gas más abundante en la atmósfera (77 %). Es un componente esencial de las proteínas y los ácidos nucleicos de todos los seres vivos. Sin embargo, únicamente las bacterias **fijadoras de nitrógeno** pueden utilizarlo. Dichas bacterias captan **nitrógeno** del aire y lo transforman en **amoníaco (NH_3);** este proceso recibe el nombre de **fijación del nitrógeno.**

A continuación, las **bacterias nitrificantes** convierten el **amoníaco** en **nitrito (NO_2^-)** y éste en **nitrato (NO_3^-).** Esta segunda fase del proceso se denomina **nitrificación.**

Los nitratos son absorbidos por las raíces de las plantas, que satisfacen de este modo su demanda de nitrógeno. Así, el nitrógeno entra en las redes alimentarias de todos los ecosistemas y es utilizado por todos los seres vivos.

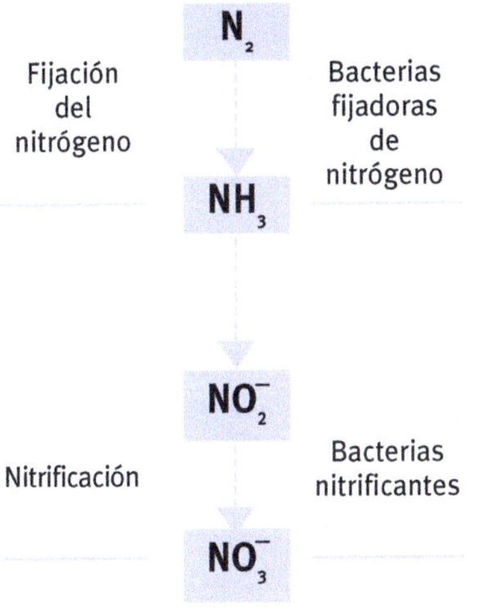

Proceso de fijación del nitrógeno y de nitrificación

2.1. *Transformación de los nutrientes: metabolismo*

A partir de los nutrientes captados tienen lugar los siguientes procesos metabólicos:

Principales procesos anabólicos

Síntesis de:

- **Almidón**, para la reserva glucídica.
- **Ácidos grasos insaturados y triacilgliceroles**, para las reservas grasas.
- **Proteínas**, para la reconstrucción celular y la regulación de funciones.

Principales procesos catabólicos

Degradación de:

- **Glucosa**, para obtener energía en la respiración.
- **Almidón** a glucosa cuando la fotosíntesis no genera suficiente glucosa.
- **Triacilgliceroles** y **ácidos grasos** en los tejidos de reserva.

En el esquema de la derecha se muestra la interrelación entre los procesos anabólicos y catabólicos. Los distintos tipos de nutrientes siguen **vías metabólicas** concretas que son reacciones encadenadas en las que se producen varias transformaciones.

El catabolismo de las proteínas y los aminoácidos en los vegetales está muy reducido. Por ello, las principales reacciones catabólicas implican sobre todo a glúcidos y lípidos.

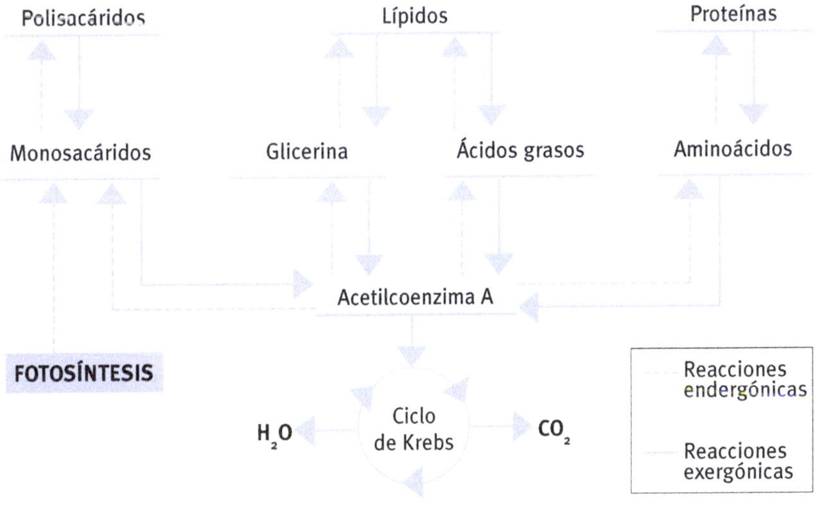

ACTIVIDADES

4. ¿Por qué las cadenas tróficas no pueden ser muy largas?
5. ¿Dónde hay más biomasa, en un bosque o en un prado? ¿Y más producción?
6. ¿Por qué un agricultor tiene que abonar cada vez más un campo conforme pasan los años? Arguméntalo tomando como referencia el ciclo del nitrógeno.

C

Según el tamaño del alimento que tomen, los animales pueden ser **micrófagos** o **macrófagos**. Los primeros se alimentan de partículas pequeñas que, normalmente, se encuentran en suspensión en el agua. Los macrófagos se alimentan de presas grandes y necesitan de apéndices bucales para apresar o para triturar el alimento.

3. La captación y la transformación de nutrientes en los animales

Todos los animales toman del exterior compuestos orgánicos y, mediante la **digestión,** los degradan parcialmente hasta obtener compuestos más sencillos que pueden ser absorbidos y utilizados por las células del organismo. Existe una gran variedad de mecanismos digestivos, que corresponden a las adaptaciones de los diferentes grupos de animales a sus condiciones de vida.

En los animales más sencillos, sin tejidos diferenciados, no existe un aparato especializado en la captación y digestión del alimento. Cada célula capta del exterior los nutrientes y los digiere. Esto implica que:

- Todas las células deben estar en **contacto con el medio** externo.
- La digestión es **intracelular.**

En los animales más complejos, la existencia de tejidos diferenciados permite la presencia de un aparato digestivo especializado en captar y digerir el alimento. Las células captan alimentos digeridos previamente. Por esta razón:

- **No** todas las células están en **contacto con el medio** exterior.
- La digestión es **extracelular.**
- Deben existir otros aparatos implicados en el **reparto de los nutrientes** a todas las células del organismo.

En un breve repaso, analizaremos los mecanismos de digestión más significativos desde el punto de vista evolutivo.

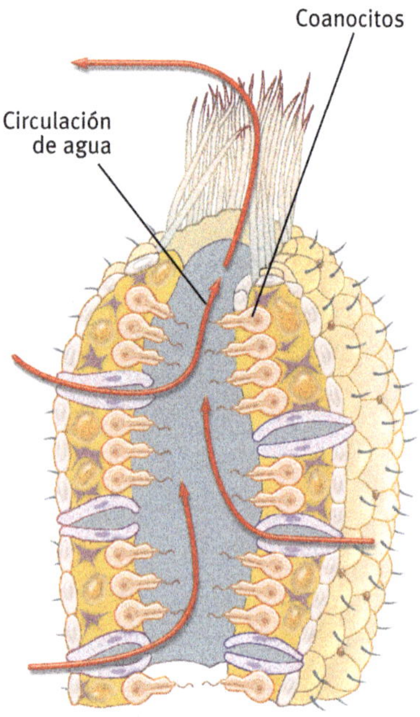

Las esponjas realizan una digestión intracelular.

3.1. *Invertebrados*

Las **esponjas** son los animales más simples. Se alimentan por **filtración**. El agua entra a través de poros inhalantes. Las sustancias alimenticias que contiene el agua son capturadas por unas células llamadas coanocitos. Este alimento será digerido en las vacuolas digestivas.

Los **cnidarios**, en cambio capturan su alimento a partir de unas células urticantes, llamadas cnidoblastos, muy abundantes sobretodo en los tentáculos. Los cnidarios son, evolutivamente, el primer grupo que presenta tubo digestivo. Este, tiene forma de saco, se llama **cavidad gastrovascular** y presenta un solo orificio de entrada y salida. La digestón se realiza en esta cavidad, por lo que se trata de una digestión extracelular.

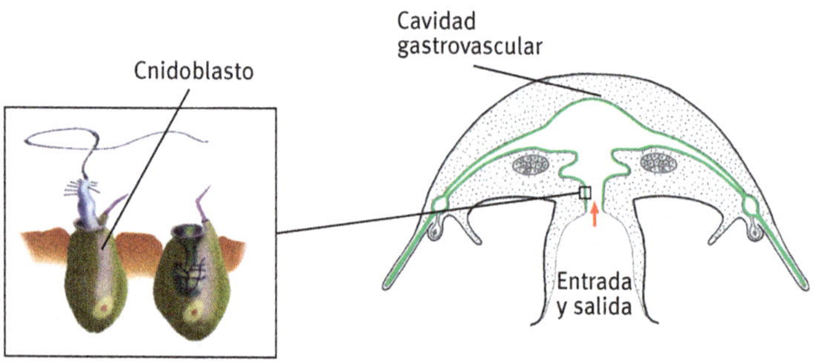

Los **cnidoblastos** poseen un cnidocilio sensible al contacto y una vesícula llena de líquido urticante, donde se encuentra un filamento enrollado. Cuando la posible víctima roza el cnidocilio, el cnidoblasto despliega su filamento, inyectándole el líquido urticante, que paralizará a la presa.

La aparición del celoma determina la existencia de un sistema digestivo especializado. En los animales *celomados* aparece ya una **cavidad digestiva** que, con distintas adaptaciones propias de cada grupo, responde a las siguientes características básicas:

- Es un *tubo abierto* al exterior por dos orificios: la **boca,** para la entrada o **ingestión** de *alimentos,* y el **ano,** para la salida o **egestión** de *residuos* no absorbidos.
- A lo largo del tubo digestivo, se distinguen distintos **tramos especializados** en procesos diferentes. Relacionadas con la **boca,** aparecen estructuras que favorecen la captura y trituración del alimento.
- Existen **glándulas anexas** al tubo, especializadas en la secreción de sustancias que facilitan la digestión.

La **digestión** es **extracelular,** porque se realiza dentro del tubo digestivo, pero fuera de las células.

Los **nutrientes** obtenidos son **absorbidos** y **repartidos** a las células del resto del organismo mediante un sistema circulatorio de transporte.

En los **anélidos,** y otros grupos afines, el tubo digestivo es **alargado.** Consta de *boca, faringe, esófago, estómago, intestino* y *ano,* y posee *glándulas anexas* repartidas a lo largo de todo el recorrido.

En los **moluscos** aparece en la boca un órgano especial para raspar el alimento, la **rádula**. El intestino es largo y enrollado. Los moluscos contienen una víscera única y voluminosa, el **hepatopáncreas**.

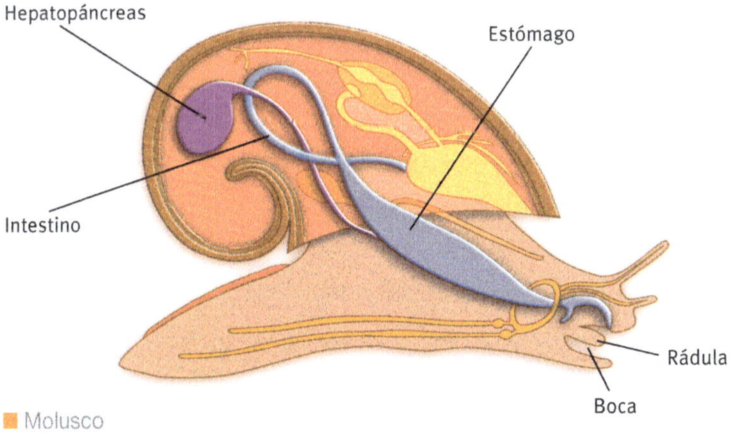

Molusco

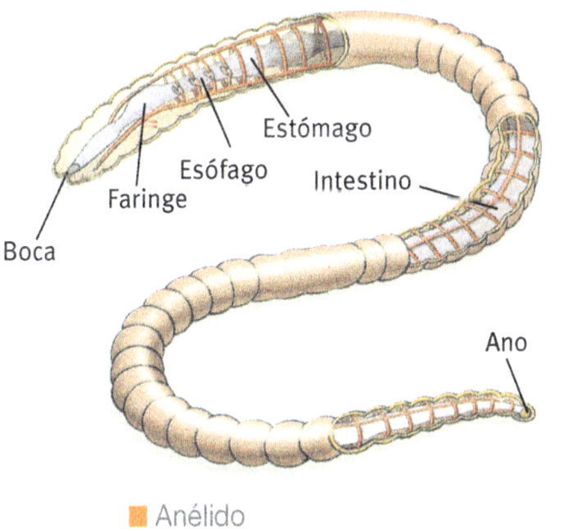

Anélido

En los **artrópodos,** especialmente en los insectos adaptados al medio terrestre, el tubo digestivo alcanza un mayor grado de especialización. Ello es una consecuencia de la necesidad de economizar el agua, ligada al desarrollo de la vida en el medio terrestre.

- Todos los artrópodos macrófagos tienen digestión en el **estómago.**
- Disponen de numerosas **piezas bucales** para masticar y triturar.
- Tienen **absorción intestinal** de nutrientes y, en su parte terminal, **reabsorción de agua** y formación de excrementos.

Los *arácnidos* son especiales, ya que no mastican y segregan enzimas digestivos sobre la presa. La **digestión** es parcialmente **externa;** luego absorben los líquidos resultantes de esta digestión previa.

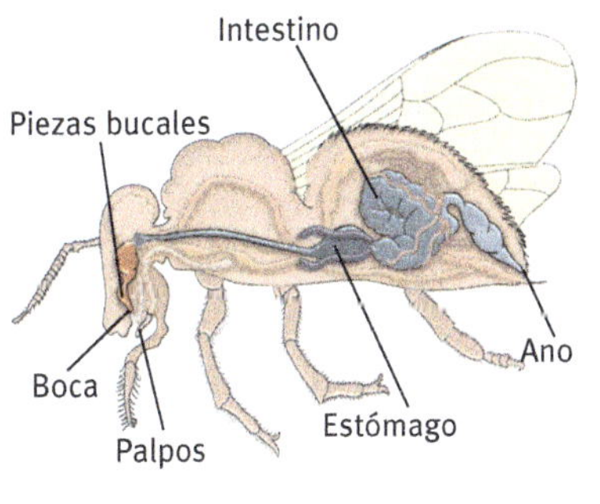

Artrópodo

3.2. Vertebrados

Los distintos grupos presentan una estructura del aparato digestivo muy similar. Las diferencias fundamentales se encuentran en las características de la boca, las glándulas anexas y el intestino grueso.

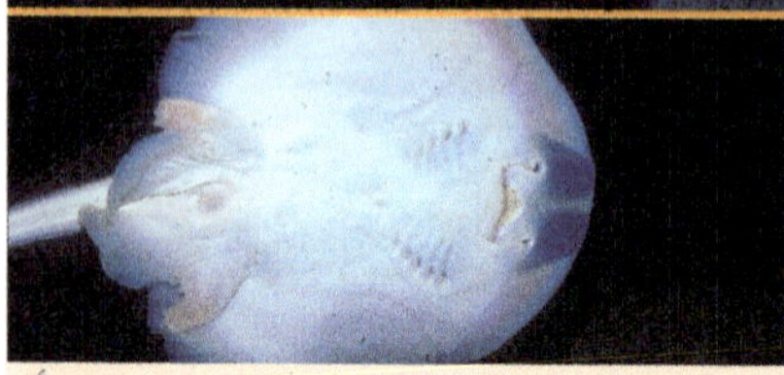

En los **osteictios**, los dientes sólo sujetan el alimento y la boca está en posición terminal. En los **condrictios**, la boca se sitúa en posición ventral, y tiene dientes muy cortantes, que son adaptaciones propias de grandes carnívoros. En ambos grupos el hígado está muy desarrollado y separado del páncreas.

La presencia de glándulas salivales y lengua es una adaptación a la vida terrestre. Con ellas se humedecen los alimentos y se facilita su circulación por el sistema digestivo. En los **anfibios** la lengua se inserta en la parte anterior de la boca para proyectarse hacia fuera al capturar las presas. Se inicia la diferenciación entre el intestino delgado y el grueso.

En los **reptiles** ofidios, la lengua es bífida y la boca y el esófago son dilatables, para tragar grandes presas. Los dientes no trituran, sólo sujetan.

En general, los reptiles presentan intestino delgado, intestino grueso y ciego totalmente diferenciados.

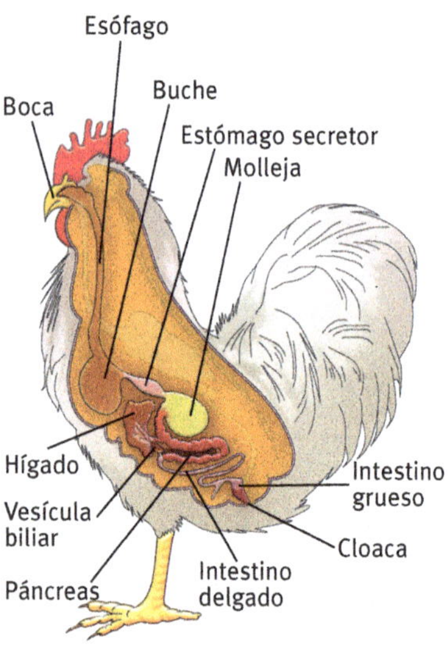

Aparato digestivo de un ave

En las **aves,** el tipo de alimentación es muy variado. Existen especies frugívoras, granívoras y la mayoría, además, completa su dieta con la caza de insectos, larvas y otros invertebrados según la estación del año y la abundancia de alimentos.

Por este motivo, su proceso digestivo es laborioso, lo cual se pone de manifiesto en las características anatómicas de su aparato digestivo.

- El tubo digestivo es largo y en él se distinguen:
  - **Buche,** para el reblandecimiento de granos y semillas.
  - **Estómago secretor** de enzimas digestivos.
  - **Estómago musculoso** o **molleja,** para la digestión mecánica o trituración de los alimentos.
- Aparece la **vesícula biliar,** para la acumulación de jugos procedentes del hígado y la regulación de su secreción.

En los **mamíferos,** el aparato digestivo alcanza la máxima complejidad. En general el tubo digestivo presenta:

- La boca con **labios** para succionar y mamar. Está separada de las fosas nasales por el **paladar,** lo que les permite respirar y masticar al mismo tiempo.
- La **lengua** es muy móvil y facilita la **deglución** del bolo alimenticio insalivado.
- El número de **piezas dentarias,** su forma e implantación en los maxilares dependen del tipo de alimentación.

- El **cardias** y el **píloro** son válvulas situadas a la entrada y la salida del estómago, respectivamente, e impiden que el bolo retroceda.
- La forma y el tamaño del estómago dependen del tipo de alimentación. La máxima especialización se encuentra en los **rumiantes,** que se han adaptado a la **digestión de la celulosa.**
- En el intestino delgado se diferencian: duodeno, yeyuno e íleon, cuya parte terminal desemboca en el **ciego** del intestino grueso.

Sobre estas características generales, se presentan adaptaciones, debidas a las diferencias entre nutrición *herbívora* o *carnívora.*

Los mamíferos no segregan enzimas para digerir la celulosa, pero pueden utilizarla si se encuentra en su intestino un tipo de bacterias simbiontes que la degradan por fermentación. En función de su capacidad para utilizar la celulosa, distinguimos tres tipos de mamíferos:

- Los **carnívoros,** como el perro, y los **omnívoros,** como el ser humano, no digieren la celulosa, debido al reducido tamaño del ciego y el colon, órganos en los que se encuentran las bacterias que degradan estas sustancias.
- Los **herbívoros no rumiantes,** como el caballo, aprovechan sólo una parte de la celulosa que ingieren, ya que aunque el ciego y el colon están bastante desarrollados, el tiempo de permanencia del bolo alimenticio en ellos es relativamente corto.
- Los **herbívoros rumiantes,** como la vaca, tienen una parte de su estómago, denominada **panza,** especializada en alojar a las bacterias que degradan la celulosa. En ella se hace la digestión e, incluso, se inicia la absorción, y continúa en el resto del tubo digestivo.

En los dibujos inferiores se pueden observar algunas características morfológicas determinadas por el tipo de alimentación:

- El tubo digestivo es **corto en los carnívoros** y **largo en los herbívoros rumiantes.** Así, se facilita en estos últimos la digestión total de la celulosa y su absorción.
- Los herbívoros **no rumiantes** tienen el **ciego muy desarrollado,** mientras que en los **rumiantes** está **muy reducido,** ya que la digestión de la celulosa se ha completado prácticamente en su estómago.
- El **hígado** es de tamaño **reducido en herbívoros** y **muy desarrollado en carnívoros,** pues estos últimos ingieren muchas proteínas que requieren, para su digestión, enzimas específicos.

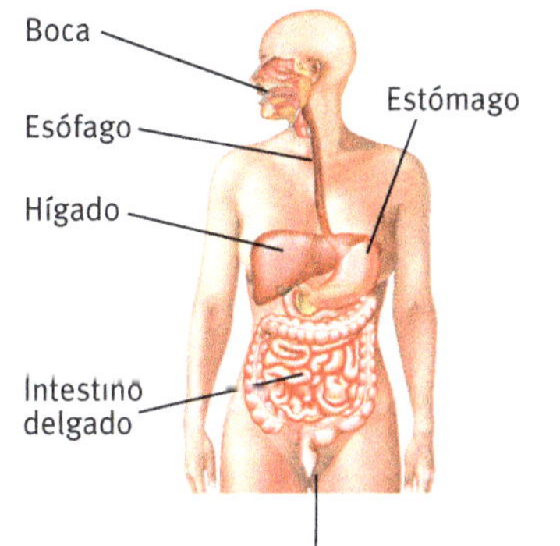

Omnívoro

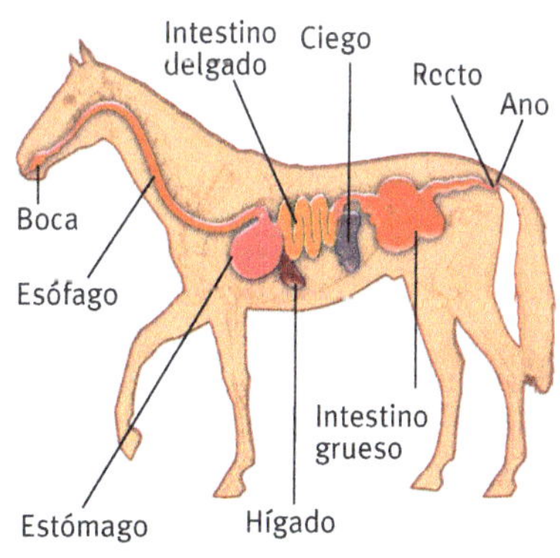

Herbívoro no rumiante

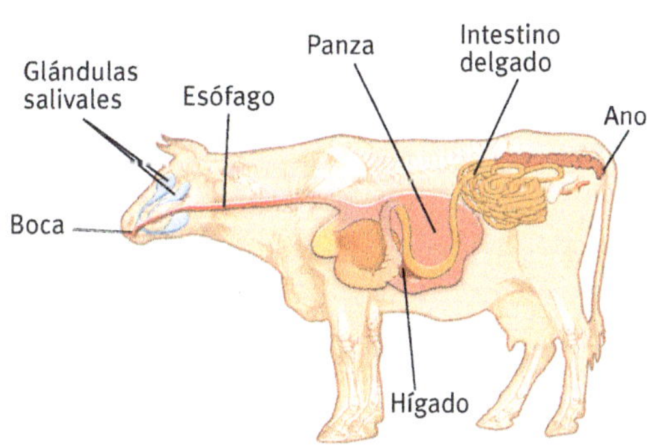

Herbívoro rumiante

3.3. *Digestión*

Es el proceso de transformación de las sustancias alimenticias que se ingieren en nutrientes asimilables por las células del organismo. En el caso del ser humano, el proceso se desarrolla en las siguientes fases:

— **Ingestión.** Es la entrada de los alimentos en la boca.

— **Trituración.** Se realiza mediante la masticación con los dientes y mediante los movimientos del estómago.

— **Transporte.** Tiene lugar gracias a movimientos *peristálticos,* o contracciones musculosas de las paredes del tubo, que hacen avanzar el bolo alimenticio desde la boca hasta el ano.

— **Digestión química.** Mediante los jugos digestivos segregados por las glándulas anexas, los polímeros se transforman en monómeros:

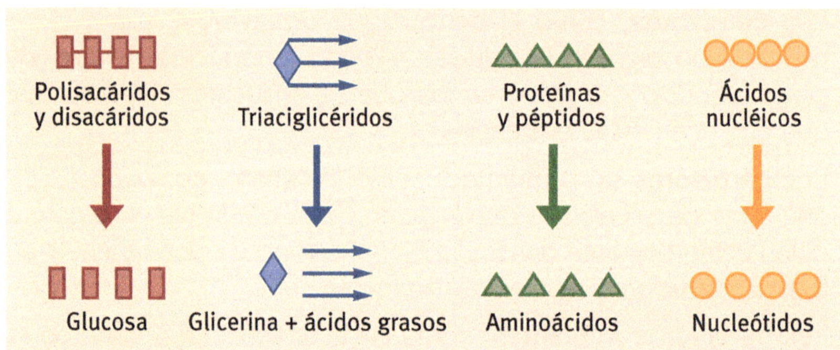

Las sustancias obtenidas pueden cruzar las membranas celulares.

— **Absorción de nutrientes.** Los productos obtenidos de la digestión pasan a la sangre a través de las microvellosidades del **intestino delgado.** La absorción puede realizarse:

- Por **transporte activo,** con gasto energético, como los monosacáridos.
- Por **transporte pasivo** o difusión simple, como sucede con el agua y algunos elementos minerales.
- Por **endocitosis,** como algunas proteínas y lípidos.

Una vez que los nutrientes pasan a la sangre, ésta los transporta y distribuye por todas las células del organismo.

— **Reabsorción de agua y absorción de vitaminas**. Tiene lugar en el **intestino grueso.** El agua pasa la pared del intestino por difusión simple, mientras que las vitaminas lo hacen por transporte activo. Este proceso espesa y compacta las heces.

■ Principales glándulas y enzimas que intervienen en la digestión química. Podemos ver los sustratos sobre los que actúan los enzimas y también los productos que generan.

GLÁNDULAS	ENZIMAS	SUSTRATOS	PRODUCTOS
Salivales	amilasa salival	almidón	Maltosa
Estómago (jugo gástrico)	Pepsina	proteínas	polipéptidos
	lipasa	grasas	ácidos grasos y glicerina
Intestino delgado (jugo intestinal)	maltasa	maltosa	glucosa
	lactasa	lactosa	glucosa y galactosa
	lipasa	grasas	ácidos grasos y glicerina
Páncreas (jugo pancreático)	amilasa	almidón	maltosa
	lipasa	grasas	ácidos grasos y glicerina
	tripsina	polipéptidos	aminoácidos

3.4. *Transformación de los nutrientes: metabolismo*

Las moléculas orgánicas sencillas obtenidas de la degradación de los alimentos, absorbidas y repartidas a las células del organismo ponen en marcha los procesos metabólicos.

PRINCIPALES PROCESOS ANABÓLICOS	PRINCIPALES PROCESOS CATABÓLICOS
Síntesis de: • **Glucógeno,** para constituir reserva glucídica. • **Triacilgliceroles saturados,** para la reserva lipídica. • **Proteínas** a partir de aminoácidos, para la reconstrucción celular y la regulación de funciones.	Degradación de **polímeros** a monómeros (glucógeno a glucosa, triacilgliceroles a ácidos grasos, proteínas a aminoácidos). A continuación, los monómeros son degradados: • **Glucosa,** para obtener energía en la respiración. • **Ácidos grasos,** para obtener energía complementaria. • **Aminoácidos,** para obtener energía en casos de desnutrición extrema (falta de glucosa y de ácidos grasos).

Las distintas sustancias pueden participar tanto en procesos anabólicos como catabólicos. El cuadro siguiente muestra las principales vías metabólicas de glúcidos, lípidos y proteínas en el caso concreto de la especie humana.

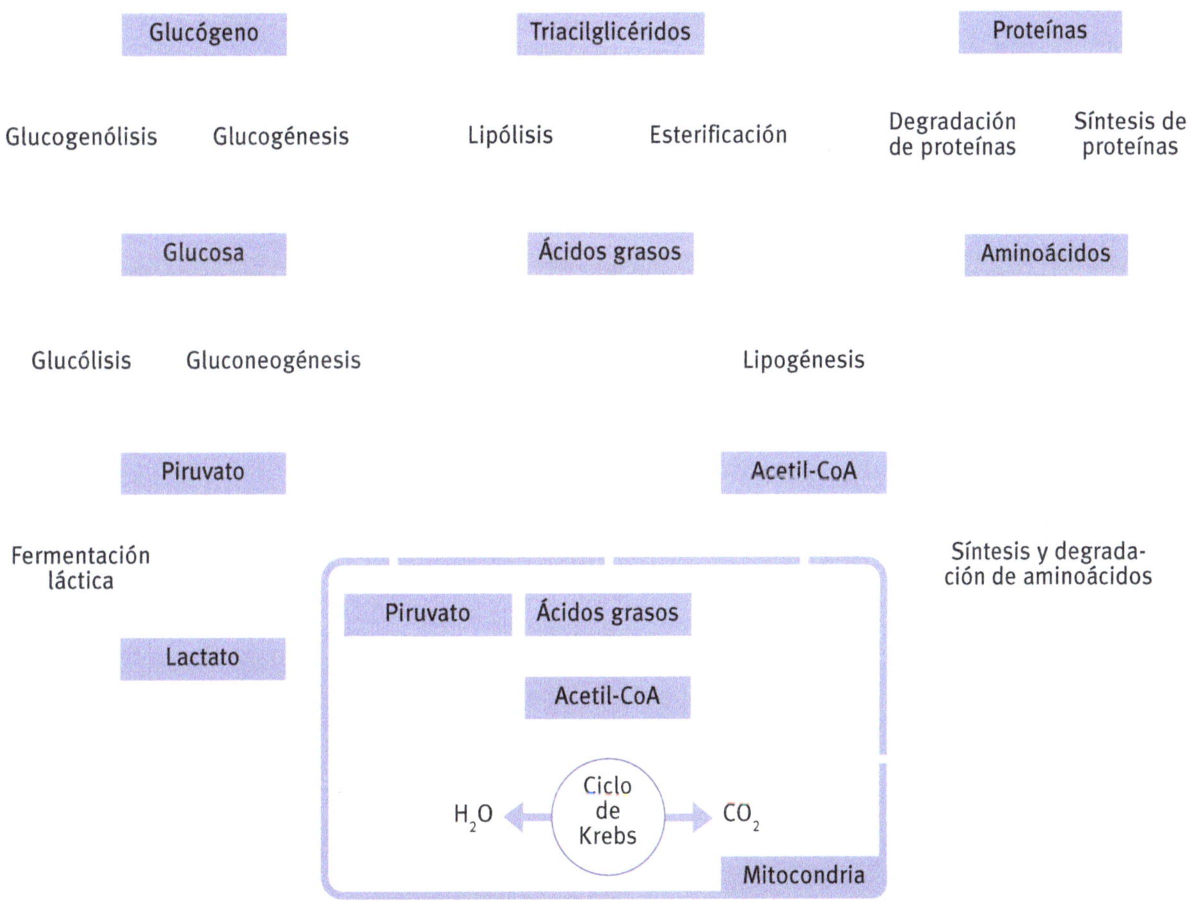

ACTIVIDADES

7. Explica qué entendemos por digestión intracelular y extracelular. Pon ejemplos de organismos que presenten estos tipos de digestión.
8. Compara el aparato digestivo de un rumiante con un herbívoro no rumiante.
9. ¿Qué tipo de digestión tienen los arácnidos? Justifica la respuesta.

3.5. *Metabolismo del ser humano*

El mantenimiento de las funciones vitales y el desarrollo de las actividades cotidianas supone un gasto energético.

El **metabolismo basal** es la cantidad mínima de energía que se necesita para mantener en funcionamiento el organismo, permaneciendo despierto y en reposo absoluto. Este valor se encuentra en torno a los 23 kcal/kg de peso y día en las mujeres, y 24 en los hombres.

El **metabolismo energético** o de **actividad** es la cantidad de energía necesaria para satisfacer todas las actividades propias de cada individuo en un día. Este valor depende de la edad, la talla, el peso, el sexo y las actividades realizadas. En general, cualquier factor que aumente la actividad química de las células incrementa el metabolismo energético. Del valor del metabolismo energético depende la dieta que debe seguirse.

La **dieta** es la cantidad y el tipo de alimentos que se toman en un período de tiempo determinado. Una **dieta equilibrada** es aquélla capaz de proporcionar las cantidades necesarias de nutrientes energéticos (glúcidos y lípidos), estructurales (proteínas) y reguladores del metabolismo (vitaminas y minerales), evitando desequilibrios y excesos nocivos.

Para conseguir una alimentación equilibrada se deben respetar, diariamente, los porcentajes siguientes de nutrientes

NUTRIENTE	% CALÓRICO
Proteínas	10% al 15%
Hidratos de carbono	45% al 50%
Lípidos totales	30% al 35%
monoinsaturados	15% al 20%
poliinsaturados	<7%
saturados	<10%

Los alimentos se clasifican en siete grupos y se representan mediante la **pirámide alimenticia**. Los alimentos que se deben tomar de forma moderada se situan en la parte de arriba de la pirámide, de manera que en la base se situan los alimentos que debemos ingerir en cantidades mayores.

Grupo 1: leche y derivados (queso, yogur...)
Grupo 2: carnes, pescados y huevos
Grupo 3: legumbres, patatas, frutos secos...
Grupo 4: verduras y hortalizas
Grupo 5: frutas
Grupo 6: pan, harina, dulces...
Grupo 7: aceites, mantequilla, manteca...

METABOLISMO ENERGÉTICO SEGÚN DIFERENTES ACTIVIDADES (EN KCAL/KG/DÍA)

- **Actividad física ligera:** trabajo de oficina, enseñanza, trabajo del hogar con aparatos mecánicos...
  — Hombres: 42
  — Mujeres: 36
- **Actividad física media:** trabajo en la construcción, agricultura, pesca, industria ligera, comercio, trabajo del hogar sin aparatos mecánicos, estudiantes...
  — Hombres: 46
  — Mujeres: 40
- **Actividad física intensa:** industria metalúrgica, minería...
  — Hombres: 54
  — Mujeres: 47

C

Valor energético de los nutrientes

1 g de glúcidos .... 3,74 kcal
1 g de lípidos .... 9,30 kcal
1 g de proteínas .... 4,20 kcal

La combinación de alimentos de los distintos grupos en un número determinado de raciones diarias permite establecer una dieta equilibrada. Respecto a los alimentos del grupo 7, se considera que el aceite de una ensalada, la salsa de un guiso o la mantequilla para tres tostadas equivalen, en cada caso, a 10 g.

Para la elaboración de una dieta equilibrada puedes seguir el proceso que exponemos a continuación:

— Observa la tabla inferior y determina, según tu peso normal, la cantidad aproximada de raciones diarias de cada grupo de alimentos que debes tomar.

Peso	Grupo 1	Grupo 2	Grupos 3 y 6	Grupo 4	Grupo 5	Grupo 7
50 kg	1,5	1,5	4	1,5	1,5	40-60 g
60 kg	1,5	1,5	6	1,5	1,5	40-60 g
70 kg	2	2	6	2	2	40-60 g
80 kg	2	2	7	2	2	40-60 g

— Elige los alimentos que te apetezca tomar en un día, de modo que se respeten las raciones diarias recomendadas de cada grupo.

— Repártelos a lo largo del día. Es conveniente realizar cuatro comidas: desayuno, comida, merienda y cena.

— Para el siguiente día realiza un nuevo reparto; elige las raciones evitando repetir los alimentos que tomaste el día anterior.

ALGUNAS RECOMENDACIONES MÁS

- Procura cocer los alimentos en aceite de semillas (oliva, girasol, maíz...).
- Los alimentos deben conservarse en el frigorífico.
- Debe transcurrir el mínimo tiempo posible entre la preparación de un alimento y su consumo.
- La congelación es una buena técnica de conservación que reduce muy poco el poder nutritivo de los alimentos. Utilízala siempre que sea preciso, pero recuerda que no se puede volver a congelar un alimento que ya ha sido descongelado.
- En la cocina, todos los utensilios y superficies deben desinfectarse diariamente con jabón o lejía, para evitar la proliferación de bacterias y prevenir las infecciones alimentarias.

ALIMENTO	RACIONES
Grupo 1: leche y derivados	
leche	200 cc
yogur	250 g
queso fresco	100 g
queso curado	40 g
Grupo 2: carnes, pescados y huevos	
ternera	100 g
pollo	250 g (con huesos)
merluza	120 g
huevos	100 g (sin cáscara)
jamón dulce	100 g
Grupo 3: legumbres, patatas, frutos secos...	
lentejas	60 g
patatas	350 g
Grupo 4: verduras y hortalizas	
ensalada	200 g
judías verdes	100 g
zanahoria	100 g
Grupo 5: frutas	
fruta	130 g
Grupo 6: pan, harina, dulces...	
pan, pastas...	60 g
Grupo 7: aceites, mantequillas, manteca...	
grasas	60 a 80 g

ACTIVIDADES

10. Calcula las calorías que proporciona un plato combinado formado por un bistec de 150 g (suponemos que 100 g son enteramente proteínas), y 120 g de patatas fritas (supondremos que 20 g son de aceite y el resto de glúcidos).

11. Organiza el menú de 4 días para 2 personas de forma que sea totalmente equilibrado sabiendo que en la despensa y en la nevera de casa hay:
3 cartones de leche, un pollo entero, 4 bistecs, media docena de huevos, 2 caballas, 4 melocotones, 250 g de fresas, 4 plátanos, 3 kg de patatas, 3 paquetes de kilo de pasta, pan de molde, 6 yogures, un paquete de alubias, un chorizo, una lechuga, 1 kg de tomates, 0,5 kg de zanahorias, un paquete de madalenas, mantequilla y aceite.

12. Elabora un listado de los alimentos que consumiste ayer indicando la cantidad o el peso aproximado. Relaciona cada alimento de la lista con el grupo al cual pertenece y el número de raciones consumidas. Valora si tu dieta de ayer era equilibrada o no y el porqué.

4. La excreción en los vegetales

Como resultado del procesamiento y la utilización de nutrientes, se producen sustancias diversas que deben ser eliminadas del organismo. Su excreción constituye una fase del proceso de nutrición.

A diferencia de lo que ocurre con los animales, muchas sustancias excretadas en los vegetales no son claramente perjudiciales para ellos. Los productos de excreción de los vegetales se acumulan en vacuolas o en el citoplasma, o bien se expulsan al exterior mediante pelos secretores (como ocurre en las ortigas), conductos especiales como los resiníferos del pino o en bolsas de esencia como en los frutos cítricos.

Los productos de excreción vegetales son el *agua*, las *sustancias nitrogenadas* y las *sustancias no nitrogenadas*.

+

En condiciones de disponibilidad de agua, un árbol de envergadura media puede perder más de 150 l de agua al día.

Una planta de maíz excreta de 2 a 3 litros de agua al día.

Un cactus grande, en una zona desértica, elimina tan sólo unos 25 cm^3 de agua al día.

Agua. Se excreta por dos mecanismos: la *transpiración* y la *gutación*.

La **transpiración** es la eliminación de **vapor de agua** a través de los estomas de las hojas.

- Depende de la apertura o el cierre de los **estomas;** por ello, tal como vimos en la página 214, está directamente relacionada con la intensidad luminosa y la temperatura.
- Cumple en la planta **funciones** diversas:
  - — **Elimina el agua** producida durante la respiración celular.
  - — Asegura el **transporte** de agua y nutrientes desde las raíces hasta las partes verdes de la planta.
  - — **Aumenta la concentración** de los nutrientes en las hojas, ya que se captan del suelo muy diluidos.
  - — Permite la **regulación térmica,** pues una parte de la energía solar que reciben las hojas se consume en la transpiración, con lo que se evita el aumento de temperatura de las hojas.

Gutación

La **gutación** es la eliminación de agua en forma de **gotas,** a causa de la elevada presión de empuje que puede alcanzarse dentro de los vasos conductores ascendentes.

- Se produce en plantas de porte pequeño, cuando la **humedad atmosférica** es muy **elevada**.
- El agua sale por unos **estomas acuíferos** especiales situados en el extremo de los grandes nervios de las hojas.

Sustancias nitrogenadas. Los vegetales, debido a su reducido metabolismo catabólico de proteínas y aminoácidos, a diferencia de los animales no necesitan órganos diferenciados para eliminar estas sustancias.

SUSTANCIAS NITROGENADAS	**Alantoína**	Se almacena en tallos y raíces para ser utilizada nuevamente en la síntesis de proteínas.
	Arginina	Se transformará en **glutamina** y **asparagina,** que se almacenarán para la síntesis proteica.
	Alcaloides	Se acumulan en las vacuolas de las células. Ejemplos: **cafeína, nicotina, codeína** o **morfina.**

Sustancias no nitrogenadas. Muchas de ellas se producen y se acumulan en frutos, hojas, tallos o raíces.

SUSTANCIAS NO NITROGENADAS	**Pigmentos**	La coloración roja, azul o morada de las flores y de los frutos se debe a la secreción de **antocianos.**
	Terpenos	Se trata de lípidos que pueden ser esencias volátiles que dan olor característico a la planta como el **mentol, geraniol, limoneno...** Por otro lado pueden ser sustancias viscosas como la **resina** de las coníferas.
	Sales	En las plantas halófilas (es decir, plantas que pueden vivir en terrenos con una elevada concentración de sales) encontramos glándulas secretoras de sal que pueden expulsar el cloruro sódico en forma de pequeños cristales, los cuales acabarán recubriendo las hojas.

Resina en una conífera

Coloración azul debida a los antocianos

5. La excreción en los animales

Como resultado final del metabolismo celular, se produce un conjunto de sustancias que, debido a su toxicidad o a su elevada concentración, tienen que ser eliminadas. En el caso de los animales, el catabolismo de las proteínas y de los aminoácidos genera una elevada cantidad de residuos nitrogenados, los cuales son muy perjudiciales para el organismo.

Una de las sustancias más tóxicas es el **amoníaco,** NH_3, que procede de la desaminación oxidativa de los aminoácidos. El modo como lo eliminan los distintos grupos de animales depende de la cantidad de agua disponible en el medio en el que se desenvuelven.

Según el sistema de eliminación que emplean para excretar los residuos, los animales se pueden clasificar en:

- **Amoniotélicos.** Excretan directamente **amoníaco.** Son animales acuáticos, como los peces osteictios o peces óseos.
- **Uricotélicos.** Excretan **ácido úrico,** en forma casi sólida. Son animales propios de lugares con poca disponibilidad de agua, como insectos, reptiles terrestres y aves.
- **Ureotélicos.** Excretan **urea.** Es el caso de los mamíferos.

  La urea se sintetiza en el hígado a partir del NH_3 y su excreción requiere poca agua, porque es una sustancia muy soluble.

Los animales presentan diferentes adaptaciones para la excreción. Éstas dependen de las diversas sustancias que excretan y de su medio ambiente. Vamos a conocer los sistemas excretores de los principales grupos de seres vivos.

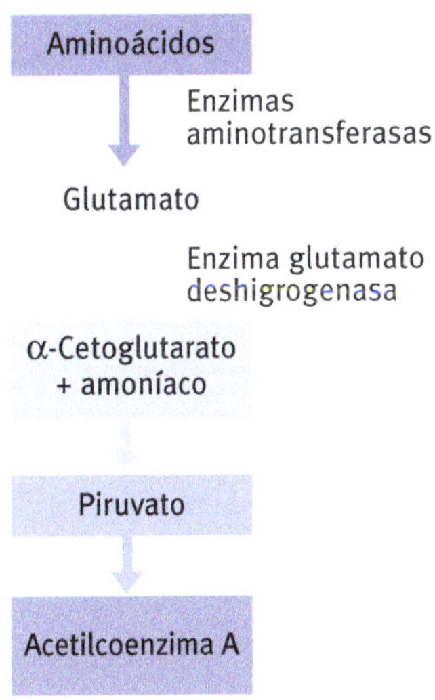

Reacción de la desaminación oxidativa de los aminoácidos

5.1. *Invertebrados*

Los animales más sencillos, en los que las células están poco especializadas y en contacto con el medio exterior, realizan la excreción de sustancias tóxicas por difusión simple. Es el caso de las esponjas y los cnidarios.

La aparición de tejidos especializados provoca la necesidad de recoger y eliminar al exterior las sustancias de excreción que producen las células del organismo.

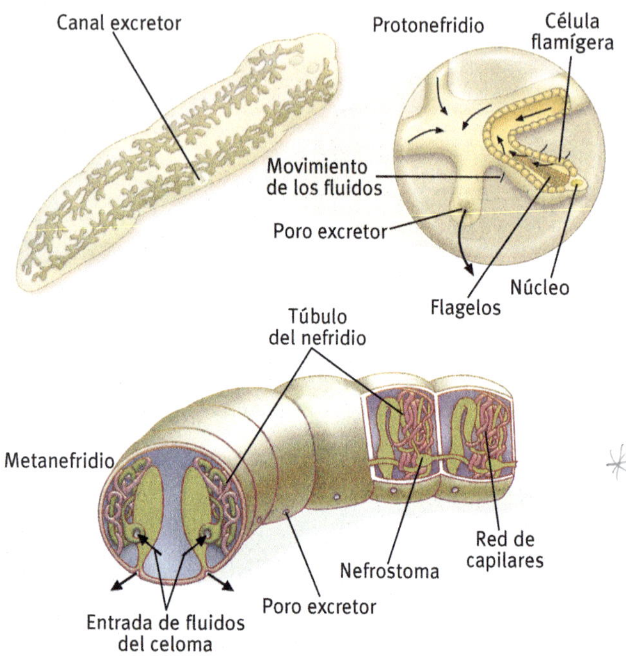

Los **nefridios** son órganos excretores formados por una serie de túbulos que vierten los productos de excreción al exterior del cuerpo, a través de poros excretores. Desde un punto de vista evolutivo, los protonefridios de los platelmintos son anteriores a los metanefridios de los anélidos:

- Los **protonefridios** son tubos cerrados por un lado, que están llenos de células flamígeras, que se mueven en forma de llama, de manera que arrastran los desechos hacia el conducto excretor.
- En los **metanefridios** los tubos están abiertos por ambos lados y se sitúan dos en cada anillo, uno a cada lado. Cada metanefridio consta de un embudo ciliado que recoge el líquido celomático. Los productos de desecho salen a través del poro excretor.

C

Los **arácnidos** tienen glándulas coxales y tubos de Malpighi.

En los **crustáceos** a lo largo del ciclo vital se alternan la presencia de glándulas antenales y maxilares.

En los **insectos** se encuentran, únicamente, tubos de Malpighi.

En los *artrópodos* se combinan dos sistemas de excreción:

- Mediante órganos glandulares de secreción que desembocan junto a algún apéndice, como las **glándulas coxales** en *arácnidos,* y las **glándulas antenales** en *crustáceos.*
- Mediante **tubos** ciegos, llamados **de Malpighi.** Éstos filtran el medio interno recogiendo los residuos disueltos en él, que posteriormente verterán al interior del tubo digestivo. Es el caso de los *insectos.*

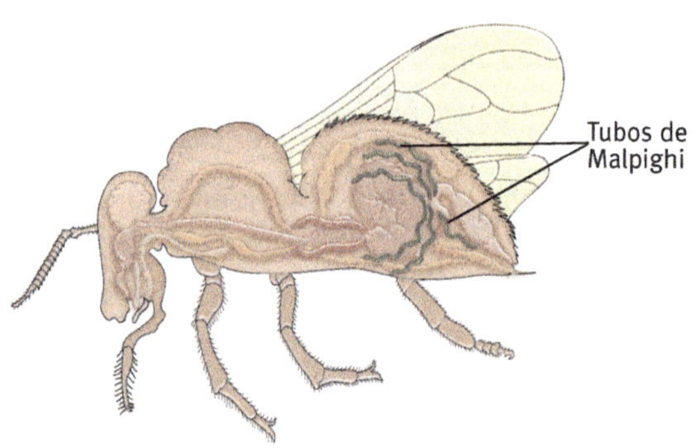

Insecto

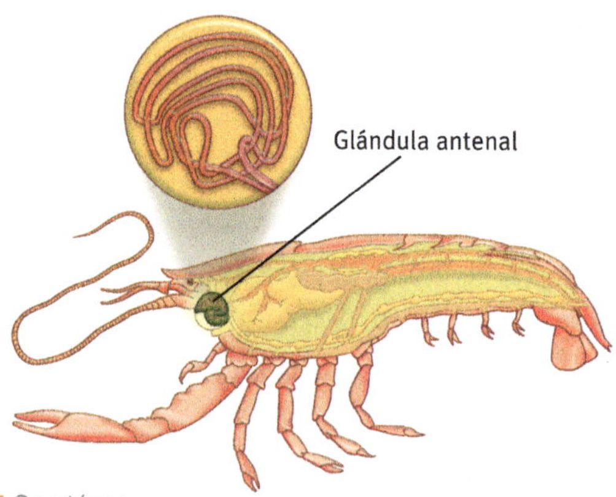

Crustáceo

5.2. *Vertebrados*

La excreción en los vertebrados tiene lugar en los **riñones.** Cada riñón consta de multitud de *nefronas* que están en contacto íntimo con el sistema circulatorio.

La **nefrona** es la unidad funcional del riñón de los vertebrados. En la especie humana, se encuentran un millón de nefronas en cada riñón. Tomando como ejemplo nuestra especie, cada nefrona consta de:

- El **corpúsculo renal,** que se localiza en la *corteza renal* y consta del **glomérulo,** u ovillo de capilares sanguíneos, y la **cápsula de Bowman** que lo engloba a modo de cáliz.
- El **túbulo renal,** que se sitúa en la zona de la **médula renal.** Es un conducto tortuoso, de diámetro variable, que recibe nombres diversos en los distintos tramos de su recorrido: *túbulo contorneado proximal, asa de Henle, túbulo contorneado distal* y *túbulo colector.*

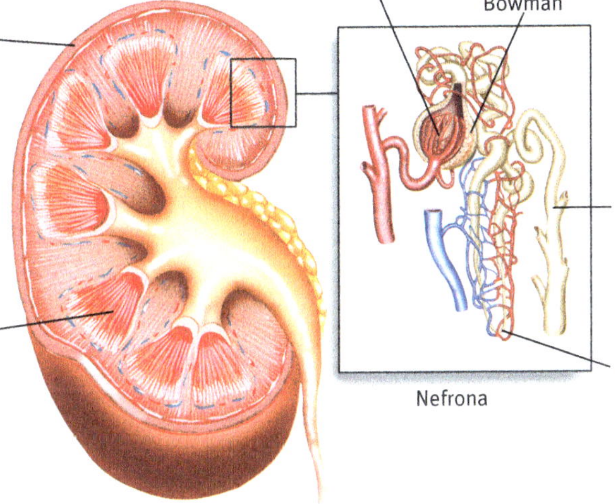

La **función** de la nefrona se desarrolla en las siguientes etapas:

— **Filtración.** Pasan sustancias desde la sangre contenida en los capilares que forman el glomérulo **hacia** el interior de la **cápsula de Bowman.** El líquido que se filtra tiene una composición muy *similar al plasma sanguíneo, pero sin proteínas.* En él se encuentra disuelta la *urea*. En un adulto se forman unos 180 litros de filtrado glomerular al día.

— **Reabsorción.** El líquido filtrado circula hacia el **túbulo renal.** En él se reabsorben, es decir, **pasan de nuevo a la sangre,** *glucosa, aminoácidos* y *sodio,* por transporte activo y, pasivamente, *agua.*

— **Secreción.** Desde los capilares sanguíneos se **excretan directamente** a los túbulos de la nefrona, en especial en su zona distal, mediante transporte activo, K^+, Na^+ y productos nitrogenados, como *urea* y *ácido úrico.*

El resultado final son unos 1500 cm^3 de orina al día, (un 1 % del volumen filtrado inicialmente), que contiene un 90 % de agua y un 10 % de solutos: urea, sodio, potasio...

C

La **excreción** en los vertebrados no se realiza exclusivamente a través de los riñones, sinó que intervienen otros órganos o estructuras:

Riñón ⟶ orina

Aparato digestivo ⟶ heces

Piel ⟶ sudor

Pulmones ⟶ vapor de agua y CO_2

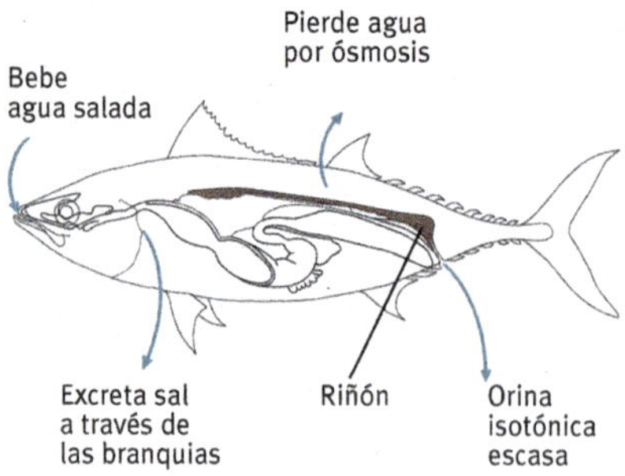

■ Pez de agua salada óseo

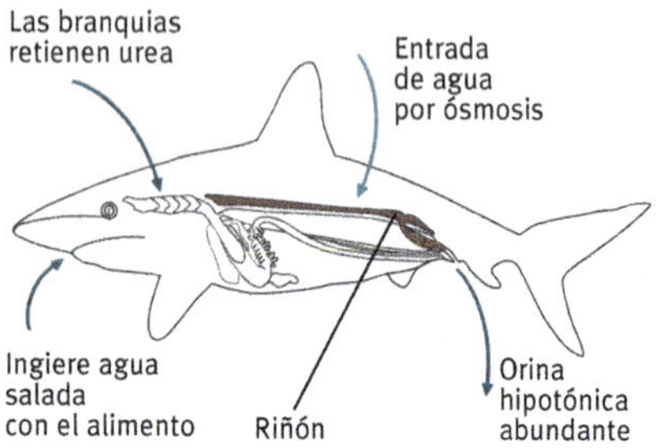

■ Pez de agua salada cartilaginoso

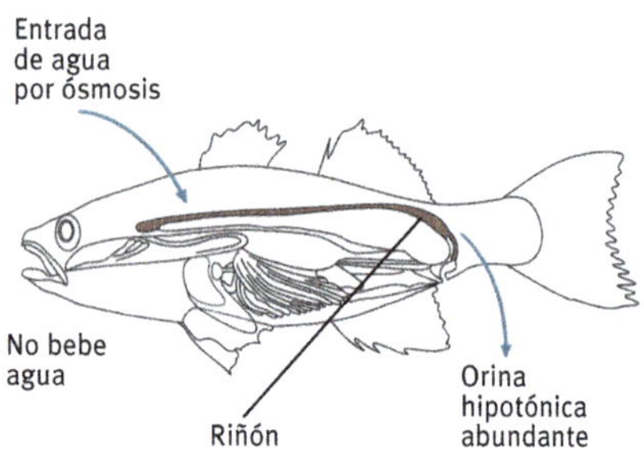

■ Pez de agua dulce

El funcionamiento de la nefrona **regula el volumen** y la **composición química de la sangre** y mantiene el equilibrio del medio interno.

Tomando como base la nefrona, los distintos grupos de vertebrados presentan adaptaciones a las condiciones de su medio ambiente.

Los **peces**, en general, presentan dos riñones grandes, alargados y, en parte, fusionados entre ellos. De ellos parte un conducto que se ensancha y desemboca en el orificio urogenital.

Encontramos diferentes adaptaciones según se trate de peces de agua salada o dulce:

- Los **peces de agua salada** son hipotónicos respecto el medio. Presentan nefronas simples, sin glomérulo, para evitar una excesiva pérdida de agua. Estos peces pueden ser óseos o cartilaginosos:
  - — Los peces **óseos** ingieren agua salada. La sal la excretan a través de las branquias.
  - — Los peces **cartilaginosos**, como el tiburón, convierten su medio hipotónico en hipertónico gracias a una acumulación de urea. Poseen al final del intestino una glándula secretora de sal.
- Los **peces de agua dulce** tienen un medio interno hipertónico, por lo que el agua tiende a entrar por ósmosis. Estos peces excretan abundante orina, por lo tanto sus nefronas son completas.

Los peces de agua salada presentan nefronas simples, sin glomérulo, para evitar una excesiva pérdida de agua, debido a que son hipotónicos respecto al medio. En los de agua dulce las nefronas son completas.

La estructura básica del aparato excretor en el resto de los vertebrados tiene características similares.

Los **anfibios** presentan riñones alargados, semejantes a los de los peces; en ellos aparece la vejiga urinaria.

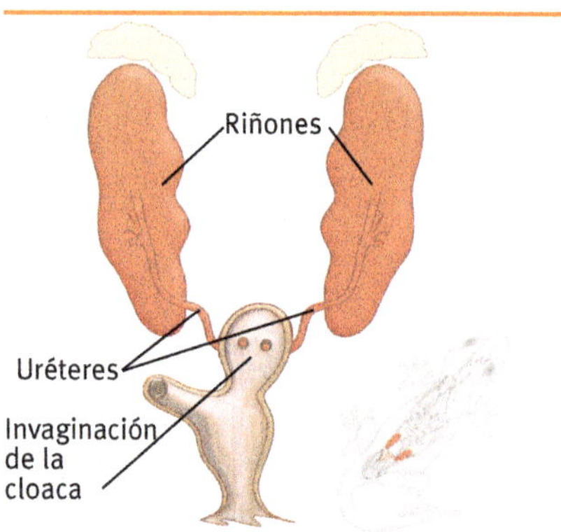

Los **reptiles** tienen riñones pequeños, pero muy especializados y eficientes, con uréteres, y un inicio de vejiga urinaria formada a partir de una invaginación de la cloaca.

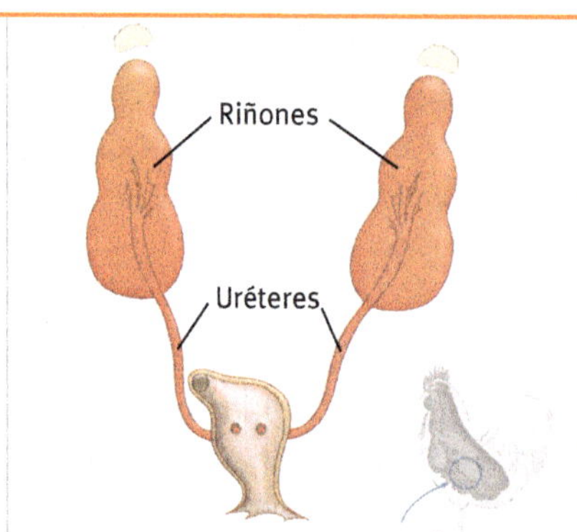

Las **aves** no presentan vejiga urinaria y los uréteres desembocan directamente en la cloaca.

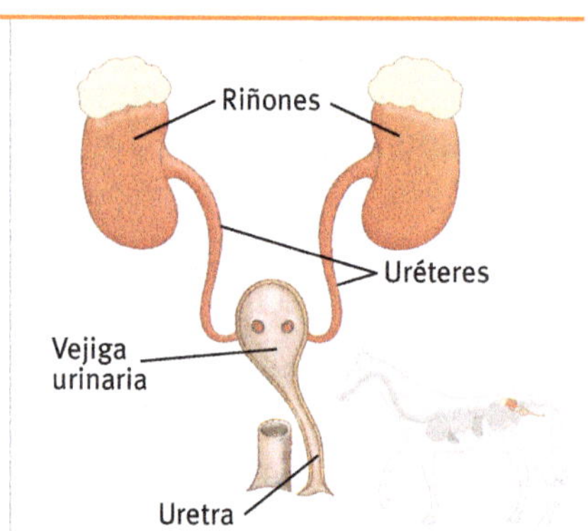

Los **mamíferos** tienen dos riñones, dos uréteres, una vejiga urinaria y una uretra. Ésta, en las hembras es independiente del aparato reproductor, mientras que en los machos presenta función excretora y reproductora.

13 Lee el sigiente texto y responde las preguntas que se plantean a continuación:

El clembuterol es el producto más usado para el engorde ilegal de las carnes que ingiere el ser humano. La frecuencia con que se dan a conocer manipulaciones fraudulentas de los alimentos, cada vez más escandalosas, ha cortado de cuajo el apetito de los consumidores.

Es en la carne donde más fácilmente se producen todo tipo de actos delictivos. Al fin y al cabo son estos alimentos, cuya producción puede controlarse mejor, donde estas prácticas obtienen mayores beneficios. No en vano, y según datos del Ministerio de Agricultura, Pesca y Alimentación, la cantidad de carne media comprada por cada español es de 65,9 kilos al año

El uso de sustancias prohibidas, como el clembuterol, es una constante en la ganadería de todos los países industrializados. España no es una excepción. Y si quiere buscarse una sustancia fraudulenta que destaque sobre todas las demás, es ésta.

Este esteroide alcanzó su mayor fama cuando se descubrió su empleo por el velocista canadiense Ben Johnson. Produce un aumento de la masa muscular, ya que desencadena una mayor síntesis proteíca, con pérdida de materia grasa. Las reses tratadas con clembuterol producen una carne mejor moldeada, prieta y con menos grasa. El pero es su calidad mucho inferior. Al hacerse en menos tiempo, es más dura, absorbe más agua y pierde sabor por su menor concentración de grasas. Da filetes más bonitos, pero luego se quedan en la sartén. El abusivo empleo del clembuterol se ve favorecido por el llamado plazo de supresión: en 15 días desaparece cualquier resto en la carne de los animales tratados. Entonces, sólo se localiza en el hígado y las vísceras.

Si los ecologistas hacen hincapié en los productos contaminantes que, accidental o intencionadamente, se introducen en los alimentos, las organizaciones de consumidores luchan contra los aditivos. "Muchos se usan para ocultar la calidad de los productos. De manera especial, los cárnicos.

El E-124, por ejemplo, es uno de los más abundantes. Este llamativo colorante rojo se emplea en todos aquellos alimentos que cuando están en buen estado, muestran un color rojizo o rosado. Es, sobre todo, el caso de las carnes rojas, pero no el único. Salchichas, chorizos, embutidos, batidos, yogures, galletas, huevos, incluso chicles y caramelos lo llevan en su composición. Carentes de cualquier justificación tecnológica, solamente sirven para engañar al consumidor sobre las cualidades reales del producto que compran. Su empleo más frecuente es para enmascarar la falta de frescura de los alimentos o el uso de materias primas de baja calidad.

Los potenciadores de sabor son otros de los denunciados. Sus principales efectos en el hombre van desde dolores de cabeza hasta diarreas, ardores de estómago, vómitos y falta de apetito. Se emplean de forma generalizada, encontrándose incluso en platos precocinados, como las pizzas congeladas.

Los expertos consideran que la solución del problema pasa por la creación de un organismo regulador europeo. Inspirado en el modelo de la Food and Drugs Administration norteamericana (FDA), debe poseer la entidad suficiente para centralizar y homogeneizar todas las disposiciones.

Adaptado de: Pero, ¿qué comemos? de A. Merino. www.elmundo.es

— Explica las principales ideas que se tratan en el texto.

— Explica, a partir de la información del artículo, qué es el clembuterol y qué efectos causa en el animal.

— En el artículo, se habla además del clembuterol de otros productos utilizados para adulterar o alterar características de los alimentos. Explica cuáles son estos productos y qué efectos producen.

— Consulta las siguientes direcciones electrónicas que hablan sobre los aditivos y responde las preguntas que tienes a continuación:

http://vm.cfsan.fda.gov/~mow/sfoodadd.html

http://www.saludalia.com/Saludalia/web_saludalia/vivir_sano/doc/nutricion/doc/aditivos.htm

a) ¿Qué son los aditivos? ¿Qué es un aditivo directo? ¿Y uno indirecto?

b) ¿Qué es la FDA? ¿ Qué función desempeña?

c) ¿Cuáles son los principales tipos de aditivos?

d) ¿Qué son los números E?

e) ¿Existe algún problema conocido de intolerancia a los aditivos o de enfermedades producidas por éstos?

14 Observa la siguiente figura que muestra el flujo de energía en un ecosistema terrestre (energía expresada en kcal·m^{-2}·día)

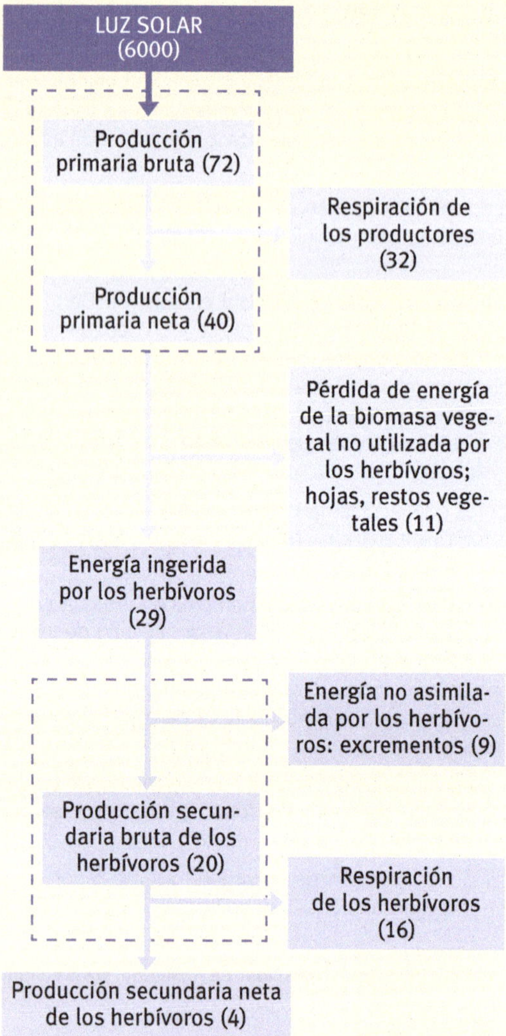

— Comparando los valores de producción primaria bruta y producción primaria neta y observando el esquema, ¿cómo podríamos definir la producción primaria neta?

— Mirando el esquema, ¿podrías decir cómo se pierde la energía entre un nivel trófico y el siguiente?

— Justifica, con la ayuda de la información que te ofrece el esquema, el hecho de que el número de niveles tróficos sea limitado.

— Explica qué camino seguirá en el ecosistema la biomasa vegetal y la materia contenida en los excrementos de los herbívoros. ¿Qué organismos participan?

— Explica porqué la energía no describe un ciclo sino un flujo tomando como ejemplo el esquema de esta actividad.

15 Para estudiar el crecimiento de dos especies de plantas en suelos con nitrógeno, se cultivaron algunos ejemplares de cada especie en parcelas con diferentes cantidades de nitrógeno. Al cabo de una semana se midió la biomasa de las plantas en cada parcela. Los resultados se reflejan en la tabla siguiente:

Cantidad de nitrógeno en el suelo (mg/kg)	Biomasa media de las plantas de la especie 1 (g)	Biomasa media de las plantas de la especie 2 (g)
1	0,05	0,2
5	0,25	0,5
10	0,5	1,2
30	1,2	2,5
100	1,25	4,0
200	1,3	5,0

— Construye el gráfico de las curvas de crecimiento de las dos especies de plantas.

— Identifica las variables dependiente e independiente.

— ¿Cuál presenta mayor crecimiento? ¿Por qué?

16 El nitrógeno es un factor limitante de la producción primaria. ¿Por qué?

17 Esta gráfica representa la pirámide de biomasa de un ecosistema. Explica cómo es posible la sostenibilidad del ecosistema.

18 La franja ecuatorial se caracteriza por la constancia de sus condiciones climáticas: a lo largo de todo el año la temperatura y la humedad son elevadas y eso permite el desarrollo de las selvas. En cambio, en las latitudes medias de climas templados, donde el ciclo estacional es muy importante, predominan los bosques templados.

— Teniendo en cuenta que el reciclaje de los nutrientes se debe básicamente a los microorganismos descomponedores, razona dónde será más rápido el reciclaje, en la selva tropical o en el bosque templado.

— En cada uno de los casos (selva o bosque) justifica si los nutrientes se encuentran mayoritariamente en el suelo o en los árboles.

19 El gráfico siguiente representa las mediciones de transpiración y absorción de agua efectuadas durante un día a un grupo de fresnos.

— Explica qué relación se deduce entre la absorción de agua por parte de las raíces y la transpiración, así como el momento del día en que ambos fenómenos se producen.

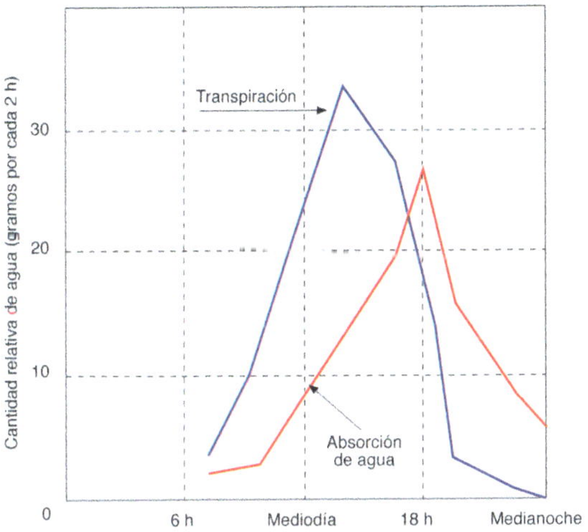

20 Contesta las siguientes preguntas:

— ¿Cómo penetran los nutrientes en las algas y en las briófitas?

— Según el modo de obtención de los nutrientes y de la energía, ¿qué tipo de nutrición tienen las plantas? ¿Y los animales? ¿Y los hongos?

21 ¿De qué se alimentan los hongos? ¿Qué tipo de digestión llevan a cabo, intracelular o extracelular?

22 Observa esta figura que representa dos soluciones separadas por una membrana. El medio externo tiene agua, iones cloruro e iones sodio. El medio interno, el citoplasma, presenta la misma composición pero no contiene tanta concentración de iones cloruro y sodio.

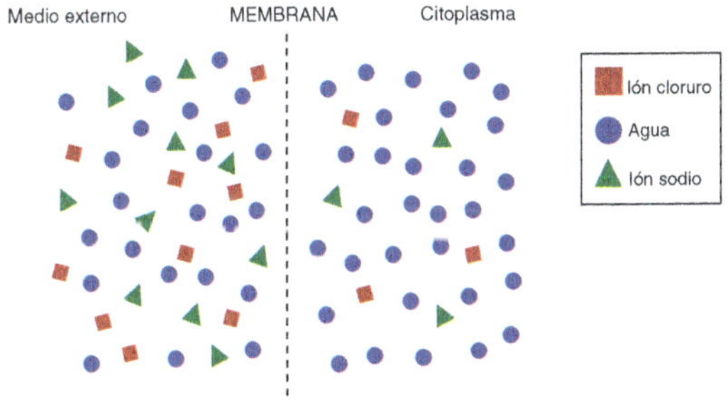

— Indica qué moléculas actúan como disolvente y cuáles como soluto.

— Explica qué es la osmosis e indica mediante flechas hacia dónde se desplazará el agua.

— La ósmosis, ¿es un fenómeno pasivo o activo?

23 En este esquema se representan dos soluciones de glucosa separadas por una membrana.

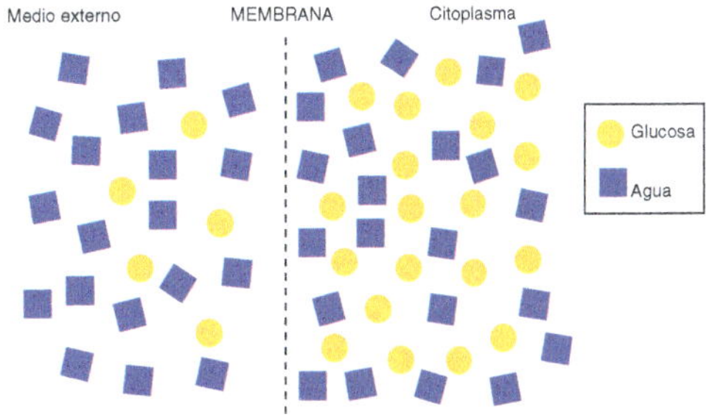

— Qué tipo de transporte se realizará para que ambos medios tengan la misma concentración de glucosa?

— Si en el citoplasma de la célula hay 18 moléculas de glucosa y en el exterior sólo 5 pero la célula necesita 20 moléculas de glucosa, ¿cómo penetrarán las moléculas que le faltan? ¿Qué tipo de transporte se llevará a cabo?

24 ¿Por qué las plantas no necesitan un sistema excretor como los animales?

— ¿Las esponjas tienen sistema excretor?

25 ¿Cuál es el sistema excretor más rudimentario? Qué organismos lo presentan? ¿Y el más sofisticado?

— ¿Puede considerarse el sistema excretor como un regulador hídrico en los animales? Razona tu respuesta.

26 ¿Cuáles son los procesos involucrados en la formación de la orina?

— ¿Cuál es la composición química de la orina?

27 Además de la nefrona, existen otras estructuras y órganos que también funcionan como sistemas de excreción. Explica cuáles son y qué función realizan.

TIC

28 En esta página web encontrarás información sobre el sistema digestivo de los rumiantes. Consúltala y contesta las siguientes preguntas:

http://www.viaganadera.com/sabiasque/pagina1.asp

— Explica qué quiere decir que los rumiantes tienen 4 estómagos.

— ¿Dónde se dirige el alimento tras la deglución?

¿Dónde se produce la digestión mecánica? ¿Y la química?

— ¿A qué compartimento se dirige el bolo alimenticio generado en la rumia?

— ¿Cuánto tiempo dedica un rumiante a comer? Y a rumiar?

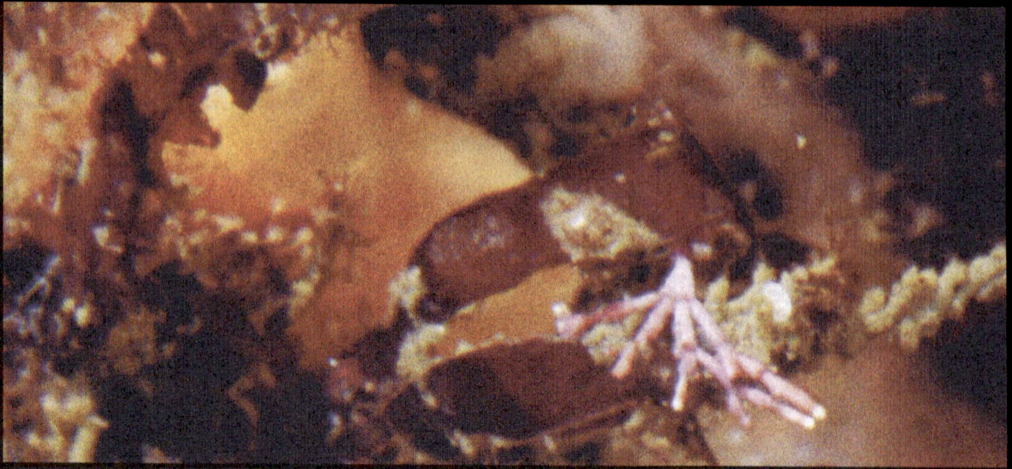

10

La nutrición: intercambio de gases

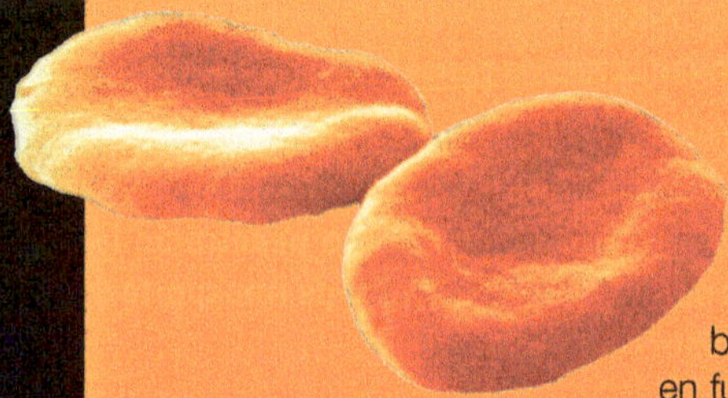

Los seres vivos han desarrollado distintos mecanismos de intercambio de gases en función de sus necesidades y de las características del medio en el que viven. Como resultado de ello, es posible describir muchas estrategias respiratorias distintas que implican una gran diversidad de órganos.

Habitualmente, a mayor actividad metabólica aumentan las necesidades de intercambio de gases y en consecuencia la complejidad del sistema respiratorio.

De igual forma, aquellas especies con mayores requerimientos respiratorios también estarán dotadas de un sistema circulatorio suficientemente desarrollado con el que distribuir por todo el organismo el elevado volumen de gases intercambiados.

Así pues, en general, en aquellos organismos de mayor tamaño, de elevada actividad metabólica y que viven en hábitats más extremos podremos observar los sistemas respiratorios y circulatorios más complejos.

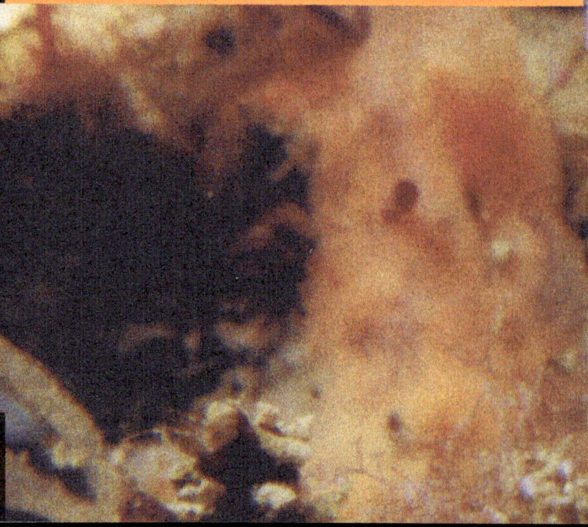

circulación

■ Branquias de poliqueto. *Bispira bolutacornis.*

ORGANIZACIÓN DE LOS CONTENIDOS

1. La respiración en los vegetales
- 1.1. Regulación del intercambio de gases

2. La respiración en los animales
- 2.1. Invertebrados
- 2.2. Vertebrados
- 2.3. La respiración en el ser humano

3. El transporte de sustancias en los vegetales
- 3.1. Vías de conducción ascendentes
- 3.2. Vías de conducción de moléculas orgánicas

4. La circulación en los animales
- 4.1. Invertebrados
- 4.2. Vertebrados
- 4.3. El corazón humano y las enfermedades cardiovasculares

OBJETIVOS

- Comprender la necesidad de un sistema de respiración y un sistema de circulación para el desarrollo de la función de nutrición en los seres vivos.
- Identificar los mecanismos de circulación en los vegetales y su relación con la respiración y la elaboración y el transporte de nutrientes.
- Conocer los distintos tipos de aparatos respiratorios y circulatorios en los animales.
- Comprender el funcionamiento del corazón y la causa principal de las enfermedades coronarias.

En la unidad anterior, hemos analizado cómo obtienen las sustancias nutritivas los vegetales, así como las características de los sistemas de digestión y excreción de diferentes grupos de animales. En ésta estudiaremos los mecanismos de respiración y circulación en diversos grupos de seres vivos; de este modo, completamos el análisis de la función de nutrición, ya que la mayoría de los seres pluricelulares necesita del trabajo coordinado de los cuatro aparatos o sistemas indicados.

La máxima cantidad de energía que las células pueden obtener, tal como vimos en la unidad 7, se consigue mediante la oxidación total de compuestos orgánicos como la glucosa utilizando oxígeno; dicha reacción desprende dióxido de carbono. De ahí que existan mecanismos de captación de oxígeno del exterior y, debido a su toxicidad, también de eliminación del dióxido de carbono.

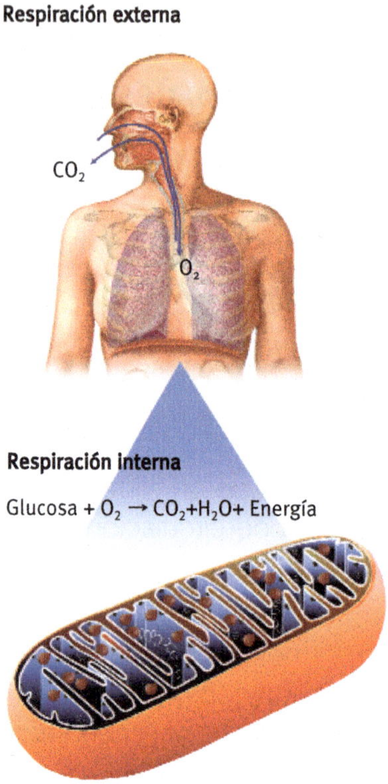

Mitocondria

Podemos distinguir entre respiración interna y respiración externa.

- La **respiración interna** tiene lugar en las células. Durante ella:
  - Se produce la **reacción química:**

    $$\text{Glucosa} + O_2 \rightarrow CO_2 + H_2O + \text{Energía}$$

  - Se **intercambian gases** a través de la membrana plasmática: entra oxígeno y sale dióxido de carbono.

    Estas sustancias, como ya sabes, son solubles en lípidos y atraviesan las membranas plasmáticas por **difusión simple,** a favor del gradiente de concentración.

- La **respiración externa o ventilación** es el intercambio de gases entre el medio interno de un ser vivo y su medio ambiente.

  En los animales, el proceso suele estar relacionado con un sistema de conducción constituido por un **líquido circulante** que, a menudo, contiene **moléculas** o **pigmentos,** especializados en unirse a los gases respiratorios para facilitar el transporte de éstos.

Mientras la respiración interna es un proceso que tiene lugar del mismo modo en numerosos y variados grupos de seres vivos, la respiración externa se realiza de modo muy diferente, según las características de los distintos grupos de seres vivos.

Vamos a analizar algunos casos representativos, dentro de los reinos vegetal y animal, siguiendo los progresos que han favorecido la adaptación a diversos ambientes.

1. La respiración en los vegetales

Tal como hemos visto en la unidad 7, las células de los vegetales también respiran; por tanto, en ellas podemos observar:

- La **respiración interna** en las células de la planta.
  - La glucosa reacciona con el oxígeno produciéndose dióxido de carbono, agua y energía.
  - Los gases se intercambian a través de la membrana plasmática y la pared celular.

- La **respiración externa** o **ventilación** es el intercambio de gases con el medio ambiente.
  - — El O_2 del aire penetra a través de los **estomas**. También se capta una pequeña parte de oxígeno por los **pelos absorbentes** de la raíz.
  - — Los tejidos adyacentes a la epidermis de la hoja tienen **espacios intercelulares** por los que circulan los gases.
    - El **parénquima clorofílico lagunar** en contacto con el envés de la hoja posee de un 50 a un 80 % de volumen de aire.
    - El **parénquima clorofílico en empalizada** en el haz de las hojas tiene entre un 10 y un 40 % de su volumen ocupado por aire.

Dado que la respiración externa o ventilación es cualquier intercambio de gases con el medio, incluye otros intercambios de gases relacionados con la fotosíntesis y la transpiración.

Respecto a la **fotosíntesis:**

— Los vegetales toman $\mathbf{CO_2}$ del aire atmosférico o CO_2 disuelto en el agua, en el caso de plantas acuáticas.

— Desprenden $\mathbf{O_2}$ a la atmósfera o al agua.

Respecto a la **transpiración:**

— Las plantas pierden hasta el 90 % del agua absorbida por las raíces en forma de **vapor de agua,** a través de los estomas.

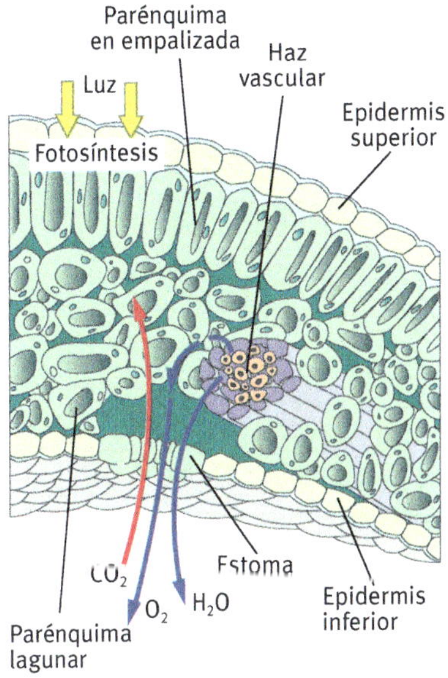

Corte transversal de una hoja: intercambio de gases en la fotosíntesis

C

El intercambio de gases se realiza a través de los estomas. Este intercambio de gases, entre los espacios con aire y las células, se produce siempre por **difusión simple,** a favor del gradiente de concentración.

1.1. *Regulación del intercambio de gases*

Las plantas regulan el intercambio de gases (O_2, CO_2 y vapor de H_2O) mediante la apertura y el cierre de los estomas. Los orificios estomáticos se encuentran delimitados por dos células oclusivas o estomáticas que tienen capacidad para regular la apertura.

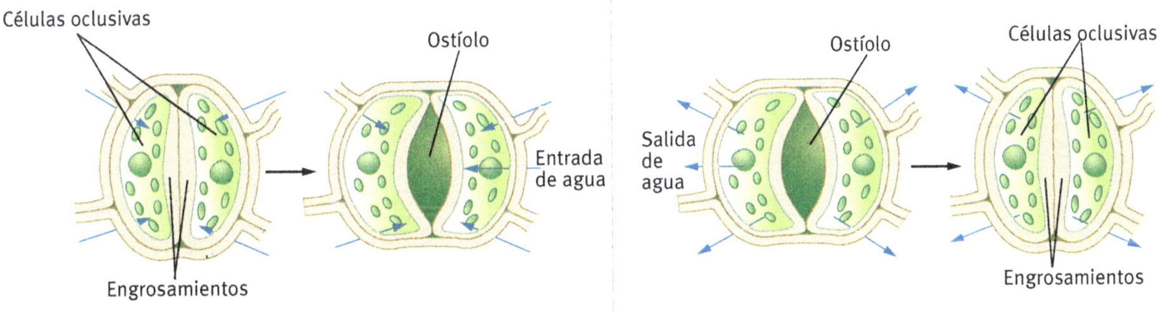

Las células oclusivas tienen unos engrosamientos asimétricos en sus paredes celulares. Cuando aumentan su contenido en agua, toman una forma arqueada y dejan entre ellas un orificio central denominado **ostíolo.**

Si las células oclusivas reducen su contenido en agua adoptan una forma alargada y ambas células contactan por el centro, cerrando el ostíolo.

La entrada y salida de agua de las células oclusivas depende de las concentraciones de potasio y de sodio. Éstas, a su vez, están controladas por el *ácido abscísico,* una hormona vegetal que modifica la permeabilidad de las membranas celulares para la entrada de iones.

2. La respiración en los animales

También en este caso distinguimos:

- Una **respiración interna** que tiene lugar en el interior de las células y que se produce según la reacción química vista anteriormente.
- Una **respiración externa** o **ventilación** que se produce entre los individuos y su medio ambiente.

Las características de la respiración de los animales dependen de la estructura del cuerpo y de su medio ambiente.

- La **respiración directa** es aquella en la que no existen estructuras especializadas para la respiración: el oxígeno del agua pasa por difusión al interior de todas las células. Éstas, a su vez, liberan el CO_2 por difusión al agua. Se presenta por tanto en organismos acuáticos sencillos, como los poríferos, los celenterados, los platelmintos y los nemátodos.
- La **respiración indirecta** es aquella que requiere la presencia de un órgano respiratorio, capaz de realizar el intercambio gaseoso. Este órgano respiratorio se caracteriza por presentar un epitelio delgado y muy vascularizado (con muchos vasos sanguíneos). Destaca también la presencia de líquidos y pigmentos que circulan por el organismo para hacer más efectivo el transporte de gases. Es propia del resto de invertebrados y de todos los vertebrados.

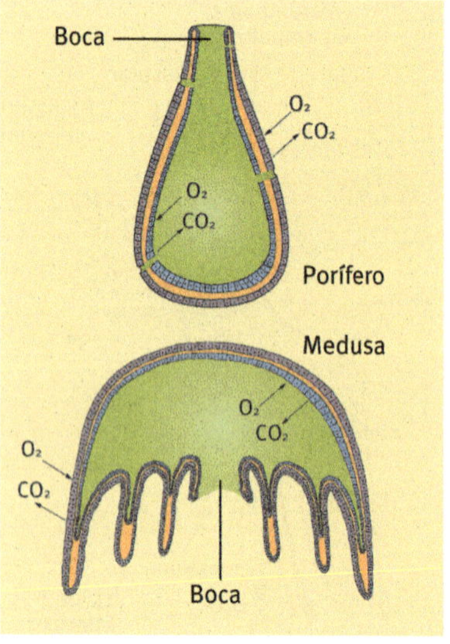

Intercambio de gases en un animal diblástico

C

En los animales, tanto acuáticos como terrestres...

- La región donde se intercambian los gases está separada del medio por un sistema de membranas.
- Los gases que atraviesan el sistema de membranas (entrada de O_2 y salida de CO_2) están siempre disueltos en agua.
- El sistema de membranas debe estar siempre húmedo para ser funcional.
- Los gases atraviesan las membranas por difusión (sin gasto energético).

Aunque forman parte del sistema circulatorio, que se trata a fondo en la segunda parte de esta unidad, en la descripción de los mecanismos respiratorios se explican, muy a menudo, las características del líquido circulante así como de los pigmentos respiratorios, debido a la íntima relación que existe entre respiración externa, o ventilación, y circulación, o transporte.

Los órganos especializados en la respiración externa dependen sobre todo de las características del medio con el cual se deben realizar los intercambios.

En la atmósfera, la concentración de oxígeno es de unos 210 ml de oxígeno por litro de aire. En el agua esta concentración es muy inferior: entre 5 y 8 ml de oxígeno por litro.

- En los animales que toman el oxígeno del aire, la **velocidad de difusión** de los gases a través de las membranas respiratorias es muy **elevada.** Sin embargo, se ven amenazados por el **peligro de desecación,** que evitan encerrando las membranas en cavidades especiales y dotándolas de mecanismos que las mantengan húmedas.
- En el medio acuático, los animales fuerzan el movimiento del agua en torno a las branquias, para mejorar así la **velocidad de difusión,** que de otro modo sería muy **baja** o incluso **inexistente**. Por otro lado, **no** corren el **peligro de desecación** que mencionábamos para los animales terrestres.

Vamos a conocer los cuatro tipos de respiración indirecta: *cutánea, traqueal, branquial* y *pulmonar*.

Respiración cutánea

- Propia de animales que habitan **ambientes terrestres y húmedos.**
- El oxígeno y el dióxido de carbono se difunden a través de la **piel.**

  Ambos gases son transportados por pigmentos respiratorios.

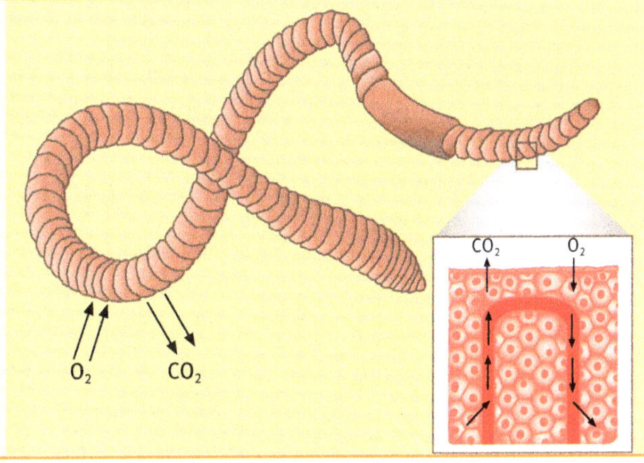

Respiración traqueal

- Es propia de los **artrópodos.**
- Las **tráqueas** son una red de tubos que recorren el cuerpo del animal y que comunican con el exterior a través de un espiráculo.
- Las tráqueas, que están impermeabilizadas para evitar la pérdida de agua, se dividen en **traqueolas.**
- Las traqueolas llegan a todas las células del cuerpo, por lo que no es necesario un sistema de transporte de gases ni pigmentos respiratorios.

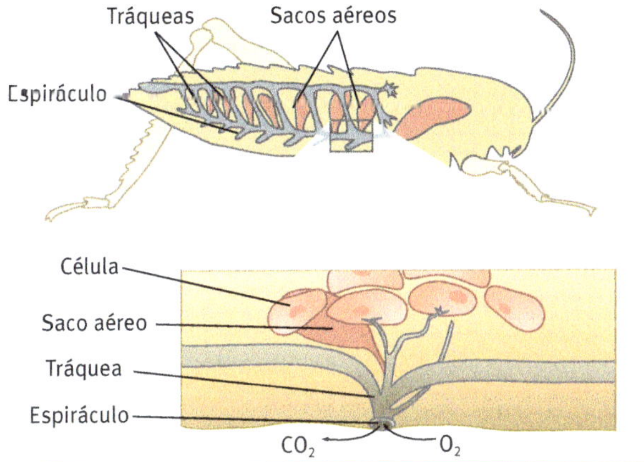

Respiración branquial

- Se presenta en **animales acuáticos** pero también en algunos **terrestres.**
- Las **branquias** están formadas por membranas especializadas en captar el O_2 disuelto en el agua.
- Existen dos tipos de branquias, las externas, presentes en larvas de anfibios, y las internas, en peces y artrópodos.
- El oxígeno captado es transferido al líquido circulante; el CO_2 sale de las células y es transportado tambien por ese líquido.
- Ambos gases viajan unidos a pigmentos o proteínas transportadoras hasta su destino.

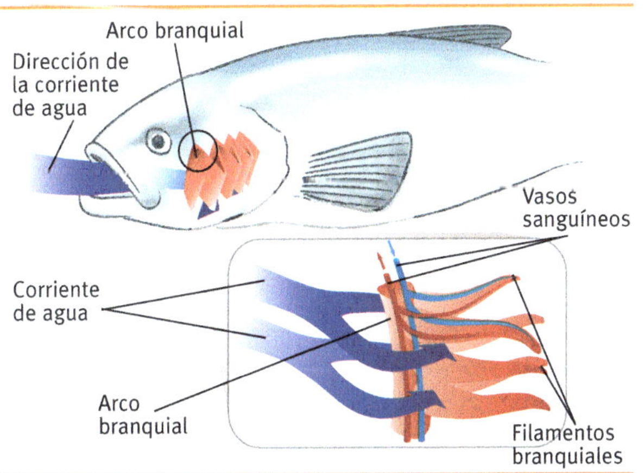

Respiración pulmonar

- Es propia de los **animales terrestres.**
- Los **pulmones** son bolsas membranosas situadas en el interior del cuerpo en las que tiene lugar el intercambio de gases.
- Suele haber un par de pulmones, aunque existen ejemplos, como las serpientes que presentan uno solo.
- El oxígeno se transfiere desde los pulmones al líquido circulante; el dióxido de carbono realiza el paso inverso.
- Existen pigmentos especializados en el transporte de ambos gases.

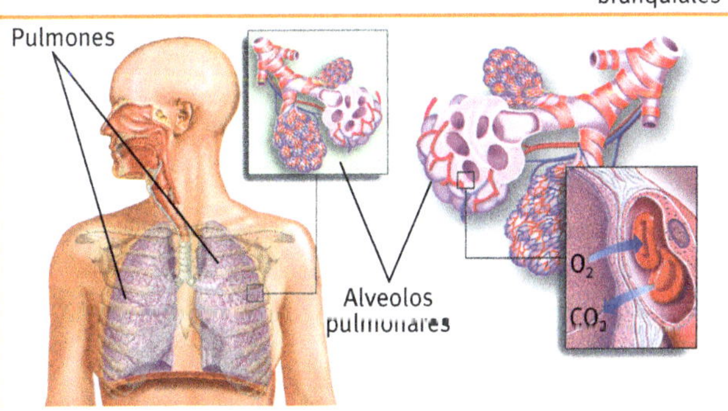

Las branquias y los pulmones son los órganos más eficaces para el intercambio gaseoso porque tienen una superficie de intercambio mucho mayor que las tráqueas o la piel.

2.1. *Invertebrados*

Los **anélidos** son un grupo variado en el que encontramos representantes acuáticos y terrestres:

- Los *terrestres* viven en ambientes muy húmedos y tienen respiración **cutánea,** es el caso de la lombriz de tierra.
- Los *acuáticos* tienen respiración **branquial,** como es el caso de la sanguijuela.

Los pigmentos respiratorios pueden ser la *hemoglobina,* de color rojo, la *hemeritrina*, de color rosado o la *clorocruorina,* de color verde, que confieren un color específico al líquido circulante.

En los **moluscos,** los órganos respiratorios se encuentran en el interior de la cavidad paleal y reciben el nombre de **ctenidio,** por tener un aspecto plumoso que sirve para incrementar la superficie de intercambio de gases. En los moluscos acuáticos se corresponde con las **branquias,** y en los terrestres con una cámara de función similar al **pulmón,** ya que esta rodeada de vasos sanguíneos que favorecen el intercambio gaseoso.

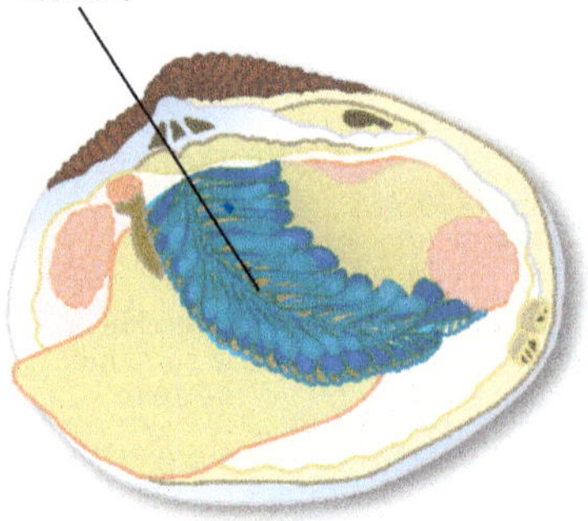

Molusco acuático. Bivalvo.

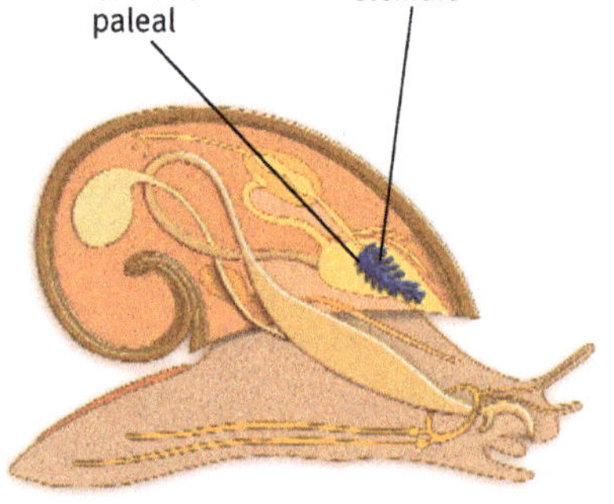

Molusco terrestre. Gasterópodo.

Para aumentar la eficacia del intercambio de gases en las branquias, en los moluscos acuáticos el líquido circulante se desplaza en sentido contrario al flujo del agua. En los cefalópodos, este intercambio también se ve favorecido por la circulación del agua provocada por los movimientos de propulsión del cuerpo.

El pigmento respiratorio de los moluscos es la *hemocianina,* que tiene una tonalidad azulada.

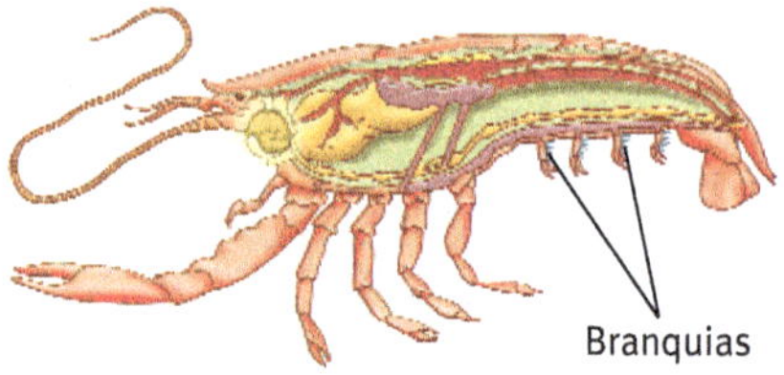

Artrópodo acuático. Crustáceo.

En el caso de los **artrópodos** encontramos especies:

- Acuáticas: respiran por **branquias** situadas en la base de sus apéndices. El movimiento de éstos asegura una mayor circulación del agua por cada branquia. El pigmento respiratorio es la **hemocianina.** La mayoría de los crustáceos respira de este modo.
- Terrestres: desarrollan un sistema respiratorio específico, las **tráqueas.** Este mecanismo respiratorio es característico de los insectos y los arácnidos. Se distinguen dos tipos de tráqueas: las *filotráqueas* y las *dendrotráqueas.*

Las **filotráqueas** son unas invaginaciones globosas que se encuentran en el interior del cuerpo y tienen la pared interna recubierta de láminas donde se intercambian los gases. El líquido circulante se denomina **hemolinfa** y el pigmento es la *hemocianina.* Este tipo de respiración es característico de los *arácnidos.*

Las **dendrotráqueas** son invaginaciones en forma de tubos reforzados por quitina. Se abren al exterior por los *espiráculos* y se ramifican por el interior del cuerpo. En este caso, el transporte de gases por el cuerpo no requiere de la ayuda de la hemolinfa. Se presentan en los *insectos.*

Arácnido

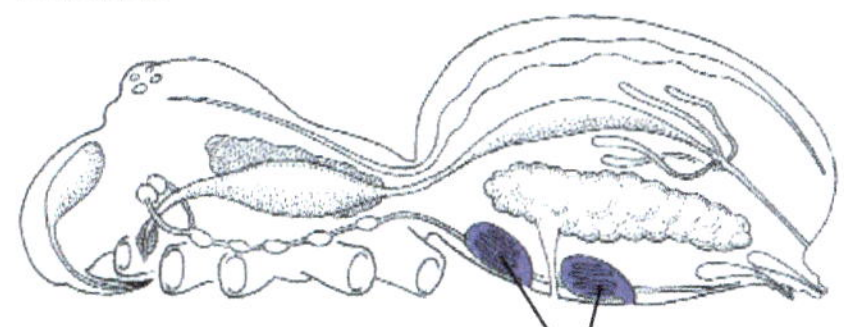

Insecto

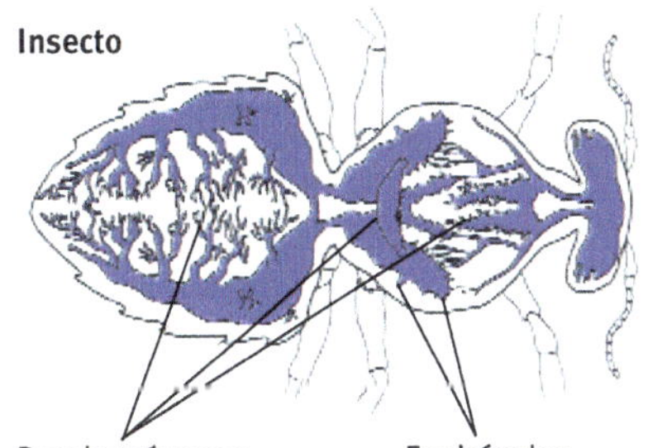

ACTIVIDADES

1. En la respiración se obtiene CO_2. ¿Podrías decir cuántas moléculas de CO_2 se obtienen a partir de una molécula de glucosa? ¿Dónde se produce esta reacción? ¿Con qué nombre se conoce este tipo de respiración?
2. Si el porcentaje de oxígeno presente en el agua es mucho menor que en la atmósfera, ¿cómo consiguen captarlo con eficiencia los organismos acuáticos?
3. Elabora una relación de los diferentes pigmentos respiratorios de los invertebrados y di el color que los caracteriza.

2.2. *Vertebrados*

Atendiendo a la respiración, existen dos grandes grupos de vertebrados: los *peces,* que están adaptados a la vida acuática, y los vertebrados *tetrápodos* (llamados así porque presentan cuatro extremidades), que están adaptados a la vida terrestre. En este último grupo, incluimos a los mamíferos que conquistaron el medio acuático después de su adaptación al medio terrestre, como es el caso de las ballenas, las orcas y los delfines.

Todos los vertebrados se caracterizan por:

- La presencia de un pigmento transportador denominado *hemoglobina*, que se encuentra en el interior de células especializadas llamadas *glóbulos rojos.*
- La existencia de un líquido circulante, llamado *sangre.*

Los **peces** respiran por **branquias.**

- Están formadas por una doble lámina, sostenida por un arco branquial. Cada lámina está integrada por una hilera de filamentos muy vascularizados.
- Las branquias presentan una elevada eficacia, ya que captan el 80 % del oxígeno disuelto en el agua.

En los **peces cartilaginosos** o **condrictios** el agua entra por el espiráculo (que en el tiburon es lateral) y se dirige a las hendiduras branquiales, por donde sale.

Estos peces no pueden impulsar el agua desde el espiráculo a las hendiduras branquiales; el propio movimiento del pez es el que permite la circulación del agua.

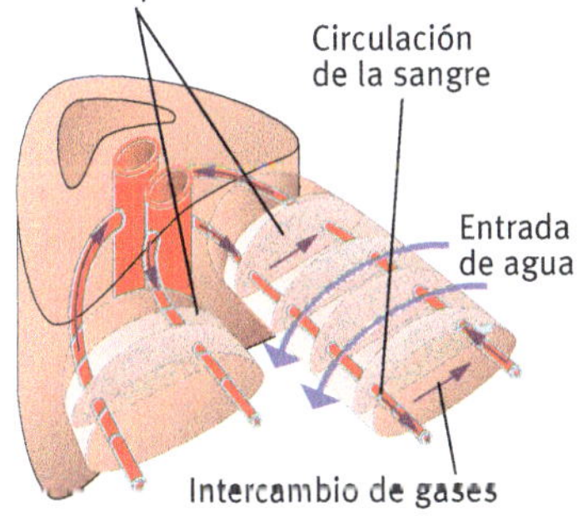

La dirección que sigue el agua a través de las branquias es perpendicular al movimiento de la sangre: esto permite que el intercambio de gases sea muy eficaz.

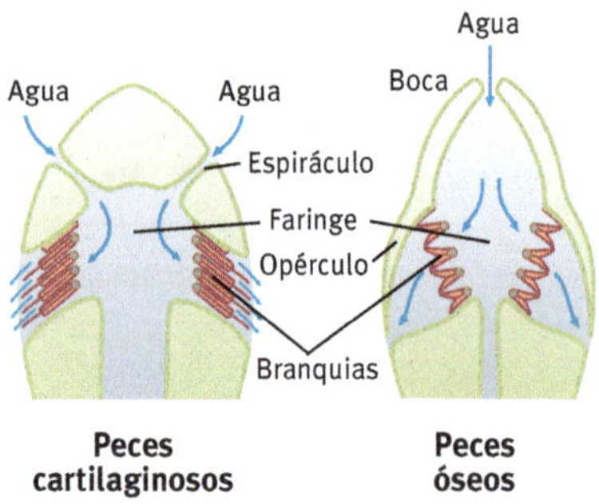

En los **peces óseos** u **osteíctios** existen normalmente 4 pares de branquias repartidos a ambos lados de la cabeza y cubiertos por un opérculo.

— En la inspiracion el pez abre la boca, el agua entra en la cavidad bucofaríngea y queda retenida en la cavidad branquial porque el opérculo la cierra.

— En la espiración, el pez cierra la boca, se contraen las paredes de la cavidad bucofaríngea y se levanta el opérculo, por donde sale el agua.

Los **vertebrados tetrápodos** (anfibios, reptiles, aves y mamíferos) tienen respiración pulmonar, con un aparato respiratorio en el que pueden observarse órganos comunes a todos los grupos.

En los **anfibios,** puede comprobarse claramente que el tipo de respiración de un animal es el que mejor se adapta al medio en que vive.

Las **larvas**, que son acuáticas, respiran por **branquias.** En el caso de la larva de tritón, ésta posee branquias externas.

Los **adultos** combinan la respiración **cutánea** con la **pulmonar.** Los pulmones son pequeños y con pocos alveolos, por lo tanto la superficie para el intercambio de gases es reducida.

El proceso es el siguiente:

— El aire entra por las fosas nasales, que se abren a la cavidad bucal.

— Los movimientos de deglución (hundimiento y elevación del suelo de la boca) favorecen el paso de aire desde la boca a la faringe y a los pulmones.

— Existen otras dos zonas de intercambio de gases: el interior de la boca y la faringe, que están muy vascularizadas, y la piel desnuda, muy vascularizada y siempre húmeda.

Aunque algunos **reptiles** desarrollan su vida en estrecha relación con los ambientes acuáticos, todos los representantes de este grupo tienen los órganos respiratorios adaptados a la vida terrestre.

- La respiración es exclusivamente **pulmonar,** ya que tienen una piel impermeable que impide cualquier intercambio de gases. Los pulmones son simples, pero suficientes para captar el oxígeno necesario para la actividad de estos animales. A pesar de ello, el desprendimiento de dióxido de carbono es bajo y poco eficiente, con lo cual soportan un alto contenido de este gas en los tejidos y en la sangre.
- Aparece un **paladar** no óseo que permite separar la cavidad bucal de las vías respiratorias; pueden masticar y respirar al mismo tiempo.
- En algunas especies, la **laringe** puede producir sonidos. Un ejemplo de ello son los silbidos que emiten los ofidios o serpientes.

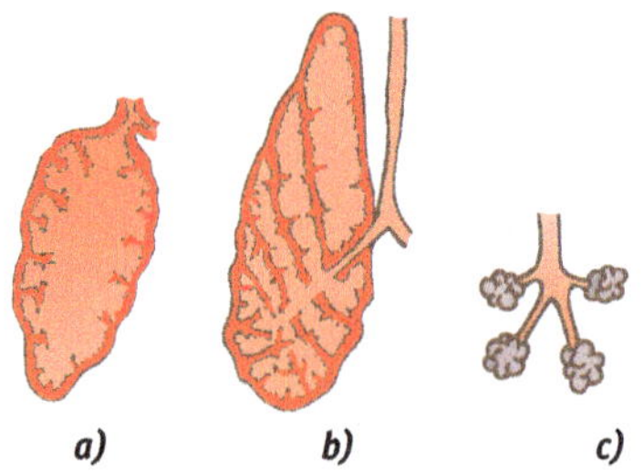

Evolución de la superficie de absorción pulmonar: *a)* pulmón de anfibio *b)* pulmón de reptil y *c)* alveolo de pulmón de mamífero

Las **aves** exigen una gran eficacia respiratoria, puesto que el vuelo supone un gran trabajo muscular, y el proceso de obtención de la energía para desarrollarlo necesita un gran aporte de oxígeno.

Sin embargo, su morfología aerodinámica no puede dar cabida a grandes pulmones. Por ello, además de las estructuras comunes a todos los vertebrados tetrápodos, presentan algunas adaptaciones a su modo de vida.

- Presentan **sacos aéreos** membranosos en conexión con los pulmones y los **huesos neumatizados,** huesos especializados en los que la médula ha sido reemplazada por aire, para aligerar el peso del animal.

  Los sacos aéreos contribuyen a aligerar el peso del animal y completan la función pulmonar como reserva de aire. Sin sobrecargar el peso, dedican una parte importante del volumen de su cuerpo al aparato respiratorio.

- En la zona de unión entre la tráquea y los bronquios, la inmensa mayoría de las aves presenta la **siringe,** órgano fonador que les permite el canto.

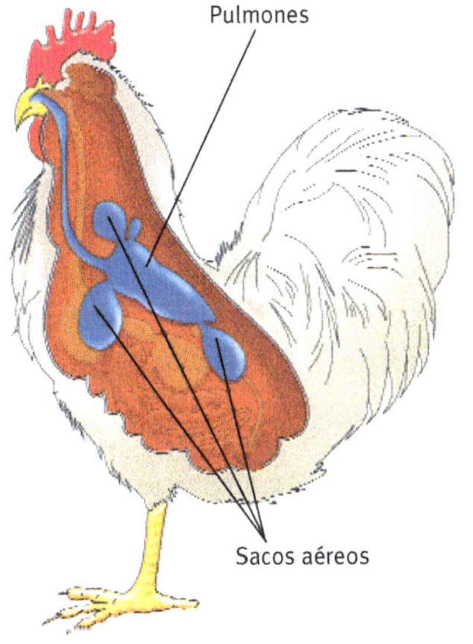

Los **mamíferos** presentan un aparato respiratorio similar al de todos los vertebrados tetrápodos, pero con algunas adquisiciones bastante ventajosas:

- El **gran desarrollo pulmonar,** con numerosas ramificaciones en los bronquiolos y, por tanto, gran número de alveolos. Esto supone una gran superficie para el intercambio de gases.
- La formación de dos **pleuras** o membranas protectoras que recubren los pulmones y la pared interna de la caja torácica. Entre ellas se encuentra el *líquido pleural* con función lubricante. El conjunto protege y facilita el desplazamiento de los pulmones al compás del movimiento de la caja torácica.
- La formación del **diafragma,** un músculo esencial en los movimientos ventilatorios, que aísla los pulmones de todas las vísceras de la cavidad abdominal.
- La aparición de **paladar óseo.**

Como ejemplo de respiración en los mamíferos, en el subapartado siguiente vamos a describir el caso del ser humano.

ACTIVIDADES

4. Los peces y los poríferos son animales acuáticos. ¿Puedes explicar qué diferencias existen entre ellos en la manera de obtener el oxígeno disuelto en el agua?

5. ¿Por qué se dice que el órgano respiratorio terrestre más evolucionado y adecuado son los pulmones?

6. ¿Qué son las branquias externas? ¿Y las internas? Razona cuáles son más evolucionadas.

7. Explica los siguientes conceptos:

a) ctenidio　　b) arco branquial

c) huesos neumatizados　　d) líquido pleural

2.3. *La respiración en el ser humano*

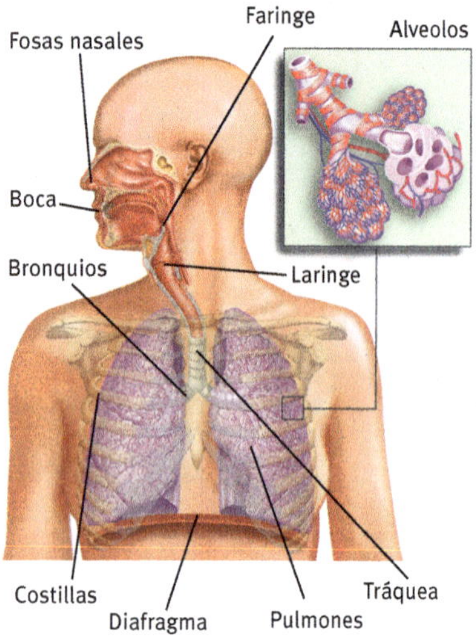

En el aparato respiratorio humano las *vías respiratorias* conducen el aire hacia los *pulmones,* los órganos en los que se produce el intercambio de gases.

Los pulmones son elásticos y están adheridos a las paredes de la caja torácica por la doble membrana pleural. La compresión y expansión de la caja torácica, mediante los *músculos intercostales* y el *diafragma,* provoca los movimientos de entrada y salida de aire.

La **inspiración,** o entrada de aire del exterior hacia los pulmones, es un mecanismo **activo:**

— El diafragma se contrae y desciende; los músculos intercostales también se contraen y elevan las costillas.

— Como consecuencia de la actividad muscular anterior, el volumen de la caja torácica aumenta y disminuye la presión del aire en su interior. La presión es ligeramente negativa respecto al aire exterior.

— El aire penetra como si fuese succionado desde el exterior hasta los alveolos pulmonares.

La **espiración,** o salida de aire alveolar hacia el exterior, es **pasiva:**

— El diafragma y los músculos intercostales se relajan. El primero asciende y las costillas se deprimen.

— Disminuye el volumen de la caja torácica, la presión del aire en su interior aumenta y se hace ligeramente superior a la del exterior.

— El aire se ve obligado a salir por las vías respiratorias.

La finísima membrana que forma los alveolos pulmonares está siempre húmeda; durante la espiración, las paredes de los alveolos podrían pegarse entre sí y provocar su colapso. Para evitarlo, los alveolos segregan una sustancia proteica que actúa como un *tensioactivo,* es decir, reduce la tensión de estas superficies y evita el colapso.

Podemos definir unos valores que nos dan idea de la actividad de nuestro aparato respiratorio.

- La **capacidad pulmonar total** es el volumen máximo que pueden contener los pulmones después de una inspiración forzada. Corresponde a unos 6 l en hombres y 4,2 l en mujeres.
- La **capacidad vital** es la cantidad máxima de aire que se mueve a través del aparato respiratorio en condiciones de máximo esfuerzo respiratorio. Corresponde a unos 4,5 l en hombres y 3,2 l en mujeres.
- El **volumen residual** es la diferencia entre la capacidad pulmonar total y la capacidad vital, y corresponde al aire que, después de una espiración, permanece en las vías respiratorias y los alveolos. Suele ser 1 l de aire.
- El **volumen de ventilación** es la cantidad de aire que penetra en los pulmones durante una inspiración normal, y corresponde a unos 500 ml. De esta cantidad, sólo 350 ml de aire llegan a los alveolos, ya que 150 ml quedan retenidos en las vías respiratorias.

Difusión de gases entre los alveolos y los capilares

La composición del aire que se encuentra en los alveolos pulmonares no es igual a la del aire atmosférico, debido a que:

- El aire que penetra por las vías respiratorias se humedece y, por tanto, se enriquece en **vapor de agua.**
- Constantemente se absorbe **oxígeno** y se desprende **dióxido de carbono.**

La energía cinética de las moléculas de los gases hace que éstas choquen contra las membranas de los compartimentos del cuerpo donde se encuentran encerradas. Los choques constituyen la **presión parcial del gas.**

Los gases atraviesan la membrana del alveolo y la pared de los capilares. El paso se hace por difusión simple y a favor del gradiente de presión de cada gas:

- La presión parcial de O_2 en el alveolo es de 100 mmHg, mientras que en la sangre, antes del intercambio, es de 40; el gradiente de presión favorece el paso del oxígeno desde el alveolo **hacia la sangre.**
- La presión parcial de CO_2 en la sangre antes del intercambio es de 46 mmHg, mientras que en el alveolo de 40; el gradiente de presión favorece el paso de CO_2 **desde la sangre al alveolo.** Aunque el gradiente es pequeño, la solubilidad del CO_2 en los lípidos hace que su intercambio, que tiene lugar a través de las membranas de cada compartimento, sea muy eficaz.

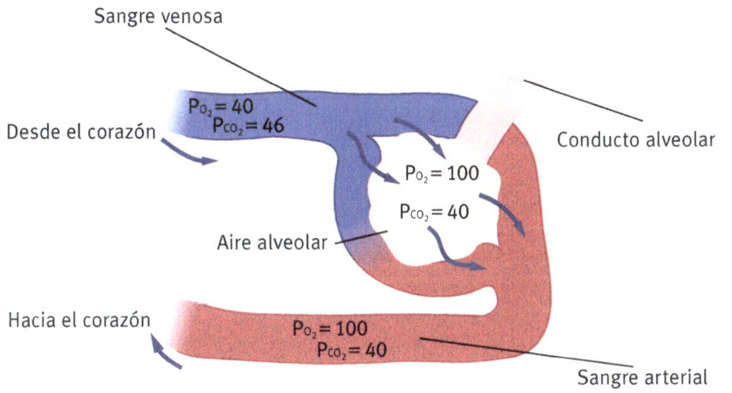

Difusión de gases de los capilares sanguíneos a las células del cuerpo

El proceso tiene lugar del mismo modo que el intercambio de gases en los alveolos pulmonares.

- La presión parcial de O_2 en la sangre es mayor que la del interior de las células; por ello, el oxígeno penetra **hacia el citoplasma** por difusión, a través de la membrana plasmática, y es utilizado para la respiración celular.
- La presión parcial de CO_2 en la sangre es menor que en el interior de la célula, a causa de la respiración mitocondrial. Por tanto, el CO_2 **sale de la célula,** por difusión, y penetra en los capilares sanguíneos.

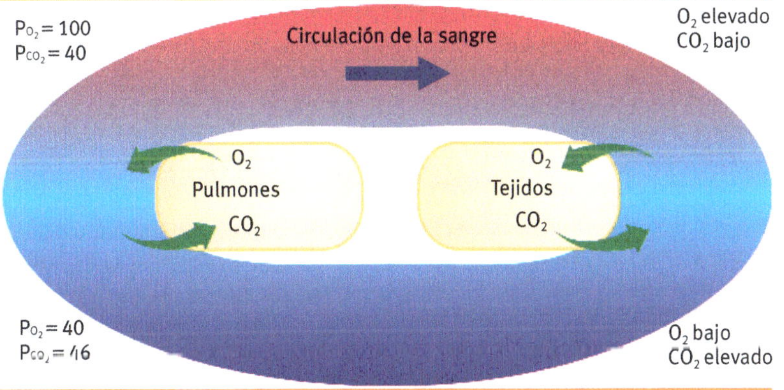

Transporte de gases por la sangre

El plasma es el responsable mayoritario de transportar moléculas de CO_2 en forma de ácido carbónico.

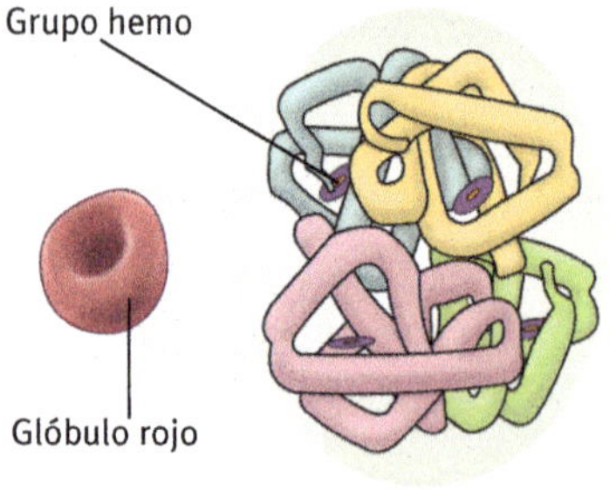

La hemoglobina es un pigmento respiratorio. Se encuentra en el interior de los glóbulos rojos y contiene 4 átomos de hierro. Cada átomo de hierro se une a una molécula de oxígeno. La hemoglobina oxidada recibe el nombre de oxihemoglobina y transporta el 97% del oxígeno.

+

Reacción de transformación del 70 % del CO_2

$CO_2 + H_2O \rightarrow H_2CO_3 \rightarrow H^+ + HCO_3^-$

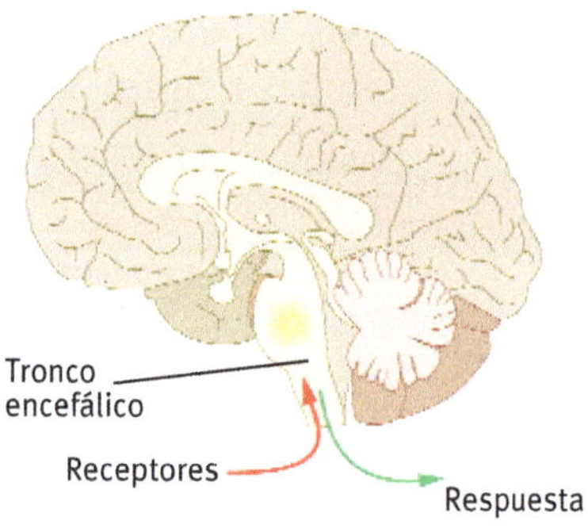

+

Las infecciones respiratorias como las neumonías, los traumatismos torácicos provocados por un accidente o las enfermedades pulmonares como el asma pueden producir una insuficiencia respiratoria.

Ésta tiene lugar cuando la presión parcial de O_2 en la sangre desciende por debajo de 60 mm de mercurio, independientemente de la cantidad de CO_2 transportado.

En estos casos, puede ser necesario tratar a la persona con oxigenoterapia; es decir, administrando oxígeno, para aliviar sus problemas ventilatorios.

El plasma es el líquido en el que se encuentran los glóbulos rojos y el resto de los elementos celulares de la sangre. La mayoria del CO_2 producido en la respiración celular difunde hacia los capilares y entra en los glóbulos rojos, donde se transforma en ácido carbónico. Tan sólo un 20% del CO_2 se combina con la hemoglobina (carboxihemoglobina).

El CO_2 restante que penetra en los glóbulos rojos pero que no se combina con la hemoglobina reacciona con el agua y forma ácido carbónico. Éste se disocia debido a su inestabilidad en iones bicarbonato y iones hidrógeno que vuelven a salir hacia el plasma. Cuando la sangre llega a los capilares del alveolo, los iones bicarbonato vuelven a entrar hacia los glóbulos rojos, donde se combinarán con iones de hidrógeno, generándose de nuevo ácido carbónico. Ahora, el ácido carbónico se disocia en CO_2 y H_2O. Por la diferencia de presiones el CO_2 difunde hacia el alveolo para ser expulsado al exterior y el agua es eliminada en forma de vapor.

Regulación de la respiración

El sistema nervioso central ajusta el ritmo de la respiración a las necesidades del cuerpo. De este modo, contribuye al mantenimiento de la regulación del medio interno. La respiración está regulada por dos factores: el centro respiratorio del bulbo raquideo y por un control químico.

- El **centro respiratorio** está formado por una serie de nervios que se dirigen hacia las paredes del diafragma y de los músculos intercostales. Durante la inspiración los pulmones se inflan, de modo que los nervios sensoriales de las paredes pulmonares se estimulan y envían señales al centro nervioso, que se inhibe. Su inhibición provoca la espiración, los pulmones se relajan, los nervios sensoriales dejan de estimularse y el centro respiratorio del bulbo raquídeo no recibe ninguna señal de inhibición, por lo que vuelve a empezar la inspiración.
- La **presión parcial de O_2 y CO_2** circulantes deben ser constantes. Cuando hacemos ejercicio, la concentración de CO_2 aumenta en sangre y respiramos más rápido, para poderlo eliminar a través de los pulmones. Los quimioreceptores son receptores químicos específicos de estos dos gases; si los receptores detectan un descenso de la presión parcial de oxígeno, los quimioreceptores envían el mensaje hacia el centro respiratorio para que se aumente la actividad respiratoria.

ACTIVIDADES

8. Las personas que viven a gran altitud tienen más glóbulos rojos que la media y la red de capilares sanguíneos es más extensa. Razona el porqué.
9. ¿Por qué si voluntariamente detenemos la respiración llega un momento en que ésta se realiza de forma involuntaria?
10. Corrige las frases que sean falsas:
    a) Todo el volumen de ventilación llega a los alveolos.
    b) La composición del aire que se encuentra en los alveolos es la misma que la del aire atmosférico.
    c) La hemoglobina puede transportar tanto O_2 como CO_2.
    d) El centro respiratorio modifica el ritmo de contracción de los músculos respiratorios según la información que recibe de los receptores.

3. El transporte de sustancias en los vegetales

La obtención de nutrientes útiles, la excreción de productos tóxicos y los intercambios de gases son procesos que tienen lugar en zonas concretas del organismo, tanto en los vegetales como en los animales. Por ello, se requieren mecanismos de transporte que comuniquen estas zonas con el resto de las células del organismo.

Vamos a describir los mecanismos de circulación en los vegetales, teniendo en cuenta que en las plantas los líquidos circulantes no transportan los gases respiratorios (CO_2 y O_2), sino las sustancias necesarias para que tenga lugar la fotosíntesis y las que se obtienen como resultado de ésta. Nos referiremos exclusivamente a las plantas cormófitas, ya que sólo en ellas se da la especialización en tejidos que hace necesarios y, a la vez, permite estos transportes.

Normalmente, la raíz es el órgano de absorción, situado en el extremo opuesto a las hojas, que tienen actividad fotosintética. Ambos órganos están comunicados por un sistema de circulación, de abajo arriba, a través de las *vías conductoras ascendentes*. Desde las hojas y otras partes verdes se reparten, mediante las *vías de conducción de sustancias orgánicas,* las sustancias orgánicas que se obtienen en la fotosíntesis.

3.1. *Vías de conducción ascendentes*

El conjunto de vías conductoras ascendentes recibe el nombre de **vasos leñosos** o **xilema.**

- El transporte tiene lugar entre los pelos absorbentes de la **raíz** y las zonas verdes fotosintéticas, en especial las **hojas.**
- El líquido circulante en sentido ascendente se suele denominar **savia bruta,** compuesta por agua y sustancias minerales disueltas.
- Todos los vasos leñosos son aproximadamente del mismo calibre. Están dispuestos en haces paralelos que discurren por la raíz, ascienden por el tallo y disminuyen de grosor a medida que algunos se desvían, para llegar a las hojas, donde forman sus nerviaciones. Pueden ser tráqueas o traqueidas:
  - Las **tráqueas** son vasos rígidos que forman un tubo continuo. Se han formado a partir de células dispuestas una a continuación de otra, que han lignificado sus paredes celulares. Ello provoca la degeneración de los orgánulos citoplasmáticos y la desaparición de los tabiques de separación, hasta constituir un tubo continuo de **células muertas.**
  - Las **traqueidas** son más finas; están formadas por **células muertas** con las paredes **lignificadas,** pero independientes entre ellas, ya que se conservan los tabiques de separación que presentan un gran número de orificios.

Las traqueidas son los vasos conductores más primitivos, propios de las plantas *pteridófitas,* como los helechos, y también de las plantas fanerógamas más antiguas evolutivamente, como las *gimnospermas.*

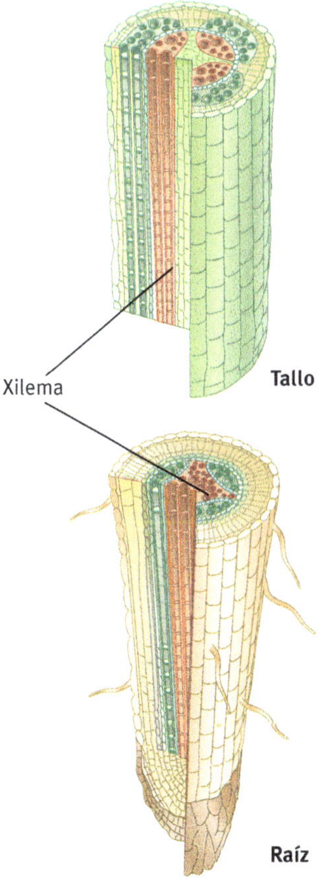

Las plantas fanerógamas más recientes desde el punto de vista evolutivo, es decir, las *angiospermas,* tienen el xilema formado mayoritariamente por tráqueas, que se consideran vasos conductores más perfectos y evolucionados que las traqueidas. En un árbol se produce xilema periódicamente. El conjunto de vasos leñosos más recientes recibe el nombre de **albura** y son vasos funcionalmente conductores. La parte de xilema más vieja, con menor o nula función conductora, recibe el nombre de **duramen,** y participa en el sostén del vegetal.

Mecanismos de transporte por el xilema

Algunas especies de árboles pueden alcanzar alturas superiores a los 100 m, como secuoyas y eucaliptos. En estos casos, el transporte ascendente a través del xilema supone un reto considerable. Es también notable la velocidad de ascenso de la savia bruta detectada en algunas especies: 1 m/h en coníferas, unos 4 m/h en hayas y entre 20 y 45 m/h en algunos robles. Aunque no está totalmente aclarado, en la actualidad se acepta que el ascenso se debe a un mecanismo de **tensión-cohesión,** ya apuntado en la unidad 9, y que se detalla a continuación:

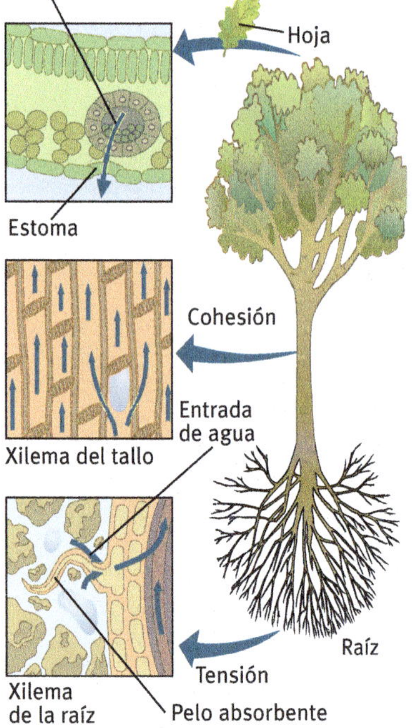

El transporte a través de la planta funciona mediante un mecanismo de tensión-cohesión, y gracias a la fuerza de la aspiración generada por la transpiración.

— Los **pelos absorbentes** captan agua por ósmosis y minerales que penetran en los tubos conductores del **xilema** por transporte activo.

— La entrada continua de agua por la raíz hacia los vasos conductores provoca una **presión hidrostática** que empuja el líquido en sentido ascendente en contra de la gravedad. Este hecho se ve favorecido por la longitud y el calibre de los vasos xilemáticos, ya que les hacen funcionar como verdaderos capilares.

— Las moléculas de agua se comportan como **dipolos:** establecen fuerzas de cohesión muy fuertes entre ellas que impiden la interrupción de las finas **columnas de líquido.**

— A través de los estomas, tiene lugar la pérdida de vapor de agua mediante la transpiración; ésta crea una **fuerza continua de aspiración** a lo largo de los vasos leñosos que causa la entrada de agua en la raíz. De hecho, todo el mecanismo depende de la energía solar que motiva la transpiración.

3.2. *Vías de conducción de moléculas orgánicas*

Como resultado de la fotosíntesis, las plantas obtienen monosacáridos, como la glucosa, que luego pueden transformar en moléculas más complejas, como disacáridos, polisacáridos y aminoácidos.

Las moléculas orgánicas son trasladadas y repartidas a otras partes del vegetal disueltas en agua. Este proceso recibe el nombre de **translocación** y tiene lugar mediante los **vasos liberianos** o **floema.**

- El transporte se produce desde las hojas y otras partes verdes hacia el tallo y las raíces.

  También puede haber translocación desde los tejidos de reserva (semillas, bulbos, tubérculos...) a otras partes de la planta que lo necesiten. El almidón contenido en estos tejidos se degrada a sacarosa, que es la molécula utilizada para la movilización de la glucosa por la planta. Al llegar al órgano receptor, la sacarosa se degrada a glucosa para ser utilizada.

El desplazamiento puede ser en sentido descendente o ascendente, dependiendo de la posición del órgano fotosintetizador o de reserva respecto al receptor.

- El líquido que se desplaza se ha denominado **savia elaborada.** Contiene agua y solutos, de los cuales un 90 % es sacarosa; el resto son aminoácidos, ácidos grasos, vitaminas, hormonas e iones como K^+.
- Los vasos liberianos son **conductos finos** que discurren por el tallo y la raíz. Están formados por **células vivas,** con tabiques de separación oblicuos y la pared celular sin lignina. Al madurar, pierden el núcleo, los orgánulos citoplasmáticos se disponen periféricamente y se forman **poros** en las paredes, que confieren a éstas el aspecto de una criba. Por ello, también se denominan *vasos* o *tubos cribosos.*

  Conectadas a las células de los tubos cribosos se encuentran **células acompañantes,** muy activas metabólicamente.
- La velocidad de transporte depende de la época del año y puede variar desde 30 cm/h a 200 cm/h en algunos casos.
  - En invierno, la mayoría de las cribas se encuentra taponada por **calosa** y la circulación está muy **reducida.**
  - En primavera, se reabsorbe la calosa y se **activa la circulación.**
  - En un gran número de especies vegetales, los vasos liberianos sólo son funcionales durante un año y son reemplazados por otros nuevos a partir del cámbium.

En plantas *pteridófitas* y *gimnospermas* no hay células acompañantes y los tabiques perforados de las células cribosas son muy inclinados. En las *angiospermas* se encuentran células acompañantes y tubos liberianos con placas cribosas muy perpendiculares.

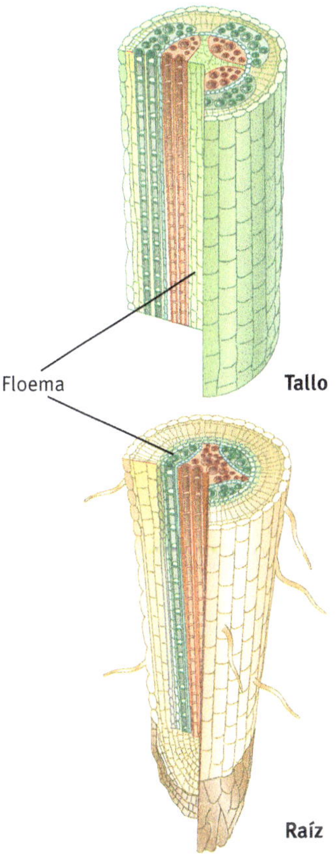

El transporte por el floema

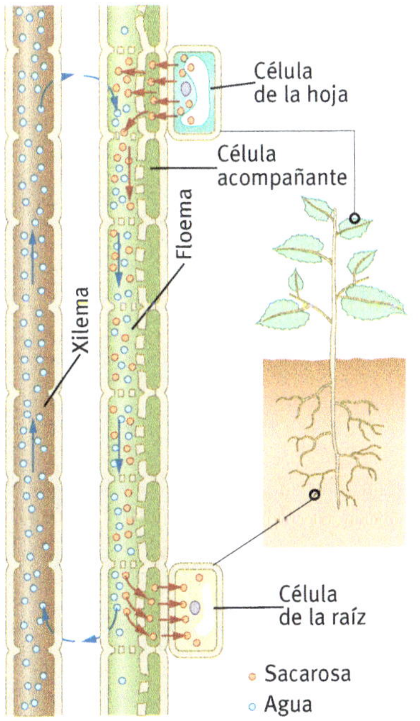

Mecanismos de transporte por el floema

La hipótesis del **flujo en masa** es la explicación más plausible de este proceso. Se basa en los puntos siguientes:

- En los órganos donde se produce la fotosíntesis, los glúcidos **salen** del citoplasma de las células en forma de **sacarosa.**
- Ésta **entra** en los vasos liberianos por **transporte activo,** contra gradiente de concentración. El proceso se ve facilitado por la actividad de las **células acompañantes.**
- La entrada de sacarosa en los tubos cribosos provoca la **entrada de agua,** por ósmosis, procedente **del xilema.**
- Cuando el órgano consumidor **capta sacarosa** y ésta abandona el vaso liberiano, disminuye la concentración de soluto en su interior.
- El **agua** también sale del tubo criboso, por el mismo proceso osmótico, y **vuelve** a los vasos del **xilema.**

La diferencia de presión del agua dentro del tubo floemático entre el órgano fotosintetizador y el receptor crea la corriente de flujo.

4. La circulación en los animales

Hemos visto en la unidad 8 que en los animales la presencia del medio interno surge como una necesidad para mantener un ambiente para las células con unas características constantes. Como sabes, los dos componentes del medio interno son:

- El **líquido intersticial** donde viven inmersas todas las células.
- El **líquido circulante** que se desplaza por el organismo como vehículo de transporte.

Al mismo tiempo que aumenta la complejidad de la organización celular y el tamaño en los distintos grupos de animales, se plantea la necesidad de vasos conductores para el transporte del líquido circulante, así como un órgano impulsor que lo bombee. Por ello presentan un **aparato circulatorio,** con las siguientes funciones:

1: La sangre recoge los nutrientes y los pone al alcance de las células.
2: La sangre facilita el O2 a las células y recoge el CO2 procedente de la respiración.
3: La sangre recoge los residuos metabólicos.
4: La sangre transporta hormonas, anticuerpos y células fagocitarias (leucocitos).
5: En animales homeotermos, la sangre colabora en la regulación de la temperatura.

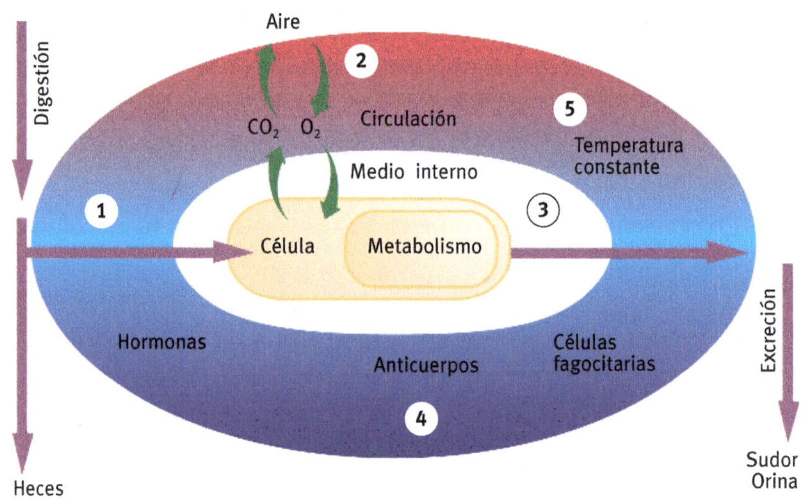

4.1. Invertebrados

Los invertebrados pueden presentar dos tipos de circulación: **abierta,** si, en un momento de su recorrido, el líquido circulante sale de los vasos que lo conducen y se extiende por los tejidos; o **cerrada,** cuando el líquido circulante siempre se desplaza contenido en los vasos.

En algunos grupos, no existe un líquido circulante claramente diferenciado; es el caso de las esponjas y los equinodermos.

Las **esponjas** tienen como líquido circulante **agua exterior** que ponen en movimiento por el interior del cuerpo.

— El animal **succiona** el agua que penetra por los poros inhalantes y la impulsa hacia numerosos conductos y cavidades.

— Las células que revisten los conductos y cavidades **captan** el O_2 y los nutrientes disueltos en el agua, y expulsan a ella el CO_2 y los metabolitos residuales.

— Finalmente, el agua **sale** por el *ósculo* u orificio principal.

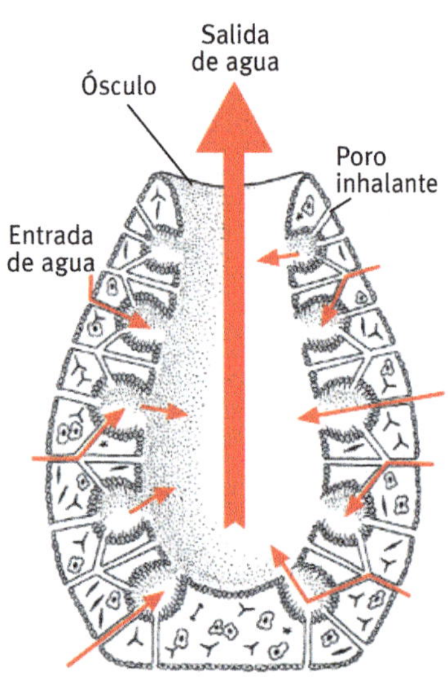

Circulación del agua en una esponja

En el sistema circulatorio de los **anélidos** el líquido circulante se denomina **hemolinfa** y toma el color del pigmento respiratorio disuelto según las especies: *hemoglobina* (roja), *hemeritrina* (rosa) o *clorocruorina* (verde).

La **circulación** es **cerrada,** ya que la hemolinfa discurre siempre por el interior de un sistema de vasos conductores.

— Un gran **vaso dorsal** reparte la sangre desde atrás hacia delante, gracias a la contracción de sus paredes musculosas. A partir de él, surgen ramificaciones hacia todos los órganos del cuerpo.

— De los órganos salen pequeños vasos que se reagrupan hasta reunirse **ventralmente** y formar **vasos** de mayor calibre.

Los **moluscos** pueden presentar un **sistema** circulatorio **abierto** (gasterópodos y bivalvos) o **cerrado** (cefalópodos). La hemolinfa suele presentar como pigmento respiratorio la hemocianina, que le da color azul.

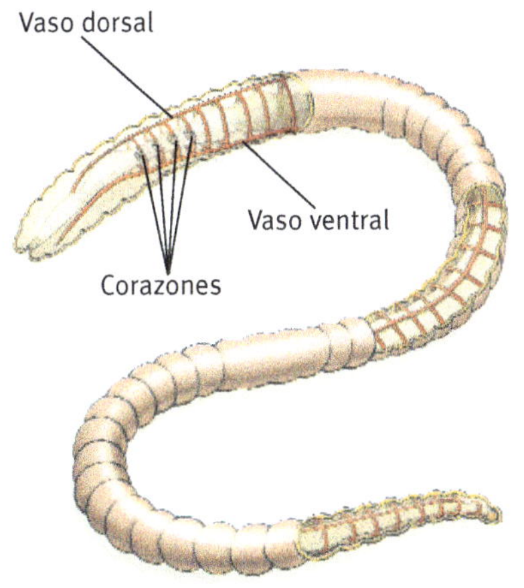

Anélido

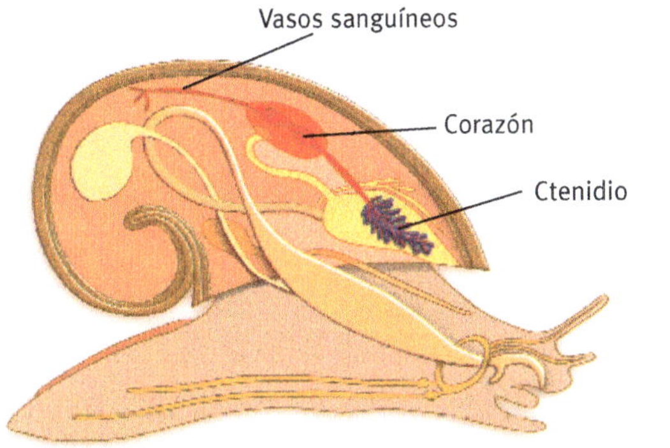

En **gasterópodos** y **bivalvos** la hemofinfa es bombeada por el corazón, protegido por una cubierta membranosa o pericardio. Del corazón salen vasos que van al hemoceloma, que baña todas las células del cuerpo. Mediante los vasos sanguíneos, la hemolinfa retorna al corazón.

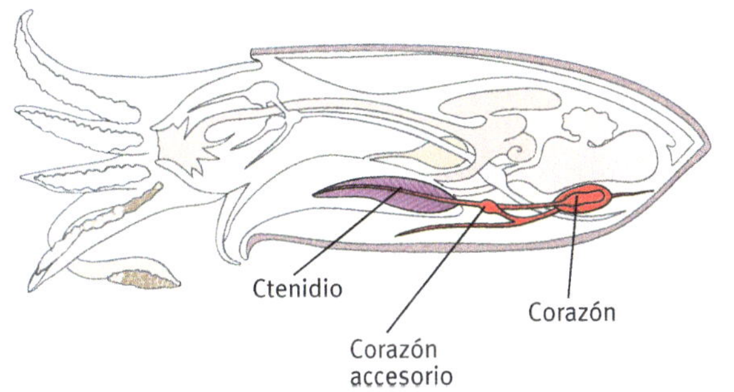

En **cefalópodos,** además del corazón principal, existe un sistema de corazones situados a nivel de las branquias, para aumentar el volumen de intercambio de gases.

En los **artrópodos,** el líquido circulante es la **hemolinfa** y los pigmentos suelen ser *hemoglobinas* (rojas) y *hemocianinas* (azules).

La circulación es **abierta.**

— Un gran **vaso dorsal,** formado por diversas cavidades separadas por válvulas, impulsa la hemolinfa de modo similar a un corazón.

— Cuando las cavidades del vaso se contraen, el líquido es repartido por las arterias hasta las **lagunas hemocélicas,** lugares donde se acumula la hemolinfa. Desde ellas, la hemolinfa baña directamente los órganos.

— Cuando las cavidades del vaso se relajan, la hemolinfa penetra en ellas.

Las especies que respiran por **filotráqueas** tienen un sistema de circulación **más reducido** que las que respiran por **dendrotráqueas** o **branquias.** Ello se debe a que las primeras son muy eficientes en la distribución de los gases respiratorios y el sistema circulatorio sólo se precisa para el transporte de las sustancias resultantes de la digestión de los alimentos.

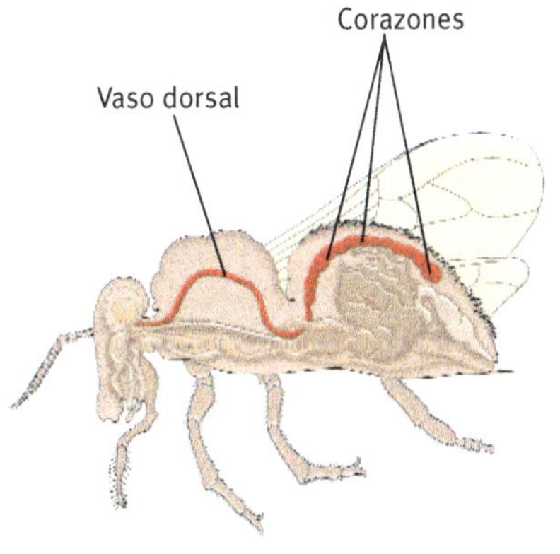

Insecto

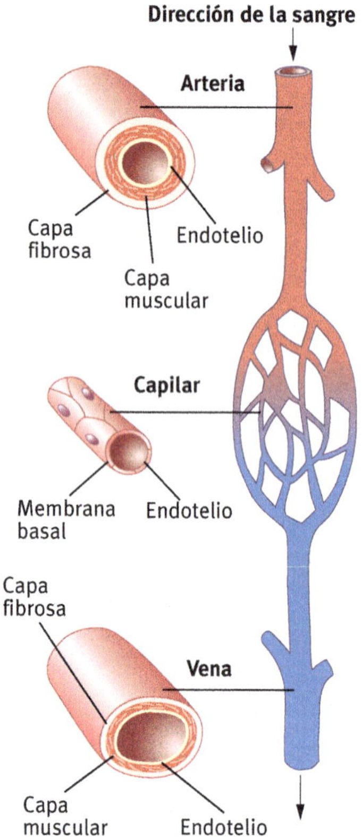

4.2. Vertebrados

En todos los vertebrados, el sistema de circulación es **cerrado.**

- La **sangre** es el líquido circulante; el pigmento respiratorio es la **hemoglobina.** Ésta es de color rojo y está contenida en los *hematíes o glóbulos rojos.* Otros elementos celulares sanguíneos son los *leucocitos*, con función defensiva, y fragmentos celulares denominados *plaquetas,* que intervienen en los procesos de coagulación.

  El **plasma** es el líquido en el que se encuentran inmersos todos los componentes de la sangre.

- El aparato circulatorio consta de un **corazón** y un conjunto extenso de vasos sanguíneos.

  — El **corazón** es el órgano impulsor de la sangre; presenta unas cavidades, aurículas y ventrículos, que se contraen y relajan rítmicamente.

  — Los vasos sanguíneos reciben el nombre de **venas** cuando ellos, o sus prolongaciones, devuelven la sangre al corazón, independientemente de la cantidad de oxígeno que contiene esta sangre.

  — Las **arterias** son los vasos que salen del corazón para distribuir la sangre por el cuerpo a través de sus ramificaciones, independientemente de la cantidad de oxígeno contenida en dicha sangre.

  Arterias y venas tienen una membrana interna de recubrimiento o **endotelio,** una capa media **muscular** y una capa externa **fibrosa.**

  — Conectando las arterias y las venas existe una red de vasos de diámetro pequeño. Son los **capilares,** formados por una capa interna o **endotelio** y una capa externa de tejido conjuntivo o **membrana basal.**

  Las **arterias,** que reciben de lleno el empuje y la presión de la sangre bombeada por los ventrículos, tienen una musculatura potente y sus paredes son **muy elásticas.**

  Las **venas** tienen las paredes más **flácidas** y las de calibre medio poseen válvulas, para impedir el retroceso de la sangre.

- Como sistema circulatorio auxiliar se encuentra el **sistema linfático.** Por él, mediante un conjunto de ganglios y vasos distintos de los sanguíneos, circula la **linfa,** un líquido que procede del drenaje del líquido intersticial y contiene las células que participan en la respuesta inmunitaria.

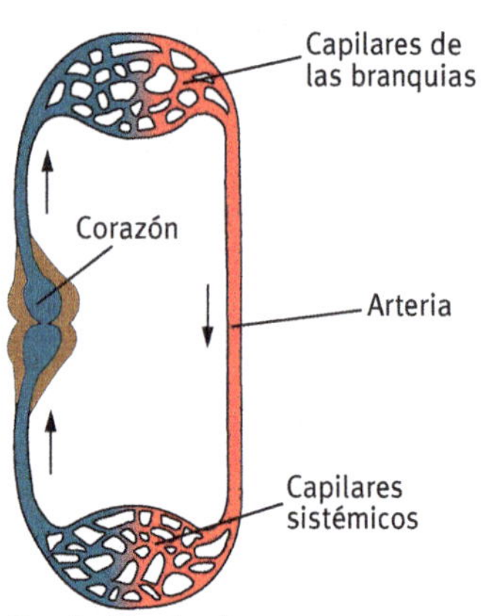

Circulación simple

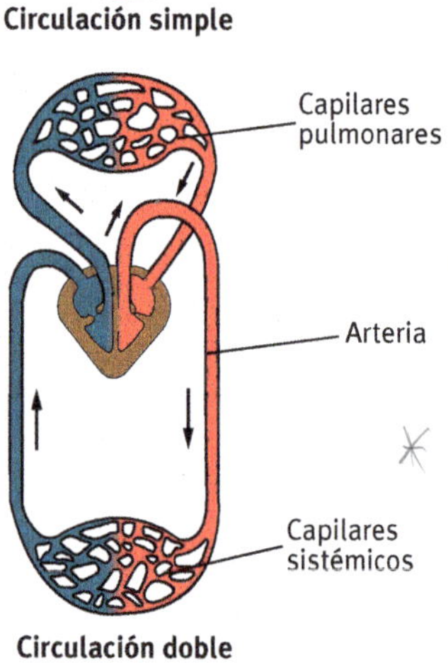

Circulación doble

En los vertebrados, la circulación puede ser:

- **Simple:** existe un solo circuito por el que se mueve la sangre para recorrer todo el cuerpo, la sangre pasa una sola vez por el corazón.
- **Doble:** hay dos circuitos diferenciados y la sangre debe pasar dos veces por el corazón para completar su recorrido por el cuerpo.
- **Incompleta:** se produce una mezcla de sangre rica en oxígeno y sangre pobre en este gas.
- **Completa:** no hay mezcla de los dos tipos de sangre.

Los **peces** tienen un corazón, con una aurícula y un ventrículo, situado en posición ventral y muy cerca de las branquias.

- Del corazón sale la arteria **aorta ventral,** que se ramifica en arterias branquiales.
- Estas arterias penetran en las branquias, donde la hemoglobina se carga de O_2. La sangre sale por la **aorta dorsal,** que reparte la sangre oxigenada y los nutrientes por el cuerpo a través de los capilares.
- La sangre empobrecida en O_2 y nutrientes es recogida por las *venas* y transportada a la aurícula del corazón.

En los peces la sangre recorre todo el cuerpo, pasando una sola vez por las cavidades cardíacas y nunca se mezcla sangre rica en O_2 con sangre pobre en dicho gas.

Por tanto, la circulación de los peces es **simple** y **completa.**

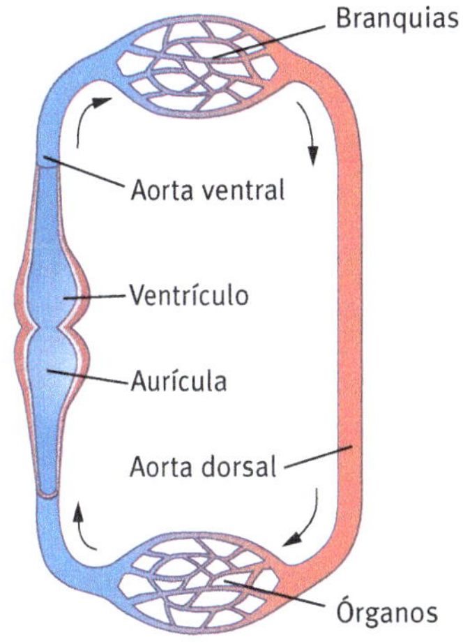

Sistema circulatorio de un pez

Todos los **vertebrados tetrápodos** tienen un sistema de circulación **doble,** ya que se distingue un circuito menor y un circuito mayor.

- El circuito menor conecta el aparato circulatorio con el aparato respiratorio haciendo posible el intercambio de gases.
  - — La sangre sale del ventrículo y, mediante las arterias pulmonares, llega a los pulmones para cargarse de O_2.
  - — Luego regresa a la aurícula izquierda del corazón mediante las venas pulmonares.
- El circuito mayor distribuye la sangre por todo el cuerpo y, una vez realizado todo el recorrido, la devuelve al corazón.
  - — La sangre enriquecida en oxígeno parte del ventrículo, a través de la arteria aorta, y recorre todos los órganos del cuerpo mediante la red de capilares arteriales.
  - — Los capilares venosos se encuentran a la salida de los órganos y se agrupan formando vasos de mayor calibre: las venas. Éstas desembocan en las grandes venas cavas, que entran en la aurícula derecha.

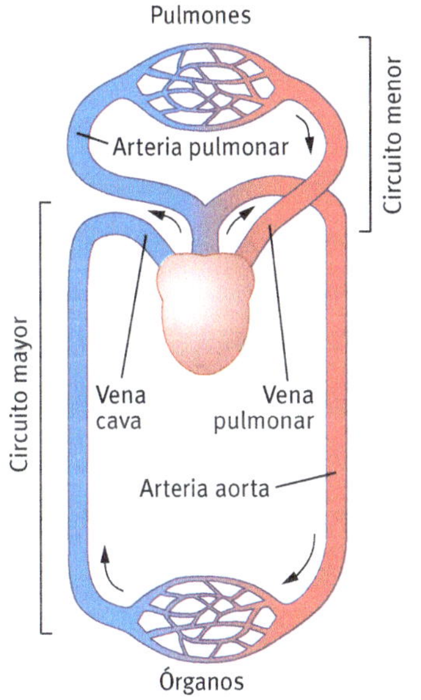

Sistema circulatorio de un vertebrado tetrápodo

Al analizar los distintos grupos de vertebrados tetrápodos, observamos las siguientes características:

En **anfibios** adultos, el corazón presenta dos aurículas, que no se contraen al mismo tiempo, y un solo ventrículo.

— La sangre sale del ventrículo y capta O_2, no sólo de los pulmones, sino también a través de la red de capilares que se extiende por debajo de la piel.

— Luego regresa a la aurícula izquierda mediante las venas pulmonares.

— De allí pasa al ventrículo, de donde parte para recorrer el cuerpo y, finalmente, regresar a la aurícula derecha.

— Aunque las dos aurículas no se contraen al mismo tiempo, en el ventrículo se produce una mezcla de sangre rica en oxígeno y de sangre pobre en este gas.

La circulación en los anfibios es **doble** e **incompleta.**

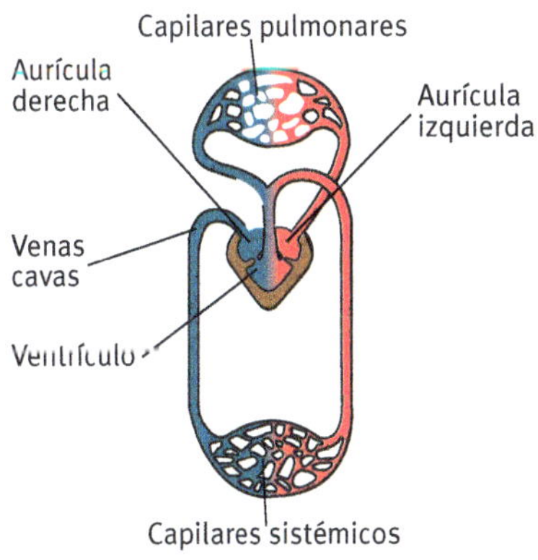

Sistema circulatorio de un anfibio

Ciclo cardíaco

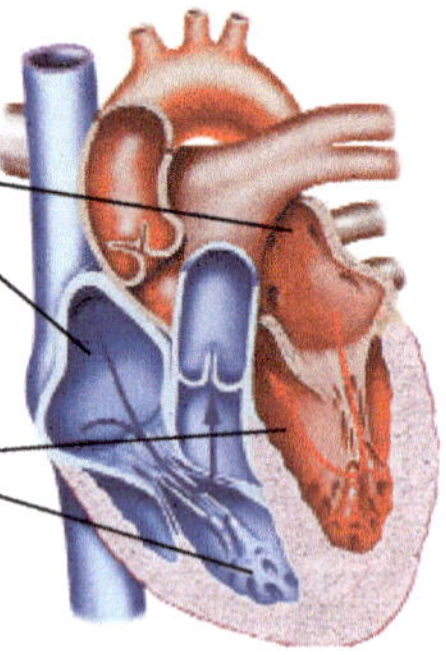

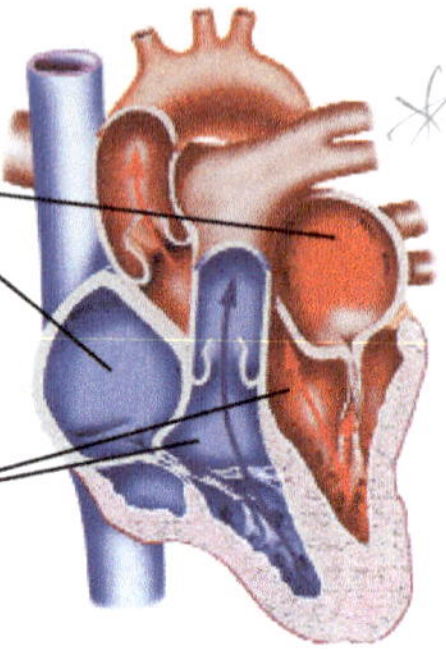

+

William Harvey (1578-1567), médico inglés, fue el primero en describir el movimiento de la sangre. A través del método científico pudo comprobar que la sangre daba alimento, calor y fuerza a todas las células del cuerpo. Describió la diferencia entre venas y arterias. Confirmó que los ventrículos obligan a la sangre a dirigirse hacia los pulmones y que desde ellos vuelve al corazón para dirigirse hacia todo el cuerpo. También estableció que el recorrido de la sangre no siempre se realiza a la misma velocidad, sino que depende de diversos factores como la edad, el reposo, estado de ánimo... Estos descubrimientos siguen siendo válidos 400 años más tarde.

En los **reptiles,** el corazón presenta dos aurículas y un ventrículo. Esta cavidad ventricular se encuentra parcialmente tabicada, por lo cual la mezcla de sangre rica en O_2 procedente de la aurícula izquierda y sangre pobre en O_2 que llega por la aurícula derecha es mínima.

Debido a ello, la circulación es **doble** y casi completa en reptiles y totalmente **completa** en los cocodrilos.

Las **aves** y los **mamíferos** tienen el corazón dividido en dos aurículas y dos ventrículos totalmente independientes. Entre las aurículas y los ventrículos existen válvulas que impiden que la sangre retroceda por el interior del corazón. Su circulación es **doble** y **completa.**

4.3. *El corazón humano y las enfermedades cardiovasculares*

El corazón está formado por tres capas de tejido.

- El **pericardio** constituye la parte más externa y tiene función protectora de las restantes.
- El **miocardio** es la capa muscular que forma el corazón; sus fibras son de tipo estriado. Es más grueso en los ventrículos que en las aurículas: en el primer caso, alcanza hasta 2 cm de grosor, mientras en las aurículas puede tener sólo unos milímetros. Entre las fibras musculares que forman el miocardio, cabe destacar las del sistema de *conducción eléctrica.* A partir de ellas, se generan impulsos eléctricos que se dirigen a las aurículas y a los ventrículos, dando lugar al **ciclo cardíaco,** que puedes observar en la ilustración de la izquierda.
- El **endocardio** es una capa delgada de tejido epitelial que recubre el interior de las cavidades del corazón y las válvulas que las separan.

El funcionamiento del corazón depende de un suministro de nutrientes y oxígeno para la contracción del miocardio y de su adecuada irrigación.

Las **arterias coronarias** parten de la aorta y se ramifican recubriendo la superficie del corazón, la sangre que contienen proporciona oxígeno y nutrientes al miocardio. Después del intercambio con este tejido, las **venas cardíacas** devuelven la sangre al interior del corazón.

Una reducción del flujo de sangre por las arterias coronarias provoca un sufrimiento cardíaco que desencadena una *angina de pecho* o un *infarto.* Los síntomas de estos trastornos se manifiestan en forma de dolor en el pecho y el brazo izquierdo.

Estas enfermedades se incluyen dentro de las cardiovasculares, que son la principal causa de muerte en los países desarrollados. El flujo coronario se reduce, sobre todo, como consecuencia de la **aterosclerosis,** es decir, la formación de placas de ateroma, compuestas básicamente de lípidos, que obstruyen el interior de la arteria. Algunos de los factores que favorecen la formación de estas placas son el tabaco, la hipertensión arterial y el exceso de lípidos en sangre, como el colesterol. Para prevenirlas se recomienda seguir hábitos saludables: no fumar, controlar la tensión arterial y reducir en la dieta el nivel de colesterol.

11 Vamos a estudiar los gases que intervienen en la respiración de los organismos. Para estudiar qué gases contiene el aire que respiramos haremos una experiencia para la que necesitaremos el siguiente material:

— dos tubos de ensayo,

— azul de bromotimol,

— una pajita,

— una mancha de bicicletas.

El proceso a seguir es el siguiente:

— Cogemos dos tubos limpios (A y B), los llenamos de agua (10 ml) y añadimos a cada uno 3 gotas de azul de bromotimol. El azul de bromotimol es un indicador de pH; con un pH basico, este indicador es de color azul mientras que si el pH es acido, menor a 6,5, presentará color amarillo.

— Soplamos con la ayuda de una pajita dentro del tubo A. Insuflamos aire con la ayuda de una mancha para bicicletas en el tubo B.

Observamos los resultados y a continuación completamos la tabla de datos siguiente:

	COLOR DEL TUBO ANTES DE INYECTAR AIRE	COLOR DEL TUBO TRAS INYECTAR AIRE
Tubo A		
Tubo B		

A partir de los datos observados responde las siguientes cuestiones:

a) Compara los resultados de los tubos antes y después de inyectar aire y emite una hipótesis sobre lo que crees que ha sucedido.

La tabla siguiente muestra el % de gases contenidos en la inspiración y en la espiración:

	% aire inspirado	% aire espirado
O_2	20	16
N_2	79	79
CO_2	0.03	4.03
Otros gases	0.97	0.97
Vapor de agua	variable	saturado

b) ¿Por qué el contenido gaseoso de la inspiración y espiración no es el mismo?

c) ¿Qué gases no intervienen en el intercambio gaseoso?

d) ¿De dónde proviene el vapor de agua en el aire espirado?

e) Compara los datos que te damos en la tabla con la hipótesis que has emitido en la experiencia de los tubos. ¿Piensas que la hipótesis que has dado es válida? ¿Por qué?

En el caso de que tu hipótesis no sea válida, emite otra que creas que esté de acuerdo con tus razonamientos a partir de los experimentos realizados.

f) Explica detalladamente de dónde procede la sustancia que hace cambiar el pH del tubo A, cómo es transportado por la sangre hasta los alveolos pulmonares y de qué manera se libera.

12 Explica qué quiere decir que los gases atraviesan la membrana plasmática por difusión simple.

13 Explica cómo es la respiración externa de una planta.

— ¿Podemos considerar la fotosíntesis como un tipo de respiración externa? ¿Y la transpiración? ¿Por qué?

14 Observa este dibujo. Di de qué se trata, que función desempeña e indica el nombre de las estructuras señaladas.

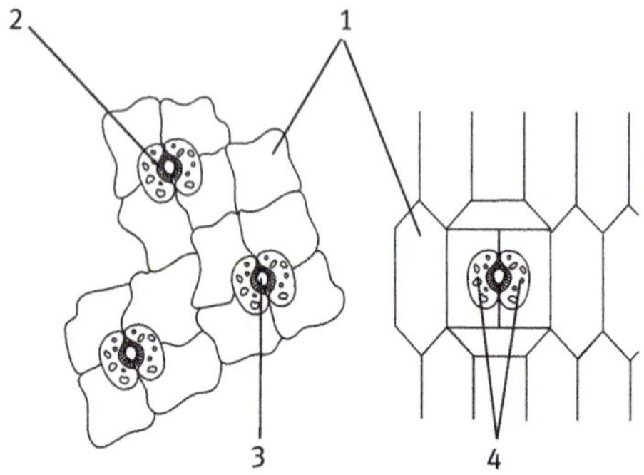

15 Las graficas muestran el rendimiento fotosintético de una planta bajo diferentes condiciones ambientales. Contesta las siguientes preguntas:

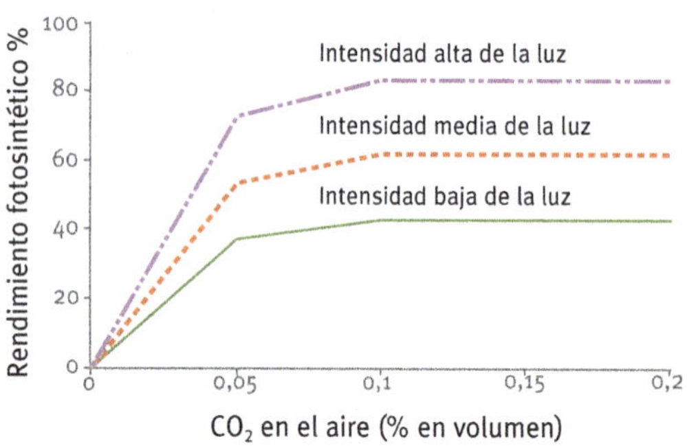

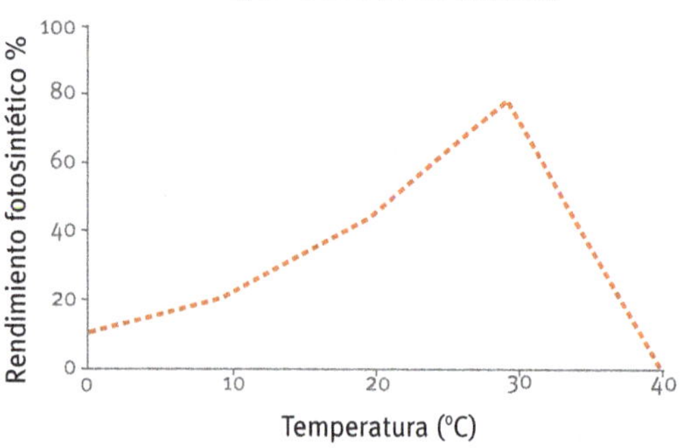

— ¿Qué efecto tiene la concentración de CO_2 sobre la fotosíntesis?

— ¿Qué efecto tiene la temperatura?

— ¿Cuál es el efecto de la intensidad luminosa?

— ¿Qué condiciones optimas de temperatura, concentración de CO_2 e intensidad de luz necesita una planta para tener el máximo rendimiento fotosintético?

16 ¿Qué diferencia existe entre una ventilación pulmonar y respiración?

17 ¿Por qué las superficies respiratorias en los animales terrestres son internas?

18 ¿Qué dificultades presentan los organismos acuáticos frente los terrestres respecto la respiración? ¿Y los terrestres frente los acuáticos?

19 Observa el dibujo:

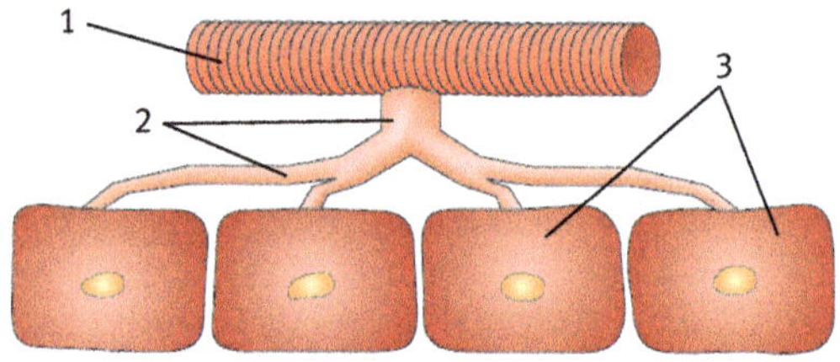

— ¿Qué tipo de respiración representa?

— ¿Qué animales presentan estas estructuras?

— Indica los nombres de las 3 estructuras señaladas.

— Explica porqué este tipo de respiración no necesita pigmentos respiratorios.

20 ¿Qué sistemas de respiración poseen los siguientes seres vivos?

lombriz, rana, lagartija, manta, mejillón, pulpo

21 Razona porqué algunas especies de tiburones para poder respirar no pueden dejar de nadar.

22 ¿Qué grupo animal presenta unos pulmones como a? ¿Y como b? ¿Y como c? Estos dibujos estan ordenados según la escala evolutiva. ¿Qué podemos decir respecto la evolución pulmonar?

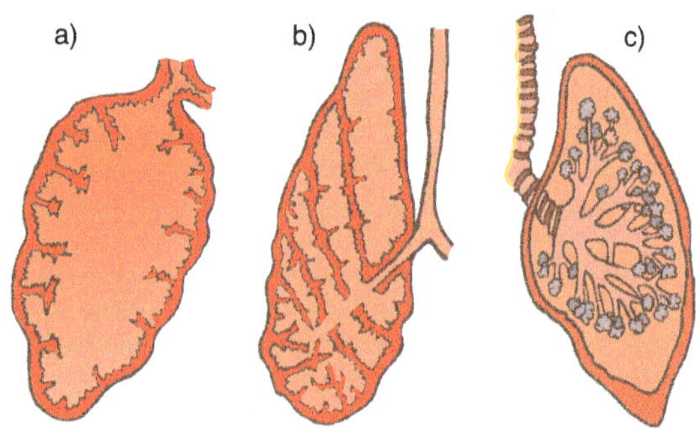

23 En el siguiente dibujo se representa el intercambio gaseoso a nivel alveolar. Intenta asignar un nombre a cada estructura señalada con un número del 1 al 9. Recuerda que en el pulmón entran las arterias pulmonares y salen las venas pulmonares.

— Haz una copia del dibujo y pinta de rojo la sangre oxigenada y de azul la sangre rica en CO_2.

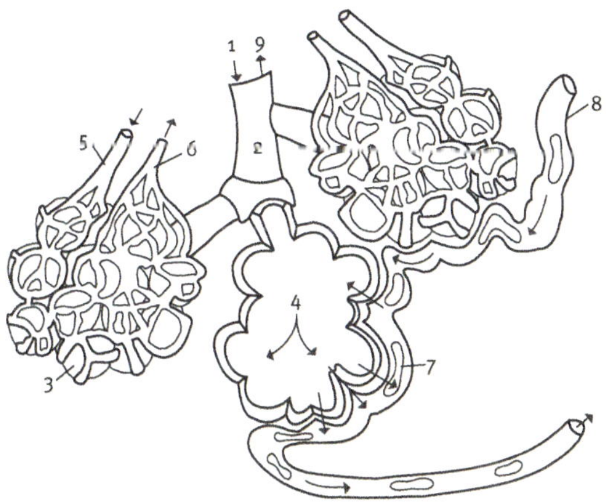

24 En este gráfico se representa el ciclo respiratorio humano.

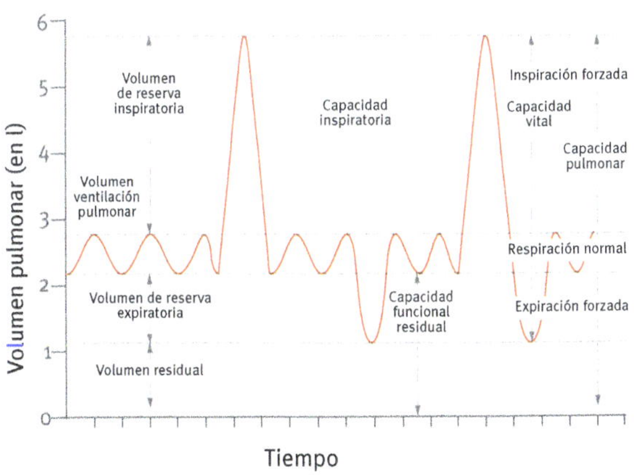

— ¿Cuál es el volumen máximo y mínimo que pueden contener los pulmones? ¿Qué nombre recibe cada uno de estos valores?

— Define según el grafico que es la capacidad pulmonar.

— ¿Qué volumen de aire entre y sale de los pulmones en la respiracion normal?

— Cuál es el volumen de reserva expiratoria? Y el volumen de reserva inspiratoria?

— Después de hacer 10 minutos de flexiones, deduce cómo cambiará el gráfico y dibújalo.

25 Explica gracias a qué mecanismos la sabia bruta asciende a través del xilema.

26 Explica cuando un tubérculo como la patata puede considerarse:

— un órgano productor

— un órgano consumidor

TIC

27 En esta dirección electrónica encontrarás la explicación interactiva del movimiento del agua a traves de las plantas. En ella verás como tus dudas sobre la transpiración se aclaran. Después de visitarla contesta las preguntas que te presentamos.

http://croptechnology.unl.edu/animationOut.cgi?anim_name=transpiration_esp.swf

— ¿Qué es la capa límite? Explica cómo su grosor influye en la transpiración y qué factores la hacen aumentar de grosor.

— Si la humedad relativa del ambiente es alta explica porqué disminuye la transpiración.

— Explica porqué una mayor temperatura aumenta la tasa de transpiración.

— Explica en qué condiciones se cierran los estomas.

— Explica las 3 rutas posibles del agua para llegar desde los pelos radicales al xilema.

28 Explica cómo se relaciona el sistema circulatorio con la respiración en los animales pluricelulares.

29 Atribuye cada dibujo del sistema circulatorio al grupo animal al que pertenece. Razona tu respuesta.

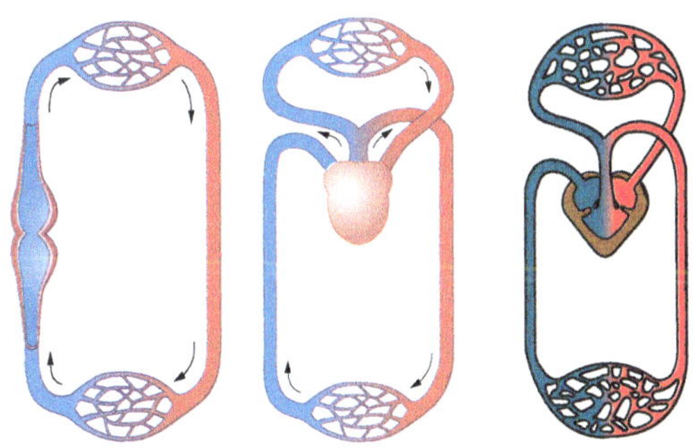

30 Explica que diferencias se dan entre:

— circulación abierta y cerrada.

— circulación doble y simple.

— circulación completa e incompleta.

— arterias y venas.

11

La relación: la respuesta motora

Cuando una partícula extraña entra en nuestras fosas nasales estornudamos. Esta acción en apariencia sencilla implica un mecanismo complejo en el que el sistema nervioso primero recibe, la información de su entorno (*contacto con la partícula extraña*), interpreta la información (*no es beneficiosa y por tanto se deben tomar medidas para expulsarla*) y elabora una respuesta adecuada (*contrae diversos músculos que llevan a la expulsión brusca de aire o estornudo*).

Todos los seres vivos presentan sistemas más o menos complejos que les permiten interpretar su entorno y elaborar la respuesta adecuada que asegure su seguridad y supervivencia. Cuando la respuesta implica de alguna manera al aparato locomotor (esqueleto y músculos) hablamos de respuesta motora.

Un ejemplo de respuesta motora es la huída: todos los animales, al detectar un peligro elaboran una respuesta que comúnmente implica el alejamiento de la zona de peligro; muchas veces, de ello depende su vida.

ORGANIZACIÓN DE LOS CONTENIDOS

1. La función de relación

2. El sistema nervioso

3. El impulso nervioso

3.1. La sinapsis

4. Captación de los estímulos: los receptores

4.1. Los órganos de los sentidos

5. Integración de la información y elaboración de respuestas

5.1. El sistema nervioso de los vertebrados

5.2. Emisión de la respuesta motora

5.3. La sinapsis neuromuscular

6. Los efectores del movimiento: el aparato locomotor

6.1. El aparato locomotor en los invertebrados

6.2. El aparato locomotor en los vertebrados

OBJETIVOS

- Comprender la fisiología de la neurona y la propagación del impulso nervioso.
- Conocer la función de los receptores sensoriales de los distintos grupos de animales según el tipo de estímulo que captan.
- Comprender el funcionamiento de los centros nerviosos como áreas de integración de los estímulos, de análisis de dichos estímulos y de elaboración de las respuestas.
- Conocer la distinta morfología del aparato locomotor según los grupos taxonómicos y analizar su estructura.

1. La función de relación

Los seres vivos están en contacto con el medio que les rodea y necesitan reaccionar ante las variaciones de éste para poder sobrevivir. Al mismo tiempo, perciben las condiciones de su medio interno, lo que contribuye a mantener el equilibrio necesario para el desarrollo de sus funciones vitales.

La función de **relación** es un conjunto de procesos mediante los cuales los seres vivos captan los cambios del medio externo e interno, y responden de forma adecuada.

En la función de relación se distinguen las siguientes fases:

- **Presencia** de un **estímulo,** que es un cambio que se puede producir tanto en el interior del organismo como en el exterior.
- **Captación** del estímulo por medio de unos **receptores.**
- **Producción** de una **respuesta.**

Almendros en flor

Las fases anteriores son comunes a todos los seres vivos; sin embargo, existen distintos mecanismos que varían según el grupo taxonómico o incluso según la especie. En la siguiente tabla podemos ver algunos ejemplos:

ORGANISMO	ESTÍMULO	CAPTACIÓN DEL ESTÍMULO	RESPUESTA
Almendro	Las horas de luz solar se alargan (empieza la primavera)	Por medio de sustancias químicas	Comienza la floración
Ostra	Un cuerpo extraño, por ejemplo un grano de arena, se introduce en el cuerpo del molusco	El cuerpo extraño provoca una irritación en el molusco que es captada por las células nerviosas	Las células epiteliales del manto empiezan a secretar nácar que irá rodeando y aislando el cuerpo extraño
Humano	Calor excesivo cuando acercamos la mano al fuego	Mediante receptores del calor de la mano	Nuestro centro nervioso ordena que apartemos la mano del fuego

Ostra perlífera

En los animales, el desarrollo de la función de relación alcanza un nivel de complejidad muy elevado. Esto es posible gracias al tejido nervioso, que es el responsable tanto de la captación de los estímulos, como de su transmisión y análisis, y de la elaboración de las respuestas necesarias.

Estímulo → **Análisis** → **Respuesta**

En esta unidad vamos a describir la actividad del sistema nervioso en distintos grupos de seres vivos.

2. El sistema nervioso

El sistema nervioso está formado por dos tipos de células: las *neuronas* y las *células gliales.*

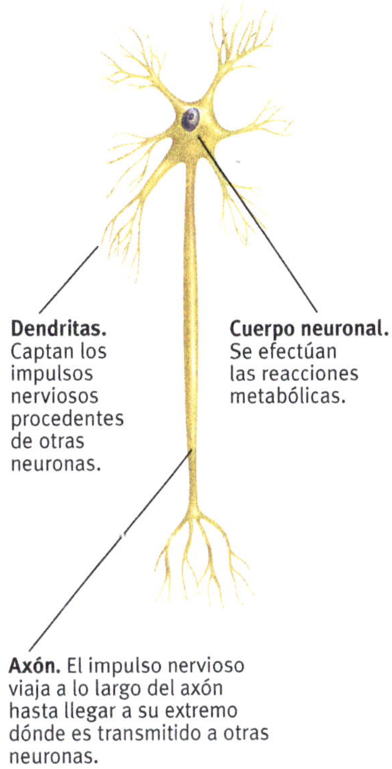

- Las **neuronas** están especializadas en transmitir los estímulos en forma de pequeñas corrientes eléctricas, que reciben el nombre de **impulsos nerviosos.**

  Las neuronas se clasifican según su función:

  - **Neuronas sensitivas.** Captan estímulos y los transmiten a los centros donde son analizados. La asociación de varias neuronas sensitivas forma un **nervio sensitivo.**
  - **Neuronas de asociación.** Constituyen los centros de recepción y análisis de los estímulos o **centros nerviosos.**
  - **Neuronas motoras.** Reciben los impulsos procedentes de los centros nerviosos y los transmiten a los órganos donde debe producirse una respuesta. La asociación de varias neuronas motoras forma un **nervio motor.**

Neuronas de asociación (centros nerviosos) ← Neuronas sensitivas ← Estímulo

Neuronas de asociación (centros nerviosos) → Neuronas motoras → Respuesta

- Las **células gliales** rodean a las neuronas. Como hemos visto en la unidad 8, pueden ser de tres tipos: *astrocitos,* que actúan como soporte y aportan sustancias nutritivas a las neuronas; *microglía,* con función de defensa; y *células de Schwann,* que favorecen la transmisión del impulso nervioso.

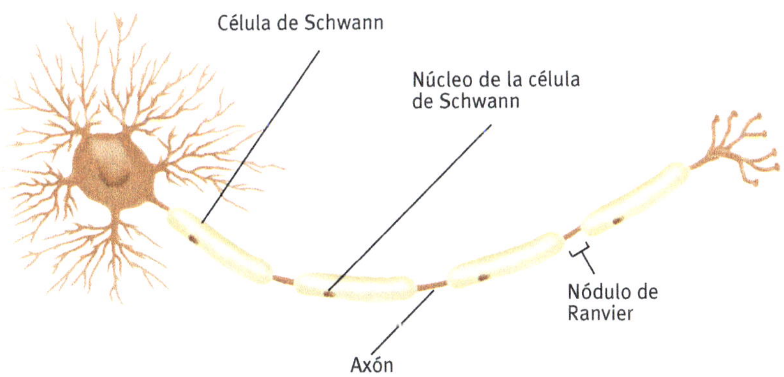

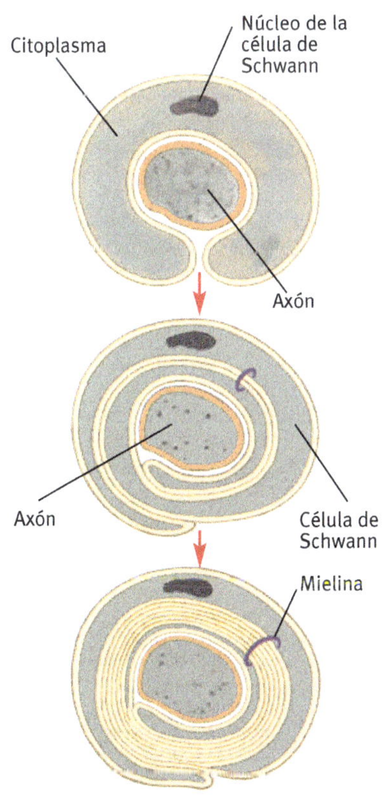

■ En su membrana plasmática, las **células de Schwann** contienen una sustancia aislante denominada mielina. Estas células crecen en torno a los axones de las neuronas: a lo largo de un axón se distinguen varias células de Schwann que lo envuelven, formando **vainas de mielina.** Entre una célula y otra quedan pequeñas separaciones denominadas **nódulos de Ranvier.** Por ello, en un axón alternan zonas recubiertas con mielina y zonas sin este recubrimiento.

3. El impulso nervioso

La función del sistema nervioso es transmitir impulsos nerviosos. Esto es posible gracias a que las neuronas tienen un potencial eléctrico generado por las diferencias de concentración de determinados iones entre el interior y el exterior de la membrana plasmática de la neurona.

La membrana neuronal es impermeable a las moléculas polares y a los iones, por lo que el transporte de éstas se realiza por difusión facilitada a través de **canales iónicos** (un tipo de transporte pasivo regulado por voltaje), o a través de la **bomba Na^+/ K^+**, un tipo de transporte activo.

Las diferencias de concentración de iones entre el interior y el exterior de la neurona generan una **diferencia de potencial** entre los dos lados. La diferencia de potencial, y los mecanismos que la generan, variarán según el estado de excitación de la neurona, como vamos a ver a continuación:

POTENCIAL EN REPOSO

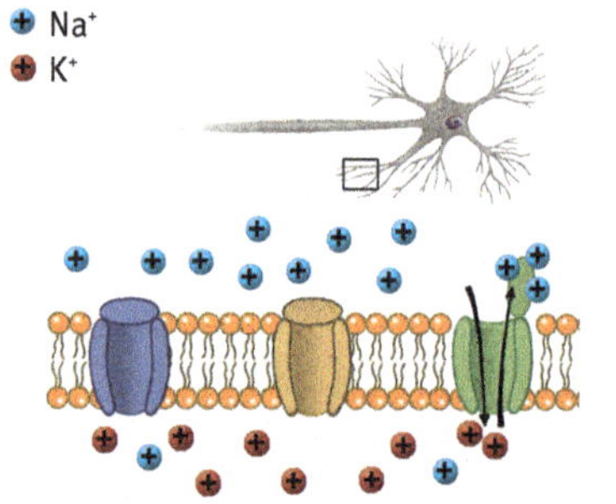

Cuando la neurona está **en reposo,** es decir, cuando no transmite un impulso nervioso, su citoplasma tiene carga negativa debido a una mayor presencia de aniones que de cationes, mientras que en el exterior presenta una carga eléctrica positiva. Esto genera una **diferencia de potencial** entre citoplasma y medio exterior de –70 mV. Este valor es el **potencial de reposo.** En estas condiciones las concentraciones de Na^+ y K^+ del citoplasma y del medio externo se mantienen muy diferentes:

— La bomba Na^+/K^+ funciona para mantener la diferencia de potencial de –70mV, bombeando 3 iones de Na^+ hacia el exterior del axón por cada 2 iones de K^+ que entran hacia el citoplasma.

— Los canales iónicos de Na^+ y K^+ están cerrados.

DESPOLARIZACIÓN Y POTENCIAL DE ACCIÓN

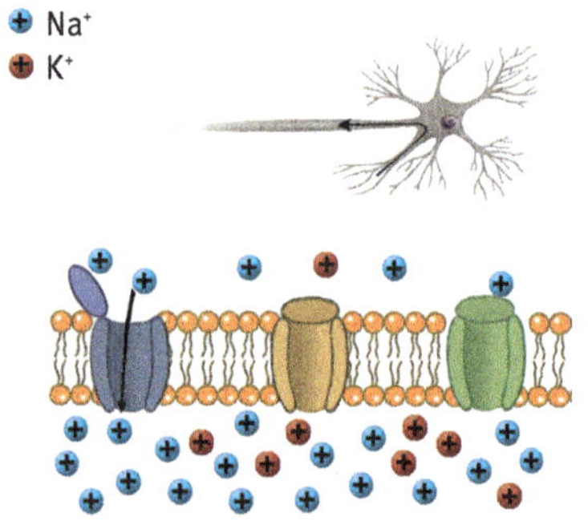

Con la llegada del estímulo, la membrana neuronal se **despolariza** (se intentan equilibrar las diferencias de carga entre exterior y citoplasma del axón).

— Deja de funcionar la bomba Na^+/K^+.

— Se modifica la conformación de las proteínas de canal de Na^+ de manera que se produce el paso de este ión hacia el interior de la neurona.

Con la entrada del Na^+, el potencial eléctrico de la célula pasa a tener un valor positivo de 40 y 50 mV. Este valor corresponde al **potencial de acción** que se propaga a lo largo de la membrana, desde las dendritas hasta el axón.

REPOLARIZACIÓN

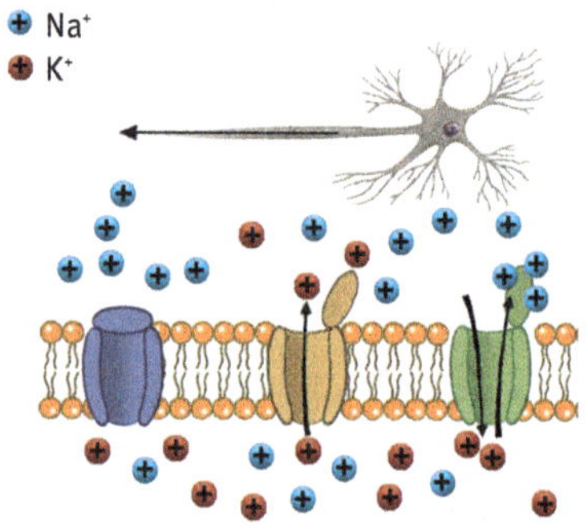

Cada zona de la membrana que ha alcanzado el potencial de acción vuelve nuevamente al potencial de reposo. Este proceso es la **repolarización.** Para ello, se produce otro intercambio de iones a través de membrana:

— Cambia de nuevo la conformación de las proteínas transportadoras de Na^+, y se cierra el paso a este ión.

— Otras proteínas permiten el paso de iones K^+ fuera de la neurona.

— La bomba Na^+/K^+vuelve a funcionar para restablecer el potencial de reposo.

Durante un período de 0,5–2 mseg la fibra no se puede volver a excitar de nuevo; es el llamado **período refractario absoluto.** En un período de 2-3 mseg la neurona recupera su excitabilidad aunque se necesitará un estímulo mayor, es el llamado **período refractario relativo.**

La velocidad con que se propaga el impulso nervioso depende de la presencia de vainas de mielina en los axones.

AXÓN SIN VAINA DE MIELINA

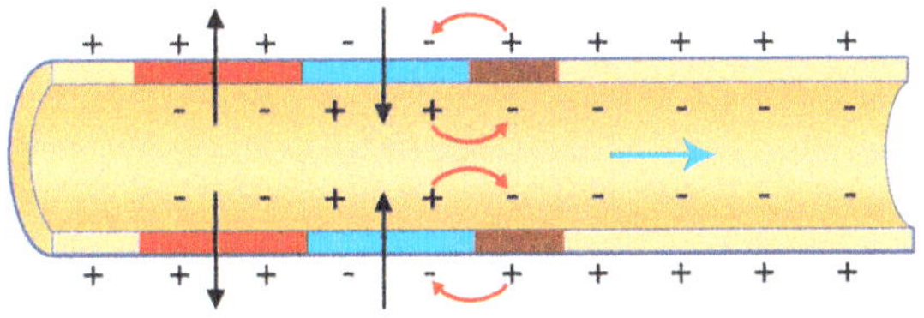

— Los iones Na^+ que han entrado son atraídos por la carga negativa del segmento de neurona más próximo, lo que provoca de nuevo un potencial de acción. Posteriormente se repolariza esta zona.

— Este proceso se va repitiendo a lo largo del axón, segmento a segmento.

Las neuronas transmiten los impulsos nerviosos a velocidades de unos pocos metros por segundo.

AXÓN CON VAINA DE MIELINA

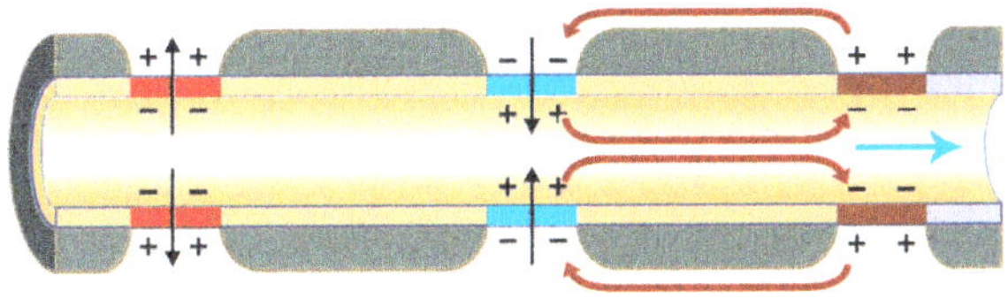

En los axones con vaina de mielina, el intercambio de iones se produce sólo en los nódulos de Ranvier. En estas zonas, la membrana está en contacto con el líquido extracelular, y se localizan las proteínas que transportan el Na^+ al interior de la célula.

— El impulso nervioso se transmite por saltación de un nódulo a otro.

Las neuronas mielinizadas conducen el impulso nervioso a velocidades de hasta 200 metros por segundo.

3.1. *La sinapsis*

Cuando el impulso nervioso llega al extremo del axón, se transmite a las dendritas de otra neurona. Esta transmisión se produce en el espacio que separa ambas neuronas y que, como ya sabes, se denomina **sinapsis.**

El paso del impulso nervioso de una neurona a otra es un proceso que se lleva a cabo por medio de unas moléculas sintetizadas por la propia neurona, denominadas **neurotransmisores.**

— En el extremo presináptico se encuentran las vesículas con los neurotransmisores.

— Al llegar el potencial de acción, los neurotransmisores se vierten al espacio sináptico.

— Los neurotransmisores liberados se unen a unos receptores situados en las dendritas de la neurona siguiente y originan un impulso nervioso que se transmite a lo largo de esta neurona.

— Los neurotransmisores se vuelven a liberar al espacio sináptico, y son en parte reabsorbidos por la célula productora, y en parte degradados.

+

Para ver una animación de la propagación del impulso nervioso puedes visitar la siguiente dirección web:

http://www.blackwellpublishing.com/matthews/actionp.html

Axón

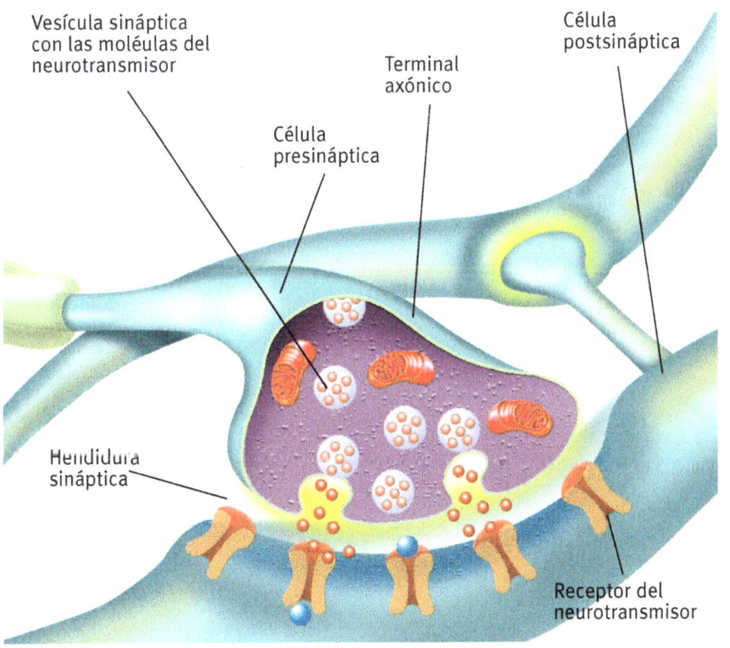

+

Sir **Bernard Katz** (1911-2003) nació en Leipzig, Alemania, y trabajó en Inglaterra después de doctorarse en medicina. Es reconocido por su trabajo sobre la bioquímica del nervio. Compartió el Premio Nobel de medicina en 1970 con Julius Axelrod y Ulf von Euler.

Estudió como se generaba y conducía el impulso nervioso sirviéndose del axón gigante del calamar. Entre sus principales trabajos destaca la manera en que el impulso nervioso se transmite entre células como el nervio y el músculo. Otras investigaciones revelaban que existían comunicaciones entre el nervio y el músculo a través de un transmisor llamado acetilcolina, que se fijaba en los receptores musculares. Katz, junto con Paul Fatt, estableció que la acetilcolina, cuando se une a los receptores de las células musculares, induce la abertura de canales iónicos y por lo tanto la creación de corriente eléctrica, provocando la contracción muscular.

Algunos de los neurotransmisores más abundantes son la *dopamina,* la *acetilcolina,* el *ácido gammaaminobutírico (GABA)* o la *serotonina.* No son todos de la misma naturaleza, y pueden tener distintos efectos sobre la célula posináptica: **excitar,** es decir, iniciar un potencial de acción, o **inhibir,** bloquear el potencial de acción que ha iniciado otra neurona. Las neuronas pueden contener un tipo u otro de neurotransmisor. Veamos un ejemplo de cada uno.

- La **acetilcolina** es un neurotransmisor liberado por los axones de determinadas neuronas, cuya función es excitar, es decir, despolarizar la neurona receptora para iniciar un potencial de acción. Puede liberarse también a la musculatura esquelética y provocar su contracción.
- El **GABA** (ácido gammaaminobutírico) es un neurotransmisor presente en el sistema nervioso central que tiene función inhibidora; las neuronas que liberan GABA eliminan la acción excitadora que han transmitido otras células nerviosas.

Para el buen funcionamiento del sistema nervioso, es necesaria la acción conjunta de neurotransmisores con función inhibidora y con función excitadora. El exceso o la falta de alguno puede originar enfermedades muy graves. Es el caso de la **corea de Huntington,** debida principalmente a una deficiencia de **GABA,** y que se manifiesta por convulsiones de la cara y las extremidades, o la **enfermedad de Parkinson,** producida por la destrucción de neuronas sintetizadoras de dopamina, lo que causa temblores al enfermo que la padece.

ACTIVIDADES

1. En la función de relación se distinguen 3 fases: presencia del estímulo, captación del estímulo y producción de una respuesta. Especifica, para cada una de las frases siguientes, cuál es el estímulo, cómo se capta y qué respuesta se desencadena:
   - Una brisa marina hace que la piel se ponga de gallina.
   - La planta crece en dirección a la luz solar.
   - La presencia de una presa para una medusa hace que los tentáculos urticantes disparen veneno para inmovilizar la presa.
   - Un perro comienza a segregar saliva al oler comida.
2. Contesta las siguientes preguntas sobre la transmisión del impulso nervioso:
   - ¿Qué tipo de estructura es la bomba Na^+/K^+? ¿Cuál es su función? ¿En qué fase del impulso nervioso está activa? ¿Y en cuál inactiva?
   - ¿Qué entendemos por despolarización? ¿A qué se debe que el axón se despolarice?
   - ¿Qué es la repolarización? ¿Cómo se produce?
3. Explica que procesos se llevan a cabo en cada una de estas partes que participan en la sinapsis:
   - Neurona presináptica.
   - Neurona postsináptica.
   - Hendidura sináptica.

4. Captación de estímulos: los receptores

La percepción de la luz, el calor, el contacto... es posible gracias a la existencia de células nerviosas especializadas, que transforman estas manifestaciones de energía en impulsos nerviosos.

Estas células, o receptores sensoriales, presentan diversas formas en los distintos grupos de animales. Los animales más sencillos, como las esponjas, presentan células nerviosas dispersas que son a la vez receptoras y emisoras de respuestas. Al avanzar en la escala evolutiva, las células receptoras pueden estar especializadas en la captación de un estímulo concreto, como es el caso de las que captan la temperatura, que están distribuidas por la piel de los mamíferos. En otras ocasiones, se agrupan y constituyen un órgano especializado con estructuras añadidas para facilitar la recepción del estímulo, como es el caso del ojo. Según su localización, distinguimos dos tipos de receptores:

- **Externos.** Se encuentran localizados en la periferia del organismo y reciben estímulos del exterior. Son externos los receptores del tacto y los órganos de los sentidos.
- **Internos.** Están situados en el interior del organismo y detectan los cambios del medio interno. Es el caso de los receptores situados en músculos y tendones en los vertebrados, que reciben información sobre la postura del cuerpo.

Además, según el tipo de estímulo, los receptores pueden ser:

Mecanorreceptores	Detectan estímulos mecánicos, como el tacto, la presión, la gravedad y las ondas sonoras.
Termorreceptores	Captan aumentos y descensos de las temperaturas.
Quimiorreceptores	Son sensibles a sustancias químicas, como los gustativos y olorosos.
Fotorreceptores	Son sensibles a la energía lumínica.
Algorreceptores	Captan estímulos, ya sean mecánicos, térmicos o químicos, que causan dolor.

A continuación, describimos el mecanismo de acción de las células sensitivas receptoras:

— Cuando el estímulo llega a la célula receptora, provoca un cambio de conformación de las proteínas de la membrana citoplasmática transportadoras de sodio, lo que facilita la entrada de iones Na^+ a través de ésta. Se produce un potencial de acción que en este caso recibe el nombre de **potencial receptor.**

— Como consecuencia de este potencial, las células receptoras segregan neurotransmisores, los cuales provocan el inicio del impulso nervioso en las neuronas que están en contacto con ellas.

— El impulso nervioso prosigue su camino por las neuronas, hasta llegar a los centros nerviosos, en el caso de que los haya, donde se analiza la información recibida.

C

Para que los receptores sensoriales capten un estímulo, es necesario que éste tenga una determinada intensidad mínima que se denomina **umbral.** Por ejemplo, el oído humano es sensible a sonidos de determinada intensidad, y por debajo de ésta no podemos oírlos; si dos personas se hablan al oído, una tercera que esté cerca no puede percibir lo que dicen.

Muchos receptores sensitivos tienen la capacidad de habituarse ante un estímulo continuo y dejar de captar dicha sensación. Los receptores del tacto, dispersos sobre la piel, se acostumbran al contacto continuo de la ropa y dejan de percibir el estímulo táctil.

C

INICIO DEL POTENCIAL RECEPTOR EN UNA CÉLULA QUIMIORRECEPTORA

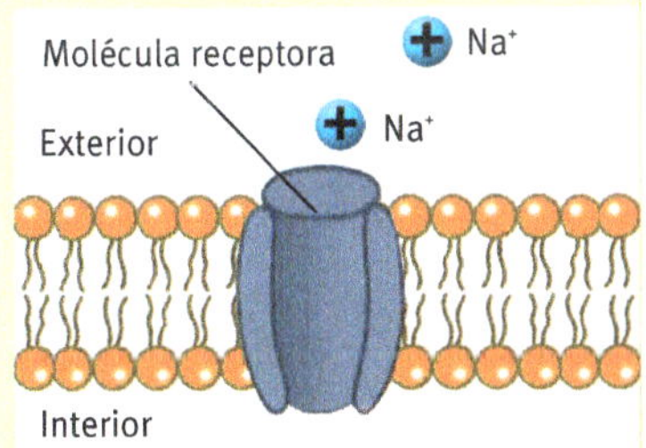

Las proteínas de membrana que permiten el paso de Na^+ se mantienen cerradas gracias a una molécula receptora, que puede reaccionar con sustancias químicas.

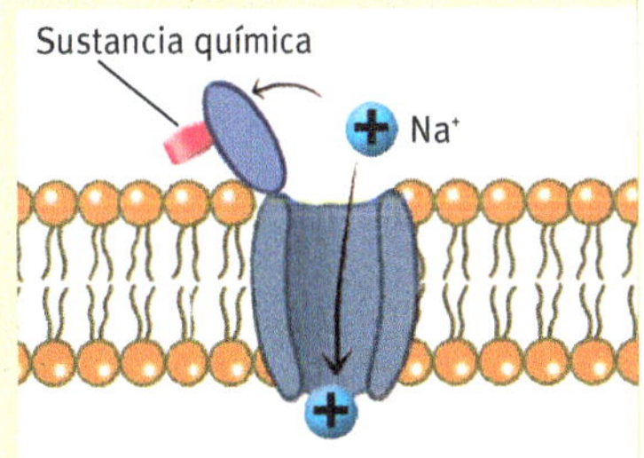

Cuando una sustancia química específica llega a la membrana de la célula sensitiva, reacciona con la molécula receptora. Ésta produce el cambio de conformación de la proteína de membrana y los iones Na^+ entran en la célula.

4.1. Los órganos de los sentidos

Se constituyen mediante la agrupación de células receptoras junto con estructuras anexas destinadas a facilitar la percepción de las células sensoriales. Su existencia permite tener una mayor sensibilidad y discriminación de los estímulos y, en algunos casos, como la vista y el oído, orientar la sensibilidad hacia la procedencia de éstos.

+

Lo que un artrópodo ve es la suma de la visión de todos sus omatidios. Este tipo de visión se denomina **visión en mosaico.**

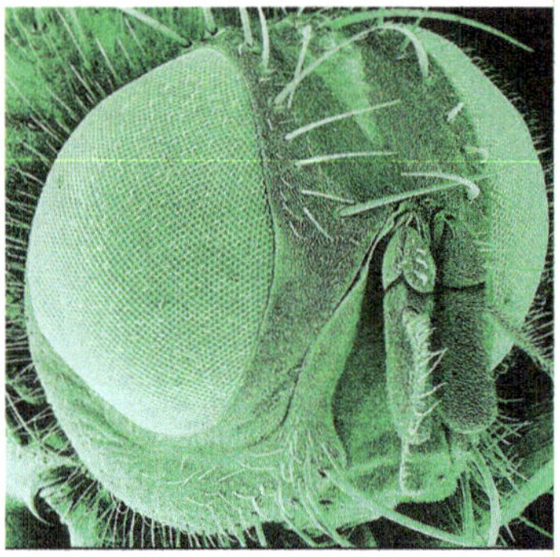

Receptores de la luz: los ojos

La unión de fotorreceptores da lugar a los órganos de la visión, los ojos. En algunos casos, simplemente perciben la dirección del foco luminoso, y en los casos más complejos, permiten diferenciar formas, colores, localización y velocidad del movimiento.

A lo largo del proceso evolutivo han aparecido diversos tipos de órganos para la visión:

- En los cnidarios, la mayor parte de los moluscos y los anélidos, las células fotosensibles captan la intensidad de la luz, y se agrupan en una depresión del cuerpo del animal. Reciben el nombre de **ocelos.**
- El ojo propio de los artrópodos es el **ojo compuesto,** constituido por numerosas unidades estructurales que reciben el nombre de *omatidios.* Exteriormente se aprecian como numerosas superficies poligonales denominadas *facetas.*
- El **ojo** característico de los vertebrados y de los cefalópodos consta de un globo ocular en cuyo interior se localizan los fotorreceptores.

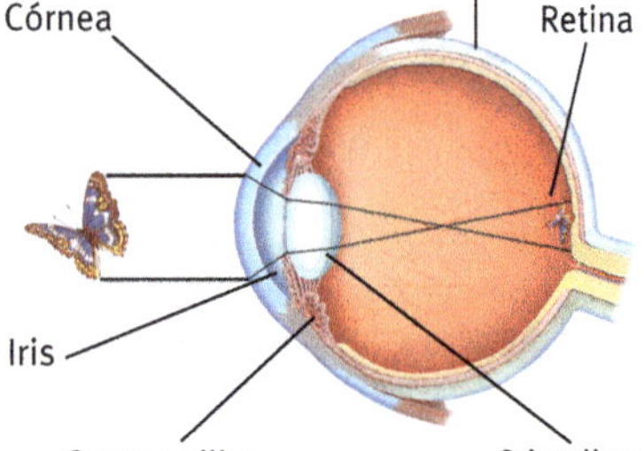

Mecanismo de la visión

— El estímulo visual entra a través de la córnea.

— La entrada de luz es regulada por el iris.

— La función del cristalino es enfocar el objeto que se mira; el cristalino puede abombarse o aplanarse mediante la acción del cuerpo ciliar, según el objeto esté cerca o lejos.

— Las imágenes se proyectan en la retina, formada por células receptoras que contienen pigmentos fotosensibles y células nerviosas.

— Las prolongaciones de las células nerviosas constituyen el nervio óptico, que hace llegar los estímulos al sistema nervioso central.

La retina presenta varias capas de células. Una de estas capas está formada por conos y bastones que son las células fotorreceptoras:

- Los **conos** perciben detalles de las imágenes, y permiten la visión en color; son sensibles a la luz intensa.
- Los **bastones** captan el negro, el blanco y matices de gris, debido a su gran sensibilidad a la luz, aunque ésta sea tenue.

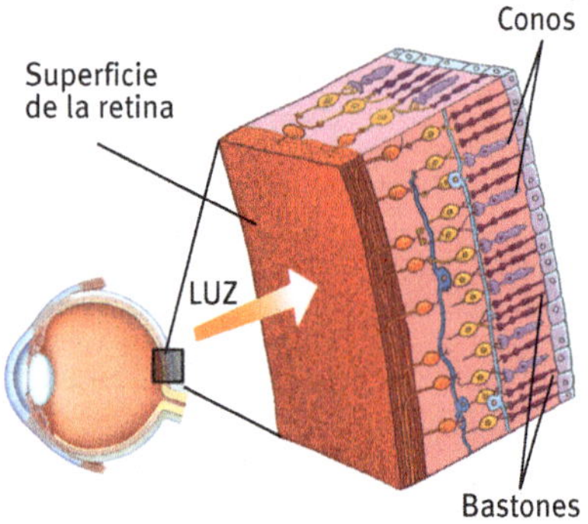

La *mancha amarilla* o *fóvea* es la zona de la retina donde se localizan más conos, y es la que proporciona una imagen más detallada.

Los animales nocturnos tienen retinas formadas principalmente por bastones, mientras que los diurnos poseen casi exclusivamente conos. En el ser humano predominan los conos sobre los bastones.

Receptores del sonido: los órganos de la audición

Los órganos receptores de las ondas sonoras pueden ser de dos tipos:

- **Órgano timpánico.** Lo poseen los insectos, y es una modificación del sistema traqueal usado para la respiración. Se puede localizar en antenas (moscas), tórax (polillas), abdomen (escarabajos) o patas (saltamontes). El mecanismo en estos casos es el siguiente:
  - Las ondas sonoras producen vibraciones en la membrana timpánica.
  - Las vibraciones son detectadas por las células receptoras y enviadas a las neuronas.

  Los insectos pueden localizar los sonidos que produce otro individuo de su propia especie, y determinar la distancia desde la que se emiten. El órgano timpánico también les permite distinguir la proximidad de los depredadores, como los murciélagos.

- **Oído.** Es propio de los vertebrados y alcanza su máximo desarrollo en aves y mamíferos. El oído presenta distinta anatomía según el medio en que habitan los animales, y alcanza mayor perfección conforme se avanza en la escala evolutiva. Así, los peces sólo poseen *oído interno,* mediante el que detectan las vibraciones del agua; los anfibios y reptiles, como adaptación al medio terrestre, presentan *oído medio*, que tiene la función de amplificar las ondas sonoras transmitidas por el aire. El *oído externo* aparece en el grupo de los cocodrilos y en las aves, que presentan *conducto auditivo externo.* En los mamíferos, el oído externo se completa con un *pabellón auricular* muy desarrollado, que es móvil en algunas especies. El mecanismo de la audición en estos casos es el siguiente:
  - El oído externo recoge las ondas sonoras y las conduce hasta el tímpano, que vibra.
  - El oído medio transmite las vibraciones por medio de la cadena de huesecillos hasta la ventana oval.
  - La membrana conecta con el oído interno y hace vibrar el líquido del caracol, produciendo el movimiento de los cilios del órgano de Corti, que transmiten la señal a las neuronas del nervio auditivo.

Oído

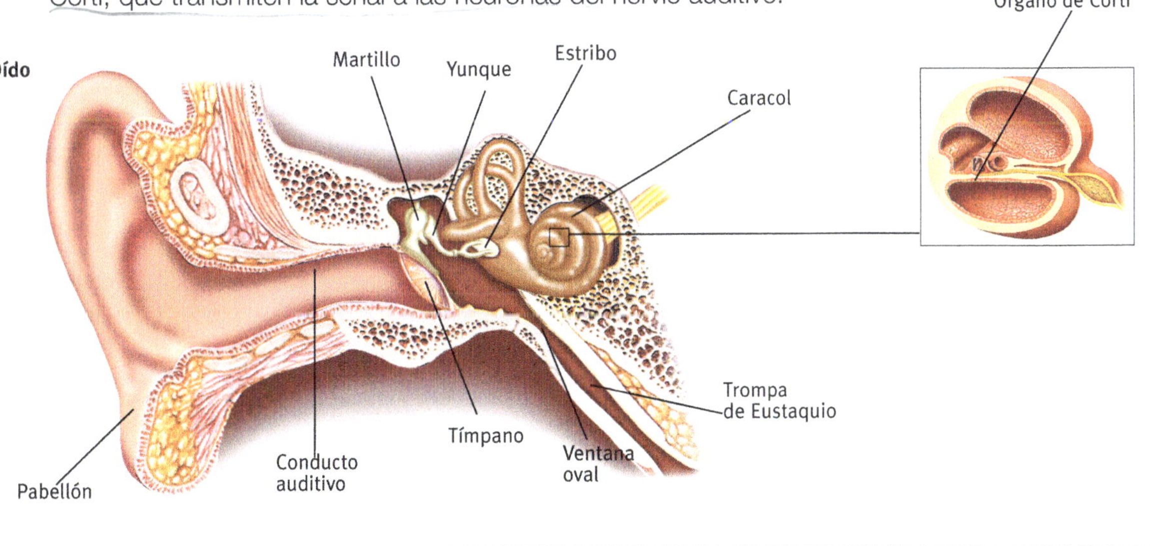

Receptores de la posición: órganos del equilibrio

Es importante que el ser vivo obtenga información precisa acerca de la posición y el movimiento de su cuerpo para que pueda orientarse y desplazarse.

En los animales encontramos distintos órganos del equilibrio:

ESTATOCISTOS

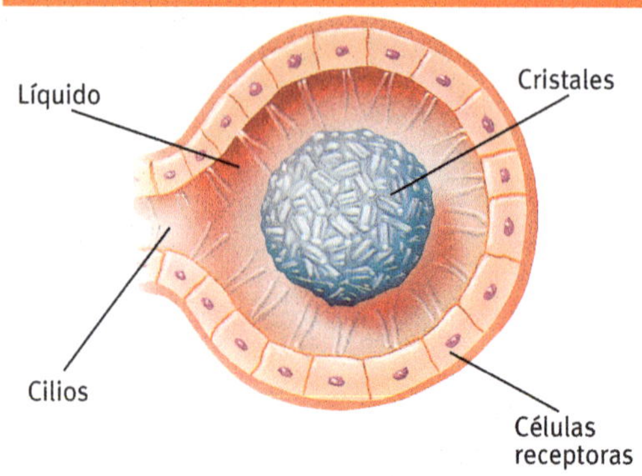

Los **estatocistos** están presentes en los invertebrados. Son sensibles a la fuerza de la gravedad. Cuando el estatocisto se inclina, a causa del movimiento del animal, el peso de los cristales hace que los cilios se doblen. El movimiento de los cilios inicia una serie de impulsos en las células receptoras que se transmiten a las células nerviosas.

En los cnidarios, como la medusa, los estatocistos se sitúan en el borde de la umbrela. En los moluscos, como el caracol, están cerca de los ojos.

En los crustáceos, del grupo de los artrópodos, se sitúan en las antenas. El estatocisto contiene una masa de granos de arena pegados, el estatolito, que al moverse estimulan las células ciliadas. Los granos de arena se cambian por otros nuevos durante la muda.

LÍNEA LATERAL

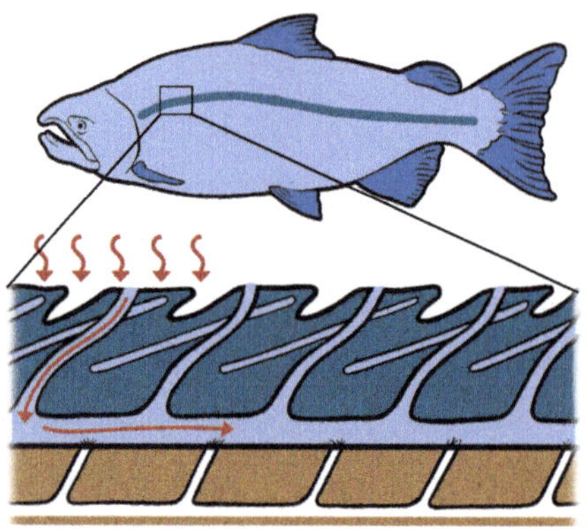

La **línea lateral** la poseen los peces y las larvas de anfibios. Consiste en una línea de células ciliadas que se agrupan a lo largo del cuerpo del animal, formando una línea longitudinal que se ramifica al llegar a la cabeza. Los movimientos del agua estimulan los cilios, que transmiten a los centros nerviosos la posición del animal.

APARATO VESTIBULAR

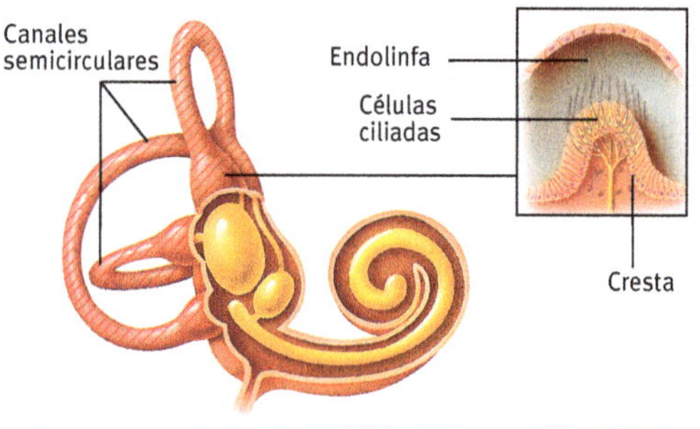

El **aparato vestibular** se encuentra en los mamíferos, en una cavidad del hueso temporal. Está compuesto por los *canales semicirculares* y los *órganos otolíticos*.

Canales semicirculares

Detectan los *movimientos de giro* de la cabeza.

- Al mover la cabeza, la **endolinfa** se desplaza al menos en uno o dos **canales.**
- Esto provoca que los cilios de las células receptoras de la **cresta** se doblen produciendo un impulso.

ÓRGANOS OTOLÍTICOS: SÁCULO Y UTRÍCULO

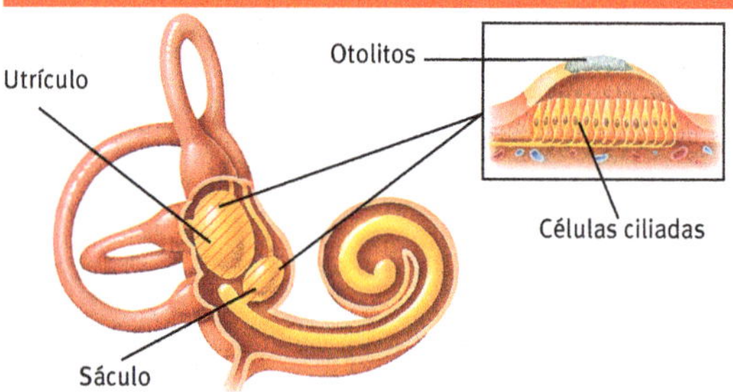

Órganos otolíticos

Transmiten a los centros nerviosos información sobre los *movimientos lineales*, e intervienen en el *equilibrio estático*.

- Al mover la cabeza, la endolinfa produce el desplazamiento de los **otolitos,** que son gránulos calcáreos.
- Los cilios de las células receptoras se doblan y transmiten los impulsos nerviosos.

Receptores químicos: órganos del gusto y del olfato

Es posible detectar el gusto de las sustancias debido a que sus partículas pueden disolverse en el medio, como las sales en el agua. El olor puede captarse gracias a que algunas partículas se suspenden en el medio, cómo ocurre con el néctar de las flores en el aire.

- El **gusto** se detecta mediante quimiorreceptores que se estimulan con el contacto de determinadas sustancias en disolución. Dichos quimiorreceptores se sitúan próximos a la boca y están asociados a respuestas relacionadas con la alimentación.
- La sensación del **olor** se produce cuando determinadas sustancias que se encuentran en suspensión estimulan receptores específicos, asociados tanto a la alimentación como a estímulos ligados al comportamiento: la reproducción, el marcaje de un territorio, la caza, etc.

La existencia de quimiorreceptores es común a todos los animales. En los más sencillos, como los cnidarios, se encuentran dispersos.

Los quimiorreceptores más conocidos en los invertebrados son los **pelos gustativos** y los **receptores olfatorios** de los insectos.

En los vertebrados, existe una gran interrelación entre la capacidad de percibir olores y sabores. Así, en los animales terrestres, la percepción del gusto en la lengua se ve completada y amplificada por la percepción del olor en las fosas nasales.

- Los **órganos del gusto** en los vertebrados son los **botones gustativos.** Éstos tienen distinta localización según el grupo de vertebrados. En los peces, están situados en la boca, la faringe, la piel de las aletas y en las barbas. El medio acuático en el que viven facilita el transporte de las sustancias que estimulan el gusto, por ello es posible la localización de los receptores en el exterior de la boca.

  En los vertebrados terrestres, los botones se sitúan en la boca y la lengua, donde se mantienen siempre húmedos gracias a la saliva. En ellos el mecanismo de percepción del gusto es el siguiente:

+

Algunos animales, como los insectos sociales o los mamíferos, segregan sustancias olorosas denominadas **feromonas,** que provocan determinadas respuestas en los individuos de su especie.

Mecanismo de percepción del gusto

— Las moléculas disueltas en el agua o la saliva entran en contacto con los botones gustativos, situados en las papilas. Los botones gustativos son agrupaciones de neuronas en forma de racimo.

— Dichas moléculas estimulan a las células receptoras.

— Esta interacción origina un impulso que es transmitido a los centros nerviosos.

Existe la capacidad de percibir cuatro sabores: dulce, salado, amargo y ácido (o agrio), que se localizan en lugares concretos de la lengua. La combinación de estos cuatro sabores da lugar a la gama de gustos que se detectan al comer.

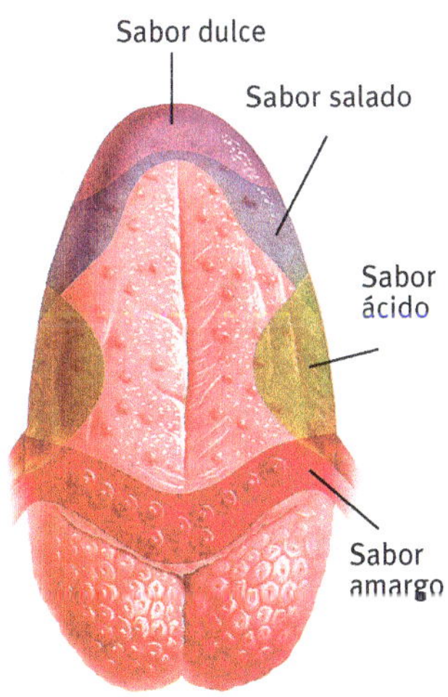

Zonas de la lengua donde predomina la percepción de un determinado gusto

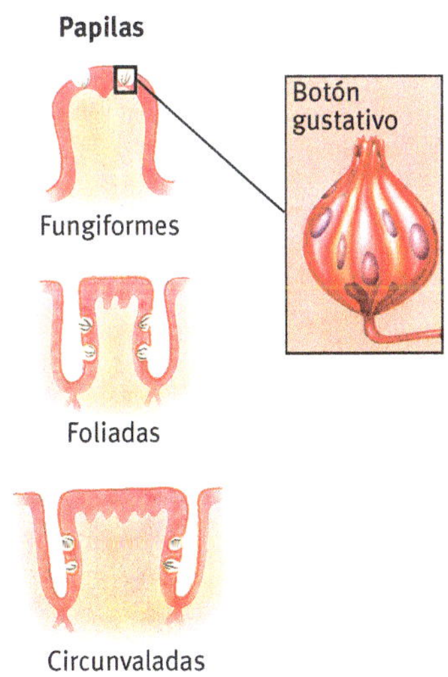

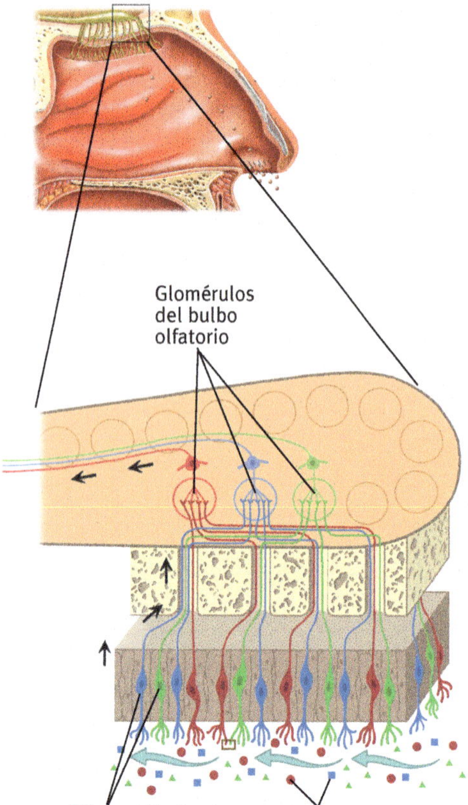

- El **órgano del olfato** en los vertebrados está situado en el interior de la **cavidad nasal.** En los peces esta cavidad está recubierta de quimiorreceptores olfatorios y se comunica con el exterior por medio de unos poros.

En los vertebrados terrestres la cavidad tiene doble función: vía respiratoria y órgano olfativo. Está formada por las **fosas nasales,** que presentan dos orificios para la entrada de aire y comunican con la cavidad bucal. En el interior de las fosas nasales, los **cornetes nasales** originan una serie de invaginaciones, tapizadas por una mucosa que tiene la función de filtrar, humedecer y calentar el aire. En la parte superior de esta mucosa se localizan los quimiorreceptores sensibles al olfato; por ello recibe el nombre de **mucosa olfatoria.**

Mecanismo del olfato

— Las moléculas transportadas por el aire entran en contacto con la mucosa olfatoria. El **mucus** pone en contacto las moléculas hasta los receptores situados en las **células ciliadas.**

— Cuando los receptores olfatorios se activan, envían señales eléctricas hacia los glomérulos del **bulbo olfatorio.**

— Las señales se transmiten hacia los centros nerviosos cerebrales.

Pueden distinguirse siete olores básicos: **floral, alcanforado, almizclado, etéreo, mentolado, pútrido** y **picante.**

Receptores del tacto, la temperatura y el dolor

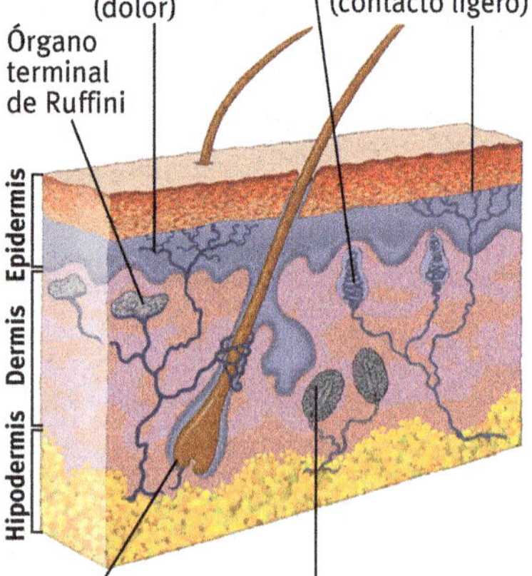

Los órganos de los sentidos están localizados en el lugar del cuertpo más adecuado para captar los estímulos con la máxima eficacia. Así, la mayor parte de los órganos de los sentidos se localiza en la cabeza (ojos, oídos, receptores del gusto y del olfato), ya que la parte cefálica es la que primero recibe determinados estímulos.

Los receptores del tacto, la temperatura y el dolor, sin embargo, están repartidos por el organismo, puesto que ello es necesario para detectar este tipo de cambios en el medio interno y externo.

En los invertebrados, los receptores más frecuentes son neuronas poco diferenciadas, en las cuales diferentes estímulos provocan un impulso nervioso. En los vertebrados, estos receptores se localizan en el exterior y en el interior del cuerpo:

- En el exterior, se encuentran receptores de **sensibilidad cutánea,** que detectan estímulos externos. La sensibilidad cutánea depende de *mecanorreceptores, termorreceptores* y *algorreceptores.* La dispersión de estos receptores en el cuerpo no es homogénea, sino que existe una mayor concentración de éstos en determinadas zonas. En el ser humano, las células sensibles al tacto se encuentran sobre todo en los dedos de las manos, con las que manipulamos los objetos, y en los labios, para tener contacto previo con los alimentos.

- En el interior del cuerpo, hay receptores de **sensibilidad propioceptiva** y **nociceptiva,** que captan estímulos internos. La sensibilidad propioceptiva es posible gracias a *terminaciones nerviosas*, en conexión con músculos y tendones, que son sensibles a la postura y los movimientos de locomoción. La sensibilidad nociceptiva viene dada también por *terminaciones nerviosas,* receptoras de estímulos dolorosos, situadas en los vasos circulatorios y las vísceras.

  También existen *quimiorreceptores* especializados en la captación de las características químicas del medio interno: su pH y la concentración de oxígeno y de dióxido de carbono.

ACTIVIDADES

4. Señala las afirmaciones incorrectas y reformúlalas para que sean verdaderas.
   - Los algorreceptores detectan estímulos causados por la gravedad o ondas sonoras.
   - Los órganos de la visión están constituidos por la asociación de fotorreceptores.
   - Las ondas sonoras hacen vibrarla cadena de huesecillo, y estos, a su vez, la membrana timpánica, que comunica con la ventana oval
   - Al mover la cabeza, se produce el desplazamiento de los otolitos, hecho que detectan los cilios de las células que se encuentran en el sáculo y utrículo.
   - Papilas gustativas es sinónimo de botones gustativos.
   - En la piel encontramos tres tipos de receptores: mecanorreceptores, termorreceptores y algorreceptores.
5. Define los siguientes conceptos:
   - Receptor
   - Omatidio
   - Órgano timpánico
   - Estatocisto
   - Otolito
6. Describe el mecanismo de la audición.

5. Integración de la información y elaboración de respuestas

La interpretación de la información que captan los receptores no depende únicamente de ellos: en los animales que presentan centros nerviosos, son éstos los que interpretan los estímulos y elaboran las respuestas adecuadas. Así, vemos realmente un objeto cuando el estímulo visual llega a la parte del cerebro que recibe la visión.

En la ilustración de la derecha puedes observar la importancia que tiene la interpretación de la imagen representada que lleva a cabo el cerebro. Las áreas negras aparecen como el perfil de dos caras, mientras que las blancas representan la silueta de un jarrón. A pesar de ser una sola imagen, el cerebro puede hacer dos interpretaciones diferentes.

Así, la **percepción** no es la recepción pasiva de las señales sensoriales, sino la interpretación activa para parte de los centros nerviosos de los estímulos que éstos reciben.

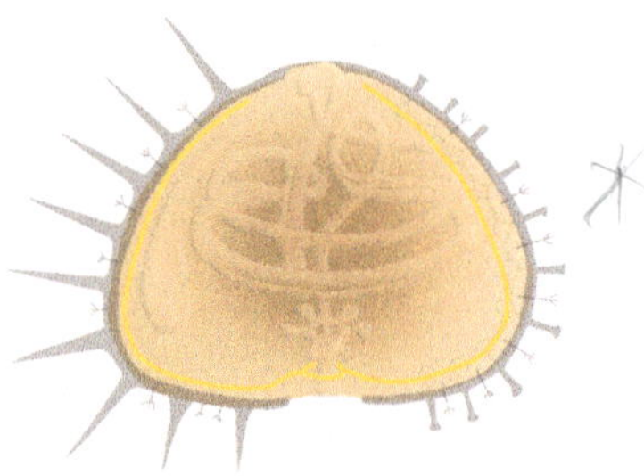

Sistema nervioso de un equinodermo

Sistema nervioso de un anélido

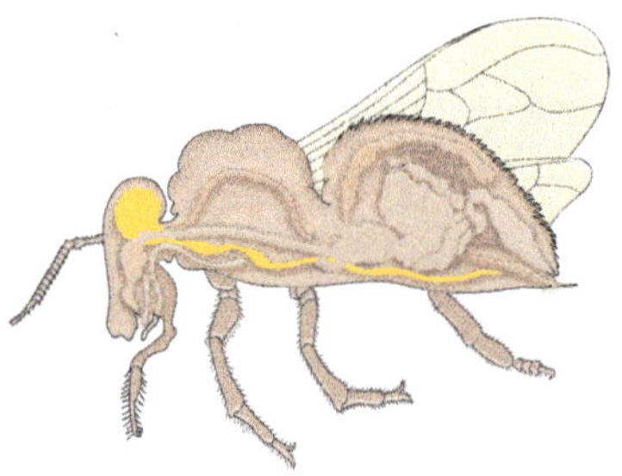

Sistema nervioso de un artrópodo

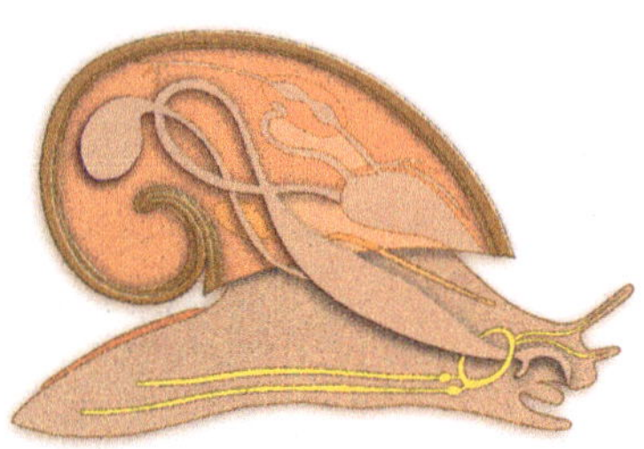

Sistema nervioso de un molusco gasterópodo

El sistema nervioso adquiere una mayor complejidad a medida que se avanza en la escala evolutiva. La siguiente tabla resume las características del sistema nervioso de algunos grupos de invertebrados:

Sin sistema nervioso	Los **poríferos** sólo poseen **células nerviosas dispersas.**	
Redes nerviosas	En los **cnidarios** las neuronas están dispersas pero conectadas formando una red nerviosa. Los impulsos nerviosos viajan así en todas direcciones.	
Anillos nerviosos	Propios de los **equinodermos.** El sistema nervioso está formado por una serie de **anillos nerviosos** de los que parten los nervios radiales.	
Sistemas nerviosos bilaterales	Los animales con simetría bilateral suelen tener un mayor número de células nerviosas que se suelen concentrar: — En un extremo del animal, normalmente en posición ventral y frontal. — Formando cordones que se pueden transformar en ganglios, nervios o centros nerviosos.	**Anélidos:** Las neuronas se agrupan en **ganglios,** que se distinguen en cada segmento del cuerpo. Existe un **cerebro** que centraliza toda la información. **Artrópodos:** Los **ganglios** se fusionan en la región frontal formando un **cerebro.** **Moluscos:** De este grupo, los bivalvos y gasterópodos poseen un centro nervioso en forma de **anillo ganglionar** mientras que los cefalópodos poseen un **cerebro.**

5.1. *El sistema nervioso de los vertebrados*

En los vertebrados, el sistema nervioso alcanza un mayor desarrollo. En él se distinguen las siguientes estructuras:

- **Centros nerviosos,** formados por el **encéfalo** y la **médula espinal.** Reciben la información sensorial, la procesan y emiten respuestas.
- **Ganglios nerviosos,** formados por la agrupación de cuerpos neuronales. Se encuentran en el trayecto de algunos nervios.
- **Fibras nerviosas,** que son los axones de las neuronas, a través de los que se transmiten los estímulos o las respuestas a éstos.
- **Nervios,** formados por la unión de fibras nerviosas.

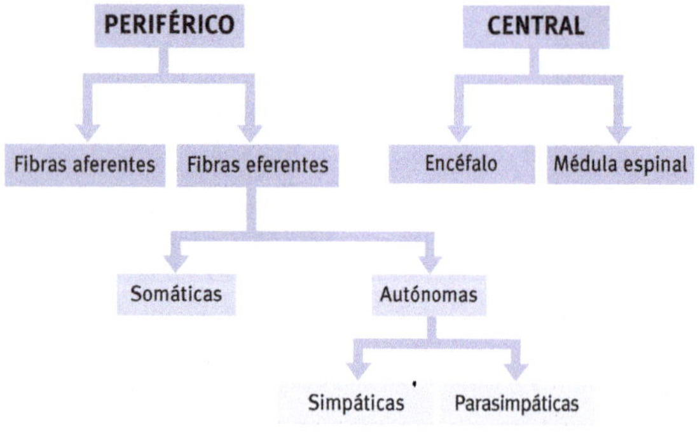

Estas estructuras se organizan formando el *sistema nervioso central* y el *sistema nervioso periférico.*

El sistema nervioso central

Está formado por un gran número de neuronas agrupadas y conectadas entre sí. Los cuerpos de las neuronas tienen color grisáceo, mientras que las fibras cubiertas de mielina son blancas, debido al color de las células de Schwann. Por ello, a las zonas del sistema nervioso con acúmulo de cuerpos neuronales se les denomina **sustancia grias,** y a las zonas donde se encuentran haces de axones con sus respectrivas vainas se les llama **sustancia blanca.** El sistema nervioso central está constituido por el *encéfalo* y la *médula espinal.*

- El **encéfalo** controla toda la actividad del organismo, en respuesta a las percepciones del medio. Se encuentra protegido por los huesos del cráneo, y consta de *cerebro, cerebelo* y *tronco encefálico.*
  - — El **cerebro** es el centro donde se integra y analiza la información que proviene de los receptores sensoriales, y se procesa la respuesta a esta información en forma de movimientos del cuerpo o secreción de glándulas. La sustancia gris se sitúa en la parte exterior, formando la **corteza cerebral,** y la blanca en la parte interior. A lo largo de la escala evolutiva, se presenta un desarrollo gradual del cerebro, ligado al perfeccionamiento de sus funciones. En el ser humano se desarrolla la capacidad de razonar y elaborar un lenguaje para comunicarse.
  - — El **cerebelo** coordina los movimientos que permiten el mantenimiento del equilibrio del cuerpo. La distribución de la sustancia gris y la blanca es igual que en el cerebro.
  - — El **tronco encefálico** controla el latido cardíaco y la respiración. La sustancia gris está en el interior y la blanca en el exterior.
- La **médula espinal** es la vía de conexión entre el encéfalo y el resto del cuerpo y se localiza en el interior de la columna vertebral. A veces actúa como un centro emisor de respuestas elaborando *actos reflejos.*

  Está constituida por un cilindro delgado, en cuya parte central se localizan los cuerpos de las neuronas; en la parte externa están los axones. Por este motivo, tiene color gris en su interior y blanco en el exterior. Los axones forman largos nervios sensitivos, que ascienden hasta el cerebro, y motores, que se dirigen a distintas partes del cuerpo.

El sistema nervioso periférico

Está constituido por nervios, que pueden ser de dos tipos:

- **Nervios sensitivos.** Recogen los estímulos captados por los órganos de los sentidos y los canalizan hacia el sistema nervioso central mediante impulsos nerviosos. Reciben el nombre de **fibras aferentes.**
- **Nervios motores.** Proceden del sistema nervioso central y trasmiten los impulsos nervisosos hacia los órganos que elaborarán la respuesta (músculos o glándulas). Se denominan **fibras eferentes.**

  Los nervios que salen del encéfalo son los *nervios craneales,* mientras que los que parten de la médula espinal se llaman *nervios espinales.*

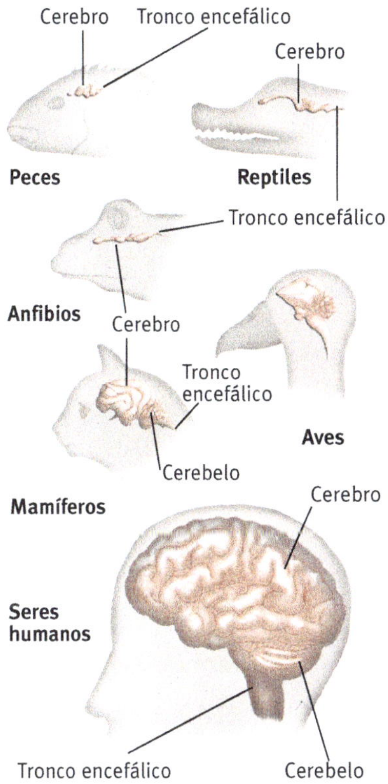

Esquema del encéfalo de los distintos grupos de vertebrados.

C

El sistema nervioso central está rodeado por tres membranas llamadas **meninges,** separadas entre sí por el **líquido cefalorraquídeo.**

C

El hecho de retirar la mano cuando tocamos un objeto muy caliente es un acto reflejo que evita que suframos una quemadura grave.

Un **acto reflejo** es una respuesta involuntaria y tiene lugar cuando un estímulo es conducido a la médula espinal y ésta envía un impulso nervioso que produce el movimiento rápido de la parte afectada.

El recorrido del impulso nervioso desde que se capta el estímulo hasta que se produce el movimiento es más corto que en los actos voluntarios, controlados por el cerebro, por lo que la respuesta es más rápida.

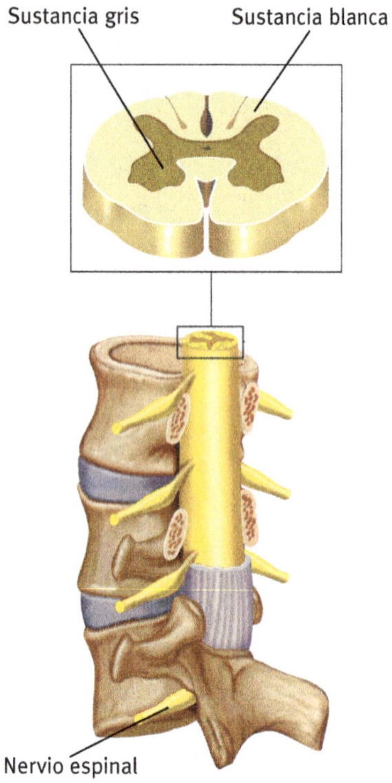

Médula espinal y sección de la misma

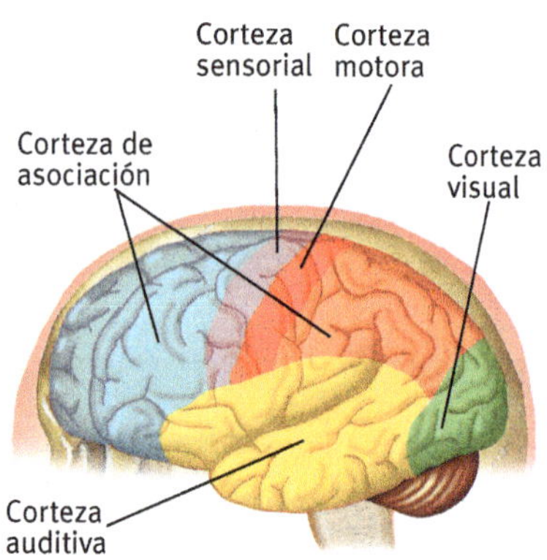

Áreas de la corteza

- Los **nervios craneales** son de tres tipos:
  - — Nervios craneales sensitivos, que reciben los estímulos de los receptores y los envían al sistema nervisoso central, como los nervios olfatorios y los ópticos.
  - — Nervios craneales motores, que envían las órdenes del sistema nervioso central ahcia la musculatura, como los nervios motores oculares.
  - — Nervios craneales mixtos, es decir, con fibras sensitivas y motoras, como el nervio trigémino, que inerva la cara.
- Los **nervios espinales** salen del espacio existente entre dos vértebras. Todos ellos son mixtos, es decir, contienen fibras sensitivas y motoras.

El cerebro humano

A lo largo del proceso evolutivo, el cerebro humano ha alcanzado un elevado grado de desarrollo que le ha permitido un mayor perfeccionamiento de sus funciones.

El cerebro humano consta de dos **hemisferios cerebrales,** derecho e izquierdo, divididos por el **cuerpo calloso,** que es una densa masa de fibras nerviosas.

La **corteza cerebral** es la parte del cerebro donde se integra la información que llega a éste, se elaboran las respuestas y se llevan a cabo los procesos de la memoria y el razonamiento. Por esta razón, existe una relación directa entre la superficie de corteza cerebral y el grado de perfeccionamiento de las funciones cerebrales. Para aumentar esta superficie se han desarrollado unos repliegues, denominados **circunvoluciones cerebrales,** que se aparecen en los mamíferos. El gran número de éstas en el ser humano hace que su corteza cerebral tenga una superficie de 2 500 cm^2. La corteza está dividida en **lóbulos,** que son zonas limitadas por unos surcos y que reciben el nombre del hueso del cráneo más cercano: *formal, parietal, temporal* y *occipital.*

Existen distintas áreas de la corteza, las cuales desarrollan funciones especializadas:

- **Corteza motora:** controla el movimiento de los músculos esqueléticos. Se localiza en el lóbulo frontal.
- **Corteza sensorial:** recibe los estímulos del tacto, la temperatura, el dolor y el gusto. Se sitúa en la parte superior del lóbulo pariental.
- **Corteza visual:** a ella llegan las sensaciones visuales; ocupa el lóbulo occipital.
- **Corteza auditiva:** recibe los estímulos sonoros y se encuentra en la zona más próxima al oído, en el lóbulo temporal.
- **Corteza de asociación:** existen otras áreas de la corteza cuya función es relacionar la información que recibe el cerebro acerca del medio con las emociones, la memoria y el aprendizaje. Estas zonas se localizan en el lóbulo frontal, el temporal y el occipital, aunque el mecanismo de los procesos que tiene lugar en ellas está todavía en estudio.

En la parte central del encéfalo se encuentran dos estructuras formadas por sustancia gris, el *tálamo* y el *hipotálamo,* que participan en los procesos de integración y procesamiento de los estímulos.

- El **tálamo** interviene en la conexión de las zonas sensitivas y motoras de la corteza.
- El **hipotálamo** recibe información sobre el estado general del organismo, y regula procesos metabólicos a través del sistema endocrino, que describiremos en la próxima unidad, o bien a través de impulsos nerviosos que causan sensaciones como la sed o el hambre.

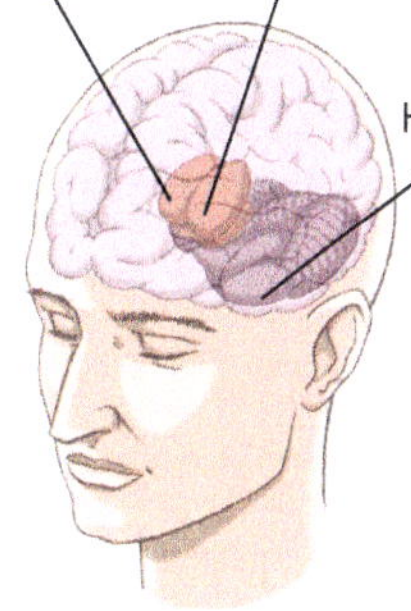

La **inteligencia** es la facultad de pensar, conocer y comprender la información que recibimos del medio, y se manifiesta en los cerebros más desarrollados, como el del ser humano. Se basa en la capacidad de memorizar y, por tanto, de aprender.

La **memoria** se desarrolla a partir de una serie de conexiones entre neuronas, localizadas básicamente en la corteza de asociación, que almacenan los estímulos que llegan al cerebro. Comprende tres fases distintas:

- La **memoria inmediata,** que es la información que se acaba de recibir y se mantiene unos segundos en las redes de la corteza sensorial.
- La **memoria reciente,** que retiene hechos y pensamientos localizados en un período de tiempo de unas horas.
- La **memoria remota,** que comprende los hechos y pensamientos más antiguos.

Cualquier nuevo estímulo que llegue al cerebro, ya sea visual, sonoro, táctil.., es comparado con una experiencia similar anterior, de este modo se desarrolla el proceso del **aprendizaje.**

El ser humano tiene la capacidad de comunicarse mediante el **lenguaje,** que es el código empleado para que el individuo pueda comunicarse con otros seres humanos. Se aprende durante la infancia, desde que el niño es capaz de emitir sonidos y comienza a asociar algunos sonidos con determinados objetos, desarrollando el lenguaje oral; el lenguaje escrito se aprende en una fase de maduración posterior de los niños. Así, la memoria es la base que permite el aprendizaje del lenguaje.

El proceso del **pensamiento** se basa en la capacidad de comunicarse por medio del lenguaje. Permite asociar, además de estímulos concretos, ideas almacenadas en la memoria, con lo que se llega a desarrollar el **razonamiento,** y a tener capacidad de **juicio.**

Las **emociones** dependen de la actividad que se lleva a cabo en neuronas situadas en la base del cerebro, que están conectadas con al corteza cerebral y el hipotálamo. Esta conexión permite que, tanto un determinado estímulo externo como un pensamiento, produzcan reacciones en el organismo, como la aceleración del pulso que provoca el hecho de estar ante un peligro, ya sea real o producido por la imaginación.

La capacidad de desarrollar las funciones cerebrales se transmite de forma hereditaria, pero su desarrollo real depende del proceso de aprendizaje de cada individuo, que formará su inteligencia y su personalidad.

5.2. *Emisión de la respuesta motora*

Para responder a los estímulos, desde el sistema nervioso central se emiten las órdenes hacia distintos puntos del organismo. Esta respuesta puede dirigirse a los músculos, donde se producrá un movimiento, o bien a las glándulas endocrinas a través del hipotálamo, para dar lugar a una secreción hormonal.

Las respuestas viajan desde el sistema nervioso central hasta la zona donde se produce la respuesta del mismo modo que lo hacen los estímulos: por impulsos nerviosos. En esta unidad nos centraremos en el estudio de las respuestas relacionadas con los músculos.

Los cuerpos de las neuronas de los centros nerviosos emiten las respuestas, que viajan por los **nervios motores** y hasta los músculos efectores. Según la respuesta que transportan, los nervios motores forman:

- La red nerviosa **somática,** que controla los músculos que pueden moverse de forma voluntaria, la musculatura esquelética. El impulso comienza en el área motora de la corteza cerebral o en el cerebelo, sigue por el tronco encefálico, la médula espinal, y de ésta salen los nervios motores a distintos niveles hacia los músculos.
- La red nerviosa **autónoma,** que controla los músculos que se mueven de forma involuntaria, es decir, la musculatura lisa y la cardíaca. El impulso puede comenzar en el tronco encefálico o en la médula espinal. Cada nervio motor sale de la médula espinal, conecta con un ganglio nerviosos y desde éste parten las conexiones nerviosas hacia los diferente órganos.

Por su parte, la respuesta autónoma se realiza de dos modos:

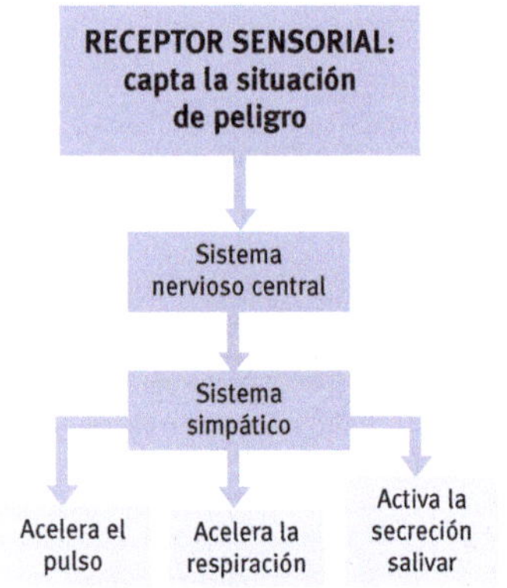

RESPUESTA DEL SISTEMA NERVIOSO SIMPÁTICO

El **sistema nervioso simpático** se activa en situaciones de estrés: miedo, confusión, ira, huida, lucha, ejercicio físico. Su activación comporta la liberación adrenalina y noradrenalina, neurotransmisores que tienen los siguientes efectos sobre el organismo:

— Dilatación de pupilas

— Aumento de la frecuencia cardíaca y de la contracción cardíaca y, como consecuencia, aumento de la presión arterial

— Broncodilatación para permitir la entrada de aire

— Vasodilatación de corazón hígado, músculos y tejido graso

— Glucogenolisis en el hígado y lipólisis del tejido adiposo

— Aumento de la glucosa sanguínea

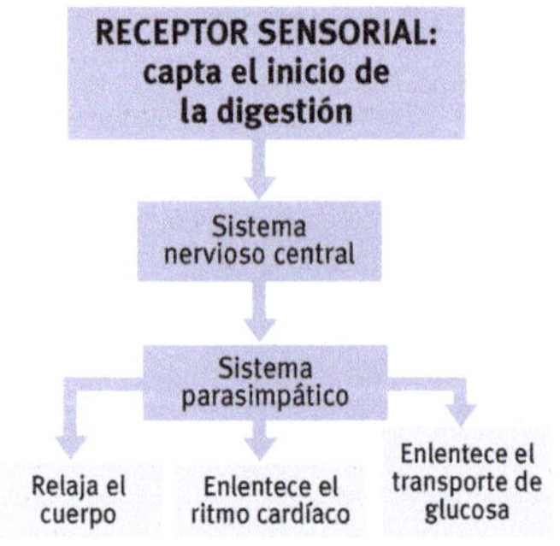

RESPUESTA DEL SISTEMA NERVIOSO PARASIMPÁTICO

El **sistema nervioso parasimpático** relaja el cuerpo y por lo tanto ejerce una función contraria al simpático. Su actuación conlleva la liberación, entre otros neurotransmisores, de acetilcolina, que tiene los siguientes efectos sobre el organismo:

— Reduce el ritmo respiratorio y el cardíaco

— Estimula la secreción salivar

— Estimula los músculos involuntarios del sistema digestivo para favorecer la digestión

— Estimula la producción de orina

— Retarda el transporte de glucosa hacia órganos no implicados en la digestión.

La mayoría de los órganos recibe una doble inervación: simpática, estimuladora, y parasimpática, relajante. La alternancia de estos dos tipos de estímulos es un mecanismo fundamental en la regualción de las funciones vitales. La siguiente tabla resume los efectos de la inervación simpática y parasimpática sobre diferentes órganos:

ÓRGANO	INERVACIÓN PARASIMPÁTICA	INERVACIÓN SIMPÁTICA
Pupila	Contracción	Dilatación
Corazón	Bradicardia	Taquicardia
Bronquios	Contracción	Dilatación
Glándulas sudoríparas	Secreción generalizada	Secreción local
Glándulas salivares	Secreción acuosa	Secreción viscosa
Vejiga	Contracción incompleta	Mayor contracción
Tubo digestivo	Aumento del peristaltismo	Disminución del peristaltismo

5.3. *La sinapsis neuromuscular*

Los nervios motores transportan las respuestas del sistema nervioso central en forma de estímulos, que se van transmitiendo a lo largo de las fibras nerviosas hasta llegar a los músculos. El impulso nervioso que llega al músculo produce una excitación de las fibras de éste, las cuales se contraen. La contracción del músculo esquelético en los vertebrados es la más estudiada. En los músculos se produce una sinapsis entre la neurona y la musculatura, de forma similar a la existente entre dos células nerviosas. La unión entre una neurona y una fibra muscular se denomina **sinapsis neuromuscular,** y la estructura obtenida es la **placa motora.**

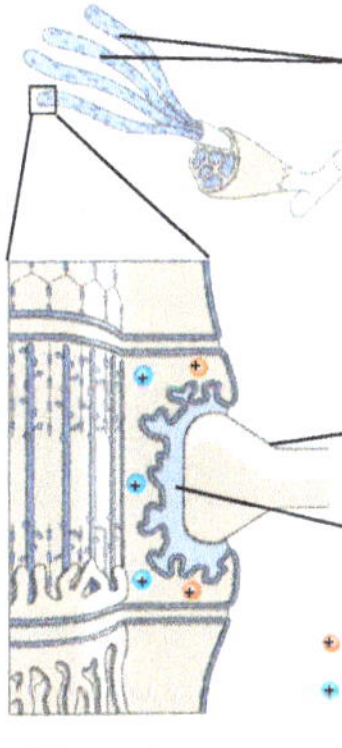

Sinapsis neuromuscular

— En el extremo del axón se produce una secreción de acetilcolina.

— Esta sustancia es captada por los receptores de la membrana plasmática de la fibra muscular. Esto provoca un cambio en la concentración de iones Na^{+} que genera un potencial eléctrico.

— El cambio de potencial produce la secreción de iones Ca^{2+} desde el retículo endoplasmático al citoplasma, que provocan la contracción de la fibra muscular.

A diferencia de la sinapsis entre neuronas, la sinapsis neuromuscular es siempre excitadora, produce siempre una contracción muscular.

ACTIVIDADES

7. ¿Qué centro de control nervioso se considera el más primitivo? ¿Qué grupo de animales lo presenta?

8. ¿Qué son los nervios? ¿En qué se diferencian de los ganglios?

9. Explica qué es la sustancia gris y la sustancia blanca. Indica, a continuación, cómo se distribuyen las dos sustancias en cerebro, cerebelo y médula espinal.

10. Define y localiza los siguientes conceptos:

a) cuerpo calloso
b) corteza motora
d) tronco encefálico
e) hipotálamo

11. ¿Qué es un acto reflejo? Pon un ejemplo y explica porqué la respuesta se da más rápidamente que la respuesta somática.

C

El **esqueleto hidrostático,** característico de los anélidos y los equinodermos, está formado por un fluido que llena los espacios celomáticos en el interior del cuerpo, y actúa como soporte para la acción de la musculatura, que produce el movimiento.

6. Los efectores del movimiento: el aparato locomotor

El movimiento en los animales es posible gracias a la transmisión nerviosa y a la existencia de un **aparato locomotor.** Está compuesto por:

- **Sistema muscular,** formado por los músculos. Es el responsable de los movimientos, debido a las contracciones de los músculos.
- **Sistema esquelético,** que actúa como soporte y anclaje de la musculatura, y como protección de las partes delicadas. Está constituido por estructuras sólidas (huesos, placas...) que se mueven por efecto de la contracción y la relajación de los músculos, que tiran de ellas. Según los grupos de animales, puede ser:
  - — **Exoesqueleto,** o esqueleto externo. Lo presentan los invertebrados, y está formado por placas o piezas duras de sales de calcio o quitina.
  - — **Endoesqueleto,** o esqueleto interno. Lo presentan algunos invertebrados, como los poríferos y los equinodermos, y todos los vertebrados. Está formado por espículas (estructuras con forma de aguja) de sílice o huesos de sales de calcio.

El exoesqueleto confiere mayor protección al organismo que el interno, aunque el endoesqueleto es más ligero y permite mayor movilidad.

6.1. *El aparato locomotor en los invertebrados*

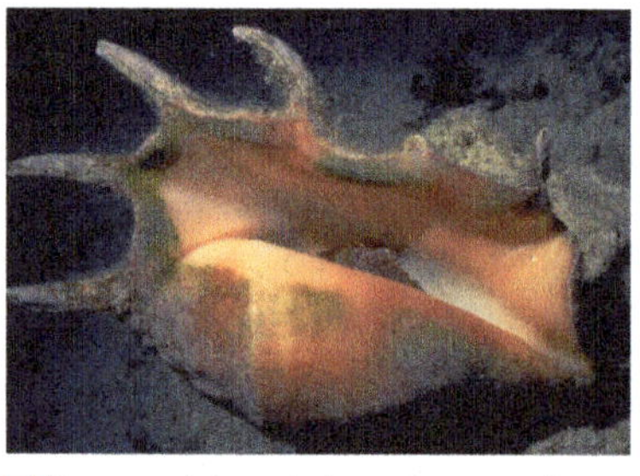

Exosqueleto calcáreo de un molusco

En los diferentes grupos de invertebrados existen distintas adaptaciones que afectan al aparato locomotor. En los poríferos, que tienen la organización más sencilla, el esqueleto es interno y está compuesto por fibras de proteína y espículas de sílice o de calcio. Los cnidarios que presentan formas coloniales, como los corales, poseen un exoesqueleto formado por placas calcáreas; mientras que las medusas no poseen estructuras duras y se impulsan con contracciones de su cuerpo.

GRUPO	ESQUELETO	DESPLAZAMIENTO
Anélidos	Interno hidrostático	Acción de la musculatura sobre el esqueleto hidrostático
Antrópodos	Externo de quitina	Acción de los músculos sobre los apéndices articulados
Equinodermos	Interno calcáreo y esqueleto hidrostático: aparato ambulacral	Por medio del aparato ambulacral
Moluscos	Externo calcáreo (concha) en algunos grupos	Mediante saltos (bivalvos), reptación (gasterópodos), o por contracciones que provocan un bombeo de agua (cefalópodos)

Exosqueleto de quitina de un artrópodo

6.2. *El aparato locomotor en los vertebrados*

El aparato locomotor está constituido por un conjunto de músculos estriados y un esqueleto interno que funcionan de forma coordinada.

- Los **músculos esqueléticos** están formados por tejido muscular y constituyen **husos** de fibras insertados a los huesos mediante **tendones** de tejido conjuntivo.
- El **esqueleto interno,** en las primeras fases del desarrollo de los vertebrados está formado por tejido cartilaginoso. En la mayoría de los casos, el cartílago se va osificando durante el crecimiento del individuo, debido a la incorporación de sales de calcio y fósforo, que van formando el tejido óseo, propio del adulto.

  La unión entre dos huesos recibe el nombre de **articulación.** La superficie de contacto entre los dos huesos está recubierta por cartílago, y su función es evitar el roce y el desgaste óseo en los movimientos. El movimiento se consigue mediante un **sistema de palancas.**

  Los elementos de una palanca son:

  - Punto de apoyo (A).
  - Fuerza motriz (F). Es la fuerza que hace posible el movimiento.
  - Fuerza de resistencia (R). Es la que hay que vencer para ejecutar el movimiento.

  Flexión del antebrazo:

  - El punto de apoyo (A) lo constituye la articulación del codo.
  - La fuerza motriz (F) es producida por el músculo biceps en el punto de inserción del antebrazo.
  - La fuerza resistencia (R) es el peso del antebrazo.

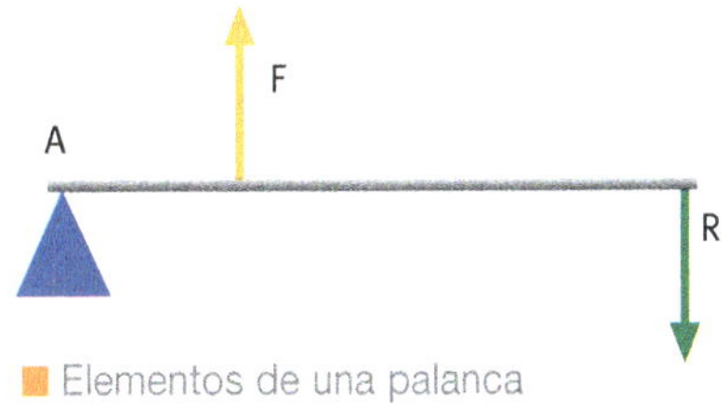

Elementos de una palanca

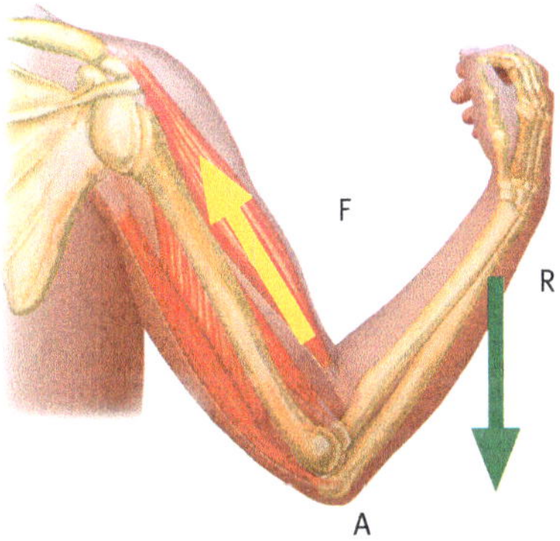

Flexión del antebrazo

Las partes del esqueleto comunes a la mayoría de los vertebrados son:

- La **columna vertebral** es el eje del esqueleto y está formada por vértebras, pequeñas piezas de tejido óseo y cartilaginoso, con un orificio central por el que pasa la médula espinal. Algunas vértebras se articulan con las costillas.
- El **cráneo,** constituido por una serie de huesos que protegen el encéfalo.
- Las **cinturas,** que son arcos óseos que permiten la articulación de las extremidades con la columna vertebral. La **cintura escapular** articula las extremidades anteriores; la **cintura pélvica** articula las extremidades posteriores.
- En la mayoría de los casos se distinguen las **extremidades,** que son cuatro apéndices articulados, unidos a las cinturas escapular y pélvica. Se utilizan para el desplazamiento, y presentan distintas modificaciones, en función de la adaptación al medio más apropiada; algunas especies de reptiles carecen de extremidades y se desplazan por reptación.

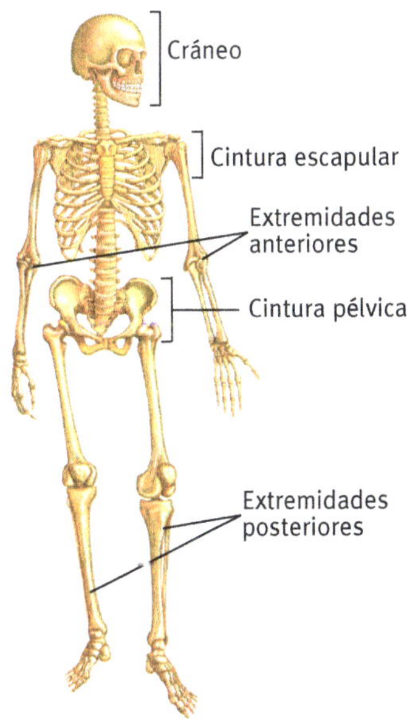

Esqueleto del ser humano

Esqueleto de un pez

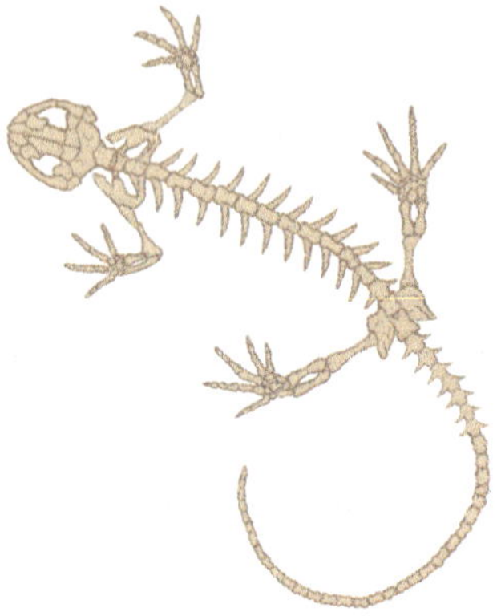

Esqueleto de un anfibio

Esqueleto de un ave

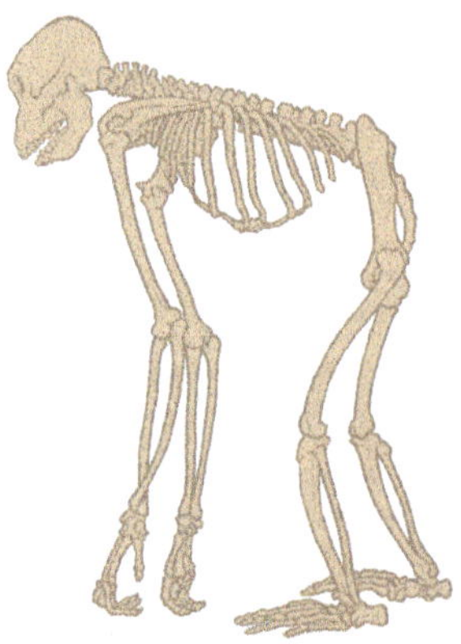

Esqueleto de un mamífero

La musculatura es similar en todos los grupos, y está organizada para posibilitar el movimiento mediante el sistema de palancas. A pesar de que en todos los grupos se distingue una estructura básica común, las adaptaciones a la vida en diferentes medios se han visto favorecidas por modificaciones en la estructura del esqueleto.

- Los **peces** presentan un esqueleto cartilaginoso u óseo, con expansiones membranosas que constituyen las aletas. Se desplazan mediante movimientos ondulatorios laterales de la musuculatura y la propulsión de las aletas.
- En el grupo de los **anfibios** el esqueleto es siempre óseo, y aparecen las cuatro extremidades, necesarias para el desplazamiento en el suelo. Ya presentan un cuello formado por una sola vértebra. El desplazamiento se realiza en el agua, nadando, y en el medio terrestre, mediante saltos.
- Los **reptiles** tienen un esqueleto óseo resistente, con costillas que se cierran por delante en el esternón, y cuello con varias vértebras. Se desplazan por el medio terrestre o acuático; los reptiles sin patas se impulsan mediante la reptación, producida por movimientos ondulatorios laterales de su musculatura.
- Las **aves** presentan un esqueleto adaptado al vuelo, con extremidades anteriores modificadas en forma de alas. Es un esqueleto ligero debido a que posee huesos huecos, cráneo reducido y a la ausencia de dientes.
- En los **mamíferos,** el esqueleto está provisto de columna vertebral que permite un movimiento de flexión hacia delante y hacia atrás. Las cinturas y las articulaciones están bien desarrolladas.

El desplazamiento es ágil y rápido, para lo cual disponen de una potente musculatura y distintas adaptaciones a la locomoción. Así, los gamos caminan sobre pezuñas o los gatos se desplazan sobre las puntas de los dedos, para conseguir mayor velocidad.

El esqueleto de los mamíferos marinos está adaptado al medio acuático, especialmente el de los cetáceos, como la ballena, el delfín, la orca... que tienen una apariencia externa muy parecida a la de los peces, aunque el funcionamiento de su organismo es el propio de los mamíferos.

El desplazamiento en el medio aéreo es poco común en los mamíferos, a excepción de los murciélagos, cuyas extremidades anteriores presentan un alargamiento de los dedos entre los que se encuentra un recubrimiento membranoso que les permite el vuelo.

Los primates, grupo que incluye al ser humano, tienen como característica más relevante la capacidad de manipular objetos, gracias a que su pulgar puede oponerse al resto de los dedos de la mano.

En el ser humano, el cráneo ha experimentado un aumento de tamaño, sobre todo de la zona frontal, donde se localizan las áreas de asociación de la corteza cerebral. La locomoción se caracteriza por la posición bípeda, es decir, sobre dos piernas, hecho que ha provocado un desarrollo de la cintura pélvica, para poder soportar el peso de las vísceras del abdomen y del feto, en el caso de las mujeres embarazadas.

12 Lee el siguiente texto y responde a las preguntas que encontrarás a continuación:

En 1929, **Hans Berger,** un psiquiatra austríaco, descubrió que el cerebro produce campos eléctricos. Colocando electrodos sobre la cabeza de diversos sujetos, descubrió que estos campos eléctricos se pueden registrar. A este registro lo denominó **electroencefalograma** o **EEG.**

Cuando una neurona se activa, se produce un movimiento de cargas eléctricas y, como una neurona contacta con otras, este movimiento de cargas se va transmitiendo por todo el cráneo y cuero cabelludo, produciéndose campos eléctricos.

Berger observó que la actividad eléctrica generaba una diferentes tipos de ondas, unas más pequeñas y rápidas otras más grandes y lentas... El investigador las estudió y clasificó según su frecuencia, bautizándolas con letras del alfabeto griegos: alfa, beta, delta y theta.

Cuando las neuronas no están procesando información, por ejemplo en una situación de calma, donde la persona está sin pensar en nada, con los ojos cerrados, las neuronas oscilan de una manera sincronizada, produciendo ondas lentas y de gran tamaño, llamadas alfa, que tienen una frecuencia de 8 a 12 ondas por segundo. En cambio si la persona está realizando un esfuerzo mental, las neuronas de la corteza dejan de ir sincronizadas, cada neurona se activa independientemente de las demás, y esto produce una serie de ondas más rápidas y pequeñas. Se han dejado de generar las **ondas alfa** para dar paso a las **ondas beta.**

Cuando nos dormimos aparecen otro tipo de ondas que son las **theta** y las **delta,** siendo estas últimas las más grandes y lentas.

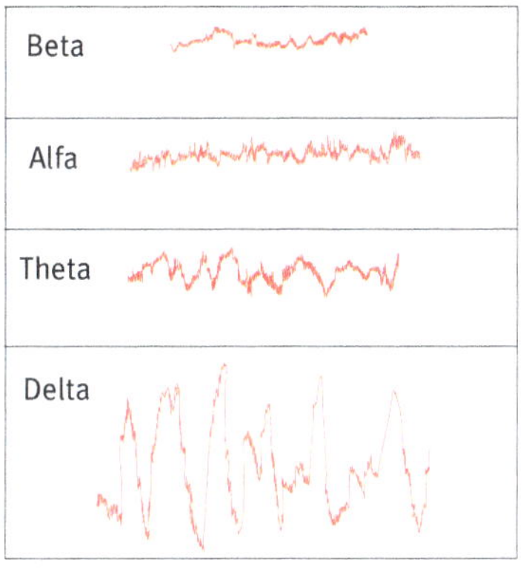

El sueño tiene diferentes fases, primero se entra en un sueño lento para continuar con otro tipo de sueño llamado fase REM. En una noche se suelen dar 5 alternanzas entre sueño lento y fase REM. Es en la fase REM donde soñamos, donde el sueño es profundo y donde generalmente cuesta más de despertar una persona y, si esta estaba soñando, suele recordar el sueño. Esta fase se caracteriza por tener un registro de ondas en el electroencefalograma muy similar a cuando se está despierto (ondas alfa y beta) ya que el cerebro se encentra en una intensa actividad, por ejemplo, los glóbulos oculares se mueven rápidamente debajo de los párpados (REM en inglés son las siglas de "rapid eyes movements"), el gasto de energía es muy elevado así como el gasto de oxígeno por parte del cerebro.

A continuación resuelve las cuestiones acerca del siguiente gráfico:

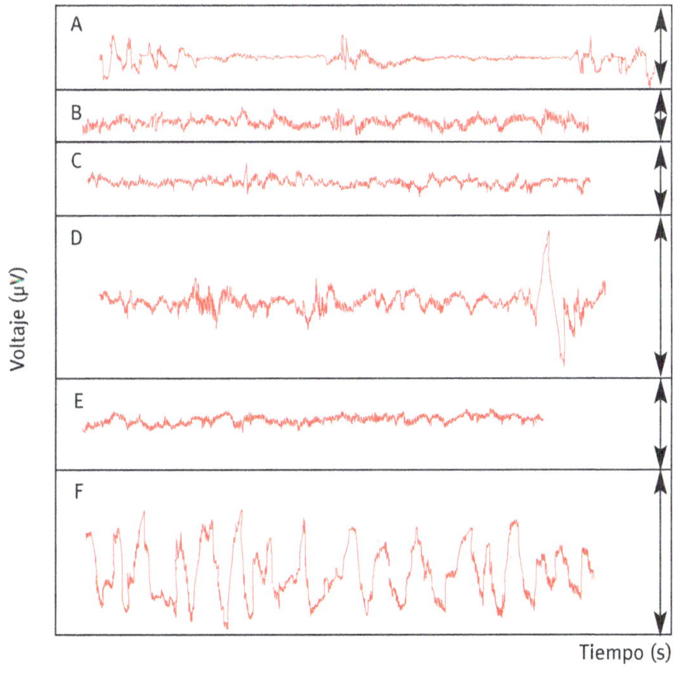

— Uno de estos EEG pertenece a un bebé que presenta convulsiones. ¿Puedes identificar cuál es?

— Ordena los 5 EEG restantes desde que una persona está activa hasta que está en la fase de sueño lento, no profundo (no está representada la fase REM).

— Fíjate que las ondas, contra más relajados estamos, más se elevan. ¿A qué crees que se debe este fenómeno? (Recuerda que la frecuencia son las ondas registradas por segundo.)

— ¿Por qué los electrodos para hacer un EEG se sitúan a nivel de la corteza cerebral?

— ¿Es verdad la afirmación que durante el sueño el cerebro descansa en el sentido que está inactivo?

13 Explica como se da la captación, transmisión , análisis y respuesta del siguiente estímulo: un perro huele su comida. Ayúdate del esquema de la página 259.

14 Asigna a cada definición el tipo de neurona que le corresponde: sensitiva, motora o de asociación:

— Reciben estímulos del medio interno y externo.

— Controlan las glándulas y las fibras musculares.

— Establecen conexiones entre neuronas, formando así circuitos.

15 ¿Qué es un nervio? Los nervios pueden ser de 3 tipos. Explica cuáles son.

16 La siguiente tabla muestra la composición iónica intracelular y extracelular de una neurona.

IÓN	CONCENTRACIÓN CITOPLASMÁTICA (mM)	CONCENTRACIÓN EXTRACELULAR (mM)
K^+	400	20
Na^+	50	440
Cl^-	52	500

— ¿Qué iones tendrían tendencia a entrar al citoplasma? ¿Y a salir?

17 Explica los siguientes conceptos:

— Diferencia de potencial.

— Polaridad de la membrana.

— Despolarización de la membrana del axón.

— Repolarización de la membrana axónica.

TIC **18** Observa el siguiente gráfico y anota el estado de los canales de membrana del axón en cada uno de los puntos señalados:

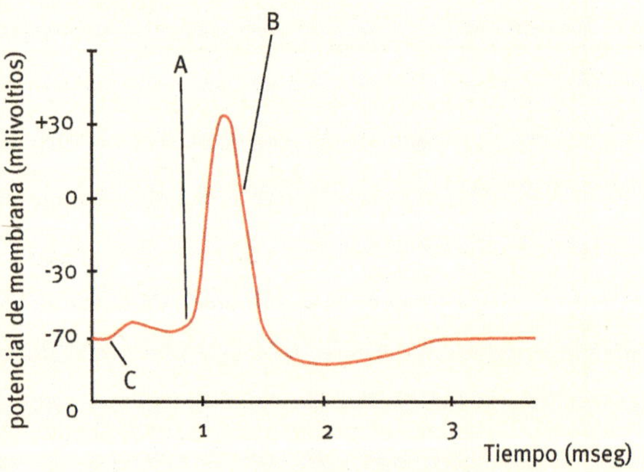

— ¿A partir de qué momento la célula se despolariza?

— ¿A partir de qué momento se cierran los canales de Na^+ y se abren los de K^+?

— ¿A partir de qué momento la célula recupera su excitabilidad, aunque necesitaría un estímulo mayor al normal? ¿Cómo se llama ese momento?

La información que necesitas la encontrarás en la siguiente dirección web:

http://www.blackwellpublishing.com/matthews/channel.html

19 Las siguientes afirmaciones están relacionadas con alguna fase del impulso nervioso. Determina si pertenece al potencial de reposo, al potencial de acción o a la repolarización. Ordena después las frases por orden de ejecución.

a. El interior de la membrana está cargado negativamente con respecto al exterior.

b. La bomba Na^+/K^+ bombea 3 iones Na^+ hacia el espacio extracellular por cada 2 iones K^+ bombeados hacia el citoplasma neuronal.

c. Los iones Na^+ se extraen rápidamente del axón gracias a la bomba Na^+/K^+.

d. Cuando se abren los canales de Na^+, los iones pasan precipitadamente al interior celular y la polaridad de la membrana se invierte. A continuación, se abren los canales de K^+ regulados por voltaje y se inactivan los de Na^+ regulados por voltaje.

e. El incremento resultante en la concentración interna de iones Na^+ despolariza la siguiente área contigua de la membrana, haciendo que sus canales iónicos de Na^+ se abran y permitiendo que el proceso se repita.

f. El impulso nervioso se mueve en una sola dirección porqué el segmento del axón anterior al lugar donde se produjo el potencial de acción tiene un período refractario breve durante el cual sus canales iónicos de Na^+ no se abren.

20 El dibujo muestra el perfil de un potencial de acción. Las 4 etapas principales están marcadas con las letras A, B, C y D:

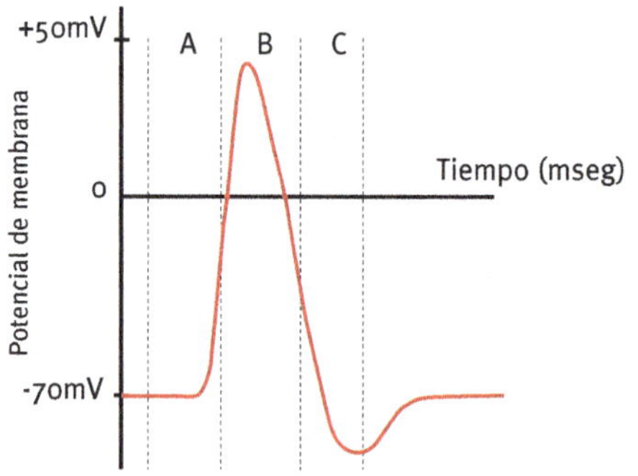

Elabora una tabla como la siguiente describiendo cada una de las etapas señaladas en el gráfico así como el estado de los canales Na^+ y K^+ regulados por voltaje.

	DESCRIPCIÓN DE LA ETAPA	CANAL DE NA^+ (abierto/cerrado)	CANAL DE K^+ (abierto/cerrado)
A			
B			
C			

21 ¿Qué pasaría si la célula del ejercicio anterior recibe otro estímulo de igual magnitud, pero esta vez, en el momento D?

22 ¿Cuál es la diferencia entre la propagación de un potencial de acción en un nervio mielinizado y en uno no mielinizado?

23 El siguiente dibujo representa una sinapsis. Asigna a cada número, la estructura correspondiente y explica brevemente los procesos que se llevan a cabo.

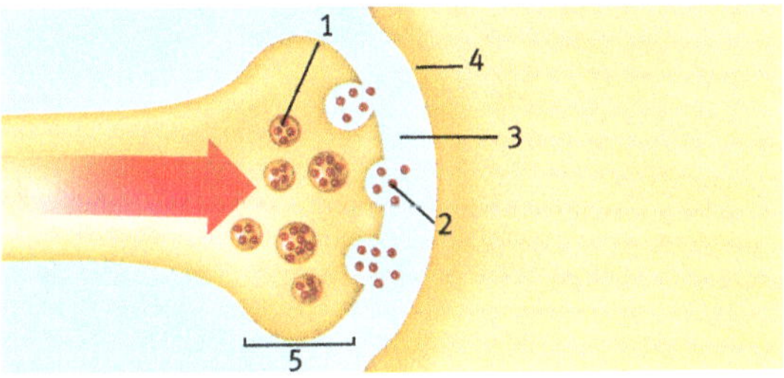

24 Di el nombre de las estructuras señaladas en cada una de las siguientes ilustraciones:

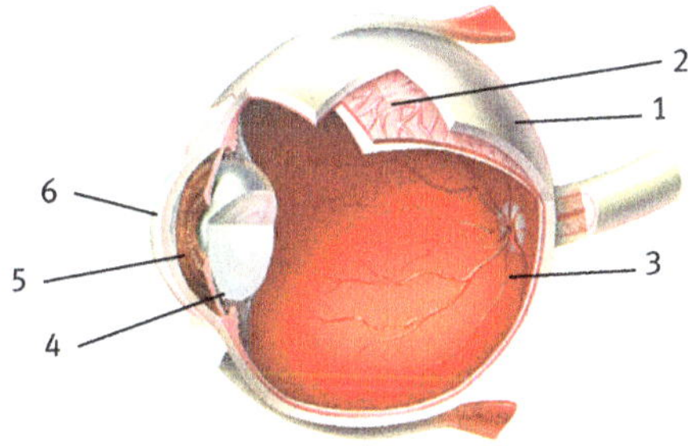

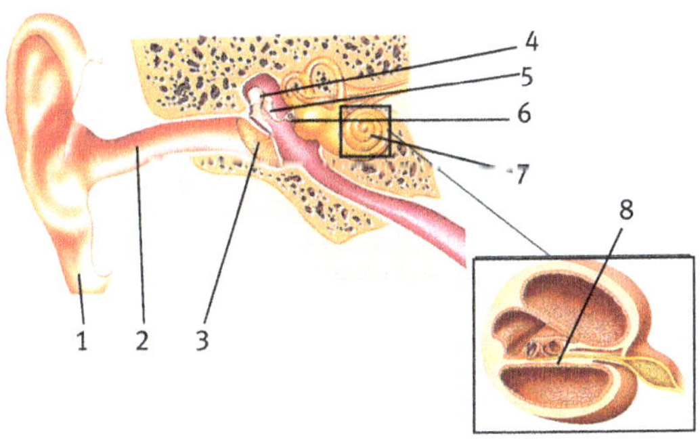

25 Empezando por la parte de fuera, con la punta del lápiz, sigue la espiral hasta llegar al centro.

Describe la ilustración antes y después de pasar el lápiz ¿A qué se debe la diferencia?

26 ¿Qué consecuencia negativa produciría una lesión en el bulbo raquídeo?¿Y en la hipófisis?

27 En una situación de estrés, ¿cuáles son los nervios autónomos que se activan? ¿Qué efecto se consigue al activar estos órganos? Incluye en tu explicación al menos dos órganos.

28 ¿Por qué se dice que las fibras musculares son células excitables? ¿Qué neurotransmisor provoca su contracción?¿Qué dos iones principales participan en la contracción?

29 Elabora un esquema sencillo del funcionamiento de un arco reflejo. En éste deben aparecer: terminales nerviosas sensitivas, terminales nerviosas motoras, neurona sensorial, neurona motora, médula espinal, placa motora.

30 Dibuja el esquema del sistema de palancas de los siguientes movimientos:

a. El brazo de un deportista que levanta una mancuerna con una mano.

b. La cabeza de una persona que alza la vista para ver el techo.

TIC

31 El Parkinson es una enfemedad neurodegenerativa que pese a ser descubierta en 1817 no se conoció su causa hasta 1960. Responde las siguientes preguntas sobre esta enfermedad:

a. ¿ Por qué síntomas se caracteriza la enfermedad del Parkinson?

b. ¿Esta enfermedad afecta más a una edad que a otra? ¿Cuáles son los factores de riesgo?

c. ¿Cuál es el mecanismo de actuación de los medicamentos contra el Parkinson?

La información que necesitas la encontrarás en la siguiente web:
http://www.elmundo.es/elmundosalud/especiales/2005/09/parkinson/que_es.html

12

La relación: la respuesta hormonal

En los seres vivos encontramos numerosos procesos que pese a estar controlados por el sistema nervioso, no dependen de una respuesta motora. Estos procesos son fruto del equilibrio en la actuación de unas sustancias, llamadas hormonas, que son liberadas al medio interno y que pueden llegar a actuar lejos del lugar de su liberación. Es la llamada respuesta hormonal.

Algunos ejemplos de este tipo de respuesta son: el movimiento de los girasoles siguiendo al Sol, el crecimiento vertical de las plantas, la caída de las hojas de los árboles caducifolios en otoño, el sudor cuando hace calor, el inicio de la secreción gástrica durante la alimentación, la maduración de los órganos sexuales, o muchos de los procesos asociados al crecimiento y al mantenimiento de la temperatura corporal en los animales homeotermos.

La disfunción en la secreción de una hormona puede conllevar graves perjuicios al organismo que la padece.

ORGANIZACIÓN DE LOS CONTENIDOS

1. Las hormonas

2. Las hormonas vegetales

2.1. Funciones y actividad de las fitohormonas

2.2. Movimientos de las plantas

3. Respuesta hormonal en los animales: el sistema endocrino

3.1. El sistema endocrino de los invertebrados

3.2. El sistema endocrino de los vertebrados

4. El sistema endocrino humano

4.1. Hipotálamo

4.2. Hipófisis

4.3. Glándulas endocrinas

4.4. Hormonas tisulares

4.5. Mecanismos de acción hormonal

5. El crecimiento en el ser humano

5.1. Anomalías en la secreción hormonal

OBJETIVOS

- Entender el concepto de hormona y su mecanismo de acción.
- Conocer las hormonas vegetales y su acción sobre las plantas.
- Comprender la respuesta hormonal en los animales invertebrados y vertebrados.
- Conocer las principales hormonas del sistema endocrino humano, así como las glándulas y los tejidos que las segregan.
- Entender que el conocimiento de los mecanismos metabólicos del organismo es la vía para vencer las enfermedades y mejorar la calidad de vida humana.

En la unidad anterior hemos descrito el modo en que los animales perciben los estímulos del ambiente; esta información se procesa y genera una respuesta motora. Esto es posible gracias a la transmisión eléctrica de los impulsos nerviosos a través de las neuronas. La respuesta motora la llevan a cabo los músculos y se traduce en un movimiento.

Existe además otro tipo de respuestas a los estímulos que se llevan a cabo mediante sustancias químicas: las **hormonas.** En esta unidad vamos a estudiar las principales hormonas, tanto vegetales como animales, y sus mecanismos de acción.

1. Las hormonas

Las **hormonas** son sustancias cuya función consiste en regular el desarrollo de reacciones químicas. Son las responsables de cambios fisiológicos y morfológicos, como el crecimiento, el desarrollo de órganos reproductores, la reproducción y el envejecimiento.

+

El término **hormona** viene del griego *hormon*, que significa 'estimular'. Sin embargo, las hormonas pueden estimular o inhibir los procesos, por lo que, en realidad, son reguladores químicos.

Las hormonas se producen en pequeñas cantidades y su efecto es más duradero que el impulso nervioso. Algunas se sintetizan en el mismo lugar donde van a actuar, mientras que otras son transportadas desde el lugar de síntesis al lugar de acción, a través de la savia en las plantas, y de la sangre en los animales, o bien por el aire si son compuestos volátiles.

El mecanismo de acción de las hormonas es más lento que el impulso nervioso pero su efecto es más duradero. Por ejemplo, la **insulina** regula la asimilación de glucosa en los vertebrados; se segrega cuando el nivel de glucosa en sangre lo precisa y se mantiene en la cantidad adecuada hasta que se recuperan los niveles normales.

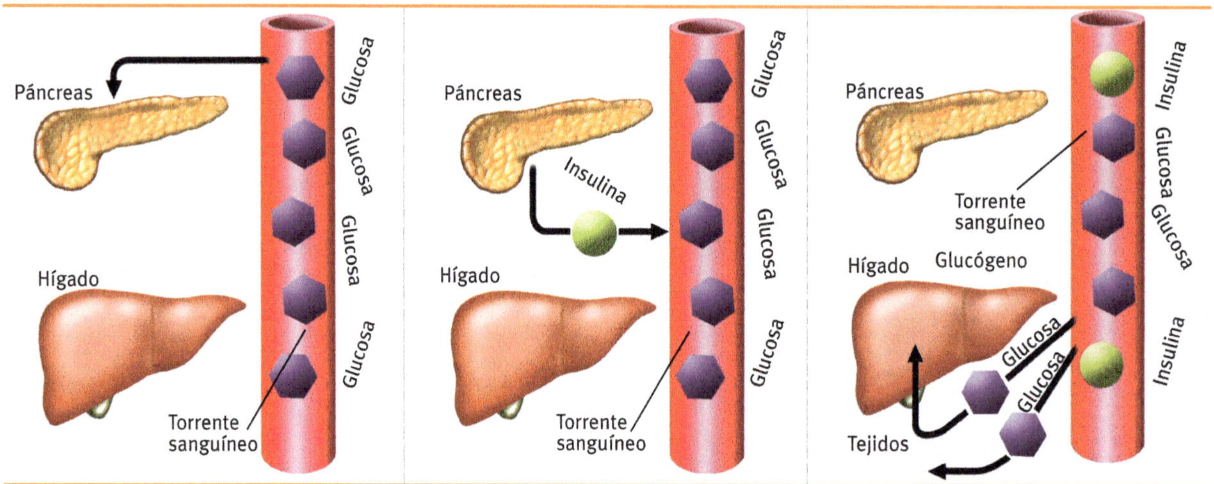

La asimilación de los alimentos produce un aumento de la concentración de glucosa en la sangre que es detectado por el páncreas.

El páncreas segrega la hormona insulina.

La insulina reduce el nivel de glucosa en la sangre, facilitando su almacenamiento en el hígado y su metabolización en los tejidos.

En caso contrario, cuando el nivel de glucosa en sangre es bajo, el páncreas segrega otra hormona, el **glucagón**, que promueve el paso a sangre de la glucosa almacenada en los tejidos.

2. Las hormonas vegetales

Las **hormonas vegetales** o **fitohormonas** son los compuestos orgánicos responsables del crecimiento y desarrollo de las plantas. Se sintetizan en una parte de la planta y actúan normalmente en otra parte donde, a muy bajas concentraciones, producen una respuesta fisiológica. No se sabe con exactitud a qué estímulos responden ni cómo se sintetizan; no obstante, los principales estímulos que afectan al desarrollo de los vegetales son el *grado de humedad*, la *temperatura*, la *gravedad*, las *horas de luz* y el *contacto con los objetos*.

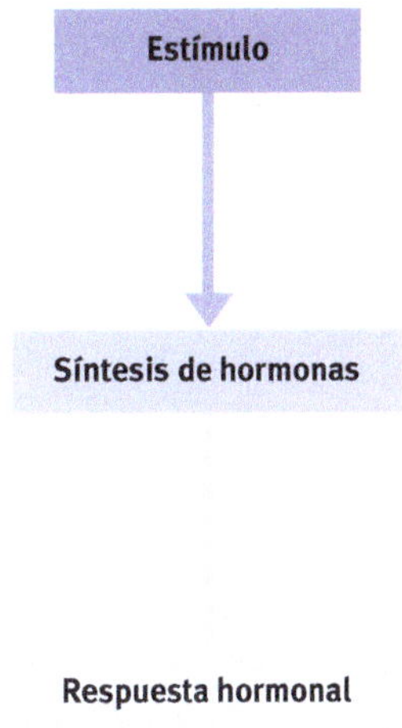

A diferencia de los animales, en las plantas no se han determinado receptores de estímulos concretos, ni órganos diferenciados de percepción ni órganos de procesado de la información. A pesar de ello se han reconocido algunos receptores especializados en captar determinados estímulos. Entre los receptores más conocidos destacan el **fitocromo**, los **estatocitos** y las **células sensibles al tacto**.

- El **fitocromo** es un pigmento sensible a la luz roja y a la infrarroja, y permite a las plantas detectar la duración del día y la noche. Induce la síntesis de hormonas que dan lugar a procesos como la floración. Según la época de floración distinguimos:
  - — Las **plantas de día largo** florecen cuando hay más horas de luz y las noches son más cortas. Son ejemplos la lechuga, la patata, las espinacas y muchas plantas ornamentales como el clavel.
  - — Las **plantas de día corto** florecen al principio de la primavera o el otoño, cuando las noches son largas. A principios de primavera florecen, por ejemplo, las fresas o el botón de oro; a principios de otoño, entre otros, los crisantemos o las ponsetias.
  - — Las **plantas de día neutro** florecen durante todo el año, como los geranios.
- Los **estatocitos** son células especializadas en detectar la gravedad. Contienen **amiloplastos,** orgánulos llenos de almidón que actúan de modo similar a los estatolitos de los artrópodos; gracias a ellos, las plantas detectan la posición del tallo y la raíz, y se desencadena una respuesta hormonal que dirige su crecimiento en el sentido correcto.
- **Células sensibles al tacto:** las células de las hojas de algunas especies captan estímulos táctiles que desencadenan respuestas hormonales o movimientos rápidos.

La venus atrapamoscas *(Dionaea muscipula)* es una planta carnívora que, al estímulo del contacto de la presa, responde cerrándose.

Además de su papel regulador, los especialistas suponen que algunas hormonas vegetales podrían actuar también como receptores, ya que parece que responden directamente a determinados estímulos.

2.1. *Funciones y actividad de las fitohormonas*

Son compuestos que se sintetizan en pequenas cantidades como resultado de un estímulo, y regulan los procesos de crecimiento y desarrollo. Circulan por los vasos conductores del floema y el xilema, o de célula a célula por difusión o transporte activo. Algunas de las hormonas vegetales más conocidas son: *auxinas, citoquininas, etileno, ácido abscísico* y *giberelinas.*

■ Principales zonas donde se sintetizan las hormonas vegetales

HORMONAS	LUGAR DE SÍNTESIS	FUNCIONES PRINCIPALES	FACTORES QUE INFLUYEN EN SU ACTIVIDAD
Auxinas	Los brotes apicales	Regulan el crecimiento del tallo y la raíz.	La luz activa el mecanismo de acción.
Citoquininas	Las raíces	Favorecen el crecimiento de las ramas laterales. Evitan el envejecimiento de las hojas porque retrasan la inactivación del ADN y estimulan la síntesis de clorofila.	Son activadas por la luz roja y la temperatura alta.
Etileno	Los frutos	Es un gas que se libera al exterior de la planta y actúa sobre los frutos cercanos, favoreciendo su maduración.	Es activado por las auxinas, el déficit hídrico y las heridas; y lo inhibe el ácido abscísico.
Ácido abscísico	En todo tipo de tejidos: hojas, tallos, frutos, semillas y raíces	Controla la apertura y el cierre de los estomas. Inhibe el crecimiento de la planta y el desarrollo de las semillas y los frutos.	Aumentan la concentración la sequía, el encharcamiento de las raíces, el frío y la luz roja.
Giberelinas	Tallos y hojas jóvenes	Germinación de las semillas	Activan su acción los fotoperiodos largos

C

La **turgencia celular** consiste en un aumento del volumen de la célula vegetal, de modo que el citoplasma ejerce una presión sobre la pared celular.

Cuando el medio que envuelve a las células tiene menor concentración de iones que el interior celular, el agua penetra en la vacuola, por el proceso de ósmosis. La célula vegetal se hincha y el citoplasma queda presionado sobre la pared celular.

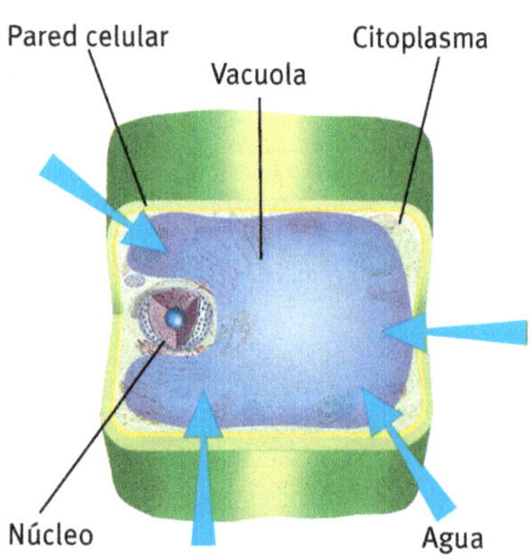

La actividad de las fitohormonas está todavía en estudio. Uno de los procesos mejor conocidos es el **mecanismo de acción de las auxinas,** que describimos a continuación:

— Las auxinas se producen en el extremo del tallo y provocan su alargamiento, respondiendo a los estímulos luminosos. El efecto de las auxinas permite la curvatura del tallo en dirección a la luz, sea cual sea la localización del foco luminoso (vertical o lateral).

Este alargamiento de las células recibe el nombre de **crecimiento ácido,** y es posible gracias a un rápido aumento de la turgencia celular, que permite a la planta cambiar la dirección de su crecimiento en unas pocas horas.

— Desde el tallo, las auxinas migran hasta la raíz, favoreciendo su crecimiento si se encuentran en bajas concentraciones, o inhibiéndolo si se encuentran a altas concentraciones. Las auxinas también son las responsables del crecimiento de raíces adventicias en un tallo cortado.

A continuación vamos a seguir los pasos del mecanismo de acción de las auxinas durante el crecimiento ácido del ápice del tallo, observando aquellos fenómenos que le ocurren a una planta que crece en sentido vertical después de cambiar su ubicación haciendo que la luz pase a incidirle lateralmente.

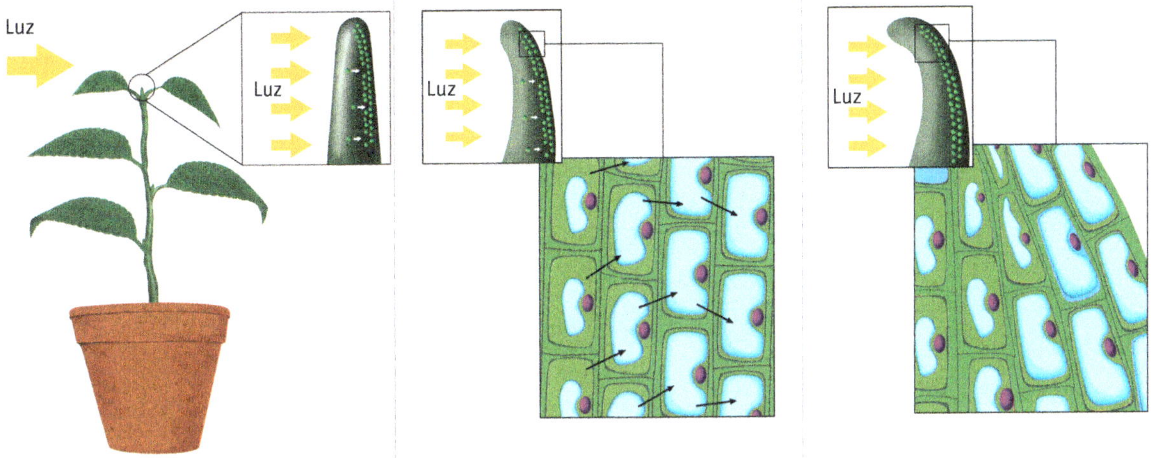

Las auxinas presentes en el tallo, por la presencia de la luz, se desplazan en sentido opuesto al estímulo luminoso. En estas células se activan las proteínas transportadoras de H^+, que transportarán los H^+ del citoplasma a la pared celular.

La acidez de la pared celular activa una enzima que rompe los enlaces entre las cadenas de celulosa de la pared, dándole mayor plasticidad. El agua penetra en la vacuola, aumentando la turgencia celular de forma que la célula se alarga.

Las células de la zona afectada por las auxinas se alargan, mientras que el resto de las células no se modifica y el tallo crece curvado hacia la luz.

2.2. *Movimientos de las plantas*

Las plantas presentan movimientos que responden a estímulos del medio, y que son posibles gracias a la actividad de las hormonas. Así, disponen de mecanismos para mover sus órganos y orientarlos en la posición más favorable.

Los movimientos se pueden basar en dos tipos de mecanismos: el crecimiento en extensión y alargamiento de una parte de la planta, como ocurre en los *tropismos* y las *nutaciones*; o bien, cambios reversibles de la turgencia de un grupo de células que producen el movimiento de un órgano de la planta, como es el caso de las *nastias*.

Los tropismos

Son movimientos de crecimiento dirigidos por estímulos externos. Pueden producirse en dirección al estímulo, y se denominan tropismos positivos, o bien ser opuestos a éste; en este segundo caso se habla de tropismos negativos. Las auxinas son las responsables de estos fenómenos, mediante el proceso de alargamiento celular que hemos descrito. Los más conocidos son el *fototropismo* y el *geotropismo*.

- **Fototropismo.** El crecimiento se produce como respuesta a un estímulo luminoso. El tallo presenta fototropismo positivo, las raíces tienen fototropismo negativo.
- **Geotropismo.** Es el crecimiento en función de la fuerza de la gravedad. El tallo tiene geotropismo negativo y la raíz, geotropismo positivo. Los estatocitos son las células receptoras. En ellas se encuentran los amiloplastos, que contienen almidón y detectan la posición de la planta. De este modo activan a las auxinas, cuya acción provoca el crecimiento de tallos y raíces en la dirección correcta.

+

El fototropismo positivo de los tallos fue descrito por primera vez por Charles Darwin en 1881.

Darwin observó la curvatura del ápice de plantas en sus primeras fases de desarrollo, en dirección a la luz.

Si cubría la punta de la plantita con papel de estaño y la sometía a luz lateral, no se producía la curvatura hacia la luz.

De ello concluyó que en el ápice de la planta había alguna sustancia (que más tarde se identificó como auxina) que se transmitía hacia la parte inferior y producía su curvatura.

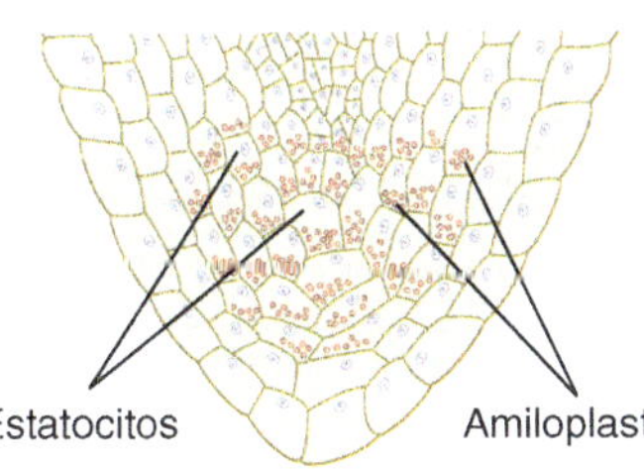

Situación de los estatocitos en el extremo de la raíz

Como describimos en las páginas 286 y 287, la acumulación de las auxinas en unas capas celulares u otras, en función de la dirección del estímulo, provoca el crecimiento ácido de las células que causa la curvatura del tallo o la raíz.

Veamos los procesos de fototropismo positivo del tallo y geotropismo positivo de la raíz, cuando una planta se coloca en posición invertida.

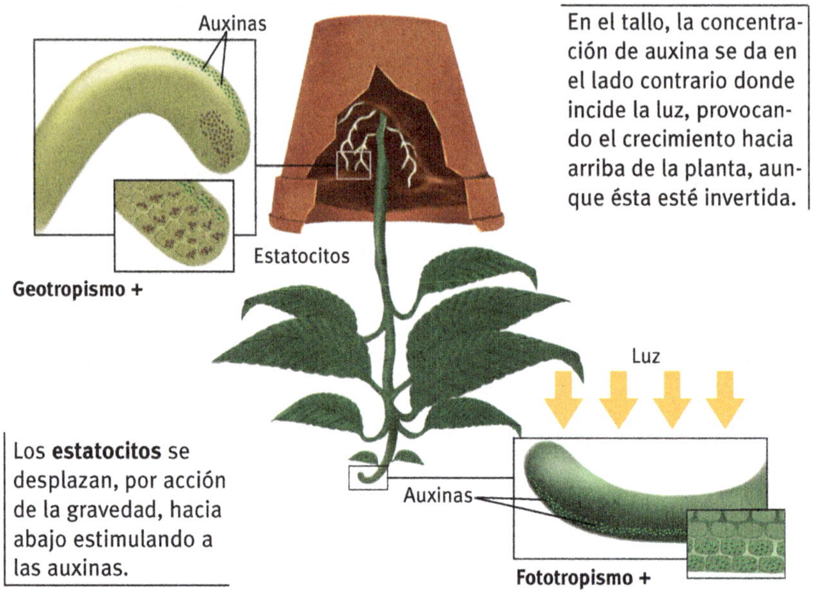

Las nutaciones

Consisten en un crecimiento desigual del tallo o de las hojas, que produce una forma helicoidal de estos órganos. Este fenómeno se da en las plantas trepadoras. Existen dos tipos básicos de nutaciones: los **tallos volubles** y los **zarcillos**.

Tallo voluble

Los **tallos volubles** crecen en espiral, y cuando encuentran un soporte se enrollan sobre él.

Los poseen plantas como el jazmín y el lúpulo.

Zarcillos del guisante

Los **zarcillos** son hojas modificadas que, al entrar en contacto con un objeto, se enroscan a su alrededor. La respuesta de los zarcillos es más rápida que la de los tallos volubles, ya que en unos minutos pueden enroscarse sobre un soporte.

Poseen zarcillos plantas como la calabaza o el guisante.

Los tallos y las hojas poseen en su epidermis células que actúan como receptores, percibiendo el contacto con el soporte sobre el que se produce el crecimiento.

Las nastias

Son movimientos rápidos de una parte de la planta que, a diferencia de los casos anteriores, puede volver a la posición inicial cuando el estímulo que ha originado el movimiento cesa. Las más conocidas son las *fotonastias* y las *seismonastias.*

- **Fotonastias.** Están provocadas por las variaciones de la luz entre el día y la noche; es el caso de los **movimientos nictinásticos**, que presentan algunas especies de hojas compuestas, como la acederilla y la mimosa. Los folíolos que constituyen las hojas se abren por la mañana y se cierran al anochecer. Estos movimientos están inducidos por el fitocromo, que actúa como receptor de los estímulos luminosos.
- **Seismonastias.** El movimiento brusco o el simple contacto con un objeto produce el cierre de las hojas. La mimosa posee células sensibles al tacto que captan los movimientos bruscos. Así, cuando se agita una rama, ésta cierra sus hojas, desarrollando, de este modo, un sistema de defensa ante los depredadores.

Las nastias son posibles gracias a unos grupos de células de la base del pecíolo denominados **pulvínulos,** que actúan a modo de bisagra, y mediante un aumento o disminución de su turgencia provocan que unos folíolos se doblen sobre otros y luego se vuelvan a separar.

El mecanismo de acción, tanto en las fotonastias como en las seismonastias, se basa en un cambio de permeabilidad para los iones K^+ de las membranas plasmáticas de las células que intervienen en estos movimientos. Al entrar K^+ en una célula también entra agua y se produce la turgencia celular.

Las hojas de la acederilla presentan movimientos nictinásticos.

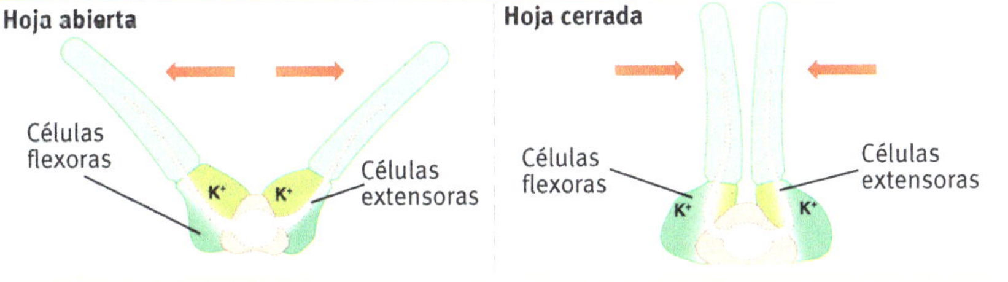

Las células extensoras experimentan una turgencia, debido a la entrada de K^+ a través de la membrana. Las células flexoras pierden la turgencia debido a la salida de K^+.

El mecanismo es inverso: hay una entrada de K^+ en las células flexoras, por lo que aumenta la turgencia; en las células extensoras disminuye, lo que produce el cierre.

ACTIVIDADES

1. Explica qué diferencias encontramos entre las hormonas vegetales y las animales.
2. Los centros de integración y coordinación de la información están constituidos por el sistema nervioso y el endocrino. ¿Qué diferencias hay entre ellos?
3. Explica qué hormona vegetal está implicada en los siguientes procesos:

a) Aceleración del crecimiento de las raíces en un esqueje que acabamos de plantar.

b) Activación de la germinación de las semillas de avena.

c) Aceleración de la maduración de un plátano situado junto a otros plátanos maduros.

d) Inhibición de la germinación de las semillas.

Claude Bernard (Saint Julien 1813-París 1893) está considerado uno de los padres de la fisiología. Entre sus descubrimientos destacan el mecanismo de actuación de medicamentos y tóxicos, la descripción de los nervios vasomotores y la función del páncreas, así como que la glucosa se almacena en el hígado en forma de glucógeno.

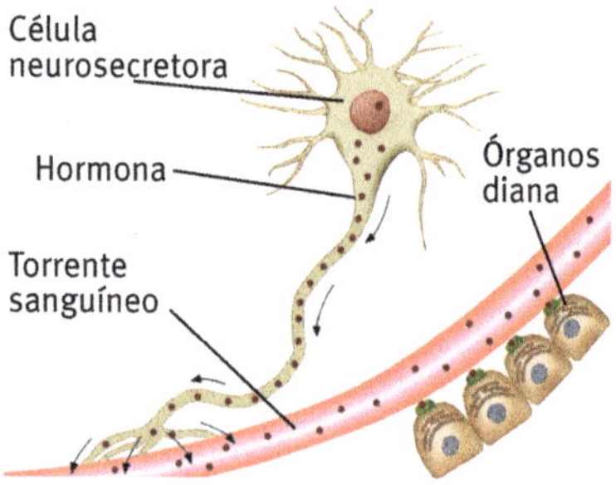

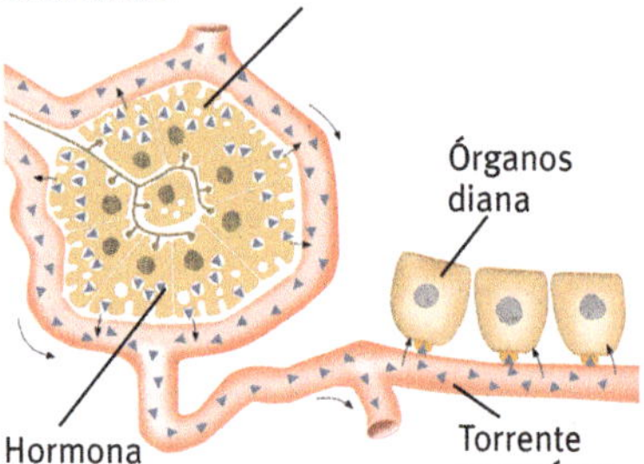

Tejidos secretores

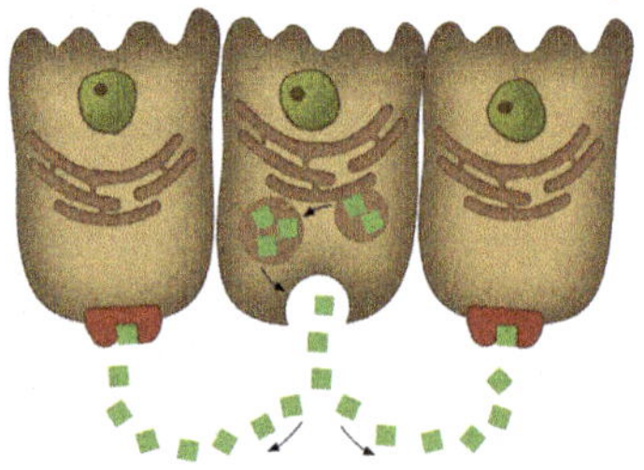

3. Respuesta hormonal en los animales: el sistema endocrino

La función de relación en los animales está controlada por el sistema nervioso, que recibe los estímulos tanto externos como internos, los procesa de forma más o menos compleja, según el grupo taxonómico, y emite una respuesta. Esta respuesta puede ser motora o bien hormonal.

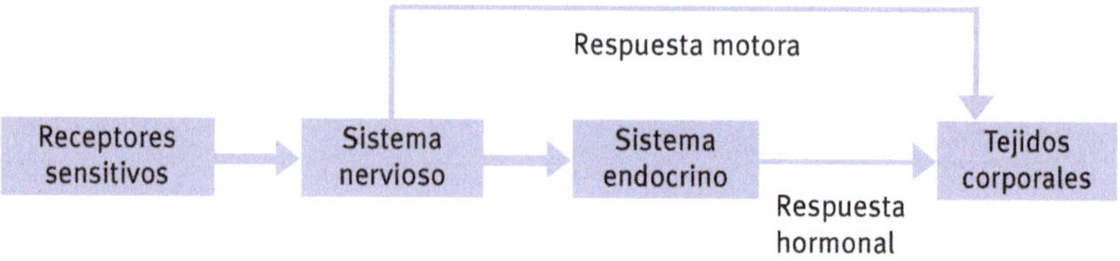

Como indicábamos al inicio de la unidad, las hormonas son sustancias sintetizadas en determinados tejidos, que pueden actuar localmente o ser transportadas por la sangre hasta un órgano en el que ejercen su acción. Ésta puede ser activadora o inhibidora. Su actividad es absolutamente necesaria para el buen funcionamiento de los órganos y el desarrollo de las distintas etapas del ciclo vital de los animales.

El conjunto de tejidos y glándulas secretoras y las hormonas que sintetizan recibe el nombre de **sistema endocrino** u **hormonal.** Los tejidos u órganos sobre los que las hormonas ejercen su acción son los **efectores de la respuesta hormonal.** Las hormonas pueden sintetizarse en *células nerviosas neurosecretoras, glándulas* o *tejidos secretores.*

- Las **células nerviosas neurosecretoras** son neuronas del sistema nervioso central que sintetizan unas sustancias denominadas **neurohormonas** y las liberan a los capilares sanguíneos. Éste es el caso de la *oxitocina,* que se forma en el hipotálamo e interviene en el proceso del parto y en la producción de leche durante la lactancia.
- Las **glándulas de secreción interna** o **endocrinas** se encuentran repartidas en distintas partes del cuerpo. Se llaman así porque no poseen conductos y vierten sus productos a la sangre. El sistema circulatorio transporta las hormonas hasta zonas alejadas del lugar de síntesis, donde ejercen su acción; son los **órganos diana.** Por ejemplo, las *glándulas suprarrenales*, situadas sobre los riñones, segregan *adrenalina,* sustancia que produce un aumento del ritmo cardíaco y dilatación de las pupilas.
- Las células de determinados **tejidos** pueden sintetizar **hormonas** que ejercen su acción en el mismo lugar en que se produce la síntesis. Un ejemplo de ésta, en los vertebrados, es la hormona *gastrina,* segregada por células de la mucosa gástrica del estómago, que estimula la secreción del jugo gástrico.

A continuación, vamos a estudiar el funcionamiento del sistema endocrino de los animales invertebrados y los vertebrados.

3.1. *El sistema endocrino de los invertebrados*

A partir de los cnidarios, todos los grupos de invertebrados poseen un sistema endocrino, ya bien desarrollado en los artrópodos, que poseen células nerviosas neurosecretoras y glándulas endocrinas, cuyas secreciones actúan sobre distintos órganos, regulando el desarrollo.

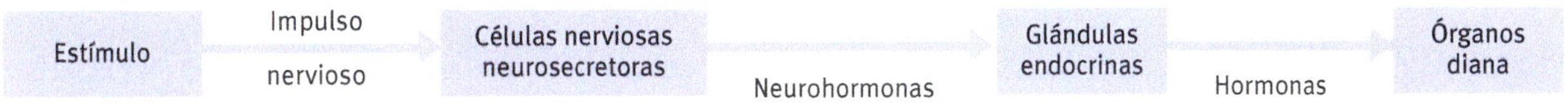

Los artrópodos han sido muy estudiados, debido al interés que despiertan las fases de su desarrollo y el comportamiento en los insectos sociales. Las células nerviosas neurosecretoras se encuentran en los ganglios cerebrales o en puntos concretos de los ganglios ventrales, y regulan procesos fisiológicos, tales como:

- El crecimiento y la **muda** del exoesqueleto. La muda hace posible el crecimiento del animal porque provoca la renovación de las piezas del exoesqueleto, de forma que se desprenden las viejas y crecen otras nuevas, a medida que el tamaño del animal aumenta.
- La **metamorfosis.** Es un cambio que experimentan algunos animales durante su desarrollo, hasta alcanzar la edad adulta. Se manifiesta en la variación de la forma y las funciones de su organismo.
- La maduración del aparato reproductor y los cambios morfológicos sexuales.

En los **insectos,** las células neurosecretoras regulan el funcionamiento de diversas glándulas endocrinas, como:

- Los **cuerpos alados,** que sintetizan la **hormona juvenil,** responsable de mantener el aspecto de larva mientras dura esta fase del desarrollo.
- Las **glándulas protorácicas,** que sintetizan la hormona **ecdisona,** cuya función es favorecer el paso de larva a pupa cuando disminuye la concentración de la hormona juvenil y de pupa a adulto cuando desaparece la hormona juvenil.

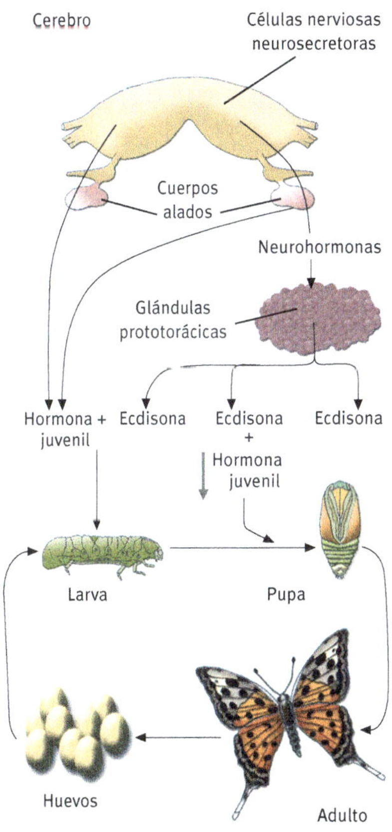

La acción equilibrada de la **hormona juvenil** y la **ecdisona** regula el proceso complejo de metamorfosis en los insectos.

Es también muy importante la existencia de **feromonas,** moléculas orgánicas olorosas que se liberan al medio en forma de partículas volátiles. Las feromonas constituyen un método químico de comunicación entre individuos de la misma especie. El animal receptor de la feromona modifica su conducta en respuesta a este estímulo. Algunos insectos, como las mariposas, utilizan feromonas como atracción sexual. En insectos sociales, como las abejas, sirven para marcar rutas que deben seguir otros individuos para llegar a un campo de flores, mantener el grupo unido o advertir de un peligro.

+

El conocimiento de las feromonas tiene aplicaciones en el **control de las plagas** de insectos perjudiciales para la agricultura.

Por ejemplo, la procesionaria del pino se combate aplicando un tratamiento a las hembras para que no produzcan las feromonas que atraen a los machos. De este modo, se evita la reproducción y se puede controlar la plaga.

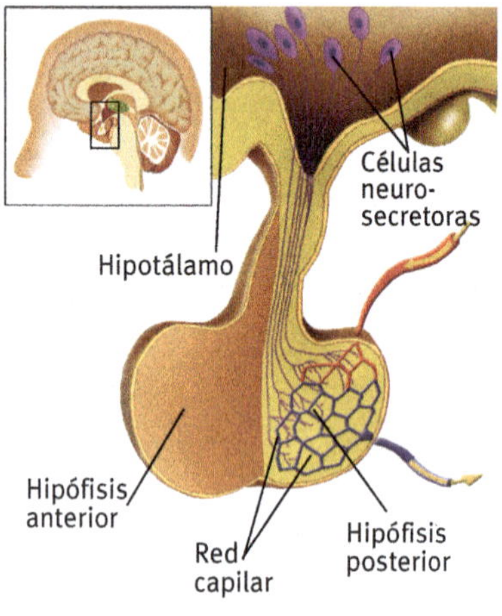

Glándulas y tejidos secretores

3.2. *El sistema endocrino de los vertebrados*

En los vertebrados el sistema endocrino está bien desarrollado. El mecanismo de producción hormonal está jerarquizado y controlado por el sistema nervioso central. Las partes del encéfalo implicadas son el hipotálamo y la hipófisis.

- El **hipotálamo,** como ya hemos visto en la unidad anterior, es una agrupación de neuronas situada en la parte inferior del cerebro. Recibe informaciones del medio externo e interno, las procesa y emite respuestas en forma de neurohormonas, que se dirigen a la hipófisis.
- La **hipófisis** es una pequeña estructura glandular situada debajo del hipotálamo y dividida en dos lóbulos: hipófisis anterior e hipófisis posterior. Las secreciones enviadas por el hipotálamo activan la secreción de hormonas en la hipófisis, hormonas que actúan sobre otras glándulas y tejidos secretores.

De este modo, el sistema nervioso central regula el funcionamiento de todo el sistema endocrino.

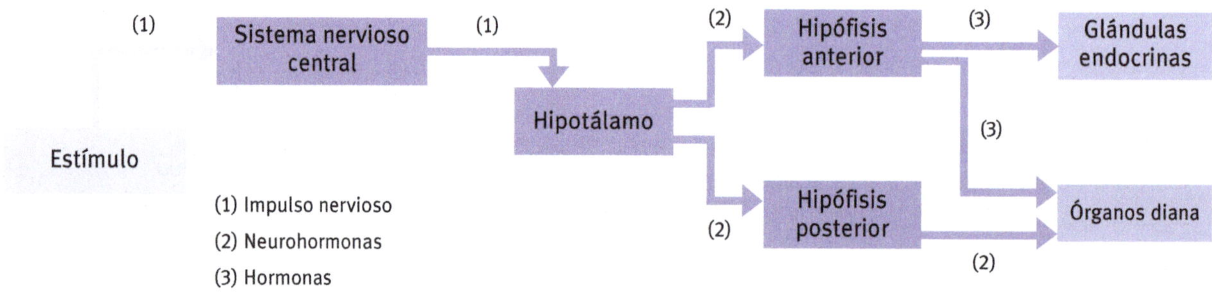

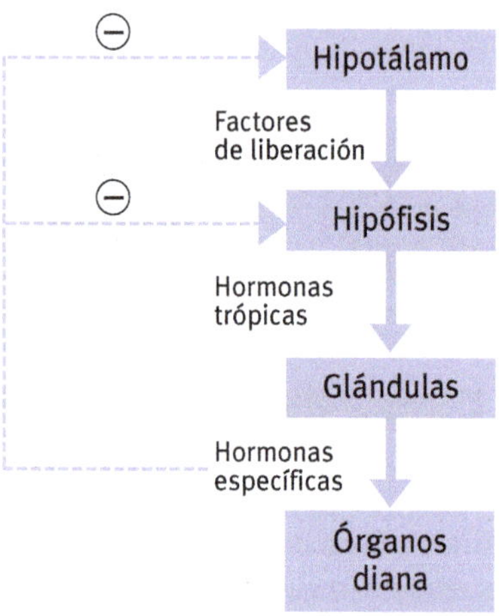

La retroalimentación permite mantener los niveles de hormonas adecuados para todas las funciones que dependen del control endocrino

La acción de las hormonas se regula a sí misma por un mecanismo denominado **retroalimentación** o ***feed-back.*** Está controlado por el hipotálamo y permite mantener unos niveles hormonales adecuados en la sangre mientras dura el estímulo que ha provocado la secreción. El proceso es el siguiente:

— Cuando el estímulo es percibido por el sistema nervioso central, el **hipotálamo** libera neurohormonas (**factores liberadores**) que se dirigen a la hipófisis.

— En la **hipófisis** se activa la liberación de las **hormonas trópicas,** que por la sangre van a las glándulas.

— En las **glándulas** se activa la síntesis de **hormonas específicas** que, también vía sangre, llegan y actúan sobre los órganos diana.

— La presencia de las hormonas específicas en sangre hace que se inhiba la liberación de hormonas trópicas por parte de la hipófisis y de factores de liberación por parte del hipotálamo.

El mal funcionamiento del sistema endocrino puede dar lugar a un déficit o un exceso de una hormona. Esto provoca anomalías importantes en el organismo.

En el cuadro siguiente constan las principales glándulas endocrinas y los tejidos secretores de hormonas en peces, anfibios, reptiles y aves. Se puede observar un progresivo desarrollo del sistema endocrino, relacionado con la creciente complejidad de los procesos vitales de estos grupos.

GRUPO TAXONÓMICO	SISTEMA ENDOCRINO
Peces	El hipotálamo controla la acción de la hipófisis. Las neurohormonas segregadas por la hipófisis estimulan el desarrollo y funcionamiento de las gónadas: ovarios y testículos. Estas glándulas son responsables de: • La formación de gametos: óvulos y espermatozoides. • Los caracteres sexuales secundarios, que diferencian a los machos de las hembras.
Anfibios	Además del hipotálamo, la hipófisis y las glándulas sexuales, presentan: • La glándula tiroides, que segrega la hormona tiroidea. Ésta regula el proceso de metamorfosis de la larva al individuo adulto. • Las glándulas paratiroides, cuya hormona regula la concentración de calcio en la sangre. • Glándulas suprarrenales, que producen adrenalina.
Reptiles	Presentan las mismas glándulas que los anfibios, y también: • Células pancreáticas productoras de insulina, que regula la entrada de glucosa en el citoplasma celular.
Aves	La glándula pineal o epífisis detecta las horas diarias de luz solar **(fotoperíodo)** y envía la información al hipotálamo, el cual regula procesos como: • La conducta reproductora ante el sexo opuesto. • La muda, cambio parcial o completo del plumaje, que suele producirse después de la nidificación y antes de la migración. • La migración, desplazamiento periódico de una población de aves en función de los cambios estacionales.

Los alevines de doncellas *(Coris julis)* son hembras y se distinguen por sus colores discretos; al crecer cambian de sexo y su coloración pasa a ser más vistosa.

ACTIVIDADES

4. Un ejemplo de neurohormona es la oxitocina. ¿Dónde se sintetiza? ¿Dónde actúa? ¿Qué funciones desempeña? Organiza la información en una tabla.

5. Un ejemplo de hormona secretada por una glándula endocrina es la adrenalina. ¿Qué glándula la secreta? ¿Sobre qué órganos actúa? ¿Qué efectos produce? Organiza la información en una tabla.

6. Respecto al sistema endocrino de los invertebrados, contesta:
   a) ¿Qué son los cuerpos alados?
   b) ¿Qué es y qué función desempeña la ecdisona? ¿Dónde se produce?
   c) ¿Qué es la metamorfosis? ¿Cómo se dirige el proceso de la metamorfosis?
   d) ¿Cómo se puede controlar una plaga de procesionaria?

7. Contesta las siguientes preguntas:
   a) ¿Cómo llega un estímulo al hipotálamo?
   b) ¿De qué manera responde el hipotálamo a un impulso nervioso?
   c) ¿De qué manera responde la hipófisis al estímulo del hipotálamo?

8. Explica en qué consiste el mecanismo de retroalimentación.

4. El sistema endocrino humano

La producción de hormonas en el ser humano es similar al resto de los mamíferos; diversas glándulas y tejidos secretores intervienen en la regulación de los procesos fisiológicos del organismo.

En la siguiente ilustración podemos apreciar la localización de los principales componentes del sistema endocrino humano.

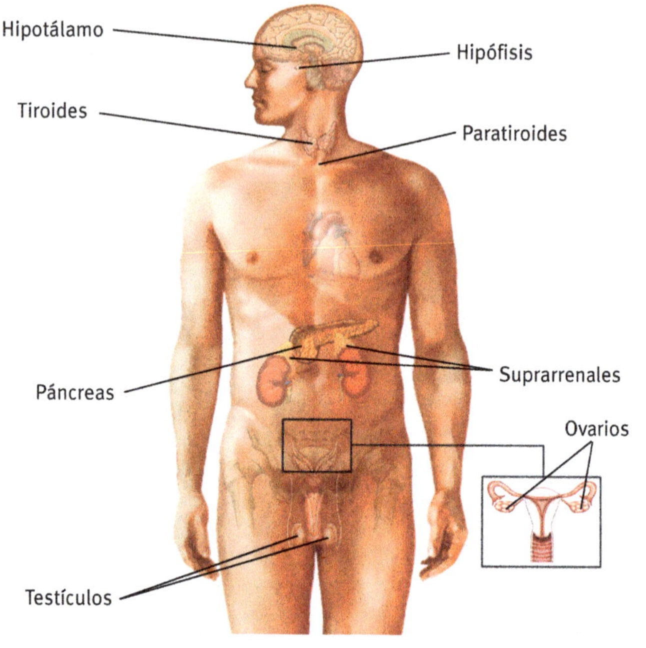

A continuación describimos los órganos secretores y la función de las distintas hormonas que intervienen en el metabolismo humano.

4.1. *Hipotálamo*

Está formado por tejido nervioso, cuyas neuronas reciben información procedente de la corteza cerebral acerca de los estímulos externos o internos.

El hipotálamo sintetiza neurohormonas que envía a la glándula hipófisis y pueden ser de dos tipos:

- Neurohormonas de función estimuladora o inhibidora sobre la hipófisis anterior.
- Otras neurohormonas pasan a la hipófisis posterior, donde son almacenadas, y desde allí se vierten a la sangre para alcanzar los órganos diana. Son la *oxitocina* y la *hormona antidiurética* (ADH).
  - — La **oxitocina** estimula las contracciones de la musculatura del útero durante el parto y después de la expulsión del feto, para favorecer la recuperación de su volumen normal. También provoca la salida de la leche de las células glandulares de la mama.
  - — La **hormona antidiurética (ADH)** estimula la reabsorción de agua en las nefronas de los riñones. Así se mantienen constantes los valores de la presión sanguínea.

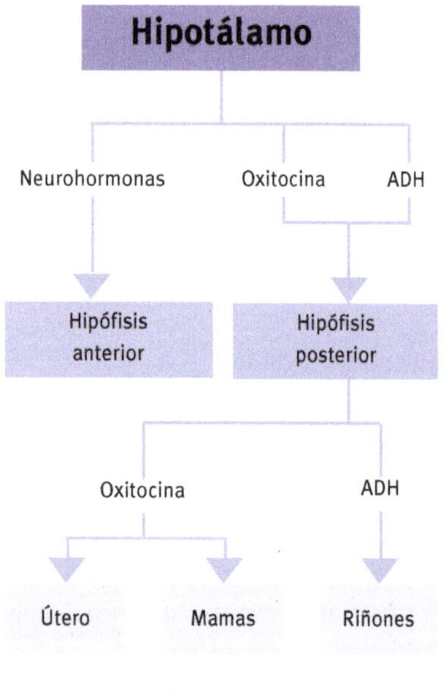

4.2. *Hipófisis*

En la hipófisis distinguimos:

- **Hipófisis anterior** o **adenohipófisis.** Tiene función glandular y segrega hormonas que pasan a la sangre.
- **Hipófisis posterior** o **neurohipófisis.** Almacena las neurosecreciones del hipotálamo y las libera al torrente sanguíneo.

La hipófisis anterior segrega seis hormonas diferentes, que se pueden clasificar en dos grupos:

- **Hormonas que actúan directamente sobre órganos diana.** Son la *hormona del crecimiento* o *somatotropina* y la *prolactina.*
  - La **hormona del crecimiento** o **somatotropina (GH)** actúa sobre los tejidos aumentando la síntesis de proteínas y estimula el crecimiento de los huesos.
  - La **prolactina (PRL)** estimula la síntesis de leche en las células glandulares de la mama después del parto. Es activada por la succión del recién nacido.
- **Hormonas que estimulan glándulas endocrinas.** Pertenecen a este grupo la *hormona estimulante del tiroides (TSH)*, la *hormona adrenocorticotrópica (ACTH),* la *hormona foliculoestimulante (FSH)* y la *hormona luteinizante (LH).*
  - La **hormona estimulante del tiroides (TSH)** estimula las células de la glándula tiroides para que aumente su producción de tiroxina.
  - La **hormona adrenocorticotrópica (ACTH)** estimula la producción de cortisol en la corteza de las glándulas suprarrenales.
  - La **hormona foliculoestimulante (FSH)** actúa sobre los testículos, activando la producción de espermatozoides, y estimula la maduración de los ovocitos (células que darán lugar a los óvulos).
  - La **hormona luteinizante (LH)** estimula la producción de testosterona por parte de las células intersticiales de los testículos y activa la maduración de los ovocitos a óvulos y su expulsión hacia las trompas de Falopio.

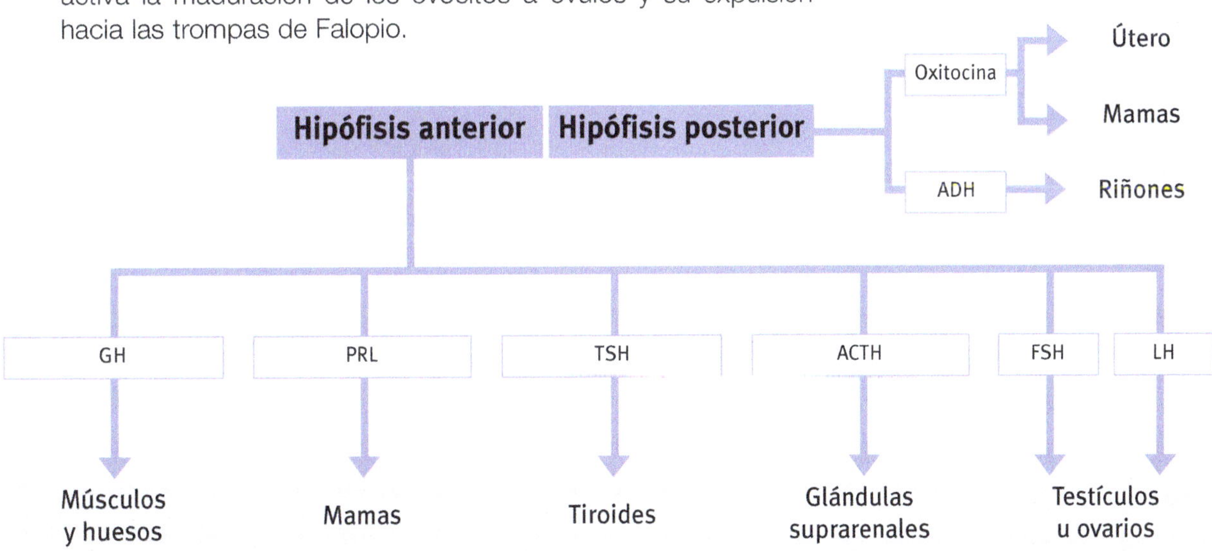

4.3. *Glándulas endocrinas*

Los centros de control hormonal, el hipotálamo y la hipófisis, regulan la síntesis y la secreción de hormonas en distintas partes del organismo.

A continuación, resumimos en una tabla las principales glándulas endocrinas, las hormonas que sintetizan y el mecanismo que activa su secreción.

GLÁNDULA		HORMONA	ACCIÓN	MECANISMO ACTIVADOR
Tiroides Situada en la base del cuello, junto a la tráquea.		**Tiroxina**	Acelera el metabolismo celular y, por tanto, el crecimiento del individuo.	Hormona estimulante del tiroides (TSH); segregada por la hipófisis anterior.
		Calcitonina	Inhibe la liberación de calcio de los huesos.	El aumento de iones Ca^{2+} en la sangre.
Paratiroides Situadas en la parte posterior del tiroides.		**Hormona paratiroidea** o **parathormona**	Estimula la liberación de calcio de los huesos.	La disminución de iones Ca^{2+} en la sangre.
Los **islotes de Langerhans,** grupos de células especializadas que ocupan una parte del páncreas. Éste está situado a la altura de la cintura, en el lado izquierdo del cuerpo.		**Insulina** Sintetizada por las células beta de los islotes de Langerhans.	Favorece la absorción y utilización celular de la glucosa. A partir de la glucosa se sintetiza el glucógeno, que se almacena en el hígado.	La concentración alta de glucosa en la sangre.
		Glucagón Sintetizada por las células alfa de los islotes de Langerhans.	Estimula la degradación del glucógeno del hígado a glucosa.	La baja concentración de glucosa en la sangre.
Glándulas suprarrenales, situadas sobre los riñones.	• **Corteza suprarrenal** (parte externa)	**Cortisol**	Degradación de proteínas y grasas a glucosa.	La hormona adrenocorticotrópica.
		Aldosterona	Activa la absorción de los iones Na^{+} y la expulsión de K^{+} en las nefronas. Mantiene la presión sanguínea.	La pérdida de iones Na^{+} por la orina.
	• **Médula suprarrenal** (parte interna)	**Adrenalina** **Noradrenalina**	Contribuyen a reforzar la actividad del sistema simpático.	El sistema simpático.
Testículos		**Andrógenos,** (Ej.: testosterona)	Estimulan la formación de espermatozoides, y caracteres sexuales masculinos.	La hormona luteinizante.
Ovarios		**Estrógenos** y **progesterona**	Intervienen en el ciclo menstrual y el embarazo.	La hormona luteinizante.

4.4. *Hormonas tisulares*

Una parte de las hormonas de nuestro organismo se sintetiza en tejidos no glandulares. Estas hormonas no son vertidas a la sangre, sino que actúan localmente. La gastrina y la histamina son dos de las más conocidas.

- La **gastrina** es segregada por la mucosa del estómago, en la zona del píloro. Estimula la secreción del jugo gástrico y del pancreático, y modifica la motilidad gástrica y esofágica.
- La **histamina** se encuentra en los mastocitos, que son un tipo de glóbulos blancos. Actúa como un potente dilatador de los capilares y de los vasos sanguíneos, y produce la contracción de la musculatura lisa.

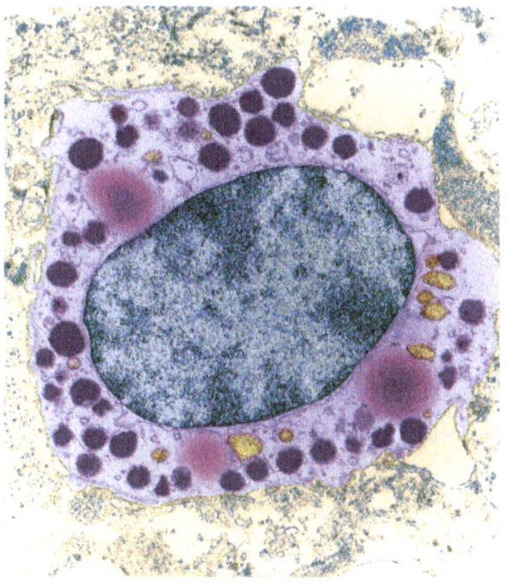

Microfotografía de un mastocito con diversos gránulos de histamina

4.5. *Mecanismos de acción hormonal*

Cuando las hormonas llegan a los órganos efectores, intervienen en las reacciones del metabolismo celular de distintos modos, dependiendo de la naturaleza química de la hormona.

Según su composición, las hormonas pueden clasificarse en *hormonas lipídicas* u *hormonas peptídicas.*

Hormonas lipídicas: Conjunto de hormonas liposolubles que pueden atravesar con facilidad la membrana plasmática. Muchas de ellas son derivadas del colesterol.

Hormonas peptídicas: Conjunto de hormonas compuestas por una cadena más o menos larga de aminoácidos. Se trata de moléculas hidrosolubles y por tanto no pueden atravesar la membrana plasmática por sí solas.

A continuación, mostramos dos ejemplos, que pertenecen a cada uno de estos grupos: la *testosterona* y la *insulina*.

HORMONAS LIPÍDICAS
Cortisol
Aldosterona
Andrógenos
Estrógenos
Progesterona

HORMONAS PEPTÍDICAS	
Insulina	Oxitocina
Tiroxina	Parthormona
Calcitonina	Hormona antidiurética
Adrenalina	
Gastrina	

Hormona lipídica: testosterona

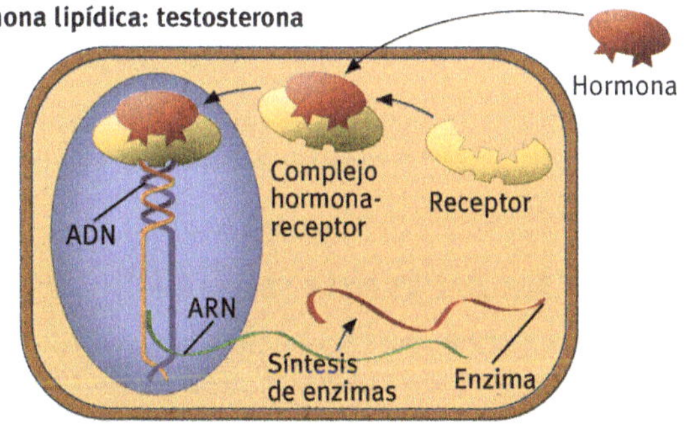

— La hormona lipídica atraviesa la membrana y se une al receptor que está en el citoplasma.

— El complejo hormona-receptor pasa al interior del núcleo celular y se induce la síntesis del ARNm.

— El ARNm se traduce, generándose una proteína.

— La proteína estimula la formación de espermatozoides y la aparición de los caracteres sexuales secundarios.

Hormona peptídica: insulina

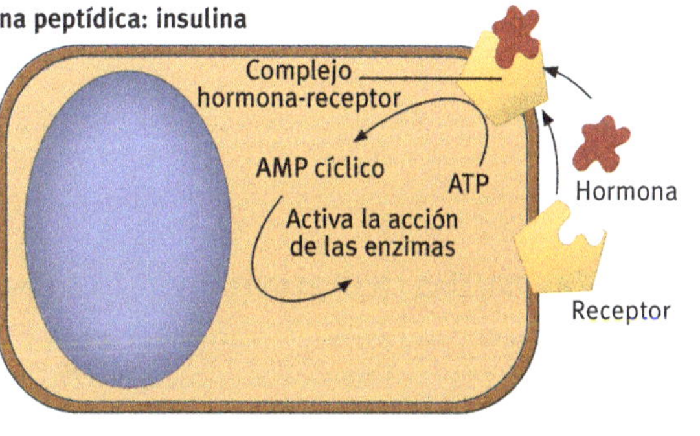

— La insulina se une a un receptor situado en la membrana plasmática.

— El complejo hormona-receptor induce la síntesis de adenosín monofostato cíclico (AMP cíclico) a partir de ATP.

— El AMP cíclico activa la acción de los enzimas que participan en el paso de glucosa al interior de la célula y en la síntesis de glucógeno.

5. El crecimiento en el ser humano

Hemos descrito la actividad de las distintas hormonas, pero éstas no actúan de forma aislada, sino que la acción conjunta de todo el sistema endocrino hace posible el desarrollo de los procesos vitales.

Un ejemplo de la acción conjunta de las hormonas es el crecimiento del individuo, proceso que se inicia con el nacimiento y cesa al llegar a la edad adulta. Se debe a la acción de la *hormona del crecimiento* y de la *tiroxina*.

CONTROL HORMONAL DEL CRECIMIENTO HUMANO

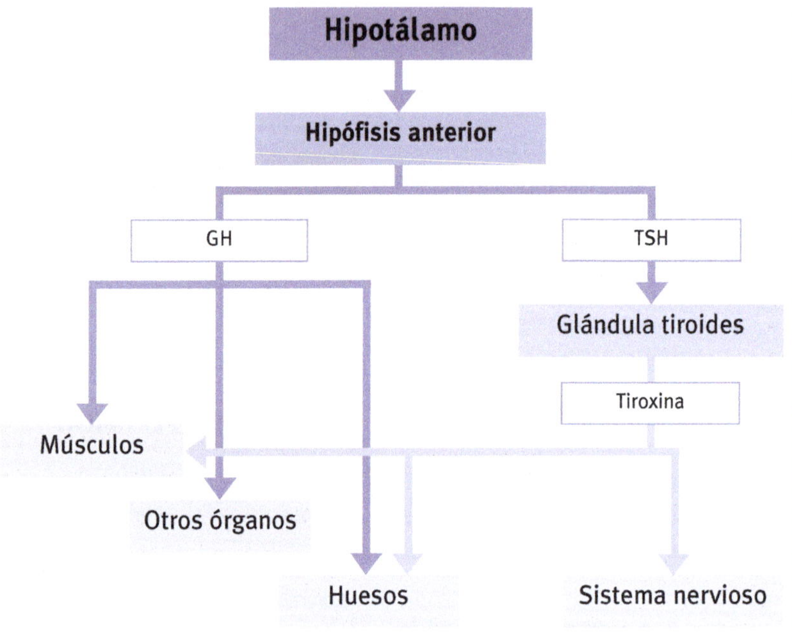

— El **hipotálamo** controla el buen funcionamiento de todo el proceso. Segrega *neurohormonas* que activan a la hipófisis.

— La **hipófisis anterior** segrega la *hormona estimulante del tiroides* (TSH), que actuará sobre esta glándula, y la *hormona del crecimiento* (GH), cuyos órganos diana son los músculos y los huesos, principalmente.

La hormona del crecimiento favorece la utilización de aminoácidos para sintetizar proteínas estructurales, por lo que produce un aumento de la masa muscular, y un alargamiento y calcificación de los huesos.

— La glándula **tiroides** segrega la *tiroxina,* que activa los procesos de obtención de la energía necesaria para las reacciones de síntesis de las moléculas que forman los músculos y los huesos. También interviene en la maduración del sistema nervioso.

5.1. *Anomalías en la secreción hormonal*

Las disfunciones en la secreción hormonal producen diversos trastornos en el organismo. En el siguiente cuadro resumimos dos de las anomalías hormonales más estudiadas:

ANOMALÍAS EN LA SECRECIÓN DE TIROXINA	ANOMALÍAS EN LA SECRECIÓN DE GH
Hipotiroidismo: Se produce cuando la síntesis y secreción de Tiroxina o TSH es insuficiente. Sus síntomas principales son un cansancio excesivo, sensación de frío, pérdida de memoria, dificultad de movimientos y aumento de peso.	**Enanismo y gigantismo:** Son las enfermedades relacionadas con el defecto o el exceso de secreción de GH durante la época de crecimiento. Se trata de personas pequeñas o muy grandes de cuerpo proporcionado y con plenas facultades mentales.
Hipertiroidismo: Se da cuando el funcionamiento de la tiroides es excesivo y se caracteriza por un incremento de la actividad, intolerancia al calor, nerviosismo, irritabilidad, aumento del ritmo cardíaco y pérdida de peso.	**Acromegalia:** Se produce cuando la secreción de GH es excesiva durante la etapa adulta. Se caracteriza por un crecimiento desmesurado de las extremidades, así como de los huesos de la cara.
En ambos casos aparece un abultamiento debajo del mentón, el bocio, producido por el aumento de la tiroides, para compensar el mal funcionamiento hormonal.	

9 Lee el siguiente texto y responde a las preguntas que encontrarás a continuación:

La diabetes es una enfermedad que se caracteriza porque los niveles de glucosa en sangre (glucemia) son más elevados de lo normal. En un individuo no diabético los niveles de glucosa antes de las comidas está entre 90-130 mg/dl y después de dos horas de las comidas el nivel es menor a los 180 mg/dl. Un paciente diabético en ayunas suele tener un nivel de glucemia mayor a 126 mg/dl. La hormona que se encarga de regular los niveles del azúcar en sangre es la insulina. Esta hormona, producida en el páncreas, se encarga de introducir la glucosa en el citoplasma de las células, sobre todo del hígado, músculos y corazón. Distinguimos básicamente 3 tipos de diabetes:

- El Tipo 1 se diagnostica cuando el páncreas no produce insulina.
- El Tipo 2 es la forma más común de diabetes; se caracteriza por una producción insuficiente de insulina o bien por una resistencia a la acción de ésta.
- La diabetes gestacional aparece durante el embarazo y generalmente desaparece en cuanto nace el bebé.

Los síntomas dependen del tipo de diabetes pero suelen ser sed, aumento de las micciones y del apetito, presencia de picores, de infecciones...

La diabetes se diagnostica midiendo, en situación de ayunas, los niveles de glucemia en sangre venosa. A esta cifra se la denomina glucemia basal.

Cuando los niveles de glucemia basal son dudosos se hace una curva de glucemia para determinar la existencia de diabetes. Esta prueba se llama SOG (Sobrecarga Oral de Glucosa) y consiste en medir la glucosa en sangre cuando al paciente en ayunas se le suministra una sobrecarga de glucosa de 75 g. El paciente, días antes de la prueba, debe mantener una dieta alta en carbohidratos. La prueba dura 2 horas y cada media hora se suele hacer una extracción de sangre para determinar el nivel de glucosa.

a. ¿Qué es la glucemia? ¿Cómo se mide?

b. El texto también nos habla de glucemia basal. Define este concepto.

c. Hay un tipo de diabetes que se llama insulino-dependiente. ¿A qué tipo de diabetes crees que corresponde? Razona tu respuesta.

d. Explica mediante un esquema cómo se regula el nivel de glucosa en sangre.

e. ¿Qué células del páncreas producen insulina? ¿Y glucagón?

f. ¿Qué es una curva de glucemia?

g. La tabla muestra los valores de glucemia después de administrar 75 g de glucosa en ayunas a dos pacientes, uno no diabético y otro diabético:

	NIVEL GLUCOSA (mg/100 ml)	
Tiempo (min)	**Paciente 1**	**Paciente 2**
T = 0	80	160
T = 30	110	210
T = 60	160	290
T = 90	120	330
T = 120	90	325
T = 150	85	315

— Representa gráficamente los valores de la tabla.

— Comenta las diferencias en la absorción de glucosa por parte de los tejidos entre las dos curvas.

— ¿Cuál de los pacientes crees que és diabético? ¿Cómo lo has deducido?

ACTIVIDADES DE SÍNTESIS

10 En 1926 Friz Went realizó el siguiente experimento:

a) Cortó el ápice de una planta en crecimiento y puso dicho ápice sobre un bloque de agar duante 4 horas. Después colocó el bloque de agar centrado sobre el coleóptilo cortado tal y como se muestra en la siguiente ilustración:

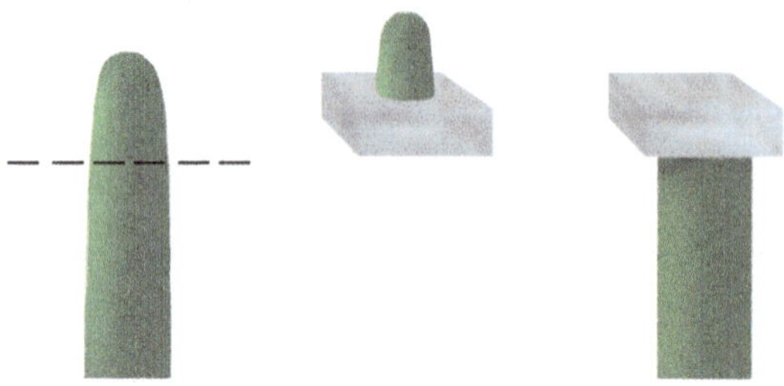

b) Observó que la planta recuperaba su crecimiento.

Contesta las siguientes preguntas:

— ¿Por qué el ápice se deja un tiempo encima de un bloque de agar?

— ¿Qué hormona es la responsable del resultado del experimento?

— Si colocáramos el bloque de agar ladeado en lugar de centrado, ¿la planta crecería según la opción A o según la opción B? Razona tu respuesta.

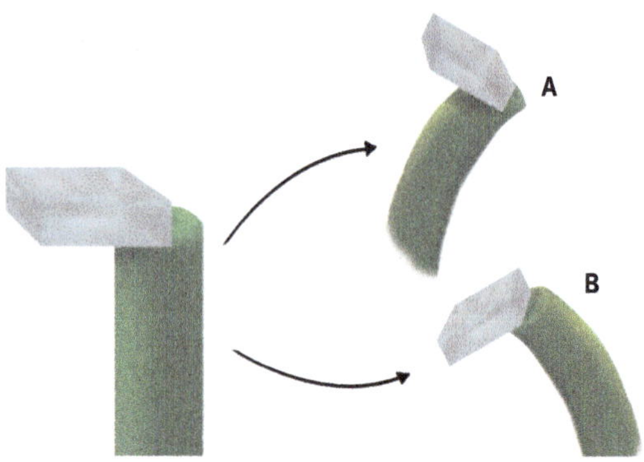

11 Asigna a la función descrita a la hormona vegetal que la lleva a cabo:

— Inhibición del desarrollo de las semillas.

— Estimulación de la síntesis de clorofila.

— Retardo en el envejecimiento vegetal.

— Aceleración de la maduración de los frutos.

— Inhibición del envejecimiento de las hojas.

— Regulación del crecimiento del tallo de la planta.

12 Explica el significado de los siguientes conceptos: fotoperíodo, geotropismo negativo, fotonastia y nutación.

13 En el siguiente esquema se representa la multiplicación vegetal por esquejes:

— ¿A qué hormona corresponde el tratamiento del tercer paso?

— Describe el proceso que permite a una planta el crecimiento de sus raíces.

14 Observa las siguientes fotografías. Comenta qué movimiento de la plantas se representan y explica en qué consiste dicho movimiento.

15 ¿Qué diferencia hay entre excreción y secreción?

16 ¿Qué es el hipotálamo? ¿Qué relación existe entre hipotálamo e hipófisis?

17 Lee atentamente el siguiente texto:

"La hipófisis es una pequeña glándula, no más grande que una judía, controlada por el hipotálamo, y que regula a las glándulas endocrinas. A su vez, la hipófisis también está regulada por las hormonas que secretan dichas glándulas endocrinas".

— ¿A qué mecanismo hace referencia el texto?

— Aclara su significado mediante un ejemplo donde se vea claramente los diferentes niveles de regulación de la hipófisis.

18 El hipotálamo está unido a la hipófisis por un tallo recorrido internamente por neuronas secretoras. La hipófisis es de hecho dos glándulas de origen embriológico diferente: la adenohipófisis o hipófisis anterior y la neurohipófisis o hipófisis posterior.

ACTIVIDADES DE SÍNTESIS

— ¿Cuál de ellas almacena las secreciones provinentes del hipotálamo? ¿Es la misma que tiene función glandular?

— ¿De dónde provienen las hormonas secretadas por el lóbulo posterior?

— ¿Qué dos hormonas secreta la hipófisis anterior?

19 ¿Qué diferencias hay entre neurohormona y hormona? Pon un ejemplo de neurohormona explicando su lugar de síntesis, su lugar de actuación, su función y una hormona resultado de su estimulación.

20 Nombra y sitúa las glándulas endocrinas que aparecen en la siguiente ilustración:

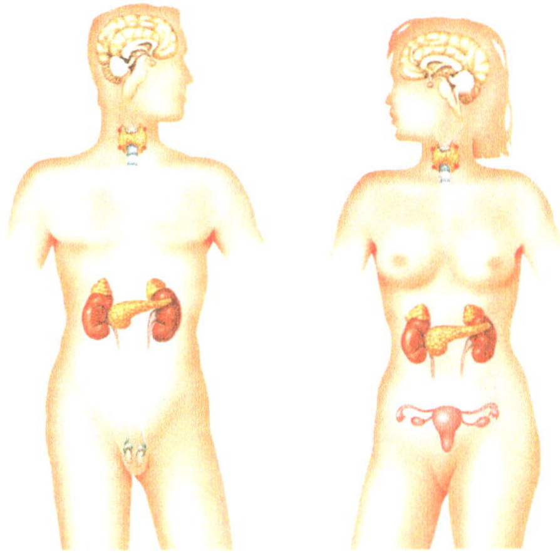

21 Observa el siguiente esquema y contesta las preguntas que encontrarás a continuación:

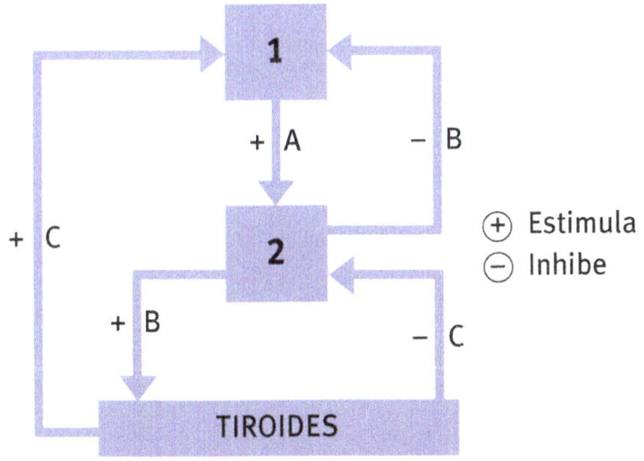

— Pon los nombres correspondientes a 1 y 2.

— ¿A qué corresponden las letras A, B y C?

— Si aumenta la concentración de C, ¿inhibirá o estimulará a 1?

— Si disminuye la concentración de B, ¿inhibirá o estimulará la 1? ¿Y a C?

— ¿Qué anomalía se produce por un defecto en la secreción de B?

22 La hormona luteinizante (LH) actúa a nivel de ovarios estimulando la producción de estrógenos y progesterona, y la hormona foliculoestimulante (FSH) estimula la producción de los ovocitos.

— Representa esquemáticamente el sistema de retroalimentación de la LH y la FSH.

23 Busca información de la enfermedad de Cushing y el síndrome de Cushing en enciclopedias o bien en buscadores de internet y contesta las siguientes preguntas:

— ¿Qué hormona es la responsable del síndrome?

— ¿Qué hormona es la responsable de la enfermedad?

— ¿Qué relación hay entre estas dos hormonas?

Dibuja el sistema de retroalimentación implicado en una persona sana.

24 Elabora una tabla como la siguiente completando los espacios vacíos:

HORMONA	LUGAR DE PRODUCCIÓN	ACCIÓN
		Intervienen en el ciclo menstrual y en el embarazo
Calcitonina		
	Células alfa del páncreas	
		Degradación de proteínas y grasas a glucosa
	Parte interna de la médula suprarrenal	
Testosterona		
		Favorece la absorción de la glucosa

13

La función de reproducción

La función de reproducción asegura la supervivencia de la especie y es por ello que una parte importante de la vida de los seres vivos está destinada a llevarla a cabo.

Mediante la función de reproducción, los descendientes adquieren características de los progenitores, manteniendo así un flujo de genes entre generaciones que permite una cierta uniformidad de la especie.

No obstante, a lo largo de la evolución las especies han ido adquiriendo diferentes estrategias de reproducción dependiendo de su entorno.

Incluso existen especies que varían su estrategia dependiendo del ambiente concreto en el momento de reproducirse.

El conocimiento profundo de los mecanismos reproductivos por parte del ser humano ha conllevado múltiples beneficios relacionados con la producción agrícola y ganadera así como al avance en las ciencias biomédicas.

ORGANIZACIÓN DE LOS CONTENIDOS

1. Reproducción asexual
- 1.2. Reproducción asexual vegetativa
- 1.2. Reproducción asexual por esporas

2. Reproducción sexual
- 2.1. Gametogénesis
- 2.2. Fecundación
- 2.3. Desarrollo embrionario

3. Ciclos biológicos

4. Comparación entre la reproducción asexual y la sexual
- 4.1. Reproducción alternante

5. Técnicas de reproducción
- 5.1. Técnicas aplicadas a las plantas
- 5.2. Técnicas aplicadas a los animales

OBJETIVOS

- Conocer los principales métodos de reproducción.
- Relacionar los tipos de reproducción con las condiciones en las que viven las plantas y los animales.
- Comprender las distintas etapas de la reproducción sexual en los organismos pluricelulares.
- Interpretar los distintos tipos de ciclos biológicos e identificar en qué momento se producen la meiosis y la fecundación.
- Analizar las ventajas y los inconvenientes de la reproducción asexual y la sexual.

Los seres vivos intercambian materia y energía con el medio para llevar a cabo la función de nutrición; mediante la función de relación captan la información del entorno, la procesan y elaboran respuestas; y con la reproducción, se caracterizan por la formación de nuevos individuos.

La **función de reproducción** consiste en la formación de nuevos organismos semejantes a sus progenitores. Los descendientes compensan las pérdidas producidas por la muerte de individuos de la especie; por tanto, aunque la reproducción no es indispensable para la vida de un organismo, lo es para asegurar la supervivencia de su especie a través del tiempo. Los seres vivos han desarrollado diversos métodos de reproducción.

- En los **organismos unicelulares** todo el ser participa en la reproducción; su única célula se divide para formar dos células hijas.
- En los **organismos pluricelulares** la división de cada una de sus células permite al individuo crecer, renovar y reparar sus tejidos. Pero precisan estrategias más complejas que los unicelulares para la formación de un nuevo ser vivo. Existen dos tipos de reproducción:
  — **Reproducción asexual:** a partir de un fragmento del progenitor o de unas células de éste **(espora)** se genera el nuevo organismo.
  — **Reproducción sexual:** para la formación del nuevo individuo se necesita la unión de dos células especializadas llamadas **células reproductoras** o **gametos.** El resto de las células que integran los organismos con reproducción sexual son las **células somáticas** o **vegetativas,** las cuales forman los distintos tejidos.

A continuación vamos a describir los dos tipos de reproducción que tienen lugar en los organismos pluricelulares.

Gemación de una hidra

1. Reproducción asexual

Se basa en el desarrollo de un nuevo individuo a partir de una o varias células del progenitor; el descendiente es un individuo completo idéntico al progenitor. Existen dos tipos de reproducción asexual: la **reproducción asexual vegetativa** y la **reproducción asexual por esporas.**

1.1. *Reproducción asexual vegetativa*

En la reproducción asexual vegetativa, el descendiente se genera a partir de un grupo de células del progenitor. Comprende diferentes estrategias reproductivas:

En los escifozoos, como las medusas, la escisión recibe el nombre de estrobilación

- **Gemación:** En este tipo de reproducción un conjunto de células se diferencia sobre la superficie del organismo como un abultamiento y crece por repetidas divisiones celulares hasta formar otro individuo. El nuevo ser puede permanecer unido al progenitor formando una colonia o bien desprenderse.
- **Escisión:** En esta estrategia el individuo adulto se fragmenta longitudinalmente o transversalmente, dando lugar, al menos, a dos individuos. Se puede dar en cnidarios, en equinodermos y en anélidos. En los escifozoos, la escisión recibe el nombre de estrobilación.

- **Fragmentación:** Se basa en la generación de nuevos individuos a partir de un fragmento y se da en vegetales. En jardinería se utiliza la multiplicación por esquejes. A continuación presentamos las formas más importantes de fragmentación:

ESTOLONES	Son ramas que debido a su crecimiento llegan a tocar el suelo, generando un nuevo individuo al enraizar.
BULBOS	Son tallos subterráneos de forma cónica que almacenan sustancias. En las hojas mas cercanas al bulbo se originan nuevos bulbos, capaces de originar una planta nueva.
RIZOMAS	Son tallos alargados que crecen bajo el suelo en forma horizontal generando cada cierta distancia un nuevo individuo de crecimiento vertical.
TUBÉRCULOS	Son tallos subterráneos que adoptan una forma algo esférica y almacenan en su interior sustancias de reserva. En su superficie desarrollan yemas, capaces de originar una nueva planta.

Estolón

Bulbo

Rizoma

Tubérculo

En ocasiones, estos mecanismos tienen otro tipo de finalidad además de la reproducción:

La formación de yemas no siempre implica la reproducción del individuo. En las plantas, las yemas están constituidas por meristemos y son responsables del crecimiento en longitud del tallo y del desarrollo de ramas, hojas o flores.

- Por su parte, la escisión permite la regeneración de órganos o tejidos y en estos casos no tiene finalidad reproductiva. A partir de fragmentos de raíces, tallos u hojas, algunas plantas pueden regenerar un organismo completo. Estos fragmentos reciben el nombre de **esquejes.** Aunque de modo natural no se considere una técnica reproductiva, esta capacidad ha sido utilizada por el ser humano para la multiplicación de plantas.

- **Regeneración:** No se considera un proceso de reproducción del individuo entero porque sólo se regenera una parte de los tejidos perdidos por los animales cuando se encuentran bajo una situación de peligro. Por ejemplo, los artrópodos pueden regenerar las patas, las antenas, los reptiles, la cola...

1.2. *Reproducción asexual por esporas o esporulación*

La espora es una estructura resistente, que puede ser haploide o diploide. En el caso de la esporulación asexual las esporas se generan por mitosis y se llaman mitósporas. Las mitósporas pueden generar un nuevo individuo, idéntico al progenitor, por divisiones celulares sucesivas. Este tipo de reproducción se da en algunas algas, en pteridófitos, en briófitos y en algunos hongos.

ORGA-NISMOS	DOTACIÓN DIPLOIDE	DOTACIÓN HAPLOIDE
Maíz	20	10
Pino	24	12
Patata	48	24
Perro	78	39
Mosquito	6	3
Ser humano	46	23

2. Reproducción sexual

Es la forma de reproducción más frecuente en los organismos pluricelulares. Se desarrolla en las siguientes etapas:

— **Gametogénesis:** es el proceso de formación de las células reproductoras o gametos, que se caracterizan por contener la mitad de cromosomas que una célula somática.

Si el número de cromosomas de una célula somática se denomina dotación **diploide** y se representa **2n,** el de un gameto recibe el nombre de **haploide** y se representa **n.** El número de cromosomas es característico de cada especie.

— **Fecundación:** es la unión de gametos para formar una célula nuevamente diploide (2n) que recibe el nombre de célula huevo o **zigoto.**

— **Desarrollo embrionario:** son las mitosis sucesivas del zigoto hasta transformarse en un individuo semejante a los progenitores.

2.1. *Gametogénesis*

Es el proceso de formación de las células reproductoras o gametos, que contienen la mitad de cromosomas que una célula somática.

Grupo taxonómico	Gametos femeninos	Gametos masculinos
Briófitos y pteridófitos	Ovocélula	Espermatozoide
Espermatófitos	Ovocélula	Célula espermática
Animales	Óvulo	Espermatozoide

■ Nombre de los gametos en los grupos taxonómicos que presentan oogamia

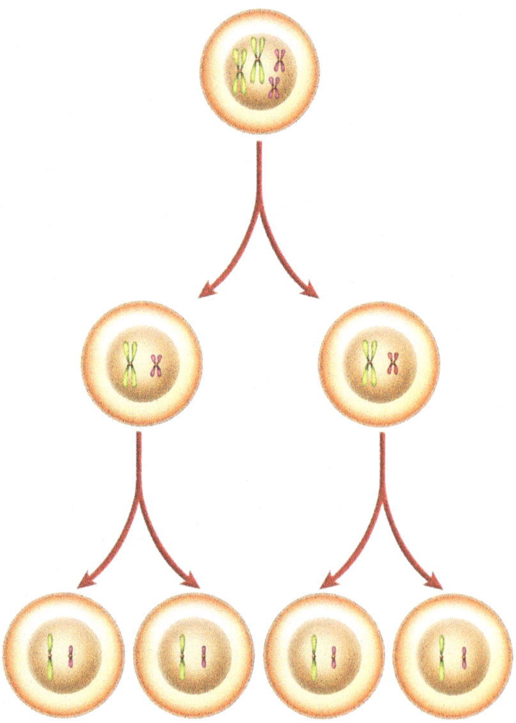

Hablamos de **isogamia** cuando todos los gametos generados por los individuos de una especie son iguales entre sí. En la **anisogamia** los gametos presentan diferente tamaño: los femeninos son mayores y se llaman macrogametos, y los masculinos son menores y se llaman microgametos. Finalmente, en la **oogamia** los gametos presentan diferente tamaño y forma; su nombre varía según se trate de especies vegetales o animales.

La gametogénesis tiene lugar a partir de células precursoras presentes en los órganos sexuales; estas células, llamadas **células madre** o **precursoras de gametos**, entran en división por mitosis y dan lugar a los gametos por **meiosis.**

La meiosis es un proceso de división celular reduccional, ya que las células hijas tienen la mitad de la dotación cromosómica de la célula madre: pasan de células diploides a células haploides. Es una fuente de variabilidad genética.

La meiosis consta de dos divisiones: en la primera se separan los cromosomas homólogos, reduciéndose el número de cromosomas a la mitad; la segunda división es muy similar a la mitosis.

Los organismos que sólo producen un tipo de gametos reciben el nombre de **unisexuales.** Se diferencia entre machos y hembras, que pueden ser morfológicamente parecidos o presentar diferencias externas considerables; en este último caso, se habla de dimorfismo sexual.

Los organismos que producen ambos tipos de gametos son llamados **hermafroditas.**

A continuación, describimos el proceso de gametogénesis en los vertebrados, el cual se realiza con ligeras diferencias en testículos y en ovarios.

ESPERMATOGÉNESIS MASCULINA EN LOS VERTEBRADOS

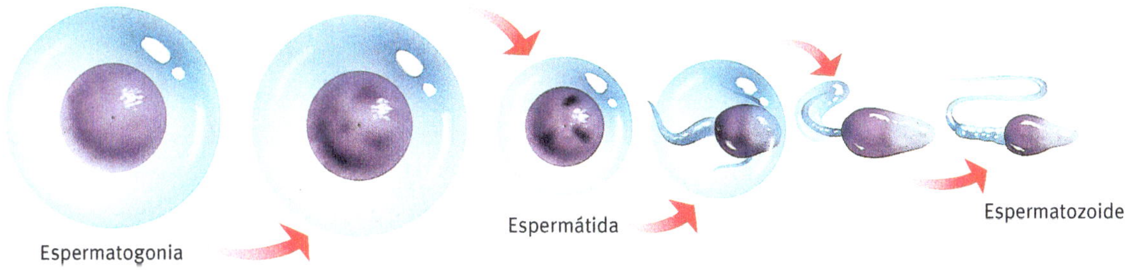

— Las células precursoras se denominan **espermatogonias** y son diploides (2n).

— Después de la meiosis obtenemos cuatro **espermátidas** que son haploides.

— Cada espermátida dará lugar a un espermatozoide después de un proceso de cambio morfológico llamado diferenciación.

— Por tanto, de cada espermatogonia se obtienen cuatro espermatozoides.

GAMETOGÉNESIS FEMENINA EN LOS VERTEBRADOS

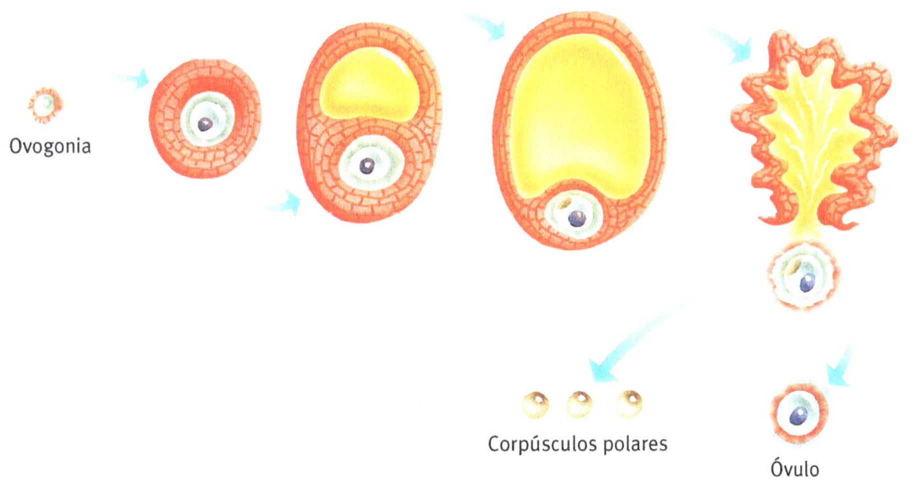

— Las células precursoras se denominan **ovogonias** (2n).

— Cada ovogonia se divide por meiosis originando unas células intermediarias denominadas ovocitos.

— Los **ovocitos** dan lugar al **óvulo** (n), que contiene la mayor parte del citoplasma, y otras tres células más pequeñas denominadas **corpúsculos polares** (n), que normalmente degeneran.

En los procesos de gametogénesis de los invertebrados encontramos algunas diferencias.

Por ejemplo, en algunas especies, durante la gametogénesis femenina las ovogonias dan lugar a los ovocitos y a unas células denominadas **células nutritivas,** que envuelven a los ovocitos. La función de estas células es sintetizar proteínas, ácidos nucleicos, etc. para proporcionarlos al ovocito.

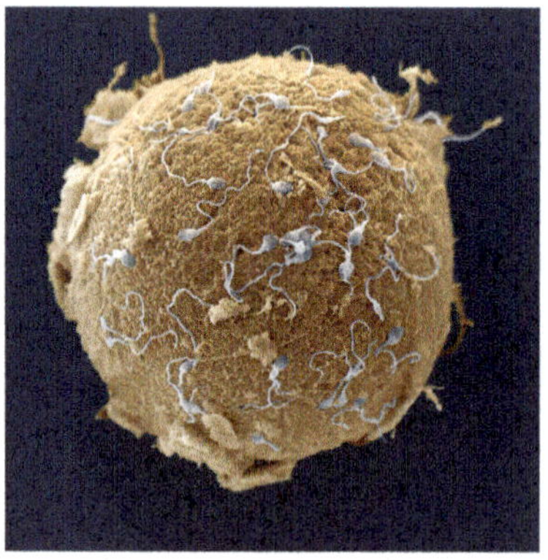

Óvulo humano rodeado de espermatozoides visto al microscopio electrónico de rastreo

2.2. *Fecundación*

Es la fusión del gameto femenino con el gameto masculino y da lugar a la célula huevo o **zigoto,** el cual es diploide debido a la fusión de los dos núcleos haploides.

La fecundación se denomina **cruzada** cuando se unen un gameto femenino y uno masculino procedentes de dos individuos distintos. Para que se pueda producir la fecundación cruzada en los organismos hermafroditas, éstos poseen mecanismos que evitan la **autofecundación** como sucede en los siguientes casos:

- Los órganos productores de gametos maduran en momentos distintos. Por ejemplo, en los lirios los granos de polen maduran antes que los sacos embrionarios.
- En otros casos, la situación de los órganos sexuales impide la autofecundación. Así ocurre en algunos anélidos que tienen los órganos reproductores masculinos y femeninos en segmentos sucesivos.

Respecto a los animales distinguimos entre **fecundación externa** o **interna** según el lugar donde se produzca la unión de los gametos masculinos y los femeninos.

- En la **fecundación externa,** óvulos y espermatozoides son liberados en el agua y se requiere la formación de miles de gametos para asegurar que unos pocos se fecunden. Es un tipo de fecundación propio de invertebrados acuáticos y de los peces.

  Los anfibios, a pesar de presentar fecundación externa, realizan acoplamiento, para favorecer la sincronización en la expulsión de los gametos; de esta forma se asegura el mayor número posible de fecundaciones.

- La **fecundación interna** supone la fusión de los gametos en el interior del cuerpo de un individuo progenitor, frecuentemente la hembra. En este caso el macho deposita los espermatozoides en las vías genitales de la hembra, proceso que tiene lugar durante la cópula.

  La fecundación interna es el recurso más utilizado por los animales adaptados al medio terrestre, donde la falta de agua no posibilita ni la dispersión ni el encuentro de los gametos, y provoca su desecación.

  En algunos moluscos y artrópodos no se produce la cópula y el macho deposita sus espermatozoides dentro de un estuche protector o **espermatóforo,** que la hembra recoge e introduce en su cuerpo.

La fecundación interna tiene más ventajas que la fecundación externa y, por tanto, ofrece mayores posibilidades de éxito reproductivo.

Las ventajas que presenta son las siguientes:

- La producción de pocos gametos, en comparación con los que se producen en la fecundación externa, supone un ahorro de energía.
- La supervivencia de estos gametos es mayor, ya que no están expuestos a condiciones ambientales desfavorables o a depredadores.
- Las probabilidades de encuentro de los gametos y, por tanto, de que se produzca la fecundación son muy altas.

2.3. *Desarrollo embrionario*

Una vez que se ha llevado a cabo la fecundación, el zigoto inicia una serie de divisiones mitóticas y de posteriores diferenciaciones celulares hasta constituir un organismo pluricelular similar a sus progenitores, como hemos visto en la unidad 8.

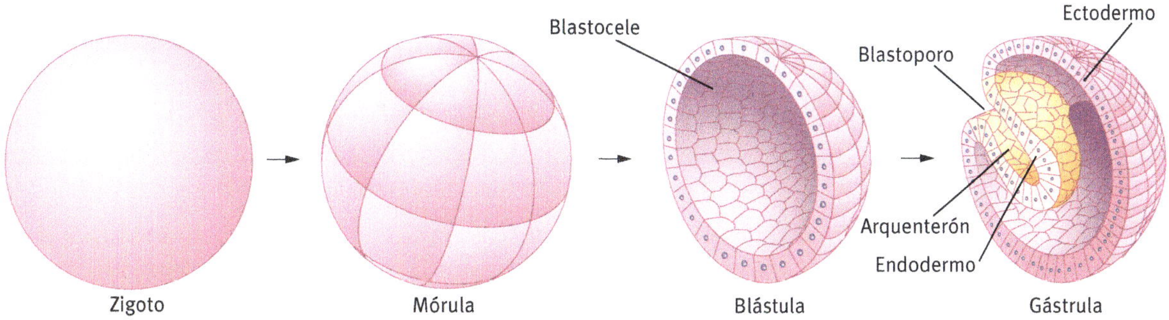

En las primeras fases después de la fecundación, las sustancias contenidas en el citoplasma del óvulo que se ha fecundado nutren a las células descendientes de éste, hasta el momento en que se implanten en la estructura encargada de proteger y alimentar al embrión.

Por esta razón, en la mayoría de las especies el óvulo contiene una gran cantidad de sustancias nutritivas y es una célula de tamaño mucho mayor que el espermatozoide.

Según dónde tiene lugar el desarrollo del embrión, los animales se clasifican en:

TIPOS	CARACTERÍSTICAS
Ovíparos	• El desarrollo embrionario se produce en el interior de un huevo gracias a las sustancias nutritivas que éste contiene y que forman el vitelo nutritivo. Estas sustancias nutren al embrión durante su desarrollo. • En el caso de los animales acuáticos, para evitar que los huevos sean desplazados por el agua, éstos poseen mecanismos de flotación, o bien, filamentos. Los huevos depositados en el medio terrestre están recubiertos de una envoltura protectora para evitar su desecación. • Son ovíparos los insectos, algunos peces, los anfibios, algunos reptiles, las aves y un único mamífero: el ornitorrinco.
Vivíparos	• El embrión se desarrolla en el interior del útero materno, donde la placenta le proporciona protección y alimento. La placenta es un tejido formado durante el embarazo y sirve para intercambiar nutrientes y material de desecho entre la sangre materna y la fetal. • Son vivíparos ciertos reptiles, algunos peces cartilaginosos y todos los mamíferos con excepción del ornitorrinco.
Ovovivíparos	• El desarrollo embrionario se produce dentro de un huevo, que a su vez es protegido en el cuerpo de la madre. • Son ovovivíparos algunos peces y algunos reptiles como la culebra.

Existen algunos ovíparos, como los insectos y los anfibios, cuyos huevos no contienen suficiente vitelo nutritivo para alimentar al embrión durante todo el proceso de desarrollo. Por ello, el embrión abandona el huevo antes de completar dicho proceso y nace en **estadio larvario.** La larva se alimenta de sustancias del medio para seguir su proceso de metamorfosis hasta convertirse en adulto.

3. Ciclos biológicos

Un ciclo biológico es el conjunto de acontecimientos que tienen lugar en un organismo desde la formación del zigoto hasta que se vuelve a reproducir. Los ciclos biológicos implican, excepto en la mayoría de los animales, estrategias dobles de reproducción. Es decir que, en su ciclo biológico, el organismo alterna la reproducción sexual con la asexual.

En la reproducción sexual intervienen siempre dos procesos fundamentales, la **meiosis** y la **fecundación.** La meiosis reduce a la mitad (n) el número de cromosomas y la fecundación reúne nuevamente el número de cromosomas diploide (2n). Ello implica que durante el ciclo biológico se presenten tanto células haploides como diploides.

Según la dotación cromosómica de los individuos adultos y el momento en el que tienen lugar la meiosis y la fecundación, se distinguen tres tipos de ciclo biológico:

- **Ciclo haplonte.** El organismo adulto está constituido por células haploides y sólo el zigoto es diploide. La meiosis ocurre después de la fecundación, una vez formado el zigoto, y por ello se denomina **meiosis zigótica**.
- **Ciclo diplonte.** El individuo adulto está constituido por células diploides y sólo realiza la meiosis para formar gametos haploides destinados a participar en la fecundación. Es una **meiosis gametogénica**.
- **Ciclo diplo-haplonte.** La especie presenta, en su ciclo de vida, dos individuos adultos morfológicamente diferentes, uno haploide y otro diploide. La meiosis se da en el individuo diploide, para generar las **esporas**. Se trata por tanto de una **meiosis esporogénica**.

CICLO HAPLONTE DEL ALGA *CHLAMYDOMONAS*

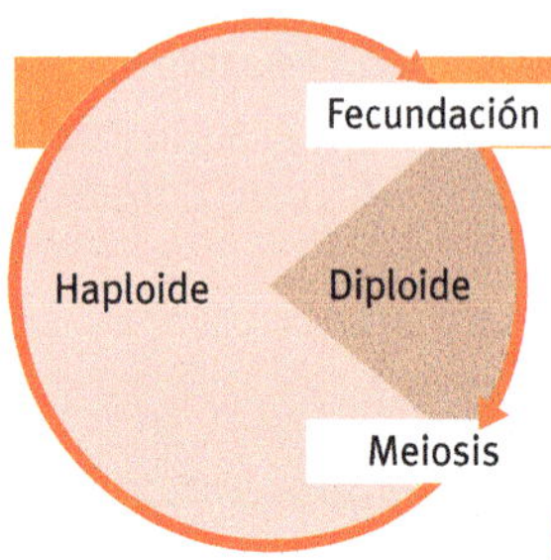

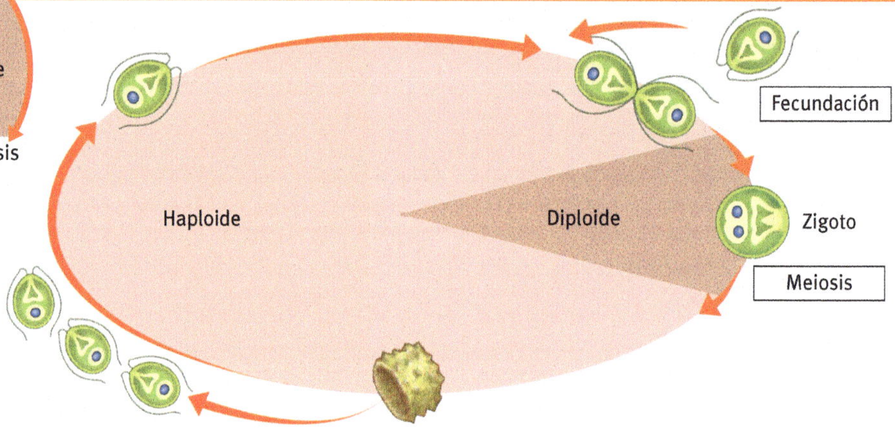

— El organismo es **haploide** y se reproduce asexualmente por divisiones mitóticas.

— En ocasiones, dos células se unen y se produce la **fecundación.**

— El **zigoto diploide** presenta una cubierta muy gruesa que le permite sobrevivir si las condiciones no son favorables.

— En este zigoto tiene lugar la **meiosis** y se originan cuatro células **haploides** que constituyen cuatro individuos adultos y cierran el ciclo.

Presentan este tipo de ciclo los protoctistas, algunas algas y algunos hongos.

CICLO DIPLONTE DEL CNIDARIO *AURELIA*

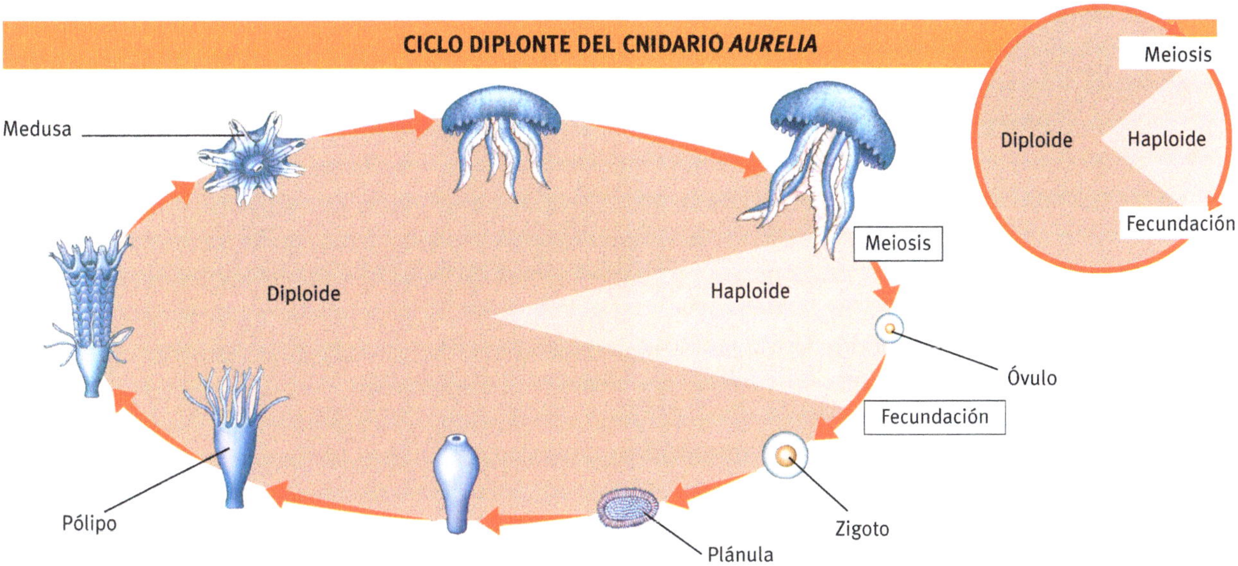

— El pólipo es un organismo **diploide** que se reproduce asexualmente alternando la gemación y la escisión, y da lugar a medusas que se apilan encima de él. Una vez que se desprenden, crecen hasta convertirse en adultas.

— Las medusas adultas producen por **meiosis** los gametos **haploides,** óvulos y espermatozoides según sean medusas hembra o macho, y los liberan en el agua.

— La **fecundación** origina un **zigoto diploide,** del que se desarrolla una larva denominada plánula.

— Posteriormente la plánula, tras sufrir una serie de cambios, se transforma en un pólipo.

Este ciclo se da en algas, hongos y animales invertebrados.

El ciclo que mostramos a continuación es un ciclo diplonte como el anterior pero existe una gran diferencia entre ambos: en el primero se produce una alternancia entre la reproducción sexual y asexual, mientras que en el segundo sólo tiene lugar la reproducción sexual:

CICLO DIPLONTE DEL MAMÍFERO *MUS MUSCULUS*

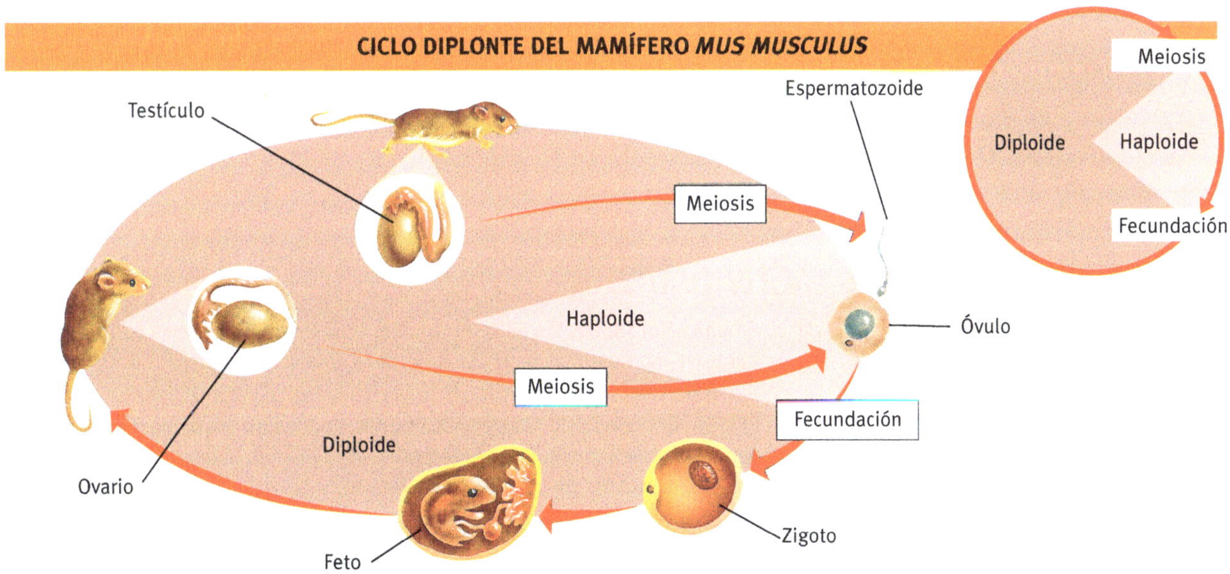

— El organismo es **diploide.** Cuando alcanza la madurez sexual tiene lugar la **meiosis** y produce los gametos haploides.

— Tras la **fecundación,** el **zigoto diploide** origina un ratón hembra o macho que se reproducirá sexualmente cuando sea adulto.

Se da en animales invertebrados y vertebrados.

CICLO DIPLO-HAPLONTE DEL HELECHO *PTERIDIUM AQUILINUM*

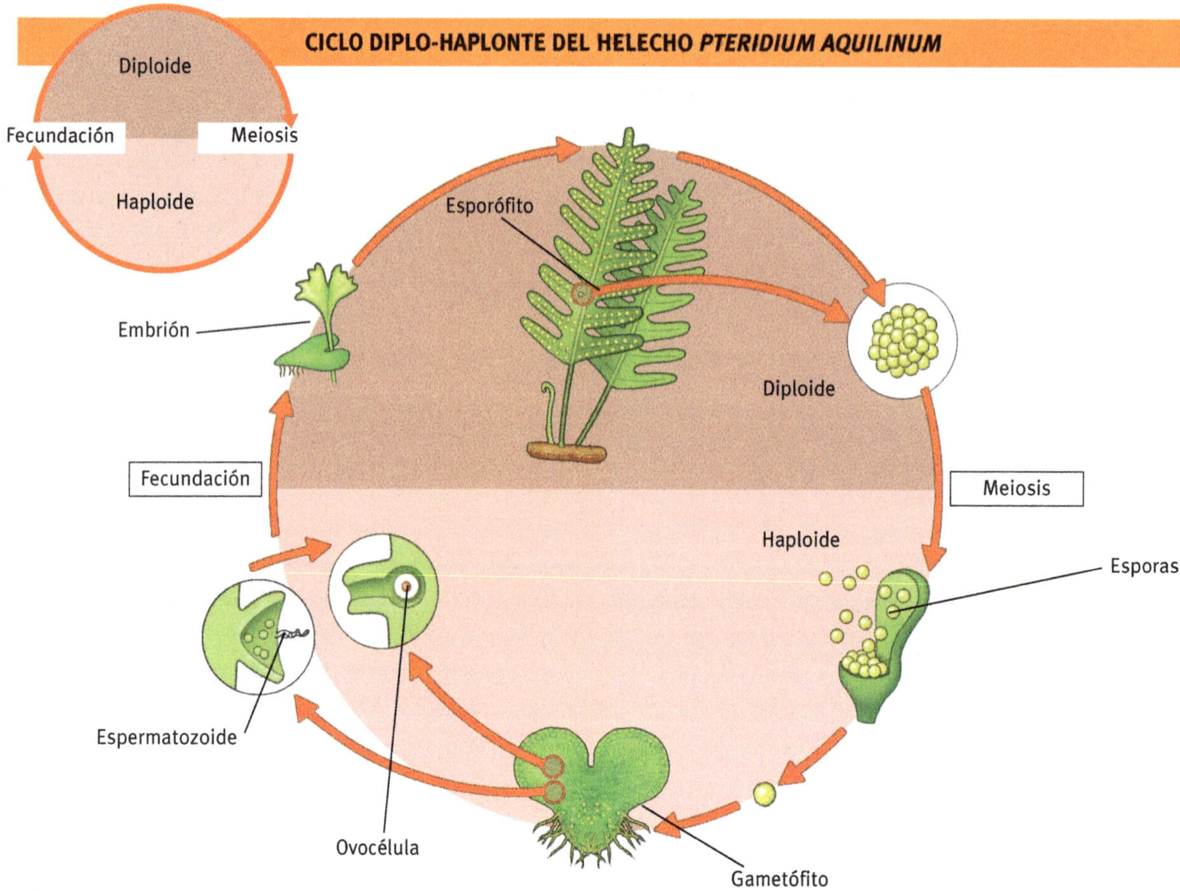

— Existen dos tipos de individuos adultos: el **gametófito** y el **esporófito**.

— El **gametófito** es haploide y produce gametos por mitosis en el **gametangio**.

— Dos gametos procedentes de dos gametófitos diferentes se fusionarán durante la fecundación, dando lugar a un zigoto diploide.

— Al crecer, el zigoto dará lugar al **esporófito**, que es diploide y produce esporas haploides mediante la **meiosis esporogénica**.

— Las esporas son estructuras de dispersión que al germinar y crecer darán lugar a un nuevo **gametófito**.

Existen ejemplos de este tipo de ciclo en algas, plantas y hongos.

	Plantas criptógamas	**Plantas fanerógamas**
Gametangio masculino	Anteridio	Grano de polen
Gameto masculino	Anterozoides	Célula espermática
Gametangio femenino	Arquegonio	Saco embrionario
Gameto femenino	Ovocélula	Ovocélula

ACTIVIDADES

1. En una célula somática del ser humano, ¿cuántos cromosomas hay? ¿Cuántos pares de cromosomas hay? ¿Es, por tanto, una célula diploide? Razona la respuesta.
2. El ciclo biológico de una alga es haplonte. Realiza un esquema donde aparezcan los conceptos siguientes: gametos, alga juvenil, alga adulta, fase haploide, fase diploide, meiosis, fecundación, (n) y (2n).
3. Observa los ciclos biológicos de *Chlamydomonas*, *Aurelia*, *Mus musculus* y *Pteridium aquilinum*. Elabora un informe que trate los siguientes puntos:
   - Predominio de la fase haploide o de la diploide.
   - Momento en que se produce la meiosis.
   - Momento en que se produce la fecundación.
   - Si se utiliza o no una estrategia doble de reproducción, y, si es así, qué tipos de reproducción asexual implica.

4. Comparación entre la reproducción asexual y la sexual

Ambas estrategias reproductivas presentan características diferentes.

REPRODUCCIÓN ASEXUAL	REPRODUCCIÓN SEXUAL
• En el proceso de reproducción participa un solo individuo.	• Participan dos individuos diferentes de la misma especie.
• No se requieren células especializadas, únicamente células capaces de dividirse activamente para formar un ser pluricelular.	• Se requiere la participación de células especializadas, llamadas gametos, que se unen mediante la fecundación para dar lugar al zigoto.
• Todas las divisiones celulares se realizan mediante mitosis; por tanto, todas las células resultantes son genéticamente iguales y el nuevo individuo es genéticamente idéntico a su progenitor.	• En algún momento del ciclo biológico se lleva a cabo la meiosis. El individuo descendiente es genéticamente diferente a los progenitores ya que la mitad de sus cromosomas proceden de un progenitor y la otra mitad de cromosomas del otro.

En el caso de los animales, la reproducción asexual es frecuente en aquellos que viven fijos al sustrato o tienen pocas posibilidades de encuentro con otros individuos de su misma especie.

A partir de la comparación entre ambas estrategias de reproducción, observamos que la **reproducción asexual** es un mecanismo más sencillo que la reproducción sexual, ya que implica un solo individuo y, por tanto, supone un gasto de energía y de tiempo menor.

La población se incrementa rápidamente, siempre y cuando el ambiente sea favorable y de características estables.

Por su parte, la **reproducción sexual** es más compleja que la reproducción asexual, ya que intervienen dos individuos de la misma especie y de sexos opuestos, que previamente se han de encontrar. Por tanto, implica un gasto de energía y de tiempo mayor.

En este caso la población crece muy despacio en comparación con las poblaciones con reproducción asexual, pero en cada generación se producen nuevas combinaciones del material hereditario que aumentan las posibilidades de adaptación de los individuos a distintos ambientes.

La reproducción sexual conlleva un alto gasto energético en parte debido al tiempo requerido para que dos individuos de la misma especie pero de sexo opuesto coincidan.

ACTIVIDADES

4. Respecto la reproducción sexual contesta:
   - ¿Cómo la definirías?
   - ¿Todos los gametos se originan por meiosis?
   - ¿Un gameto sin fecundar puede originar un individuo? ¿Cómo se llama esta forma de reproducción?
   - Define los siguientes conceptos: isogamia, ovocélula, gametófito, gónada y esporangio.
5. Razona qué tipo de reproducción es más ventajosa en un ambiente en proceso constante de cambio.
6. ¿Por qué la reproducción asexual supone menos gasto de energía y de tiempo que la sexual?
7. Hay especies que se pueden reproducir tanto asexual como sexualmente. ¿De qué crees que depende que lo hagan de una u otra manera?

4.1. *Reproducción alternante*

Las dos estrategias de reproducción presentan ventajas o inconvenientes en función de las características de cada especie y de las condiciones ambientales. Como has visto en el apartado de los ciclos biológicos, algunos organismos alternan ambos tipos de reproducción. Un caso especialmente interesante es el ciclo biológico de las abejas.

Estos insectos viven en sociedad y, en condiciones normales, la colonia o enjambre consta de una reina que es la hembra reproductora, diploide; de miles de obreras o hembras no reproductoras, diploides; y de centenares de zánganos o machos, haploides.

A continuación se muestra el ciclo biológico, en el que intervienen los tres tipos de individuos de un enjambre:

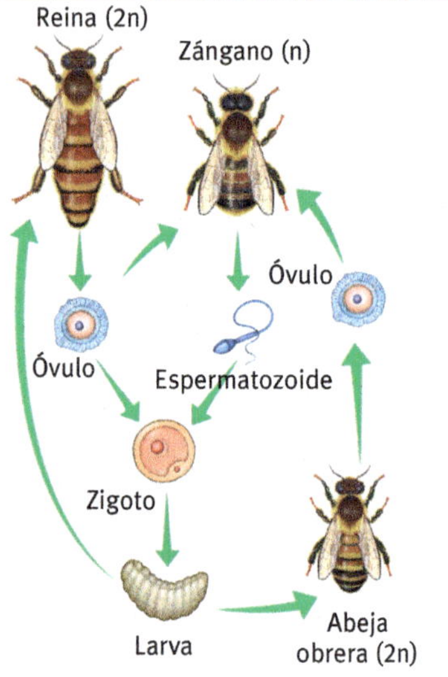

CICLO BIOLÓGICO DE LAS ABEJAS

— La reina se aparea con un **zángano** de otra colmena y almacena los espermatozoides, que son haploides, en la espermateca, una cavidad de su cuerpo encargada de recibir y almacenar estos gametos.

— Los óvulos, también haploides, son fecundados diariamente a medida que pasan por la espermateca, y dan lugar a **zigotos** diploides.

— Por sucesivas divisiones mitóticas, los zigotos originan las **larvas,** que son diploides.

— Las larvas son alimentadas por las obreras hasta el momento en que sufren la metamorfosis. Ésta dura unos 12 días, después de los cuales se convierten en adultos.

— Los adultos serán una **reina** o a una **abeja obrera** según la alimentación que reciban durante la fase larvaria.

— No todos los óvulos de la reina son fecundados; de los óvulos no fecundados surgen los **zánganos.**

— Las **abejas obreras** presentan los órganos reproductores inhibidos, pero en ciertas ocasiones producen óvulos, aunque la anatomía de su aparato reproductor no permite la cópula. En situaciones excepcionales, el desarrollo de estos óvulos también puede dar lugar a **zánganos.**

■ Mazorca de maíz no modificada genéticamente.

■ Mazorca de maíz modificada genéticamente.

En las abejas, como en otros invertebrados, se produce la **partenogénesis,** es decir, el desarrollo de un individuo adulto a partir de una célula sexual no fecundada. La combinación de esta variante reproductiva con la reproducción sexual es la causa de la determinación del sexo.

5. Técnicas de reproducción

Desde hace miles de años, el ser humano ha mostrado gran interés por obtener el máximo rendimiento de las especies que cultivaba y criaba.

Por ello ha intentado conseguir variedades de plantas de crecimiento rápido, que producen frutos grandes o con un contenido nutritivo mayor, y de la misma manera ha intentado obtener especies ganaderas más dóciles o que proporcionan más leche o mejor carne.

Así pues, la intervención humana en la reproducción de las especies que tienen interés económico es muy antigua, pero en los últimos años se han desarrollado intensamente diversas técnicas para aumentar el rendimiento de dichas especies.

5.1. *Técnicas aplicadas a las plantas*

Como ya vimos al tratar la reproducción asexual, a diferencia de la mayoría de los animales, las células somáticas de una planta pueden regenerar la planta entera.

Una de las técnicas más utilizadas en la agricultura moderna es la **micropropagación,** que consiste en obtener miles de ejemplares idénticos a partir de una sola planta de interés frutícola u ornamental. Veamos cómo llevarla a cabo:

— Se obtienen muestras de tejidos a partir de una planta.

— Estos tejidos se cultivan en un medio estéril con sales minerales y otros compuestos necesarios para la nutrición de las células.

— El cultivo es tratado con hormonas para estimular su crecimiento hasta desarrollar una plántula.

— La plántula se traslada a un invernadero en condiciones ideales para que complete su desarrollo.

En la micropropagación se extrae una parte del tejido de la planta y se introduce en un medio de cultivo.

Actualmente también se trabaja intensamente en los **cultivos transgénicos,** obtenidos a partir de la manipulación del material genético de sus células para que presenten unas características determinadas.

— Se obtienen cultivos de células somáticas.

— Se inyecta a los núcleos de estas células fragmentos de ADN procedentes de otra especie o variedad, que contienen información sobre resistencia a insectos parásitos, a sustancias herbicidas, a enfermedades víricas, etc.

— Se estimula el crecimiento y se facilita el desarrollo de las plantas como en el caso anterior.

Si, por ejemplo, el fragmento de ADN proporciona a la planta resistencia a un herbicida, es posible aplicar esta sustancia tanto al cultivo como a otras plantas no deseadas que se encuentran junto a él, de manera que sólo éstas se vean afectadas.

Hoy en día, ya se trabaja con más de cien especies vegetales manipuladas mediante cultivos transgénicos: arroz, berenjenas, centeno, colza, fresas, maíz, manzanas, patatas, soja, tabaco, tomate, trigo, etc.

C

En la actualidad, los organismos transgénicos generan un gran debate en la comunidad científica, ya que se desconocen los efectos que pueden producir los organismos modificados genéticamente sobre los sistemas naturales.

5.2. *Técnicas aplicadas a los animales*

Desde la Antigüedad, el ser humano ha obtenido provecho de los animales domésticos utilizando su capacidad de trabajo, la carne, los huevos, la leche y otros productos derivados.

Para incrementar este provecho, se han **seleccionado razas** y desde hace algunos años se aplican en ganadería las **técnicas de reproducción asistida,** que son además un buen método de investigación y de avance científico.

Estas técnicas se aplican a mamíferos. En especial, es muy frecuente entre el ganado bovino la **inseminación artificial,** que consiste en fecundar a las vacas con el semen de machos previamente seleccionados.

John Gurdon (Reino Unido 1933), es un destacado biólogo británico especialista en el campo de la embriología. Revolucionó la comunidad científica cuando en 1962 anunció que había usado el núcleo de una célula intestinal para clonar una rana *(Xenopus laevis)*. Fue el primero en demostrar en animales que el núcleo de una célula diferenciada todavía tiene el potencial para transformarse en otros tipos de células.

Las técnicas que usó para transferir núcleos se siguen utilizando hoy en día.

Con este proceso se consigue por un lado, aumentar el número de descendientes y por otro, que éstos presenten las características por las cuales fueron seleccionados sus progenitores.

Otra técnica muy desarrollada es la **transferencia de embriones.** Consiste en utilizar unas hembras para la fecundación y otras para la gestación.

— Una vez seleccionadas las hembras por sus características genéticas, se estimulan hormonalmente para que tengan numerosas ovulaciones.

— Los óvulos son fecundados mediante inseminación artificial y a los pocos días se extraen los embriones por vía vaginal.

— Cada embrión es transferido al útero de una hembra con características genéticas menos destacadas, que hará de madre adoptiva durante la gestación.

Con esta técnica se evita el período de gestación de las hembras de las que se han obtenido los embriones y así pueden ser de nuevo fecundadas, incrementando el número de descendientes seleccionados.

En muchas ocasiones, se manipulan los embriones antes de ser transferidos; las **manipulaciones embrionarias** más frecuentes son:

- *Clonación de embriones*. Esta técnica se basa en la separación de las células de un embrión cuando todavía no se han diferenciado. Estas células conservan la capacidad de desarrollar un individuo completo y son transferidas a una madre adoptiva, lo que origina el nacimiento de varios individuos idénticos.
- *Modificación del material genético.* En este caso se combina el material genético de varios embriones de especies distintas para obtener animales con características de las diferentes especies.
- *Animales transgénicos.* La técnica es la misma que se aplica en el caso de los vegetales, aunque aquí la introducción de fragmentos de ADN se produce en las células de los embriones. De este modo se obtienen individuos con períodos de ovulación más cortos, ritmos de crecimiento más rápidos, resistencia a enfermedades, etc.

Muchas de estas manipulaciones incluyen además la *congelación de los embriones* para su conservación, o bien, la *selección* de éstos según diversas características. Si lo que se busca es la producción de leche, sólo se transfieren los embriones hembra, mientras que si el objetivo es la producción de carne, se transfieren los machos.

Las técnicas de intervención en la reproducción han originado una fuerte polémica en los últimos años. Se contraponen, por un lado, el interés económico que tiene aumentar la productividad y el rendimiento del ganado, y por otro, la ética de estas prácticas que someten a los animales a una explotación considerada abusiva por algunos. También existen opiniones contrarias basadas en el riesgo que puede suponer el consumir productos derivados de especies transgénicas.

Por ello es preciso llevar a cabo una valoración científica y ética de estos procedimientos, que ayude a preservar la dignidad de los animales y la salud de las personas.

8 Lee el siguiente texto y responde a las preguntas que encontrarás a continuación:

Vacas que producen proteínas de interés farmacológico

En octubre del año 2003 una empresa argentina presentó a Mansa, la primera vaca capaz de producir leche con hormona del crecimiento. La hormona podía ser recuperada de la leche mediante procesos químicos diversos y sirve para el tratamiento del enanismo hipofisario.

Años más tarde, en abril del 2007, la misma empresa presentó a otra vaca, Patagonia. En este caso la proteína humana presente en la leche de esta vaca era la insulina, hormona que regula la absorción de glucosa por las células.

En ambos casos la hormona presente en la leche de la vaca era idéntica a la humana ya que se trata de vacas transgénicas que expresan el gen humano de cada una de las hormonas. Las vacas se obtuvieron a partir de un óvulo de vaca sin núcleo y una célula fetal de vaca Jersey (buena productora de leche) a la que se le había añadido (transfección) una copia del gen de la hormona que se quería expresar.

Otra de los retos era conseguir que las hormonas sólo se produjeran en las glándulas mamarias para que no resultaran tóxicas para la propia vaca. El objetivo se consiguió añadiéndole un marcador de expresión al gen de forma que sólo "funcionase" en la producción de leche.

Antes de que apareciesen Mansa y Patagonia, ya se disponía de la hormona de crecimiento y de insulina humanas. Pero los métodos de producción eran muy costosos y se producían en mucha menor cantidad. Todo ello encarecía el coste de los fármacos. La producción y posterior aislamiento a partir de la leche puede permitir la reducción de hasta un tercio en el precio de la hormona de crecimiento y de la insulina.

A continuación podemos ver un esquema del proceso de obtención de estas vacas transgénicas:

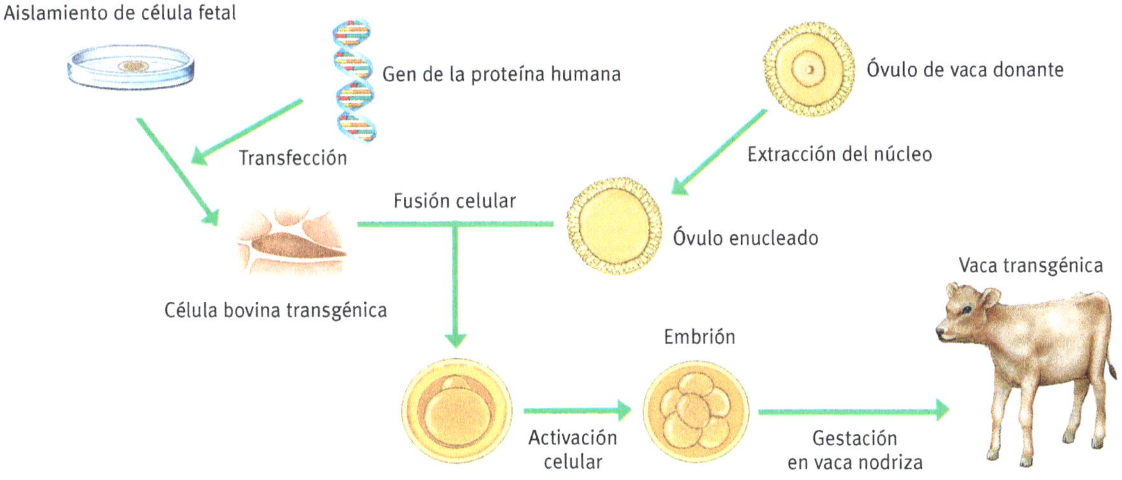

- Enumera ordenadamente las ideas principales del texto.
- Explica el proceso a seguir para obtener una vaca transgénica como Mansa o Patagonia.
- Elabora un informe resumido sobre la hormona de crecimiento y la insulina en el que debe constar: el órgano y el lugar de producción de cada una de ellas, su función y forma de actuación y las consecuencias de su carencia.
- Según el texto y el esquema anteriores contesta:
  - ¿Por qué se utilizan células fetales de vaca de raza Jersey?
  - ¿De qué especie debe proceder el gen transfectado en la célula fetal de la vaca?
  - ¿Todas las células de la vaca son portadoras del gen de la hormona humana? ¿Por qué está presente sólo en la leche de la vaca?
  - ¿La leche de estas vacas está destinada al consumo directo de las personas afectadas por la ausencia de estas hormonas? Razona tu respuesta.
- Elaborad por parejas una tabla con argumentos a favor y en contra de la utilización de animales transgénicos. Podéis obtener información adicional en la siguientes páginas web:
  - http://www.ctv.es/USERS/infomed/op/4.htm
  - http://www.todo-ciencia.com/biologia/0i87881600d990196098.php

9 Elabora una tabla con las principales diferencias entre la reproducción sexual y la asexual.

10 De las siguientes afirmaciones sobre las dotaciones cromosómicas, di cuáles son falsas y cuáles verdaderas. Razona la respuesta:

— Una célula diploide (2n) tiene siempre un número par de cromosomas.

— Una célula haploide (n) tiene siempre un número impar de cromosomas.

— En una misma especie, los macrogametos tienen una dotación cromosómica mayor que los microgametos.

— Los gametos tienen siempre un número impar de cromosomas.

— Los gametos de cualquier especie son siempre haploides.

11 Observa la ilustración y responde a las preguntas que encontrarás a continuación:

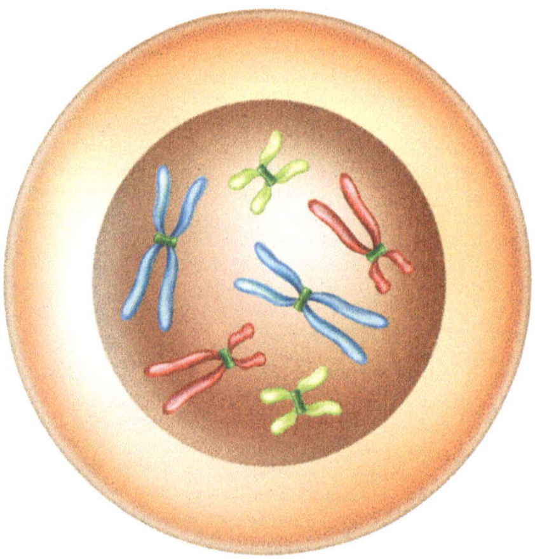

— ¿Cuántos pares de cromosomas tiene esta célula?

— ¿Cúal es su dotación cromosómica?

— ¿Cuál es el valor de n?

— ¿Crees que es una célula somática? Razona tu respuesta.

12 Asigna a cada frase el tipo de reproducción asexual correspondiente y pon un ejemplo de organismo que la lleve a cabo.

— Las ramas en crecimiento tocan el suelo, enraizan y generan un nuevo individuo.

— Ciertos tallos crecen horizontales en el interior del suelo y pueden generar nuevos individuos que crecerán verticalmente.

— Ciertos tallos subterráneos acumulan en su interior substancias de reserva y pueden originar nuevos individuos a partir de yemas.

— Células provenientes del zigoto se separan generando cada una de ellas un organismo.

— Un adulto se divide transversalmente originando uno o más un individuos.

13 Durante la gametogénesis, ¿cuántos óvulos se generan por cada ovogonia?¿cuántos espermatozoides se generan por cada espermátida? Busca información sobre la gametogénesis en vertebrados y explica qué posible significado biológico tiene esta diferencia.

14 Observa los datos de la siguiente tabla:

	CÉLULA A	CÉLULA B
Cantidad de ADN (picogramos/célula)	3,65	7,30
Dotación cromosómica		
Tipo celular		

— Completa la tabla teniendo en cuenta que una de ellas es una espermatogonia y la otra una espermátida.

15 Elabora una tabla con las diferencias entre la fecundación externa y la fecundación interna, así como las ventajas de cada una de ellas.

16 ¿Por qué los organismos hermafroditas acostuman a presentar mecanismos que impidan la autofecundación?

17 ¿Qué ventajas y qué desventajas tienen el oviparismo, el viviparismo y el ovoviviparismo?

18 Representa mediante un esquema el ciclo de un organismo vegetal que se pueda multiplicar tanto sexual como asexualmente. En el esquema deben aparecer los siguientes conceptos: gametos, zigoto, fase haploide, fase diploide, forma adulta, fecundación, meiosis, n, 2n y forma juvenil.

19 Completa la siguiente tabla sobre los tres tipos de ciclo biológico:

CICLO BIOLÓGICO	DOTACIÓN ADULTO	TIPO DE MEIOSIS	ORGANISMOS EJEMPLO
	n		
		Gametogénica	

20 Di qué tipo de dotación cromosómica presentan las siguientes estructuras referentes al ciclo diplohaplonte: esporangio, gametófito, anterozoide, espora, esporófito. Razona tu respuesta.

21 Observa los siguientes ciclos biológicos y responde las preguntas que encontrarás a continuación:

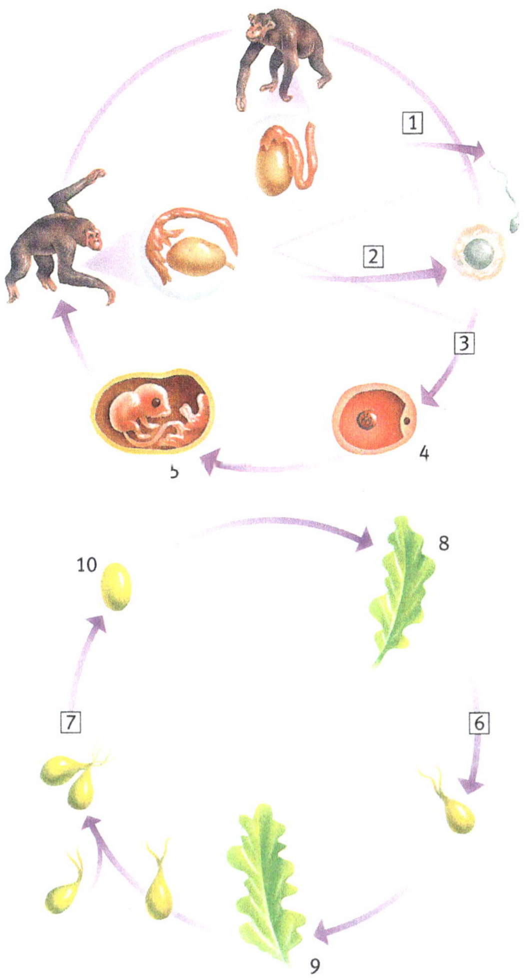

— ¿A qué tipo de organismo pertenece cada uno?

— Los números que aparecen, ¿a qué procesos (con recuadro) o estructuras (sin recuadro) corresponden?

— Separa la parte del ciclo haploide de la diploide en cada uno de ellos.

22 Lee atentamente el siguiente texto:

"El ciclo de vida del mosquito *Culex* sp. varía entre 10 y 14 días. Los adultos (machos y hembras) generan gametos. De la fecundación se originan los huevos, que se depositan en la superficie del agua. Las larvas emergen de los huevos al cabo de 24-48 h. Éstas viven en el agua unos 5 días hasta que se convierten en pupas, etapa intermedia de la metamosfosis durante la cual la larva se transforma en adulto."

Haz el esquema del ciclo biológico de este mosquito, donde se vea perfectamente el lugar donde se da la meiosis, la fecundación, la fase haploide y la fase diploide.

23 Algunas especies, como el alga unicelular *Chlamydomonas,* pueden reproducirse tanto sexualmente como asexualmente, según las condiciones del medio. Cuando las condiciones del medio son estables, se reproduce asexualmente y, al contrario, si las condiciones del medio son inestables, lo hace sexualmente. Intenta explicar el porqué de esta afirmación.

24 En febrero de 1997 se anunció el nacimiento de la oveja Dolly, el primer mamífero clonado. El método utilizado fue la transferencia nuclear y está simplificado en el siguiente esquema:

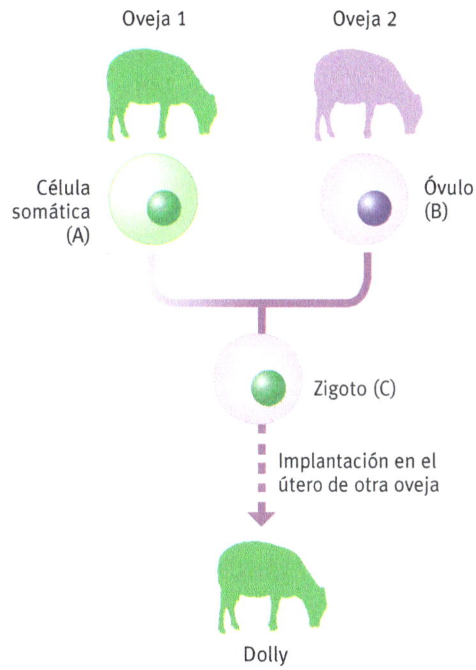

— Indica el tipo de dotación cromosómica tienen las estructuras A, B y C.

— ¿Qué significa la palabra clon? ¿De cuál de las dos ovejas iniciales es clon Dolly?

— Fíjate bien en el proceso de clonación y responde: ¿todo el ADN de Dolly proviene de la misma oveja? Razona tu respuesta.

TIC

25 Consulta la siguiente dirección web y contesta las preguntas:

http://www.elmundo.es/elmundo/2001/graficos/noviembre/semana4/clon/clonacion.swf.

— ¿Qué es la clonación terapéutica?

— ¿Qué diferencia básica existe respecto a la clonación de Dolly? ¿Qué ventajas presenta?

14

La reproducción y la sexualidad

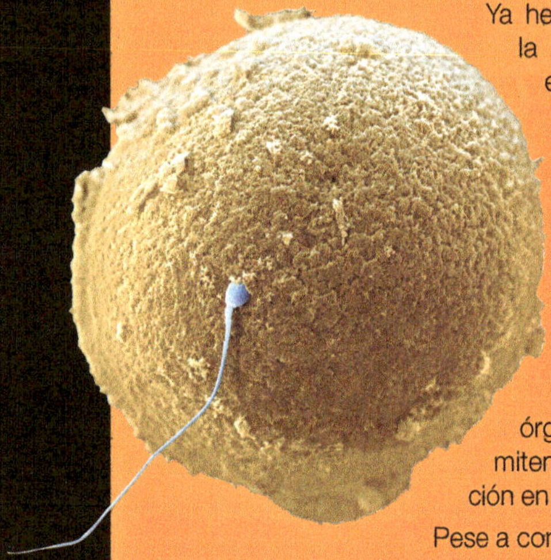

Ya hemos visto que la reproducción es la función vital que permite la supervivencia de una especie.

En esta unidad vamos a profundizar en el estudio de los órganos que permiten la reproducción en el ser humano.

Pese a compartir muchas características con la reproducción del resto de mamíferos placentarios, debemos tener en cuenta que en la especie humana la función reproductiva va más allá del mero fenómeno biológico: incluye factores como la afectividad, el cariño, la responsabilidad, la educación... Es por ello por lo que la sexualidad es una característica inherente a nuestra especie.

Asociadas a la función reproductiva, existen las llamadas enfermedades de transmisión sexual. Utilizando los medios preventivos adecuados, todas ellas son evitables.

umana

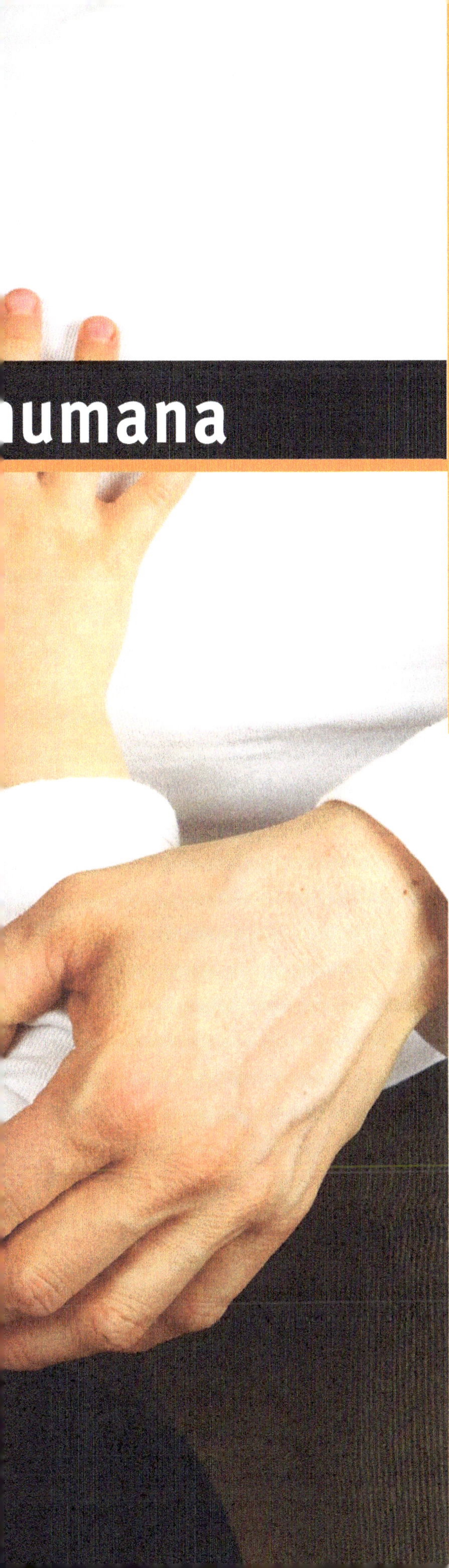

ORGANIZACIÓN DE LOS CONTENIDOS

1. El aparato reproductor masculino

1.1. Desarrollo de las características del sexo masculino

2. El aparato reproductor femenino

2.1. Desarrollo de las características del sexo femenino

2.2. Ciclo ovárico y ciclo menstrual

3. Fecundación, embarazo y parto

3.1. El embarazo

3.2. El parto

3.3. La lactancia

3.4. Planificación familiar

4. La sexualidad

5. Enfermedades de transmisión sexual

5.1. El sida

OBJETIVOS

- Conocer la anatomía y el funcionamiento de los aparatos reproductores masculino y femenino.
- Analizar el control hormonal en el desarrollo del sistema reproductor y en los procesos de espermatogénesis y ovogénesis.
- Conocer las fases de la reproducción humana y la lactancia, así como los mecanismos hormonales que las regulan.
- Comprender la sexualidad como aspecto inherente a la condición humana.
- Conocer las principales enfermedades de transmisión sexual y su prevención.

En la unidad anterior hemos descrito la reproducción en los distintos grupos de seres vivos. Ahora, vamos a centrarnos en el ser humano.

Los mamíferos presentan fecundación interna y desarrollo de los embriones también interno, lo que favorece la supervivencia de los descendientes. Por ello, en algunos grupos de mamíferos, entre los que se encuentra el ser humano, los embarazos suelen ser de un número bajo de individuos, normalmente sólo uno, suficiente para garantizar el éxito reproductivo de la especie.

A pesar de estas semejanzas, el ser humano presenta diferencias respecto a ellos, relacionadas con un aspecto propio de nuestra especie: la sexualidad. Ésta viene determinada por aspectos no sólo biológicos, sino también por cuestiones psicológicas, culturales y emocionales.

Iniciamos la descripción de la función de reproducción humana con un estudio de los sistemas reproductor masculino y femenino.

1. El aparato reproductor masculino

Está formado por los órganos siguientes:

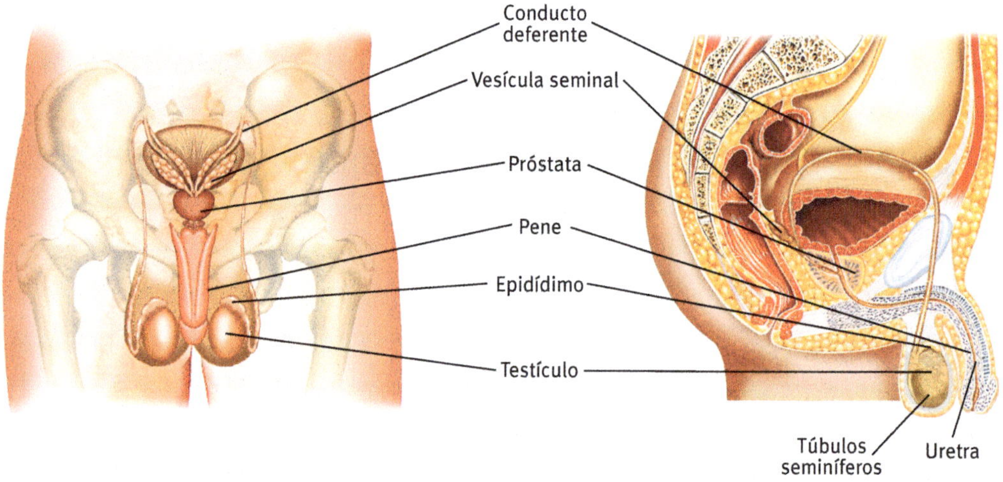

Estos órganos tienen como función la formación de espermatozoides y la incorporación de éstos al aparato reproductor femenino, para fecundar a los óvulos.

El proceso que da lugar a los espermatozoides recibe el nombre de **espermatogénesis.** Ésta se lleva a cabo en los **testículos,** que son las dos gónadas masculinas.

En el interior de los testículos se encuentran los **túbulos seminíferos,** donde se forman los espermatozoides. El tejido conjuntivo situado entre los túbulos contiene las **células de Leydig,** secretoras de hormonas sexuales.

Los testículos están rodeados por una envoltura epitelial denominada **escroto,** que contiene en su interior unas capas musculares y fibrosas con función protectora.

La espermatogénesis se desarrolla como sigue:

— Los túbulos seminíferos contienen **espermatogonias,** células precursoras de los espermatozoides.

— Mediante la meiosis, cada espermatogonia diploide (46 cromosomas) da lugar a cuatro espermatozoides haploides (23 cromosomas); este proceso dura unos 70 días.

— Los espermatozoides son impulsados hacia los **epidídimos,** donde maduran. Posteriormente pasan a los **conductos deferentes.** El interior de éstos está tapizado con células ciliadas que posibilitan el avance de los espermatozoides hasta la **uretra.** Las **vesículas seminales** y la **próstata** incorporan las secreciones que nutren a los espermatozoides, y que, junto a ellos forman el **semen** o **esperma.**

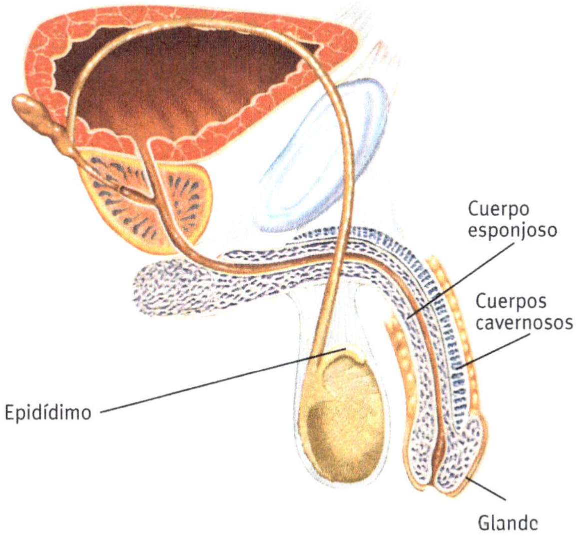

Si hay actividad sexual, el semen es eyaculado a través del orificio de la uretra. Si no es expulsado, se almacena en las vías genitales, donde en parte se reabsorbe, y en parte es expulsado por la orina. Durante la eyaculación, las **glándulas de Cowper** segregan un líquido lubricante que también se expulsa por la uretra.

El **pene** es un órgano adaptado para el coito, el acto por el cual el pene erecto se introduce en la vagina de la mujer. Posee un engrosamiento en su extremo, denominado **glande,** que está recubierto por el prepucio. Éste es un repliegue de piel que se puede desplazar sobre la superficie del glande y dejarlo al descubierto. El interior del pene está constituido por el cuerpo esponjoso, que rodea la uretra, y dos cuerpos cavernosos, que se engrosan llenándose de sangre durante la estimulación sexual y producen la erección necesaria para llevar a cabo el coito.

1 Espermatogonia
(2n=46)

Meiosis

4 Espermátidas
(n=23)

Diferenciación

4 Espermatozoides
(n=23)

1.1. *Desarrollo de las características del sexo masculino*

El aparato reproductor se forma durante el desarrollo embrionario.

FORMACIÓN DEL APARATO REPRODUCTOR MASCULINO

— Hasta la séptima semana de gestación, las estructuras precursoras del aparato reproductor son las mismas para ambos sexos, y están en contacto con el aparato excretor. A partir de ese momento, se inicia la diferenciación del aparato reproductor. Si el embrión es masculino, se forman los testículos en el interior de la cavidad abdominal.

— Los testículos empiezan a segregar hormonas, los **andrógenos,** que originan la diferenciación del aparato reproductor masculino, el cual mantiene una parte en común con el excretor: la uretra.

— A partir del tercer mes, los testículos comienzan a descender, hasta que, al final del embarazo se sitúan en el exterior de la cavidad abdominal.

— En el momento del nacimiento, los órganos reproductores están totalmente formados.

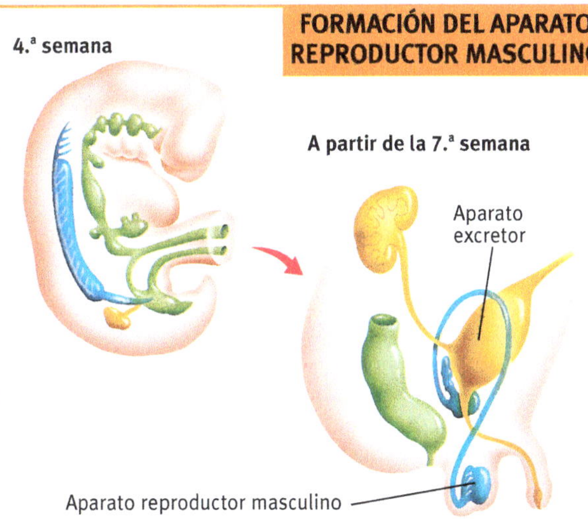

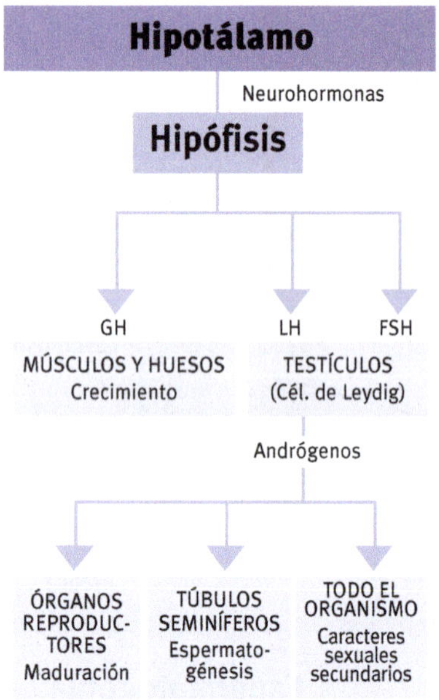

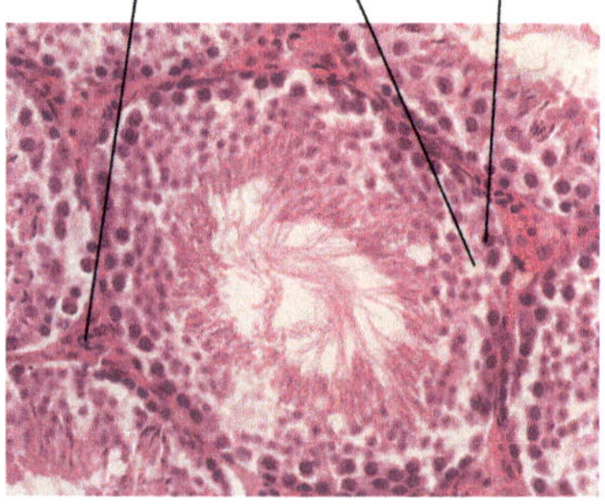

Las células de Sertoli protegen y nutren a las espermatogonias; las células de Leydig sintetizan andrógenos.

Pubertad

En esta etapa se inicia una serie de cambios que llevan a la adquisición de las características adultas: se *completa el crecimiento*, y se lleva a cabo la *maduración del sistema reproductor*, a la vez que el desarrollo de los *caracteres sexuales secundarios*. Se inicia aproximadamente a los 14 años y dura hasta los 17 ó 20. Los cambios que tienen lugar se deben a la secreción hormonal, que está regida por el hipotálamo.

- **Crecimiento.** El hipotálamo segrega neurohormonas que actúan sobre la hipófisis, donde se activa la producción de la **hormona del crecimiento (GH)**. Ésta actúa sobre numerosos tejidos y es responsable del aumento de la musculatura y de completar el proceso de osificación.
- **Maduración del sistema reproductor.** La hipófisis segrega también las hormonas **foliculoestimulante (FSH)** y **luteinizante (LH).**

  Estas hormonas activan el desarrollo de las células de Leydig en los testículos, que inician la síntesis de las hormonas sexuales masculinas: los **andrógenos,** principalmente la **testosterona.**

  Los andrógenos actúan sobre el sistema reproductor masculino produciendo:

  — La maduración y el crecimiento de los órganos reproductores.

  — El inicio de la espermatogénesis.

  Una vez alcanzada la madurez de los órganos reproductores, el proceso de la espermatogénesis se produce de forma constante. A partir de los 50 años comienza a descender el nivel de producción de espermatozoides.
- **Caracteres sexuales secundarios.** La actuación de los **andrógenos** produce cambios en el resto del organismo, que determinan las características físicas del hombre adulto. Algunos de los principales caracteres sexuales secundarios son:

  — Aumento de la musculatura.

  — Crecimiento de vello por todo el cuerpo, sobre todo en la cara, las axilas y el pubis.

  — Ensanchamiento de los hombros.

  — Modificación de la laringe, con lo que la voz se vuelve grave.

ACTIVIDADES

1. Observa el esquema de esta página y contesta:

— ¿Qué es la hipófisis? ¿De qué partes consta?

— ¿Qué es la una neurohormona ?¿Quién las produce? ¿Dónde actúan?

— Los andrógenos estimulan la espermatogénesis. ¿Qué es la espermatogénesis? ¿En qué lugar del testículo se produce? ¿A partir de qué momento en la vida de un individuo se inicia este proceso?

2. Describe la función de cada una de las siguientes estructuras:
células de Leydig, epidídimos, vesículas seminales, glándula de Cowper, cuerpo cavernoso, células de Sertoli.

2. El aparato reproductor femenino

Consta de los siguientes órganos:

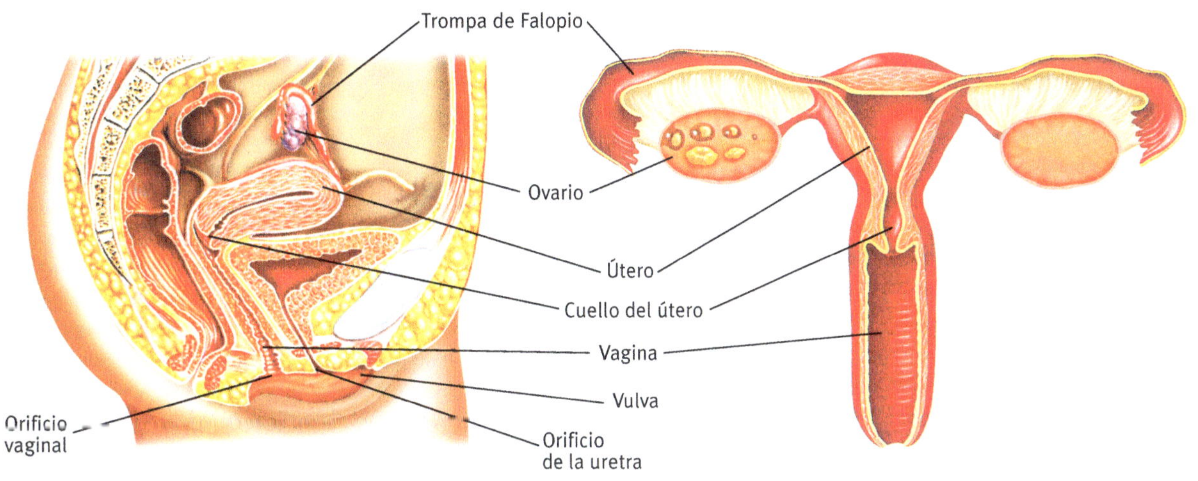

Su función básica es la producción de óvulos y, si se produce la fecundación, permitir el desarrollo del nuevo ser hasta llegar al parto.

Los **ovarios** son las dos gónadas femeninas. Contienen los **folículos primarios,** agrupaciones celulares que contienen **ovocitos,** los cuales darán lugar a los **óvulos.** Este proceso se denomina **ovogénesis.**

Normalmente, en cada ovogénesis madura un solo óvulo. Éste es transportado por una de las dos **trompas de Falopio** u **oviductos** desde el ovario hasta el **útero** o **matriz.** El epitelio interno de las trompas de Falopio posee células ciliadas, que provocan el desplazamiento del óvulo, y células secretoras de sustancias nutritivas, que le proporcionan los elementos necesarios para su supervivencia. El útero está formado por tres capas:

- El **perimetrio,** una capa externa de tejido conjuntivo.
- El **miometrio,** gruesa capa de musculatura lisa muy potente, que aumenta de tamaño durante el embarazo y se contrae en el parto.
- El **endometrio,** formado por una mucosa que se modifica, desprende y regenera periódicamente, y cuya función es proteger y nutrir al óvulo fecundado.

Si el óvulo no es fecundado se expulsa por la **vagina,** junto a las secreciones que se desprenden del endometrio. La vagina es un conducto de paredes musculosas, muy elásticas, adaptables al pene durante el coito.

La **vulva** es el conjunto de órganos genitales situados en la parte externa del organismo. Está formada por el monte de Venus, constituido por tejido adiposo y situado en la parte delantera del pubis; los labios mayores y los labios menores, formados por una serie de repliegues que rodean al **orificio de la uretra** y al **orificio vaginal;** y el clítoris, pequeño órgano eréctil muy sensible. Las **glándulas de Bartholino,** situadas en el orificio vaginal, producen una secreción lubricante.

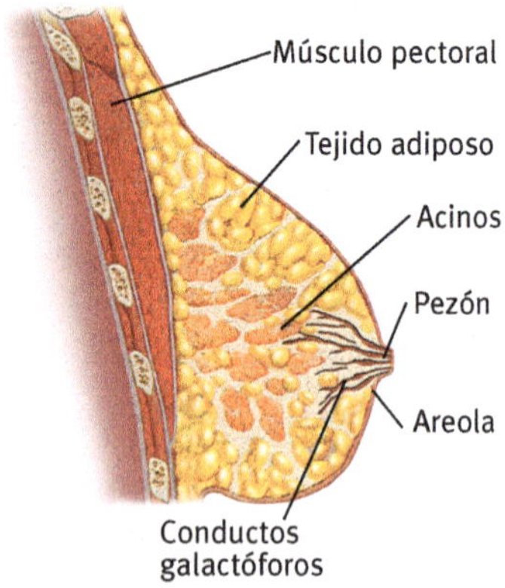

Las **mamas** tienen la función de producir y segregar leche para alimentar al recién nacido después del parto; por esta razón, pueden considerarse como integrantes del aparato reproductor femenino. Están constituidas por unas estructuras glandulares, denominadas **acinos mamarios,** y por los **conductos galactóforos,** muy ramificados, donde se vierten las secreciones de los acinos.

Los conductos galactóforos desembocan todos en el **pezón,** situado en medio de la **areola mamaria.**

2.1. *Desarrollo de las características del sexo femenino*

Al igual que el aparato reproductor masculino, los órganos reproductores femeninos se forman durante el desarrollo embrionario.

FORMACIÓN DEL APARATO REPRODUCTOR FEMENINO

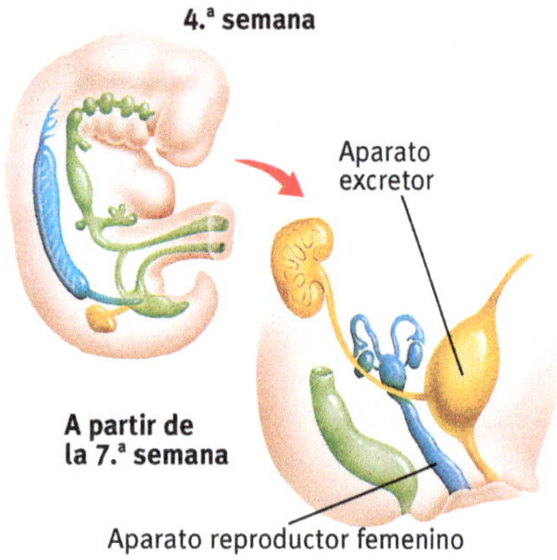

— Las estructuras precursoras del aparato reproductor son las mismas en ambos sexos.

— A partir de la séptima semana de gestación, si el embrión es femenino, se empiezan a diferenciar los ovarios. En este caso, por tanto, no hay secreción de andrógenos, lo que determina el desarrollo del aparato reproductor femenino, sin conexión con el aparato excretor.

— La **ovogénesis** se inicia ya durante el desarrollo embrionario. En los ovarios se forman los ovocitos, que se encuentran rodeados por el folículo primario, un conjunto de células nutritivas. Los ovocitos inician la meiosis para formar los gametos, pero la interrumpen y se mantienen en ese estado durante aproximadamente 12 años, momento en que comienza la pubertad. Al nacer, las niñas tienen unos 400 000 ovocitos, de los que a lo largo de la vida madurarán unos 400 ó 500.

En el momento del nacimiento, las mamas ya tienen todos los elementos necesarios para desarrollarse posteriormente y realizar su función.

Pubertad

Entre los 10 y los 14 años, y debido a la actividad hormonal, se produce la maduración del aparato reproductor femenino. El control hormonal que regula estos cambios es similar al de los chicos.

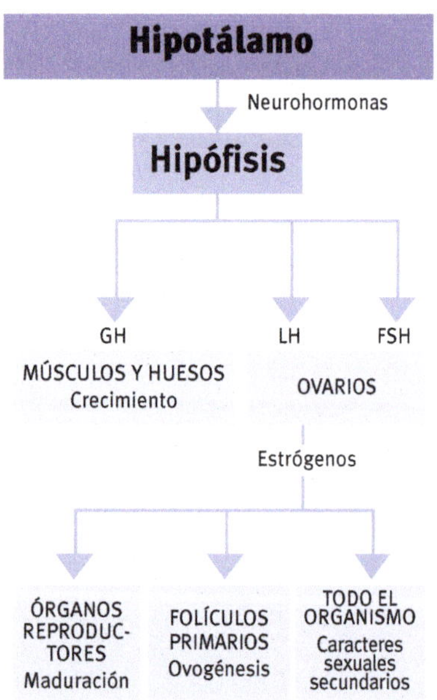

- **Crecimiento.** La hipófisis produce la **hormona del crecimiento (GH),** para completar el crecimiento propio de esta etapa.
- **Maduración del sistema reproductor.** La producción por parte de la hipófisis de hormona **foliculoestimulante (FSH)** y de hormona **luteinizante (LH)** provoca la síntesis de **estrógenos,** hormonas sexuales femeninas, por parte de los ovarios. Gracias a la producción de estrógenos, prosigue la ovogénesis iniciada durante el desarrollo embrionario. Continúa la meiosis y se forma un óvulo. También se produce un aumento del tamaño de las mamas, que alcanzarán su total desarrollo cuando la mujer embarazada llegue al final de la gestación, momento en que quedan preparadas para la producción y secreción de leche.
- **Caracteres sexuales secundarios.** Los estrógenos producen cambios como el aumento de tejido adiposo en las caderas y los pechos, el crecimiento de vello en las axilas y el pubis, y el ensanchamiento de la pelvis.

2.2. *Ciclo ovárico y ciclo menstrual*

En la pubertad se inicia el período fértil en las mujeres y se produce la primera menstruación, denominada **menarquia.** Aproximadamente entre los 40 y los 50 años, los ovarios dejan de madurar óvulos, cesan las menstruaciones y se inicia el período de la **menopausia.** La actividad del aparato reproductor da lugar a dos ciclos:

- **Ciclo ovárico.** Consiste en la maduración de un óvulo cada 28 días; se produce alternativamente en cada uno de los ovarios.
- **Ciclo menstrual.** Consiste en una serie de modificaciones del endometrio, que se engrosa con la finalidad de albergar al óvulo fecundado; si no hay embarazo, se desprende parte de la mucosa del endometrio unos 14 días después de la ovulación, originando la **menstruación** o **regla.**

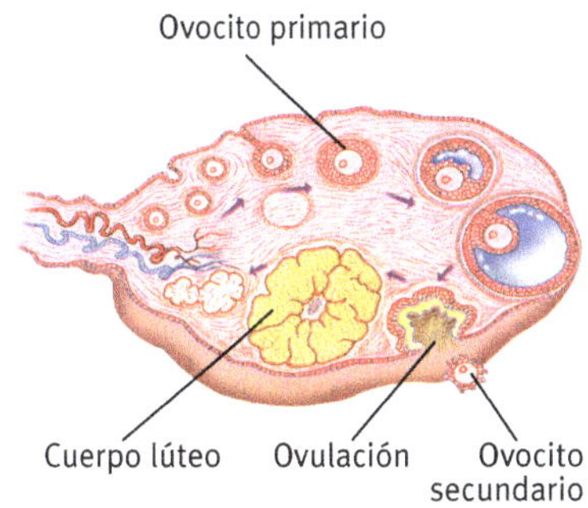

Ciclo ovárico

En el esquema siguiente se representan los cambios del endometrio durante ambos ciclos, así como la variación de las hormonas FSH, LH y progesterona:

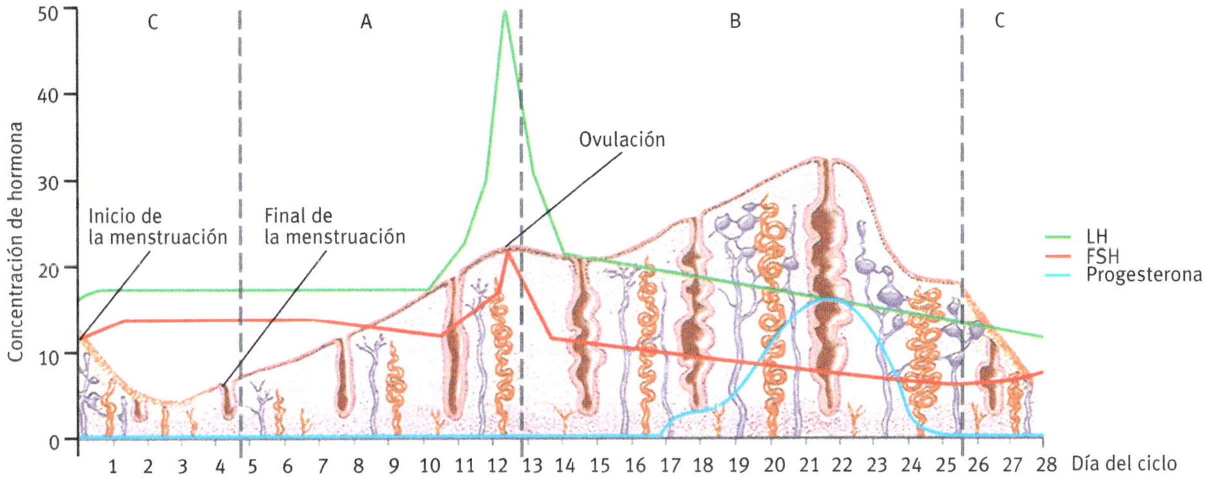

FASE	CICLO OVÁRICO	CICLO MENSTRUAL
A	Las hormonas foliculoestimulante y luteinizante actúan sobre un folículo primario, y se obtiene un folículo de Graaf. Durante este proceso, el folículo segrega estrógenos, que hacen que el ovocito prosiga la meiosis, interrumpida durante el desarrollo fetal, de modo que se produce un óvulo. La acumulación de líquido en el folículo de Graaf hace aumentar la tensión en él hasta provocar su rotura, tras la que el óvulo se libera, proceso que se denomina ovulación.	Los estrógenos producidos por el folículo de Graaf activan la división celular y el crecimiento en grosor del endometrio.
B	El folículo de Graaf roto se denomina cuerpo lúteo o amarillo, debido al color que adquiere. Éste segrega progesterona y mantiene su actividad si se produce un embarazo; si no es así, degenera. El óvulo es recogido por las trompas de Falopio, que lo transportan hasta el útero. Al no haber fecundación, el óvulo no termina la meiosis, y muere.	La progesterona hace que el endometrio se vuelva esponjoso y vascularizado, y segregue sustancias nutritivas. Si hay fecundación, el endometrio tendrá estas características durante todo el embarazo.
C	El cuerpo lúteo degenera y deja de segregar progesterona. A continuación comienza un aumento progresivo de la cantidad de hormona foliculoestimulante, que actúa sobre un folículo primario del otro ovario, y se inicia otro ciclo.	La mucosa del endometrio se expulsa por la vagina. Se llama menstruación, y dura unos cuatro días.

3. Fecundación, embarazo y parto

El óvulo sobrevive unas 24 horas después de la ovulación; por otro lado, los espermatozoides sobreviven hasta 72 horas en el aparato reproductor femenino. Así, si en el período comprendido entre tres días antes y un día después de la ovulación se produce el **coito,** puede haber fecundación. Por lo tanto, éste es el período fértil de la mujer.

En el momento de la eyaculación, centenares de millones de espermatozoides son depositados en la vagina y avanzan por el útero hacia las trompas de Falopio gracias al impulso que les proporciona el movimiento de su flagelo. Sólo unas decenas llegan a las trompas de Falopio y, si allí se encuentra un óvulo, sólo uno consigue introducir su cabeza a través de la membrana del óvulo.

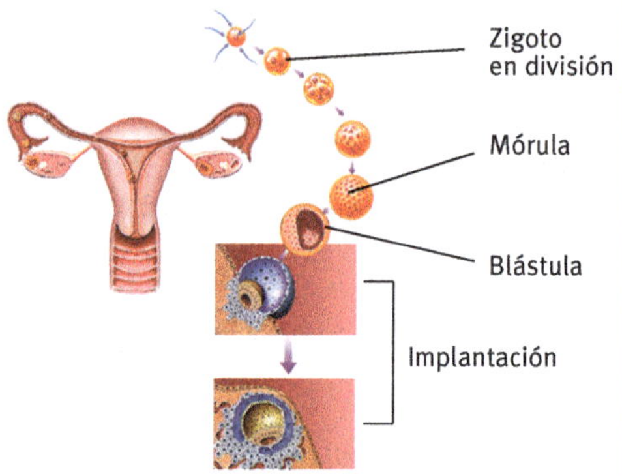

De la unión del óvulo y el espermatozoide, que son células haploides (23 cromosomas), se obtiene el **zigoto** o **célula huevo,** que es diploide (46 cromosomas).

Las sucesivas divisiones mitóticas del zigoto forman la **mórula,** que se desplaza desde las trompas de Falopio hacia el útero, mientras se sigue dividiendo.

En la fase de **blástula,** llega al útero. Una vez allí, se adhiere al endometrio, donde se desarrollará, para formar el nuevo ser.

3.1. *El embarazo*

Cuando la blástula alcanza la mucosa uterina se produce la **implantación,** mediante la cual, las células epiteliales del endometrio crecen alrededor de la blástula, que de este modo queda unida a la pared del útero. Han pasado siete días desde la fecundación y se considera que en este momento se inicia el embarazo, que durará 40 semanas.

La división y diferenciación celular sigue su curso, con la formación del **embrión** y las estructuras que lo envuelven y que se encargan de alimentarlo y protegerlo:

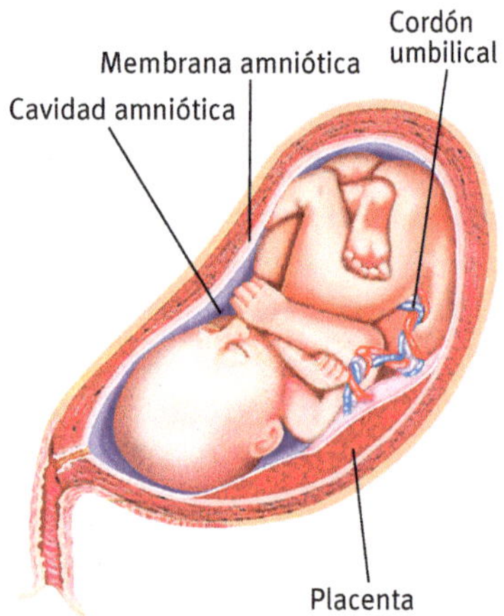

- La **cavidad amniótica** es un espacio rodeado por la **membrana amniótica** y lleno de líquido que rodea al embrión y le preserva de posibles golpes y de cambios de temperatura.
- La **placenta** es un órgano compuesto por muchos capilares sanguíneos, situado en el exterior de la cavidad amniótica. Comunica el sistema circulatorio de la madre con el del embrión. También actúa como filtro, ya que las partículas de determinado diámetro, como la mayoría de los microorganismos, no pueden atravesarla, aunque sí los medicamentos y las drogas.
- El **cordón umbilical** es un tubo largo y flexible que se origina en el ombligo del embrión, y se une a la placenta. Contiene dos arterias umbilicales, encargadas del transporte de oxígeno y sustancias nutritivas, y una vena, que transporta productos de desecho.

A lo largo del embarazo, tiene lugar una serie de cambios muy importantes:

Primer trimestre

Comienza a diferenciarse la mayor parte de los órganos. Así, se forma el sistema nervioso, los ojos con párpados, las orejas, la nariz y la boca con labios. Se distinguen los dedos de las manos y los pies. El corazón ya está formado.

Al final del primer trimestre el **embrión** pesa unos 20 gramos y mide 10 cm; tiene aspecto humano y comienza a moverse.

Segundo trimestre

Se completa el desarrollo de todos los órganos del nuevo ser, que a partir de ese momento recibe el nombre de **feto.** Los pulmones no están todavía preparados para respirar. El aparato reproductor ya está desarrollado.

La piel es muy fina, y en el cuarto mes, crece un fino vello, que recubre todo el cuerpo. Durante el quinto mes, el vello comienza a ser sustituido por pelo en las cejas y la cabeza; en el sexto mes, las glándulas sebáceas segregan una capa grasa que protege al feto de posibles roces.

El feto ya se mueve activamente y se chupa el dedo, y la madre puede sentir los movimientos.

El peso, al llegar al sexto mes de embarazo, es de unos 900 gramos. El feto tiene una talla de 32 cm.

Tercer trimestre

Se completa la maduración de los órganos, y los pulmones se preparan para respirar. El sistema nervioso tiene una madurez casi completa. A partir del séptimo mes de embarazo, si el parto se adelanta, las posibilidades de sobrevivir están entre un 60 y un 70 % de los casos. Debido al crecimiento, el espacio disponible en el útero se reduce, y el feto se mueve menos. Desaparece el vello y la piel se vuelve más gruesa.

Hacia el final del embarazo, el feto se sitúa boca abajo para encajar la cabeza entre los huesos coxales de la madre. En el momento de nacer, el peso medio es de 3000 a 3500 gramos, y la talla es de unos 50 centímetros.

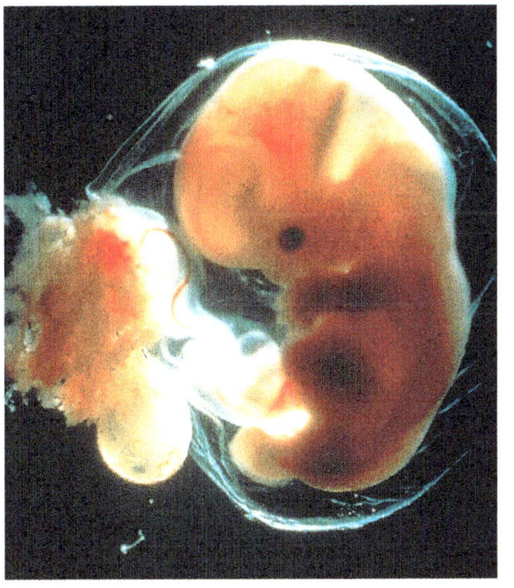

Embrión de seis semanas

PESO Y LONGITUD MEDIAS DURANTE EL DESARROLLO FETAL

	Edad	Peso (g)	Long. (cm)
Embrión	1er mes	2-3	0,75
	2º mes	5-8	3
	3er mes	18-20	15
Feto	4º mes	120	21
	5º mes	300	27
	6º mes	800-900	33
	7º mes	1300-1400	39
	8º mes	2200-2300	45
	9º mes	3200-3300	48-51

Control hormonal del embarazo

Mientras dura el período de gestación, la placenta segrega las hormonas necesarias para que el desarrollo sea el adecuado.

La placenta produce:

- **Estrógenos,** que favorecen el crecimiento del útero y la maduración de los acinos de las glándulas mamarias, donde se formará la leche.
- **Progesterona,** que mantiene las paredes del endometrio con la consistencia adecuada para el desarrollo del embarazo, y evita que haya contracciones del útero.
- **Gonadotropina coriónica (HCG),** que estimula el cuerpo lúteo del ovario, el cual seguirá produciendo progesterona durante toda la gestación.

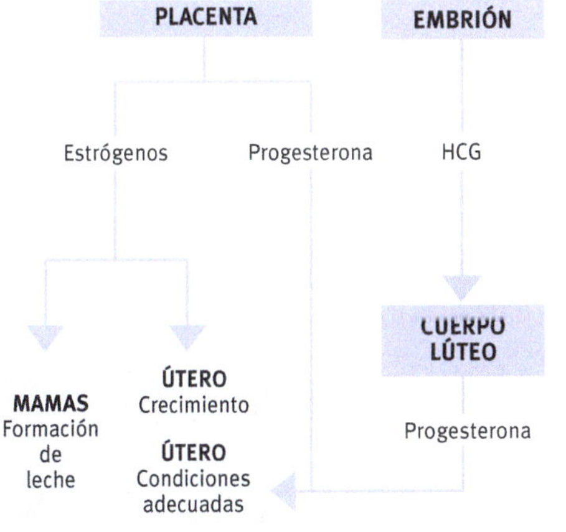

3.2. *El parto*

Al final del embarazo se produce el **parto,** que es la salida al exterior del feto y de las envolturas que lo rodean.

La fecha del parto puede calcularse del modo siguiente: se suman 10 días a la fecha de la última regla, y se avanza nueve meses del calendario. Así, si el día en que comenzó la última regla fue el 4 de enero, el día del parto será el 14 (10 + 4) de octubre (enero + 9 meses). Sin embargo, este cálculo es sólo aproximado, de modo que el parto puede producirse en el período comprendido entre 4 semanas antes y 2 semanas después de la fecha esperada. El parto suele durar entre 4 y 12 horas, y existen unos signos que indican su inicio:

- El comienzo de las **contracciones uterinas.** Son contracciones del miometrio, que se producen regularmente. Comienzan manifestándose cada cuarto de hora, o veinte minutos, y se van haciendo más fuertes e intensas.
- La **rotura de la membrana amniótica.** La presión que ejerce el feto provoca la rotura de las membranas que contienen el líquido amniótico, y la salida de éste a través de la vagina.

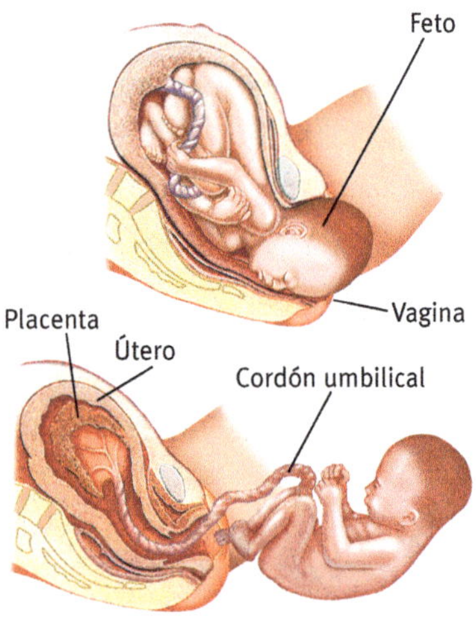

Fase de expulsión. Primero aparece la cabeza, luego el hombro y rápidamente se desliza el resto del cuerpo.

A partir de ese momento, se suceden tres fases:

- **Dilatación.** Debido a las contracciones uterinas, cada vez más frecuentes, se va ensanchando el cuello del útero. Cuando alcanza los 10 centímetros de diámetro, termina la fase de dilatación.
- **Expulsión.** Prosiguen las contracciones uterinas y, al mismo tiempo, el feto se abre paso hacia el exterior empujando con la cabeza. En el momento del nacimiento, primero aparece la cabeza, luego un hombro y, rápidamente, se desliza el resto del cuerpo. El cordón umbilical se pinza para evitar hemorragias, posteriormente se corta, y el recién nacido queda separado de la madre.
- **Alumbramiento.** Es la expulsión de la placenta, que se produce unos minutos después de la salida del feto.

Inmediatamente después del nacimiento, el niño es sometido a un primer examen médico, denominado **test de Apgar,** para comprobar su estado de salud y resolver rápidamente los posibles problemas que puedan surgir como consecuencia del parto. Posteriormente, mediante análisis de sangre y de orina, se practican las pruebas de detección de dos enfermedades hereditarias:

- El **hipotiroidismo congénito,** caracterizado por la producción insuficiente de hormona tiroidea, por lo que el sistema nervioso no madura. El hecho de que una mujer haya mantenido una dieta pobre en yodo durante el embarazo puede provocar esta enfermedad.
- La **fenilcetonuria,** debida a la falta de un enzima que degrada el aminoácido fenilalanina, lo que impide la maduración del sistema nervioso.

En el caso del hipotiroidismo congénito, el tratamiento consiste en administrar la hormona tiroxina. El tratamiento de la fenilcetonuria implica la limitación en la ingestión de alimentos que contengan fenilalanina. Ambos tratamientos se administran desde el nacimiento para evitar un retraso mental y lograr su curación total.

Control hormonal del parto

Desde el inicio hasta el final del parto, intervienen las siguientes hormonas:

- **Cortisol.** Se sintetiza en las glándulas suprarrenales del feto cuando éste alcanza la madurez suficiente, y pasa a la placenta a través del cordón umbilical. Ello produce cambios en la secreción hormonal de la placenta:
  - — Aumenta la producción de estrógenos.
  - — Disminuye la síntesis de progesterona.

  Como consecuencia de estos cambios, tienen lugar las primeras contracciones uterinas.
- **Oxitocina.** Es segregada por el hipotálamo como consecuencia de las primeras contracciones uterinas. Es la responsable de las contracciones que se producen durante el resto del parto.

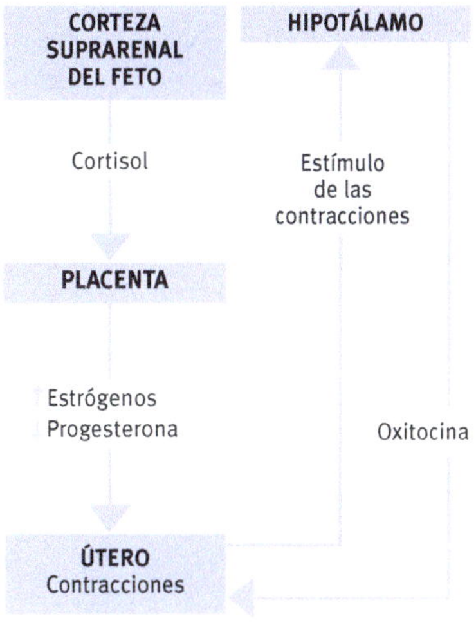

3.3. *La lactancia*

En la última fase del embarazo las mamas se preparan para segregar leche. Al nacer el niño, los acinos ya tienen preparada una primera secreción que se denomina **calostro.** Éste es el primer alimento que toma el recién nacido; contiene una gran cantidad de proteínas y anticuerpos que le servirán de defensa contra las enfermedades.

Posteriormente, varía la composición de la leche materna, ya que contiene más grasas y glúcidos.

Siempre que sea posible, es recomendable que el recién nacido se alimente de leche materna, por lo menos hasta los 4 meses de edad. En ocasiones, cuando la lactancia materna no es suficiente para la nutrición del bebé, se complementa la alimentación con leches maternizadas. Posteriormente, se irán introduciendo otros alimentos de forma progresiva.

+

La lactancia está controlada por el sistema nervioso central. Debido a ello, el estado de ánimo de la madre influye directamente en el proceso, por lo que es recomendable dar de mamar en las condiciones más tranquilas posibles.

Mecanismo y regulación hormonal de la lactancia

El mecanismo de la lactancia se basa en la succión que efectúa el recién nacido en el pezón materno. El reflejo de succión es un comportamiento innato en el ser humano, igual que en el resto de los mamíferos.

- El estímulo de la succión se transmite al sistema nervioso central, donde se procesa, iniciando la actividad del hipotálamo. Éste segrega la hormona **oxitocina,** que provoca la contracción de pequeñas fibras musculares que rodean el tejido glandular de las mamas y producen la salida de la leche al exterior.

  Al mismo tiempo, la oxitocina también origina una reducción progresiva del tamaño del útero, que había aumentado su volumen durante el embarazo.
- La hipófisis, activada por neurohormonas del hipotálamo, produce hormona **prolactina,** responsable de la formación de leche en los acinos mamarios.

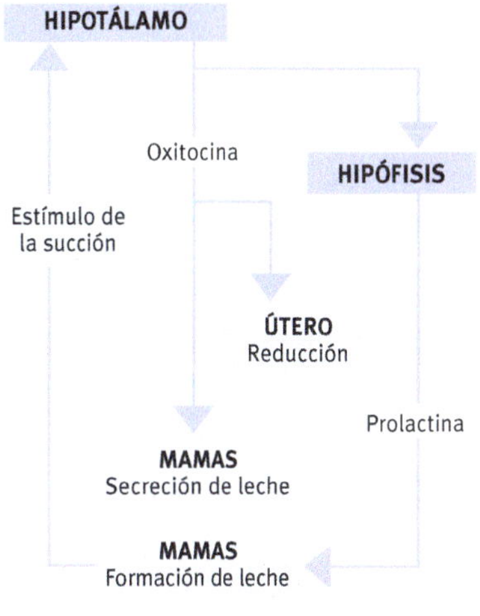

C

El primer nacimiento de un ser humano por fecundación *in vitro* (FIV) tuvo lugar en 1978. Desde entonces, un gran número de parejas con problemas reproductivos ha conseguido engendrar un hijo. Son los llamados niños probeta.

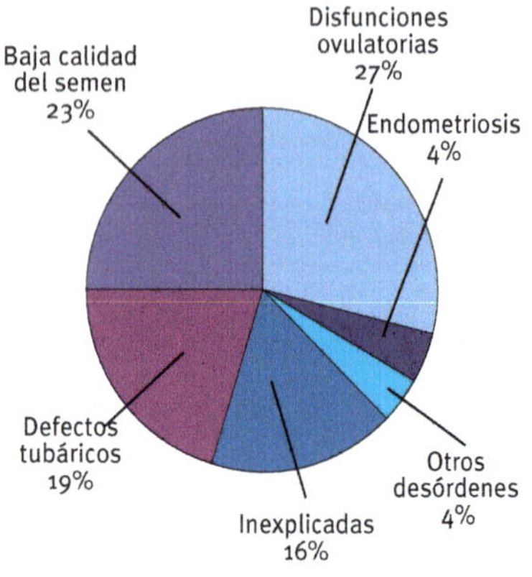

Los factores que causan la esterilidad son múltiples y pueden ser debidos tanto al funcionamiento incorrecto del aparato reproductor femenino como del masculino.

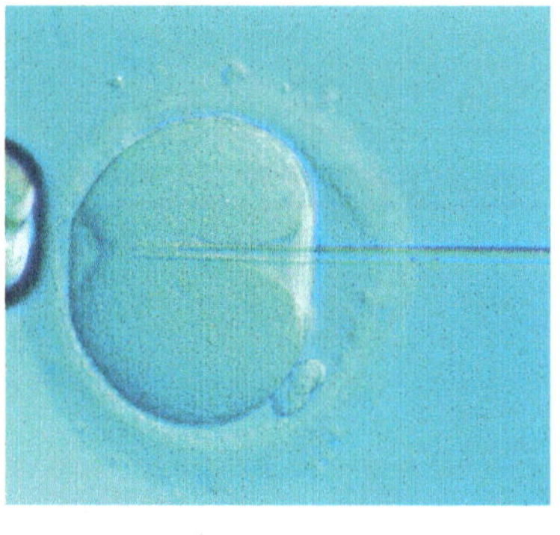

Cuando hay pocos espermatozoides o éstos tienen poca movilidad, se utiliza la microinyección bajo control microscópico, para introducir un espermatozoide en el interior del óvulo.

3.4. *Planificación familiar*

Todas las actuaciones destinadas a incrementar las posibilidades de tener un hijo o a evitar un embarazo no deseado entran en el campo de la **planificación familiar.**

En la actualidad, existen diversos centros de planificación familiar donde equipos de profesionales (ginecólogos, sexólogos, asistentes sociales...) orientan a las parejas acerca del modo más adecuado para solucionar su problema, sea éste la dificultad para tener un hijo, o bien el deseo de no concebirlo.

Técnicas de reproducción asistida

Actualmente, gracias a los avances en medicina, existen diversas técnicas para incrementar las posibilidades de tener hijos. Según diversos estudios estadísticos, se ha comprobado que un 15 % de las parejas en edad reproductiva presenta problemas de esterilidad. Los centros especializados poseen métodos para solucionar cada tipo de problema:

- El **tratamiento** mediante las **hormonas** que inducen a la ovulación se prescribe en casos de trastornos en el funcionamiento de los ovarios. También se administran para estimular la espermatogénesis.
- La **inseminación artificial** consiste en introducir semen, extraído con anterioridad del hombre, en las trompas de Falopio de la mujer. Esta técnica es la más utilizada para solucionar problemas de cantidad de espermatozoides o falta de movilidad de éstos.
- La **fecundación *in vitro*** consiste en favorecer la unión de óvulos y espermatozoides en un recipiente de laboratorio. Se utiliza principalmente en los casos de obstrucciones de las trompas de Falopio o los epidídimos. En ella se siguen estos pasos:
  - — Tratamiento hormonal de la futura madre para inducir una ovulación múltiple.
  - — Punción de los folículos de Graaf a fin de recoger los óvulos que contienen.
  - — Obtención de semen del futuro padre e inseminación, mediante la fecundación *in vitro*, de los óvulos obtenidos.
  - — Transferencia de los embriones obtenidos de la fecundación *in vitro* al útero materno. Se suelen introducir dos o tres embriones para garantizar que al menos uno se implante y siga el proceso de desarrollo.

ACTIVIDADES

3. Explica las diferencias existentes entre el proceso de la espermatogénesis y el de la ovogénesis.
4. Define los siguientes conceptos: cuerpo lúteo, folículo de Graaf, ovulación, menstruación.
5. Inseminación artificial y fecundación *in vitro* son dos técnicas de reproducción asistida. Explica qué diferencias existen entre estas dos técnicas.

Métodos anticonceptivos

Los más utilizados son:

MÉTODO	CARACTERÍSTICAS
Métodos de abstinencia periódica Efectividad: 60-77%	Se basan en el conocimiento exacto del día de ovulación para evitar así mantener relaciones sexuales durante el período fértil del ciclo menstrual. Son los métodos menos fiables ya que, en una misma mujer, el ciclo menstrual puede variar de forma irregular debido a factores como el estrés, los cambios de clima, enfermedades, etc. **Método Ogino:** La mujer debe determinar la duración de cada uno de sus ciclos menstruales durante un año y un especialista calcula el momento en el que la ovulación es más probable deduciendo así el período fértil. **Temperatura basal:** Determinación del momento de ovulación según las leves variaciones de temperatura asociadas a las diferentes fases del ciclo menstrual. **Método sincrónico:** Se basa en combinar los dos métodos anteriores. No evitan las enfermedades de transmisión sexual (ETS)
Métodos de barrera Efectividad: 88-90%	Se basan en impedir que los espermatozoides lleguen al útero y las trompas de Falopio. Los más extendidos son el diafragma, el preservativo masculino y el preservativo femenino; los dos últimos evitan las ETS. **Diafragma:** Dispositivo semiesférico de goma con un anillo elástico que adapta al cuello del útero impidiendo el paso de los espermatozoides. Sus uso debe ser prescrito por un ginecólogo. **Preservativo masculino:** recubrimiento delgado de caucho, vinilo o productos naturales que se coloca sobre el pene erecto antes de la eyaculación. Pueden tratarse con espermicida para ofrecer mayor protección. **Preservativo femenino:** recubrimiento delgado de plástico poliuretano con aros de poliuretano en cada extremo. Estos se introducen en la vagina antes del coito.
Métodos químicos	Se basan en la utilización de determinados productos químicos que pueden alterar las células implicadas en la fecundación o inhibir su formación. **Espermicida:** Productos químicos que provocan la degeneración de los espermatozoides. No se utilizan como método único sino combinados con otros métodos. **Anovulatorios:** Se basan en preparados hormonales que impiden la segregación de FSH y LH, por lo que se impide la ovulación. Poseen una efectividad del 97%. Su uso debe ser prescrito por un ginecólogo.
DIU Efectividad: 99%	Dispositivo flexible de unos 3 cm y en forma de T que se inserta en el útero alterando el endometrio y inhibiendo así la implantación. Actualmente se utilizan dos tipos: el de cobre recubierto de plástico y el liberador de hormonas. Debe ser colocado por un ginecólogo.
Métodos quirúrgicos Efectividad: 100%	Se consideran métodos de esterilización voluntaria ya que en la mayoría de los casos no son reversibles. **Vasectomía:** Es la intervención de esterilización masculina. Consiste en cortar los conductos deferentes y aislarlos de forma que los espermatozoides no puedan incorporarse al semen. **Ligadura de trompas:** Es la intervención quirúrgica femenina. Consiste en cortar las trompas de Falopio y aislar los extremos cortados, para que el óvulo no entre en contacto con los espermatozoides.

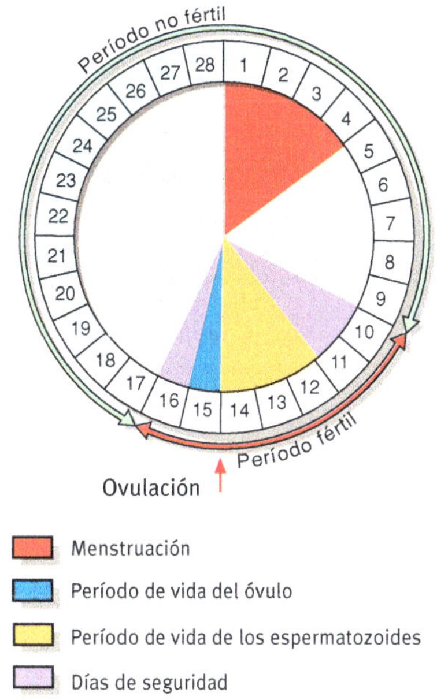

Cálculo del período fértil para una mujer con ciclos regulares de 28 días

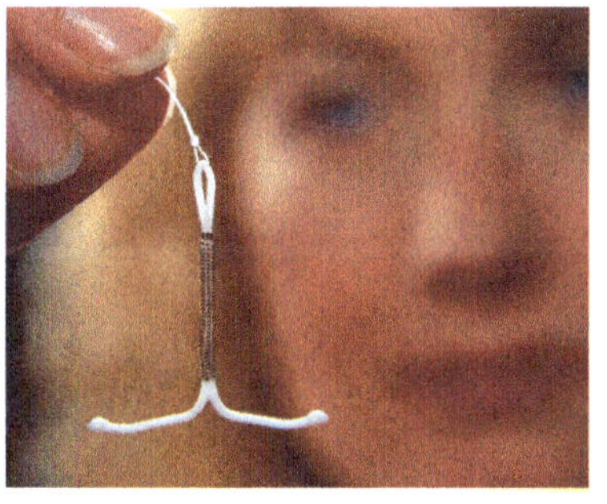

Dispositivo intrauterino (DIU)

C

El desarrollo de las técnicas para evitar un embarazo, junto con las de reproducción asistida, es un motivo de controversia en nuestra sociedad, ya que aparte de los aspectos puramente científicos, intervienen factores éticos, morales, religiosos...

4. La sexualidad

Hasta este momento, hemos tratado la función fisiológica del aparato reproductor humano. Sin embargo, las relaciones sexuales humanas no tienen como único fin la reproducción. Son una forma de comunicación entre dos personas que expresan sus sentimientos, tengan o no la intención de concebir un nuevo ser.

La relación sexual no se limita únicamente al coito, sino que entran en juego muchos estímulos (visuales, olfatorios y, sobre todo, táctiles), que producen respuestas agradables.

En la sexualidad humana, el cariño y la ternura permiten disfrutar de la relación sexual de una forma más completa. Además, son imprescindibles el acuerdo y la complicidad, ya que nunca puede llevarse a cabo una acción en contra de la voluntad de la otra persona.

Un mismo estímulo puede desencadenar respuestas distintas en diferentes personas, ya que existen factores como la capacidad imaginativa, la educación recibida o la influencia cultural, que condicionan la vida sexual de las personas.

Durante una relación sexual, tiene lugar una serie de estímulos que causan una progresiva excitación, de forma que los órganos sexuales se preparan para el coito; se produce la erección del pene y la lubricación de la vagina.

Posteriormente, si se alcanza un cierto nivel de tensión sexual, aunque no exista coito, se produce el orgasmo, que es el punto culminante de la respuesta sexual. Se caracteriza por un intensa sensación de placer que, en el caso del hombre, va acompañada de la eyaculación.

Por último, hay una fase de relajación, y el organismo regresa al estado anterior a la excitación.

C

La **identidad sexual** es el sentimiento de feminidad o masculinidad de una persona. La forma más extendida de relación sexual es la que se produce entre dos personas de distinto sexo; es la denominada relación **heterosexual**. En otras ocasiones, el sexo biológico de un individuo no corresponde con su identidad sexual, y existe una atracción hacia personas del mismo sexo. Esta relación se denomina **homosexual.**

5. Enfermedades de transmisión sexual

Se incluyen dentro de este grupo todas aquellas enfermedades que pueden transmitirse a través de una relación sexual, aunque algunas de ellas pueden contagiarse también por otras vías. Si una persona enferma mantiene relaciones sexuales con una persona sana, el riesgo de contagio de esta última es muy elevado.

A pesar de los avances médicos, estas enfermedades continúan contrayéndose en la actualidad debido, principalmente, a dos factores:

- Mayor promiscuidad de las relaciones sexuales, es decir, que éstas se producen con distintas personas y no con una sola pareja; e inicio de las relaciones en edades más tempranas.
- Falta de información acerca de los métodos preventivos.

Son enfermedades de transmisión sexual la *gonococia* y la *sífilis,* causadas por bacterias, y el *herpes genital, la hepatitis B y el sida*, provocadas por virus. Otros microorganismos, como protozoos y hongos, pueden causar dolencias que afectan a los órganos sexuales.

Éstas son algunas de las enfermedades de transmisión sexual más frecuentes. Todas se previenen utilizando el preservativo:

ENFERMEDAD	CARACTERÍSTICAS
Infección gonocócica Producida por la bacteria *Neisseria gonorrhoeae*	En los hombres, la parte infectada es la uretra, y los síntomas son la necesidad frecuente de orinar y el dolor que produce la emisión de la orina, así como una secreción amarillo-verdosa. En las mujeres, afecta a la vagina, y el único síntoma es la secreción amarillo-verdosa. Puede transmitirse al feto durante el parto y producirle una fuerte irritación en los ojos denominada **oftalmia gonocócica.** Por esta razón, para prevenir la infección, a todos los recién nacidos se les aplican unas gotas desinfectantes después del parto.
Sífilis Producida por la bacteria *Treponema pallidum*	En la zona genital se producen unas lesiones similares a las llagas denominadas chancros. Si no se trata, la enfermedad causa deterioros en los sistemas nervioso y circulatorio. Puede transmitirse al feto a través de la placenta, causando la **sífilis congénita,** que provoca graves deformaciones óseas, llagas y trastornos como la neumonía.
Herpes genital Producida por el virus del herpes simple (VHS)	En los genitales aparecen vesículas en forma de racimos, que pueden extenderse a las zonas próximas y ulcerarse. Causan un picor y dolor intensos, además de fiebre y dolor de cabeza.
Hepatitis B Producida por el virus de la hepatitis B (VHB)	Se produce una inflamación y destrucción de las células del hígado. En muchos casos no presenta síntomas. En el caso de que éstos se manifiesten son: fiebre, vómitos, dolores abdominales e ictericia, es decir, coloración amarillenta de la piel. La **hepatitis B** puede hacerse crónica, con lo que el hígado degenera, y puede desarrollarse cirrosis o cáncer. Además de la sexual, existen otras vías de transmisión de la enfermedad: • Sanguínea, a través de jeringuillas, cuchillas de afeitar y cepillos de dientes. • Materno-filial, a través de la placenta, o más probablemente en el momento del parto, cuando la sangre de la madre se pone en contacto con el feto.
Clamidiasis Producida por la bacteria *Chlamydia trachomatis*	Puede infectar la vagina, el cuello del útero, las trompas de Falopio, el ano, la uretra y los ojos. Los síntomas más habituales suelen ser: sangrado entre períodos menstruales o después del coito, dolor abdominal, fiebre, necesidad de orinar más de lo habitual y micciones dolorosas. Si no se trata, puede causar problemas serios de salud como la **enfermedad inflamatoria pélvica,** que puede causar infertilidad o embarazos ectópicos.
Verrugas genitales **Cáncer de cuello de útero** Producidos por el virus del *papiloma humano* (PVH)	El **PVH** es la causa principal del cáncer de cuello uterino, causando también las llamadas verrugas genitales. Se transmite por contacto directo piel con piel. La prueba de Papanicolau permite una detección temprana del virus aumentando las probabilidades de éxito del tratamiento. Actualmente existe una **vacuna** que está en proceso de incorporación al calendario oficial de vacunación.

Debido a la incidencia del sida y a la problemática que rodea la enfermedad, vamos a exponer sus características en el siguiente apartado.

+

Según la Organización Mundial de la Salud (OMS), cada año aparecen 250 millones de nuevos casos de personas afectadas por estas enfermedades, la mayoría de las cuales son adolescentes y jóvenes entre 15 y 35 años.

+

George Nicholas Papanicolau (1883-1962) nació en 1883 en la isla griega de Euboea. Se doctoró en zoología en Alemania. En 1913 llegó a Nueva York donde entró en la universidad de Cornell y pudo avanzar en sus estudios. Una de sus aportaciones más importantes fue el test que lleva su nombre o "Paptest". Esta técnica sigue siendo hoy en día la más utilizada y adecuada para obtener pruebas de células precursoras del cáncer de cuello uterino, diagnosticando el 95% de los cánceres cervicales en un estado temprano por lo que pueden ser tratados y la mayoría curados.

5.1. *El sida*

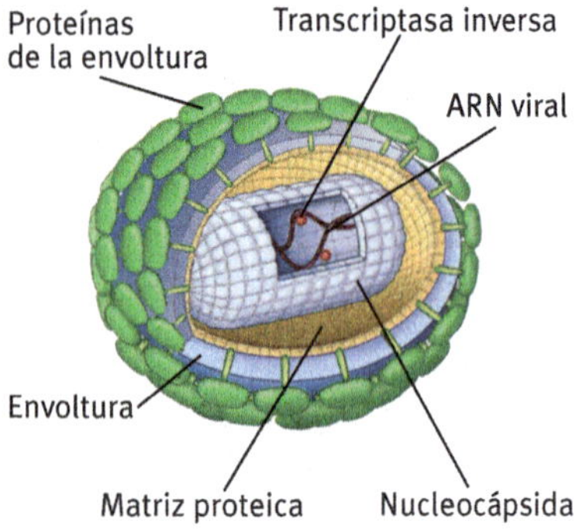

Estructura del VIH

El nombre de la enfermedad **(Síndrome de Inmunodeficiencia Adquirida)** ya define sus características. Así, es un conjunto de alteraciones, o *síndrome,* que se produce cuando el virus de la inmunodeficiencia humana (VIH) ataca al sistema *inmunitario* y provoca su funcionamiento *deficiente;* se *adquiere* por contagio.

El virus causante del sida fue descubierto en el Instituto Pasteur de París, en 1983. En el transcurso de los dos años siguientes, se pudieron conocer las principales características del virus:

- Es un **retrovirus,** es decir, su ARN se transcribe en ADN durante el proceso de infección.
- Ataca, principalmente, a los **linfocitos T4,** que forman parte del sistema inmunitario humano.

El organismo queda sin capacidad para defenderse y se producen infecciones repetidas y tumores, que son la causa de la muerte de los enfermos de sida. El mecanismo de acción del virus es el siguiente:

MECANISMO DE ACCIÓN DEL VIH

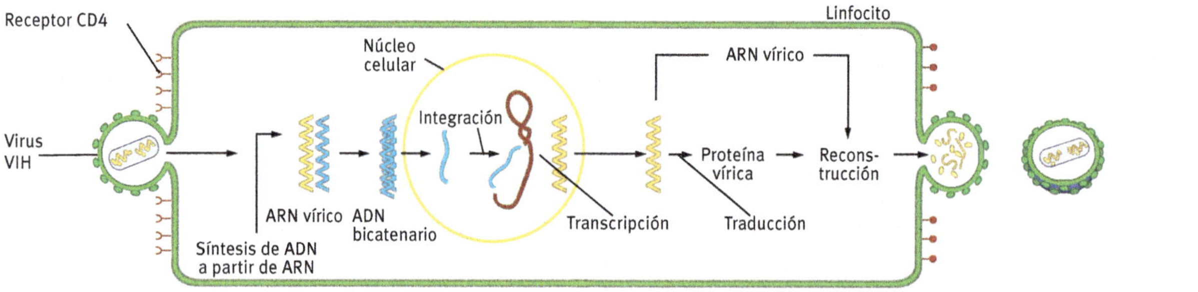

El virus se fija en los receptores CD4 del linfocito e introduce su nucleocápsida en la célula.

Mediante la transcriptasa inversa el ARN vírico se transcribe a ADN. Este ADN se dirige al núcleo del linfocito llevando a cabo un ciclo lisogénico.

Transcurrido cierto tiempo, el virus pasa a seguir un ciclo lítico: se inicia la síntesis de ARN vírico y proteínas víricas y se forman nuevos virus, que acaban rompiendo la membrana de la célula huésped cuando salen para infectar a otros linfocitos.

Para que el virus llegue a los linfocitos, existen tres **vías de contagio:**

- **Vía sanguínea.** Son vías de contagio los contactos directos con sangre contaminada, en una herida, en transfusiones y, en especial, a través de agujas y jeringuillas contaminadas. En España, ésta es la principal vía de contagio, y afecta a un 60 % de los enfermos de sida.
- **Vía sexual.** Las secreciones vaginales o el semen transmiten el virus, si se ponen en contacto con las mucosas que recubren los órganos sexuales. El riesgo es mayor si existen pequeñas heridas o úlceras en las zonas que entran en contacto. El contagio se produce tanto en relaciones homosexuales, como heterosexuales.
- **Vía materno-filial.** Cuando una mujer embarazada está infectada por el virus, puede transmitirlo al feto a través de la placenta, también en el momento del parto, a través de pequeñas heridas, o por la leche materna durante la lactancia.

C

Es muy importante destacar que, aunque la saliva y las lágrimas de los enfermos de sida contienen algunos virus, no existe la cantidad suficiente de éstos para contagiar a otra persona.

Síntomas y desarrollo de la enfermedad

Una vez que el virus ha entrado en el torrente sanguíneo, se suceden una serie de fases en la evolución de la enfermedad, que pueden variar de un individuo a otro.

— A los pocos días, aparecen unas **molestias parecidas a un resfriado,** con fiebres altas, y en ocasiones, una erupción que desaparece a las pocas semanas. Esto es debido a que el organismo comienza a desarrollar los primeros anticuerpos para combatir el virus.

— Si se practica un análisis de sangre durante los tres meses posteriores a la infección, puede ser que no se detecten anticuerpos, ya que normalmente éstos se encuentran aún en concentraciones muy pequeñas. Éste es el denominado **período ventana.**

— A partir del tercer mes pueden detectarse anticuerpos, y se dice que las personas que los presentan son **seropositivas.** Tienen la capacidad de transmitir el virus a otras personas, aunque quizá nunca presenten sida o tarden años en desarrollarlo.

— El **desarrollo de la enfermedad** varía según cada enfermo. En general, al cabo de cinco años de haber contraído la infección, el 14 % de los afectados desarrolla el sida; a los diez años, afecta al 50 %.

La destrucción de los linfocitos T4 provoca la aparición de las primeras enfermedades infecciosas y tumores. A continuación, exponemos los problemas de salud más frecuentes.

Neumonía	Es una infección del sistema respiratorio, que puede ser causada por varios microorganismos. En el caso de los enfermos de sida, la infección es producida por un protozoo poco habitual en personas sanas, *Pneumocystis carinii.*
Sarcoma de Kaposi	Es un cáncer de piel que produce manchas de color violáceo, las cuales suelen estar localizadas sobre todo en el tronco y la cabeza. Puede extenderse por debajo de la piel a otros órganos. Se manifiesta de forma más o menos leve, y no suele ser causa de muerte.
Tuberculosis	Es una infección de los pulmones que, en fases más avanzadas de la enfermedad, en las que el sistema inmunológico está más deteriorado, se extiende a otros órganos y a los huesos. Está producida por *Mycobacterium tuberculosis.*
Candidiasis	Es una infección causada por el hongo *Candida albicans.* Coloniza la boca, las comisuras de los labios y el esófago, en forma de placas blanquecinas que, cuando se desprenden, dejan una zona erosionada.
Encefalopatía	Es un conjunto de lesiones del sistema nervioso que producen fiebre, parálisis y temblores, y afectan al comportamiento. Puede ser causada por varios microorganismos, pero el más común es *Toxoplasma gondii.*
Caquexia	Es un estado de extrema delgadez causado por problemas digestivos, como las continuas diarreas, y el deterioro general causado por el síndrome. El enfermo está desnutrido, sufre anemia y presenta un bajo estado de ánimo.

— Todas estas enfermedades suelen presentarse de forma conjunta y con continuos rebrotes, causando el deterioro del cuerpo humano y, posteriormente, la muerte.

Distribución de la enfermedad

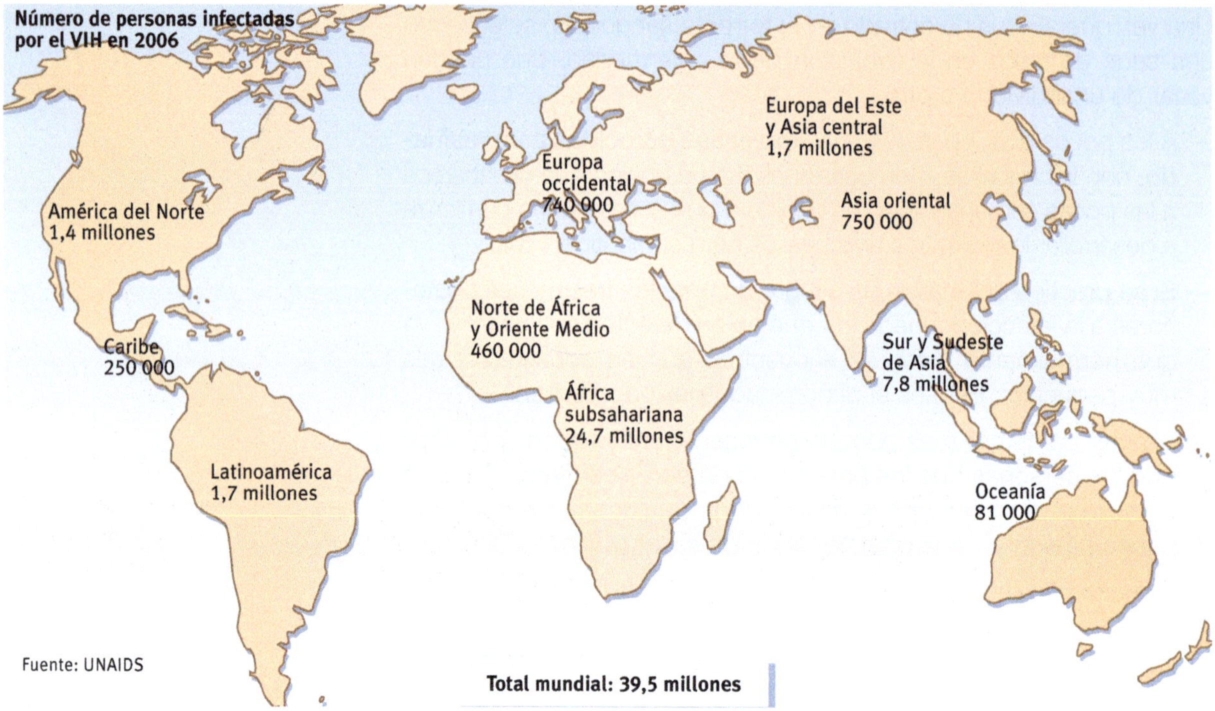

C

Endemia. Enfermedad, normalmente de tipo infeccioso, que afecta a un elevado número de habitantes de una determinada zona geográfica, y se mantiene allí permanentemente.

Epidemia. Enfermedad infecciosa que, de manera transitoria, afecta a un elevado número de personas.

Pandemia. Enfermedad epidémica que se extiende a muchos países.

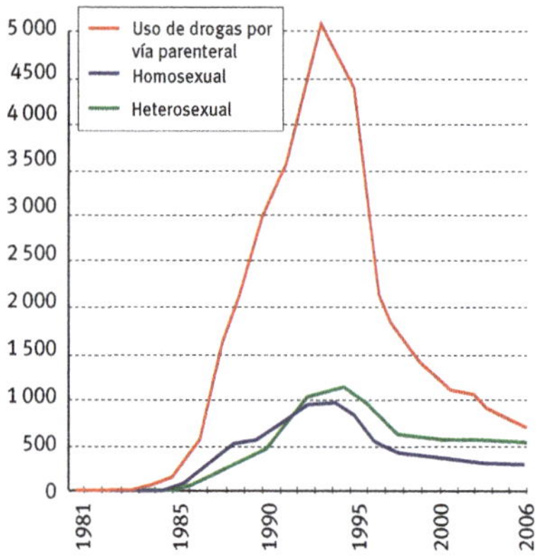

Casos nuevos de SIDA en España por categoría de transmisión (Fuente: Ministerio de Sanidad)

El sida es la enfermedad infecciosa más grave aparecida en la actualidad. Afecta a millones de personas en todo el mundo, por lo cual adquiere la denominación de pandemia. La mayoría de los afectados se concentra en el África subsahariana y en el sur y sudeste de Asia. La vía de contagio predominante varía según la zona. Así, en Europa y América del Norte la mayor parte de los enfermos son consumidores de drogas por vía intravenosa y homosexuales, mientras que en África el contagio más frecuente son las relaciones heterosexuales.

La enfermedad se da, principalmente, en los adultos jóvenes, lo que interfiere en el desarrollo social y económico de los países más afectados, que son los más pobres.

España es uno de los países europeos más castigados por esta enfermedad, a causa de las drogas. El primer caso de sida se detectó en 1981; desde entonces hasta junio de 2007 se registraron 74 885 casos. En los últimos años se ha observado una disminución de los contagios por consumo de drogas por vía intravenosa y por transmisión homosexual, acompañada de un incremento relativo de los contagios por transmisión heterosexual.

Hasta el momento no se dispone de una vacuna para la prevención ni de un tratamiento eficaz que pueda ser aplicado a todos los afectados. El esfuerzo de los científicos se centra, principalmente, en combatir el virus cuando ya ha infectado al paciente, aunque también se está trabajando para encontrar una vacuna. La única medida preventiva consiste en evitar el contagio; pero, a pesar de ello, la información sobre las medidas de prevención y el material para llevarlas a cabo no suelen llegar a grupos de población marginales o a países menos desarrollados.

Prevención

El conocimiento actual acerca del virus y sus vías de contagio permite establecer unas normas concretas y muy sencillas para evitar contraer la enfermedad.

- Utilizar preservativo cuando se mantengan relaciones con una pareja no estable o en el caso de que la pareja, a pesar de ser estable, pueda estar infectada.
- No se deben compartir jeringuillas, cuchillas de afeitar, cepillos de dientes u otros objetos que puedan haber estado en contacto con la sangre de personas infectadas.
- Las mujeres que padezcan la enfermedad, o sean portadoras del virus, deben evitar tener hijos. Si los tienen, no han de amamatarlos.

No debe temerse el contagio en un centro hospitalario, ya que el material que se utiliza para llevar a cabo las extracciones de sangre o las transfusiones es de un solo uso. Además, la sangre de los donantes es sometida a minuciosos análisis.

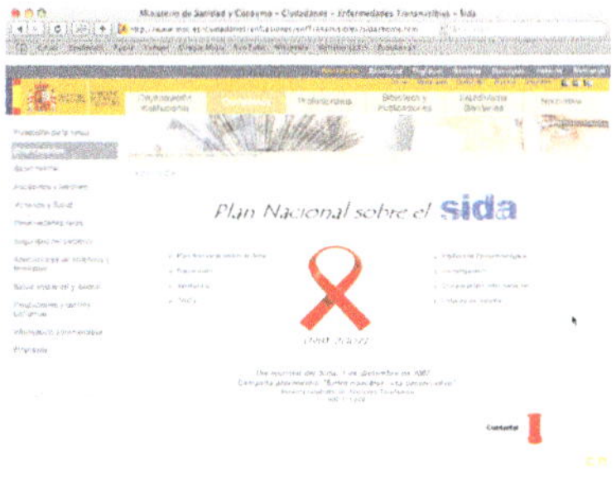

Desde distintas instituciones se realizan campañas de información para prevenir el sida.

Diagnóstico

El método más utilizado para diagnosticar el sida es la **técnica de ensayo inmunoenzimático,** que permite detectar la presencia de anticuerpos específicos contra el VIH en la sangre. Como ya hemos indicado, en la mayoría de los casos éstos ya se han desarrollado a los tres meses de la infección.

Para cuantificar el número de copias del virus por mililitro de sangre se utiliza el método denominado de la **carga viral.**

El recuento de linfocitos T4 en la sangre también es útil para conocer el grado en que ha progresado la infección. Cuanto más bajo es el nivel de linfocitos T4, más avanzada se encuentra la enfermedad y, por tanto, peor es su pronóstico.

Tratamiento

En el corto período de tiempo que ha transcurrido desde el descubrimiento del virus, el tratamiento de la enfermedad ha variado sustancialmente.

Desde el principio, se ha intentado buscar un método para destruir a los virus. Se comenzó utilizando fármacos que, en muchas ocasiones, producían efectos tóxicos debido a su actividad agresiva sobre el organismo. Actualmente se usan combinaciones de fármacos también agresivas, pero que resultan más efectivas sobre el virus, y que se adaptan a cada paciente. Se ha observado que, si se ataca al virus desde las primeras etapas de la infección, se obtienen resultados muy esperanzadores.

Los nuevos tratamientos se basan en utilizar dos o tres inhibidores de los enzimas que necesita el virus del sida para completar su ciclo vital, principalmente de la *transcriptasa inversa.* Los inhibidores evitan que este enzima desarrolle su actividad, con lo que el virus no puede replicarse.

La película Philadelphia, protagonizada por el actor Tom Hanks, refleja los problemas personales y sociales a los que se ve enfrentado un enfermo de sida.

+

Las siguientes direcciones de internet corresponden a organismos oficiales donde puedes obtener información acerca del sida.

Ministerio de Sanidad:

- www.msc.es/sida

Centro europeo de seguimiento del sida:

- www.eurohiv.org

Programa sobre sida de la ONU:

- www.unaids.org

El recuento de linfocitos T4 y el método de la carga viral sirven para determinar el tratamiento que se ha de seguir. También son útiles para averiguar si el tratamiento utilizado es efectivo, ya que si es así, la carga viral debe disminuir pasado un tiempo.

En algunas personas deja de detectarse el virus después de seguir un tratamiento durante un tiempo. Ello puede significar que se encuentra en concentraciones muy bajas, o bien que ha sido eliminado.

Este hecho supone un paso adelante; sin embargo, los tratamientos actuales presentan el inconveniente de su elevado coste. Se calcula que un enfermo de sida gasta anualmente cerca de un millón de pesetas en medicamentos. Además, es frecuente que los pacientes acudan al médico cuando la enfermedad está avanzada, lo que reduce la efectividad del tratamiento.

Podemos afirmar que, en la actualidad, la enfermedad es **tratable,** ya que se han obtenido resultados positivos en algunos pacientes. Para considerarla **curable,** es necesario que el tratamiento pueda aplicarse a la mayoría de los enfermos y permitir su curación.

Uno de los puntos de investigación actuales es la resistencia a la enfermedad que presentan algunos individuos. Parece ser que existe una característica de tipo genético que permite resistir la enfermedad.

Convivir con una persona portadora del virus VIH

Es posible que algún día nos encontremos con una persona cercana que nos confiese que es seropositivo o incluso que está enfermo. ¿Cuál será nuestra reacción?

En general, existe un temor extremo al contagio, por lo que muchas veces se toman medidas de prevención exageradas y equivocadas, como no acercarse a un enfermo ni tocarlo, o creer que beber del mismo vaso puede ser contagioso. Convivir con una persona infectada por el virus VIH no conlleva prácticamente riesgo si se toman las medidas de prevención antes citadas. Es conveniente también lavar la ropa, las sábanas y las toallas con lejía si se ensucian de sangre, semen o secreciones vaginales de estas personas.

Tener un hermano, un hijo o un compañero de clase que padezca sida no implica ningún peligro. Sin embargo, en muchos casos, el temor a contagiarse provoca que los enfermos se vean marginados de su entorno familiar, social y laboral. La soledad de estas personas se añade, muchas veces, a su sufrimiento físico y al temor a la muerte.

El desconocimiento de esta enfermedad, que hasta ahora parecía incurable, es sin duda la causa del temor que origina. La información es el mejor método para disipar dudas y plantar cara al sida; no debemos huir del problema, porque no afecta a una minoría, sino a toda la sociedad. La solidaridad, tanto individual como colectiva, es un factor imprescindible para poder combatir la enfermedad de forma efectiva y eliminar el virus. En la actualidad, el problema es especialmente grave en los países menos desarrollados, donde la enfermedad se sigue propagando sin control debido a la ignorancia y la pobreza de las personas.

6 Lee el siguiente texto y responde a las preguntas que encontrarás a continuación:

Una de cada seis parejas españolas tienen dificultad para concebir un hijo, y son cada vez más las que deciden ponerse en manos de centros especializados en **reproducción asistida.**

La edad de la mujer es uno de los factores que más influye en la dificultad de tener un hijo: en los últimos años, el porcentaje de mujeres que tienen un hijo a partir de los 35 años se ha multiplicado por dos. Entre los 35-41 años la probabilidad de concepción disminuye, ya que la producción de óvulos es menor y de peor calidad. En el caso de los espermatozoides, pierden motilidad con la edad, pese a la longevidad de su producción.

El método más simple de reproducción asistida es el de la inseminación artificial y, si éste no funciona y no da los resultados esperados se cambia de técnica: las más usuales entonces son las fecundaciones *in vitro* (FIV).

La FIV se basa en la obtención de ovocitos maduros de la mujer, los cuales se fecundan en una placa de vidrio. Después de 48-72 h de división celular, se transfieren dos o tres blástulas al útero de la mujer. Para que este proceso tenga éxito se necesitan decenas de oocitos por parte de la mujer y es por ello que la mujer debe seguir un tratamiento hormonal que estimule su producción.

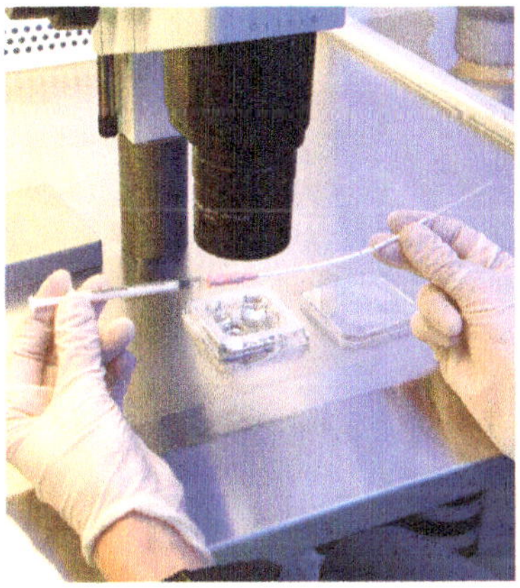

Se debe tener en cuenta que en la estimulación ovárica, no todos los ovocitos extraídos estarán en una fase de madurez adecuada para ser fecundados, que de aquellos que sean fecundados, no todos seran viables y se dividirán correctamente, y que de los dos o tres seleccionados para transferir al útero mediante una cánula, no todos se implantarán.

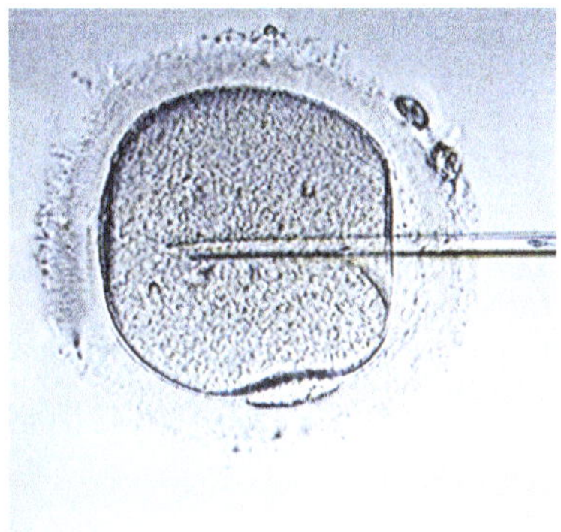

Se recurre a la *FIV* sobre todo en el caso de oligospermia u obstrucción de las trompas de Falopio. Es posible fecundar el óvulo por *microinyección directa,* en la cual se introduce mediante una aguja el espermatozoide dentro del óvulo, pero tambien se podría introducir una espermátida, en el caso que no hubiera producción de espermatozoides por parte del hombre.

Hay diferentes variantes de FIV según el problema que presente, pero todas ellas tienen la finalidad de conseguir el deseado embarazo. Actualmente la FIV tiene un éxito de un 40% de los casos.

A partir de este texto contesta las siguientes preguntas:

— Resume el contenido del texto, reflejando aquellos conceptos más importantes.

— ¿Cuáles son los principales motivos de la dificultad de concebir un hijo?

— Explica qué diferencias conceptuales hay entre la fecundación artificial y la fecundación in vitro.

— El primer paso en la FIV es estimular la ovocitación mediante una terapia hormonal. Esta terapia se basa en inyecciones intramusculares que hacen madurar a más de un ovocito por folículo y a más de un folículo. A partir de los conceptos estudiados en este tema y también en el tema 12, ¿podrías deducir qué hormonas se inyectan a la mujer para conseguir la sobreestimulación folicular?

— En el texto aparecen dos conceptos en negrita. Busca el significado de estos términos y explícalos.

— Cuando se habla de FIV, se asocian toda una serie de problemas éticos, como, por ejemplo, qué hacer con los embriones sobrantes. Busca información sobre la LEY 14/2006, de 26 de mayo, sobre técnicas de reproducción humana asistida, que encontrarás en la siguiente dirección y contesta las preguntas:

• http://nuevo.sefertilidad.com/documentos/legislacion/index.php

a. ¿Se puede experimentar con embriones humanos?

b. ¿Se pueden obtener células madre de tejidos a partir de embriones con fines terapéuticos?

c. ¿Se pueden transferir mas de tres embriones a la mujer por ciclo reproductivo?

d. ¿Una pareja puede tener un número indefinido de hijos mediante técnicas reproductivas?

e. ¿Hasta cuando se puede alargar la crioconservación de los embriones?

ACTIVIDADES DE SÍNTESIS

7 Escribe el nombre de las partes señaladas correspondientes al aparato reproductor masculino y di cuál es su función:

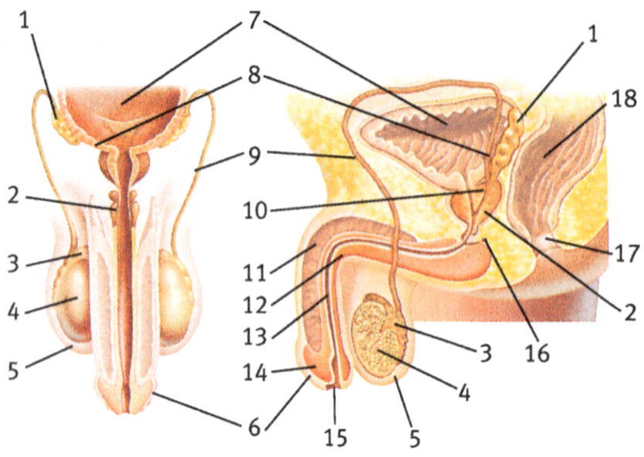

8 Indica qué partes del aparato reproductor masculino incorporan secreciones que acompañan a los espermatozoides, formando el esperma o semen.

9 ¿Qué son, dónde se localizan y qué función llevan a cabo las células de Leydig?

10 Asigna a cada número el nombre correspondiente del aparato reproductor femenino y di cuál es su función.

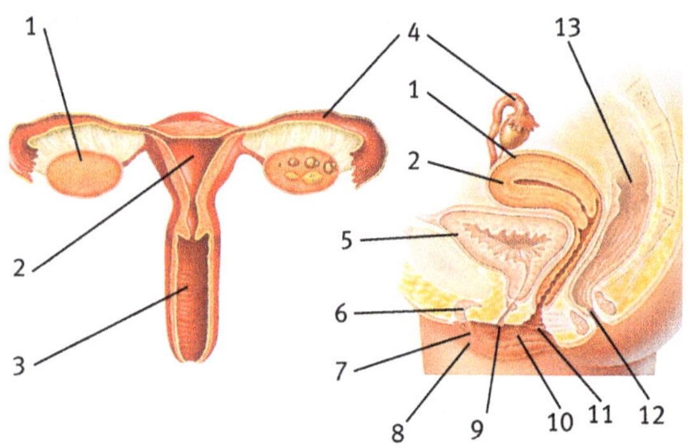

11 Elabora una tabla en la que se reflejen los caracteres sexuales secundarios masculinos y femeninos.

12 Explica los siguientes conceptos:

- Ovulación.
- Conductos galactóforos.
- Cuerpo lúteo.
- Endometrio.

13 Responde a las siguientes cuestiones sobre el esquema del ciclo menstrual que encontrarás a continuación:

- ¿Qué circunstancia del ciclo menstrual está representada en el dibujo?
- Teniendo en cuenta en qué momentos se producen la ovulación y el inicio de la menstruación, señala la fase de desarrollo del óvulo, la fase de formación del cuerpo lúteo y la fase de expulsión de la mucosa del endometrio.
- Copia la ilustración y dibuja sobre la copia la variación de los niveles de las hormonas: LH, FSH y progesterona.

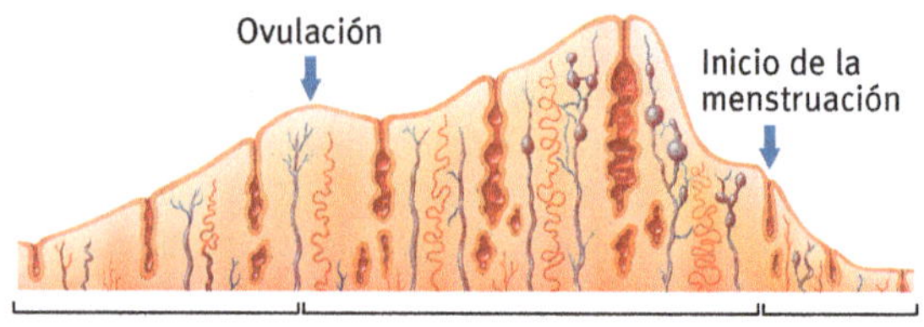

14 Observa el siguiente dibujo que representa la ovogénesis humana. Explica, basándote en los conocimientos aprendidos y en el dibujo, en qué consiste este proceso, cuándo tiene lugar y qué hormonas intervienen.

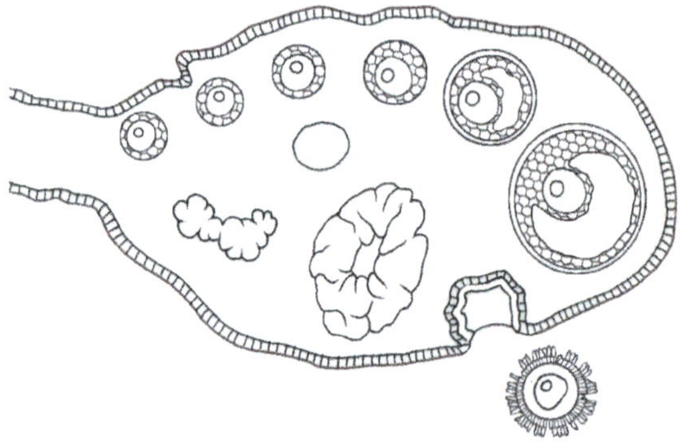

15 En el siguiente dibujo se representa el proceso de ovogénesis, oocitación, fecundación e implantación.

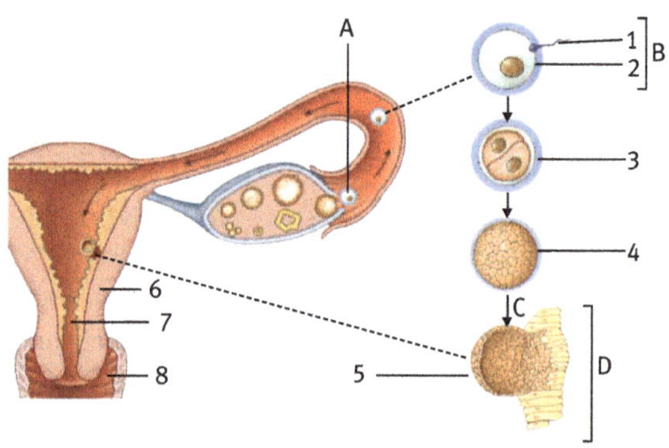

- Asigna a cada número el nombre de la estructura representada y a cada letra el nombre del proceso que se describe explicando brevemente en qué consisten.

16 La hCG es una hormona que estimula el cuerpo lúteo del ovario para mantener las secreciones de estrógenos. En la siguiente tabla están representados sus niveles a lo largo de las primeras 40 semanas del embarazo:

SEMANA DE EMBARAZO	CANTIDAD DE hCG (MUI/ml)
3	5-50
4	25-400
5	200-7000
6	1000-50 000
7-8	7000-20 0000
9-12	25 000-25 0000
13-16	15 000-25 0000
17-24	5000-15 0000
25-40	4000-10 0000

— Representa gráficamente los datos de la tabla.

— Interpreta la gráfica resultante.

1. Activación del desarrollo de las células de Leydig.
2. Activación de la síntesis de testosterona.
3. Estimulación de la síntesis de estrógenos.
4. Estimulación de la formación de la leche.
5. Inicio de la maduración de los órganos reproductivos masculinos.
6. Estimulación de la aparición de vello en las axilas.
7. Inicicación de la espermatogénesis.
8. Estimulación del crecimiento del útero durante el embarazo.
9. Estimulación del aumento de musculatura.
10. Estimulación del crecimiento de huesos y músculos.

- Estrógenos
- LH
- FSH
- GH
- Andrógenos

— Teniendo en cuenta a función de la hormona, ¿qué sentido tiene la evolución de los niveles de hCG durante el embarazo?

17 Asigna a cada proceso, la hormona responsable:

18 Explica qué son las contracciones uterinas y explica que hormonas las causan, así como quien las produce y por qué.

19 Explica qué se entiende por enfermedades de transmisión sexual. Nombra las enfermedades que aparecen en la unidad que estén causadas por virus y por bacterias.

20 Explica cuáles son las vías de contagio del VIH.

— Por qué hay mayor dificultad de controlar el sida en los países pobres?

21 Comenta las diferencias de los siguientes gráficos, en los cuales se analiza por separado los casos diagnosticados de sida en mujeres y hombres.

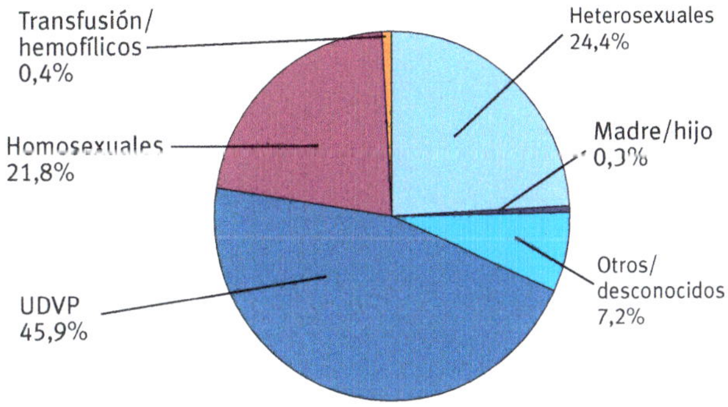

Hombres (n = 1091)

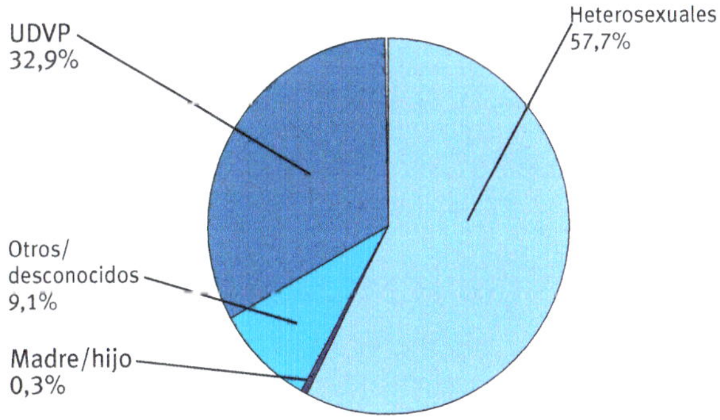

Mujeres (n = 319)

(Fuente: Ministerio de sanidad)

TIC **22** Consulta la siguiente dirección electrónica que habla sobre los métodos anticonceptivos y contesta las siguientes preguntas.

- http://elmundosalud.elmundo.es/elmundosalud/especiales/2004/12/anticonceptivos/index.html

— Explica en qué se basa el método Ogino y comenta su fiabilidad.

— Un método anticonceptivo hormonal de última generación es el anillo. Explica cómo funciona y cómo se utiliza.

— ¿Existen píldoras masculinas? Si es así, ¿en qué se basan? ¿Se encuentran en el mercado?

— Explica qué es la píldora del día después y en qué difiere con una píldora abortiva.

PRÁCTICA 4 El pulso y la tensión arterial

OBJETIVOS

- Aplicar las técnicas de medición del pulso y de la tensión arterial.
- Representar e interpretar los valores encontrados.

El **pulso** es la expansión intermitente y rítmica de una arteria producida por el impulso que la sístole ventricular da a la sangre. Se suele percibir en las arterias periféricas. La **tensión arterial** es la presión que la sangre ejerce en el interior de las arterias. La tensión arterial **sistólica** o **máxima** es causada por la sístole ventricular; la tensión arterial **diastólica** o **mínima** se produce como consecuencia de la diástole ventricular. En esta práctica se efectuará una serie de mediciones de ambos parámetros.

Material

- Cronómetro o reloj con segundera
- Esfigmomanómetro
- Papel milimetrado
- Estetoscopio

Proceso

— Por parejas, localizar la arteria radial en la muñeca izquierda con los dedos índice y medio de la mano derecha; para ello se sigue la dirección del dedo pulgar.

— Contarse mutuamente el número de pulsaciones en reposo, durante un minuto. Repetir las mediciones cada 2 minutos y anotarlas.

— Hacer ejercicio físico durante 3 o 4 minutos y tomar el pulso inmediatamente después del ejercicio. Volver a hacerlo 2, 4 y 6 minutos más tarde. Tomar nota de las pulsaciones.

— Tomar la tensión arterial del compañero en posición de reposo. La tensión arterial se mide del modo siguiente:

- Ajustar el brazalete del esfigmomanómetro al brazo izquierdo, por encima del codo, e hincharlo hasta que el indicador de presión llegue a 200 mm Hg.
- Situar el estetoscopio sobre la parte interior del codo y escuchar. Aflojar la válvula del esfigmomanómetro para que la presión ejercida sobre el brazo descienda lentamente. Cuando se oiga el latido del corazón, cerrar la válvula y observar la presión indicada.
- Abrir ligeramente la válvula y esperar a oír de nuevo el latido del corazón. En ese momento, cerrar la válvula y observar la presión que se indica. Posteriormente, aflojar la válvula totalmente y retirar el brazalete hinchable.

— Repetir la medición inmediatamente después de realizar un ejercicio físico durante unos minutos. Anotar los valores de tensión máxima y mínima en cada caso.

CUESTIONES

1. Comparar los valores del pulso y la tensión hallados en reposo con los valores que se consideran normales.
2. Construir una gráfica del número de pulsaciones por unidad de tiempo para los valores en reposo y después de la actividad física, y comentar los resultados.
3. ¿Qué diferencias se observan en los valores de tensión arterial antes y después de haber hecho ejercicio físico? ¿Qué significa *tener la tensión alta*? ¿Y *tener la tensión baja*?

PRÁCTICA 5 Ciclo biológico del grillo

Esta práctica consiste en montar un terrario para observar el ciclo biológico de los grillos.

Material

- Grillos *(Gryllus bimaculatus* o *Acheta domestica)* machos y hembras
- Dos cubetas de plástico o vidrio de dimensiones 50 x 30 x 30 cm
- Una tapa con agujeros para cada cubeta
- Tierra recogida del bosque
- Un ladrillo
- Un recipiente pequeño, de unos 5 cm de alto, y dos tubos de ensayo
- Dos bombillas de 60 o 100 W
- Lechuga y pienso para ratones

Proceso

Montar dos terrarios con las cubetas; uno para los adultos y para que las hembras hagan la puesta; en el otro eclosionarán los huevos.

— Se coloca una capa fina de tierra en el fondo de cada cubeta.

— Se llenan los dos tubos de ensayo con agua y se obturan con un algodón. Cada uno se coloca horizontalmente en un terrario y será el bebedero. Cambiar el agua y el algodón una vez por semana.

— Se sujeta un portalámparas con una bombilla, por la parte interior de cada tapa. Se tapan las cubetas y las bombillas se pueden tener encendidas todo el día.

— Además, en el terrario de los adultos se añade:

- Un ladrillo en un extremo para que puedan esconderse.
- Un ponedero, colocando algodón mojado en el fondo del recipiente pequeño, y tierra encima. Una rampa de papel permitirá que suban las hembras. Allí depositarán pequeños huevos blanquecinos. Se mantendrá húmedo para favorecer la eclosión de los huevos. Diez días después de las primeras puestas, se traslada el ponedero al otro terrario.

— Poner pienso y lechuga y renovarlos semanalmente. La lechuga debe lavarse bien para eliminar restos de insecticida.

— Observar diariamente a los grillos y anotar sus actividades. Atender al momento de la puesta para trasladar el ponedero.

OBJETIVOS

- Preparar un terrario y mantener las condiciones necesarias para la vida de los insectos y su reproducción.
- Explicar el ciclo vital de un insecto a partir de la observación del apareamiento, la puesta de huevos y el desarrollo de las larvas.

CUESTIONES

1. Fijarse en las actividades que realizan los animales, deducir cuáles son machos y cuáles hembras, y describir las diferencias entre ellos. ¿Cuál de los dos sexos emite sonidos? ¿Cuándo lo hace?
2. ¿Cómo efectúan las hembras la puesta de huevos? Describir la forma y el tamaño de éstos.
3. Describir las crías en el momento de nacer. Contar el número de mudas que hacen las crías antes de llegar al estado adulto. ¿Cuánto tiempo tardan en llegar a adultos?
4. Resumir en un informe todas las observaciones y describir el ciclo vital completo.

UNIDAD 1

1 Explica y compara los modelos de representación del universo propuestos por Aristóleles y Copérnico.

— Indica cuál de los dos es más parecido a la concepción actual de universo y en qué se ha modificado.

2 Explica las ventajas y los inconvenientes que tienen los telescopios espaciales respecto a los radiotelescopios.

3 Describe cómo se cree que se originó el universo. ¿Qué nombre recibe este momento?

— Describe en qué consisten los dos posibles modelos de evolución futura del universo e indica de qué factor depende que se produzca uno u otro.

4 Una estrella que se observa desde la Tierra emite luz de color naranja y su tamaño supera en 2,5 veces el del Sol.

— Explica qué tipo de estrella es y en qué momento de su evolución se encuentra.

— Haz un esquema y describe como serán las etapas finales de su actividad.

5 Responde las siguientes preguntas sobre el Sol:

— ¿Qué tipo de estrella es el Sol?

— ¿En qué momento de su *vida* se encuentra?

— ¿Cuál es el origen de la energía que desprendre el astro?

— ¿En qué parte del Sol tiene lugar esta generación de energía?

6 Cita los planetas que forman parte del Sistema Solar de forma ordenada según su proximidad al Sol.

— Indica bajo qué dos criterios de clasificación pueden agruparse, clasifícalos siguiendo estos criterios y explica una característica propia de cada grupo.

7 Representa un esquema de un sistema planetario que conste de los siguientes astros:

- Una estrella
- Cinco planetas
- Tres satélites
- Dos asteroides
- Un cometa

— Dibuja las orbitas los astros que correspondan e identifica con un símbolo específico cada tipo de astro representado.

8 ¿Cuándo se formó el planeta Tierra? Explica por qué las eras y los períodos que encontramos a lo largo de la historia de la Tierra no tienen la misma duración.

— Identifica dos procesos causados por seres vivos que alterasen la composición de la superficie terrestre y explica sus consecuencias.

UNIDAD 2

1 Compara los procesos de cristalización y de solidificación.

— Explica dos mecanismos de cristalización distintos.

2 ¿Qué diferencia existe entre el magma y la lava?

— Cita los dos principales tipos de magma y compara sus características.

3 Identifica los distintos tipos de rocas magmáticas que distinguimos según su profundidad de cristalización.

— Indica en cuál de estos tipos de rocas será más probable encontrar una textura granular y en cuál una textura fluida.

4 Examinando diversas muestras de rocas ígneas, reconoces en una de ellas una textura vítrea. ¿Qué puedes decir sobre el proceso de generación de dicha roca?

5 ¿Qué clase de roca sedimentaria encontramos más cerca de la roca de la que proceden los materiales sedimentarios: una roca detrítica o una roca de precipitación química? Razona tu respuesta.

— En el caso de los dos tipos de roca anteriores, explica en cuál se encontrarán más minerales alotígenos.

6 Explica cómo se produce el proceso de fosilización. Argumenta de qué tipo de roca sedimentaria suelen estar formados los fósiles.

7 El gneis es una roca metamórfica que se forma a profundidades mucho mayores que la pizarra. ¿Qué roca crees que es más densa? Razona tu respuesta.

8 Indica en qué tipo de roca se transformarán unos materiales que no alcancen la temperatura y la presión necesaria para su metamorfismo. ¿Y si, por el contrario, los valores de temperatura y presión superan ampliamente los intervalos de metamorfismo?

9 Observa las siguientes rocas y determina en qué ambiente petrogenético se ha formado cada una de ellas. Razona tu respuesta.

UNIDAD 3

1 Haz dos esquemas de las capas de la Tierra según el estado de sus materiales y según su composición química.

2 La siguiente gráfica sísmica correponde a un planeta imaginario. En el eje de abscisas se indica la profundidad en kilómetros y en el de coordenadas la velocidad de la onda en km/s.

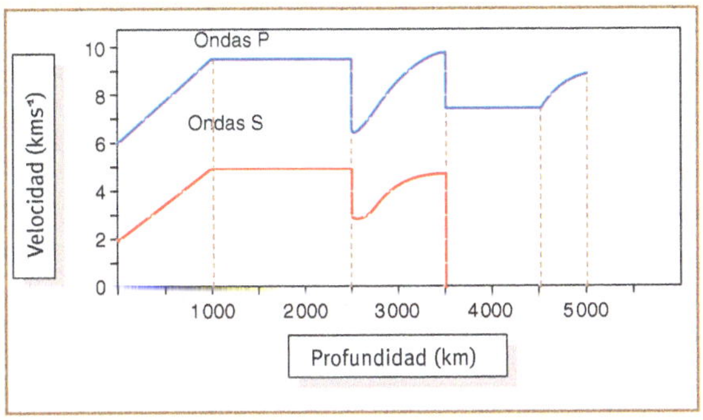

— Indica a qué profundidades se localizan las discontinuidades.

— Explica en cada uno de los intervalos si la velocidad de propagación de las ondas sísmicas aumenta o disminuye, e indica si los materiales son plásticos o rígidos.

— Representa las capas en un esquema de una sección del planeta.

3 Explica por qué coinciden en las zonas activas la distribución de los volcanes y los epicentros de los terremotos.

4 Razona por qué encontramos áreas geotérmicas en las zonas de dorsales oceánicas.

5 Los marsupiales son un grupo de vertebrados que se cree que se originó hace 125 millones de años. En la actualidad sólo se encuentran en Australia y América, principalmente en Sudamérica. Sin embargo se han encontrado fósiles de marsupiales en Europa. ¿Cómo explicarías este hallazgo si se trata de animales incapaces de nadar o volar?

6 ¿Qué es una placa litosférica? Describe los distintos tipos de límites de placa que podemos encontrar y explica una característica geológica de cada uno de ellos.

7 ¿Qué se entiende por expansión de los fondos oceánicos? ¿Por qué la superfície total de la Tierra continúa invariable a lo largo del tiempo a pesar de éste fenómeno?

8 Identifica los dos tipos de deformaciones que puede originar la actividad interna de la Tierra. ¿De qué depende que se produzca un tipo u otro?

UNIDAD 4

1 Explica los diferentes procesos geológicos externos por los que pasa una partícula desde que forma parte de una roca en lo alto de la montaña hasta que se encuentra formando parte del delta de un río.

2 Explica qué es un agente geológico externo y cita tres ejemplos.

3 ¿Crees que el ser humano puede considerarse un agente geológico? Razona tu respuesta.

— Anota en una lista las consecuencias que puede tener sobre los procesos geológicos externos, y por tanto sobre el relieve, cada una de las siguientes acciones:

- La construcción de una presa en un río.
- La construcción de espigones en un puerto.
- La construcción de una autopista.

4 Explica las diferencias entre los siguientes conceptos:

- Delta y estuario.
- Acantilado y playa.
- Casquete polar y glaciar de valle.
- Deflación y abrasión eólica.
- Erg y reg.

5 Observa las siguientes fotografías e identifica a qué formaciones del relieve corresponden así como el agente geológico que las ha formado.

6 Di en cuál de las siguientes zonas es más activo el proceso de formación del suelo. Razona tu respuesta.

- En el bosque de una zona húmeda.
- En el desierto del Sáhara.

7 ¿Qué sucede en el suelo si, después de un incedio forestal o una tala abusiva de árboles, se producen lluvias intensas?

8 Comenta la importancia que tiene el suelo para la vegetación y la importancia que tiene la vegetación para el suelo.

UNIDAD 5

1 Explica cómo se ha conformado el actual relieve de las islas Canarias.

2 Razona cómo es posible encontrar fósiles de animales marinos de más de 280 millones de años de antigüedad en algunas montañas de la península Ibérica.

3 ¿Qué papel juega la placa africana en el origen de las Cordilleras Béticas?

4 Define los siguientes conceptos:

- SIG
- Píxel
- GPS
- Teledetección
- Corte geológico
- Buzamiento

5 Observa la siguiente imagen captada por un satélite e interpreta las diferentes formaciones y estructuras geográficas que se aprecian en ella.

— Razona si la imagen que observas es tal como la capta el satélite o bien si ha sido procesada con posterioridad.

6 Observa la siguiente figura y dibuja un corte geológico del segmento marcado entre A y B.

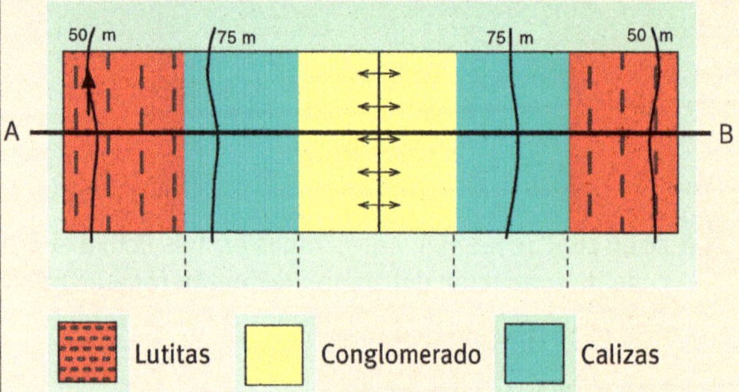

— Indica la secuencia en que se depositaron los distintos estratos.

— Explica qué es el principio de superposición de Steno.

— ¿Qué tipo de deformación ha ocurrido en este terreno?

UNIDAD 6

1 Resume las características del agua en un texto que trate los siguientes aspectos:

- Composición química
- Estructura de su molécula
- Propiedades
- Influencia en los seres vivos

2 Indica cuál es la importancia de los puentes de hidrógeno y en qué moléculas se forman.

3 Explica qué son un enlace peptídico y un enlace éster. Describe en qué casos se producen y entre qué tipo de grupos funcionales.

4 Razona qué biomoléculas se relacionan con:

- Los cromosomas
- Las membranas celulares
- La pared de las células vegetales
- El exoesqueleto de los crustáceos
- Los enzimas

5 Cita los cuatro tipos fundamentales de biomoléculas y los principales subgrupos de que constan.

— Identifica aquellos subgrupos de biomoléculas que consten de polímeros y cita el nombre del monómero que los forma.

— Indica, para cada caso, cómo se denomina el enlace que se produce entre estas subunidades.

6 Clasifica las siguientes moléculas según el grupo de biomoléculas a que pertenezcan e indica cuál es la función de cada una de ellas:

- Hemoglobina
- Celulosa
- Anticuerpo
- Cera
- ADN
- Disácarido
- Vitamina
- Ovoalbúmina

7 La aparición de ratones en depósitos húmedos de trigo y moscas en carne en putrefacción se relacionó durante siglos con un proceso de generación espontánea.

— Explica por qué crees que se podían observar estos fenómenos.

8 Según las primeras teorías del origen de la vida, ¿qué condiciones debían darse para que se formaran las primeras biomoléculas?

— ¿Crees que se dan en la actualidad?

UNIDAD 7

1 Di qué microscopio utilizarías para la oservación de: la estructura de los poros de la membrana nuclear de una célula, la disposición de las células epiteliales de la mucosa respiratoria, la distribución de los orgánulos en una célula vegetal.

2 ¿Por qué decimos que la base estructural de la membrana plasmática es una *bicapa* lipídica? Haz un dibujo sencillo de los componentes principales de la membrana plasmática y explica la función de cada uno de ellos.

3 Observa el siguiente dibujo:

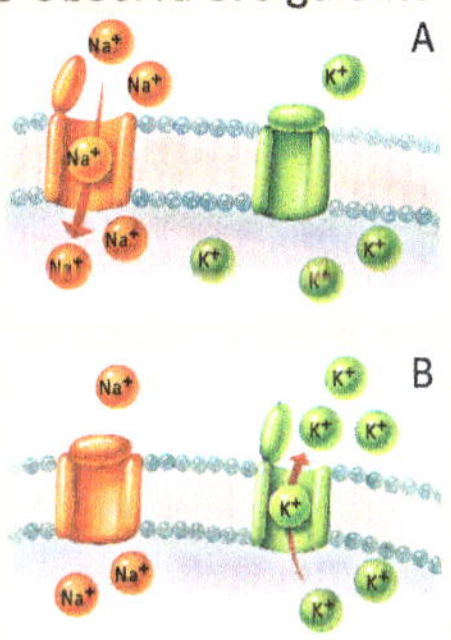

— ¿Qué tipo de estructuras permiten el paso de las sustancias?

— ¿Por qué en este caso no se consume ATP?

— ¿Qué otros tipos de transporte pasivo conoces? Explícalos brevemente.

4 Di qué estructura de la célula procariota se encarga de cada una de las siguientes funciones:

— Resistencia a la acción fagocitaria
— Síntesis de membrana
— Reserva energética de polifosfato
— Flotabilidad en medio acuático
— Síntesis de proteínas
— Movilidad

5 Enumera las estructuras comunes y no comunes de las células eucariotas vegetal y animal. Explica la estructura y función de las no comunes.

6 En conjunto, ¿las reacciones de la fotosíntesis serán endergónicas o exergónicas? Razona la respuesta.

7 Completa la siguiente tabla:

	Sustrato inicial	aerobia / anaerobia	Aceptor de electrones	Compartimento celular
Fermentación				
Respiración				

8 Haz un esquema de la mitosis y la citocinesis de una célula animal con 4 cromosomas.

— Explica los principales sucesos y estructuras de cada fase de la mitosis. ¿Qué estructura no veríamos en una mitosis vegetal?

— Enumera las diferencias entre la citocinesis en una célula vegetal y en una célula animal.

9 ¿Es efectivo tomar antibióticos para curar una gripe? ¿Y vacunarse de una enfermedad bacteriana? Justifica las respuestas.

UNIDAD 8

1 ¿Qué ventajas presenta la organización pluricelular respecto a la unicelular? ¿Y desventajas?

2 En los animales homeotermos uno de los mecanismos utilizados para mantener la temperatura del medio interno es la regulación del volumen de sangre que recorre los vasos sanguíneos más periféricos. Explica cómo crees que variará dicho volumen en función de la temperatura.

3 Define el concepto de homeostasis y enumera al menos cinco variables suceptibles de ser reguladas (*ej.: glucemia*). ¿Qué tipo de mecanismos crees que intervienen en su regulación?

4 Indica a qué tipo de tejido vegetal hacen referencia cada una de las siguientes definiciones y explica las principales características de cada uno de ellos:

— Tejido formado por células muertas lignificadas.

— Tejido que se encuentra en las partes verdes de la planta y está formado por células con abundantes cloroplastos.

— Tejido que se encarga del transporte de la savia elaborada.

5 Observa el siguiente dibujo:

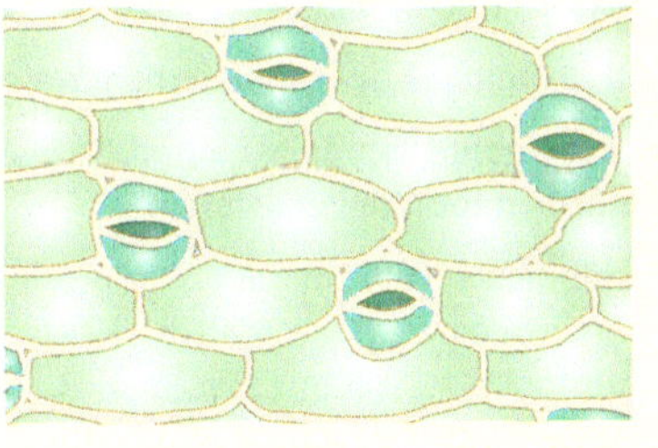

— ¿Qué estructuras están representadas?

— ¿Cuál es su función?

— ¿De qué tipo de tejido forman parte?

— Di el nombre de las células que los forman.

6 Di qué tipo de epitelio podemos encontrar en: la piel, la pared interna de la faringe, la pared interna de los vasos sanguíneos, las glándulas y las células pigmentadas de la retina.

7 Di a qué reino pertenecen los organismos que responden a las siguientes descripciones:

— Organismo pluricelular de organización especializada cuyas células no presentan pared celular.

— Tiene nutrición heterótrofa y presenta una pared celular quitinosa.

— Organismo unicelular procariota.

— Organismos autótrofos cuyas células presentan pared celular de celulosa.

— Organismo unicelular eucariota cuyas células no contienen en ningún caso quitina.

UNIDAD 9

1 Di qué tipo de relación interespecífica se establece entre los siguientes organismos:

— La garcilla bueyera se alimenta de los parásitos de la piel de los bueyes.

— La tenia vive en el tubo digestivo de los mamíferos aprovechándose de los alimentos que el animal consume.

— Las rémoras se alimentan de los restos de comida que dejan los tiburones.

2 En la siguiente tabla aparecen las biomasas relativas de todos los organismos de un ecosistema hipotético en dos años consecutivos:

	Biomasa año 1 (gC/m^2)	Biomasa año 2 (gC/m^2)
Especie A	2200	2450
Especie B	685	850
Especie C	5125	4950
Especie D	1150	1250

— Suponiendo que el ecosistema se encuentra en equilibrio, ¿a qué nivel trófico pertenece cada especie? Justifica tu respuesta

— Calcula la biomasa del ecosistema para cada año.

— Calcula la producción de cada una de las especies y la producción total del ecosistema.

— ¿Qué significa el valor negativo de producción de la **especie C**? Si este valor negativo se prolongase a lo largo de los años, ¿cómo afectaría al conjunto del ecosistema?

3 Elabora una tabla explicativa en la que se comparen la nutrición de las plantas cormófitas y la nutrición de los vertebrados teniendo en cuenta los siguientes aspectos: tipo de nutrición, tipo de nutrientes utilizados, órganos implicados en la captación de nutrientes, tipo de transformación de los nutrientes, principales procesos metabólicos, sustancias de excreción y estructuras implicadas en la excreción.

4 La siguiente tabla muestra la cantidad de nutrientes totales ingeridos por una mujer de 65 kg de peso, con una actividad física ligera, durante una semana:

	Consumo (g/semana)	Valor energético (kcal/g)	Aporte calórico recomendado (%)
Glúcidos	3600	3,74	55-65
Lípidos	4400	9,30	25-30
Proteínas	2500	4,20	10-15

— ¿Es correcto el aporte calórico de la dieta de esta mujer?(Metabolismo energético de una mujer con actividad ligera: 36 kcal/kg/día).

— ¿Es equilibrada la dieta que sigue esta mujer?

UNIDAD 10

1 ¿Qué relación existe entre la respiración interna y la respiración externa?

2 Respecto al proceso de la respiración en vegetales:

— ¿En qué consiste su repiración interna? ¿De dónde procede la glucosa que se utiliza?

— ¿De dónde procede el O_2 liberado por las plantas? ¿De dónde el CO_2? ¿Para qué captan O_2? ¿Y CO_2?

3 De las siguientes afirmaciones sobre la respiración animal, reescribe correctamente aquellas que sean erróneas explicando porqué era errónea la primera:

— En los animales terrestres la velocidad de difusión es muy elevada porqué el O_2 y el CO_2 atraviesan la membrana en forma gaseosa.

— El principal pigmento de la respiración por dendrotráqueas es la clorocruonina.

— La concentración de O_2 en el agua es inferior a la concentración en el interior de las células de las branquias que realizan el intercambio de gases.

4 Elabora una tabla en la que se comparen las características de la respiración de los anélidos, los artrópodos, los peces óseos, los peces cartilaginosos, los anfibios, los reptiles, las aves y los mamíferos.

5 Explica qué modificaciones relacionadas con la ventilación presentan las aves como adaptación al vuelo.

6 Observa el siguiente gráfico:

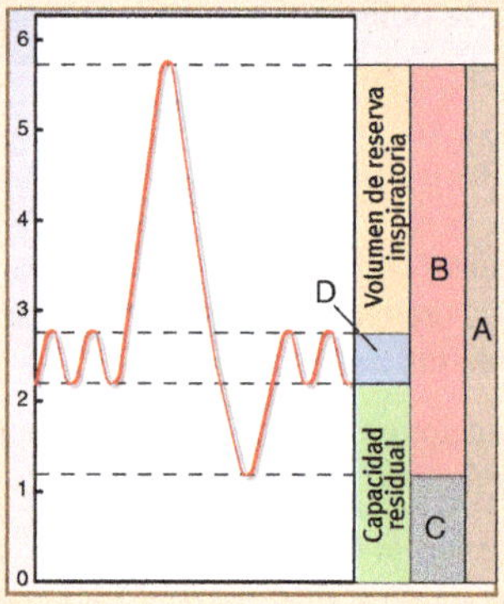

— ¿A qué parámetros corresponden las letras marcadas en el gráfico?

— Señala en el gráfico los tramos de línea correspondientes a: una inspiración normal, una inspiración forzada, una espiración normal y una espiración forzada.

7 Explica cómo varían las presiones parciales de O_2 y CO_2 en el intercambio de gases alveolar y en el de la sangre a las células del cuerpo. ¿Cómo se tranporta cada uno de estos gases en la sangre?

8 Escribe un informe resumido sobre la morfología, las características y el funcionamiento de las vías vegetales encargadas del transporte del agua y las sales minerales.

9 Explica el recorrido completo de la sangre en un mamífero y puntualiza cómo va variando la concentración de O_2 y CO_2 en cada tramo.

UNIDAD 11

1 El axón de la mayoría de nuestras neuronas presenta vaina de mielina. Explica cómo se forma esta vaina, por qué presenta discontinuidades y qué ventaja representa en la transferencia del impulso nervioso. Refuerza tu explicación con dibujos y esquemas.

2 Ordena los siguientes estadios de la neurona y di a qué fase en la transmisión del impulso nervioso (*potencial en reposo, despolarización, potencial de acción y repolarización*) corresponden:

– La bomba Na^+/K^+ no funciona, el canal iónico de Na^+ está cerrado y el canal iónico de K^+ abierto.

– Deja de funcionar la bomba Na^+/K^+ y se abren los canales iónicos de Na^+.

– La bomba Na^+/K^+ funciona y los canales iónicos de Na^+ y los de K^+ están cerrados.

– La entrada de Na^+ provoca que el potencial de membrana sea de +40 mV.

3 Explica cómo se produce una sinapsis que tiene como neurotransmisor a la acetilcolina. ¿Cómo sería si el neurotransmisor fuera el GABA?

4 Lee atentamente el siguiente texto:

"*No podía dormir debido al frío que entraba por la ventana. Fui a cerrarla y al encender la luz se me pusieron los pelos de punta al ver como una gran cucaracha corría a esconderse bajo mi cama. Era tan grande que pude oír el ruido que hacía al desplazarse.*"

– ¿Qué tipo de receptores sensoriales se han activado en el ser humano? ¿Y en la cucaracha? Explica los órganos de los sentidos implicados y su mecanismo de funcionamiento.

– ¿Qué centros nerviosos participan en la elaboración de la respuesta en el ser humano? ¿Y en la cucaracha?

– ¿En qué consiste la respuesta motora en el ser humano? ¿Y en la cucaracha? Explica el funcionamiento del aparato locomotor en ambos casos.

5 Elabora un esquema de llaves de las partes del sistema nervioso central. Expica la función de cada una de estas partes.

6 Observa la siguiente ilustración:

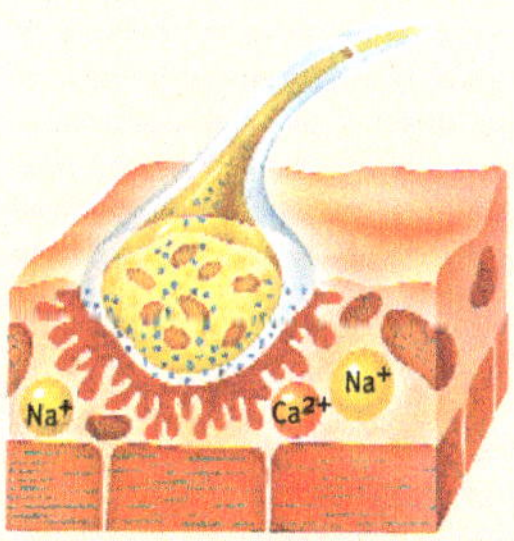

– ¿Qué tipo de interacción celular está representada? Di qué tipo de células participan y señálalas en el dibujo?

– ¿Qué neurotransmisor interviene?

– ¿Cuál es el ión característico de este tipo de interacción?

– Describe el funcionamiento de esta interacción?

UNIDAD 12

1 Enumera las principales diferencias entre el impulso nervioso y el mecanismo de acción de las hormonas.

2 Di qué hormonas vegetales son las responsables de cada una de las siguientes respuestas así como su lugar de síntesis:

– Germinación de las semillas.

– Crecimiento ácido.

– Maduración del fruto.

– Síntesis de clorofila.

– Inhibición de la maduración de las semillas.

3 Explica el posible mecanismo que permite el movimiento de los girasoles siguiendo al Sol. Incluye en tu explicación el tipo de movimiento que representa, el estímulo que lo desencadena y el receptor implicado.

4 Observa la siguiente ilustración:

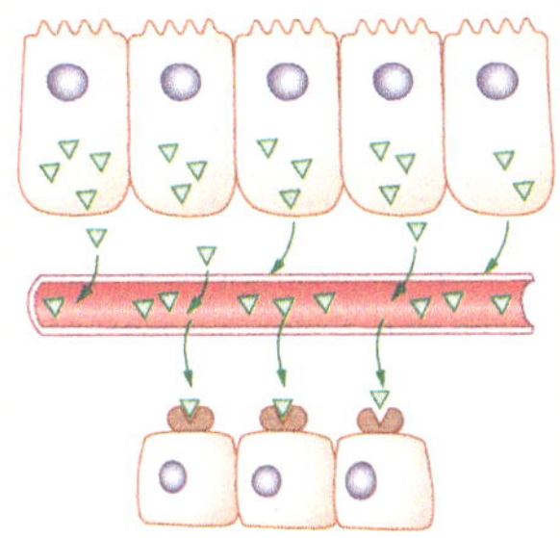

– Indica el tipo de secreción hormonal representado así como sus principales características.

– Di cuáles de las siguientes hormonas presentan este tipo de secreción: glucagón, oxitocina, insulina, hormona luteinizante, hormona antidiurética.

– ¿Cuál és el principal órgano secretor de este tipo de hormonas en el ser humano?

5 Haz un esquema del mecanismo de acción de una hormona lipídica y el de una hormona peptídica. Explica las principales diferencias entre uno y otro mecanismo.

6 Di dónde se sintetizan, la función y el mecanismo de actuación de las siguientes hormonas humanas:

– Insulina, andrógenos, estrógenos, calcitonina, tiroxina, glucagón, parathormona.

7 A la consulta de un médico llega un padre cuya hija presenta un abultamiento del cuello y que, según el padre, se cansa mucho, siempre tiene frío y pierde la memoria a menudo. ¿Qué enfermedad puede padecer la niña? ¿A qué pueden ser debidos cada uno de los síntomas de la enfermedad?

UNIDAD 13

1 ¿Qué tipo de reproducción es probable que presente un organismo adaptado a un ambiente estable pero de condicones extremas? Razona tu respuesta.

2 ¿Una célula diploide puede tener un número impar de cromosomas? ¿Una célula haploide puede tener un número par de cromosomas? Razona las respuestas

3 De las siguientes afirmaciones sobre la gametogénesis, reescribe correctamente las erróneas:

- Los espermatófitos presentan isogamia.
- Las células prescursoras de gametos dan lugar a éstos después de numerosas divisiones por mitosis.
- En la anisogamia los microgametos normalmente se corresponden con los gametos masculinos.
- Las diferencias entre un óvulo y un espermatozoide son la dotación cromosómica, la forma y el tamaño.

4 Una especie con fecundación interna, ¿puede ser ovípara? Razona tu respuesta

5 ¿Por qué en las especies oogámicas es el óvulo el gameto más grande mientras que el espermatozoide no presenta apenas citoplasma?

6 Observa la representación del siguiente ciclo biológico:

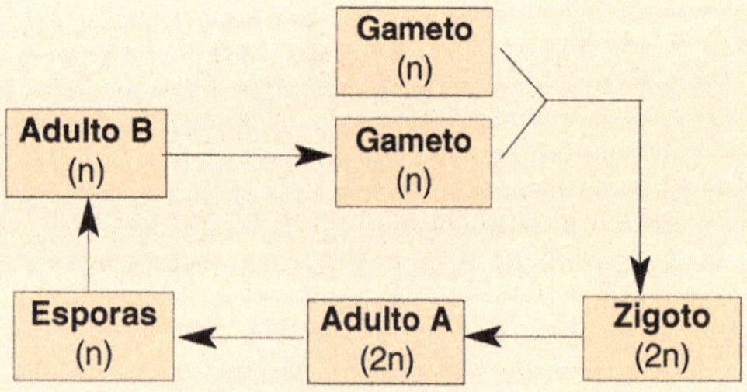

- Di de qué tipo de ciclo se trata.
- Cópialo en tu cuaderno e indica en qué momentos tiene lugar la meiosis, la fecundación y el crecimiento.
- ¿Qué tipo de meiosis es característica de este ciclo?
- ¿Qué nombre reciben los adultos A y B?
- ¿Qué organismos presentan este tipo de ciclo?

7 Elabora un esquema de texto de la reproducción alternante de las abejas, anotando la dotación cromosómica de cada individuo.

- ¿De qué depende que una larva se convierta en una abeja obrera o en una reina?
- ¿Mediante qué proceso se generan los zánganos de la colmena?
- ¿De qué individuos proceden los gametos que participan en la fecundación?

8 ¿Qué es y cómo se obtiene un individuo transgénico? Explica algunas de sus utilizaciones así como alguno de los motivos por los que su uso es controvertido.

UNIDAD 14

1 Elabora una tabla comparativa del desarrollo de los caracteres sexuales en hombres y en mujeres durante la pubertad. En la tabla deben constar:

- Período de tiempo que ocupa la pubertad, hormonas hipofisarias que participan y efecto de éstas, efecto de las hormonas sexuales en cada sexo, caracteres sexuales secundarios de cada sexo.

2 Ordena las siguientes estructuras del sistema reproductor masculino según su orden de participación en la espermatogénesis y formación del semen. Explica la fase del proceso que se da en cada una de ellas: *uretra, conductos deferentes, epidídimos, túbulos seminíferos, vesículas seminales, próstata.*

3 Explica cómo varían los niveles de FSH y LH a lo largo de los ciclos ovárico y menstrual así como los efectos asociados a dichos cambios.

4 Completa la siguiente tabla señalando las hormonas más importantes del embarazo, el parto o la lactancia y después explica el papel de cada una en las etapas en las que participa:

Hormona	Embarazo	Parto	Lactancia
Cortisol		X	
HCG			
Progesterona			
Estrógenos			
Prolactina			
Oxitocina			

5 Respecto a la fecundación *in vitro*,

- ¿Ante qué problemáticas está indicada su utilización? ¿Qué objetivo tiene el tratamiento hormonal aplicado a la mujer?
- ¿Qué se extrae de los folículos de Graaf? ¿Por qué se extrae de ahí?
- ¿Por qué son frecuentes los embarazos múltiples en los casos de fecundaciones *in vitro*?

6 Ordena los siguientes métodos anticonceptivos por orden creciente de efectividad y di cuáles de ellos protegen de las enfermedades de transmisión sexual: *vasectomía, método sincrónico, anovulatorios, preservativo, DIU.*

7 Observa el siguiente dibujo sobre el mecanismo de acción del VIH:

- Explica todos los pasos incluyendo la función de las estructuras y moléculas que participan.
- ¿Qué pasará transcurrido un tiempo?

GLOSARIO

GLOSARIO

A

abrasión. Proceso de desgaste de las rocas causado por el contacto de partículas sueltas arrastradas por el agua o el viento (pág. 83).

acuífero. Depósito de agua dulce situado por debajo de la superficie terrestre (pág. 91).

aeróbico. Proceso biológico que se produce en presencia de oxígeno (pág. 172).

aldosa. Glúcido del grupo de los monosacáridos que presenta un grupo aldehído (pág. 135).

amiloplastos. Orgánulos vegetales llenos de almidón que intervienen en la regulación de la orientación de crecimiento de tallos y raíces (pág. 285).

aminoácido. Monómero componente de las proteínas (pág. 140).

amoniotélicos. Conjunto de animales que excretan los residuos nitrogenados en forma de amoníaco (pág. 225).

anabolismo. Conjunto de reacciones bioquímicas en las células, que tienen por objeto la obtención de moléculas complejas a partir de otras más simples (pág. 168).

anaeróbico. Proceso biológico que se produce en ausencia de oxígeno, como la fermentación (pág. 172).

anatexia. Fusión de rocas de la corteza debido a condiciones extremas de presión y temperatura (pág. 36).

andrógenos. Hormonas sexuales masculinas sintetizadas por las células de Leydig en los testículos (pág. 323).

anfipático. Sustancia cuyas moléculas que presentan una parte con afín al agua y otra que la repele (pág. 138).

anticlinal. Ondulación de los estratos rocosos en forma de cubeta (pág. 72).

arco insular. Grupo de islas formadas por la actividad en un límite convergente entre placas (pág. 71).

astenosfera. Zona del interior de la Tierra que corresponde a la porción intermedia del manto superior (pág. 65).

aterosclerosis. Formación de placas de ateroma en el interior de las arterias que disminuyen su diámetro y obstaculizan la circulación (pág. 252).

ATP. Molécula con enlaces ricos en energía denominada adenosín trifosfato (pág. 160).

autótrofa. Tipo de nutrición en la que se fabrican moléculas orgánicas a partir de moléculas inorgánicas y una fuente de energía (pág. 169).

B

badlands. Grandes extensiones de terreno en zonas áridas, con numerosos barrancos de paredes verticales (pág. 89).

Big bang. Punto o singularidad inicial de temperatura y densidad infinitas a partir del cual se expandió el universo (pág. 10).

Big crunch. Proceso de contracción del universo hasta originar un nuevo punto de temperatura y densidad infinitas (pág. 11).

buzamiento. Inclinación que adquieren los estratos respecto al plano horizontal del terreno (pág 119).

C

calostro. Primera secreción de los acinos mamarios después del parto (pág. 331).

cárcavas. Surcos de poca profundidad en las laderas de las montañas (pág. 89).

cárstico, modelado. Paisaje originado por carbonatación de las rocas calizas (pág. 82).

catabolismo. Conjunto de reacciones bioquímicas en las células, que tienen por objeto la obtención de productos simples a partir de moléculas complejas (pág. 168).

células de Leydig. Células situadas en el tejido conjuntivo de los túbulos seminíferos que secretan hormonas sexuales (pág. 322).

células de Schwann. Células gliales que crecen alrededor de los axones de las neuronas constituyendo la vaina de mielina (pág. 259).

cera. Lípido complejo compuesto por ésteres de ácidos grasos de cadena larga con monoalcoholes de cadena larga (pág. 139).

charnela. Línea que une los puntos de máxima curvatura de los distintos estratos de un pliegue (pág. 72).

ciclo biológico. Conjunto de acontecimientos que tienen lugar en un organismo desde la formación del zigoto hasta su reproducción (pág. 310).

ciclo de la materia. Recorrido que sigue la materia en un ecosistema desde que, en forma inórganica, es incorporada por los organismos productores hasta que, como materia orgánica, es descompuesta por los organismos descomponedores (pág. 212).

ciclo menstrual. Modificación periódica del endometrio que origina la menstruación en la mujer, 14 días después de la ovulación, si no se produce el embarazo (pág. 327).

ciclo ovárico. Maduración de un óvulo cada 28 días, en los ovarios de la mujer (pág. 327).

ciclo de Wilson. Secuencia de fusión y ruptura de las placas litosfericas debido al conjunto de procesos geológicos internos (pág. 71).

clasto. Fragmento más o menos anguloso de una roca detrítica (pág. 43).

clorocruorina. Pigmento respiratorio de color verdoso (pág. 239).

comensalismo. Relación interespecífica nutricional entre dos organismos en la que uno de los dos sale beneficiado sin perjudicar al otro (pág. 211)

conformación. Distribución espacial de los aminoácidos de una proteína (pág. 142).

cordillera perioceánica. Serie de montañas entrelazadas entre sí, originadas por procesos de compresión entre placas (pág. 70).

cormo. Estructura vegetal formada por células diferenciadas agrupadas en tejidos (pág. 191).

corte geológico. Esquema de una sección vertical del terreno donde se muestran los accidentes geológicos, las unidades litológicas y la disposición en profundidad de los materiales (pág. 120).

covalente. Enlace químico en el que los átomos comparten electrones (pág. 132).

crioclastismo. Proceso de rotura de una roca en pequeños fragmentos debido a la formación de hielo en su superficie o en fracturas de la roca (pág. 81).

D

deflación. Tipo de erosión eólica en que las partículas que se encuentran encima de las rocas son barridas y transportadas (pág. 95).

depredación. Estrategia nutricional en la que el individuo debe cazar y matar a la presa que le sirve de alimento (pág. 211).

deriva continental. Movimiento de los continentes procedentes de la ruptura de un continente único denominado Pangea (pág. 66).

descomponedores. Organismos consumidores heterótrofos que se alimentan de la materia orgánica de los seres que han muerto (pág. 211).

diaclasa. Fractura en las rocas, que no implica un desplazamiento de los bloques resultantes (pág. 73).

diagénesis. Cambio estructural y químico que experimentan las partículas de un sedimento (pág. 43).

diploide. Dotación cromosómica de una célula somática (pág. 306).

dipolo. Molécula en la que podemos identificar dos zonas con cargas eléctricas de distinto signo e igual intensidad (pág. 132).

dorsal oceánica. Gran cordillera submarina formada por la actividad en un límite divergente entre placas (pág. 69).

disacárido. Glúcido compuesto por dos monosacáridos (pág. 135).

duna. Ondulaciones de arena que se desplazan impulsadas por el viento (pág. 95).

E

endergónica. Reacción química que libera energía (pág. 135).

endosfera. Zona del interior de la Tierra correspondiente al núcleo (pág. 65).

energía fijada. Parte de la energía solar que los organismos fotosintéticos captan y transfor-man en energía química (pág. 212).

epicentro. Punto situado en la superficie terrestre donde se registra la máxima intensidad de un terremoto (pág. 63).

era geológica. División del tiempo a partir de los acontecimientos geológicos (pág. 24).

erg. Zona de dunas en los desiertos (pág. 95).

esfingolípido. Lípido complejo compuesto por una molécula de esfingosina (o un derivado de ella), un ácido graso y un grupo de cabeza polar (pág. 139).

especie. Grupo de individuos que se reproducen entre sí, originando seres fértiles y similares a sus progenitores (pág. 200).

espermatóforo. Estructura quitinosa en la que, en algunos moluscos y artrópodos, los machos depositan sus espermatozoides para que luego los pueda recoger la hembra (pág. 308).

espematogonias. Células precursoras de los espermatozoides (pág. 323).

esquejes. En ciertas plantas, fragmento de raíces, tallos u hojas a partir de las cuáles se puede generar un nuevo individuo (pág. 305).

estatocitos. Células que participan en la regulación de la orientación de crecimiento el tallo y las raíces de una planta (pág. 285).

éster, enlace. Unión entre el carbono de un grupo hidroxilo de la glicerina con el carbono de un grupo carboxilo de un ácido graso (pág. 138).

estomas. Estructuras de la epidermis de las hojas en forma de poros que regulan en intercambio de gases y la transpiración en los vegetales (pág. 214).

estrato. Capa de sedimentos superpuestos (pág. 42).

estrógenos. Hormonas sexuales femeninas sintetizadas por los ovarios (pág. 326).

exergónica. Reacción química que consume energía (pág. 135).

F

facies metamórfica. Conjunto de minerales que indican unas condiciones ambientales determinadas bajo las que se ha producido el metamorfismo (pág. 51).

falla. Fractura en las rocas, en la que los bloques resultantes se desplazan uno respecto a otro (pág. 73).

fecundación. Unión de gametos para formar una célula diploide denominada zigoto (pág. 308).

fibras aferentes. Conjunto de neuronas que transportan el impulso nervioso de los órganos de los sentidos hacia el sistema nervioso central. (pág. 271).

fibras eferentes. Conjunto de neuronas que transportan el impulso nervioso desde el sistema nervioso central hasta los órganos efectores (pág. 271).

fitocromo. Hormona vegetal responsable de la detección del fotoperíodo y que induce a la floración en una estación concreta (pág. 285).

fitohormonas. Hormonas propias de los vegetales (pág. 285)

floema. Tejido conductor vegetal por el que se transporta la savia elaborada (pág. 193).

fosa oceánica. Depresión larga y profunda producida por el hundimiento de una placa oceánica bajo otra placa menos densa (pág. 71).

fosilización. Proceso mediante el cual la materia orgánica de un organismo muerto es sustituida por materia mineral (pág. 47).

fotótrofa. Tipo de nutrición en la que la fuente de energía es la luz (pág. 169).

G

gametogénesis. Proceso de formación de células reproductoras sexuales (pág. 306).

geocéntrico. Modelo de representación del universo en el que la Tierra es el eje a partir del cual orbitan el resto de los astros (pág. 6).

glándula. Órgano especializado en la fabricación y secreción de sustancias (pág. 187).

glándulas de Bartholino. Glándulas situadas en la vagina encargadas de secretar líquido lubricante (pág. 325).

glándulas de Cowper. Glándulas que, el sistema reproductor masculino, secretan líquido lubricante a la uretra (pág. 323).

glicerofosfolípido. Lípido complejo compuesto por una molécula de glicerina, una de ácido fosfórico y dos ácidos grasos (pág. 139).

GPS. Sistema de posicionamiento global capaz de indicar las coordenadas de un punto a partir de la información suministrada por satélites artificiales (pág. 115).

gravimetría. Disciplina científica que estudia el campo gravitatorio de la Tierra y sus anomalías. (pág. 58).

H

haloclastismo. Proceso de rotura de una roca en pequeños fragmentos debido a la formación de cristales de sal en su superficie o en fracturas de la roca (pág. 81).

haploide. Dotación cromosómica de los gametos (pág. 306).

heliocéntrico. Modelo de representación del universo en el que el Sol es el eje a partir del cual orbitan el resto de los astros (pág. 6).

hemeritrina. Pigmento respiratorio de color rosado presente en los anélidos (pág. 238).

hemocianina. Pigmento respiratorio de color azul presente en los moluscos y en los artrópodos (pág. 238).

hemoglobina. Pigmento respiratorio de color rojo presente en algunos anélidos, en muchos artrópodos y en los glóbulos rojos de los vertebrados (pág. 239).

hemolinfa. Líquido circulatorio presente de anélidos y artrópodos (pág. 248).

heteroproteína. Proteína formada por una cadena de péptidos unida a otro tipo de compuesto que se denomina grupo prostético (pág. 140).

heterótrofa. Tipo de nutrición en la que se utiliza materia orgánica procedente de otros seres vivos (pág. 169).

hidrofílico. Molécula o parte de una molécula que tiene afinidad por el agua (pág. 133).

hidrofóbico. Molécula o parte de una molécula que no tiene afinidad por el agua (pág. 133).

hipocentro. Zona del interior de la Tierra donde se produce la fractura que genera un terremoto (pág. 59).

hipófisis. Glándula característica de los vertebrados, situada debajo del hipotálamo, que segrega hormonas (pág. 292).

hipotálamo. Agrupación de neuronas que se sitúa en la parte inferior del cerebro de los vertebrados y que segrega las neurohormonas (pág. 292).

holoproteína. Proteína formada exclusivamente por cadenas de polipéptidos (pág. 140).

homeostasis. Mantenimiento del equilibrio del medio interno de un organismo (pág. 187).

hormona. Sustancia segregada por los seres vivos como respuesta a determinados estímulos y que regula el desarrollo de reacciones bioquímicas (pág. 284).

I

impulso nervioso. Pequeña corriente eléctrica que se establece entre las neuronas o entre una neurona y una fibra muscular (pág. 260).

infiltración. Movimiento de las aguas superficiales hacia el interior del terreno debido a la fuerza de la gravedad (pág. 90).

isostasia. Estado de equilibrio al que tienden los materiales de la litosfera (pág. 59).

L

linfa. Líquido circulante de composición similar al plasma sanguíneo que contiene sustancias y células especializadas en la defensa del organismo (pág. 187).

litificación. Conjunto de transformaciones que experimentan los sedimentos hasta formar una roca compacta (pág. 45).

litología. Especialidad de la geología que estudia las rocas (pág. 113).

litosfera. Zona del interior de la Tierra que comprende la corteza continental y oceánica, y los primeros 100 km del manto superior (pág. 65).

loess. Depósito de limos que, procedentes de zonas desérticas, son transportados por el viento y sedimentan en zonas húmedas (pág. 95).

M

mapa geológico. Representación de la distribución sobre el terreno de los distintos tipos de rocas que lo constituyen y de los fenómenos geológicos que lo han afectado (pág. 118).

medio interno. Conjunto de líquidos y fluidos que rodea las células de un organismo pluricelular (pág. 187).

meiosis. Proceso de división en la que a partir de una célula diploide y después de dos divisiones celulares conseutivas se obtienen cuatro células haploides (pág. 306).

menarquia. Primera menstruación de la mujer (pág. 327).

menopausia. Período que se inicia cuando cesan las menstruaciones (pág. 327).

mesosfera. Zona del interior terrestre que comprende una parte del manto superior y todo el manto inferior (pág. 65).

metabolismo. 1. Explosión violenta de una estrella (pág. 15). **2.** Conjunto de reacciones bioquímicas que se producen en el interior de las células de los seres vivos (pág. 169).

meteorito. Fragmento procedente de un asteroide o de un cometa que cae sobre un planeta u otro astro (pág. 23).

meteorización. Fragmentación de una roca o transformación química de los minerales que la forman (pág. 87).

miocardio. Capa muscular que forma el corazón (pág. 252).

mitosis. Proceso de reparto de los cromosomas en dos grupos, que tiene lugar durante la división celular (pág. 176).

monómero. Molécula pequeña que unida de forma repetitiva con otras de igual categoría constituye macromoléculas llamadas polímeros (pág. 136).

monosacárido. Glúcido sencillo formado por una cadena que tiene entre 3 y 7 átomos de carbono (pág. 135).

N

nastia. Movimiento rápido de una parte de una planta como respuesta a un estímulo externo (pág. 289).

nebulosa. Materia cósmica compuesta por gas y polvo de contorno y forma impreciso (pág. 16).

neurotransmisor. Molécula sintetizada por la neurona que transmite el impulso nervioso de una neurona a otra (pág. 261).

nivel freático. Límite o superficie de separación entre la zona en que los poros de la roca no están saturados de agua y la que sí (pág. 90).

nivel trófico. Conjunto de organismos que presentan una nutrición equivalente dentro de una cadena o de una red trófica (pág. 210).

nucleótido. Monómero de los ácidos nucleicos, compuesto por una molécula de pentosa, una de ácido fosfórico y una base nitrogenada (pág. 143).

nutación. Movimiento de crecimiento desigual de los tallos y hojas de las plantas, dirigido por estímulos externos (pág. 288).

O

O-glucosídico, enlace. Unión entre el carbono de un grupo hidroxilo de un monosacárido y el carbono del grupo aldehído o cetona de otro (pág. 135).

ocelos. Conjunto de células fotosensibles que presentan los cnidarios, la mayoría de los moluscos y los anélidos (pág. 264).

órgano. Asociación de diferentes tejidos para realizar una función determinada (pág. 186).

orgánulo. Compartimiento celular rodeado de membrana plasmática, especializado en una determinada función (pág. 165).

orogenia. Proceso de creación de relieve (pág. 107)

ósculo. Orificio principal que presentan los poríferos (pág. 248)

ósmosis. Proceso físico-químico de equilibrio de dos disoluciones de concentraciones distintas mediante el paso del disolvente a través de una membrana semipermeable que impide el paso del soluto (pág. 159).

P

Pangea. Continente hipotético formado por la unión de todas las tierras emergidas (pág. 67).

parasitismo. Relación interespecífica nutricional en la que el individuo (parásito) se alimenta de otro organismo (huésped) al cuál perjudica sin llegar a matarlo (pág. 211).

pentosa. Glúcido del grupo de los monosacáridos que presenta un grupo cetona (pág. 135).

peptídico, enlace. Unión covalente entre el grupo amino de un aminoácido y el carboxilo de otro (pág. 139).

perfil topográfico. Representación vertical del relieve, perpendicular a la superficie terrestre (pág. 119).

período geológico. División de una era según los cambios en la flora y la fauna (pág. 24).

petrogenético, ambiente. Área de la corteza terrestre con unas condiciones de temperatura, presión y composición química que posibilitan la formación de un determinado tipo de rocas (pág. 35).

pigmento respiratorio. Proteínas que transportan el oxígeno desde los órganos respiratorios hasta el resto de tejidos (pág. 238).

pirámides de tierra. Estructuras de formas peculiares producidas por una erosión diferencial (pág. 89).

piroclasto. Fragmento sólido de material volcánico, lava solidificada o rocas de la chimenea, lanzado al aire durante una erupción volcánica (pág. 62).

píxel. Cada una de las celdas o casillas que configuran una imagen digital y que contiene un número que indica el valor de su brillo (pág. 115).

placa litosférica. Cada uno de los fragmentos en los que se divide la litosfera terrestre (pág. 70).

planeta. Cuerpo gaseoso o rocoso, prácticamente esférico, que gira en torno al Sol (pág. 17).

planetesimal. Acumulación de materia, precursora de los planetas, procedente de una nebulosa (pág. 17).

plasma sanguíneo. Líquido en el que se encuentran todos los elementos celulares de la sangre (pág. 243)

pleuras. Membranas de protección situadas en el tórax; la interna se une a los pulmones y la externa a la cavidad torácica (pág. 241).

pliegue. Ondulación de los estratos rocosos (pág. 72).

polaridad. Característica de algunas moléculas en las que se distinguen dos zonas con cargas de distinto signo (pág. 132).

polímero. Macromolécula formada por la unión de varias moléculas pequeñas llamadas monómeros (pág. 136).

polisacárido. Glúcido compuesto por un gran número de monosacáridos (pág. 136).

potencial de acción. Diferencia de potencial entre el interior y el exterior de la membrana que se genera como resultado de la entrada de sodio en la célula y que se propaga de las dendritas hacia el axón (pág. 260).

potencial receptor. Potencial de acción generado en las células sensitivas receptoras (pág. 263).

protosol. Gran concentración de materia a partir de la cual se forma el Sol (pág. 17).

pubertad. Etapa del ser humano en la que se inicia una serie de cambios que llevan a la adquisición de las características adultas (pág. 324).

Q

quimiótrofa. Tipo de nutrición en la que la fuente de energía son sustancias químicas orgánicas tomadas del exterior (pág. 169).

R

radiotelescopio. Instrumentos de observación de los astros basados en la captación de ondas radioeléctricas de alta frecuencia (pág. 9).

ramblas. Pequeñas depresiones que se forman en la base de las montañas por efecto de las aguas de escorrentía (pág. 89).

receptor sensorial. Célula nerviosa especializada en percibir estímulos y transformarlos en impulsos nerviosos (pág. 263).

reg. Zona de piedras en los desiertos (pág. 95).

retroalimentación o *feed back.* Mecanismo químico de regulación en el que el resultado de un proceso estimula o inhibe el mismo proceso (pág. 292).

respiración externa (ventilación). Intercambio de gases entre un ser vivo y el medio que lo rodea (pág. 234).

respiración interna. Mecanismo de captación del oxígeno por parte de la célula, a través de la membrana plasmática (pág. 234).

retrovirus. Virus con ARN como material genético que al infectar una célula transcribe su ARN a ADN (pág. 336).

rift. Inmensa y profunda grieta originada por la separación de dos placas litosféricas (pág. 70).

S

savia. Fluido que forma el medi[illegible] interno de las plantas superior[illegible] (pág. 187).

SIG. Sistema de información g[illegible]gráfica constituido por una base [illegible] datos informática que se relaci[illegible]a con un mapa digital con locali[illegible]aciones concretas (pág. 116).

simbiosis. Relación interespecífica de dependencia entre dos organismos en la que ambos obtienen un beneficio (pág. 211).

sinapsis. Espacio que separa dos neuronas, en el que se produce la transmisión del impulso nervioso (pág. 261).

sinclinal. Pliegue con los estratos más modernos situados en su núcleo (pág. 72).

subducción. Proceso de hundimiento de una placa litosférica por debajo de otra (pág. 69).

suelo. Capa delgada de algunas zonas de la superficie terrestre, imprescindible para el desarrollo de las plantas (pág. 96).

supernova. Explosión violenta de una estrella (pág. 15).

T

talo. Estructura vegetal formada por células poco diferenciadas (pág. 191).

tectónica de placas. Teoría que propone la división de la litosfera en placas en continuo movimiento (pág. 66).

tejido. Agrupación de células semejantes que realizan una función concreta (pág. 186).

termoclastismo. Proceso de rotura de una roca en fragmentos debido a dilataciones y contracciones causadas por cambios bruscos de [illegible]emperatura (pág. 80).

[illegible]co. Organismo modificado [illegible]mente por la introducción [illegible]o o varios genes procede[illegible] otro organismo diferente (pá[illegible]315)

tráqueas (en vegetales). Vasos leñosos rígidos que forman un vaso continuo constituido por una serie de células muertas lignificadas que han perdido el citoplasma y la separación entre ellas (pág. 245).

traqueidas. Vasos leñosos rígidos de diámetro inferior a las tráqueas, constituidas por una serie de células muertas lignificadas que no han perdido la separación entre ellas (pág. 245)

triacilglicerol. Lípido complejo compuesto por una molécula de glicerina y tres ácidos grasos (pág. 138).

triplete de bases. Grupo de tres bases nitrogenadas consecutivas a cada uno de los cuales le corresponde un aminoácido según la clave genética (pág. 144).

tropismo. Movimiento de crecimiento que sufren las plantas di-rigido por estímulos externos (pág. 287).

U

ureotélicos. Conjunto de animales que excretan los residuos nitrogenados en forma de urea (pág. 225).

uricotélicos. Conjunto de animales que excretan los residuos nitrogenados en forma de ácido úrico (pág. 225).

V

vía metabólica. Reacciones metabólicas encadenadas, que son reguladas por enzimas y que siguen los pasos de la transformación de un determinado nutriente (pág. 215).

virus. Partícula sin estructura celular compuesta por ácido nucleico y proteínas, cuya supervivencia depende de que infecte una célula (pág. 179).

X

xilema. Tejido conductor vegetal por el que se transporta la savia bruta (pág. 193).

Z

zigoto. Célula resultante de la unión de dos gametos (pág. 306).